[redacted]

ORIGIN AND DISTRIBUTION OF THE ELEMENTS

[redacted]

ORIGIN AND DISTRIBUTION OF THE ELEMENTS

ACS Nuclear Chemistry and Technology Division

31 AUGUST — 4 SEPTEMBER, 1987
NEW ORLEANS, LA, USA.

Editor
G J MATHEWS

World Scientific
Singapore • New Jersey • Hong Kong

[redacted]

Published by

World Scientific Publishing Co. Pte. Ltd.
P.O. Box 128, Farrer Road, Singapore 9128

U. S. A. office: World Scientific Publishing Co., Inc.
687 Hartwell Street, Teaneck NJ 07666, USA

Library of Congress Cataloging-in-Publication data is available.

ORIGIN AND DISTRIBUTION OF THE ELEMENTS

ISBN 9971-50-455-3

Printed in Singapore by Lenco Printing Co. Pte. Ltd.

PREFACE

This Symposium on the Origin and Distribution of the Elements was held during the week of Aug. 31–Sept. 4, 1987 in New Orleans, as a part of the 194th general meeting of the Americal Chemical Society. So there, amongst the steamy wellsprings of American jazz, dixieland, and the blues, we paused to ponder the wellsprings of elemental composition. The conclusions from this symposium were the same as the conclusions from any symposium. We have done much work. We have much more work to do. But a number of exciting new ideas appear to be emerging like the first vague glimpses of a distant riverboat in the Mississippi mist. To name a few, there is the possibility that baryogenesis occured much later than the GUT's transition, and that much of what we can understand about the standard big bang now rises or falls on what one takes for the primordial ^{7}Li *abundance; Population III stars and early metallicity enrichment may now be unnecessary to understand galactic chemical evolution; SN 1987a may have answered many puzzles about the mechanism for Type II supernovae; Early main sequence stars may experience extensive mass loss; The* $^{12}C(\alpha,\gamma)^{16}O$ *reaction rate may yet change again; and those elusive hot hydrogen burning reaction rates may yet be known. In this volume are recorded some of the current wisdom on these and numerous other topics.*

Acknowledgement is made to B. Roberts and B. Voelker for help in the preparation of these proceedings and to the Donors of the Petroleum Research Fund, administered by the American Chemical Society for partial support of this Symposium.

Grant J. Mathews
University of California
Lawrence Livermore National Laboratory

TABLE OF CONTENTS

CHAPTER I

THE STANDARD AND NOT SO STANDARD BIG BANG

"... and as the engineer was about to enter the gates of heaven the third fellow tapped him on the shoulder and said 'wait a minute. You said the universe was formed from chaos. Well I'm an economist, and who do you think is responsible for all that chaos'..."

William A. Fowler
1986

The Origin of Baryons in the Universe [1]

EDWARD W. KOLB

NASA/Fermilab Astrophysics Center
Fermi National Accelerator Laboratory
Batavia, Illinois 60510
USA

ABSTRACT

In this talk I will review some recent work on the origin of the baryon asymmetry in the Universe. An asymmetry in the number of baryons compared to the number of antibaryons is crucial for nucleons to survive annihilation in the early moments of the big bang to later form nuclei, stars, planets, people, etc. The standard picture of the origin of the baryon asymmetry in the decay of a supermassive boson will be presented in detail. Some recent developments regarding effects at the electroweak phase transition, as well as possible low-temperature scenarios will be reviewed.

This symposium *The Origin and Distribution of Elements* concerns one of the most fundamental questions in science, namely "what is the Universe made of?" There are many possible answers to the question. An astronomer might answer that the Universe is made of stars or galaxies, some space dirt, and a small insignificant fraction in the form of planets. If pressed for a more fundamental answer, the astronomer might admit that stars are made of three types of elements, hydrogen, helium and metals. The abbreviated periodic table of the astronomer is often adequate for most purposes in astronomy. If pressed still further for a more fundamental answer, the astronomer might say "go see a chemist." A chemist might point to the actual periodic table and say that the Universe is made of atoms. A Universe of atoms as the fundamental building blocks might be the Universe of

[1]**Talk presented at the American Chemical Society meeting in New Orleans, Louisiana, August 31 – September 4, 1987**

the chemist, but there is a more fundamental answer. A physicist might say that the universe consists of electrons and nuclei. When asked the inevitable question the physicist will answer that electrons aren't made of anything, they are fundamental, but the nuclei are made of nucleons (neutrons and protons), which in turn are made of three quarks and some gluons added for good measure to keep the nucleons and nuclei bound.

Of course the answers of the astronomer, the chemist, and the physicist are all correct in some sense, but no single answer is the complete picture. In order to understand the Universe it is necessary to understand the structure of stars and galaxies, which requires an understanding of the fundamental particles down the entire chain from atoms to quarks. Often the best answer depends upon the scale of the problem.

Now that we have established what the Universe is made of, it is necessary to understand why it has the structure it does. This symposium is concerned with the origin of the elements. In order to understand the origin of the elements it is necessary to understand the origin of the building blocks of the elements, the nucleons. In order to understand the origin of the nucleons, it is necessary to consult that noble and rare breed of scientist, the cosmologist. The question I will address is why are there nucleons at all. A possible answer to the question would be to invoke the anthropic principle and say that if there weren't nucleons in the Universe, people wouldn't be around to hold a symposium on the subject. In my opinion, even if said with a British accent, the anthropic principle explains nothing in this instance. However, it is possible to understand the existence of nucleons on the basis of physical law. Before describing such efforts, I will quantify the question in terms of a number known as the "Baryon Number of the Universe."

It is possible to assign to quarks and antiquarks a conserved quantum number known as baryon number. The baryon number of a quark is $1/3$ and the baryon number of an antiquark is $-1/3$. Nucleons, which consist of three quarks, have a baryon number of $+1$. The baryon number is conserved in all laboratory experiments. For experiments at low energy, $E \lesssim m_N \simeq 1$ GeV, where m_N is the mass of the nucleon, the conservation of baryon number is equivalent to the conservation of mass number.

It is possible to estimate the average number density of baryons (plus antibaryons) in the Universe from the observed total mass density of the universe. It is convenient to express present energy densities in terms of a critical density, $\rho_C = 3H_0^2/8\pi G$, where H_0 is the Hubble constant (the present expansion rate of the Universe) and G is Newton's constant. In terms of $h = H_0/100$ km^{-1}sec^{-1}Mpc^{-1}, $\rho_C = 1.88 \times 10^{-29} h^2$g cm^{-3}. In terms of $\Omega_+ = (\rho_b + \rho_{\bar{b}})/\rho_C$, $n_b + n_{\bar{b}} = 1.12 \times 10^{-5} \Omega_+ \, h^2$cm^{-3}. If there are no antibaryons in the Universe (as argued below), then $n_b + n_{\bar{b}} = n_b$, and from the limit $\Omega_+ \leq 1$, it is possible to estimate the baryon charge density, $n_B = n_b - n_{\bar{b}}$. It is possible to create baryons and antibaryons so long as the difference in baryons and antibaryons is conserved. n_B should scale in expansion as inverse of the volume of the Universe. The entropy density $s = (\rho + p)/T$ also scales in the same way, so the baryon number of the Universe, $B = n_B/s$ should be constant so long as the baryon number is locally conserved and the entropy remains constant. The entropy density today is $s \simeq 2800$ cm^{-3}, so

$$B = \frac{n_B}{s} = 4 \times 10^{-9} \Omega_+ h^2. \tag{1}$$

If there is no antimatter in the Universe, $\Omega_+ = \Omega_N$, the present contribution to the total mass density due to nucleons. Limits on the TOTAL mass density of the Universe give $0.01 < \Omega_N < 1$, while limits from primordial nucleosynthesis give the much more restrictive range $0.01 < \Omega_N h^2 < 0.04$.

As mentioned above, our Universe seems to have a baryon asymmetry, more baryons than antibaryons, i.e., more matter than antimatter. Antimatter is rare on earth. It exists in "large" quantities only in the antiproton accumulators at Fermilab and CERN. Antimatter is also rare in the solar system. The fact that Neil Armstrong survived his "one small step" is evidence that the moon is matter. Planetary probes have visited seven of the nine (ten?) planets, and have shown that the solar system is matter.

Cosmic rays provide a sample of the entire galaxy (at least). Antiprotons are seen in cosmic rays at about the 10^{-4} level compared to protons. These cosmic rays are usually assumed to be secondary particles, and not primary particles from an antimatter source. The flux of antimatter nuclei is also below the 10^{-4} level compared to nuclei, and there is no clean event that signals detection of an antinucleus. Cosmic rays are solid evidence that there is a galactic asymmetry between baryons and antibaryons, and that this asymmetry is maximal, i.e., everything is baryons, and there aren't any antibaryons in the galaxy.

Our evidence on larger scales is somewhat more uncertain. There are clusters of galaxies containing intercluster gas. If both matter galaxies and antimatter galaxies exist in the same cluster there would result a large amount of annihilation which would contribute a large γ-ray flux. The absence of such a large γ-ray flux is evidence that clusters of galaxies are either all baryons or all antibaryons. There is basically no information on scales larger than clusters of galaxies ($\simeq 10^{14}\ M_\odot$).

If antimatter exists in appreciable quantities, it must be separated on scales larger than about $10^{14}\ M_\odot$. However this separation must be done before $T \simeq 1$ GeV, or the nucleons and antinucleons would have annihilated below acceptable limits. If there are equal numbers of baryons and antibaryons the evolution of the baryon number density, n_B, is described by the equation

$$\dot{n}_B + 3Hn_B = ((n_B^{eq})^2 - n_B^2)\langle\sigma_A|v|\rangle \tag{2}$$

where H is the expansion rate of the Universe, n_B^{eq} is the equilibrium number density of baryons, and $\langle\sigma_A v\rangle$ is the thermal average of the annihilation rate times relative velocity. The baryon number density tracks its equilibrium value, $n_B^{eq} \simeq (m_N T)^{3/2}\exp(-m_N/T)$, until the baryon density is so low that annihilation effectively ceases. For baryons, this happens when the density of baryons relative to the entropy density is about 10^{-19}.

There seems to be convincing arguments for a universal baryon asymmetry. Details of the arguments for a baryon asymmetry and for a symmetric Universe can be found in Ref. 1. The most reasonable conclusion is that at $T \geq 1$ GeV, there was an asymmetry between the number of baryons and the number of antibaryons. This asymmetry is characterized by the baryon number discussed above, $B = 4 \times 10^{-9}\Omega_N h^2$. Although the baryon asymmetry is maximal today (i.e., no antimatter) at $T \geq 1$ GeV, the temperature was high enough to create $N\bar{N}$ pairs, and $B \simeq 10^{-9}$ means that for every billion antibaryons there were a billion and one baryons.

The answer to the question why are there are baryons in the Universe has been replaced with another question, why is there a baryon asymmetry. The goal of this talk is to explain how such a curious (but crucial) number could arise in a Universe with symmetric, $B = 0$, or even better, random, initial conditions.

There are three basic ingredients necessary to generate a non-zero B from an initial symmetric state.[2] • *Baryon Number Violation:* There must obviously be a violation of baryon number. If baryon number is conserved in all interactions, the present baryon asymmetry must simply reflect the initial conditions. • *C and CP Violation:* Since B is odd under C and CP, they both must be violated to generate a non-zero B. • *Non-Equilibrium Conditions:* In chemical equilibrium

particle		final state	branching ratio	B
X	$\rightarrow$	qq	r	2/3
X	$\rightarrow$	$\bar{q}\bar{l}$	$1-r$	-1/3
$\bar{X}$	$\rightarrow$	$\bar{q}\bar{q}$	$\bar{r}$	-2/3
$\bar{X}$	$\rightarrow$	ql	$1-\bar{r}$	1/3

Table 1: Final states and branching ratios

the entropy is maximal when the chemical potentials associated with all non-conserved quantum numbers vanish. If baryon number is not conserved, and the Universe ever obtains chemical equilibrium, B must be zero.

To illustrate the simple model, consider a particle X which decays to final states with different baryon number (hence violating baryon number) with branching ratios given in Table 1.[3,4,5,6] Note that CP is violated if $r \neq \bar{r}$, but CPT is conserved, since the total rate for X decay is the same as the total rate for $\bar{X}$ decay.

Imagine a box containing equal amounts of X and $\bar{X}$. The baryon number produced by the decay of the X's is proportional to $B_X = r(2/3) + (1-r)(-1/3)$, and the baryon number produced by the $\bar{X}$'s is proportional to $B_{\bar{X}} = \bar{r}(-2/3) + (1-\bar{r})(1/3)$. The net asymmetry produced by an equal amount of X and $\bar{X}$'s is proportional to $\epsilon = B_X + B_{\bar{X}} = r - \bar{r}$. The baryon number vanishes, of course, if CP is conserved ($r = \bar{r}$). If there are no further baryon number violating reactions (equilibrium not maintained) a net baryon asymmetry will result. This extremely simple picture illustrates the basic idea.

Now, consider the possibility of B, C, and CP violation, and non-equilibrium conditions in the early Universe.

• *B Violation*: The existence of baryon number violation seems to be a generic feature of Grand Unified Theories (GUTs). If strong and electroweak interactions are unified, quarks and leptons typically appear as elements of a common irreducible representation of the gauge group, and gauge bosons can mediate interactions that violate B. The limit on the stability of the proton, $\tau \geq 10^{31}$ years, implies that these bosons should have masses in excess of 10^{14}GeV or so. The weaker coupling allows Higgs bosons that mediate baryon-number violation to have a somewhat smaller mass. In both cases the large mass of the intermediate bosons is responsible for the feebleness of baryon-number violation today. This suppression should have been overcome at the extremely high temperatures in the big bang, and interactions that violate baryon number should be just as common as other interactions. The requirement of B violation is easy to arrange in GUTs. A gauge or Higgs boson that violates baryon number in its decay will be generically denoted as X.

• *C and CP Violation*: C is maximally violated in the weak interactions, so C violation in the decay of the X bosons should not be a fundamental problem. CP violation is observed in the kaon system, with typical dimensionless strength of 10^{-3}. Since its origin is not well understood, it is easy to say it should also appear in the X-boson system as well.

To explicitly see how CP violation enters, consider a system with two massive particles, X and Y, with baryon number violating decays. The generalization of ϵ defined above is

$$\begin{aligned}\epsilon_X &= \sum_f B_f \frac{\Gamma(X \to f) - \Gamma(\bar{X} \to \bar{f})}{\Gamma_X} \\ \epsilon_Y &= \sum_f B_f \frac{\Gamma(Y \to f) - \Gamma(\bar{Y} \to \bar{f})}{\Gamma_Y},\end{aligned} \tag{3}$$

where the sum is over all final states f with baryon number B_f, and Γ_X (Γ_Y) is the total X (Y) decay width. For simplicity, assume there are but two final states for X and Y decay, and that the interaction Lagrangian is given by

$$\mathcal{L} = g_1 X i_2^\dagger i_1 + g_2 X i_4^\dagger i_3 + g_3 Y i_1^\dagger i_3 + g_4 Y i_2^\dagger i_4 + \text{h.c.} \tag{4}$$

The tree-level processes cannot contribute to ϵ, as they give $\Gamma(X \to \bar{i}_1 i_2) = |g_1|^2 I_X = \Gamma(\bar{X} \to i_1 \bar{i}_2) = |g_1^*|^2 I_X$ where $I_X = I_{\bar{X}}$ represents the phase space factors. The first non-zero contribution to ϵ comes from the interference of the tree-level contribution and the one-loop contributions. The interference terms contribute to the amplitude for X decay terms given by

$$\begin{aligned}\Gamma(X \to \bar{i}_1 i_2) &= g_1 g_2^* g_3 g_4^* I_{XY} + (g_1 g_2^* g_3 g_4^* I_{XY})^* \\ \Gamma(\bar{X} \to i_1 \bar{i}_2) &= g_1^* g_2 g_3^* g_4 I_{XY} + (g_1^* g_2 g_3^* g_4 I_{XY})^*,\end{aligned} \tag{5}$$

where I_{AB} now includes kinematic factors from integration over the internal momentum loop due to B exchange in A decay. If intermediate particles in the loop are kinematically allowed to propagate on shell, I_{AB} will be complex.

The difference between $X \to \bar{i}_1 i_2$ and $\bar{X} \to i_1 \bar{i}_2$ is given by

$$\begin{aligned}\Gamma(X \to \bar{i}_1 i_2) - \Gamma(\bar{X} \to i_1 \bar{i}_2) &= 2i I_{XY} \text{Im}(g_1 g_2^* g_3 g_4^*) + 2i I_{XY}^* \text{Im}(g_1^* g_2 g_3^* g_4) \\ &= 4 \text{Im} I_{XY} \text{Im}(g_1^* g_2 g_3^* g_4).\end{aligned} \tag{6}$$

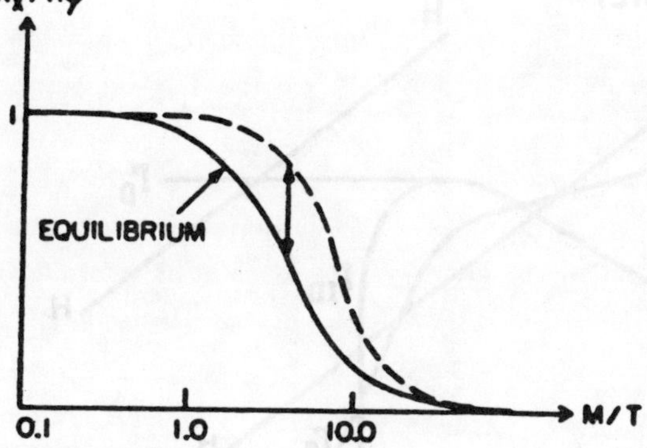

Figure 1: Departure from equilibrium denoted as the arrow between equilibrium (solid) and actual (dashed) ratio of n_X/n_γ

With a similar calculation for the other decay mode

$$\epsilon_X = \frac{4}{\Gamma_X}\text{Im}I_{XY}\text{Im}(g_1^* g_2 g_3^* g_4)\left[(B_{i_4} - B_{i_3}) - (B_{i_2} - B_{i_1})\right]. \quad (7)$$

For Y decay $\epsilon_Y = -\epsilon_X$, with $X \leftrightarrow Y$ everywhere.

There are several lessons to be learned from this tedious exercise. There must be *two* baryon number violating bosons, with masses greater than the sum of the masses in the internal loops. CP violation is manifest as complex coupling constants. The X and Y particles in the above example must not be degenerate, or the baryon number produced in X decay will cancel the baryon number produced in Y decay.

- *Non-Equilibrium Conditions*: The non-equilibrium conditions are provided by the expansion of the Universe. If the expansion rate is faster than particle interaction rates, non-equilibrium can result. Assume that at the Planck time X and $\bar{X}$ are present in equilibrium, $n_X = n_{\bar{X}} = n_\gamma$. In LTE $n_X = n_{\bar{X}} = n_\gamma$ for $T \geq m_X$, and $\ll n_\gamma$ for $T \leq m_X$. However LTE will be obtained only if the X interaction rates are greater than H. If LTE is not maintained, n_X will remain equal to n_γ and there will be an excess of X relative to the equilibrium abundance. This is illustrated in Fig. 1.

The conditions necessary for a departure from equilibrium can be quantified. The decay rate of the X, denoted by Γ_D, the inverse decay rate (i.e., X produc-

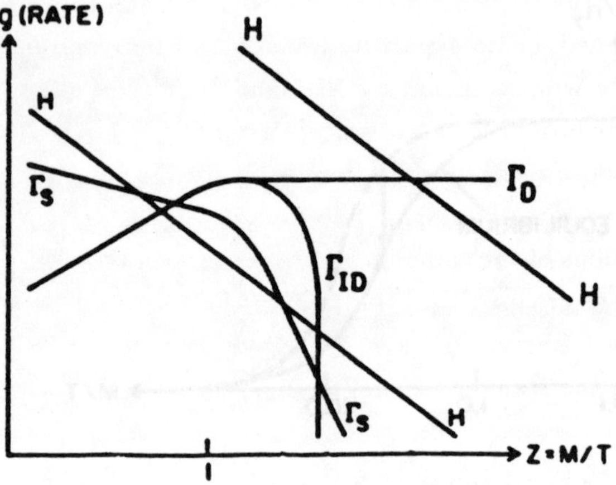

Figure 2: Rates in the early Universe

tion), denoted as Γ_{ID}, and the two-body scattering rate, denoted as Γ_S, and the expansion rate, denoted as H, are given by

$$\begin{aligned}
\Gamma_D &= \alpha m_X \begin{cases} m_X/T & m_X \leq T \\ 1 & m_X \geq T \end{cases} \\
\Gamma_{ID} &= \Gamma_D \exp(-m_X/T) \\
\Gamma_S &= n\sigma \simeq T^3 \alpha^2 \frac{T^2}{(T^2 + m_X^2)^2} \\
H &\sim g_*^{1/2} T^2 / m_{Pl}.
\end{aligned} \tag{8}$$

A comparison of the rates are given in Fig. 2. Note that all the reaction rates depend upon m_X, while H is independent of m_X. It is necessary to determine what sets the scale of H relative to the other rates. As shown in Fig. 2, H can be either large or small compared to the rates at $T = m_X$.

The relevant quantity to determine if X will be over abundant is the ratio of the decay rate to the expansion rate at $T = m_X$, given by

$$K \equiv (\Gamma_D/H)_{T=m_X} = \frac{\alpha m_{Pl}}{g_*^{1/2} m_X}. \tag{9}$$

If $K \ll 1$, equilibrium will not be maintained and the X will become over abundant. In this limit the X just drifts along and then decays. In the limit of pure drift and decay (the British limit) $B = g_*^{-1}\epsilon$. If $K \gg 1$, the X will be in good causal contact, there will be no departure from equilibrium, and baryogenesis will be thwarted. Exactly what is meant by "$\gg$" and "$\ll$" can be found by a simple model. If X is a gauge boson, then $\alpha = g^2_{\rm GUT}/4\pi \simeq 1/45$, and $K = 7\times 10^{15}{\rm GeV}/M$. If X is not a gauge boson, the effective α may be smaller, and the corresponding K may be smaller.

Detailed calculations of the complete rate equations in Grand Unified Theories based upon the gauge groups SU(5) and SO(10) have been done.[7] The scenario for the generation of a baryon asymmetry indeed can work in GUTs. The GUT scenario has become the standard model for Baryogenesis.

There have been two recent developments that might change the standard GUT scenario. The first has to do with non-perturbative effects at the electroweak

The θ-vacuum structure in electroweak theories leads to the anomalous non-conservation of baryon number.[8] There are equivalent vacua associated with different fermion numbers. Transitions between the different vacua are accompanied by a change in the baryon number. This process is usually associated with instantons that describe the tunneling between different θ-vacua. Since the process is inherently non-perturbative, the rate for baryon number non-conservation is proportional to $\exp(-4\pi/\alpha_W)$. This quantum tunneling is unimportant today, and was certainly unimportant in the early Universe.

At finite temperature the transitions between different vacua can be driven by finite temperature effects. Kuzmin, Rubakov, and Shaposhnikov (KRS) have recently shown that the tunneling at finite temperature may be appreciable.[9] In their analysis they considered an SU_2 gauge theory, which is a good approximation to the Weinberg-Salam model in the limit $\sin^2\theta_W \rightarrow 0$, α_W fixed. The Lagrangian for the SU_2 model with gauge coupling g_W is given by

$$\mathcal{L} = \frac{1}{2}D_\mu\phi^a D^\mu\phi^a - \frac{1}{4}F^a_{\mu\nu}F^{a\mu\nu} - \frac{1}{4}\lambda(\phi^a\phi^a - \sigma^2/2)^2 + \mathcal{L}_{\rm fermions}. \tag{10}$$

The basic point is that fluctuations in A_μ and ϕ caused by finite temperature effects cause transitions between different θ-vacua, with associated baryon number violation. The calculation of the transition rate is related to the calculation of the fate of the false vacuum at finite temperature. However, in this case the calculation is much more complicated. The first approximation made is to ignore the effect of the fermion fields, and to only consider A_μ and ϕ as the dynamical fields. The validity of this approximation will be discussed later. The second approximation is to replace the finite temperature "bounce" action by the free energy at the maximum of the barrier between adjacent θ-vacua.

The maximum free energy in the transition is dominated by a static configuration, given in the $A_0 = 0$ gauge by ($r^2 = \vec{x}^2$)

$$A^i_{\rm cl} = \frac{i}{g} \frac{\varepsilon_{ijk} x_j \tau_k}{r^2} f(\xi) \qquad \phi = \frac{\sigma}{\sqrt{2}} \frac{i\vec{\tau} \cdot \vec{x}}{r} \begin{bmatrix} 0 \\ 1 \end{bmatrix} h(\xi), \tag{11}$$

where $\vec{\tau}$ are the Pauli spin matrices, and $f(0) = h(0) = 0$, $f(\infty) = h(\infty) = 1$ are functions of the dimensionless parameter $\xi = r/r_0 = r g_W \sigma$. The free energy of this configuration is given by $F \simeq 2M_W/\alpha_W$, where $M_W = g_W \sigma/2$.

With the KRS assumption that the action for the tunneling rate at finite temperature is given by F, the change in the baryon number is $\dot{B} = -CBT \exp(-F/T)$, where C is a dimensionless constant of order unity, and the overall factor of T appears on dimensional grounds. Since CP is conserved, B cannot be generated by the transitions. In the early Universe g, λ, and σ are a function of temperature. The temperature dependence of g and λ is only proportional to $\ln T$ and will be ignored. The temperature dependence of σ around the critical temperature is much more important. At $T \geq T_c$, $\sigma = 0$, and as the temperature passes through the critical temperature, $\sigma \to \sigma_0 = 250$ GeV.

The rate for baryon number non-conservation is given by $\Gamma_{\Delta B} = CT \exp(-F/T)$, which is greater than the expansion rate for $T \geq T^* \simeq 200$ GeV. This would be a long enough period to eradicate any baryon number. Since there is no anomaly in $B - L$,the transitions conserve $B - L$, so more precisely, it would eradicate any baryon number with zero projection on $B - L$. The two possible solutions would be either to generate an asymmetry in $B - L$ in the same manner an asymmetry in B is generated, [2] or to generate the baryon asymmetry after the electroweak transition.

The potential impact of the KRS calculation certainly warrants more detailed calculations of the finite-temperature induced tunneling between θ-vacua. In particular, one approximation made by KRS that might be questioned is the neglect of the plasma effects on the transition. The classical field configuration are spatially large. The characteristic size of the configurations are several r_0. Within this size are $10^6 - 10^4$ particles with weak charge. Since the configuration of A_μ is pure magnetic and the plasma is a magnetic conductor, there are no plasma screening

[2] In many GUTs (in particular SO_{10}) this is easy to accomplish.

effects in the *static* configuration. However in the evolution to and from the static configuration there are time-dependent gauge fields and A_μ must have an electric component which *will* be affected by screening. The problem of screening can be addressed in a U_1 theory by calculating how long it takes to establish a U_1 field configuration with a characteristic size of several r_0. By detailed balance the time it takes to establish this unstable field configuration is the same as the time it takes for the field configuration to decay. In the limit that the conductivity of the Universe is infinite, the time is infinite, i.e., in a perfect conductor currents in the plasma are induced to oppose the establishment of the field. Of course, the Universe is not a perfect conductor, and the time for the establishment of the coherent field, Γ_A, will be proportional to the conductivity.

It is possible to map the problem at hand into one treated before. Consider the field configuration as a particle (called the Sphaleron) with mass $m = F$ and decay width $\Gamma = \Gamma_A$. Since baryon number is violated in the decay (and CP is conserved), the change in the baryon number is $\dot{n}_B + 3Hn_B = -n_B n_S^{eq}\Gamma_S$, where as usual $n_S^{eq} \simeq (mT)^{3/2}\exp(-m/T)$. Determination of Γ_S will allow a determination of the efficiency of the process in damping the baryon asymmetry.

A second development that may have implications for baryogenesis is the search for low-temperature scenarios. In the conventional GUT scenario, baryogenesis occurs at a temperature $T \simeq 10^{14}$GeV. New Inflation is a possible explanation to the origin of the large entropy in the Universe.[10] If inflation produces a large amount of entropy, it would dilute any baryon number that might exist before inflation and baryogenesis must occur after (or at least in the latter stages of) inflation. Exactly how inflation is implemented is unclear at present. However it seems that the temperature of the Universe after inflation is low, i.e. less than the GUT scale. The standard decaying particle model for baryogenesis is impossible if one starts at a temperature much less than the mass of the decaying particle. If the reheat temperature is low another mechanism for baryogenesis must be found.

One example of such a mechanism has recently been proposed by Dimopoulos and Hall.[11] They point out that in the supersymmetric extension of the standard SU(3)×SU(2)×U(1) model with explicit R-parity violation it is possible to have relatively light baryon number violating particles. In this model the proton is stable in the presence of light baryon-number violating bosons because the proton is lighter than all fermionic superpartners.

In conclusion, it is possible to understand the presence of baryons in the Universe as the result of the operation of physical laws. Whether the standard GUT scenario proves correct, or perhaps is modified by non-perturbative effects at the electroweak phase transition, or perhaps replaced by a low-temperature scenario, baryogenesis is an example of the importance of microphysics in understanding the large-scale structure of the Universe.

References

1. The evidence for the asymmetry is reviewed by G. Steigman, *Ann. Rev. Ast. Astrophys.* **14**, 339 (1976). The view that the Universe is symmetric is reviewed by F. Stecker, *Ann. NY Acad. Sci.* **375**, 69 (1981).

2. A. D. Sakharov, *JETP Lett.* **5**, 24 (1967).

3. E. W. Kolb and S. Wolfram, *Nucl. Phys.* **B172**, 224 (1980).

4. E. W. Kolb and M. S. Turner, *Ann. Rev. Nucl. Part. Sci.* **33**, 645 (1983).

5. J. N. Fry, K. A. Olive, and M. S. Turner, *Phys. Rev. D.* **22**, 2953 (1980).

6. M. Yoshimura, *Phys. Rev. Lett.* **41**, 281 (1978); D. Toussaint, S. B. Trieman, F. Wilczek, and A. Zee, *Phys. Rev. D* **19**, 1036 (1979); S. Dimopoulos and L. Susskind, *Phys. Rev. D* **18**, 4500 (1978); J. Ellis, M. K. Gaillard, and D. V. Nanopoulos, *Phys. Lett.* **B80**, 360 (1979); S. Weinberg, *Phys. Rev. Lett.* **42**, 850 (1979).

7. J. A. Harvey, E. W. Kolb, D. B. Reiss, and S. Wolfram, *Nucl. Phys.* **B201**, 16 (1982).

8. G. 't Hooft, *Phys. Rev. Lett.* **37**, 8 (1976).

9. V. A. Kuzmin, V. A. Rubakov, and M. E. Shaposhnikov, *Phys. Lett.* **155B**, 36 (1985).

10. For a review of inflation, see M. S. Turner, in *Proceedings of the 1984 Cargese School*, ed. J. Audouze and J. Tran Thanh Van.

11. S. Dimopoulos and L. Hall, LBL preprint LBL-23516 (1987).

FROM BARYON NUMBER GENERATION TO THE QUARK-HADRON PHASE TRANSITION

NEAL SNYDERMAN

Lawrence Livermore National Laboratory
University of California
Livermore, CA 94550

ABSTRACT

Between the temperature scale at which baryon number was generated and the temperature at which baryons and mesons formed, there were phase transitions associated with the Weinberg-Salam symmetry breaking, and perhaps with Peccei-Quinn symmetry breaking and supersymmetry breaking. Also, the quark-hadron phase transition may have taken place in the presence of cosmic strings. Some possible implications of these phase transitions and degrees of freedom on the evolution of the very early universe will be reviewed.

Time reversing the isentropically expanding universe implies the universe was hotter in the past in inverse proportion to the scale factor, $T \propto 1/R$. The evolution of R(t) is given, for an asymptotically flat universe by,

$$H = \frac{1}{R}\frac{dR}{dt} = \sqrt{\frac{8\pi}{3} G\rho} \quad ;$$

H, the Hubble parameter is constant throughout space (uniform expansion of space), and measures the expansion rate, which decreases with time. Newton's constant G sets the fundamental scales, the Planck length $L = \hbar/Mc = \sqrt{\hbar G/c^3} \cong 2 \times 10^{-33}$ cm, and mass $M \cong 10^{-5}$ gm $\cong 10^{28}$ eV/c^2, and Planck time $L/c \cong 10^{-43}$ sec. The explicit factors of $\hbar$ and c arise from the equation of state for the mass density $\rho = U/c^2$. The evolution of R depends on whether the energy density is dominated by matter, radiation, or by a cosmological constant,

$$R(t) \sim \begin{cases} (t/\sqrt{\hbar G/c^5})^{2/3} \;, & U \propto T^3 \;, \quad \text{matter dominated} \\ (t/\sqrt{\hbar G/c^5})^{1/2} \;, & U \propto T^4 \;, \quad \text{radiation dominated} \\ \exp(\sqrt{(8\pi/3c^2)\Lambda}\; t) \;, & U = \Lambda \;, \quad \text{cosmological constant.} \end{cases}$$

From measuring the redshift of distant galaxies, and thus their recession speed, and deducing their distances, H in this epoch of the universe is determined to be within a factor of 2 of 50 km/sec/Mpc, implying $\rho \simeq 5 \times 10^{-30}$ gm/cm^3 or $U \sim (5 \times 10^{-3}\ \text{eV})^4/(\hbar c)^3$, ($\hbar c \simeq 2 \times 10^{-3}$ eV x cm). If we assume U is now dominated by a cosmological constant, this sets the limit for it's magnitude. As we will see, this is infinitesimal compared to contributions to the vacuum energy density that existed at very early epochs. Theoretically, it is very difficult to understand how a cosmological constant could be so small without being zero.

We will extrapolate back in time to smaller R and consequently higher T to the epoch before there were nucleons, where issues of elementary particle physics must be confronted. At sufficiently high temperature the vacuum undergoes phase transitions in which the basic interactions of nature change character. We will review here some field theory issues of the phase transitions of the standard model of strong and electroweak interactions, and of the speculative axion fields. The theories of the strong and electroweak interactions are supported by a large body of experiments, although some of the parameters relevant to cosmology are either not yet theoretically calculated or experimentally accessible. The axion arises from an extra (Peccei-Quinn) symmetry[2] introduced into the standard model in order to solve the only qualitative phenomenlogical problem of the combined QCD and Weinberg-Salam-Glashow theories: that the weak interaction CP violation, observed in the neutral K-meson system, and of fundamental importance in creating the baryon number asymmetry of the universe, could otherwise induce strong interaction CP violation that would, for example, produce a large neutron static electric dipole moment - a number now measured to be less than 10^{-24} e-cm. While the strong CP problem is elegantly solved by the Peccei-Quinn symmetry, the consequent axion field theory, with great cosmological implications, is theoretically incomplete. The temperature scale associated with these phase transitions is, for the strong interactions at kT ~ 100 MeV, for the electroweak interactions at kT ~ 100 GeV, and for the axion field at kT ~ 10^{10} GeV and at the other two scales as well.

For all of these phase transitions, and for quantum field theory in general, issues of the cosmological constant arise. The resolution of this vacuum energy problem seems to require supersymmetry, a symmetry fundamentally pairing all boson and fermion degrees of freedom. The implications of supersymmetry may totally change our understanding of the origin of matter (baryon number asymmetry) and cosmological nucleosynthesis.

The cosmological implications of these phase transitions follows from quantum field theory at finite temperature. For a quantum harmonic oscillator the partition functions is,

$$Z = \sum_{n=0}^{\infty} e^{-\beta(n + 1/2)\hbar\omega} = \frac{e^{-\hbar\omega/2}}{1 - e^{-\beta\hbar\omega}} = e^{-\beta F}$$

where $\beta = 1/kT$, and F is the Hemholtz free energy,

$$F/\mathrm{Vol} = U - TS \quad ,$$

where S is the entropy per unit volume. In quantum field theory the field amplitudes can be decomposed into quantum oscillators with

$$\hbar\omega_k = \sqrt{(\hbar kc)^2 + (M(\varphi)\ c^2)^2} \quad ,$$

where the masses of the elementary particles depend, in unified gauge theories, on the value of a background field φ. The value of φ is determined from the minimum of the free energy,

$$F/\mathrm{vol} = V(\varphi) + \int \frac{d^3k}{(2\pi)^3}\left(\sum_{\text{bosons}} \left\{ \frac{1}{2}\ \hbar\omega_k + kT\ \ln\left(1 - e^{-\frac{\hbar\omega_k}{kT}}\right)\right\} - \sum_{\substack{\text{fermions}\\ \text{and anti-fermions}}} \left\{ \frac{1}{2}\ \hbar\omega_k + kT\ \ln\left(1 + e^{-\frac{\hbar\omega_k}{kT}}\right)\right\}\right) ,$$

where V is the potential for the constant background field, the sum over bosons is over all polarization and charge states of the integral spin particles, and we have included the corresponding free energy contribution of the half-integral spin fermions. If, for example, φ is a complex scalar field, the most general potential allowed by renormalizability is, (units for which $\hbar = 1$ and $c = 1$ will be assumed below unless they appear explicitly in equations),

$$V(\varphi) = \frac{\mu^2}{2}\ |\varphi|^2 + \frac{\lambda}{4}\ (|\varphi|^2)^2 + \frac{\mu^4}{4\lambda} \quad ;$$

λ must be positive so that V has a finite minimum, but μ^2 can be positive or negative. These possibilities are shown in Fig. 1. The constant $\mu^4/4\lambda$ is chosen so that for $\mu^2 < 0$, the minimum of V at $|\varphi|^2 = v^2 = \mu^2/\lambda$, is zero. There is thus an energy difference between $\mu^2 > 0$ and $\mu^2 < 0$ vacuua. While the $\mu^2 > 0$ minimum is unique, there are an infinite number of degenerate minima for $\mu^2 < 0$. V is invariant under $\varphi \to e^{i\alpha}\varphi$, where $e^{i\alpha}$ is an element of the group U(1); for $\mu^2 < 0$, $\varphi = |\varphi|\ e^{i\theta}$, and so this transformation takes one vacuum state into another. Choosing any one of these vacuua is spontaneous symmetry breaking.

In the electroweak theory the masses of the elementary particles depend on the vacuum magnitude of φ (which corresponds to the neutral component of a complex spinor; the set of minima of V is then actually a 3-sphere, and the circle of minima in Fig. 1 can be considered a circle on this sphere). Some relations are: $M_W = e\,|\varphi|/2\sin\theta_W$, $M_Z = e\,|\varphi|/2\sin\theta_W\cos\theta_W$, $m_e \simeq 2\times 10^{-6}\,|\varphi|$, $m_u \sim 2\times 10^{-5}\,|\varphi|$, $m_d \sim 4\times 10^{-5}\,|\varphi|$, and similarly for the other two generations of quarks and leptons. The mass of the W is determined from the Fermi constant (since the charged current interactions reduce to the Fermi theory at low energy), and the mass of the Z is determined from the relative strength of weak charged and neutral currents. So from experiment, v $\simeq$ 247 GeV, and $\sin^2\theta_W \simeq .22$. On the other hand, the mass of the Higgs boson, associated with oscillations in the potential well of V,

$$M_H^2 = 3\lambda\,|\varphi|^2 - \mu^2 \rightarrow \frac{4\lambda M_W^2 \sin^2\theta_W}{e^2} = \partial^2 V/\partial\varphi\partial\varphi^* \Big|_{|\varphi| = \mu^2/\sqrt{\lambda}}$$

depends on the unknown coupling parameter λ. The evolution of the universe depends dramatically on λ, or M_H.

The leading high temperature ($T \gg \mu$) contribution to F, neglecting for the moment the zero point energy, is,

$$-\,g\,\frac{\pi^2}{90}\,T^4 + \frac{e^2}{32\pi^2\sin^2\theta_W}\left(2 + \frac{1}{\cos^2\theta_W} + \frac{2}{3}\frac{m_t^2}{M_W^2}\right)T^2\,|\varphi|^2 + \frac{\lambda}{4}\left(|\varphi|^2 - \frac{\mu^2}{\lambda}\right)^2,$$

where g is the number of spin states of bosons plus 7/8 the number of spin states of fermions, and antifermions, and it is assumed that only the top quark mass is not much less than M_W. The coefficient of $|\varphi|^2$,

$$M_{eff}^2 = \frac{e^2}{32\pi^2\sin^2\theta_W}\left(2 + \frac{1}{\cos^2\theta_W} + \frac{m_t^2}{M_W^2}\right)T^2 - \frac{\mu^2}{2}$$

shows that for high temperature, $M_{eff}^2 > 0$, and there is a unique minimum for F at $|\varphi| = 0$, while for low temperature $M_{eff}^2 < 0$, and so the minimum of F is for $|\varphi| = v(T)$, giving mass to the W, Z, quarks and leptons. See Fig. 2a. The masses of these particles increase, all proportionally to v, as T decreases below T_c. See Fig. 2b. The continuous turn on of the masses is characteristic of a 2nd order phase transition. This electroweak phase transition is like that of a superconductor with respect to weak, but not electric, charge, where v is the counterpart to the gap. Since entropy is consersed in this phase transition, (no latent heat), the difference in minima of F for $|\varphi| = 0$ and $|\varphi| = v(T)$ corresponds to an energy difference between the high and low temperature vacuua. If the T = 0 vacuum with $|\varphi| = v$ has zero energy, the $T > T_c$ vacuum with $|\varphi| = 0$ has a finite energy density of $\mu^4/4\lambda = (\lambda/4)(2\sin\theta_W M_W/e)^4 \sim M_W^4$. This corresponds to a cosmological constant of $\sim (10^{11}\text{ eV})^4$, to be compared with the present energy density of the universe $\sim (5 \times 10^{-3}\text{ eV})^4$. This ratio of $\sim 6 \times 10^{-5}$ needs to be understood. In the 2nd order phase transition, this cosmological constant is small compared to the thermal energy, $(g\pi^2/30)T_c^4 >> \mu^4/4\lambda$, since $T_c^4 \sim (\pi/\alpha)^2 \mu^4$, and g ~ 100, and it is assumed that $\lambda \sim e^2$. The expansion rate therefore remains radiation dominated.

The neglect of the zero point energy is only justified if $\lambda >> e^4$. If $\lambda \sim e^4$, the zero point energy can qualitatively change the character of the potential, and consequently the phase transition. The free energy for $T >> T_c$ has a unique minimum at $|\varphi| = 0$, but as T is lowered a secondary set of local minima arise with $|\varphi| \neq 0$, that become the global minima for $T < T_c$. See Fig. 3a. The phase transition must now proceed by tunneling through an energy barrier over a finite region of space (bubble formation), and has an accompanying latent heat, $\Delta Q = T_c\Delta S$. The masses of the elementary particles now turn on discontinuously. See Fig. 3b. Because of the energy barrier there is supercooling below T_c as the Universe expands. The energy difference between the false and true vacuua acts as a cosmological constant; during the supercooling the false vacuum energy can come to dominate the thermal energy, and the universe can undergo an inflationary epoch.[4] With the completion of the phase transition the energy released reheats the universe to close to T_c, (which is much lower than for the 2nd order transition), restoring the radiation dominance of the energy density.

The zero point energy of the quantum field oscillators is divergent; while there is certainly a physical finite contribution, measured, for example, in the Lamb shift,[5] there are still infinite constants that must be subtracted. While in the usual context of quantum field theory the divergent vacuum energy does not affect physics, in the cosmological context it does. The cosmological constant is the term in the stress-energy tensor proportional to the metric tensor, and receives contributions from both V and the zero point energy,

$$\Lambda = V(\varphi) + \int d^3k/(2\pi)^3 \left(\sum_{\text{bosons}} \hbar\omega_k/2 - \sum_{\text{fermions}} \hbar\omega_k/2 \right) .$$

The divergent zero point energy has the following dependence on a momentum cutoff, $k_{max} = 1/a$, where a is an effective lattice spacing: for $k^2 \gg M^2$, expanding $\sqrt{k^2 + M^2} \sim k + M^2/2k - M^4/8k^3 + \cdots$, the divergent terms are of the form,

$$\int^{1/a} k^2 dk \sqrt{k^2 + M^2} \sim \frac{1}{4}\left(\frac{1}{a}\right)^4 + \frac{1}{4}\left(\frac{1}{a}\right)^2 M^2 - \frac{1}{8} M^4 \ln \frac{1}{a} .$$

The first term gives a cosmological constant $\Lambda = (g/8\pi^2)(1/a)^4$; this implies quantum vacuum fluctuations with wavelengths greater than 1/4 mm have as much energy as all the matter in the universe. This term must somehow be cancelled. The cutoff dependence of the other two terms can be absorbed into a renormalization of the parameters μ^2 and λ of V. That is, the magnitudes of the parameters μ^2 and λ depend on the cutoff scale 1/a. This implies the magnitude of the parameters at any distance scale depends on quantum fluctuations with all wavelengths less than that scale. The distance cutoff, a, should be chosen larger than that short distance scale for which new degrees of freedom not included in the strong and electroweak interactions become important; for example, the grand unified scale. However, this creates a fine tuning problem[6] in order to explain why $v = \mu/\sqrt{\lambda}$ (and thus the W, Z, quark and lepton masses) is not $\sim 1/a$. From the quadratically divergent part of the vacuum fluctuation energy, and for $M^2 \sim h\,|\varphi|^2$ where h is a coupling constant,

$$-\mu^2(M_W) = -\mu^2(\tfrac{1}{a}) + h \times (\tfrac{1}{a})^2 \quad ,$$

or

$$\frac{\mu^2(M_W)}{(1/a)^2} \sim \left(\frac{100\ \text{GeV}}{10^{15}\ \text{GeV}}\right)^2 = 10^{-26} = \frac{\mu^2(1/a)}{(1/a)^2} - h \quad .$$

The mass parameter μ^2 at scale $1/a$, which is sensitive to physics and distance scales smaller than a, must equal $h \times (1/a)^2$ to 1 part in 10^{-26}!

A solution to this divergent cosmological constant problem and the hierarchy problem may possibly come from supersymmetry:[7] For every polarization state of a fermion and it's associated anti-fermion there is a pair of bosons with the same quantum numbers. This fermion-boson pairing cancels the divergent parts. For exact supersymmetry, the vacuum energy is exactly zero; supersymmetry must, however, be spontaneously broken. This splits the masses of the paired fermions and bosons, and implies a finite vacuum energy. In supergravity theories, though, it is possible to spontaneously break supersymmetry with zero vacuum energy; in these theories supersymmetry becomes exact in the limit that Newton's constant goes to zero. If supersymmetry is spontaneously broken at the weak interaction scale, the hierarchy problem can be solved (if the fermionic partner of the Higgs field is massless, it can remain massless in the presence of vacuum fluctuation corrections since $\Delta m_f \propto m_f \ln 1/a$; the massless fermion protects the Higgs boson from acquiring mass as long as supersymmetry is unbroken), implying the masses of the super-partners of all the known particles are of order a coupling constant times that scale, or of order 10's of GeV.

We next consider the possible phase transitions of the strong interactions. Let us first discuss quark confinement, a property of the color field, which is a generalization of the electromagnetic field. For fixed quark sources, the vacuum at zero temperature does not allow the color electric field to spread. Color electric flux behaves like ordinary magnetic flux in real superconductors. If the temperature is raised, the superconducting ground state undergoes a phase transition to a normal phase where magnetic flux can spread. In the low temperature confining phase the interaction energy of static quark antiquark sources increases linearly with their separation. If heavy dynamical quarks are introduced into the field theory, for large enough separation of the sources (once there is enough energy in the field), quark-antiquark pairs can tunnel out of the vacuum to screen the quark sources. That is, the long flux tube breaks to form a pair of "mesons," with the quark and antiquark-source (and antiquark and quark-source) bound by a short flux tube.

The finite temperature phase transition in this theory[8] can be analyzed in terms of a "magnetization" that is non-zero at high temperature and zero at low temperature. To see this, first recognize that the partition function is the trace of the operator $\exp(-\beta H)$, which is like the time evolution operator $\exp[-(i/\hbar)tH]$ for imaginary time, $t = -i\hbar\beta$. The trace implies the initial states at $t = 0$ and final states at $t = -i\hbar\beta$ are equal. For the quantum gauge theory, this implies the vector potential is a periodic function of imaginary time, $A_\mu(\beta, \vec{x}) = A_\mu(o, \vec{x})$. The color field is described by a matrix vector potential in the Lie algebra of the group SU(3), $A_\mu = \sum A_\mu^a \lambda^a/2$, where $a = 1, \cdots, 8$, and where the λ^a are 3 x 3 traceless Hermitian matrices (the generalization of the Pauli spin matrices for SU(2)). The gauge transformations of electromagnetism,

$$A_\mu \to A_\mu + \partial_\mu \alpha = \exp(-i\alpha)\,(A_\mu - i\partial_\mu)\,\exp(i\alpha) \quad ,$$

where exp $(i\alpha)$ is a group element of U(1), is generalized to SU(3),

$$A_\mu \to \Omega^\dagger(t,\vec{x})\, A_\mu(t,\vec{x})\, \Omega(t,\vec{x}) - i\Omega^\dagger(t,\vec{x})\, \partial_\mu \Omega(t,\vec{x}) \quad ,$$

where $\Omega = \exp(i\alpha^a\lambda^a/2)$. Under this gauge transformation the following operator transforms,

$$T \exp\left[i \int_0^\beta A_0(t,\vec{x})\, dt\right] \to \Omega^\dagger(\beta,\vec{x})\, \left[T \exp i \int_0^\beta A_0(t,\vec{x})\, dt\right] \Omega(0,\vec{x}) \quad ,$$

where T refers to the (imaginary) time ordered product operation. The dynamics of the finite temperature theory is invariant not only under local gauge transformations that are strictly periodic but also under gauge transformations that are periodic up to a factor of a cube-root of unity, $\Omega(\beta,\vec{x}) = \exp(\pm 2\pi i/3)\,\Omega(0,\vec{x})$. The cube roots of unity are a sub-group of SU(3), $\{1, \exp \pm 2\pi i/3 \begin{pmatrix}1 & & \\ & 1 & \\ & & -2\end{pmatrix}\}$, and are the only SU(3) group elements that commute with all other elements (the "center" of the group). Thus, under this kind of gauge transformation,

$$M(\vec{x}) = \mathrm{tr}(T \exp i \int_0^\beta A_0(t,x)dt) \rightarrow \exp\{i(2\pi/3)\,n\}\, M(\vec{x}) \quad ,$$

where n is an integer.

The thermal expectation value of M has a simple physical meaning,

$$\langle M\rangle \propto \exp(-\beta F_{quark})$$

where F_{quark} is the free energy of the gauge field theory in the presence of a static quark source. Below a critical temperature this free energy is infinite, since the quark has an electric flux tube attached that goes off to infinity; consequently $\langle M\rangle = 0$ for $T < T_c$. Above T_c the quark free energy is finite, since it has a color coulomb field (the quarks and gluons carry different color charge, so the screening from the gluon plasma is incomplete); now $\langle M\rangle \neq 0$ for $T > T_c$.

The order of the phase transition that separates these phases is determined from an effective Hamiltonian for M.[9] As a consequence of the Z_3 transformation of the complex 3-D field $M(\vec{x})$, the effective Hamiltonian has the form,

$$H_{eff} = \int d^3x\, \{|\nabla M|^2 + \mu^2 |M|^2 + \lambda(|M|^2)^2 - a(M^3 + M^{*3})\} \quad .$$

For a = 0, H_{eff} is invariant under U(1) transformations, $M \rightarrow e^{i\alpha} M$; the cubic terms break this symmetry to Z_3, $M \rightarrow e^{\pm 2\pi i/3} M$. See Fig. 4. This effective field theory is a continuum approximation to a Z_3 magnet, (see Fig. 5) for which the spin at each lattice site can take on three orientations, the energy being lowered by parallel spins on neighboring sites. (The lattice spacing of this effective Z_3 magnet is of order a fermi.)

The free energy as a function of magnetization (in one of the three directions) is similar to that of Fig. 3a, with high and low temperature behavior interchanged. If $\mu^2 > 0$, V has a minimum for M = 0, and $\lambda |M|^4$ term insures (for $\lambda > 0$) the free energy increases for $M >> \mu$; the cubic terms, however, lower the energy for intermediate M and can produce a secondary minimum. Thus just from the symmetry of the order parameter effective field theory one can see from V, even without the fluctuation contributions that reinforce this mean field analysis, that there can be a 1st order phase transition. At high temperature a is large so the global minimum has $\langle M \rangle \neq 0$, and as the temperature is lowered, that minimum is raised until, for $T < T_c$ it becomes higher than the global minimum and true confining vacuum with $\langle M \rangle = 0$.

The strong interaction dynamics does not distinguish quark flavors. Ordinary protons are made of the 3 quarks, u u d, and neutrons are made of u d d; this independence of the strong interaction dynamics of flavor implies the charge independence of nuclear forces. This isospin symmetry of nuclear interactions is, however, only a subgroup of the flavor symmetry group of the strong interactions, chiral symmetry. The right and left handed massless quarks (recall their masses arise from the electroweak symmetry breaking - let us for the moment consider massless quarks) have independent dynamics because the strong interactions are gauge interactions. If the spinor of quark flavors is

$$\Psi = \begin{pmatrix} u \\ d \\ s \\ c \\ b \\ t \end{pmatrix},$$

the right and left chirality projections (with Dirac projections $\frac{1 \pm \gamma_5}{2}$) can be independently transformed into one another,

$$\Psi_R \rightarrow A\,\Psi_R$$

$$\Psi_L \rightarrow B\,\Psi_L ,$$

where A and B are independent (and spatially constant) $SU(N_f)$ matrices ($N_f = 6$ for the flavor spinor above). The strongly interacting particles do not fit into irreducible representatives of this large group $SU_L(N_f)$

x $SU_R(N_f)$, but only under an $SU(N_f)$ subgroup (isospin for $N_f = 2$, Gel-Mann SU(3) for $N_f = 3$, etc.). This is another example of spontaneous symmetry breaking, and the finite temperature phase transition in which the symmetry is spontaneously broken is potentially interesting for cosmology.

To analyze this phase transition, consider the generalization of "magnetization,"

$$(U^T)_{ij} = \langle \bar{\Psi}_{R_i} \Psi_{L_j} \rangle = \left\langle \begin{pmatrix} \bar{u}_R u_L & \bar{u}_R d_L & \cdots \\ \bar{d}_R u_L & \bar{d}_R d_L & \\ \vdots & & \end{pmatrix} \right\rangle$$

where the operator $\bar{\Psi}_R \Psi_L$ can create a quark-antiquark pair,

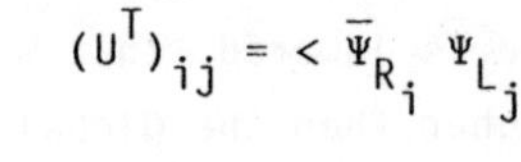

,

and the expectation value is with respect to the ground state (vacuum) of the theory. The matrix U transforms under chiral transformations of the quark fields

$$U \to B\, U\, A^\dagger \quad .$$

Since the strong interaction dynamics is invariant under this symmetry transformation, the effective potential for U must be built out of chiral group invariants,

$$V(U) = \mu^2 \,\mathrm{tr}\, U^\dagger U + \lambda \,\mathrm{tr}\, U^\dagger U U^\dagger U + \lambda'(\mathrm{tr}\, U^\dagger U)^2 - a(\det U + \det U^\dagger) + \ldots .$$

The first three terms are invariant under a larger group $U_L(N_f)$ x $U_R(N_f)$, and only the determinant terms explicitly reduce the symmetry to $SU_L(N_f)$ x $SU_R(N_f)$. For $N_f = 3$, for example, the determinant terms in the potential are cubic. For $\mu^2 > 0$, besides the local minimum for U = 0, these terms can produce an additional infinitely degenerate set of local minima. This infinite set of minima corresponding to the coset manifold $(SU_L(N_f) \times SU_R(N_f))/SU(N_f)$, are generally transformable into one another by $SU_L(N) \times SU_R(N_f)$ transformations, but any particular vacuum is invariant under $SU(N_f)$ (isospin, etc.).

At finite temperature V(U) has temperature dependent parameters. For a single direction in V(U), the free energy for different temperatures is like that of Fig. 3a. The temperature dependent coefficient a increases as $T \to 0$, eventually turning the secondary local minimum into a global minimum for $T < T_c$. This classical (mean field) analysis suggests the spontaneous chiral symmetry breaking phase transition can be 1st order. Fluctuations about this mean field analysis confirm the phase transition is 1st order and shows it is also 1st order for $N_f > 2$.[10] (Actually what is shown is there is no 2nd order transition fixed point in the ε-expansion; this kind of analysis, while not conclusive, always seems to give the correct answer.)

This analysis of the chiral phase transition assumed massless quarks. This would have been the case if the electroweak phase transition was 1st order, and the universe super-cooled below T_c^{chiral}. After the 1st order chiral transition, the electroweak phase transition would be completed,

$$\left\langle \begin{pmatrix} \bar{u}_R u_L \\ \bar{d}_R u_L \end{pmatrix} \right\rangle \sim \begin{pmatrix} \varphi^0 \\ \varphi^+ \end{pmatrix}$$

has the same quantum numbers as the Weinberg-Salam scalars, and would drive the electroweak transition.[3] From the release of electroweak vacuum energy, the universe would reheat above T_c^{chiral}; as the universe cooled again below T_c^{chiral}, there would only be 2 or 3 light quark flavors. The quark masses explicitly break chiral symmetry and induce linear terms in V(U),

$$t_r \left[\begin{pmatrix} m_u & & \\ & m_d & \\ & & m_s \end{pmatrix} U \right] + c.c. \quad ,$$

which tend to dump the well near the origin. Depending on presently unknown detailed magnitudes of parameters, the chiral transition may not be 1st order the second time around.

The axion[11,12] arises as a massless particle from a spontaneously broken U(1) symmetry, associated with the phase of a complex scalar field, $\phi = |\phi|\, e^{i\theta}$. This field is assumed to undergo a 2nd order phase transition at $T_c \sim 10^{10}$ GeV. In a 2nd order phase transition, like that of a ferromagnet, the field (magnetization) aligns over a finite region of space. The size of this region of space over which there is initially alignment (which then increases) is characterized by the Ginzberg length $\sim 1/\lambda\mu$, which is here $\sim 10^{-22}$ cm. At the intersections of these randomly oriented domains these can be trapped vortex defects.[13] The vortex is analogous to that in superfluid ^{4}He. In Fig. 6, at each point on the circle in space surrounding the vortex, the vacuum is different (the direction of the arrows refer to which minimum of V is chosen). This costs no energy from V, but there is also a term in the energy density, $|\nabla\phi|^2 = |\phi|^2 (\nabla\theta)^2 + (\nabla |\phi|)^2$. The circle in space can not be shrunk to a point keeping V at a minimum because the discontinuous change in minima ($|\phi| = f = \mu^2/\lambda$, different θ) would produce an infinity from the $f^2(\nabla\theta)^2$ term. So $|\phi| = 0$ at the center of the vortex. The intersection of the randomly oriented vacuua produce a random network of defects, with $|\phi| = 0$ along closed curves. These are axion cosmic strings. The distance over which $|\phi|$ goes from f to 0 (the core size) is $\sim 1/\mu \sim 10^{-24}$ cm.

In regions for which there is no net winding (around the circle of minima of V as one moves around a closed curve in space), the left over fluctuation in the θ-field around a particular vacuum is the axion, $a = \theta/f$, at this stage a massless particle - V = 0, and its energy goes to zero in the long wavelength limit ($\nabla\theta \to 0$).

At the electroweak phase transition the axion field changes character[14]; as the quark fields acquire mass, strong interaction (instanton) effects induce extra terms in the potential, giving the axion a mass. For $|\phi| = f$,

$$\mathcal{L} = \frac{1}{2} (\nabla a)^2 - f^2 \frac{m_a^2}{N^2} [1 - \cos(Na/f)] ,$$

where N is the (model dependent) number of quark flavors that couple to the axion, and m_a is the axion mass, proportional to the square root of the product of light quark masses. The axion mass is very temperature dependent and increases as $T \to 0$ to $m_a \sim f_\pi m_\pi/f$ or $m_a \sim 10^{-11} m_\pi \sim 10^{-3}$ ev. The set of minima of the potential is now discrete, $\theta = 2\pi/N \times$ integer. In regions of space for which θ was approximately constant before the electroweak phase transition, if θ was not in one of these minima, once the universe is old enough for the axion field to oscillate, $1/H(t) > \hbar/m_a(t) c^2$, there is a coherent classical axion field created[14] of initial amplitude $\theta = a/f \sim 1$, and consequently energy density $\sim m_a^2 f^2$. this energy density can now represent a large fraction of the energy density of the universe.

When the continuous U(1) symmetry of the axion field is reduced to the discrete Z_N the axion strings become bounded by domain walls.[15] Because of the discrete set of minima of the potential, there are 1-dimensional soliton excitations in which θ jumps by $2\pi/N$ as a domain wall of thickness $\sim \hbar/m_a c$ is crossed.

Finally, it is also possible that at the electroweak phase transition the axion strings may become superconducting wires.[16] The ordinary quarks and leptons, which are massive off the string, would be trapped as massless particles on the string. The weak interaction W and Z fields would also be massless on the string, and would mediate strong interactions on the string. The electromagnetic and color fields interact with the trapped quarks and leptons as external fields. The fascinating dynamics of axion strings, and their cosmological implications in the very early universe is still relatively unexplored.

EPOLOGUE

The title of this talk, with its implied time ordering, may have the wrong connotations. The baryon number violating processes of the electroweak theory, while of order e^{-137} at low temperature are potentially large at high temperature.[17] These baryon number violating processes could equilibrate, destroying the baryon number asymmetry generated after the grand unified phase transition. The baryon number asymmetry could then possibly be regenerated in a 1st order electroweak phase transition. It would therefore be possible for the baryon number asymmetry to be generated after the quark-hadron phase transition, at least the first time the universe went through it. Perhaps more likely, supersymmetric theories could generate a very large baryon number asymmetry[18] that would then be diluted by the electroweak baryon number violating processes. In any event, the existence of the theoretically expected particles that underlie some of these scenarios (Higgs particle, supersymmetric partners, top quark) will be experimentally tested in the foreseeable future. The results of these experiments will truly have cosmic implications.

ACKNOWLEDGMENTS

I would like to thank Grant Mathews for the opportunity to present this review.

This work was performed under the auspices of the U.S. Department of Energy by the Lawrence Livermore National Laboratory under contract number W-7405-ENG-48.

REFERENCES

1. See, for example, S. Weinberg, The First Three Minutes (1977) Bantum Books.
2. R. Peccei and H. Quinn, PRL 38, 1440 (1977).
3. E. Witten, Nucl. Phys. B 177 (1981) 477.
4. A. Guth, PRD 23 (1981) 347.
5. R. Feynman, Acta Physica Polonica, 697 (1963).
6. L. Susskind, Phys. Rep. 104 (1984) 181.
7. See, for example, H. Niles, Phys. Rep. 110 (1984) 1.
8. A. Polyakov, PL 72B (1978) 477, L. Susskind, PRD 20 (1979) 2610.
9. L. Yaffe and B. Svetitsky, PRD 26 (1982) 963.
10. R. Pisarski and F. Wilczek, PRD 29 (1984) 338.
11. S. Weinberg, PRL 40 (1978) 223, and F. Wilczek, PRL 40 (1978) 279.
12. M. Dine, W. Fishler, M. Srednicki, Phys. Lett. 140B (1981) 199.
13. T. Kibble, Phys. Rep. 67 (1980) 183.
14. J. Preskill, M. Wise, and F. Wilczek, PL 120B (1983) 127, L. Abbot and P. Sikivie, Phys. Lett. 120B (1983) 133, M. Dine and W. Fishler, Phys. Lett 120B (1983) 137.
15. P. Sikivie, PRL 48 (1982) 1156.
16. E. Witten, Nucl. Phys. B 249 557 (1985), C. Callan and J. Harvey, Nucl. Phys. B 250 (1985) 427.
17. V. Kuzmin, V. Rubakov, and M. Shaposhinikov, Phys. Lett 155B (1985) 36.
18. I. Affleck and M. Dine, Nucl. Phys. B 249 (1985) 361.

FIGURE CAPTIONS

Fig. 1 Classical potential for a complex scalar order parameter field for both $\mu^2 > 0$ (manifest vacuum symmetry) and $\mu^2 < 0$ (spontaneous symmetry breaking). Since $\varphi(\vec{x})$ is a field, a copy of this local potential sits at every point in space.

Fig. 2a Free energy as a function of order parameter. For $T > T_c$, $|\varphi| = 0$ minimizes the free energy; for $T < T_c$, $|\varphi| \neq 0$. The minimum for $T > T_c$ becomes a local maximum for $T < T_c$.

Fig. 2b The minimum $|\varphi| = v$ as a function of temperature for a 2nd order phase transition. v, and therefore all masses, turn on continuously as T falls below T_c.

Fig. 3a Free energy as a function of order parameter when vacuum fluctuation corrections are important. (For λ small, implying a light Higgs mass of ~ 10-20 GeV.) A local minimum with $|\varphi| \neq 0$ develops for $T > T_c$, and becomes the global minimum for $T < T_c$. $|\varphi| = 0$ remains a local minimum for $T < T_c$, requiring tunneling through the potential barrier.

Fig. 3b Discontinuous turn on of v after supercooling below T_c.

Fig. 4 Continuous set of minima of potential like that of Fig. 1 for $\mu^2 < 0$, become discrete due to cubic terms.

Fig. 5 Lattice spin system (Z_3 magnet) with same phase transition properties as SU(3) gauge theory.

Fig. 6 Spatial slice through cosmic axion string. Around circle in space, different minima of a potential like that of Fig. 1 for $\mu^2 < 0$, were chosen in domains as the 2nd order phase transition, as in Fig. 2, proceeded. The intersection of domains smoothed to lower the total energy. For domains with a net winding, a vortex was trapped.

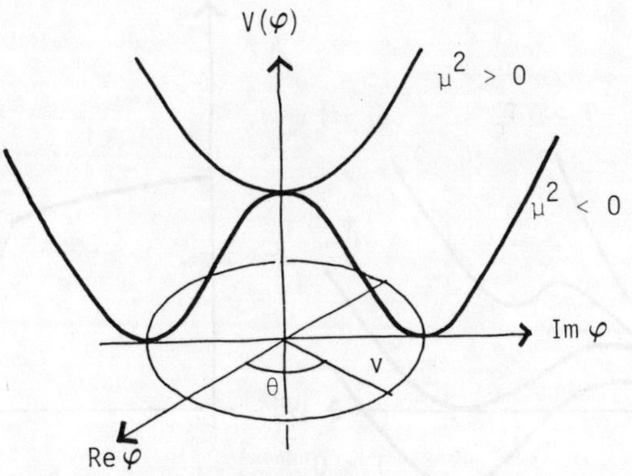

Fig. 1

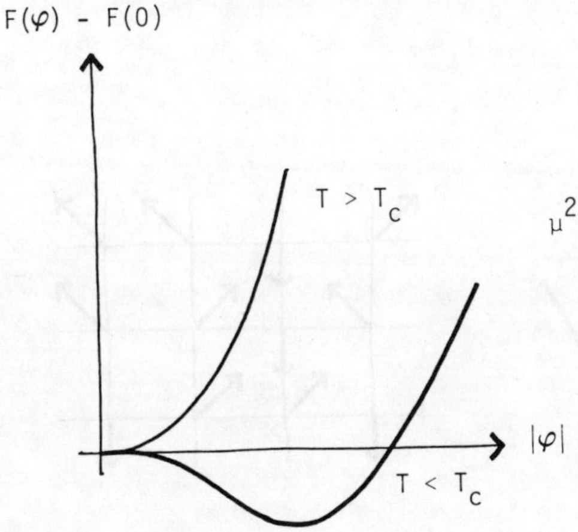

Fig. 2a

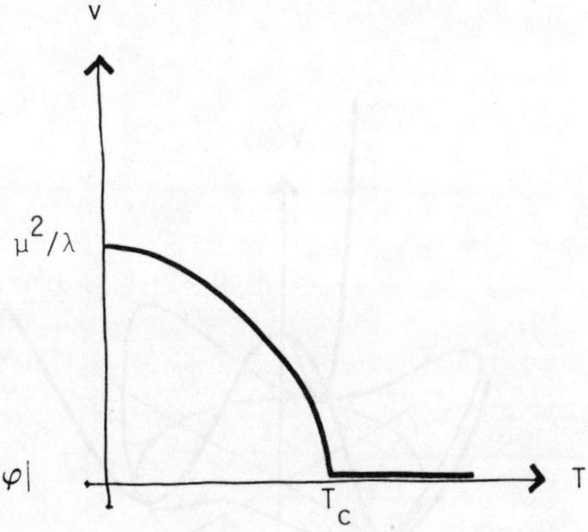

Fig. 2b

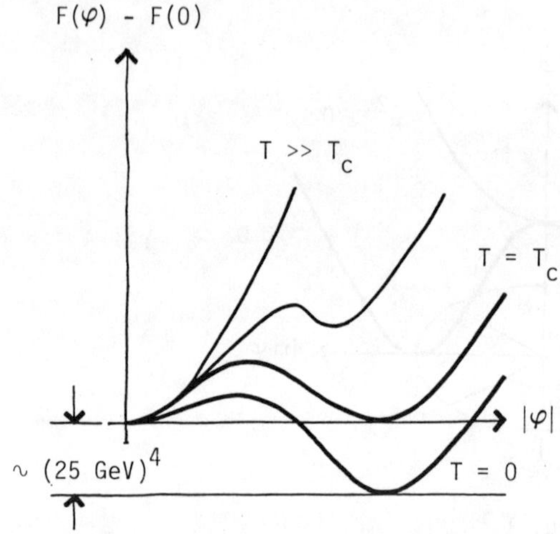

Fig. 3a

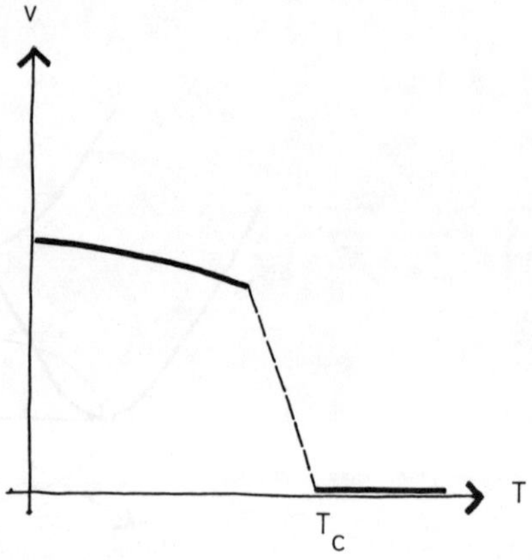

Fig. 3b

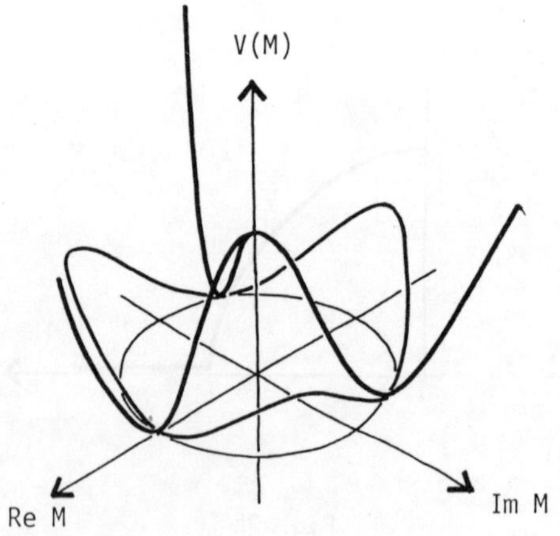

Fig. 4

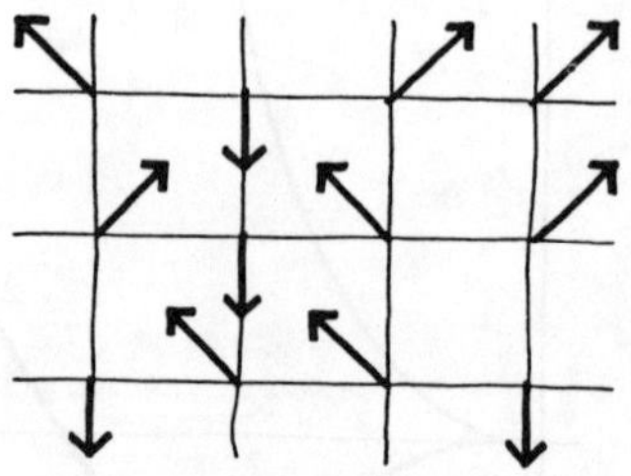

Fig. 5

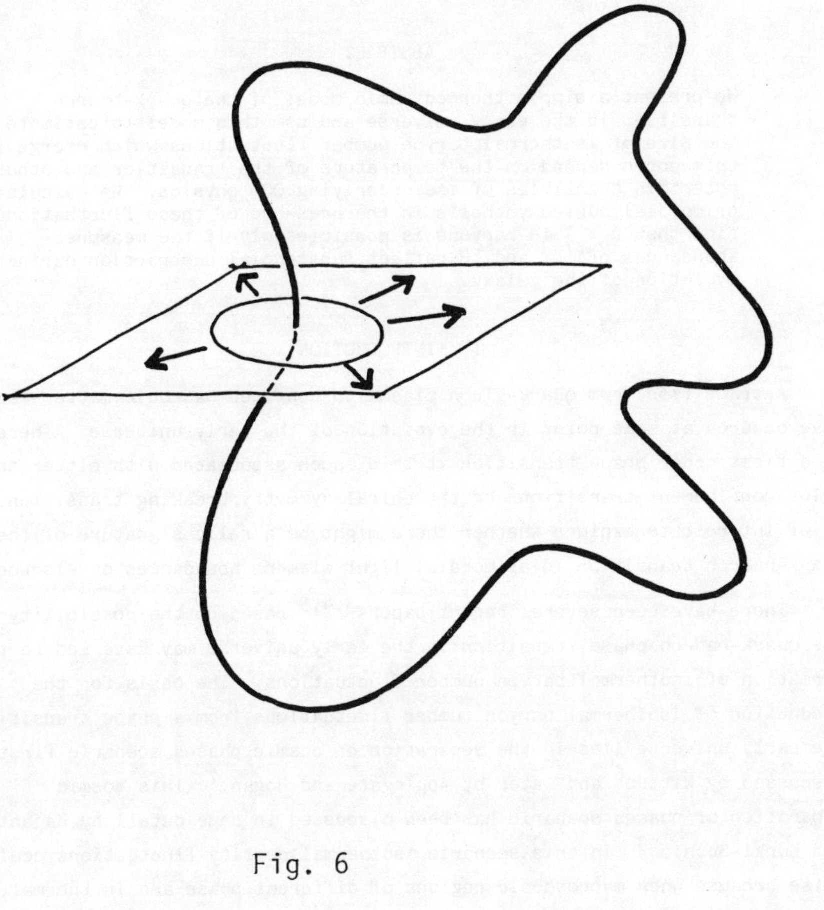

Fig. 6

DENSITY FLUCTUATIONS FROM THE QUARK-HADRON EPOCH AND PRIMORDIAL NUCLEOSYNTHESIS

G. M. Fuller, G. J. Mathews, and C. R. Alcock
Institute of Geophysics and Planetary Physics
University of California
Lawrence Livermore National Laboratory

ABSTRACT

We present a simple thermodynamic model of the quark-hadron transition in the early universe and use this model to estimate how the size of isothermal baryon number fluctuations which emerge from this epoch depend on the temperature of the transition and other uncertain quantities of the underlying QCD physics. We calculate primordial nucleosynthesis in the presence of these fluctuations and find that $\Omega = 1$ in baryons is possible only if the measured abundances of ^{7}Li and ^{2}H reflect substantial destruction during the evolution of the galaxy.

I. INTRODUCTION

A transition from quark-gluon plasma to confined hadronic matter must have occured at some point in the evolution of the early universe. There may be a first order phase transition at this epoch associated with either the color confinement transition[1] or the chiral symmetry breaking transition.[2] It is of interest to explore whether there might be a relic signature of the quark-hadron transition in primordial light element abundances or elsewhere.

There have been several recent papers[3,4,5] based on the possibility that the quark-hadron phase transition in the early universe may have led to the formation of isothermal baryon number fluctuations. The basis for the production of isothermal baryon number fluctuations from a phase transition in the early universe lies in the separation of cosmic phases scenario first discussed by Witten[6] and later by Applegate and Hogan.[7] This cosmic separation of phases scenario has been discussed in some detail by Kajantie and Kurki-Suonio.[9] In this scenario isothermal density fluctuations could arise because when macroscopic regions of different phase are in thermal, pressure and possibly, chemical equilibrium the quark-gluon plasma phase has a higher baryon concentration than does the hadron gas phase.

We will discuss the microphysics of the fluctuation generation mechanisms and we identify two scenarios. In the limit of chemical equilibrium between the two phases in coexistence there will be a different baryon concentration in each phase. In the limit where exchange of baryon number across the boundary between the two phases is not rapid enough to achieve chemical equilibrium, the fluctuations generated are always larger than in the equilibrium case, since nucleated confined vacuum contains no net initial baryon number. These scenarios for the generation of fluctuations depend on the phase transition being first-order so that latent heat is released.

The isothermal density fluctuations from the quark-hadron phase transition could be further modified at a later epoch by a diffusive separation of neutrons and protons, as pointed out by Applegate, Hogan, and Scherrer,[3] resulting in low density neutron rich regions and high density proton rich regions at the time of nucleosynthesis. The primordial nucleosynthesis resulting from a universe which is inhomogeneous in density and neutron-to-proton ratio is very different from that in the standard big bang and, in fact, may give acceptable light element abundances even with $\Omega = 1$ in baryons. This result is in contrast to previous studies of primordial nucleosynthesis with $\Omega = 1$ baryons in which not enough deuterium is produced.[10-14]

II. THERMODYNAMICS OF THE QUARK-HADRON PHASE TRANSITION

It will be most convenient for our purposes to compute the thermodynamic potential, Ω, for both the quark-gluon plasma phase and the hadron phase. We caution that we have used Ω for both the thermodynamic potential and the critical parameter for the universe. Our definition of Ω conforms with that in Landau and Lifshitz[15], so that

$$\Omega = - T \ln(Z), \qquad (1)$$

where Z is the grand partition function.

It is straightforward to compute the grand partition function, and thus Ω, if we assume that the particles are non-interacting except for an overall QCD vacuum energy in the unconfined phase. The background relativistic particles which are not strongly interacting and are in thermal equilibrium with both the confined and unconfined phases contribute (for $\mu = 0$)

$$\Omega = -V \left(g_b + \frac{7}{8} g_f\right) \frac{\pi^2}{90} T^4, \tag{2a}$$

where g_b and g_f are the statistical weights of bosons and fermions, respectively. At the epoch of the quark-hadron phase transition photons ($g_b = 2$), electrons ($g_f = 4$), muons ($g_f = 4$), and neutrinos ($g_f = 6$), yield $g = g_b + \frac{7}{8} g_f = 14.25$ so that the pressure contribution from these particles common to both phases is

$$P \simeq 1.56\ T^4. \tag{2b}$$

We treat the unconfined quark-gluon plasma as a gas of noninteracting relativistic particles plus an overall vacuum energy. In the limit of vanishing quark masses, there is a simple expression for Ω which is valid for any temperature and chemical potential:[16a-e]

$$\Omega_{qg} = \frac{-7\pi^2}{180} \cdot N_c N_f V\ T^4 \left[\ 1 + \frac{30}{7\pi^4} \left(\frac{\mu_q}{T}\right)^2 + \frac{15}{7\pi^4} \left(\frac{\mu_q}{T}\right)^4\ \right]$$
$$- \frac{\pi^2}{45} N_g V T^4 + BV \tag{3a}$$

Here N_c is the number of colors (3), N_f is the number of relativistic quark flavors (2 at lower temperatures corresponding to the u and d quarks, and 3 at higher temperatures where the strange quark becomes relativistic), and B is the QCD vacuum energy, or bag constant. The number of gluons is $N_g = 8$. The quark chemical potential is $\mu_q = \frac{1}{3} \mu_b$, where μ_b is the baryon chemical potential.

For the early universe $(\mu_b/T) \sim 10^{-8}$, so that for 2 relativistic quark flavors the contribution to the pressure in the unconfined phase from quarks and gluons is

$$P \approx 4.06\ T^4 - B \tag{3b}$$

The total pressure in the early universe in the unconfined phase includes the contributions from equations (2b) and (3b).

Note that the QCD vacuum energy contributes negatively to the pressure. The value of this vacuum energy, or bag constant, is not known. It parameterizes the temperature at which the confined and unconfined phases coexist. We emphasize that in what follows it will be this coexistence temperature T_c which will be the fundamental quantity of interest. The vacuum-energy (bag-model) approach is simply an easy and concise way to calculate the essential thermodynamics of the unconfined phase in a consistent manner.

We caution that for two reasons it is difficult to connect the value of the bag constant used in various models of hadrons with the relevant value of B in equations (3ab). First, most current research on modeling hadronic properties with the bag must address the problem of treating the bag surface.[17-22] Surface effects can be appreciable for individual hadrons. However, we expect the regions of unconfined quark gluon plasma which are in coexistence with confined hadronic matter to be of macroscopic size, so that surface effects are minimal. Second, it is dangerous to use a value of B in equations (3a) and (3b) that is derived from a particular model of hadrons because we are interested in the <u>high temperature</u> QCD vacuum energy. This vacuum energy probably comes down with increasing temperature.[18,20,22]

For the confined, or hadronic, phase we will not use the bag model but rather the spectrum[23] of known masses for baryons and mesons. This is not inconsistent with the bag model treatment of the quark-gluon phase because the isothermal baryon number fluctuations we are interested in are largest at low coexistence temperature where the neutron, proton, pion, and other light hadron masses (which do not depend on B) dominate the hadronic level density. At higher temperatures there may be an inconsistency and we discuss this limit below.

Computing Ω for mesons first (note that $\mu = 0$ for the mesons), we find for each meson that

$$\Omega = \frac{-gV\,T^4}{\pi^2} \sum_{n=1}^{\infty} \frac{1}{n^4}\, \bar{K}_2\,(nm/T), \tag{4a}$$

where m is the mass of the meson, and g is the spin plus isospin degeneracy factor, $g = (2J + 1)\,(2I + 1)$. The $\bar{K}_2$ is related to the modified Bessel function of the second order (K_2), as in Fowler and Hoyle,[24]

$$\bar{K}_2(x) \equiv \frac{x^2}{2} K_2(x) \quad (4b)$$

In the conditions we are interested in (T ~ 100-200 MeV) the pions will make the dominant contribution to the pressure in the hadronic phase; however, in the numerical calculations presented below we have summed the contributions to Ω in equation (4a) over all of the known mesons[23] and calculated all $\bar{K}_2$ numerically.

For baryons in the small chemical potential limit ($|\mu_b| < m$)

$$\Omega = \frac{-g\, T^4 V}{\pi^2} \sum_{n=1}^{\infty} \frac{(-1)^{n+1}}{n^4} \cosh\left(\frac{n\mu_b}{T}\right) \bar{K}_2\ (nm/T) \quad . \quad (5a)$$

The net baryon number density corresponding to this Ω is

$$n_b = \frac{2g\, T^3}{\pi^2} \sum_{n=1}^{\infty} \frac{(-1)^{n+1}}{n^3} \sinh\left(\frac{n\mu_b}{T}\right) \bar{K}_2(nm/T) \quad . \quad (5b)$$

Note that n_b is the number density of baryons minus that of the antibaryons. As above, we obtain the total Ω by summing over the spectrum of known baryonic masses.[23]

It is possible to make a simple analytic model of the essential statistical mechanics of the quark-hadron system and compare this to our numerical results. We define $a \equiv \pi^2/30$ and the statistical weight for the photons, neutrinos, electrons, and muons common to both phases as $g_{\nu\gamma}$ = 14.25. We then define the intrinsic statistical weight for the quark-gluon plasma as g_Q and that for the hadron soup as g_H. The total statistical weight for each phase in coexistence is then defined as

$$g_q \equiv g_Q + g_{\nu\gamma} \quad (6a)$$

$$g_h \equiv g_H + g_{\nu\gamma}. \quad (6b)$$

In the model of the quark-gluon plasma discussed above g_Q = 37 and thus $g_q \approx 51.25$. At T_c ~ 100 the pions dominate the contribution to Ω in the sum of equations (4a) and (5a), so that very roughly $g_H \approx 3$ and $g_h \approx 17.25$. The essential thermodynamics is contained in this difference in statistical weight between the phases.

The pressure, P_q, energy density, E_q, and entropy density, s_q, in the quark-gluon phase are given as

$$P_q = \frac{1}{3} g_q aT^4 - B \tag{7a}$$

$$s_q = \frac{4}{3} g_q aT^3, \tag{7b}$$

while for the hadron phase the corresponding quantities are

$$P_h = \frac{1}{3} g_h aT^4 \tag{8a}$$

We define the ratio of statistical weights in the two phases to be

$$x = \frac{g_q}{g_h} = \frac{s_q}{s_h} \tag{9}$$

Note that x in this paper is the same as r in ref. 9. Inclusion of the hadronic resonances makes g_h, and hence x, a function of temperature. Also, at high temperature, interaction corrections to g_h and x may be appreciable. In Figure 1 we present x as a function of temperature for the known mass spectrum of hadrons[23] and different numbers of relativistic quark flavors. At the lower temperature when baryon number fluctuations are expected to be largest, temperature and interaction corrections to x will be small.

The latent heat can be expressed as

$$L = T_c s_q \left(1 - \frac{1}{x}\right) \approx 4B, \tag{10}$$

where the latter approximation follows for constant x = 2.971.

Pressure equilibrium, or co-existence, between the phases, $P_h = P_q$, occurs for a temperature,

$$T_c = (g_q - g_h)^{-\frac{1}{4}} \left(\frac{3}{a}\right)^{\frac{1}{4}} B^{\frac{1}{4}} \approx 0.72\ B^{\frac{1}{4}}, \tag{11}$$

where the latter approximations assumes x = 2.971. Values of $B < (300\ \text{MeV})^4$ lead to a coexistence temperature, $T_c < 250$ MeV, in which the separation of

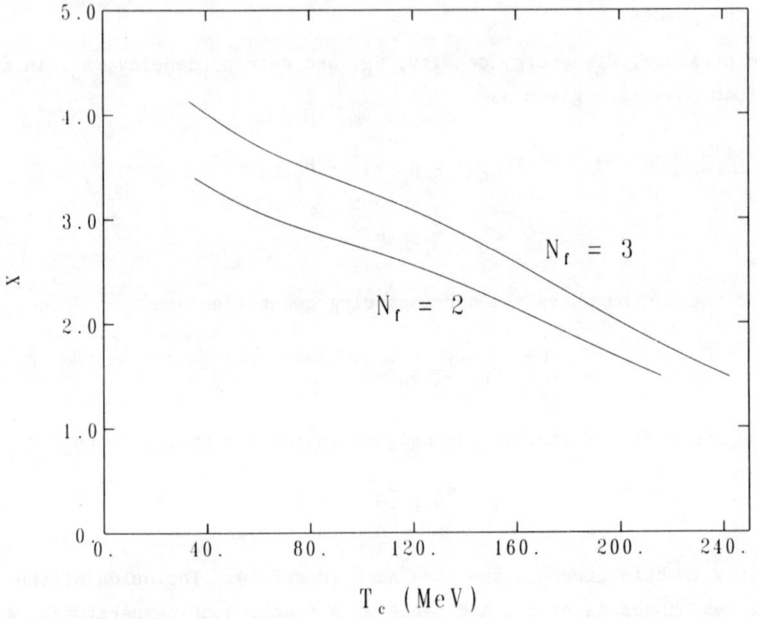

Figure 1) The ratio of the effective statistical weights of the unconfined phase to the confined phase, x, as a function of coexistence temperature, T_c. The contribution of all known hadronic resonances is included in the computation of the pressure equilibrium.

phases scenario outlined above could produce isothermal baryon number fluctuations of appreciable amplitude. If, on the other hand, the vacuum QCD energy is $B > (300\ \mathrm{MeV})^4$ then no coexistence and separation of phases is possible because the hadronic pressure always exceeds the quark-gluon plasma pressure.

Finally, even if the QCD vacuum energy is high, $B \sim 300$ MeV, it might be possible that the quark-gluon plasma phase is able to supercool well below T_c before nucleation of the hadronic phase begins. In this extreme case the QCD vacuum energy will come to dominate the pressure of the universe and a short de-Sitter exponential expansion of the universe can ensue. This miniinflation will, however, result in only a 15% increase in the scale factor.[5]

III. NUCLEATION

When a thermodynamic system is cooled (or heated) through the coexistence temperature at which a first order phase transition occurs, the new phase does not appear immediately. Significant reorganization of fundamental degrees of freedom is needed for the creation of new phase, and the thermodynamic fluctuations which can produce such reorganization have low probability. Furthermore, it is energetically unfavorable to create small volumes of new phase because of the free energy associated with the surface separating the two phases. These effects ensure that some supercooling below the coexistence temperature occurs before stable nuclei of the new phase are produced. These nuclei grow rapidly, since the old phase is metastable, releasing latent heat and thus reheating the system back to the coexistence temperature.

The physics of nucleation in the quark-hadron phase transition is less well understood than the bulk thermodynamics. However, there are three plausible mechanisms for this process. We call these mechanisms homogeneous nucleation, heterogeneous isothermal nucleation and heterogeneous non-isothermal nucleation. In what follows we will discuss only the most likely mechanism, homogeneous nucleation.[15]

If the universe is pure and isothermal, then the new phase originates through spontaneous fluctuations in the metastable phase: The nucleation rate is determined by the probability that a spontaneous fluctuation in the metastable (quark) phase will produce a "critical nucleus" of the stable (hadron) phase. This critical nucleus has a radius, r_c, determined by

$$P_h - P_q = \frac{2\sigma}{r_c} , \qquad (12)$$

where σ is the free energy per unit surface area associated with the boundary of the nucleus, which is assumed to be spherical.

New nuclei with radii less than r_c will collapse and disappear, while nuclei of radii larger than r_c will expand until a macroscopic amount of new phase is produced. The probability of a fluctuation of radius r_c is exp (-W/T) where

$$W = \frac{4\pi}{3} r_c^{\,3} \; (P_q - P_h) + 4\pi\sigma \, r_c^{\,2}. \tag{13}$$

The first term in (14) is the difference between the thermodynamic potentials of the two phases and is negative. The second term is the surface free energy of the boundary between the phases.

The rate, p(T), at which nuclei form per unit volume per unit time is given by[15]

$$p(T) = CT_c^{\,4} \; \exp\,(-W/T), \tag{14}$$

where C is a coefficient of order unity, which turns out (see below) to have no quantitative significance when we calculate the number density of nucleation sites.

The amount of supercooling is very small. Accordingly, it is convenient to expand about the coexistence temperature, T_c. We define

$$P_h - P_q = L\eta, \tag{15a}$$

$$\eta \equiv \frac{T_c - T}{T_c}, \tag{15b}$$

where L is the latent heat per unit volume of the phase transition and η is the "supercooling parameter." It follows that:

$$p(\eta) \approx CT_c^{\,4} \; \exp\,(-\frac{16\pi}{3} \frac{\sigma^3}{T_c L^2 \eta^2}). \tag{16}$$

The important characteristics of Eq. (16) are that $p(T) \to 0$ as $T \to T_c$, and that for small supercooling ($\eta \ll 1$) the argument of the expontential is large and negative, yielding a very strong increase of p with η. This means that almost all nucleation occurs at the lowest temperature achieved during the supercooling phase (i.e. just prior to reheating). Furthermore, metastable regions are reheated by the release of latent heat around individual nucleation sites, and no appreciable nucleation occurs in the reheated regions.

The important parameter for the purpose of primordial nucleosynthesis will be the mean comoving distance between nucleation sites. We estimate this distance from Eq. (16) for a given nucleation rate, using an analysis similar to that of Kajantie and Kurki-Suonio[9]. In this analysis, each expanding volume of hadronic phase acts as a piston driving a weak shock wave which expands into the quark phase at just above the sound speed $V_s \approx 3^{-\frac{1}{2}}$. These shocks reheat the quark material. No further nucleation will occur once a region has been crossed by one or more shocks. Then it follows that the fraction of the universe which is unaffected by nucleation during this period is:

$$f(t) = \exp - \int_{t_c}^{t} f(t')\, p(T')\, dt' \frac{4\pi}{3} V_s^3 \left(\frac{T}{T'}\right)^3 (t - t')^3 \tag{17}$$

The total number of nucleation sites is

$$N_n = \int_{t_c}^{\infty} f(t)\, p(T)\, dt \tag{18}$$

In these equations t is time, t_c is the time at which the universe first cools through $T = T_c$, T is the temperature at time t and T' the temperature at time t'.
An approximate (±10%) solution of for the nucleation parameter is,

$$\eta_f \approx 1.4 \frac{\sigma^{3/2}}{T_c^{1/2} L} . \tag{19}$$

The number density of nucleation sites is given by

$$N_n t^3 = \pi^2 \left(\frac{8}{3} \cdot \frac{\sigma^3}{V_s T_c L^2 \eta_f^3}\right)^3 . \tag{20}$$

The mean separation, ℓ , between nucelation sites is

$$\ell \equiv N_n^{-\frac{1}{3}} \approx 0.8 \frac{\sigma^{\frac{3}{2}} t}{T_c^{\frac{1}{2}} L} \tag{21a}$$

$$\ell \approx (5.43 \times 10^6 m) \frac{\sigma^{3/2}}{T_c^{13/2}}, \tag{21b}$$

where the second equation follows from the thermodynamic model of the last section. This length scale should be greater than the comoving proton diffusion length in order that the fluctuations not damp out entirely before the onset of nucleosynthesis. In addition, should this length turn out to be greater than the comoving neutron diffusion length, then the approximate treatment of diffusion and nucleosynthesis that is described in section VI is inadequate; however, there will still be significant modification of the nucleosynthesis yield and our broad conclusion will be qualitatively correct.

This length scale depends on two bulk properties of the phase transition (T_c and L) that may be calculated in terms of our model equations. In addition, this length scale depends on the surface tension, σ , for which we have no model calculation. Following Farhi and Jaffe[25] and Alcock and Farhi,[26] we expect the "intrinsic" contribution to σ to be small (i.e. $\sigma_I^{1/3} \ll B^{1/4}$) and the "dynamical" contribution to σ to be $\sigma_D \leq (70\ \text{MeV})^3$. If σ is near this upper limit, the length scale is long compared to the neutron diffusion length, which is ~ 30m.[3]

More important for our analysis would be a lower bound to σ since this would establish whether or not proton diffusion is important. The only strict lower bound is $\sigma > 0$, since $\sigma \leq 0$ would not result in separated phases. It is unlikely that σ is very small compared to the relative contributions of the bulk free energies of the phases, but this possibility cannot be excluded.

IV. DYNAMICS OF THE UNIVERSE DURING THE CONSTANT TEMPERATURE COEXISTENCE EPOCH

The last section described the supercooling/nucleation scenario in which the quark-gluon plasma phase is separated from the hadron phase. The duration of this nucleation epoch is short compared to the Hubble time since the fractional supercooling is small ($\eta \sim 10^{-3}$). Again, because the duration of this epoch is short, the entropy generation associated with reheating to T_c is expected to be small compared to the initial entropy. At the end of this nucleation epoch the universe is left with bubbles of hadron phase surrounded by quark gluon plasma, all in pressure equilibrium at T_c. Since the hadron bubbles were nucleated through random thermal or quantum processes they contain, on average, no net baryon number; all of the net baryon number resides in the quark-gluon plasma phase.

The subsequent evolution of this configuration of phases takes place at constant temperature, T_c.[6-9] The hadron bubbles grow with time and the fraction of the total volume of the universe which is in quark-gluon plasma, f_v, decreases from $f_v = 1.0$ to $f_v = 0$ at the end of the phase transition. The universe remains at constant temperature by trading volume from the unconfined vacuum to the confined vacuum, and thus releasing the latent heat. We summarize the dynamics and specialize our discussion to baryon-number transport in what follows.

If we denote the scale factor in the Robertson-Walker metric by A, then the Einstein equation and energy-momentum conservation can be used to solve for A(t).

$$\frac{A(t)}{A_i} = (4x)^{1/3} \left[\cos\left(\frac{3\chi\,(t - t_i)}{2\,(x-1)^{\frac{1}{2}}} + \arccos \frac{1}{2x^{\frac{1}{2}}}\right)\right]^{2/3} , \qquad (22a)$$

where the beginning of the constant-temperature epoch (end of the nucleation epoch) is taken at time t_i, corresponding to a scale factor $A(t_i) = A_i$. Similarly it can be shown that the volume fraction of the universe in quark-gluon plasma is

$$f_v = \frac{1}{4\,(x-1)} \left[\tan^2 \left(\arctan\,(4x - 1)^{\frac{1}{2}} + \frac{3\chi\,(t_i - t)}{2\,(x - 1)^{\frac{1}{2}}}\right) - 3 \right] \qquad (22b)$$

where we have assumed that $f_v \approx 1$ at the end of the nucleation phase. This approximation will be adequate for our purpose of following the baryon number transport properties.

The total proper surface area, Ξ, of the domain wall which separates the confined and unconfined phases is

$$\Xi \approx 4\pi\, r^2\, N_n\, V_o \qquad (23a)$$

where N_n ($\sim 10^{-9}$ cm^{-3}) is the number of nucleated sites per unit volume from equation (20), V_o is the proper volume of the universe (horizon volume) at the end of the nucleation epoch and r is taken to be a typical radius of a hadron bubble. Since we have approximated $f_v = 1$ at the beginning of the constant temperature epoch we must take $r = 0$. As f_v drops from 1.0 to 0.5 the radius of a typical hadron bubble increases from 0 to some maximum value r_{max}. At this point, $f_v = 0.5$, the bubbles of hadron phase percolate with the quark gluon plasma. As f_v drops from 0.5 to 0 we interpret r as the radius of a shrinking bubble of quark-gluon plasma. The assumption of spherical bubbles is strictly correct only near the beginning and the end of the phase transition where the bubble surface energy dominates the volume energy. Nevertheless, the typical bubble radius remains a useful parameterization of the total surface area in the boundary separating the phases, even if the actual surface area is larger than that given in equation (23a) because of bubble nonsphericity. During the phase transition the proper bubble radius is

$$r(t) \approx \left(\frac{4}{3}\pi N_n\right)^{-1/3} \cdot \frac{A(t)}{A_i} \cdot \tilde{f}_v \qquad (23b)$$

where $\tilde{f}_v$ is defined by

$$\tilde{f}_v = \begin{cases} 1 - f_v & \text{if } f_v \le 0.5 \\ f_v & \text{if } f_v \ge 0.5 \end{cases} \qquad (23c)$$

The mechanism of heat transport will be discussed briefly below as regards fluctuation size. We note however, that hydrodynamic heat transport is favored when the supercooling is small[7]. The neutrino mean free path is of the order ~ 1 m, so that a neutrino can random walk ~ 1 km, or 10% of the horizon distance, in a Hubble time. On the other hand, the nucleon mean free path is ~ 10^{-11} cm (where the density of mesonic and baryonic states has been estimated using equations 4a through 5a) so that a nucleon moves only ~ 10^{-3} cm in a Hubble time. In the simple model of noninteracting quarks and gluons presented here we expect the quark mean free path to be roughly comparable to that of the neutrino.

V. BARYON NUMBER TRANSPORT ACROSS THE PHASE BOUNDARY AND CHEMICAL EQUILIBRIUM

The coexistence of the two phases across a boundary is shown schematically in figure 2. Note that the latent heat transport is by the motion of the boundary wall, converting volume from one vacuum to another. This vacuum energy difference can drive particle creation. Neutrinos or photons could carry heat or entropy across the phase boundary, but baryon number is not thermally created in the hadron phase, and thus must actually flow across the boundary if the hadron phase is to have any net baryon number

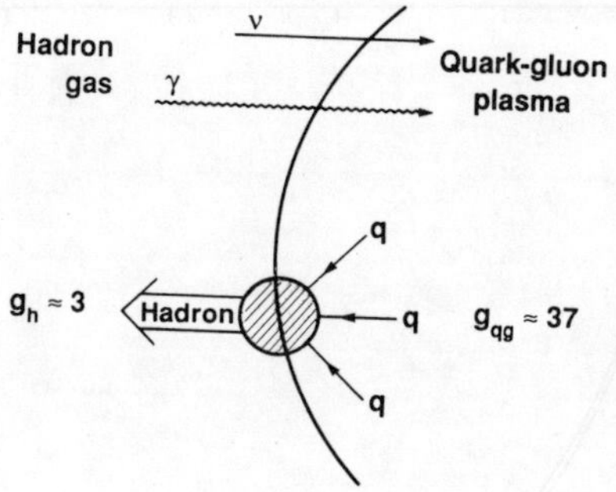

Figure 2) A schematic illustration of a phase boundary separating confined vacuum on the left from unconfined vacuum on the right. The approximate intrinsic statistical weight of relativistic degrees of freedom is shown for each phase. Entropy can be readily carried across the wall by photons, neutrinos, and other light leptons. Baryon number is transported across the boundary only by strong interaction processes. The indicated processes are reversible.

at all. Any transport of baryon number across the phase boundary must be due to strong interaction processes. We therefore consider two limits to the efficiency of baryon number transport. The first is that the baryon number is rapidly and efficiently transported across the front to achieve chemical equilibrium; and the second involves an inefficient transport process in which chemical equilibrium is not necessarily achieved on the timescale of the phase-coexistence evolution. We will show that in both of these limits isothermal baryon number fluctuations can result.

If we define,

$$R \equiv (n_b^q / n_b^h), \tag{24}$$

where the ratio of baryon number densities (n_b^q and n_b^h are the baryon densities in the quark-gluon plasma and hadron gas respectively) is to be taken immediately after decoupling of the bubbles of quark-gluon plasma, then R is a measure of the amplitude of the isothermal baryon number fluctuations. In the limit of chemical equilibrium across the front, the baryon chemical potentials

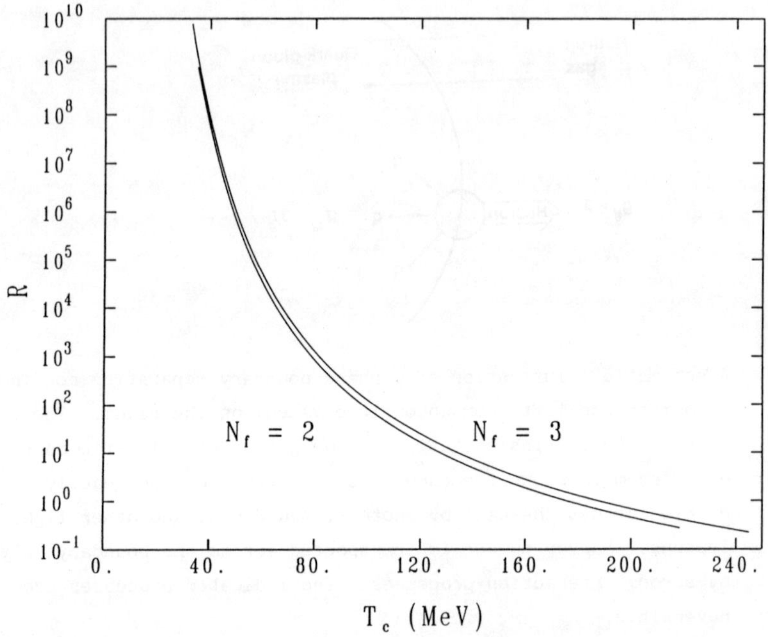

Figure 3) The equilibrium baryon number density ratio, R, as a function of coexistence temperature, T_c. Curves are given for 2 and 3 relativistic quark flavors.

are equal in the two phases and

$$R \approx \frac{2}{9} \left(\frac{\pi^3}{8}\right)^{\frac{1}{2}} \left(\frac{T_c}{m}\right)^{3/2} e^{\,m/T_c} \tag{25}$$

The sum over baryonic resonances expected to be in equilibrium in the hadron soup tends to decrease the value R by providing a higher statistical weight for baryon number in the hadron phase. This effect is smallest at low temperature where R will be largest. Figure 3 gives this equilibrium amplitude R as a function of T_c, including the hadronic resonances. As we showed in our previous paper, when $T_c < 125$ MeV, $R \geq 20$ and these flucuations will have a significant influence on nucleosynthesis. The difference in baryon number concentration in the two phases results from the difference in the mass of the particles which carry baryon number: nonrelativistic neutrons, protons, and other baryons on the one hand; and relativistic quarks on the other.

The amplitudes of isothermal baryon number fluctuations generated in the chemical equilibrium case represent, to some extent, lower limits when the actual baryon number transport processes are considered. The baryon content of the universe is initially in the shrinking bubbles of the high-temperature phase, so that unless the baryon transport rate is efficient, chemical equilibrium will never be reached.

The baryon number density in each phase can be followed in time as baryon number moves across the phase boundary. If n_b^h and n_b^q are the net baryon number densities in the hadron and quark-gluon plasma phases, respectively, then

$$\dot{n}_b^q = -n_b^q \lambda_q + n_b^h \lambda_h - n_b^q (\frac{\dot{V}}{V}) \tag{26a}$$

$$\dot{n}_b^h = + n_b^q \lambda_q - n_b^h \lambda_h - n_q^h (\frac{\dot{V}}{V}) \tag{26b}$$

where $\dot{n}_b^q$ and $\dot{n}_b^h$ are the time rates change of baryon number density in each phase, λ_q and λ_h are characteristic baryon number transfer rates from quark-gluon plasma to hadron phase and the reverse, respectively. The last terms in equations (26a) and (26b) are volume redshift factors. The total baryon number per comoving volume is constant at N_o,

$$\frac{N_o}{V} = n_b^q f_v + n_b^h (1 - f_v). \tag{27}$$

The total baryon number swept up by the wall in the quark-gluon phase, and pushed through to the hadron phase, divided by the total volume in the quark phase is

$$n_b^q \lambda_q = 4\pi r^2 \cdot N_n \cdot \frac{V_o}{f_v V} \cdot V_f \cdot F, \tag{28a}$$

where r is a typical bubble radius, N_n is as in equation (26a), V_o is the horizon volume at the end of the nucleation epoch, and F is the fraction of baryon number passed by the wall. It can be shown from simple phases space arguments that

$$F \approx (413)\ \Sigma_h\ (\frac{MeV}{T})\ \exp\ (-\frac{938\ MeV}{T})\ v_f^{-1} \tag{28b}$$

where v_f is the speed of the wall in units of c and Σ_h is the probability that the three quarks in a proton, neutron or other bound hadron will pass through the wall into the free quark-gluon plasma regime. Detailed balance can be used to show that probability for 3 quarks to recombine at the wall, Σ_q, is related to Σ_h by

$$\Sigma_q \approx (5.4 \times 10^6)\ (\frac{T}{100\ MeV})^{-10}\ \exp\ (\frac{-938\ MeV}{T})\ \Sigma_h. \tag{28c}$$

Using the above we can show that,

$$\lambda_q \simeq (1.24 \times 10^{13}\ s^{-1})\ (\frac{MeV}{T_c})\ \exp(\frac{-938\ MeV}{T_c})\ (\frac{cm}{r})\ .\ \Sigma_h. \tag{28d}$$

By reasoning similar to that employed above it can be shown that

$$n_b^h\ \lambda_h = \frac{1}{3}\ n_b^h\ V_b.\ (\frac{4\pi r^2\ .\ N_n\ .\ V_o}{f_v V})\ .\ \Sigma_h\ , \tag{29a}$$

where V_b is a typical thermal velocity for a nucleon, $V_b \approx (3T/m)^{\frac{1}{2}}$, from which we derive,

$$\lambda_h \approx (1.70 \times 10^9\ s^{-1})\ (\frac{T_c}{MeV})^{\frac{1}{2}}\ (\frac{cm}{r})\ .\ \Sigma_h\ . \tag{29b}$$

The first order differential equations for n_b^q and n_b^h are easily integrated. In the limit of a static universe, $\frac{\dot{V}}{V} = 0$, which corresponds to a static phase boundary, after a sufficiently long time $t >> \lambda_h^{-1}$ we obtain the equilibrium baryon number density ratio,

$$(\frac{n_b^q}{n_b^h})_{eq.} = \frac{\lambda_h}{\lambda_q}\quad , \tag{30}$$

which is the same ratio given in equation (28). In the expanding universe case, figure 4a and 4b present the ratio $R = n_b^q / n_b^h$ as a function of time from the onset of the constant temperature epoch to the end of the phase transition. Figure 4a corresponds to T_c = 100 MeV, while figures 4b corresponds to T_c = 200 MeV. The baryon number density ratio, R, is computed at each T_c for several values of the nucleon transmission probability: Σ_h = 1,

10^{-2}, 10^{-3}, 10^{-4}, 10^{-5}, and 10^{-6}. We note that for $\Sigma_h = 1$, baryon number chemical equilibrium is rapidly established, with R being driven to the equilibrium value on a time scale short compared to either the Hubble time or the duration of the constant temperature epoch. For $\Sigma_h << 1$ chemical equilibrium is not obtained and R can be much larger than R_{eq}.

The appropriate value of the baryon transmission probability, Σ_h, depends on the nature of the phase boundary separating the confined vacuum from the unconfined vacuum. Any calculation of Σ_h, therefore, will be highly model dependent.

The basis for isothermal baryon number fluctuation generation lies in the difference in baryon number concentration between the unconfined and confined phases; $R = n_b^q/n_b^h$. If equilibrium is not established across the phase boundary then R is larger than the equilibrium value. The amplitude of the isothermal baryon number fluctuations has been considered in our previous paper[5] and was shown to be order R.

VI. BARYON DENSITY FLUCTUATIONS AND PRIMORDIAL NUCLEOSYNTHESIS

The isothermal baryon number density fluctuations discussed in this paper can have interesting effects on primordial nucleosynthesis. There are two influences which must be considered. One is the fact that more than one baryon to photon ratio must be averaged to produce the final nucleosynthetic yield. Such an averaging by itself can significantly alter the yields from primordial nucleosynthesis.[4] The second intriguing effect[3] is that the neutron and proton components of these density fluctuations will diffuse differently after the weak reactions fall out of equilibrium. Essentially, the neutron collision mean free path is much larger (by more than an order of magnitude) than the proton mean free path since the neutron only interacts with the background plasma via nucleon scattering or via the small neutron dipole momment. The diffusion of protons, on the other hand, is slowed by the more dominant proton-electron scattering. This diffusion of neutrons leads to a filling of the low density voids between the high density fluctuations with neutron-rich material. The high-density regions are correspondingly proton-rich relative to standard big bang nucleosynthesis.[10]

In this paper we extend our previous work[5] to consider the effects of

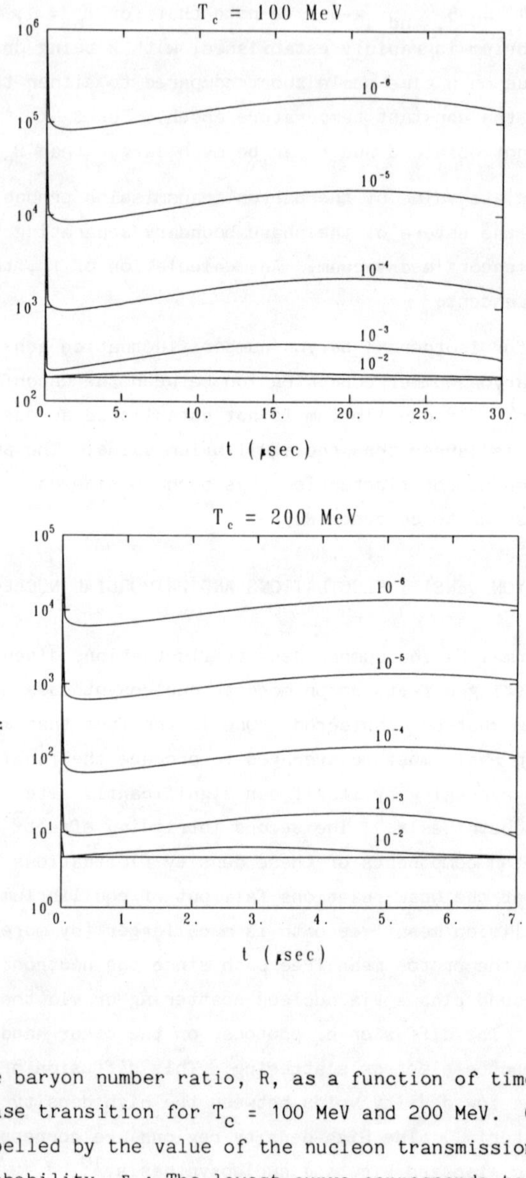

Figure 4) The baryon number ratio, R, as a function of time during the phase transition for T_c = 100 MeV and 200 MeV. Curves are labelled by the value of the nucleon transmission probability, Σ_h; The lowest curve corresponds to Σ_h = 1.

these baryon inhomogeneties in a universe with $\Omega \neq 1$, and as a function of the ratio of the baryon densities in the two phases. Our aim is to delimit the parameter space of this model which is consistent with the constraints from light element abundances.

For the purposes of this parameter study we utilize the two-zone complete diffusion scenario discussed in ref. 5 which reduces the parameter space to three quantities, the total average baryon denisty, Ω_b, the ratio of densities in the two regions, R, and the volume fraction, f_v, in the high-denisty regions just before nucleosynthesis.

We have utilized the big-bang nucleosynthesis code of Wagoner[10] with three neutrino flavors, a neutron half life of 10.6 min., and a number of nuclear reactions updated from the original version. We begin the nucleosynthesis calculations with relative baryon densities in the high density and low density regions (Ω_h or Ω_l) as a function of f_v and average Ω.

$$\Omega_h = \frac{\Omega\, R}{f_v\,(R-1) + 1} \tag{31a}$$

$$\Omega_l = \Omega_h / R \tag{31b}$$

We assume that the neutron diffusion occurs between the freezout of the weak reactions (at $T \approx 1.3 \times 10^9$ K, 0.11 MeV) and the onset of nucleosynthesis (at $T \approx 0.9 \times 10^9$ K) about 100 sec later. Note that by weak "freezeout" we mean that temperature at which free neutron decay dominates all other $n \leftrightarrow p$ reactions. The weak interaction drops out of equilibrium at a much higher temperature, $T \leq 1$ MeV, when typical weak rates become small compared to the universal expansion rate. As nucleosynthesis begins after neutron diffusion, the universe consists of a high baryon density proton-rich region, (1) with

$$\Omega^{(1)} = X_n + \Omega_h\,[1 - X_n], \tag{32}$$

$$X_n^{(1)} = X_n / \Omega^{(1)}, \tag{33}$$

where X_n is the neutron mass fraction before nucleon diffusion at $T = 1.3 \times 10^9$K, and a low baryon density neutron-rich region, 2, with

$$\Omega^{(2)} = X_n + \Omega_h [1 - X_n] \quad (34)$$

$$X_n^{(2)} = X_n/\Omega^{(2)} \quad (35)$$

The final averaged mass fractions for each nuclide are then,

$$\bar{X}_i = \frac{f_v X_i^{(1)} \Omega^{(1)} + (1 - f_v) X_i^{(2)} \Omega^{(2)}}{f_v \Omega^{(1)} + (1 - f_v) \Omega^{(2)}} \quad (36)$$

In the standard big bang, at high baryon densities, essentially all of the available neutrons are quickly converted into ^{4}He at t ~ 100 sec. In the low-density regions, the neutron mass fraction is set to close to unity when the baryon diffusion is turned on at t ~50 sec. It is then the proton mass fraction which is absorbed to make ^{4}He. Further ^{4}He production must await the decay of neutrons into protons. Thus, there is a significant neutron mass fraction until much later times. ^{2}H is therefore produced at lower densities and survives further nuclear reactions. Since ^{3}H, ^{3}He and ^{4}He are large, there is also significant production of ^{7}Li in these regions.

In figures 5a-b we show averaged nucleosynthesis yields and nucleosynthesis in the proton-rich and neutron-rich regions for Ω = 0.1 and 1.0 when R = 50. This corresponds to the ratio of baryon densities for chemical equilibrium when the coexistence temperature is 110 MeV with three relativistic quark flavors, or when T_c >110 MeV and $\Sigma_h < 1$. The intriguing feature of these results is that, for Ω_b =1, over a broad range for the parameter f_v, the main abundance constraints from primordial nucleosynthesis (i.e. ^{4}He, and ^{2}H) can be satisified. There is, perhaps, a slight under production of ^{3}He and a larger over production of ^{7}Li. This overproduction of ^{7}Li is due to the fact that ^{7}Li can be produced in significant amounts in both the neutron-rich and proton-rich regions. However, this overabundance may not be too significant due to the uncertainities[27] in the ^{7}Li abundance, the nuclear reaction rates, and stellar destruction of ^{7}Li. The overabundance appears to be a consequence of the large contribution from the ^{3}He $(\alpha, \nu)^7$ Be reaction in the high baryon-density regions and the ^{3}H$(\alpha,\gamma)^7$Li reaction in the low baryon density regions. However, diffusion during nucleosynthesis may deplete ^{7}Li.[28]

In figures 5a,b there is little dependance of the average ^{7}Li mass

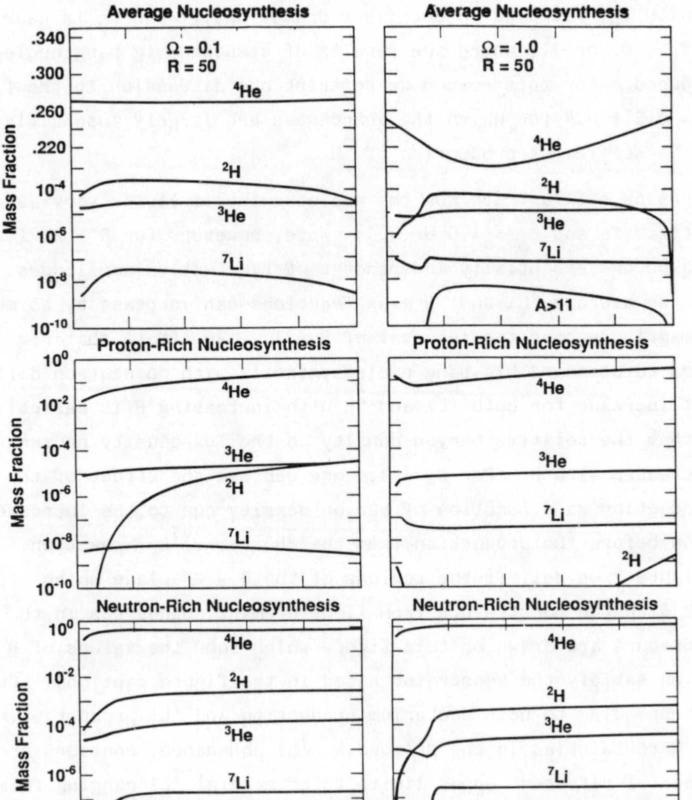

Figure 5) Primordial nucleosynthesis for the high and low density regions and their average yields as a function of volume fraction, f_v, for a fixed fluctuation amplitude, R = 50, and with: a) Ω_b = 0.1; b) Ω_b = 1.0. Note that the average yields for ^{2}H are relatively insensitive to Ω_b. Note also that the yields do not depend much on f_v for most of the range, $0.2 \le f_v \le 0.8$.

fraction with f_v, although ^2H decreases dramatically when f_v is near the extrema, f_v = 0. or 1., where the results of standard big bang nucleosynthesis are reproduced. For this reason we restrict our discussion to the f_v in the range $f_v = 0.5 \pm 0.4$ for which the abundances are largely insensitive to f_v and there is sufficient production of ^2H.

Figures 6a,b illustrate how the nucleosynthesis yields vary as a function of R for fixed f_v and Ω_b = 1.0 or 0.1. Note, however, for $R > 20$ ($1/R < .05$) the yields become essentially independent of fluctuation amplitudes. Typically the average ^7Li and ^2H mass fractions can increase by as much as an order of magnitude over the interval of R = 1. → 20. (Note that R = 1 corresponds to standard big-bang nucleosynthesis with no neutron diffusion). The slight increase for both ^7Li and ^2H with increasing R is basically due to the fact that the relative baryon density in the low-density neutron-rich region decreases with R. For $\Omega_b = 1$, one can see the effect of the minimum in ^7Li production as a function of baryon density due to the increased ^7Li destruction before ^7Be production from the $^3He(\alpha,\gamma)^7Be$ reaction.

In figure 7 we delimit the regions of the R - Ω_b plane which are consistent with the constraints from light element abundances in the two-zone model. Contours are drawn on this figure which show the values of R and Ω_b which satisfy the constraint noted in the figure captions. Curves are drawn corresponding to both deuterium production and ^7Li production. Due to the large uncertainties in the primordial ^7Li abundance, contours are shown for a number of different upper limits to primordial ^7Li ranging from the lowest values from Pop II halo stars to the highest values corresponding to a large fraction of the primordial abundance having been destroyed in stars. Two different ^2H constraints are shown to illustrate the sensitivity to the deuterium abundance constraint.

From this figure several interesting conclusions can be drawn. One is that the deuterium constraint can only be satisfied in an $\Omega = 1$ universe if the fluctuation amplitude, $R > 20$. Perhaps more useful is the observation that if the smallest upper-limits to the ^7Li abundance are correct then this constraint can only be satisfied in a universe with $R \leq 2$, which would correspond to standard big bang nucleosynthesis without fluctuations. Thus, if the Pop II or even Pop I disk ^7Li abundances are ever accepted as firm upper limits to the primordial lithium, then one could use this constraint to

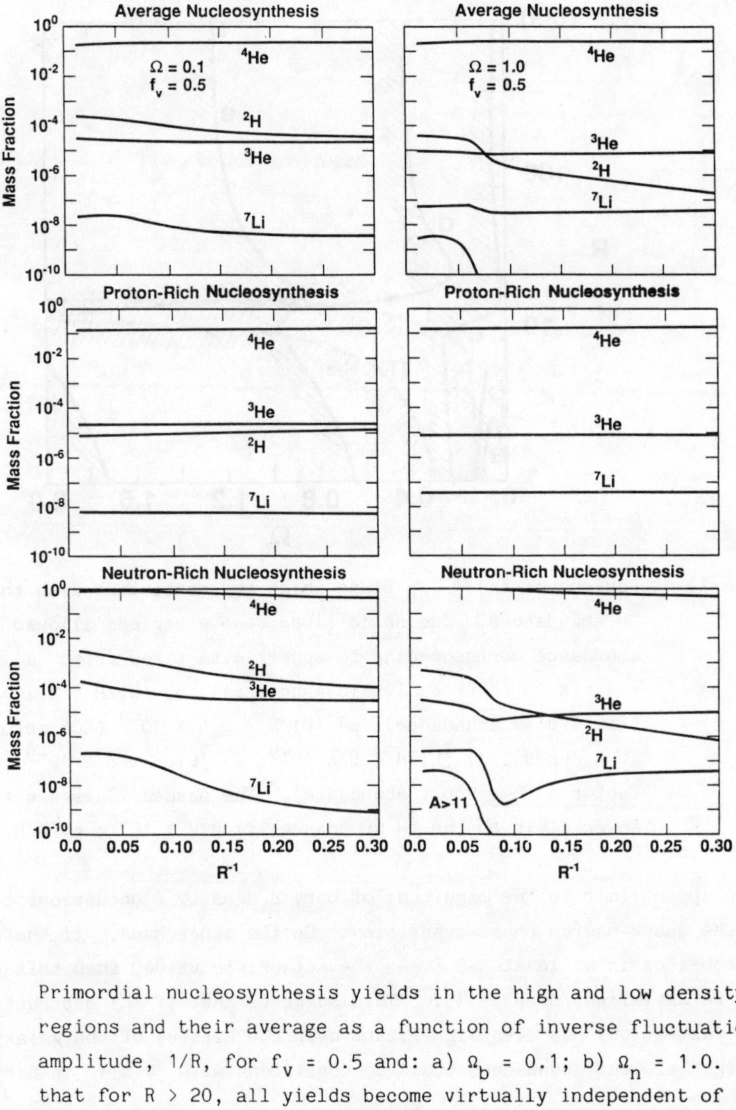

Figure 6) Primordial nucleosynthesis yields in the high and low density regions and their average as a function of inverse fluctuation amplitude, 1/R, for $f_v = 0.5$ and: a) $\Omega_b = 0.1$; b) $\Omega_n = 1.0$. Note that for R > 20, all yields become virtually independent of the fluctuation amplitude.

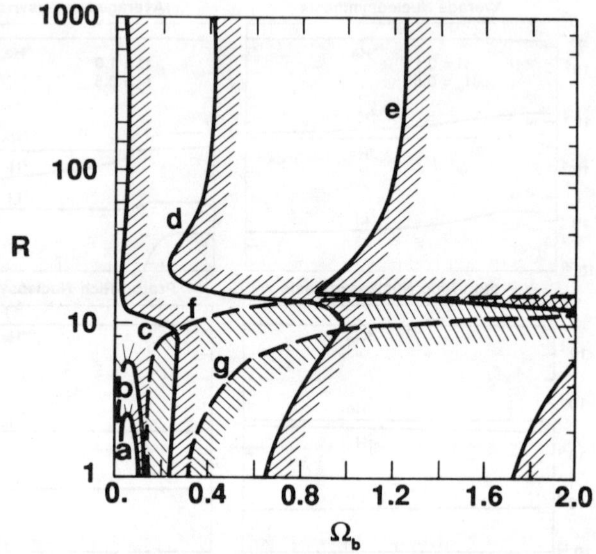

Figure 7) Regions of the R - Ω plane which are consistent with the ^{7}Li and ^{2}H abundances. The solid lines denote regions allowed by the ^{7}Li abundance corresponding to upperlimits taked from: a) ^{7}Li/H $< 1.8 \times 10^{-10}$ (Pop II halo abundance); b) ^{7}Li/H $< 8 \times 10^{-10}$ (Pop I disk abundance); c) ^{7}Li/H $< 2.6 \times 10^{-9}$ (CCI metoritic abundances); d) ^{7}Li/H $< 8 \times 10^{-9}$; e) ^{7}Li/H $< 8 \times 10^{-9}$ (Astration factor = 10 x Pop I abundance). The dashed lines are from the lower limit to the ^{2}H abundance: f) D/H $> 10^{-5}$; g) D/H $> 10^{-6}$.

set an upper limit to the magnitude of baryon density fluctuations to emerge from the quark-hadron phase transition. On the other hand, if the primordial ^{7}Li abundance is at least 3-5 times the meteoritic value, then this constraint could be satisfied for $R \geq 15$. This suggests that if ^{7}Li destruction in stars (astration) has been significant over the history of the galaxy, then the light element abundances could be consistent with $\Omega = 1$ in baryons.

If ^{7}Li has been significantly astrated then it follows that ^{2}H must also have been astrated since this nuclide is less tightly bound than ^{7}Li. It is therefore encouraging that deuterium is overproduced along with ^{7}Li in models with $R \geq 50$. It is useful to consider the constraint based upon the ratio of ^{7}Li/^{2}H in which the effects of astration cancel out to first order[29] by choosing the highest present day ^7Li abundance, and the lowest deuterium abundance a firm upper limit to this ratio can be derived ($^7Li/^2H < 3 \times 10^{-4}$). We find that for sufficiently large values of R it is possible to have Ω_b very close to unity and still satisfy this constraint.

VII. CONCLUSION

We have shown in this paper that isothermal baryon number density fluctuations are a plausible and perhaps unavoidable consequence of the transition from quark-gluon plasma to confined hadronic matter. We identify the coexistence temperature, T_c, and the nucleon phase boundary transmission probability, Σ_h, as the main determinants of fluctuation size.

We have calculated the effect of isothermal baryon number fluctuations on primordial nucleosynthesis and have shown that the light element abundance yields in a high baryon content universe could reproduce accepted primordial abundances, in a relatively parameter independent way, with the exception of ^7Li. The point is that the limit on Ω_b from primordial nucleosynthesis now rises or falls on the basis of what is taken to be the primordial ^7Li abundance. The measurements of the ^7Li abundances and the $^7Li/^6Li$ isotopic ratios, when coupled with self-consistent models of galactic chemical evolution, now seem to point to a low primordial Li abundance.[27]. If one takes the view that the measured ^7Li abundance is primordial then the present study might constrain the physics of the QCD transition. However, given the uncertainities in the primordial ^7Li abundance it is premature to exclude the study of $\Omega_b = 1$ universes.

VIII. ACKNOWLEDGMENTS

The authors acknowledge useful discussions with J. Audouze, W. A. Fowler, R. A. Malaney, L. McLerran, D. N. Schramm, N. J. Snyderman, G. Steigman, and D. Toussaint. We also acknowledge D. LaPierre for help in the preparation of this manuscript. Work performed under the auspices of the U.S. Department of Energy by Lawrence Livermore National Laboratory under contract No. W-7405-ENG-48.

IX. REFERENCES

1) L. G. Yaffe, and B. Svetitsky, Phys. Rev. **D26**, 963 (1982).

2) R. D. Pisarski and F. Wilczek, Phys. Rev. **D29**, 338, (1984).

3) J. H. Applegate, C. T. Hogan, and R. J. Sherrer, Phys. Rev. **D35**, 1151, (1987).

4) K. E. Sale and G. J. Mathews, Ap. J. **309**, L1, (1986).

5) C. R. Alcock, G. M. Fuller, and G. J. Mathews, Ap. J., **320**, 439 (1987), and G.M. Fuller, G.J. Mathews, C.R. Alcock, Preprint UCRL-97316 (1987).

6) E. Witten, Phys. Rev. **D30**, 272, (1984).

7) J. H. Applegate and C. Hogan, Phys. Rev. **D30**, 3037, (1985).

8) C. J. Hogan, Phys. Rev. Lett. **133B**, 172, (1983).

9) K. Kajantie and H. Kurki-Suonio, Phys. Rev. **D34**, 1719, (1986).

10) R. V. Wagoner, Ap. J. **179**, 343, (1973).

11) J. Yang, M. S. Turner, G. Steigman, D. N. Schramm, and K. Olive, Ap. J. **281**, 493. (1984).

12) A. M. Boesgaard and G. Steigman, Ann. Rev. Astron. Astrophys. **23**, 319 (1985); G. Steigman, private communications, (1987).

13) D. Cline, private communications, (1987).

14) D. N. Schramm, Proceedings of the XXII Recontre Moriond, March 1987.

15) L. D. Landau and E. M. Lifshitz, Statistical Physics (Pergamon Press, New York) (1969).

16) For a discussion of the analytic treatment of relativistic Fermi integrals see a) G. M. Fuller, W. A. Fowler, and M. J. Newman, Ap. J. **293**, 1 (1985); Previous analytic treatments of quark-gluon plasma in b) B. Berg, et al., Z. Phys. **C31**, 167 (1986); c) E. Suhonen, Phys. Rev. Lett. **119B**, 81 (1982); d) K. A. Olive, Nucl Phys. **B190**, 483 (1981); e) V. Dixit and E. Suhonen, Z. Phys. **C18**, 355 (1983).

17) A. Chodos, R. L. Jaffe, C. B. Thorn, and Weisskopf, Phys. Rev. **D10**, 2599 (1974).

18) A. Chodos and C. B. Thorn, Phys. Rev. **D12**, 2733 (1975).

19) A. W. Thomas in "Advances in Nuclear Physics," **13**, (eds. J. W. Negele, and E. Vogt,) pp. 1-137 (1984).

20) G. E. Brown, and M. Rho, Phys. Lett. **82B**, 177 (1979).

21) S. The'berge, A. W. Thomas, and G. Miller, Phys. Rev. **D22**, 2838, (1980).

22) L. D. McLerran and B. Sretitsky, Phys. Rev. Lett. **98B**, 199, (1981).

23) Aguilar-Benitez, et. al., Phys. Rev. Lett. **107B**, 1 (1986).

24) W. A. Fowler, and F. Hoyle, Ap. J. Suppl. **9**, 201 (1964).

25) E. Farhi and R. L. Jaffe, Phys. Rev. **D32**, 2452 (1985).

26) C. Alcock, and E. Farhi, Phys. Rev. **D32**, 1273 (1985).

27) P. Delbourgo-Salvador, C. Gry, G. Maline, and J. Audouze, Astron. and Astrophys. **150**, 53 (1985); T. Kajino, H. Toki, and S. M. Austin, Michigan State University preprint (1986); J. Audouze, and H. Reeves, in "Origin and Distribution of the Elements," G. J. Mathews, Ed., (World Scientific; Singapore) 1987 (in press).

28) R.A. Malaney and W.A. Fowler, Caltech preprint (1987).

29) G. J. Mathews and V. E. Viola, Ap. J. **228**, 375, (1979).

Big Bang Nucleosynthesis: Accelerator Tests and Can Ω_B Really be Large*

David N. Schramm

The University of Chicago

ABSTRACT

The first collider tests of cosmological theory are now underway. The number of neutrino families in nature, N_ν, plays a key role in elementary particle physics as well as in the synthesis of the light elements during the early evolution of the Universe. Standard Big Bang Nucleosynthesis argues for $N_\nu = 3 \pm 1$. Current limits on N_ν from the CERN $\bar{p}p$ collider and e^+e^- colliders are presented and compared to the cosmological bound. Supernova SN 1987A is also shown to give a limit on N_ν comparable to current accelerator bounds. All numbers are found to be small thus verifying the Big Bang model at an earlier epoch than is possible by traditional astronomical observations. Future measurements at SLC and LEP will further tighten this argument.

Another key prediction of the standard Big Bang Nucleosynthesis is that the baryon density must be small ($\Omega_B \lesssim 0.1$). Recent attempts to try to subvert this argument using homogeneities of various types are shown to run afoul of the ^7Li abundance which has now become a rather firm constraint.

---

* Prepared for the Proceedings of the American Chemical Society Meeting in New Orleans, Symposium on the Origin of the Elements.

INTRODUCTION

The interaction between cosmology and particle physics has grown at an explosive rate during the last decade. One of the first predictions to come from physics at the frontier of these fields was that Big Bang nucleosynthesis constrains[1] the number of light ($\lesssim 10MeV$) neutrino flavors, N_ν; this constraint probably limits the number of quark and charged lepton flavors as well. In fact when the cosmological limit was first proposed the particle physics limits were in the thousands, and tradition had it that new energies would lead to new particle generations. Thus the cosmological statement of small numbers seemed very risque. Such a cosmological constraint is extremely important since particle theory in general does not limit N_ν. It is therefore of great relevance that this cosmological prediction[1] is finally being tested in the laboratory by collider experiments. This first test of cosmological theory at a high energy physics facility will provide a check on the hot Big Bang model back to an earlier epoch than is feasible with any of the traditional astronomical techniques.

Steigman in his presentation here reviewed the current status of the arguments[3,4] from primordial nucleosynthesis which lead to the present bound $N_\nu \leq 4.0$ (the "standard" $N_\nu = 3$ is completely consistent with Big Bang nucleosynthesis). Next, we examine the possibilities for neutrino counting at $\bar{p}p$ and e^+e^- colliders, present the new limits from CERN and PEP and discuss the future prospects. It should be noted that cosmological and accelerator experiments do not "measure" exactly the same quantities but are quite complimentary. It is shown that the standard Big Bang model is in very good shape. At the end we turn our attention to another prediction of the standard model, namely the baryon density. We show that recent attempts to get around the constraint that $\Omega_B \lesssim 0.1$ run afoul of the lithium constraint.

NEUTRINO FAMILIES: THE QUARK-LEPTON CONNECTION

At present, three generations of quarks and leptons are known. Only the t-quark and the τ-neutrino remain to be completely confirmed. As is well known, each generation contains six quarks (3 colors of "up" quarks with $Q = +2/3$ and 3 colors of "down" quarks with $Q = -1/3$) and two leptons (a charged lepton and a neutrino). It is of fundamental importance to determine if additional families of quarks and leptons exist.

In the past, the discovery of a new generation has been made by first discovering a charged lepton. This was due largely to the lower mass of the charged lepton and/or the cleaner experimental signature (e.g.: consider the μ-lepton versus the strange or charmed quark). Today, we may be in a similar situation in that either a charged lepton from W or Z decay or, an additional neutrino flavor could provide the first evidence for another generation. In contrast, the quarks of a fourth generation, if it exists, may well be out of reach of the present colliding beam machines. In contrast, a bound to the total number of neutrino flavors may provide an upper limit to the number of quark-lepton generations.

A remarkable aspect of the hot Big Bang model for the evolution of the Universe is the dependence of primordial nucleosynthesis—in particular, the synthesis of ^4He—on N_ν, the number of light neutrino ($m_\nu \lesssim 10MeV$) flavors.[1,2,3,4] The present limit on the number of neutrino families derived from Z° decay is completely consistent with the cosmological bound. This represents the first accelerator test of cosmology. The cosmological bound limits the number of "relativistic degrees of freedom" which were present during Big Bang nucleosynthesis whereas the accelerator data limits the number of particles, which could be very massive ($\lesssim M_Z/2$) but, must couple to the Z°. That the two bounds very nearly coincide helps rule out (or, at least, constrain) the numbers of new "heavy" neutrinos and/or "light"exotic particles.

Laboratory experiments show directly that the ν_e and ν_μ are light (in the context of our discussion). Despite an heroic effort to lower the upper limit to the mass of the τ-neutrino, laboratory data[6] still permits a "heavy" ν_τ. However, by combining astrophysical and accelerator data, it has been argued that the ν_τ must be light . For convenience, we shall write $\delta N_\nu = N_\nu - 3$. Although we expect $\delta N_\nu \geq 0$, if the cosmological bound should eventually turn out to yield $\delta N_\nu < 0$, the question of the ν_τ mass might have to be reopened. In fact current best fits to astrophysical data put $N_\nu \sim 2.6$. However, uncertainties clearly allow values of 3; 4 is getting a bit marginal.

NEUTRINO COUNTING AT COLLIDERS: PRESENT RESULTS AND FUTURE PROSPECTS

The counting of neutrino families at colliders relies on the production and decay of real or virtual Z° bosons. Within the framework of the standard electroweak theory the Z^o is universally coupled to leptons and quarks. Thus, in the decay of the Z°, the branching ratio to the standard 3 neutrino species—as well as to new families—is prescribed. Basically, there are three ways to count neutrino families at colliders[5]:

A. Direct detection of $Z^\circ \to \nu_i \bar{\nu}_i$,

B. Measurement of the total width of the Z° to determine the neutrino partial width $\Gamma(Z^\circ \to \nu_i \bar{\nu}_i)$.

C. Deviation of $\frac{M^2 \omega}{M^2 cos^2 \theta \omega}$ from unity caused by radiative corrections which increase with N_ν.

Technique (A) can be used at a $\bar{p}p$ or an e^+e^- collider by detecting either

$$\bar{p}p \to Z^\circ + gluon \tag{1}$$

$$e^+e^- \to Z^\circ + photon \tag{2}$$

Recently both processes (1) and (2) have been observed and lead to limits on δN_ν, with process (2) occurring via virtual Z°'s at present since e^+e^- energies are below M_{Z° until SLC and LEP begin operations.

The measurement of the Z° width can be carried out directly or by a determination of the ratio of Z° to W widths. The direct measurement from the CERN collider experiment gives a very poor limit to δN_ν at present. For $\bar{p}p$, technique B gives the best present limit on δN_ν. This uses a directly measured ratio of the $W \to e\nu_e$ and $Z^\circ \to e^+e^-$ rates to give

$$R = \frac{R(W \to e\nu_e)}{R(Z^\circ \to e^+e^-)} = (\Gamma_Z/\Gamma_W)\left(\frac{\Gamma_{W\to e_\nu}}{\Gamma_{Z\to e^+e^-}}\right)\left(\frac{\sigma_{W^+}+\sigma_{W^-}}{\sigma_Z^\circ}\right).$$

The key idea of this technique is that $\Gamma_{W\to e_\nu}/\Gamma_{Z\to e^+e^-}$ is reliably calculated in the standard model once $\sin^2\theta_W$ is known and $(\sigma_{W^+}+\sigma_{W^-})^\sigma Z^\circ$ is determined from QCD calculations.[7] Actually the calculation of the ratio of cross sections is more reliable than the individual terms due to cancellations in the ratio. Finally, it is possible to determine Γ_W if we know all of the important decay modes of the W. The Z° and W widths are given as [including the possiblity of 4th generation charged lepton (L) and the t quark].[5,8]

$$\Gamma_Z = [2.54 + \Gamma_{Z\to t\bar{t}} + \Gamma_{Z\to L\bar{L}} + 0.17\delta N_\nu]GeV \qquad (11)$$

and

$$\Gamma_W = [2.2 + \Gamma_{W\to t\bar{b}} + \Gamma_{W\to L\nu_L}]GeV \qquad (12).$$

The present limit on the mass of a possible 4th generation lepton reduces the contribution to the Z° width to a negligible value. This is not true however for the W width. Thus, the intrinsic uncertainty in the ratio of Z to W widths comes from the t and L mass (or existence) and limits the accuracy of this technique in determining δN_ν. While it may be relatively easy to reach an uncertainty in δN_ν of ~ 2, to achieve greater accuracy will be very difficult with the ratio of widths technique. When adequate statistics are collected for processes (1) and (2) it will be possible to measure δN_ν to $\lesssim \pm 1$.

A variant on direct width measurements is to study the partial width for the process $Z^\circ \to \nu\bar{\nu}$ by

$$\gamma + Z^\circ \to \gamma + \nu\bar{\nu}$$

and look for the scattered γ in coincidence with nothing (the nondetectable neutrinos). This can be done for real Z°'s at SLC or LEP or for virtual Z°'s using heavy $q\bar{q}$ states

$$\gamma + (q\bar{q}) \rightarrow \gamma + (virtual Z^{\circ}) \rightarrow \gamma + \nu\bar{\nu}.$$

We now turn to the present measurements of, or limits to, δN_{ν}. Fig. 1 shows a comparison of the limits reached with the various techniques. The UA1 and UA2 results on the ratio of widths can be combined to give[5]

$$\delta N_{\nu} \leq 2.$$

A similar bound–using virtual Z°'s–from $e^{+}e^{-} \rightarrow \gamma\nu\bar{\nu}$ is obtained from the data of MAC and ASP at PEP and CELLO at PETRA.[9]

That δN_{ν} is small is also suggested by the data from method C which is consistent with $M_W^2 = M_Z^2(1 - \sin^2\theta_W)$. Radiative corrections due to extra, low mass neutrino flavors with corresponding new quarks would cause a deviation in this relation.[10] Limits from the data suggest $\delta N_{\nu} \lesssim 2$.

These measurements are all in excellent agreement with the cosmological results (see Fig. 1). From current experimental data we may conclude that, at most, there may be a fourth or possibly a fifth family of quarks and leptons. Given the present uncertainty ($\sim \pm 2$) in the neutrino counting techniques, the direct search for a fourth generation lepton is of great significance.

SUPERNOVA LIMITS

It is now well established that Kamioka and IMB detected[11] antielectron neutrinos from SN 1987A. Due to neutral currents, supernovae generate[12] all types of neutrinos with masses $\lesssim 10 MeV$. Since the total energy radiated in neutrinos is the neutrons for binding energy $\sim 2 \times 10^{53} ergs$, it is clear that too many additional neutrino species would dilute the number at detectable $\bar{\nu}_e$'s before the observations.[13] Plugging in numbers and uncertainties, one finds that $\delta N_{\nu} \lesssim 4$ from this technique.[14] Thus giving a completely independent argument that δN_{ν} is small as predicted by cosmology.

LI AND Ω_B

With the increasing success of the N_ν prediction, confidence continues to grow on the standard cosmological model. One important astrophysical constraint of the standard model[15] is $\Omega_B \lesssim 0.1$. Recently possible loopholes in this constraint have received publicity. These loopholes come from the possibility that the quark-hadron transition can produce variations in n/p ratios in the early Universe and that mixtures of such variable n/p ratios can fit D, ^3He, and ^4He abundances[16] for $\Omega_B \sim 1$. These models do indeed manage to fit D/H with a high Ω_B and D/H was the original reason used for limiting Ω_B. However any model with large variations seems to inevitably over produce ^7Li. This occurs because Li observations fall exactly at the Li production minimum in the standard model. Thus mixing of larger or smaller density values moves away from the minimum and is incapable of giving such small values. In particular, as Alcock et al.[16] show, models which have enough n/p variations to yield $\Omega_B \sim 1$ produce more ^7Li than is observed in Pop I which in turn is ~ 10 times the probable primordial Pop II abundance.

Although Li abundances were less of a constraint a few years ago, it has now been shown by three independent groups that the Pop II abundances are really at Li/H $\sim 10^{-10}$ which precisely hits the minimum of the standard model.[17] These new measurements range over Pop II metallicities, and thus cannot be fit by any standard convective depletion of Pop I abundances. In addition Baade[18] has found that Li in the LMC is consistent with the Pop II value. Since the LMC is metal deficient relative to Pop I (yet metal rich compared to Pop II) this shows that the ^7Li of Pop I does indeed seem to be an enhancement over Pop II rather than Pop II being some special depleted case. The enhancement mechanism is presumably

$$^3HE + \ ^4He \rightarrow \ ^7Be \rightarrow \ ^7Li$$

in convective zones. Such Li production is seen in some red giants so we know ^7Li is enhanced in the galaxy. In addition we know that ^6Li is not made in stars nor in the standard Big Bang but must come from cosmic ray spollation over the history of the galaxy. Any large scale Li destruction mechanism would preferentially destroy ^{6}Li. Thus it would be difficult to lower the excess ^7Li produced in these variable n/p models even down to Pop I abundances and still have the observed ^6Li found in the solar system.

Admittedly the details of Li production and destruction in stellar convective zones is only qualitatively not quantitatively understood. However considering the difficulties of quantitatively understanding turbulence, convection, etc. (weather) everything really fits together quite well. The detailed quantitative weather next week in New Orleans might be hard to predict but the qualitative statement that it will be warmer in New Orleans than at the North Pole is hard to argue against. Similarly the lack of a full quantitative understanding of stellar convection does not really enable one to overthrow all of the well established lithium trends. As Kawano et al.[17] argue, lithium has come of age as a cosmological problem and it's not deuterium that has now become the strongest argument for $\Omega_B \lesssim 0.1$.

ACKNOWLEDGEMENTS

I wish to thank my collaborators Gary Steigman and Dave Cline for their help in earlier versions of this presentation.

REFERENCES

1. G. Steigman, D. Schramm and J. Gunn, Phys. Lett. B66, 502 (1977).
2. G. Steigman, in this volume (1987).
3. J. Yang, M.S. Turner, G. Steigman, D. Schramm and K. A. Olive, Ap. J. 281, 493 (1984); A.M. Boesgaard and G. Steigman, Ann. Rev. Astron. Astrophysics 23, 319 (1985).
4. G. Steigman, K. A. Olive, D. Schramm, M. S. Turner, Phys. Lett. B76, 33 (1986).
5. D. Cline, D. Schramm and G. Steigman, Comments on Nuclear and Particle Physics (and references therein) (1987).
6. Proc. Recontre de Moriond on Neutrino Masses, ed. Tran Thanh Van. (1986).

7. D. Cline and J. Rohlf, UA1 technical note TN 83-40 (1983), unpublished; N.B. Deshpande et al., Phys. Rev. Lett. 54,1757 (1985); N. Cabibbo, Proc. Third Topical Workshop on $p\bar{p}$ Collider Physics (Rome, January 1983) p. 567 (Geneva 1983); F. Halzen and K. Mursula, Phys. Rev. Lett. 51 (1983).
8. D. Schramm and G. Steigman, Phys. Lett. 141B, 337 (1984).
9. T. Levine, Proc. of the SLAC Summer Institute 1986; U. of Wisc. Preprint No. WISC-EX-86/275 and U. of Wisc. PhD Thesis.
10. G. Altarelli, Ecole d'Ete de Physique des Particules 1986 (as referenced by H. Reeves, 1986 Saclay Preprint).
11. Bionta, R. et al., Phys. Rev. Lett. 58, 1494 (1987); Hirata, K., et al. Phys. Rev. Lett. 58, 1490 (1987).
12. D. Schramm and W.D. Arnett, Ap. J. (1975).
13. D. Schramm, Proc. Recontre de Moriond (1987).
14. D. Schramm, Comments Nuc. Part. Phys. (1987); J. Ellis and K. Olive, Phys. Lett. (1987).
15. J. Gott, J. Gunn, D. Schramm and B. Tinsley, Ap. J. and references therein (1974).
16. C. Alcock, G. Fuller and G. Mathews, Ap. J. (1987); J. Applegate, C. Hogan and R. Scherrer, Phys. Rev. D35, 115 (1987).
17. L. Kawano, D. Schramm and G. Steigman, Fermilab preprint (and references therein) (1987).
18. D. Baade, Proc. ESO workshop on SN 1987A.

FIGURE CAPTIONS

1. Present limits to N_ν from cosmology, $p\bar{p}$ and e^+e^- experiments (from ref. 5).
2. Mass fraction of helium -4, Y_p versus the baryon-to-photon ratio, N, for $N_\nu = 2, 3$ and 4. The three curves at each N_ν correspond to neutron half-lives of 10.4, 10.5 and 10.6 min. The horizontal dotted line is the current 3τ upper limit on ^4He from observations. While it is not impossible that systematic errors could raise this to 0.25 and thus marginally allow $N_\nu = 4$, higher values seem well excluded. However it is far more difficult to exclude systematic errors that could give lower values for He. Thus no lower bounds have been drawn. The vertical dotted lines correspond to upper and lower bounds on the baryon-to-photon ratio coming from ^3He, D and ^7Li. It is IMPORTANT to note that ^7Li independently supports the independent limits previously derived from ^3He and D. The current central observed abundance values of Y_p and N are 0.235 and 4×10^{-10} which yield $N_\nu \sim 2.5$. Obviously $N_\nu = 3$ is a perfectly allowed value and a good fit to the data once error bars are included.

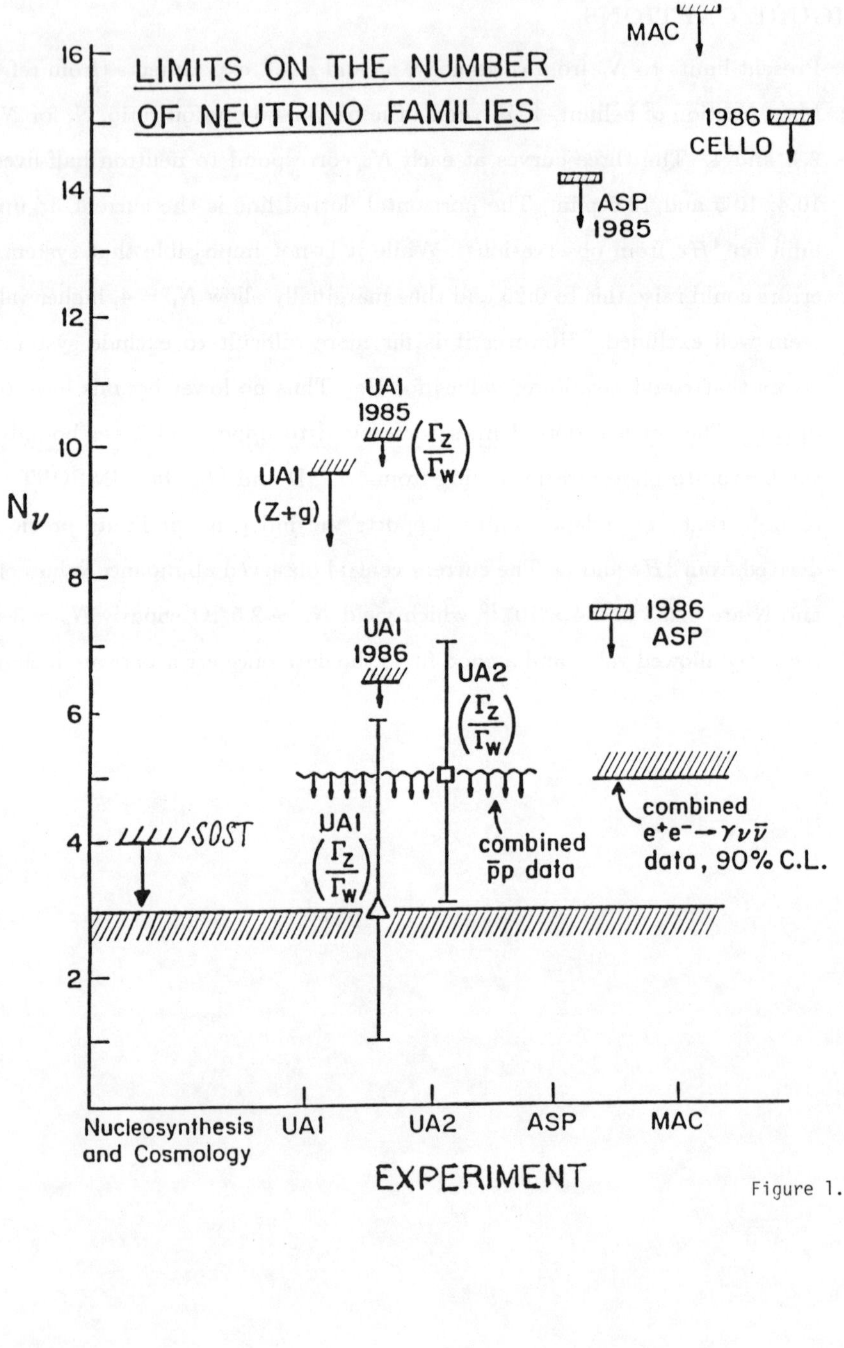
LIMITS ON THE NUMBER
OF NEUTRINO FAMILIES
MAC
1986
CELLO
ASP
1985
UA1
1985
UA1
(Z+g)
1986
ASP
UA1
1986
UA2
SOST
UA1
combined
p̄p data
combined
e⁺e⁻ → γνν̄
data, 90% C.L.
Nν
16
14
12
10
8
6
4
2
Nucleosynthesis
and Cosmology
UA1
UA2
ASP
MAC
EXPERIMENT

Figure 1.

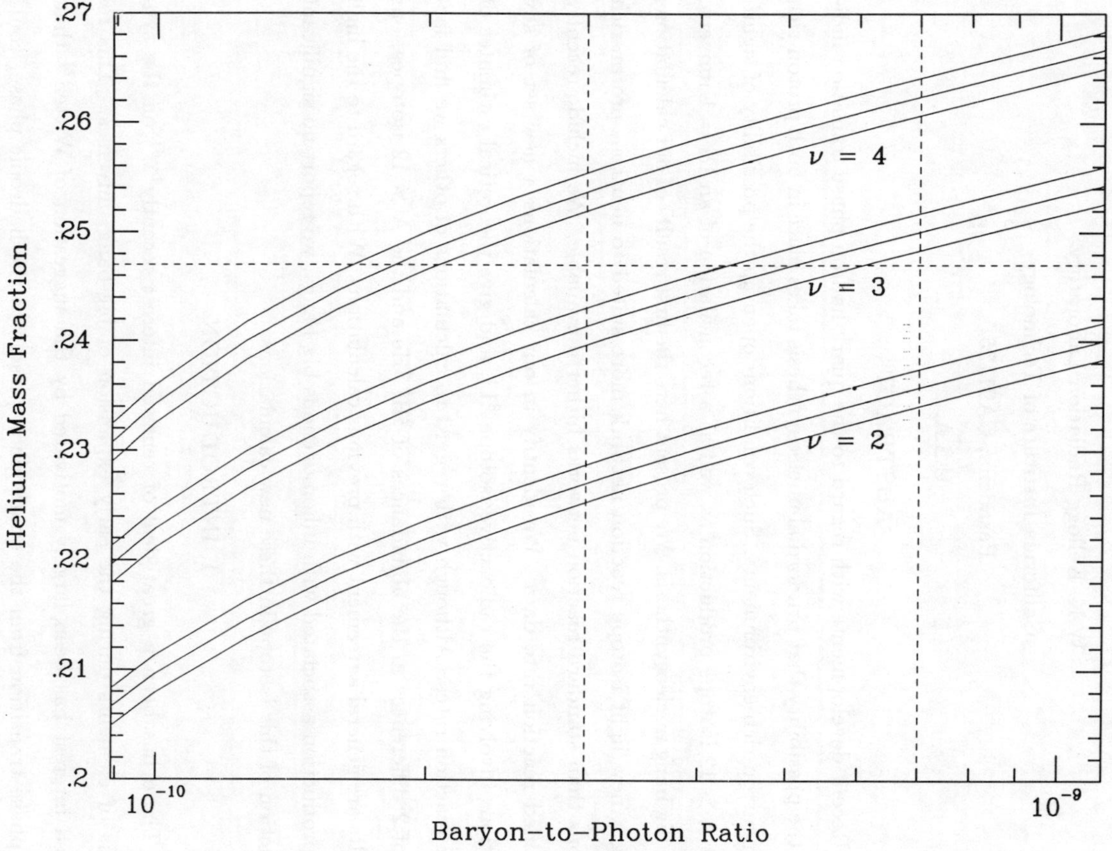

.27
.26
.25
.24
.23
.22
.21
.2
Helium Mass Fraction
$\nu = 4$
$\nu = 3$
$\nu = 2$
10^{-10}
10^{-9}
Baryon–to–Photon Ratio

Figure 2.

The Consequences of Baryonic Diffusion on Primordial Nucleosynthesis

Robert A. Malaney and William A. Fowler

W.K. Kellogg Radiation Laboratory

California Institute of Technology

Pasadena, CA91125

U.S.A.

ABSTRACT

Recent developments with regard to the quark-hadron phase transition indicate the possibility that primordial nucleosynthesis takes place in both proton-rich and neutron-rich environments. Such conditions open up the possibility of significant A > 12 isotope production as well as a reconcilation of an $\Omega_b = 1$ universe with big bang nucleosynthesis. We present here the first results of our calculations using a new light isotope reaction network more suited to neutron, proton-rich regions than standard reaction networks hitherto available. We include recently obtained reaction rate data. We identify in our calculations a new set of key reactions involving the radioactive isotope ^{8}Li, and give theoretical estimates of their reaction rates. Although with regard to calculations of others we find factor of 2 differences in the abundances of ^{2}H, ^{3}He and the A > 12 isotopes, our results are general agreement with previous calculations. With regard to the high ^{7}Li production associated with inhomogenous big bangs, we report no significant reduction of this isotope in these new calculations.

I. INTRODUCTION

There has been a great deal of renewed interest recently[1−6] on the synthesis of elements during the early expansion of big-bang universes. This renewed interest has been largely motivated by the suggestion of Witten[7] that, the phase transition from the quark-gluon plasma to the hadronic phase, could give rise to isothermal baryonic density fluctuations, or high density "bubbles", in the early universe.[1−7] It was later suggested[1−2] the characteristic size of such bubbles could be of such dimensions that, protons, which interact via coulomb scattering, could not diffuse across the bubbles before the onset of nucleosynthesis, whereas neutrons, which scatter more weakly via their magnetic dipole moment and nucleon-nucleon interaction, can. The neutrons could subsequently escape from the bubble into lower density regions before the onset of nucleosynthesis. The resulting scenario then, is one far removed from standard big bang nucleosynthesis in which it is implicitly assumed that during the epoch of nucleosynthesis the baryonic density and chemical composition are homogenous. In this non-standard big bang possessing baryonic density fluctuations, the universe at the the epoch of nucleosynthesis consists of high density proton-rich regions coupled with low density neutron-rich regions. In such circumstances it may be possible to reach two conclusions very different than those arrived at from the standard big bang. The first of these concerns the reconcilation of the observed light isotope abundances with an $\Omega_b = 1$ universe[8-9] (see below for definition of Ω_b), and the second concerns the production of heavy elements (atomic mass, A > 12) in observable quantities.[1-4]

Due to these important developments we feel it is appropriate at this time to construct a new light-element nucleosynthesis reaction network more suited, than standard networks, to the neutron-rich, proton-rich environment of a non-standard big bang. This new network contains many additional nuclear reactions and the most up-to-date reaction rate data. We present here our first results from these calculations.

II. NUCLEOSYNTHESIS

In what is to follow it will be useful to define the following quantities. The parameter Ω_b is given by

$$\Omega_b = \frac{\rho_b}{\rho_c}, \tag{1}$$

where ρ_b is the baryonic density and ρ_c is the critical density required to close the universe[8-9]. In terms of h_o (the Hubble constant in units of 100 km s^{-1} Mpc^{-1}),

ρ_c is given by

$$\rho_c = 1.9 \times 10^{-29} h_o^2 \text{ g cm}^3 . \tag{2}$$

For an adiabatic universal expansion, ρ_b is given by

$$\rho_b = a T_9^3 , \tag{3}$$

where a is the baryonic density parameter, and T_9 is the temperature in units of 10^9 K. The density parameter can be written in terms of Ω_b using

$$a = \frac{1.9 \times 10^{-29} \Omega_b h_o^2}{T_9^3} . \tag{4}$$

Using the relation

$$\Omega_b h_o^2 = 3.5 \times 10^7 \eta \left(\frac{T}{2.7}\right)^3 , \tag{5}$$

where η is the baryon to photon ratio $\frac{n_b}{n_\gamma}$, we have

$$a \approx 3.4 \times 10^4 \eta . \tag{6}$$

The calculations presented here assume three neutrino flavours and a neutron half-life of 10.6 min. Figure 1a displays the time behaviour for some of the light isotopes for the *idealised* case of an initially pure neutron-rich region. This calculation is characterised by a neutron to photon ratio of $\eta = 10^{-8}$ (this density and initial neutron abundance are not related, as yet, to any quantitive calculation, and are simply chosen at this point in order to illustrate the different nucleosynthesis aspects of a non-standard big bang). Figure 1b shows a similar calculation for the same value of η only this time assuming a neutron to baryon ratio of 1/6 at the onset of nucleosynthesis (i.e., a standard big bang). The critical difference between these calculations as far as the nucleosynthesis is concerned is the low proton abundance and high ^{2}H/^{1}H ratio of the non-standard big bang relative to the standard big bang. In fact it can be seen from figure 1a that the ^{2}H/^{1}H ratio can be greater than 1, a situation far removed from the calculation of figure 1b.

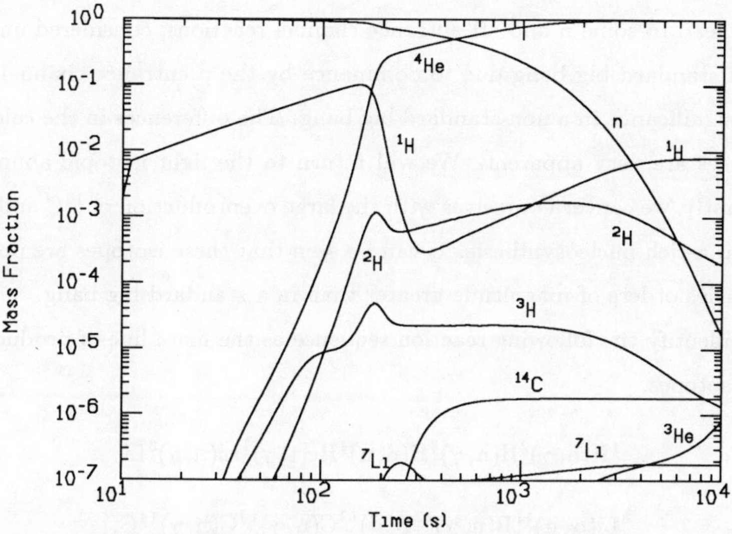

Figure 1a. Element Production for $\eta = 10^{-8}$ and $X_n^i = 1$

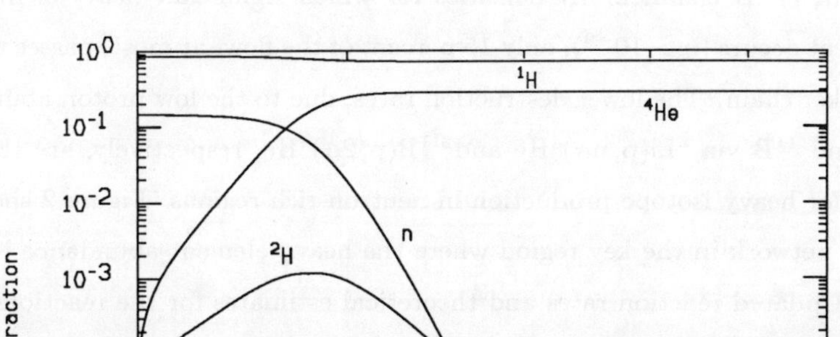

Figure 1b. Element Production for $\eta = 10^{-8}$ and $X_n^i = 0.16$

This can lead to some n and ^{2}H entrance channel reactions, considered unimportant in a standard big bang due to dominance by the p entrance channel, being of great significance in a non-standard big bang. The differences in the calculated abundances are very apparent. We will return to the light isotopic abundances later. Firstly, we concern ourselves with the large oveproduction of ^{12}C and ^{14}C in the neutron-rich nucleosynthesis. It can be seen that these isotopes are produced in quantities orders of magnitude greater than in a standard big bang.

We identify the following reaction sequence as the main line of production of A > 12 isotopes,

$$^{1}\mathrm{H}(\mathrm{n},\gamma)^{2}\mathrm{H}(\mathrm{n},\gamma)^{3}\mathrm{H}(\mathrm{d},\mathrm{n})^{4}\mathrm{He}(\mathrm{t},\gamma)^{7}\mathrm{Li}(\mathrm{n},\gamma)^{8}\mathrm{Li}$$

$$^{8}\mathrm{Li}(\alpha,\mathrm{n})^{11}\mathrm{B}(\mathrm{n},\gamma)^{12}\mathrm{B}(\beta)^{12}\mathrm{C}(\mathrm{n},\gamma)^{13}\mathrm{C}(\mathrm{n},\gamma)^{14}\mathrm{C},$$

the ^{14}C produced subsequently decaying to ^{14}N. A weaker chain passes through the $^{7}\mathrm{Li}(\alpha,\gamma)^{11}\mathrm{B}$ channel. At densities for which significant heavy-element enhancement occurs ($\eta > 10^{-9}$), only 15 percent of the flow, at most, passes through this weaker chain. The lower destruction rates, due to the low proton abundance, of ^{8}Li and ^{11}B via $^{8}\mathrm{Li}(\mathrm{p},\mathrm{n}\alpha)^{4}\mathrm{He}$ and $^{11}\mathrm{B}(\mathrm{p},2\alpha)^{4}\mathrm{He}$, respectively, are the main reasons for heavy isotope production in neutron-rich regions. Figure 2 shows the reaction network in the key region where the heavy element abundance is determined. Updated reaction rates and theoretical estimates for the reactions of figure 2 are given in the appendix. The inclusion of the $^{8}\mathrm{Li}(\mathrm{n},\gamma)^{9}\mathrm{Li}$ and $^{8}\mathrm{Li}(\mathrm{d},\mathrm{n})^{9}\mathrm{Be}$ reactions leads to a "leak" from the main chain producing the heavy isotopes. This leak can lead to a 50 percent reduction in the amount of heavy elements produced. However, we caution here that this reduction in the heavy element production is sensitive to the reaction rates of the above reaction sequence, some of which are poorly known. A full error analyses of the heavy element production will be presented in a forthcoming paper. Although not shown in figure 2, it should be noted that, the excited states of ^{9}Be (formed by the β decay of ^{9}Li) lie above the neutron emission threshold, and as such there is a $\sim$ 35 percent branching through the

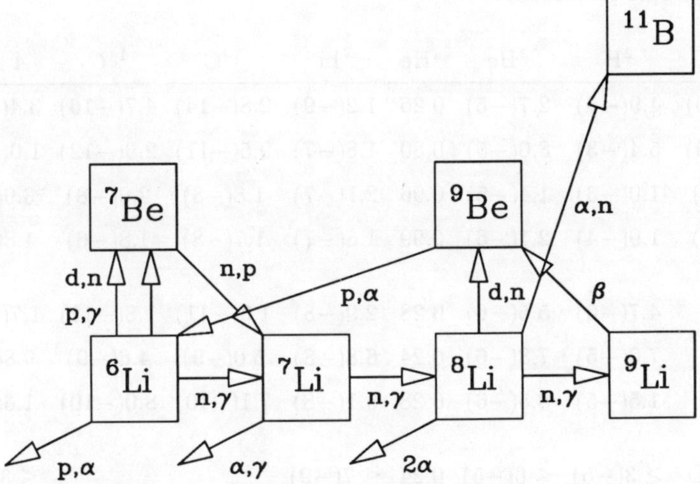

Figure 2. Reaction network in region of ^{8}Li

^{9}Be$(\gamma, n)^8$Be channel,[10] with the ^{8}Be formed subsequently decaying to two alpha particles.

Table 1 lists the mass fractions of the isotopes produced in regions possessing different values of the baryon to photon ratio, η, and the initial neutron mass fraction, X_n^i. The calculation listed with $X_n^i = 0.16$ and $\eta = 5\times10^{-10}$ corresponds to the standard big bang. The observed primordial abundances of the isotopes are also listed.[2] The mass fractions are calculated after 10^4s and assuming all the ^{3}H has decayed to ^{3}He and all the ^{7}Be to ^{7}Li. Recently, Alcock, Fuller and Mathews[5−6] have attempted to make quantitive predictions of the baryonic density fluctuations arising from the quark-hadron phase transition. Using their notation, a value of $\Omega_b = 1$, and the calculated ratio of the baryon density in the quark phase to that in the hadronic phase (for a bag model[11]) constant of $B = 145$ MeV^4), we also list in table 1 the total mass fraction of the isotopes for different values of f_v, the volume fraction of the universe in the high baryon

Table 1. Isotopic Abundances

X_i^n	η	^{2}H	^{3}He	^{4}He	^{7}Li	^{12}C	^{14}C	$A > 12$
0.16	5(−10)	4.9(−5)	2.7(−5)	0.26	1.2(−9)	2.8(−14)	4.7(−16)	3.4(−14)
1.0	1(−10)	5.4(−3)	8.0(−5)	0.80	1.8(−7)	9.5(−11)	2.9(−12)	1.0(−10)
1.0	1(−9)	1.0(−3)	1.4(−5)	0.96	2.1(−7)	1.8(−8)	2.0(−8)	3.9(−8)
1.0	1(−8)	1.9(−4)	2.1(−6)	0.99	1.8(−7)	1.7(−8)	1.8(−6)	1.8(−6)
$f_v = 0.1$		4.7(−6)	5.5(−6)	0.28	2.9(−8)	1.3(−11)	1.8(−12)	1.7(−11)
$f_v = 0.5$		7.0(−5)	7.8(−6)	0.24	6.8(−8)	5.0(−9)	4.6(−9)	9.8(−9)
$f_v = 0.9$		1.5(−5)	8.5(−6)	0.28	5.1(−8)	7.1(−10)	8.0(−10)	1.6(−9)
OBSERVED		$\gtrsim$ 3(−5)	$\leq$ 6(−5)	0.24	$\leq$ 7(−9)			$\leq$ 3(−7)

density regions just before the onset of nucleosynthesis.

With regard to previous calculations,[1−6] the calculated abundances of ^{2}H given in table 1 are roughly a factor of 2 higher, and the abundances of ^{3}He roughly a factor of 2 lower. This reduction in the ^{3}He abundance results in a slight underproduction of this isotope, relative to the observed values. The heavy isotope production is found to be either a factor of 2 lower or higher depending on the density of the neutron-rich region. However, overall there remains general agreement between the different sets of results. As can be seen from table 1, it is possible to reconcile an $\Omega_b = 1$ universe with the observed abundances of ^{2}H, ^{3}He and ^{4}He. We also note that, assuming $\Omega_b = 1$, and presently available predictions of the baryon density fluctuations at the quark-hadron phase transition,[5−6] the maximum mass fraction of heavy elements produced is $\sim 10^{-8}$. This is over an order of magnitude lower than the abundances of the most metal-poor star known.[12] However, it is has been suggested[3−4] that this level of ^{14}C production is a large enough basis from which to build an r-process which could account for the r-process abundances observed in metal-poor stars.

Throughout these and previous calculations,[1-6] there exists the persistent problem of an overproduction of ^{7}Li. We find here, relative to earlier calculations, a reduction in the ^{7}Li production of 10 percent at most. Even assuming the high ^{7}Li mass fraction of Population I stars[13] to be the relevant measurement rather than the lower determination from Population II stars,[14-15] this overproduction is at best a factor of 4 too large. Even taking into account the uncertainties in the primordial abundance determination of ^{7}Li, this overproduction can clearly be taken as evidence against any significant baryon density fluctuations arising from the quark-hadron phase transition. A reduction of the ^{7}Li abundance due to some error in the reaction rates seems an unlikely answer to this problem as ^{7}Li is overproduced via different reaction sequences in both the proton and neutron-rich regions (via ^{3}He$(\alpha,\gamma)^7$Be$(\beta)^7$Li in the former, and ^{3}H$(\alpha,\gamma)^7$Li in the latter region). The minimum change in the reaction rates required to reduce the ^{7}Li to the observed value occurs in the narrow range of parameter space where the production of ^{7}Li is at its lowest in the neutron-rich region ($f_v \sim 0.1$), where we find a reduction by a factor of 2 in the reaction rates of both the d(p,γ)^{3}He and ^{3}He$(\alpha,\gamma)^7$Be rates removes the ^{7}Li discrepancy. The present uncertainty of these two reaction rates[16] would be incompatible, however, with such a solution to this problem.

It has recently been suggested[6] that an astration of the universal ^{7}Li abundance in stellar interiors could account for the low observed ^{7}Li. Such a suggestion would be supported by recent evidence[17] that early main-sequence stars undergo extensive mass loss. ^{7}Li and ^{2}H in a large fraction of the stellar interior is destroyed as convective zones process material through the high temperature regions of the star. If the unprocessed material near the stellar surface, containing the primordial ^{7}Li and ^{2}H, is lost to the interstellar medium via mass loss, then, a substantial reduction of the ^{7}Li presently observed in stellar atmospheres would indeed be anticipated.

III. CONCLUSIONS

By employing a new nuclear reaction network more suited to a non-homogenous early universe than standard networks hitherto available, we confirm the general conclusions found by earlier calculations. We caution, however, that the production of heavy isotopes is very sensitive to the theoretically determined rates of the reaction sequence we identify as responsible for the heavy isotope build-up. No experimental cross section information is yet available on the key isotope of this sequence, namely ^{8}Li, due to its short half-life of 0.85 seconds. However, as new techniques of measuring nuclear properties of short-lived radioactive isotopes become available, the rate determination of the heavy-isotope producing reaction $^8\mathrm{Li}(\alpha,\gamma)^{11}\mathrm{B}$ and the competing $^8\mathrm{Li}(\mathrm{n},\gamma)^9\mathrm{Li}$ and $^8\mathrm{Li}(\mathrm{d},\mathrm{n})^9\mathrm{Be}$ reactions (both of which to serve to recycle ^{8}Li back to alpha particles), will be one of the most important challanges in experimental nuclear astrophysics. Such measurements will be fundamental in determining the significance of primordial heavy element production.

The authors would like to thank C. A. Barnes and G. J. Mathews for useful discussions. An allocation of computer time at the San Diego Supercomputing Center is gratefully acknowledged. This work was partly supported by National Science Foundation Grant PHY85-05682. RAM acknowledges the financial support of an SERC/NATO Fellowship (U.K.).

APPENDIX

Rates for Key Heavy-Isotope Producing Reactions (in sec^{-1})

$$^7\mathrm{Li}(\mathrm{n},\gamma)^8\mathrm{Li} = 6015\rho_b + 1.141\times 10^4 \rho_b T_9^{-\frac{3}{2}} \exp\left(-\frac{2.756}{T_9}\right)$$

$$^8\mathrm{Li}(\mathrm{n},\gamma)^9\mathrm{Li} = 4.294\times 10^4\rho_b + 6.047\times 10^4 \rho_b T_9^{-\frac{3}{2}} \exp\left(-\frac{2.866}{T_9}\right)$$

$$^8\mathrm{Li}(\mathrm{d},\mathrm{n})^9\mathrm{Be} = 2.92\times 10^{11} \rho_b T^{-\frac{2}{3}} \exp\left(-\frac{10.259}{T_9^{\frac{1}{3}}}\right)$$

$$^{8}\mathrm{Li}(\alpha,\mathrm{n})^{11}\mathrm{B} = 8.62\times 10^{13}\rho_b\left(\frac{T_{9A}^{\frac{5}{6}}}{T_9^{\frac{3}{2}}}\right)\exp\left(-\frac{19.461}{T_{9A}^{\frac{1}{3}}}\right)$$

where

$$T_{9A} = \frac{T_9}{1+(\frac{T_9}{15.1})}$$

Note: The ^{8}Li(d, n)^{9}Be rate is set equal to the ^{7}Li(d, n)^{8}Be rate. This is plausible since both reactions can be regarded as high Q stripping reactions in which both ^{7}Li and ^{8}Li capture the extra proton required to close their proton shell from the incoming deuteron.

REFERENCES

1) Applegate, J. H. and Hogan, C. J., *Phys. Rev.*, **D31**, 3037 (1985).
2) Applegate, J. H. and Hogan, C. J., Scherrer, R. J., it Phys. Rev., **D35**, 1151 (1987).
3) Applegate, J. H., to appear in *Proceedings of the Conference on Supernova in Honor of Hans Bethe's 80th Birthday*, ed. G. E. Brown, (1987).
4) Applegate, J. H. and Hogan, C. J., Scherrer, R. J., *Ap. J.*, submitted, (1987).
5) Alcock, C. R., Fuller, G. M. and Mathews, G. J., *Ap. J.*, in press, (1987).
6) Fuller, G. M., Mathews, G. J. and Alcock, C. R., preprint, (1987).
7) Witten, E., *Phys. Rev.*, **D30** , 272 (1984).
8) Guth, A. H., *Phys. Rev.*, **D23** , 347 (1981).
9) Guth, A. H. and Steinhardt, P. J., *Sci. Am.*, **250** , 116 (1981).
10) Ajzenberg-Selove, F., *Nuc. Phys.*, **A413**, 1 (1984).
11) Thomas, A. W., in *Advances in Nuclear Physics*, **13**, 1, (1984).
12) Bessel, M. S. and Norris, J., *Ap.*, J. **285**, 622 (1984)
13) Boesgaard, A. M. and Steigmann, G., *Ann. Rev. Astron. Astrophys.*, **23**, 319 (1985).
14) Spite, F. and Spite, M., *Astron. Astrophys.*, **115**, 357 (1982).
15) Spite, M., Maillard, J. P. and Spite, F., *Astron. Astrophys.*, **141**, 56 (1984).
16) Caughlan, G. R., Fowler, W. A., Harris, M. J. and Zimmerman, B. A., *At. Data Nuc. Data Tables*, **32**, 197 (1985).
17) Willson, L. A., Bowen, G. H. and Struck-Marcell, C., ***Comments on Astrophysics***, **12**, 17 (1987).

Note Added in Proof: Recently, Malaney and Fowler (1987, submitted to Ap.J.) have shown a possible way of resolving the ^{7}Li overproduction problem. By including the late-time diffusion of neutrons, they show how an $\Omega_b = 1$ universe can reproduce *all* of the observed primordial isotopes.

Early nucleosynthesis, particle physics and the quark–lithium connection

Jean Audouze[1,2], Pascale Delbourgo–Salvador[3], Hubert Reeves[4,1] and Pierre Salati [5,6].

1 Institut d'Astrophysique du CNRS, Paris 75014, France
2 Laboratoire René Bernas, Orsay 91405, France
3 CEA–DAM Bruyères le Chatel, 91680, France
4 Service d'Astrophysique CEA–IRF, Saclay 91191, France
5 LAPP, Annecy 74019, France
6 Dept of Physics and Astronomy, U.C. Berkeley, USA

ABSTRACT

Three questions relevant to the primordial nucleosynthesis of the very light elements are discussed in this contribution:

1- It is argued that the " standard " Big Bang nucleosynthesis models are strenghtened if D is destroyed thoroughly during the galatic history. This can be achieved by specific models of chemical evolution of galaxies like those assuming a rate of star formation varying with time.

2- The existence of non baryonic particles such as massive neutrinos or supersymetric particles (gravitinos, photinos ...) might affect this early nucleosynthesis.

If they are massive ($\geq$500 MeV) and long lived enough ($\geq 10^5$sec) the energetic photons released by their possible decay might affect the relative abundances of the light elements. In the case of the photinos, which are the supersymetric particles and which might be experimentally detectable in a near future,this possible effect can be used as to constrain the predictions on their physical properties (mass, lifetime...).

3- The early nucleosynthesis can be affected by the inhomogeneities triggered by the quark-hadron phase transition. It is argued that the primordial abundance of ^{7}Li limits very severely this possibility. As in the case of photinos the relation between the early synthesis of ^{7}Li and the characteristics of this quark-hadron phase transition may provide interesting constraints on some important physical parameters such as the constant B of the quark-bag model.

1. INTRODUCTION

The importance of the early nucleosynthesis (responsible of the formation of the lightest elements D,^{3}He,^{4}He and ^{7}Li) both in cosmology and in particle physics has been emphasized very often in the current literature (see eg Boesgaard and Steigman (1985), Audouze (1987): The Standard Big Bang nucleosynthesis model (BBN) provides constraints on the baryonic density of the Universe such that ($\Omega_B \leq$0.1) and on the maximum number of families of relativistic particles (neutrinos) such that $N\nu \leq$ 3-4.The merits of BBN and a comprehensive review of the primordial abundances of the lightest elements are presented by G. Steigman in this book. In this contribution, we aim to address three questions dealing with the success (or the limitations) of BBN. The first question is related to the validity of the "canonical" (standard) model of primordial nucleosynthesis. The relatively low abundance of ^{4}He favours galactic evolution models leading to a thorough destruction during the galactic history (section 2). The second question concerns the possible existence of massive, unstable elementary particles such as massive neutrinos or particles like photinos and gravitinos predicted by supersymetric (SUSY) theories. These particles might decay in releasing high energy photons which could photodisintegrate the light elements. If Ω_B is larger than 0.1–0.2,^{4}He and ^{7}Li would be overproduced while D and ^{3}He would be underproduced. One could imagine that if such particles exist and release such photons they could transform in part ^{4}He and ^{7}Li into D and ^{3}He (Audouze et al. 1985, Salati et al. 1987). In the case of photinos i.e. the SUSY particles which are the more likely to be determined if they exist we have shown (Salati et al. 1987) that BBN can be used to constrain the physical parameters (lifetime, mass, decay modes...) of such particles (section 3).

Finally, we have also considered the effect of the quark–hadron (Q–H) phase transition on BBN. After the pionering work of Applegate et al. (1987) this question has become very fashionable among the BBN specialists (see eg Fuller et al. and Malaney and Fowler, this book) . The Q–H phase transition can indeed induce some inhomogeneous BBN and many workers (Applegate et al.,1987, Alcock etal,1987) believed that they found a way to reconcile the early production of the very light elements with Ω_B=1. In section 4, we argue that the ^{7}Li abundances and nucleosynthesis are so constraining that this claim is hopeless. As in the case of photinos the physics of which can be constrained by BBN it is shown in section 4 that the physics of the Q–H phase transition can be constrained by the primordial abundance of ^{7}Li.

2. THE MERITS OF THE CANONICAL BBN MODELS

It has been argued in many papers (Yang etal, 1984, Boesgaard and Steigman, 1985, Steigman, this conference) that the comparison between the primordial abundances of D, ^{3}He, ^{4}He and ^{7}Li and standard (canonical) BBN model has two very important consequences : the first is cosmological : the baryon density is limited to a value such that $\Omega_B \sim 0.1$ (i.e.10% of the density predicted by inflationary models); the second is related to particle physics : the maximum number of neutrino (lepton) families is 4 and even 3. This means that particle physicists should discover at most one more lepton family if standard BBN models apply .As expressed by D.Sciama (1983), " standard BBN model is a triumph in cosmology" and one of us,HR (see eg Reeves, 1974) shares this enthusiasm. The three other authors of this contribution such as Pagel (1986) notice that the agreement between the primordial abundances of the light elements and the predictions coming from the standard BBN model requires some careful "fine turning". As shown in figure 1, adapted from Yang etal (1984) and discussed at more length in Audouze 1987, the only value of $\eta = n_B/n\gamma$ consistent with X(^{4}He)=Y<0.25 and (D+^{3}He)/H$\leq 10^{-4}$ is $\eta = 3\ 10^{-10}$ corresponding to $0.01 \leq \Omega_B \leq 0.04$ (the width of this range is only due to the uncertainty on the Hubble constant H_o: $50 \leq H_o \leq 100$ km s^{-1} Mpc^{-1}!

In the frame of the standard BBN model,and given the fact that $Y_p < 0.25$ (Kunth,1986) the only way to enlarge the possible range of η and therefore Ω is to assume that (D+^{3}He)/$H_{primordial}$ can be larger than 10^{-4}. In that case this implies that D is significantly destroyed during the galactic history. Our group

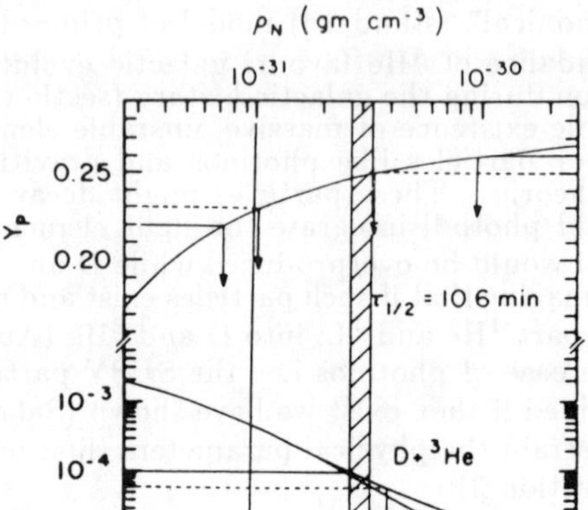

Figure 1. Calculations of the light element abundances made in the frame of the standard BBN model by Yang et al. (1984). The dashed domain indicates the η range ($\eta = n_B/n_\gamma$ i.e. the baryonic density relative to the photon density) compatible both with Y_p=0.24±0.01 (Kunth, 1986) and with the Yang et al. 1984 prescription on D/H<10^{-4}. One sees that the resulting is quite narrow η=3.2±0.2 making such comparison farily contrived.

(Delbourgo–Salvador etal, 1985, Vangioni–Flam and Audouze, 1987) have considered different possibilities to achieve that goal. There are (i) infall/outflow of processed material in the galactic disk, (ii) strong stellar mass losses occuring during the pre main sequence phase (Delbourgo– Salvador etal, 1985), (iii) a promising model is the chemical evolution scheme in which the rate of star formation is assumed to be varying. Figure 2 displays a comparison of the D destruction between such a model (called model II) in which the rate of star formation is constant for $t \leq 1$ ou 2 Gyr and proportional to the density of interstellar gas for $t \geq 1$ ou 2 Gyr. Models III and IV with bimodal rate of star formation (see Wyse and Silk, 1987) are not satisfactory in that respect (see Vangioni–Flam and Audouze, 1987, for more details). This model of varying star formation has not only the merit to account for a thorough destruction of D during the galactic history but also for the evolution of heavier elements such as O,N,Fe and the s and the r process elements (see eg Andreani etal, 1988).

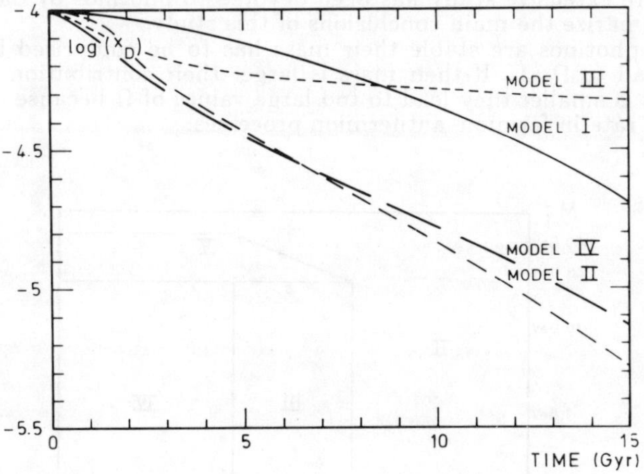

Figure 2. Evolution of the D abundance during the galactic history . Models I to IV refers to various scenarios described in Vangioni–Flam and Audouze (1987). Model II assuming a time varying rate of star formation is the most successful to account both for a thorough D destruction ($D_{present}/D_{primordial} \leq 0.1$) and a metallicity consistent with the observations

In conclusion one has still the choice either to adopt the simplest models of galactic evolution to the expense of a very narrow possible range of baryonic densities ; or to assume more complex galactic evolution models leading to a thorough D destruction which would enlarge the range of baryonic values such that $6\ 10^{-3} \leq \Omega_B \leq 6\ 10^{-2}$ (Delbourgo–Salvador etal, 1985).

3. PHOTON DECAY OF NON BARYONIC PARTICLE

In this section and in the following we examine two scenarios which attempt to reconcile the possibility of large baryonic densities Ω_b up to 1 (remember that inflationary models predict Ω=1) with the results of BBN. The first possibility is to invoke partial photodisintegration processes induced by high energy photons coming themselves from the decay of massive unstable particles. These photodisintegration processes could transform part of ^4He (and ^7Li) into D and ^3He. An important feature concerning the occurence of such processes has been established by Lindley (1985). Such photodisintegration processes can only take place if the product $E_{HBB} < 1/50(MeV)^2$ where E_H is the energy of the high energy photons and E_{BB} the energy of the thermal photons. Given the fact that $E_H \geq$20–50MeV to be able to induce such photodisintegration processes $E_{BB} \leq 4\ 10^{-4}$–10^{-3}MeV which means that $t \geq 10^5$–10^6sec.

As a consequence, particles like massive neutrinos and gravitinos with masses of a few hundred MeV and lifetimes of 10^5–10^6sec might decay into energetic photons able to partly photodisintegrate ^4He and ^7Li (see eg Audouze et al., 1985)

A more extensive study has been devoted to photinos by Salati et al. (1987) let us summarize the main conclusions of this study.

(i) If photinos are stable their mass has to be comprised between 5 and 8 GeV to lead to Ω=1. If their mass is larger their contribution is negligible . If their mass is smaller they lead to too large values of Ω because of their too large anhilation rate in fermion–antifermion processes.

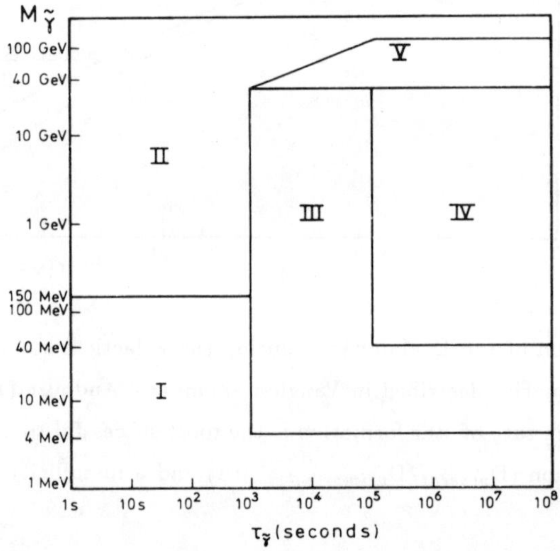

Figure 3. Photino mass $M_{\tilde{\gamma}}$ and lifetime $\tau_{\tilde{\gamma}}$ domains defined by the nucleosynthesis constraints for Ω=1. Each domain is discussed in the text. The only allowed domain is domain II (low lifetime–high-mass) (from Salati et al., 1987).

(ii) The situation is especially interesting if photinos are unstable. Figure 3 displays the constraints on their mass/lifetime coming from the early nucleosynthesis. Region II i.e. photinos massive enough with relatively low lifetime,corresponds to the physical conditions which respect the constraints coming from the early nucleosynthesis. Region I (light photinos with long lifetime) corresponds to the conditions which make photinos looking as another family of neutrinos which would increase the ^{4}He abundance. Regions III, IV and V correspond to physical conditions where photinos by their photon decay may affect significantly the abundances of D, ^{3}He and ^{4}He and are therefore forbidden by the astronomical observations. In order to fulfill the conditions set up by the early nucleosynthesis, photinos have to be relatively massive and unstable.

When the different decay modes are envisaged for the photinos these constraints on the lifetime induce further constraints on different parameters such as the supersymetric breaking scale, the mixing parameter describing the lepton number violation or the vacuum expectation value of SUSY particles (Salati etal, 1987).

To sum up the decay of particles predicted by the supersymetric theories can effect the outcome of the primordial nucleosynthesis. The observed abundances of the very light elements can be used to constrain the physical characteristics (mass, lifetime...) of such particles.

4. QUARK–HADRON PHASE TRANSITION AND EARLY NUCLEOSYNTHESIS

The possible influence of the quark–hadron phase transition on the early nucleosynthesis has been envisaged in many recent papers (Applegate etal, 1987, Alcock et al, 1987, Audouze et al, 1987, Reeves, 1987, Fuller et al, 1987). Around a temperature of 100 MeV the quarks confine and transform themselves into hadrons. As in a first order phase transition during this very short period where $T_{BB} \sim 100$ MeV there is the possibility of coexistence between the quark–gluon phase and the hadron phase. Because of the difference of the statistical weight between these two phases (the relativistic quarks have indeed a larger statistical weight than the non relativistic hadron), a natural segregation between a high density phase (the quark–gluon phase) and a low density phase (the hadron phase) takes place.

The contrast of density (between these two phases) is simply given by :

$$c = \frac{n_b^q}{n_b^h} = \frac{\sqrt{\Pi}^{3/2}}{9} = \frac{e^Z}{Z^{3/2}}$$

with $Z = m_N/T_c$ and T_c is the temperature of the phase transition T_c is related to a term B called the bag constant which expresses the negative pressure term coming from the quark confirment if $T_c \sim 106$ MeV , c=120 (the phase density contrast is large for a small bag constant predicted by the MIT model) if kT= 240 MeV (large bag constant predicted by the chiral model) c=3.

This quark–hadron phase transition induces therefore inhomogeneities : the quark zone which is a baryon rich zone which will become a proton rich zone while the hadron zone which is a baryon poor zone will become a neutron rich zone... This is because the neutrons have a diffusion mean free path much larger than the protons : the neutron density is constant while the overall baryon density is large in the former quark zone.

There are two basic parameters which fix the degree of inhomogeneity induced by this phase transition a) the baryon density contrast c, and b) the fractional mass on the high density region between the two zones such that $\Omega_T = f_v \Omega_q + (1-f_v)\Omega_H$, where Ω_T, Ω_q and Ω_H are the baryon cosmological parameters of the whole universe, the quark rich zone and the hadron rich zone. Reeves (1987) and Reeves et al. (1987) consider also the possibility to vary the neutron diffusion.

In the case of a total neutron diffusion the resulting abundances X_{el} of the light elements are given by Audouze et al., 1987.

$$X_{e\ell} = X^1_{e\ell} f_v \frac{c + X_n(c-1)(fv-1)}{1 + fv(c-1)} + X^2_{e\ell}(1 - fv)\frac{1 + X_n fv(c-1)}{1 + fv(c-1)}$$

where X_n is the n/p ratio when the weak interactions stop and "freeze" this ratio. $X^1_{e\ell}$ and $X^2_{e\ell}$ are the resulting abundances in the high density regions(1) and the low density region(2). These abundances depend on their relative densities Ω^1_B and Ω^2_B which are such that

$$\Omega^1_B = \Omega_T \frac{c + X_n(c-1)(fv-1)}{1 + fv(c-1)}$$

and

$$\Omega^2_B = -\Omega_T \frac{1 + X_n fv(c-1)}{1 + fv(c-1)}$$

To give an example, in the case of Ω_T=0.01 and c=10 for f_v=0.2, Ω^1=0.035 and Ω^2=0.005 (60% of the light elements come from the first zone) for f_v=0.8, Ω^1=0.012 and Ω^2=2.5 10^{-3} (98% of the light elements come from the first zone).

A first series of calculations have been reported by Audouze et al. (1987). They clearly show that ^{7}Li is significantly overproduced as soon as Ω >0.05 and c>5. The abundance of D is also significantly decreased with respect to the standard case. All calculations performed so far in the frame of this model (Alcock et al., 1987, Applegate et al., 1987, Fuller et al., 1987, Malaney and Fowler, 1987) exhibit this conspicuous ^{7}Li overabundance). A graphical explanation of such a striking behaviour is currently proposed by Reeves (1987) and Reeves et al. (1987).

The basis of this explanation comes from the fact that ^{7}Li is overproduced at low densities by the (T, ^{4}He)^{7}Li reaction and at high densities by the (^{3}He^{4}He) ^{7}Be reaction. Figures 4 and 5 show respectively the ^{7}Li "trench" and the quark hadron "butterflies" : the "butterfly" is delineated by the three parameters related to the quark- hadron (c, f_v and Ω) while the ^{7}Li trench is due to two different origins at low and at high density of this element.

The reader is referred to Reeves (1987), Reeves et al. (1987)and Audouze et al. (1987) for more details : the outcome of their analysis is that the ^{7}Li abundance constitutes the most stringent constraint against models where the contrast c is higher than 3–5. The inhomogeneity of the Universe due to the quark–hadron is strongly limited by this abundance. As a consequence the high values of the bag constant $(250 MeV)^4$ predicted by the chiral model are favoured compared to the low values predicted by the MIT bag model.

Instead of seeking a universe with Ω_B=1 which is not consistent with the observed abundances of ^{7}Li it seems to us more appropriate to use the observed abundances of this element to constrain the physics of the quark–hadron phase transition and possibly that of quarks through the determination of the bag constant.

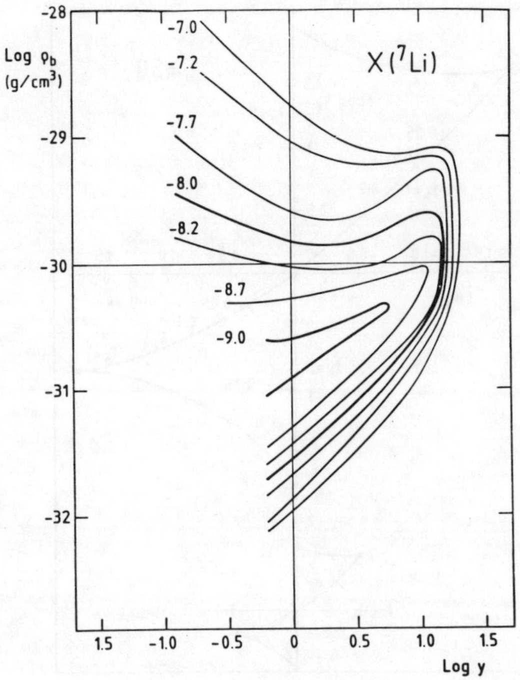

Figure 4. <u>The ^{7}Li trench</u>. The BB yield of ^{7}Li is calculated as a function of the present baryonic density in g/cm^3 (in ordinates) and the n/p ratio, in units of the standard (homogeneous density) BB nucleosynthesis n/p ratio (in abscissa).

The isoabundance lines are marked with their calculated values. They form a "trench" with steep slopes, especially in the neutron–rich side. The population II star observations correspond to the deepest part of the trench ($X_7=10^{-9}$). The strong upper limit to the BB contribution is ($X_7=10^{-8}$).

As described in the text, only the small–size "butterflies" can be "adjusted" to this trench, giving an upper limit of c=2 (or a strong upper limit of c=4) to the density contrast emerging from the quark–hadron phase transition. The range of the mean baryonic densities of the universe is essentially the same as obtained from the standard homogeneous density universe (from Reeves <u>et al.</u>, 1987.)

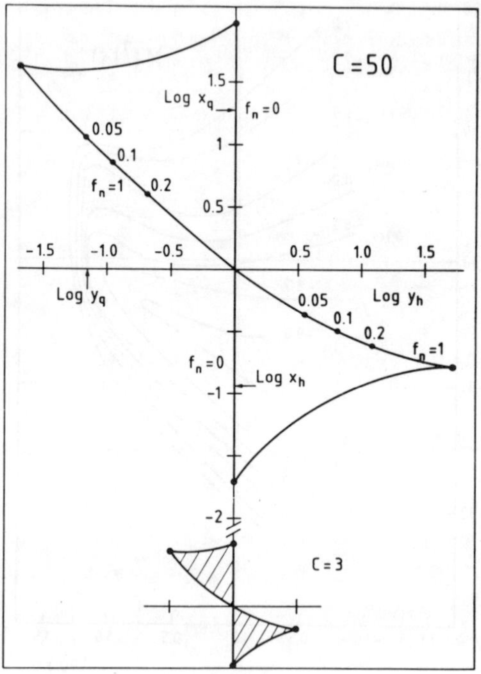

Figure 5. The quark–hadron "butterfly"

In ordinate the parameter space associated to the quark–hadron phase transition x_q and x_h are the baryonic densities in the high density phase (q : former quark sea) and the low density phase (h : nucleated hadronic bubbles), normalized to the mean baryonic density. In abscissa, y_q and y_h are the neutron to proton ratio in these phases, normalized to the standard BB (homogeneous density) n/p ratio.

Each scenario is described by 1) a choice of $c=x_q/x_h$, the contrast between the two phases 2) a choice of f_q, the fractional volume in the q– phase and 3) a choice of f_n the fractional amount of neutron diffusion between the two phases at the moment of nucleosynthesis.

The set of all these points takes the form of a "butterfly" with a wing span proportional to c. The circled numbers on the main diagonal are the assumed (paired) values of f_q in the case of complete neutron diffusion (f_n=1). The ordinate corresponds to f_n=0. Intermediate values of f_n span the area of the wings.

The large butterfly corresponds to a contrast c=50 ; the small one to c=3. (from Reeves et al., 1987)

5. CONCLUSIONS

Our current analyses of the Big Bang nucleosynthesis which have been summarized in this short presentation lead us to propose the following conclusions :

a) The standard BBN model predicting $\Omega_B \leq 0.1$ and at most 3–4 families of leptons remains the scenario which provides the best agreement with the observed abundances of the very light elements (D, ^{3}He , ^{4}He and ^{7}Li). The only point of contention is to adopt or not a model predicting a large destruction of D during the galactic history. A few observational tests regarding the D abundance in quasars or in different regions of our galaxy will favour either a low D destruction scheme such as this suggested by Yang et al., 1984, or a large D destruction scheme proposed by some of us, Delbourgo–Salvador etal., 1985, Vangioni–Flam and Audouze, 1987.

b) It seems therefore useless to attempt to day to build up a BBN scenario such that Ω_B=1. The early ^{7}Li abundance constitutes a difficulty which seems today impossible to overcome. If Ω=1 because of the necessity of considering an inflationary universe, the density of the Universe should come from non baryonic particles.

c) The Big Bang nucleosynthesis BBN can be used instead to provide useful constraints on different aspects of particle physics and cosmology. For instance, it is easier to reconcile the BBN predictions with the existence of the photinos if those particles are massive and have a relatively low lifetime. Concerning the quark–hadron phase transition it seems to us unavoidable to restrict the degree of inhomogeneity coming this transition to fulfill the ^{7}Li abundance requirements. This restriction may have the advantage to restrict the range of possible values for the quark bag constant. One can be impressed by the longevity of the standard (simplest) BBN model and by its predictive value not only regarding the baryonic density but also many different aspects of cosmology and particle physics such as the number of leptons, the physical properties of some supersymetric particles and the quark –hadron phase transition.

ACKNOWLEDGEMENTS

This work reported in this presentation has been supported in part by PICS n^{0}18. We are grateful to Elisabeth Vangioni–Flam for a careful reading of the manuscript and Marie–Claude Pantalacci for her skillful typing of it. One of us (JA) is indebted to Grant Mathews for having invited him to such an outstanding conference, for his patience and his help.

REFERENCES

Alcock,C.R., Fuller,G.M. and Mathews,G.J., 1987, ApJ. (15th september issue)
Andreani,P., Vangioni–Flam,E. and Audouze,J., 1988, ApJ.,in preparation
Applegate,J.H., Hogan,C.J. and Scherrer,R.J., 1987, Phys.Rev.,D 35, 1151
Audouze,J., 1987, in Observational Cosmology, eds A.Hewitt, G.R.Burbidge and Li Zhi Fang, Reidel Dordrecht, p.89.
Audouze,J., Delbourgo–Salvador,P. and Salati,P., 1987, in Confrontation between theories and experiments in cosmology, eds.J.Audouze and F.Melchiorri, Varenna Summer School proceedings (in press)
Audouze,J., Lindley,D. and Silk,J., 1985, Astrophys.J.Letters, 293, L
Boesgaard,A. and Steigman,G. (1985), Ann. Rev. Astron. Astroph., 23, 319
Delbourgo–Salvador,P.,Gry,C., Malinie,G. and Audouze,J. (1985), Astron. Astroph., 150,53
Fuller,G.M.,Mathews,G.J. and Alcock,C.R., preprint
Kunth,D., 1986, P.A.S.P., 98, 984
Maleney,R. and Fowler,W.A., (1987), this volume
Pagel,B.E.J., 1986, in Material Content of the Universe(published by the Royal Astron. Society)
Reeves,H., 1974, Ann. Rev. Astron. Astroph., 12, 437
Reeves,H., in Confrontation between theories and experiments in cosmology, eds J.Audouze and F.Melchiorri,Varenna Summer School proceedinga (in press)
Salati,P., Delbourgo–Salvador,P. and Audouze J., 1987, Astron. Astroph., 173,1
Sciama,D.W., 1984, in Large Scale Structure of the Universe, cosmology and fundamental physics, eds G.Setti and L.Van Hove, First ESO–CERN symposium p. 3
Vangioni–Flam,E. and Audouze,J.,1987, Astron. Astroph. (in press)
Wyse R. and Silk,J., 1987, ApJ. Letters, 313, L 11
Yang,J., Turner,M.S., Steigman,G., Schramm,D.N., Olive K.A., 1984, Astrophys. J., 281, 493

Cosmology with Decaying Vacuum Energy*

KATHERINE FREESE[1], FRED ADAMS[2],
JOSHUA FRIEMAN[3] AND EMIL MOTTOLA[1]

[1] *Institute for Theoretical Physics, University of California*
Santa Barbara, California 93106

[2] *Departments of Physics and Astronomy*
University of California, Berkeley, California 94720

[3] *Stanford Linear Accelerator Center, Stanford University*
Stanford, California 94305

ABSTRACT

Motivated by recent attempts to solve the cosmological constant problem, we examine the observational consequences of a vacuum energy density which decays in time. For all times later than $t \sim 1$ sec, the ratio of the vacuum to the total energy density of the universe must be small. Although the vacuum cannot provide the "missing mass" required to close the universe today, its presence earlier in the history of the universe could have important consequences. We discuss restrictions on the vacuum energy arising from primordial nucleosynthesis, the microwave and gamma ray background spectra, and galaxy formation. A small vacuum component at the era of nucleosynthesis, $0.01 < \rho_{vac}/\rho_{rad} < 0.1$, increases the number of allowed neutrino species to $N_\nu > 5$, but in some cases would severely distort the microwave spectrum.

---

* Invited talk presented by J. Frieman at the ACS Symposium on "The Origin and Distribution of the Elements," New Orleans, 31 August-4 September 1987.

The cosmological constant has alternately been the most maligned and the most neglected of the constants of Nature. The seeds of this mistreatment lay in its early history: it is well known that Einstein introduced it purely to obtain steady state cosmological solutions to general relativity. When Hubble subsequently discovered the universe was expanding, the original motivation for the cosmological term was removed, and Einstein recommended with embarrassment that it be dropped as an ugly blemish on his theory. Except for occasional aberrations, as a rule, cosmologists have been happy to follow Einstein in forgetting about it. With the advent of unified gauge theories, this is no longer possible.

In the standard cosmological model, the early universe is thought to have passed through a series of symmetry breaking phase transitions at various energy scales M_x. As the temperature drops below M_x, the vacuum energy density associated with the order parameter (e.g., a Higgs field) changes by $O(M_x^4)$. It is therefore puzzling that the upper bound on the present value of the vacuum energy density, $\rho_{vac} < (0.004 \text{ eV})^4$, is much smaller than any of the energy scales associated with particle physics. Even if such a cancellation can be arranged classically, there is at present no known low energy symmetry which prevents quantum corrections from inducing a large value for ρ_{vac}. Since the stress energy of the vacuum is $T_{\mu\nu} = \rho_{vac} g_{\mu\nu}$, i.e., $p_{vac} = -\rho_{vac}$, it enters Einstein's equations precisely as a cosmological constant $\Lambda = 8\pi G \rho_{vac}$. The upper bound above, which is equivalent to $\Omega_{vac} = \rho_{vac}/\rho_{crit} \lesssim 3$, comes from the limit on the cosmological constant from measurements of the Hubble constant and the age of the universe[1], $\Lambda \lesssim 9H_0^2 \simeq 10^{-56} \text{ cm}^2$, or $\Lambda/m_{Pl}^2 \lesssim 10^{-119}$ in gravitational units. Thus, the problem: the cosmological constant is known to be tiny in any natural scale of units, but in the context of particle physics it does not appear to be a naturally small parameter.

The hope is sometimes voiced that a fundamental quantum theory of gravity will require $\rho_{vac} = 0$, but such a theory must in fact give rise to a cosmological term (say, at the Planck scale) which is precisely cancelled by all lower energy contributions (e.g., at the electroweak scale and below) to one part in 10^{119}!

Thus, the cosmological constant problem is essentially a difficulty of physics at very 'low' energies, suggesting that its solution will come instead from new physics which is manifest at low energies, or large distance- and timescales.

Along this line, several authors have recently discussed mechanisms for dynamically reducing ρ_{vac} to a very small value over cosmological timescales[2,3]. The simplest example[2] is a classical scalar field with a potential which depends on the spacetime curvature, $V(\phi) = V_0 - \xi R\phi^2$, where $\rho_{vac} = V_0 = \Lambda_0/8\pi G$ is the initial vacuum energy density and $\xi > 0$. Neglecting the scalar field kinetic energy, Einstein's equations give $R \sim 8\pi G V(\phi)$. If the field starts near the origin, then initially $R \sim \Lambda_0$ and the field begins to roll down the potential exponentially fast. This reduces both $V(\phi)$ and, by Einstein's equation, the Ricci scalar R over time. But this implies that the slope of the scalar field potential is also decreased, so the field slows down. Asymptotically, $\phi \sim t$ and the effective vacuum energy redshifts away, $\rho_{vac} \sim t^{-2}$. This simple 'feedback' model is indicative of the classical relaxation mechanisms which have been proposed[2]. It is also possible that the dynamical effects of *quantum* fields may render de Sitter space (the spacetime dominated by a cosmological constant) unstable to conformal perturbations [3]. At present, the significance for cosmology of such an instability is unclear, since it is not known how the system would evolve away from the initial de Sitter solution.

These ideas suggest the intriguing possibility that the universe evolves to a state in which the effective cosmological term (ρ_{vac}) is small and continues to decrease with time. In this talk, I summarize the consequences for observational cosmology of such a continuously *decaying* vacuum energy density[4,5]. Such scenarios are interesting because a redshifting vacuum energy can have effects over many expansion times, while a constant vacuum density (the usual case) could only have become dynamically important at very recent epochs. Our major conclusion is that dynamical models of decaying vacuum energy of a rather general variety are consistent with observational cosmology; however, the deviation from the standard model must be small.

Suppose that in addition to ordinary matter and radiation, there is an energy density associated with the vacuum which is time dependent, $\rho_v(t)$. This additional component enters the Einstein equation for the Robertson-Walker scale factor $a(t)$,

$$\left(\frac{1}{a}\frac{da}{dt}\right)^2 \equiv \left(\frac{\dot{a}}{a}\right)^2 = \frac{8\pi G}{3}(\rho_m + \rho_r + \rho_v), \tag{1}$$

where we have assumed flat spatial sections ($k = 0$). The pressure of the vacuum is assumed to have the usual form $p_v = -\rho_v$, while $p_m = 0$, $p_r = \rho_r/3$.

The conservation equation for energy-momentum takes the form

$$\dot{\rho}_v + \frac{1}{a^3}\frac{d}{dt}(\rho_m a^3) + \frac{1}{a^4}\frac{d}{dt}(\rho_r a^4) = 0 \tag{2}$$

We notice immediately that if $\dot{\rho}_v \neq 0$, at least one of the ordinary adiabatic relations $\rho_r \sim a^{-4}$, $\rho_m \sim a^{-3}$ ceases to be valid. If $\dot{\rho}_v < 0$ then entropy or matter *must* be generated in the expansion. In this talk, I focus on the radiation-dominated epoch ($\rho_m < \rho_r$), in which case the second term in Eqn.(2) is small and can be neglected. That is, we are only concerned with the coupling of the vacuum to massless radiation, and shall further assume the coupling to massive particles is suppressed, e.g. due to threshold effects. (In the dynamical decay scenario of refs.2,3, the only lengthscale in the problem appears to be the Hubble radius or possibly the Compton wavelength associated with a very small mass, so we expect the radiation emitted to be peaked at very long wavelengths or ultralow energies. For a discussion of vacuum decay to massive particles, see ref.[4].)

It is useful to define a new parameter which characterizes these models,

$$x = \rho_v/((\rho_r + \rho_v) \tag{3}$$

From Eqn.(2), we obtain[5] the evolution equation

$$\frac{\dot{x}}{x} = \frac{\dot{\rho}_v}{\rho_v} + 4\frac{\dot{a}}{a}(1 - x) \tag{4}$$

There are three possibilities for the behavior of $x(t)$ at large t: (i) for $x \to 1$, the vacuum term dominates, and the universe becomes de-Sitter-like as the radiation is redshifted away. This case is ruled out at the level of the bounds of Ref. 1; (ii) the vacuum density falls more rapidly than the radiation density, *i.e.*, $x \to 0$, and we recover the standard cosmological model; (iii) the only genuinely new cosmology is obtained if x approaches a non-zero constant betwen 0 and 1, which corresponds to the vacuum and radiation densities redshifting at the same rate. If the vacuum and the radiation are coupled by particle creation or if the relaxation mechanism is of the 'feedback' form discussed above (i.e., the scalar field responds to the total curvature, which gets a contribution from the radiation), one may expect behavior of the form (iii) to be generic. We consider only this case in the following.

From Eqn.(4), we find

$$\rho_v(t) = \left(\frac{x}{1-x}\right)\rho_r \sim a^{-4(1-x)}. \tag{5}$$

As expected, the radiation density drops more slowly as a function of the scale factor than in the standard cosmology, whereas the matter density approximately redshifts in the usual way, $\rho_m \sim a^{-3}$, if matter creation is negligble. With this scaling of the two components, the constraint that an early radiation epoch be followed by a matter dominated era requires that $x < \frac{1}{4}$ in both the matter and radiation epochs. Eqns.(1) and (5) are easily solved to yield

$$a \sim t^{\frac{1}{2(1-x)}} \;,\; \rho_v = 3x/32\pi G(1-x)^2 t^2 \tag{6}$$

Increasing x towards unity speeds up the expansion rate of the universe in the radiation era.

As long as the created radiation reaches thermal equlibrium, it can be characterized by its temperature, with $\rho_r = \frac{\pi^2}{30} g_{\rm eff} T^4$; here $g_{\rm eff}$ is the number of relativistic degrees of freedom. In this case, from Eqns.(5,6), we find

$$T(t) = \left[\frac{16\pi^3 G\, g_{\rm eff}(1-x)}{45}\right]^{-\frac{1}{4}} t^{-\frac{1}{2}}, \tag{7}$$

for the radiation temperature as a function of time in the radiation-dominated epoch. The electromagnetic radiation created by vacuum decay thermalizes completely up to at least a time $t_T \approx 10^5$ sec (and, in particular, throughout primordial nucleosynthesis); thus it makes sense to describe the radiation by a Planck spectrum with a temperature given by (7). The assumption of thermal equilibrium determines how the radiation number density and energy per particle change with time. From Eqn.(5), as long as the photons remain in thermal equilibrium, we have $T \sim a^{x-1}$. To maintain a Planck distribution, the energy per particle must redshift like the temperature, so $E_\gamma \sim a^{x-1}$ as well. From Eqn.(5), this implies the photon number density scales as $n_\gamma \sim a^{-3(1-x)}$.

The assumption that photons are created in vacuum decay also implies that the baryon to photon ratio, n_B/n_γ, decreases as the universe expands. Since baryons are not created, $n_B \sim a^{-3}$, and the baryon to photon ratio thus scales as

$$\eta = \frac{n_B}{n_\gamma} \sim a^{-3x} \sim T^{\frac{3x}{1-x}} \tag{8}$$

at least up to t_T. (We have not considered the case where the vacuum decays, in whole or in part, into noninteracting, nonthermal particles such as gravitons, or shadow photons but our treatment does include massless neutrino production.)

i) Nucleosynthesis

Since a non-zero vacuum component changes both the expansion rate through Eqn.(6), the temperature-time relation, Eqn.(7), and the baryon to photon ratio (8), it can alter the delicate balance with nuclear reaction rates at the time of helium and deuterium synthesis that holds in the standard cosmology. (For a review of standard big bang nucleosynthesis, see the talk by Steigman in this volume and references therein.) At the high temperatures in the early universe, the ratio of neutrons to protons is determined by its thermal equilibrium value,

$$n/p = e^{-Q/kT}, \quad T \geq T_F \tag{9}$$

where the neutron-proton mass difference $Q = 1.293$ MeV and k is Boltzmann's constant. Neutrons drop out of equilibrium below a freeze-out temperature T_F, where the weak interaction rates can no longer keep up with the expansion of the universe. Below T_F the n/p ratio continues to fall due to β-decay on the time scale of the neutron half-life τ_n. In the standard model, nucleosynthesis begins at a temperature approximately given by

$$T_D = \frac{2.2 \text{ MeV}}{-\ln \eta}. \tag{10}$$

Once T_D is reached, deuterium becomes stable against photodissociation and nucleosynthesis takes place very rapidly, efficiently converting essentially all of the available neutrons into ^{4}He. In this approximation, the primordial helium abundance Y_p is given by

$$Y_p = \left(\frac{2n}{n+p}\right)_D = \left(\frac{2n}{n+p}\right)_F \exp\left[-\Gamma(t_D - t_F)\right] \approx \frac{2e^{-\Gamma t_D}}{1 + \exp\left[Q/kT_F\right]} \tag{11}$$

where the final approximation is valid since $\Gamma^{-1} = \tau_n / \ln 2 \gg t_F \sim 1$ sec.

In the presence of a small vacuum component $x \ll 1$ (we will see from the numerical results that x must be less than 0.1), we can illustrate heuristically the deviation from the standard model. Recall that freeze-out occurs when a typical $n \Leftrightarrow p$ weak interaction rate $\Gamma \sim G_F^2 T_F^5$ is equal to the expansion rate $H \sim [G(\rho_r + \rho_v)]^{\frac{1}{2}}$. Since $\rho_v + \rho_r = \rho_r/(1-x)$ and $\rho_r \sim T^4$, we have $H \sim T^2/(1-x)^{-1/2}$. Equating $\Gamma \simeq H$, we obtain

$$T_F = \bar{T}_F (1-x)^{-\frac{1}{6}} \tag{12}$$

where an overbar indicates the standard model value ($x = 0$). By itself this would tend to increase Y_p by increasing the n/p ratio at freeze out. However, Eqns.(7)

and (10) indicate that

$$t_D = \bar{t}_D(1-x)^{-\frac{1}{2}}, \tag{13}$$

which increases the available time for neutrons to β-decay. This turns out to be the larger effect in the domain of interest, so that Y_p is a *decreasing* function of x.

The numerical analysis, using Wagoner's[6] nucleosynthesis code modified to include a nonzero vacuum term, does indeed follow this general trend, although the results differ quantitatively. One of the most dramatic changes from the standard model (not included in the discussion above) is the behavior of the entropy, which in the standard model is constant throughout and after nucleosynthesis (except for the infusion of e^+e^- pairs). In the decaying vacuum model the entropy per baryon can change drastically through nucleosynthesis and continues to change afterward according to Eqn.(8).

We present our results graphically for ^{4}He and D abundances in Figs.1-3 for $N_\nu = 3, 4$, and 5 neutrino species. For comparison with observation we have chosen a temperature $T_N = 10^8$ K to signal the end of nucleosynthesis; after this time, element abundances from the code no longer change significantly even for nonzero x, although the entropy continues to drop according to Eqn.(8). Requiring that $0.22 \leq Y_p \leq 0.26$ and $10^{-5} \leq D/H \leq 10^{-4}$[7], we find the (η, x) plane at T_N is restricted as shown. For fixed number of light neutrino species N_ν, the constraints on the vacuum component are: $x < 0.08(N_\nu = 3)$, $x < 0.09(N_\nu = 4)$, and $x < 0.10(N_\nu = 5)$. Although at most four neutrino (or equivalent numbers of light) species can be accommodated in the standard model, for $x \geq 0.01$ five neutrinos (or more) are consistent with the observed element abundances (see Fig.3).

We have also confirmed consistency with observation of the ^{7}Li abundance obtained from the code for this range of parameters. The abundances of ^{4}He, D, and ^{7}Li are all lower than in the standard model, whereas the H abundance is slightly higher. The constraints on $\eta(T_N)$ become more restrictive in the presence of a nonzero x, but remain within the same range as in the standard model ($10^{-10} \leq \eta(T_N) \leq 10^{-9}$). Thus we reach these conclusions regarding nucleosynthesis:

1) Primordial nucleosynthesis in the presence of a vacuum component with $x \leq 0.1$ is consistent with observations of abundances of ^{4}He, D and the other light nuclei.

2) If $x > 0$, the preferred values of η at nucleosynthesis lie within the same range as in the standard model.

3) If $x > 0$, Y_p decreases. Observational estimates of the ^{4}He abundance in HII regions have decreased recently, and may make even 3 light neutrinos uncomfortable for the standard nucleosynthesis model (See the contribution by Pagel in these proceedings.) If these observations hold up and if $N_\nu > 3$ neutrinos are found at the SLC (or if $N_\nu > 4$ neutrinos are found, regardless), the standard model would be in difficulty. This could be resolved by the presence of a small vacuum component at nucleosynthesis. We also note that non-standard scenarios (with baryon diffusion or late hadronic decays) with a large baryon density, $\Omega_B \simeq 1$, rely on reducing an initial overproduction of ^{4}He. Since our model underproduces ^{4}He, the constraint on the vacuum component cannot be evaded by invoking such a scenario.

ii) Microwave Background Distortions

An interesting feature of models with $x =$const. and $\rho_v \sim 1/t^2$ is that some fraction of the microwave background photons in the present universe was created by the decay of the vacuum. The spectrum of radiation emitted by the decaying ρ_v is model-dependent: in general it may be quite different from the Planck distribution appropriate for fully equilibrated radiation. If this is the case, and if the processes involved in the relaxation of the injected photon spectrum toward equilibrium are not 100% efficient, then distortions of the Planck spectrum may arise[8].

In this section we explicitly assume that the vacuum does not decay into photons fully equilibrated to a Planck spectrum. As mentioned above, the most likely possibility seems to be that the emission is peaked at long wavelengths ($E_\gamma < kT$). In that case the photons would be efficiently absorbed by the free electron plasma via inverse bremsstrahlung, since the cross section for this process rises like $1/\omega^3$. At frequencies lower than the plasma frequency, any electromagnetic radiation produced by the decaying vacuum is rapidly damped and its energy transferred to the plasma by ohmic heating. (However, a zero-frequency magnetic component may survive; one can speculate that vacuum decay may produce large-scale primordial magnetic fields.) The result in either case would be to increase the electron energy density relative to the radiation density. However, at very early times, *i.e.*, for redshifts greater than $z_T = 6.3 \times 10^4 (\Omega_B h^2)^{-\frac{6}{5}}$, the injected energy is completely thermalized by double Compton and bremsstrahlung process: no distortions survive. [In this and all the following we take $H_O = 100h$ km/sec Mpc^{-1} for the present value of the Hubble parameter, $T_O = 2.7\,^\circ$K for the present radiation temperature and we assume 3 massless neutrino species;Ω_B is the density parameter for the ionized gas. We also neglect the small x-dependent factors in all redshifts defined in this section.]

When $z < z_T$, the injected radiation energy heats the electron plasma, and photon production by the electron gas continues in the far Rayleigh-Jeans region ($\hbar\omega \ll kT$). At higher energies, however, photon production by the hot electrons becomes inefficient, and Compton scattering cannot redistribute the excess low energy radiation toward the peak. Thus, except at the very low end of the spectrum, we have $T_e > T_r$, and multiple scattering off the electrons shifts the background radiation to higher frequencies without changing the total number of photons. The spectrum then takes on a Bose-Einstein form, with a nonzero chemical potential μ. If $\frac{T_e - T_r}{T_r} \ll 1$, the resulting value of μ (also small compared to kT_r) depends only on the total amount of energy injected into photons, $\Delta\rho_r$, and is *independent* of their initial frequency distribution:

$$\mu = 1.4\ kT_r \frac{\Delta\rho_r}{\rho_r} \tag{14}$$

For a continuous injection of photons by vacuum decay, $\frac{\Delta\rho_r}{\rho_r}$ is the time integral of $-\frac{\dot{\rho}_v}{\rho_r}$ (the contribution of neutrinos to ρ_r cancels from the integral),

$$\frac{\Delta\rho_r}{\rho_r} = -\int_{z_i}^{z_f} \frac{\dot{\rho}_v}{\rho_r}\frac{dt}{dz}dz = \frac{x}{1-x}\ln\left(\frac{\rho_r(z_i)}{\rho_r(z_f)}\right)$$

$$= 4x\ln\left(\frac{z_i}{z_f}\right) \tag{15}$$

by Eqn.(5). Thermalization to a Bose-Einstein spectrum requires multiple photon scattering, but the average number of scatterings per photon decreases as the universe expands. For times later than $t_1 \cong 10^{11}$ sec (or redshifts less than $z_1 = 8.5\times10^3(\Omega_B h^2)^{-\frac{1}{2}}$), multiple scattering becomes inefficient, and the background subsequently evolves too slowly to relax to a Bose-Einstein spectrum. So we take $z_i = z_T$ and $z_f = z_1$ for these μ distortions. Since observations of the microwave background spectrum require $\mu < 0.01kT_r$ [9] we obtain the bound on x:

$$x < 4\times10^{-4} \tag{16}$$

where we have taken $\Omega_B h^2 = 2.5\times10^{-2}$ here and below.

At later times $t > t_1$, energy injected and efficiently absorbed by the electron plasma produces a different distortion of the microwave background spectrum. Compton scattering shifts the photons to higher energies, creating an excess in the Wien region and a shortage in the Rayleigh-Jeans part of the spectrum. The resulting spectrum is parametrized by a new variable y, which can again be related to the total energy injected:

$$y = \frac{1}{4}\frac{\Delta\rho_r}{\rho_r} = x\ \ln\left(\frac{z_i}{z_f}\right) \tag{17}$$

by Eqn.(15).

If we assume that the injected photons are too low in energy to reionize the gas after it has recombined, then z_f for this distortion should be taken to be of the order of the redshift of recombination, $z_2 \cong 10^3$. For $t > t_2$, energy injection will continue to raise the temperature of the residual ionized gas and heat the intergalactic medium, without distorting the microwave spectrum.

Taking $z_i = z_1$ and $z_f = z_2$ in Eqn.(17) and using the observational bound $y < 0.02$ [9] yields the following bound on x:

$$x < 5 \times 10^{-3} \tag{18}$$

so that the μ bound is the most stringent constraint on x we have obtained. These constraints are so severe because the background radiation is being subjected to the injection of energy over many expansion times, when the processes responsible for restoring equilibrium are inefficient. We reiterate that the key assumptions used in deriving these bounds are (i) that the vacuum produces photons which are out of equilibrium with the pre-existing radiation and (ii) that essentially all of the energy injected by the vacuum decay goes into heating the electron gas to $T_e > T_r$. If the vacuum decays into some non-interacting form of dark matter instead, then (ii) need not be true and we would again lose the very stringent bounds of Eqns.(16,18).

iii) Entropy Generation

Since entropy is produced by the decay of the vacuum, η decreases with temperature according to Eqn.(8). The vacuum energy density can be constrained by the evolution of η after nucleosynthesis. The most stringent bound is obtained if the vacuum decays to a thermal spectrum of radiation for all time. In this case, Eqn.(7) applies through the present epoch,

$$\eta(2.7\text{ K}) = \eta(T_N) \left(\frac{2.7\text{ K}}{T_N} \right)^{\frac{3x}{1-x}} . \tag{19}$$

Taking $\eta(2.7\text{ K}) \gtrsim 2 \times 10^{-11}$ [7] for a conservative lower bound today and $\eta(T_N) \leq$

10^{-9} at nucleosynthesis, we find

$$x \leq 0.07 \tag{20}$$

On the other hand, if, as considered above, the radiation produced by vacuum decay only thermalizes up to a time t_T, then η remains constant for $T < T_T \simeq \left(3.3 \times 10^5 (\Omega h^2)^{-\frac{6}{5}}\right)^{1-x} 2.7$ K. In this case, we obtain the less stringent bound

$$x < 0.15. \tag{21}$$

Additional model-dependent bounds on x from the evolution of η arise from consideration of big bang baryogenesis; the limits are comparable to those above.

We have investigated the cosmological constraints on and consequences of a vacuum energy density which dynamically decays in time. We conclude that such a scenario can be consistent, but the universe cannot be vacuum-dominated for times later than about $t \sim 1 sec$. For vacuum decay to a non-thermal radiation distribution, the microwave background spectrum provides the strongest constraint, $x < 4 \times 10^{-4}$. On the other hand, if the radiation produced by the vacuum retains a Planck spectrum for all time, the requirement that the baryon-to-photon ratio not drop too low after nucleosynthesis gives the strongest bound, $x < 0.07$. Vacuum decay appears to be a promising framework for solving the cosmological constant problem, but more work needs to be done in constructing realistic particle physics models. For example, the simplest 'feedback' relaxation models suggest values of $x \sim 1$. It would be interesting to see if more sophisticated models can naturally generate the requisite smaller values of x reported here, without fine tuning.

We thank Bill Hiscock, Rocky Kolb, Joe Silk, Albert Stebbins, Mike Turner and Bob Wagoner for helpful discussions and Hardy Hodges for aid in running the nucleosynthesis code. We especially thank Albert Stebbins for enlightening discussions about microwave background distortions. K.F. especially thanks

Michael Turner for discussions and for hospitality at Fermilab, where part of this work was carried out. This research was supported in part by the National Science Foundation under Grant No.PHY82-17853, supplemented by funds from the National Aeronautics and Space Administration, at Santa Barbara and by the Department of Energy, contract DE-AC03-76SF00515 at SLAC.

REFERENCES

1. P.J.E. Peebles, *Astrophys.Jour.* **284** (1984), 439 ; J.E. Gunn and B.M. Tinsley, *Nature* **257** (1975), 454 ; J.R. Gott, J.E. Gunn, D.N. Schramm and B.M. Tinsley, *Astrophys.Jour.* **194** (1974), 543 . This bound applies to a vacuum density which has remained constant over the last several expansion times of the universe.

2. A. Dolgov, in *The Very Early Universe*, ed. by G. Gibbons, S.W. Hawking, and S.T.C. Siklos (Cambridge Univ. Press, Cambridge 1983). E. Witten, in *Shelter Island II:Proc.*, ed. by R. Jackiw, N. Khuri, S. Weinberg and E. Witten (MIT Press, Cambridge 1985). F. Wilczek, 1983 Erice Lectures. L. Abbott, *Phys.Lett.* **150B** (1985), 427; lectures in *Bariloche SILARG Symposium*, 1985. T. Banks, *Phys.Rev.Lett.* **52** (1984), 1461, *Nucl.Phys.* **B249** (1985), 332 ; S. Barr, Brookhaven preprint BNL 38423 (1986). L. H. Ford,, *Phys.Rev.* **D35** (1987), 2339.

3. A. Starobinsky, *Phys.Lett.* **91B** (1980), 99 ; L.H. Ford, *Phys.Rev.* **D31** (1985), 704 ; C. Hill and J. Traschen, *Phys.Rev.* **D33** (1986), 3519; E. Mottola, *Phys.Rev.* **D31** (1985), 754, **D33** (1986), 1616, **D33** (1986), 2136 ; P. Mazur and E. Mottola, *Nucl.Phys.B*, to appear.

4. K. Freese, F. Adams, J. Frieman and E. Mottola,, *Nucl.Phys.* **B287** (1987), 797.

5. W.A. Hiscock, *Phys.Lett.* **166B** (1986), 285.

6. R.V. Wagoner, *Astrophys.Jour.* **179** (1973), 343.

7. A.M. Boesgaard and G. Steigman, *Ann. Rev. Astron. Astrophys.* **23** (1985), 319 ; D.N. Schramm and R.V. Wagoner, *Ann. Rev. Nucl. Science* **27** (1977), 37. K.A. Olive, D.N. Schramm, G. Steigman, M.S. Turner and J. Yang, *Astrophys.Jour.* **246** (1981), 557 ; J. Yang, M.S. Turner, G. Steigman, D.N. Schramm and K. Olive, *Astrophys.Jour.* **281** (1984), 493 .

8. Ya.B. Zeldovich and R.A. Sunyaev, *Astophys.Sp.Sci.* **4** (1969), 301 . For reviews, see R.A. Sunyaev and Ya.B. Zeldovich, *Ann. Rev. Astr. Ap.* **18** (1980), 537 , and L. Danese and G. De Zotti, *Rev. del Nuovo Cim.* **7** (1977), 277 .

9. J.B. Peterson, P.L. Richards, and T. Timusk, *Phys.Rev. Lett.* **55** (1985), 332 ; G.F. Smoot, *etal.*, *Astrophys.Jour.* **291** (1985), L23 .

FIGURE CAPTIONS

1) Element abundances as a function of the vacuum energy density parameter x and the baryon-to-photon ratio at T_N, $\eta = \eta_{-10}10^{-10}$ for $N_\nu = 3$ neutrino species. The primordial ^{4}He abundance satisfies $0.22 < Y_p < 0.26$ and the ratio $10^{-5} < D/H < 10^{-4}$. Cross-hatching indicates the allowed region.

2) Same as Fig.1 for $N_\nu = 4$ neutrino species.

3) Same as Fig.1 for $N_\nu = 5$ neutrino species. Note that models with $x > 0$ can accomodate $N_\nu > 4$.

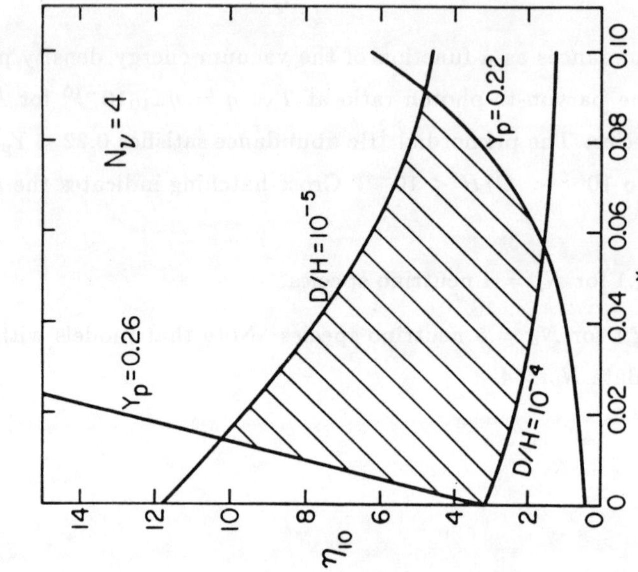

$N_\nu = 4$
$Y_p = 0.26$
$Y_p = 0.22$
$D/H = 10^{-5}$
$D/H = 10^{-4}$
η_{10}
x
0
2
4
6
8
10
12
14
0.02
0.04
0.06
0.08
0.10

Fig. 2

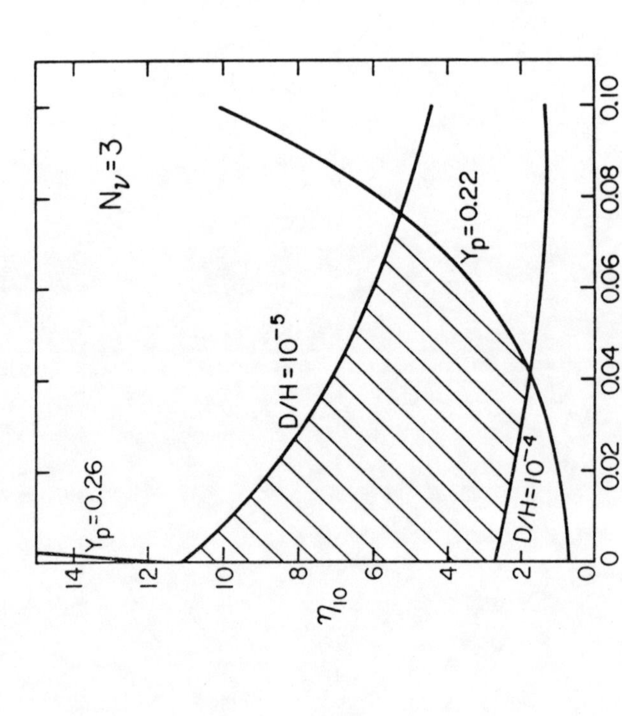

$N_\nu = 3$
$Y_p = 0.26$
$Y_p = 0.22$
$D/H = 10^{-5}$
$D/H = 10^{-4}$
η_{10}
x
0
2
4
6
8
10
12
14
0.02
0.04
0.06
0.08
0.10

Fig. 1

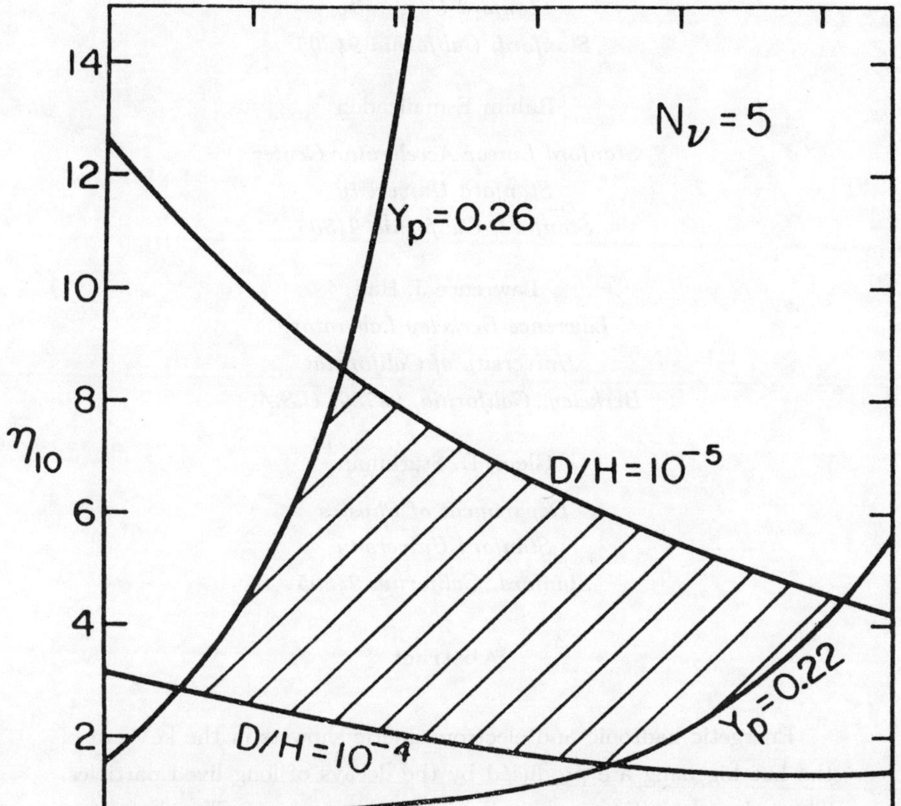

14
12
10
8
6
4
2
0
η_{10}
0
0.02
0.04
0.06
0.08
0.10
x
$N_\nu = 5$
$Y_p = 0.26$
$D/H = 10^{-5}$
$Y_p = 0.22$
$D/H = 10^{-4}$

Fig. 3

Primordial Nucleosynthesis During The keV Era *

Savas Dimopoulos

Department of Physics
Stanford University
Stanford, California 94305

Rahim Esmailzadeh

Stanford Linear Accelerator Center
Stanford University
Stanford, California 94305

Lawrence J. Hall
Lawrence Berkeley Laboratory
University of California
Berkeley, California, 94720, U.S.A.

Glenn D. Starkman

Department of Physics
Stanford University
Stanford, California 94305

Abstract

Energetic hadronic and electromagnetic showers in the keV era of the hot big bang are produced by the decays of long lived particles. These showers initiate a new phase of nucleosynthesis. The abundance ratios of D, ^3He, ^6Li and ^7Li are given by fixed points of rate equations, which are determined by nuclear physics not by the nature of the decaying particle. The fixed points are independent of prior abundances, so that constraints from the MeV era of nucleosynthesis evaporate, except for a requirement that ^4He not be underproduced. For example, $\Omega_B = 1$ and many more than four neutrino species are both

*Talk presented by Lawrence Hall at the American Chemical Society Symposium "Origin and Distribution of the Elements", New Orleans Aug. 31 –Sept. 4 1987.

possible. Within the accuracy of our calculation (there are uncertainties of at least a factor of three), the abundances agree with those inferred from observations. Considerable ^6Li is produced and must be depleted in both population II halo stars and in the galactic disk. We predict ^6Li, ^3He and D abundances in primordial material which are higher than conventional nucleosynthesis.

In a recent paper[1] we gave a detailed analysis of how the decays of long–lived massive particles can lead to resynthesis of the light nuclei. In this talk we review the analytic derivation of these results and explore some of their consequences.

Consider a particle X of mass m_X anywhere from a few GeV to hundreds of TeV and lifetime[2] τ_X between $10^4 - 10^6$ sec. When the X decays the universe is a hot plasma of photons, electrons, protons and light nuclei at a temperature of several keV. The interactions of the decay products of the X with the background plasma cause electromagnetic[3] and baryonic showers.[1] The high energy photons of the electromagnetic shower are removed by pair production off the abundant thermal photons $\gamma\gamma \rightarrow e^+e^-$. This produces[3] a sharp upper cutoff to the photon spectrum given by

$$E_{max} \simeq \frac{m_e^2}{25T} \ln \frac{\eta}{5.10^{-10}} \tag{1}$$

$\eta = n_B/n_\gamma$ is the baryon to photon number density ratio. As the universe cools, E_{max} increases so that successively more and more tightly bound nuclear species become vulnerable to photodissociation. The resulting photon spectrum per X–decay is given by[1]

$$\xi_{\gamma^*}(E) = \begin{matrix} \frac{M_X}{2E_{max}^{1/2}} & \frac{1}{E^{3/2}} & E < E_{max} \\ 0 & & E > E_{max} \end{matrix} \tag{2}$$

These photons interact predominantly with electrons except for an occasional photodissociation of a light nucleus.

Baryonic showers occur when the X decays into an energetic baryon.[1] An energetic baryon, whether a neutron <u>or</u> a proton, loses very little energy to the plasma of photons and electrons as long as its kinetic energy exceeds a GeV[1]. Therefore, it slows down by sequentially colliding and sharing its energy with several protons and nuclei and exciting them to large kinetic energies. These "heated" protons and nuclei collide with each other and undergo a new phase of nucleosynthesis resulting in the production of ξ_i nuclei of type i per baryonic X decay. These ξ_i's were calculated in reference [1].

We proceed with an approximate analytic treatment of simplified rate equations for the nuclear abundances. This treatment stresses the physics of the abundance evolution, and gives results which are in excellent agreement with the numerical solutions of the complete rate equations given in reference [1].

Taking τ_X small enough so that the ^4He reduction occurs predominantly by baryo destruction rather than photodestruction[4]

$$\dot{f}_4 = +f_X^o \Gamma_X e^{-\Gamma_X t} r_B^* \xi_4 \tag{3}$$

where ξ_4 is negative, r_B^* is the effective branching ratio of X to final states containing baryons as precisely defined in ref. [1], $\Gamma_X = \tau_X^{-1}$, f_A is the number density of A divided by the thermal photon number density and f_A^o is the initial abundance. The fractional destruction of ^4He therefore determines[1] $\frac{f_X^o}{f_B} r_B^*$,

$$\frac{f_X^o}{f_B} r_B^* \simeq 1.15.10^{-3} - \frac{(Y - .24)}{40} + 2.2.10^{-4} \ln \Omega_B h_o^2 + 7.10^{-4}(N_{100} - 3.36). \tag{4}$$

where N_T is the effective number of degrees of freedom contributing to ρ at temperature T keV:

$$\rho(T) = \frac{\pi^2}{30} N_T T^4. \tag{5}$$

$\Omega_B h_o^2$ and N_{100} appear because they determine the amount of ^4He produced during the conventional nucleosynthesis era, Y is the final value of the primordial ^4He mass fraction.

The equations for the abundances of $D, {}^3He, {}^6Li$, and ^7Li have a common feature: the dominant production occurs inside the hadronic showers and involves the relevant ξ_i, and the dominant destruction is by photodissociation.

$$\dot{f}_i = f_X^o \Gamma_X e^{-\Gamma_X t} r_B^* \xi_i - n_i \int_{Q_i}^{E_{max}(t)} f_{\gamma^*}(E) \sigma_{\gamma i}(E) dE \tag{6}$$

where Q_i is the binding energy, n_i the number density and $\sigma_{\gamma i}$ the photodissociation cross–section of the nuclear species i, $E_{max}(t)$ is given by (1), together with the time temperature relation

$$\frac{t}{sec} = \frac{2.4}{\sqrt{N_T}} \left(\frac{MeV}{T}\right)^2. \tag{7}$$

The non–thermal photon spectrum is given by its fixed point value,[1]

$$f_{\gamma^*}(E) = \frac{\xi_{\gamma^*}(E) f_X^{\circ} \Gamma_X e^{-\Gamma_X t}}{n_e \sigma_c(E)}. \tag{8}$$

where $\sigma_c(E)$ is the Compton scattering cross–section averaged over non–forward angles, and $\xi_{\gamma^*}(E)$ is given in equation (2).

If equations (6) reach their fixed points ($\dot{f}_i = 0$) the abundances of D. ^3He, ^6Li, and ^7Li, after the keV era nucleosynthesis, are given by

$$\frac{f_i}{f_B} \simeq 30 \xi_i Q_i^{1/2} E_{max}^{1/2}(t_{fi}) \frac{r_B^*}{M_X}. \tag{9}$$

We have taken $\sigma_{\gamma i}(E)/\sigma_c(E) = 1/30$. At the freeze–out time t_{fi}, the photodissociation rate for species i drops below the expansion rate of the universe and the baryoproduction of species i is negligable because sufficient X's have decayed. In practice this occurs while the equations are at their fixed points. In order that the abundances reach the fixed point values of equation (9), $E_{max}(t_{fi}) > Q_i$, which translates into a condition on τ_X

$$\tau_X \left(\frac{N_1}{3.36}\right)^{1/2} \gtrsim 4 \cdot 10^5 sec \tag{10a}$$

where we take 1 keV to be the characteristic temperature for the keV era of nucleosynthesis.

The constraint that little ^4He is photodissociated,[4] in particular that the resulting ^3He production is negligible, gives

$$\tau_X \left(\frac{N_1}{3.36}\right)^{1/2} \lesssim 8 \cdot 10^5 sec \tag{10b}$$

The fixed point result of equation (9) is of great importance. The keV era nucleosynthesis erases any previous abundance of $D,^3He,^6Li$ and ^7Li. Furthermore the ratios

$$D : {}^3He :{}^6 Li : {}^7Li \simeq 1 : 1 : 10^{-5} : 10^{-6} \tag{11}$$

are given predominantly by the ratios of ξ_i since the remaining i dependence is weak.[1] The absolute value of these abundances determines M_X/r_B^*:

$$\frac{M_X}{r_B^*} \simeq (2 - 10) \cdot 10^4 GeV. \tag{12}$$

The abundances (9) were obtained from a simple analytic approximation; they agree well with a more detailed numerical calculation.[1] However, the ξ_i have not been computed numerically, and contain uncertainties arising from a lack of nuclear cross–section data. The detailed calculation of the ξ_i in reference 1 could easily be off by a factor of 3. This would affect the results (11) and (12).

Consider the processes leading to ξ_i. For a given X mass it is first necessary to estimate the average number of $\bar{p}p$ or $\bar{n}n$ pairs per X decay and their spectra. For certain values of M_X e^+e^- data is available, but an extrapolation is required for most masses. Fortunately, within our uncertainties this only affects the magnitudes of all the ξ_i and not their ratios. For ξ_2 and ξ_3 the number of D, ^3H or ^3He produced in a N^4He collision is computed (N represents a primary $n, \bar{n}, p$ or $\bar{p}$). Next one must find out what fraction of these D, ^3H or ^3He are destroyed before they are slowed to thermal equilibrium. Finally the process must be iterated: high energy p and n undergo many nuclear scatterings before they thermalize.

Calculations of ξ_6 and ξ_7 are much lengthier. The dominant production mechanisms are $^3H^*\ ^4He \to\ ^6Li\ n$ and $^4He^*\ ^4He \to\ ^7Li\ p$, $^7Be\ n$. To calculate these one must know the spectra of $^3H^*$ and $^4He^*$ produced in high energy N^4He collisions. One must then compute the fraction of $^3H^*$ and $^4He^*$ which produce ^6Li and ^7Li. A major uncertainty here is that, as far as we know, the ^3H and ^4He spectra and the ^6Li and ^7Li production cross–sections for these processes have not been measured for all energies and scattering angles of interest to us. Hence there are uncertainties in the ratio ξ_6/ξ_7 as well as in the ratios of ξ_6 or ξ_7 to ξ_2 or ξ_3. However, it is clear why ξ_6 and ξ_7 are much smaller than ξ_2 and ξ_3: there are extra stages in the processes of building up ^6Li and ^7Li and the probability of each stage occurring is small because there are many other possible reactions which do not produce ^6Li or ^7Li.

Our scenario leads to significant primordial ^6Li production, typically $f_6/f_7 \simeq 10$. Is this a problem? Since f_6/f_7 ratios $\leq \cdot 1$ are seen in both population II halo stars[5] and in the galactic disk[6] substantial ^6Li depletion must occur in both cases. For the halo stars this is probable, the rate for ^6Li destruction is about a hundred times that for ^7Li destruction at any point in the convective zone of the star. Hence even if the ^7Li depletion is very small, a large ^6Li depletion can occur.

Depletion of ^6Li in the the galactic disk seems to require significant stellar processing of the disk material. One then expects that if stellar processing depletes

^6Li by a factor 1–A (the astration factor), D will also be changed by $1-A$, and ^3He by $1-(1-g)A$ (where g is expected to be between 1/4 and a 1/2)[7]. Hence we should choose our X parameters, M_X/r_B^* and τ_X, such that our primordially produced ^6Li, D and ^3He are all overabundant. For example for $M_X/r_B^* = 3.10^4$ GeV and $\tau_X = 6.10^5$ sec. we find primordial abundances of 1.5×10^{-10} for ^7Li, 2×10^{-9} for ^6Li, 2.5×10^{-4} for ^3He and 9×10^{-5} for D, all relative to H. The ^7Li abundance agrees well with the population II stars, and, as in the standard scenario, essentially all the ^7Li observed in the disk (10^{-9} abundance) must be produced during the evolution of the galaxy. If the astration factor $A = .9$ (90% of disk material processed), then the disk abundances are 2×10^{-10} for ^6Li, 1×10^{-4} for ^3He (using $g = 1/3$) and 9×10^{-6} for D.

These numbers agree well with observations. This is highly significant: it may have been that the abundance ratio predictions were off by many orders of magnitude. However, this success must be qualified. Until we have better data on the $N^4He \rightarrow^4 He$, ^3H and $^3H\ ^4He \rightarrow^6 Li\ n$ reactions, we cannot have confidence in our values of ξ_6 and ξ_7. Our present uncertainties are at least a factor of 3. The ranges quoted for τ_X and M_X/r_B^* in (10) and (12) assume a factor of 3 uncertainty. When these uncertainties are resolved, these allowed ranges may increase or they may disappear altogether. Our present central values for the ξ_i require about 90% stellar processing in the disk. With more accurate values for the ξ_i this may also change substantially; for example lower ^6Li could mean that 50% or less astration is needed. Fifty percent stellar processing is required in the standard scenario to explain the observed abundances of heavy elements. Some authors claim that these abundances are consistent with over 90% processing.

Our theory of nucleosynthesis has several implications for both particle physics and cosmology. In sharp contrast to the standard theory, it works for a very broad range of N_{100}, N_1 and η (or $\Omega_B h_o^2$). This implies that we can afford to allow major changes to the structure of the theory at 100 keV or 1 keV or its baryon to photon density without reaching a contradiction with the observed light element abundances.

Our theory gives us the freedom to explore new avenues in both cosmology and particle physics. Ordinary nucleosynthesis conflicts with any physics that creates a ^4He overabundance, whereas our theory is consistent with any such excess. The first application of these ideas[1] was to show that it is possible to have a completely baryonic $\Omega = \Omega_B = 1$ universe. In fact it is possible to have a universe with any proportion of weakly interacting and baryonic dark matter in the allowed range $0.03 < \Omega_B h_o^2 < 1.1$.[8] A second application is that we now have the freedom to have more than three light neutrino species (i.e., $N_1 = N_{100} = N_\nu$).

To conclude, we have discovered that within the accuracy of our calculations a late decaying particle satisfying (4), (10) and (12) can account for the observed light nuclear abundances. The D : ^3He : ^6Li : ^7Li ratio is determined by fixed point behavior, which erases prior abundances and yields predictions which depend only on nuclear physics. The D : ^3He : ^7Li ratios are successful. We predict that primordial material should have higher ^6Li, ^3He and D abundances than would be expected with standard nucleosgnthesis. This appears to be a viable possibility which needs further investigation.

Acknowledgments

It is a pleasure to thank David Schramm, Mike Turner and Bob Wagoner for very valuable discussions and suggestions.

S.D. And G.D.S. acknowledge support by the National Science Foundation under Grant NSF–PHY–86–12280 to Stanford University. G.D.S. is a National Science and Engineering Research Council of Canada Postgraduate Scholar, R.E. acknowledges support by the D.O.E. under contract DE–AC03–76SF00515 at SLAC. L.J.H. acknowledges support by the D.O.E. under contract DE–AC03–7600093 and in part by NSF under Grant PHY–85–15857. L.J.H. also acknowledges partial support by a Sloan Foundation Fellowship and a Presidential Young Investigator

References

1. S. Dimopoulos, R. Esmailzadeh, L. Hall and G. Starkman "Is the Universe Closed by Baryons?" SLAC preprint SLAC preprint SLAC–PUB–4356, June 1987.

2. Lifetimes in this range can result from dimension five operators at the Planck or Grand scales (e.g., the gravitino) or by dimension six operators at the intermediate scale.

3. M.L. Burns and R.V.E. Lovelace, AP.J. 262 (1982) 87.
F.A. Aharonian and V.V. Vardanian, Yerevan Physics Institute preprint

827(54)-85 (1985).
D. Lindley, Ap. J. 294(1985) 1.

4. In reference (1) this is shown to be a necessity. Photodestruction of ^4He leads to too large a ^3He abundance.

5. E. Maurice, F. Spite and M. Spite, Astrophys. 132 278 (1984).

6. P.A. Vander Bout, R. L. Snell, S. S. Vogt and R. G. Tull, Ap. J. 221, 598 (1978).

7. J. Yang, M. S. Turner, G. Steigman, D. N. Schramm and K. A. Olive, Ap. J. 281 493 (1984).

8. Such theories have recently been drafted to explain the observed large scale structure of the universe. P.J.E. Peebles, "Yet Another Scenario fore Galaxy Formation" Princeton university preprint 1987 and "Origin of the Large Scale Galaxy Peculiar Velocity Field: A Minimal Isocurvature Model" Princeton University preprint, January 1987.
G.R. Blumenthal, B.A. Dekel and J. R. Primack, SCIPP 87/81.

Primordial Nucleosynthesis in a Universe with Nonlinear Inhomogeneities

Richard A. Matzner

Relativity Center
The University of Texas at Austin
Austin, Texas 78712

Joan M. Centrella

Department of Physics and Atmospheric Sciences
Drexel University
Philadelphia, Pennsylvania 19104

Tony Rothman

Relativity Center
The University of Texas at Austin
Austin, Texas 78712

James R. Wilson

Lawrence Livermore National Laboratories
Livermore, California 94550

ABSTRACT

A completely relativistic, nonlinear plane-symmetric hydrodynamics and general relativity code is used to give the geometrical evolution within which we compute cosmic nucleosynthesis of light isotopes through ^4He. We present models of two different classes.

In the first class, either dynamically important constituents of matter, or dynamically important amounts of gravitational wave energy density are strongly inhomogeneous on scales comparable to the horizon size at $2 \times 10^{10} K$, at the initialization of the nucleosynthesis computation. Local enhancement or depression of the element abundance is then found, which may persist to the present. Average abundances found are not identical to the predictions of homogeneous cosmologies, and local ^4He abundances fluctuate in the range 22% - 27% for models whose homogeneous counterpart predicts $X(^4He) \sim 0.24$. These inhomogeneous models are consistent with all observational constraints, in particular, with the observed isotropy of the microwave radiation background.

In the second class, the geometry is taken to be an exactly homogeneous isotropic (Robertson-Walker) background, but the baryon density is taken to be inhomogeneous on scales 0.0005 to 0.05 times the horizon scale at $2 \times 10^{10} K$. This kind of baryon inhomogeneity may arise in some models of the quark-hadron phase transition. Because neutron electromagnetic interactions are only via their magnetic moment, substantial diffusion of neutrons into the baryon scarce regions is possible, leading to unusual neutron-rich evolution. We present examples of such evolution, which show shifts in the final averaged-abundances.

INTRODUCTION

The universe we live in is verifiably lumpy. On a large enough scale, however, one can make the argument that there is some sort of average smoothness, and observations certainly provide a strong basis for this assertion. For instance, the isotropy in the microwave background to about one part in 10^5 seriously restricts possible models of the cosmos, and argues strongly for a smooth universe (Uson and Wilkinson, 1984).

Other observables of the universe we inhabit are the nuclear abundances. These are measured with relatively high accuracy, and their values can be compared to the theoretical prediction in a homogenous, isotropic (i.e., completely regular) cosmology. In general they agree remarkably well with those predictions, giving further credence to the idea of a very regular cosmology (Boesgaard and Steigman, 1985).

Nevertheless, there are definite reasons to explore, as we do, the possibility of inhomogeneous cosmologies. One reason is to find out if the resulting abundances are sensitive to the inhomogeneities. They are, and this provides a further restriction on the spectrum of inhomogeneities. This is of some interest if one is restricting inhomogeneities that are not otherwise restricted and are essentially just reflections of the initial state of the universe. It is of substantially more interest if the inhomogeneity is predicted by some theory. In fact, we shall discuss applications to both of these points. Whatever restrictions are developed then raise the challenge of consistency with the observed large scale structure in the universe.

THE COSMOLOGICAL MODEL

The computer code used to carry out these simulations is a fully general relativistic 1-dimensionally inhomogeneous code which simultaneously evolves the geometry, fluid sources, and the abundances of the nuclear isotopes. In this paper we concentrate principally on the evolution of the inhomogeneities in the abundances. The nuclear chain is an abbreviated one, including reactions only up to ^4He. The rates are taken from Wagoner (1973). An improved and extended nucleosynthesis program is being developed with H. Kurki-Suonio. The geometrical evolution is described extensively in Centrella and Wilson (1983,1984); the code including the nuclear evolution is described by Centrella, Matzner, Rothman and Wilson (1986).

In summary, the code solves the first order evolution equations for the single dynamical variable of the geometry, and for its associated momentum. All other geometrical variables are solved for via the constraint equations which are elliptic equations defined at a particular time. The choice of time variable is based on the constant trace of extrinsic curvature (= constant average Hubble parameter) definition. This introduces another elliptic equation into the system, that for the **lapse function** that implements this time choice. One also solves elliptic equations for the **shift vectors** that maintain a simple, diagonal, two element equal, form for the metric. This class of metrics is general enough to admit general dynamical inhomogeneities in matter, in energy density, and in gravitational wave density. A typical evolution shows the development of localized density enhancements, together with propogating gravitational, sound and shock waves.

The hydrodynamics describing the matter content of the universe is carried out in the formulation developed by Wilson (1979). Energy density and baryon density transport are independently possible.

No provision is made in this code for "free" or almost "free" streaming of neutrinos; they are treated as part of the $p = \rho/3$ radiation fluid.

The nuclear evolution follows the baryon matter streamlines. At nucleosynthesis temperatures (with observationally consistent entropy/baryon) the baryon matter is a dynamically unimportant constituent of the cosmic fluid, but of course provides the basic stuff of which the eventual measurements are made. Abundance diffusion is important only on relatively short length scales, but has been included to handle the neutron diffusion phenomenon described below. The nuclear evolution is carried out using an implicit scheme implementing the technique described by Wagoner, Fowler and Hoyle (1967) to linearize the nonlinear nuclear evolution equations.

LARGE SCALE PRIMORDIAL INHOMOGENEITIES

In this section we consider the evolution of inhomogeneities on scales large enough that abundance diffusion may be neglected. We will consider dynamically important energy density and gravitational-wave inhomogeneities.

The inhomogeneities we consider here, $\delta e/e \sim 5 - 6$, are of the scale of the horizon size ct at the time the simulation begins ($T \sim 2.0 \times 10^{10} K$; $t \sim 1$ second). It should be realized that the horizon scale indicates the largest scale fluctuations that can be in causal contact. Fluctuations on larger scales can grow; those on smaller scales can be dissipated by causal process.

There arises the question of whether inhomogeneities as we now consider can have any relevance at the present cosmological epoch. We must of course consider the conservative results; these computed abundances lead to certain nonstandard mass-averaged final results. On the other hand, one may ask whether localized enhancements or suppressions may be substantial enough to survive to the present. The baryon mass contained in a sphere of horizon size at nucleosynthesis is only a few solar masses and spatial variations of this size would not be detectable. But our models are plane symmetric, and formally our inhomogeneities represent infinite amounts of matter. Planarity arises naturally and a collapsing pancake, for instance, may have two faces within causal contact while extending far more than a horizon size in the two transverse directions. Further, inflow in the transverse directions can be important and can continue long after the 'short' dimension comes within its horizon, because the plane initial conditions mean pressure gradients are small in the transverse directions. Hence the inhomogeneities we find in nuclear abundances may turn out to be persistent inhomogeneities of substantial mass, that could survive as anamolous-abundance regions to the present.

COMPUTATION

Our numerical experiments encompass a range of model inhomogeneities, and initial choices for the matter parameters (the entropy per baryon, the number of light neutrino flavors). In all cases we are more interested in, and more confident of, the effect in our results of the changes from the standard model, than in trying to produce a very precise result equal to the observed abundances.

We will present several different models. In this section, all computation is done on a 50-spatial zone grid and the space is periodically identified; zone 1 is equivalent to the hypothetical zone 51, etc. The evolution is carried through from a rather high initial temperature ($T \sim 2 \times 10^{10} K$) to low temperatures ($\sim 10^8 K$).

Inhomogeneities are introduced by giving a series of amplitudes of cosine contributions to the variables of interest. For instance, the energy density variable E is written

$$E = \sum_{n=o}^{6} E_n cos(2\pi n z / z_{max}) \tag{1}$$

with essentially arbitrarily chosen E_i. (Here z_{max} is the identification length.) Perturbations in the gravitational wave energy labelled by a wave momentum variable K_1, in the baryon density, or in the fluid momentum, are handled in exactly the same way. Once the appropriate Fourier harmonics are chosen, the program uses the field variables so defined to solve the initial value problem; a set of elliptic equations defined on one constant-time slice. This process changes the values of some initially specified quantities to achieve consistency, but typically the resultant data are close to that specified.

As already mentioned, the baryon matter is dynamically unimportant during nucleosynthesis. After the initial value problem is solved, *then* the baryon matter is specified. There are a number of ways that the baryon density can be specified. The simplest is to give a baryon density which increases where the energy density does, with $\rho_{bar} \propto \epsilon^{\frac{3}{4}}$. This 'adiabatic' variation has a constant ratio of photons per baryon. The second approach, and one we report here, is to set exactly constant baryon density on our defined time=constant space after solving the initial value problem, regardless of varying energy density. A third possibility is the so-called isothermal case, where the baryon density varies while the photon density is constant. Once the time slicing is chosen, it can be seen that the constant baryon density situation can be viewed as a linear combination of the adiabatic and isothermal fluctuations, *at least in perturbation theory.*

RESULTS

We present in Table 1 a list of the large-scale inhomogeneous computations we have carried out. We present figures obtained from five of the computations. Run #1, Figs. 1a-1d, is an adiabatic variation from a Robertson-Walker isotropic homogeneous evolution. At the beginning of the computation, the temperature exceeded $2 \times 10^{10} K$ everywhere and the entire computational arena covered only about 0.9 horizon sizes. This is a 2-neutrino flavor model with a baryon to photon ratio $\eta = 6 \times 10^{-10}$ corresponding to a baryon density now (when $T_\gamma = 2.7K$) of $\rho_{bar} = 4 \times 10^{-31} gm/cm^3$. We use certain arbitrary standard inhomogeneities in the energy density:

$$\begin{aligned} E_1 &= 0.6E_o, \\ E_2 &= 0.3E_o, \\ E_3 &= 0.15E_o, \\ E_4 &= 0.075E_o, \\ E_5 &= 0.0375E_o, \\ E_6 &= 0, \end{aligned} \tag{2}$$

cf Eq. (1); and the momentum density $S_{zi} = E_o v_{zi}$, with

$$\begin{aligned} &v_{20} = 0 \\ &v_{z1} = 0.8 \qquad && v_{z5} = 0.05 \\ &v_{z2} = 0.4 \qquad && v_{z6} = 0.025 \\ &v_{z3} = 0.2 \\ &v_{z4} = 0.1 \end{aligned} \tag{3}$$

Table 1

run #	HORIZ at $T = 2\times 10^{10}K$	(0 = adiabatic)	flavors	$\eta \times 10^{10}$	Inhomogeneity	^{4}He	$\langle He^4 \rangle$	$\langle ^2H \rangle$	$\langle ^3H \rangle$
1	0.9	0	2	6	Standard $\delta E, \delta S$	0.226-0.267	0.250	4.08×10^{-5}	1.90×10^{-5}
2	0.9	1	2	6	Standard $\delta E, \delta S$	0.225-0.258	0.240	7.45×10^{-5}	2.42×10^{-5}
3	0.9	0	3	6	Standard $\delta E, \delta S$	0.241-0.282	0.264	4.63×10^{-5}	1.99×10^{-5}
4	0.4	0	2	6	Standard $\delta E, \delta S$	0.229-0.273	0.243	4.52×10^{-5}	2.00×10^{-5}
5	1.3	0	2	6	Standard $\delta E, \delta S$		No	nucleosynthesis	
6	0.9	0	2	6	$\delta E = 0,\ \delta S_z = 0$	0.251-0.264	0.257	4.24×10^{-5}	1.92×10^{-5}
					$K_1 = \sum_{i=1}^{5} K_{1i}$				
					$\times \cos((2i\pi z)/z_{\max})$				
					$K_{1i} = 1, 0.5, 0.25, 0, 0$				
7	0.9	1	2	6	$K_{1i} = 1, 0.5, 0.25, 0, 0$	0.250-0.265	0.257	4.22×10^{-5}	1.92×10^{-5}
8	0.9	0	2	6	$K_i = 1, 0.5, 0, 0, 0.5, 0.5$	0.252-0.262	0.257	4.18×10^{-5}	1.91×10^{-5}
9	0.9	0	2	6	as 6 plus std $\delta E, \delta S_z$	0.239-0.285	0.259	5.28×10^{-5}	2.11×10^{-5}
10	0.9	1	2	6	as 7 plus std $\delta E, \delta S_z$	0.233-0.286	0.256	7.17×10^{-5}	2.36×10^{-5}
11	0.9	0	2	6	as 8 plus std $\delta E, \delta S_z$	0.238-0.280	0.258	5.03×10^{-5}	2.06×10^{-5}
12	0.9	0	2	1.5	std $\delta E,\ \delta S_z$	0.213-0.246	0.233	3.76×10^{-4}	5.25×10^{-5}
13	0.9	0	2	1.5	$\delta E = 0, \delta S_z = 0$	0.232-0.249	0.240	3.97×10^{-4}	5.28×10^{-5}
					$K_{1i} = 1, 0.5, 0.25, 0, 0$				
14	2.5	0	2	6	standard $\delta E,\ \delta S_z$	0.225-0.262	0.246	3.83×10^{-5}	1.83×10^{-5}
15	2.5	0	2	6	standard $\delta E, \delta S_z$	0.227-0.289	0.248	4.50×10^{-5}	2.01×10^{-5}
16	1.4	0	3	1.5	standard $\delta E, \delta S_z$	0.225-0.255	0.241	3.94×10^{-4}	5.27×10^{-5}
17	2.5	0	3	1.5	standard $\delta E, \delta S_z$	0.227-0.262	0.239	4.51×10^{-4}	5.56×10^{-5}

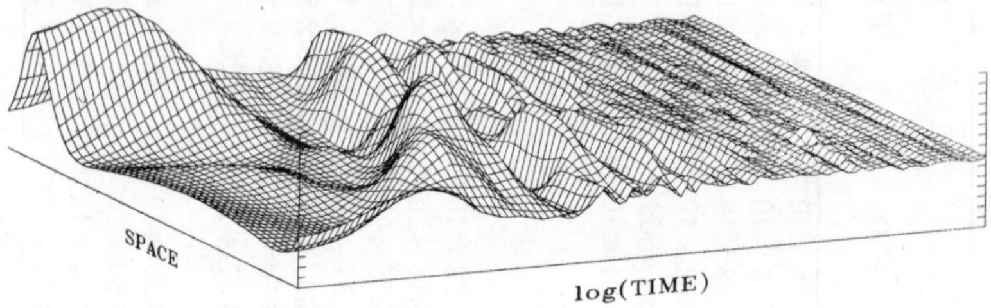

RUN 1
E/TRK**2

SCALE	0.5000E-01		
MAX E	0.3954E+02	MAX E	0.7135E-09
MIN E	0.5963E+01	MIN E	0.6479E-09
(INITIAL SLICE)		(FINAL SLICE)	

Fig. 1a

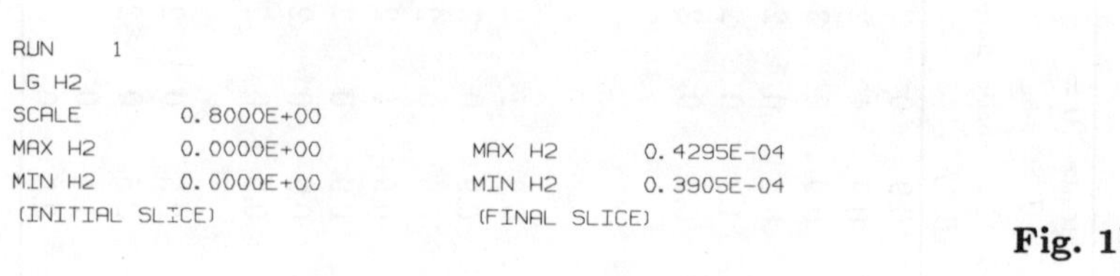

RUN 1
LG H2

SCALE	0.8000E+00		
MAX H2	0.0000E+00	MAX H2	0.4295E-04
MIN H2	0.0000E+00	MIN H2	0.3905E-04
(INITIAL SLICE)		(FINAL SLICE)	

Fig. 1b

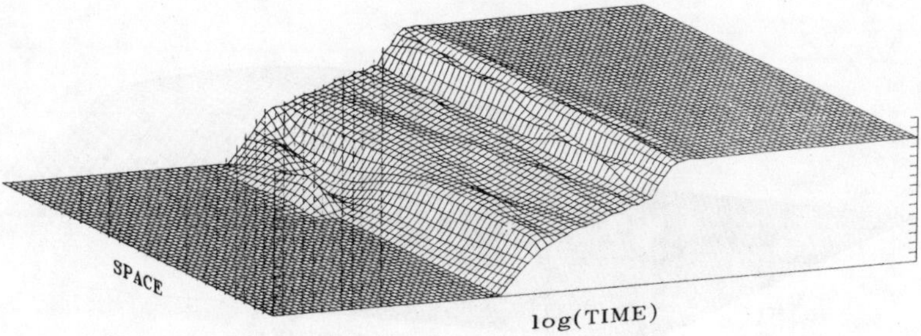

RUN 1
LG HE3
SCALE 0.8000E+00
MAX HE3 0.0000E+00
MIN HE3 0.0000E+00
(INITIAL SLICE)

MAX HE3 0.1942E-04
MIN HE3 0.1878E-04
(FINAL SLICE)

Fig. 1c

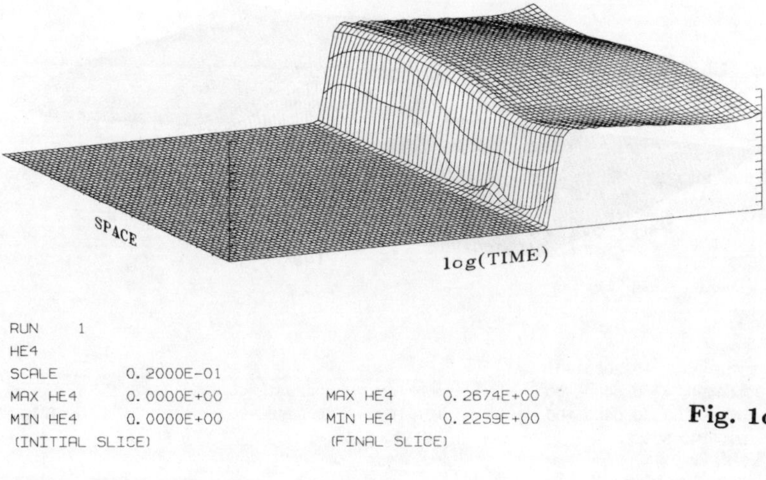

RUN 1
HE4
SCALE 0.2000E-01
MAX HE4 0.0000E+00
MIN HE4 0.0000E+00
(INITIAL SLICE)

MAX HE4 0.2674E+00
MIN HE4 0.2259E+00
(FINAL SLICE)

Fig. 1d

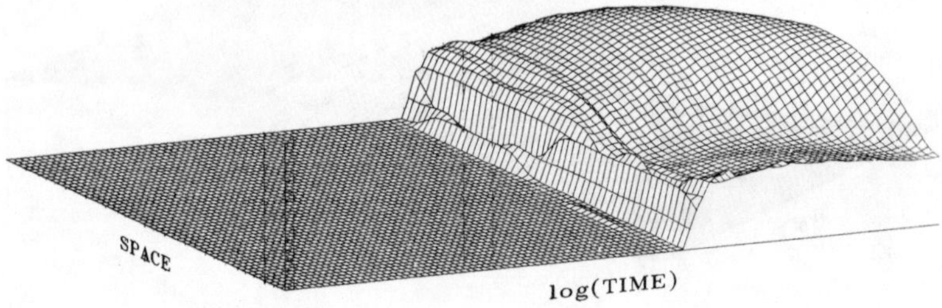

RUN 2
HE3
SCALE 0.2500E-05
MAX HE3 0.0000E+00 MAX HE3 0.3565E-04
MIN HE3 0.0000E+00 MIN HE3 0.1713E-04
(INITIAL SLICE) (FINAL SLICE)

Fig. 2a

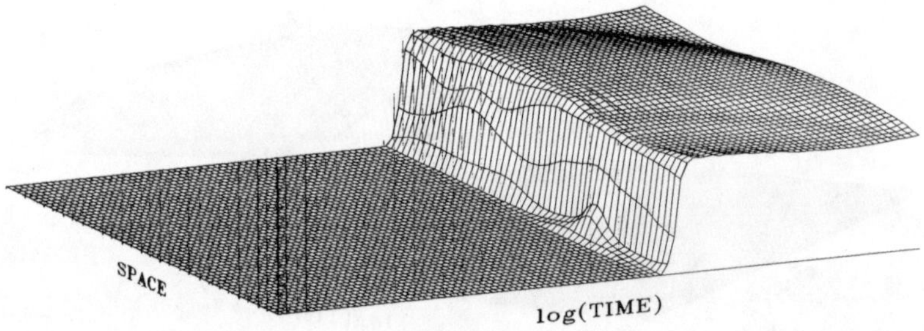

RUN 2
HE4
SCALE 0.2000E-01
MAX HE4 0.0000E+00 MAX HE4 0.2581E+00
MIN HE4 0.0000E+00 MIN HE4 0.2248E+00
(INITIAL SLICE) (FINAL SLICE)

Fig. 2b

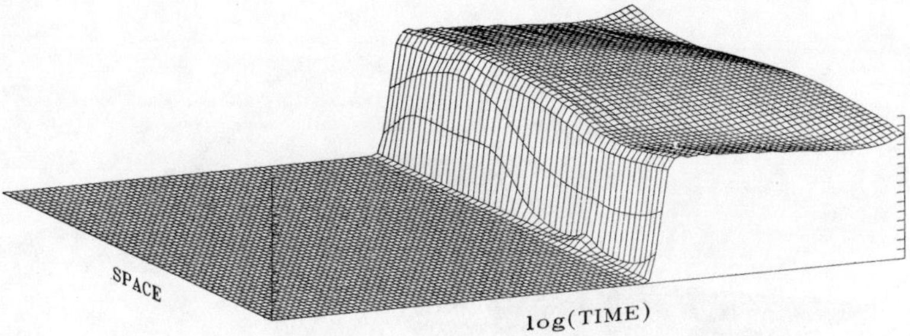

RUN 3
HE4
SCALE 0.2000E-01

MAX HE4	0.0000E+00	MAX HE4	0.2817E+00
MIN HE4	0.0000E+00	MIN HE4	0.2406E+00
(INITIAL SLICE)		(FINAL SLICE)	

Fig. 3

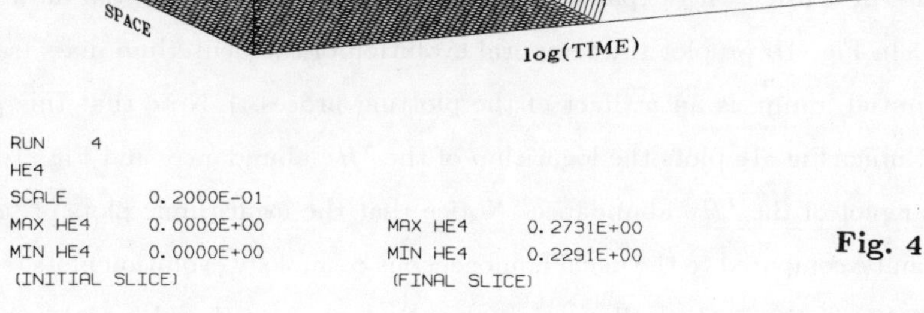

RUN 4
HE4
SCALE 0.2000E-01

MAX HE4	0.0000E+00	MAX HE4	0.2731E+00
MIN HE4	0.0000E+00	MIN HE4	0.2291E+00
(INITIAL SLICE)		(FINAL SLICE)	

Fig. 4

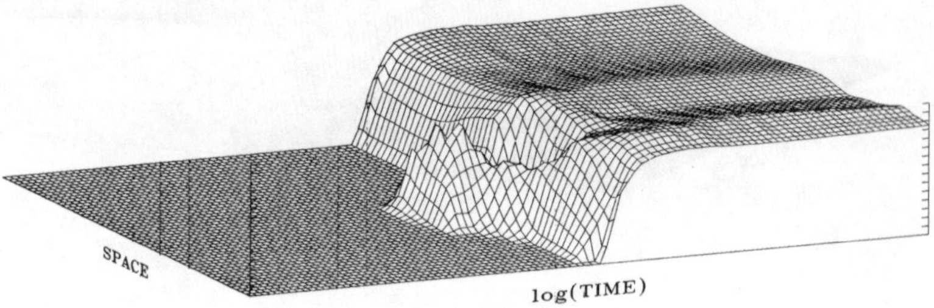

RUN 17
HE4
SCALE 0.2000E-01
MAX HE4 0.0000E+00 MAX HE4 0.2619E+00
MIN HE4 0.0000E+00 MIN HE4 0.2271E+00
(INITIAL SLICE) (FINAL SLICE)

Fig. 5

Run #1 is set with this data.

To show that this is a genuinely inhomogeneous model, we give in Fig. 1a the evolution of an energy density variable E, normalized to remove the overall decrease in density with expansion. The inhomogeneity in the initial data is 5 ~ 6 to 1. In Fig. 1b we plot the temporal evolution of the deuterium mass fraction. (The initial 'ramp' is an artifact of the plotting process.) Note that this plot is logarithmic. Fig. 1c plots the logarithm of the ^3He abundance, and Fig. 1d gives a linear plot of the ^4He abundance. Notice that the logarithmic plots of ^2H and ^3He can be compared to the usual homogeneous cosmology evolution plots (see, eg. Wagoner et al. 1967; Centrella et al. 1986). By looking at the edge of the plot, the reader can see that the evolution is quite similar to the evolution in homogeneous cosmologies. On the other hand, all these plots show that there is inhomogeneity in the time development of the isotopes and in their final abundance. The final ^4He abundance is inhomogeneous, with mass fraction varying 0.226 to 0.267.

Run #2 is similar to run #1 except that it is not an **adiabatic** variation of the homogeneous case. Instead, the baryon density is chosen homogeneous on the first slice, with the baryon to photon ratio specified at one particular point in the space. The point of specification is grid zone **2**, and the result of this choice is to provide a smaller average baryon to photon ratio than was the case in Run #1. The resulting ^2H and ^3He are dramatically more inhomogeneous, and the final ^4He is decreased, all in accord with the effect of lowered baryon-to photon ratio at certain places in the grid. Figures 2a and 2b give the final ^3He and the final ^4He, respectively. Figure 3 gives the final ^4He abundance for Run #3, which is the 3-neutrino flavor model otherwise identical to Run #1. It can be seen that ^4He is raised by approximately 0.015 in the mass fraction, almost exactly the same difference in ^4He found in homogeneous models when an extra neutrino flavor is added.

Figure 4 gives the final ^4He abundance in an adiabatic variation from the homogeneous isotropic standard model. This model has the same sort of inhomogeneity as that of Run #1, but it is compressed into only ~0.45 horizon scales initially. The interesting feature is the dramatic isolated peak in ^4He abundance.

Figure 5 shows the result of Run #17. This is a low density, large spatial scale ($\sim$ 2.5 horizon size at $T = 20 \times 10^9 K$) evolution. In this model, there are three neutrino flavors, but the baryon density is adjusted downward to lower the average ^4He abundance back to a 'standard' value. The ^4He evolution is dramatically inhomogeneous. However, the final inhomogeneity is not as great as in Run #1, and the averaged abundances in Run #17 are quite close to 'standard' values.

In investigating the behavior of these inhomogeneous models, for instance, the behavior of Run #2, it is of interest to compare their evolution to some sort of "averaged" equivalent. We present in Table 2 such a comparison of the results

Table 2

run #	$\langle {}^2H \rangle$	Homo ^2H	Homo with K_1 ^2H	$\langle {}^3He \rangle$	Homo ^3He	Homo with K_1 ^3He	$\langle {}^4He \rangle$	Homo ^4He	Homo with K_1 ^4He
1	4.08×10^{-5}	3.69×10^{-5}		1.90×10^{-5}	1.84×10^{-5}		0.250	0.242	
2	7.45×10^{-5}	6.07×10^{-5}		2.42×10^{-5}	2.30×10^{-5}		0.240	0.239	
3	4.63×10^{-5}	4.18×10^{-5}		1.99×10^{-5}	1.91×10^{-5}		0.264	0.256	
4	4.52×10^{-5}	3.69×10^{-5}		2.00×10^{-5}	1.84×10^{-5}		0.243	0.242	
6	4.24×10^{-5}	3.69×10^{-5}	4.52×10^{-5}	1.92×10^{-5}	1.87×10^{-5}	1.97×10^{-5}	0.257	0.242	0.265
7	4.22×10^{-5}	3.54×10^{-5}	4.34×10^{-5}	1.92×10^{-5}	1.80×10^{-5}	1.93×10^{-5}	0.257	0.242	0.265
8	4.18×10^{-5}	3.69×10^{-5}	4.54×10^{-5}	1.91×10^{-5}	1.87×10^{-5}	1.97×10^{-5}	0.257	0.242	0.265
9	5.28×10^{-5}	3.69×10^{-5}	4.46×10^{-5}	2.11×10^{-5}	1.87×10^{-5}	1.96×10^{-5}	0.259	0.242	0.263
10	7.17×10^{-5}	5.27×10^{-5}	3.54×10^{-5}	2.36×10^{-5}	2.15×10^{-5}	1.80×10^{-5}	0.256	0.240	0.242
11	5.03×10^{-5}	3.69×10^{-5}	3.65×10^{-5}	2.06×10^{-5}	1.87×10^{-5}	1.82×10^{-5}	0.258	0.242	0.242
12	3.76×10^{-4}	3.39×10^{-4}		5.25×10^{-5}	5.08×10^{-5}		0.233	0.226	
13	3.97×10^{-4}	3.39×10^{-4}		5.28×10^{-5}	5.08×10^{-5}		0.240	0.226	
14	3.83×10^{-5}	3.69×10^{-5}		1.83×10^{-5}	1.87×10^{-5}		0.246	0.242	
15	4.50×10^{-5}	3.69×10^{-5}		2.01×10^{-5}	1.87×10^{-5}		0.248	0.242	
16	3.94×10^{-4}	3.92×10^{-4}		5.27×10^{-5}	5.29×10^{-5}		0.241	0.240	
17	4.51×10^{-4}	3.92×10^{-4}		5.56×10^{-5}	5.29×10^{-5}		0.239	0.240	

for all our runs. We give for each run the mass averaged abundances (from Table 1) and then the results from homogeneous models starting with values of the parameters equal to the average of the inhomogeneous parameters. In those cases (Runs ##6-11) where there is significant wave amplitude in the data, there are two possible approaches to this presentation. The average of the wave variable K_1 is zero. But K_1^2 acts as an effective energy density driving the expansion, and has a nonzero average. Hence we should logically include the average of K_1^2 in the total energy density. In cases where it is significant, we give both averaged comparisons.

In considering these large-scale computations, it is necessary finally to decide how well the resultant nuclear abundances correspond to observations. There is a standard way to handle the light isotope abundances, which removes the difficult-to-determine specific entropy (i.e., the baryon to photon ratio) from the problem, by plotting the ^4He abundance vs some other isotope abundance. There are astrophysical reasons to plot the mass fraction of ^4He against the number density of ^3He plus that of ^2H (the ^2H can be burned in stars, but tends to end up in ^3He), with these number densities standardized to the density of hydrogen. If one takes a standard (isotropic homogeneous) model, varying the baryon to photon ratio η produces a single curve (parametrized by η) for a particular number ν of massless neutrinos in the $X(^4He) - (N(^3He) + N(^2H))$ plane. Increasing η moves the point along the constant ν curve toward increasing $X(^4He)$. We plot our averaged abundances here on similar such plots (cf Boesgaard and Steigman, 1985).

Figure 6 gives the results. Only Runs ## 3, 16, 17 are 3-neutrino runs. They differ in their matter parameters in that Run #3 is a high density $\eta = 6 \times 10^{-10}(\rho_{bar} = 4 \times 10^{-31}$ now) 3-neutrino run. It is set up identically to Run #1 (a 2-neutrino flavor run) except for the difference in neutrino numbers. As we have already seen from Figs. 1 and 3, and is also apparent in Fig. 6, the effect of this extra neutrino is increase the $X(^4He)$ by about 0.015. This is the same magnitude as in the increase in going from 2 to 3 neutrino flavors the homogeneous case.

Of the 2-neutrino models, Runs ##2, 4, 14 lie slightly below the curve for the isotropic situation. Run #2 is a non-adiabatic (constant density on the first time slice) model. Runs ##4 and 14 have adiabatic irregularities but differ from Run #1 by having different irregularity scale compared to the horizon. All the other 2-flavor models computed, however, show enhanced or, for those models with substantial wave contributions (Runs ## 6-11), strongly enhanced ^4He for a specified $(N(^2H)+N(^3He))/N(H)$. Adding wave energy is like adding a (perhaps fractional) neutrino species.

The graph also shows the behavior of the 3-neutrino flavor models. Run #3, which differs from Run #1 by simply the addition of the third neutrino, shows substantial enhancement ^4He compared to the isotropic homogeneous case. The

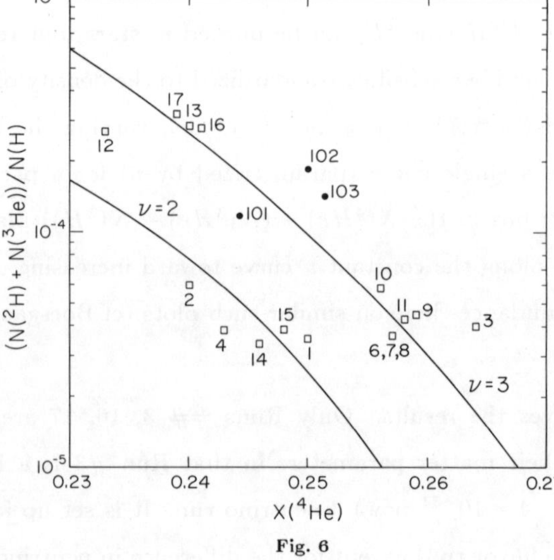

Fig. 6

two low density ($\eta = 1.5 \times 10^{-10}, \rho_{bar} = 10^{-31} gm/cm^3$ now) 3-neutrino runs ##16,17 differ also by having perturbation scales much larger than the horizon scale. In these latter cases, the scale of the grid size in Run #16 equals 1.5 times the horizon at $T \sim 20 \times 10^9 K$, falling to within the horizon size at $T \sim 3.5 \times 10^9$). In Run #17 the grid is originally 2.5 times the horizon size, falling within the horizon at $T \sim 2 \times 10^9$. In contrast, the grid in Run #3 is essentially equal to the horizon at $T \sim 20 \times 10^9$.

The conclusion appears to be that shorter wavelength perturbations lead to enhancements to the ^4He at a given $(N(^3He)+N(^2H))/N(H)$, while longer wavelength perturbations (those that are comparable to the horizon size a later times lower temperatures), tend to produce averaged results very close to the isotropic-homogeneous results appropriate to the specified horizon entropy ratio. Long wavelength adiabatic fluctuations evolve identically to a homogeneous cosmology at each point, as expected. It is only the short wavelength perturbations where causal efforts can influence the evolution that this relationship is broken. A further complication is the presence of gravitational wave energy, which tends to move a model horizontally to the right in Fig. 6, as additional neutrino flavors would do. We will discuss below the points marked 101, 102, 103 on Fig. 6.

SMALL SCALE INHOMOGENEITIES ARISING FROM THE QUARK/HADRON PHASE TRANSITION

The inhomogeneities discussed in previous sections, which have scales typical of the horizon size at nucleosynthesis, have to be postulated a priori as constituents of the initial universe. While they are certainly admitted by observations of the present state of the universe, they are not predicted, and in fact may be counter-indicated if the universe passed through an early inflationary epoch. It has recently been noted, in contrast, that there is a possible rather natural mechanism, which can produce horizon and abundance gradients over scales of meters to perhaps tens of kilometers at temperatures $\sim 2 \times 10^{12} K$, corresponding to hundreds of meters to thousands of kilometers at the nucleosynthesis starting temperature of $2 \times 10^{10} K$, about 10^{-2} to 10^{-6} horizon sizes at $2 \times 10^{10} K$.

This small scale would of course be thoroughly mixed in the subsequent evolution of the universe, but the nonstandard nuclear evolution point by point in the universe can lead to nonstandard abundances after averaging these abundances. In particular, one makes use of the fact that with some choices of the quark-hadron parameters, the baryon density can be quite inhomogeneous. At temperatures below about $5 \times 10^9 K$, neutron-proton interconversion via the weak interaction becomes impossible, and the neutrons diffuse through the plasma impeded only by their strong collisions with protons, and by their electromagnetic interactions via the magnetic moment. The protons remain locked in the electron-photon plasma until temperatures about a factor of 20 lower, when enough electrons have annihilated to allow them to stream also, relatively freely.

Recent investigations by Kurki-Suonio (1985), (1987) and by Kurki-Suonio and Kajantie (1986) have suggested that large spatial scale baryon inhomogeneities are unlikely in the quark-hadron phase transition. Applegate, Hogan and Scherrer (1987) suggest that inhomogenieties of $\sim$ 100 meter scales are possible at the quark hadron phase transition, $T \sim 2 \times 10^{12} K$. We consider such possiblities, and have carried out a number of simulations under the assumption that the baryon number can vary by an arbitrarily large factor on small scales. From the viewpoint of the most interesting nucleosynthesis, one investigates the situation in which there are regions totally devoid of nucleons.

Our plane symmetric model uses the simple diffusion coefficients for neutrons and protons that were obtained by Applegate, Hogan and Scherrer (1987). They are:

$$D_{ne} = \frac{\pi}{16}\left[\frac{M}{m_e}\right]^2\left[\frac{\hbar}{m_e}\right]\frac{1}{(\alpha\kappa)^2}\frac{e^{m_e c^2/T}}{(T/m_e c^2) + 3(T/m_e c^2)^2 + 3(T/m_3 c^2)^3}, \quad (4)$$

the coefficient for neutron diffusion based on neutron-electron electromagnetic scattering; and

$$D_{np} = \tfrac{1}{3} v/(n_p \sigma) \ , \tag{5}$$

the coefficient for strong-force scattering of neutrons by protons. D_n, the neutron diffusion coefficient is given by $D_n = (D_{nd}^{-1} + D_{np}^{-1})^{-1}$. In Eq. (4), M is the neutron mass, m_e is the electron mass, $\hbar$ is Planck's constant, $\kappa = -1.91$ is the anomolous magnetic moment of neutrons, and $\alpha = \frac{1}{137}$ is the fine structure constant. In Eq. (5), σ is the neutron-proton cross section and $v = \sqrt{3T/M}$ is the thermal neutron velocity.

Applegate, Hogan and Scherrer (1987) also give the coefficient for proton diffusion based on scattering on electrons:

$$D_{pe} = \frac{3\pi}{8\alpha^2 \Lambda} \left[\frac{\hbar}{m_e} \right] \frac{T/(m_e c^2) e^{m_e c^2/T}}{1 + 2(T/m_e c^2) + 2(T/m_e c^2)^2} \tag{6}$$

In Eq. (6), Λ is a 'Coulomb logarithm' cutoff taken as $\Lambda = 5$. We include a more or less heuristic term that prevents protons from diffusing faster than neutrons at low temperature:

$$D_p = (D_{pe}^{-1} + D_{np}^{-1})^{-1} \tag{7}$$

This last term, the additional D_{np}^{-1}, has very little effect on the mass- averaged final abundances because by the time D_{pe} becomes large, the nucleosynthesis has almost completed. For purposes of numerical stability the presence of such a term is crucial, so we include it.

In all our simulations we take a fiducial 3-neutrino standard density model. This model has $\eta = 3 \times 10^{-10}$, $\rho_{bar} = 2 \times 10^{-31} gm/cm^3$ now and gives in the homogeneous case mass fractions: $X(^2H) = 1.25 \times 10^{-4}$; $X(^3He) = 3.2 \times 10^{-5}$; $X(^4He) = 0.249$. The average baryon density is held fixed as we consider small-scale variations of this quantity.

At high enough temperatures, $T \gtrsim 5 \times 10^9 K$, the neutron- proton interconversion means that the neutrons cannot diffuse; it also means that the n/p ratio is maintained at essentially its thermal ratio. At high temperatures, the evolution proceeds as it does in any cosmology, and in particular, in the low-baryon density regions the onset of nucleosynthesis occurs at a lower temperature than it does in the high-baryon density regions.

In any region, the onset of nucleosynthesis occurs rather abruptly as the photon temperature falls to the binding energy of deuterium. In the usual model, the number density of neutrons is always less than that of protons. The onset of nucleosynthesis very quickly removes the free neutrons and leaves essentially only protons and heavier isotopes. In contrast, in these small scale baryon inhomogeneous models, it is at least in principle possible that more neutrons could diffuse into regions than the number of protons there initially. In such a case the nuclear evolution could be dramatically nonstandard. In the simulations to be presented here, there is one example in which the evolution is more or less standard (a neutron deficit after nucleosynthesis begins) and several which are anti-standard in that they have a proton deficit after nucleosynthesis begins.

The results presented here should be definitely taken as preliminary because of the very short nuclear chain implemented. This code gives good agreement with other codes in standard homogeneous, isotopic situations. However, as Mathews (1987) has pointed out, one must rethink the inclusion or exclusion of reactions in the network when the evolution may occur in a neutron-rich environment. We (with H. Kurki-Suonio) have an enhancement program underway to expand and improve the nuclear network in this code.

The diffusion equation is treated in the following way. Firstly, the evolution of the nuclear abundances is done via a variant of the Wagoner et al. (1967) scheme, wherein the evolved quantities $Y_i (i = 1 \text{ to } 6, \text{ here})$ are related to the mass fraction of the isotope by

$$a_i Y_i = \rho_i / \rho_{bar} \quad ; \tag{8}$$

the atomic mass a_i of the i_{th} isotope multiplied by the evolved quantity Y_i gives the mass fractional density of the i_{th} isotope. (As before, ρ_{bar} is the total baryon density).

Consistently with the "operator splitting" approach of the code, the diffusion equation is applied to the neutron ($i = 1$) and proton ($i = 2$) abundances by:

a) Evolve normally the abundances Y_i (with inhomogeneous ρ_{bar})in a Robertson-Walker radiation dominated background to obtain a density for each isotope.

b) multiply each Y_i by $a_i\rho_{bar}$: $X_i = a_i Y_i \rho_{bar}$

c) Evolve the additional change in X_i, $i = 1, 2$ via

$$\frac{\partial X_i}{\partial t} - D_i(\nabla^2 X_i) = 0 \tag{9}$$

where the diffusion coefficients D_i are evaluated as indicated above; there is no sum over i in the second term.

d) With the new X_i, resum to obtain the baryon density:

$$\rho_{bar} = \sum_i X_i \ . \tag{10}$$

e) Obtain the diffused values of the Y_i by

$$Y_i = a_i^{-1} X_i / \rho_{bar} \tag{11}$$

The diffusion equation in step c) involves unit lapse, and zero shift vector because the evolution is carried out in a Robertson-Walker dynamical background. Because the background is so simple, we are spared some of the paradoxes of a relativistic formulation of diffusion. (Glaviano and Raymond, 1981). Such difficulties and paradoxes arise if the diffusion velocities ($\sim F_i/X_i$ with F_i defined by Eq. (13) below) approach the speed of light so we demand they stay small. A more technical numerical problem is the fact that regardless of the speed of light criterion, instabilities arise if the diffusion velocities are such that they imply diffusion across one zone in one time step (If the speed of light condition is met, then this second problem can be corrected by different, implicit, numerical techniques;

see Press, Flannery Teukolsky, Vetterling 1986). In our simulations the diffusion velocities are sufficiently slow that neither problem arises.

The Laplacian in the diffusion equation is expanded in the Robertson-Walker spatial metric: $d\gamma^2 = a^2(t)(dx^2 + dy^2 + dz^2)$ as:

$$\nabla^2 X = a^{-3}(a^{-2}X_{,i}\, a^3)_{,i} = a^{-2}X_{,ii} \tag{12}$$

for this spatially constant metric. Differencing proceeds by defining an abundance flux (here we have only z-dependence):

$$F_i = -a^{-2}D_i\frac{\partial X_i}{\partial z} \tag{13}$$

(no sum on i on the right side of this equation) and evolving this part of the abundance time development as a transport problem,

$$\frac{\partial X_i}{\partial t} + \frac{\partial}{\partial z}F_i = 0 \tag{14}$$

using Barton's monotonicity algorithm (Centrella and Wilson 1983, 1984). Plane symmetry here is perhaps more difficult to justify than for the adiabatic fluctuation discussed earlier. Nevertheless, we believe these results are indicative.

The expected spatial scale of the inhomogeneities arising from the quark-hadron transition is substantially smaller than the horizon size at nucleosynthesis, and in fact, just amounts to scales of meters to kilometers at the transition temperature. The overall expansion extends the scale of these inhomogenieties at nucleosynthesis by at least a factor $\sim$ 300. The neutron diffusion rates are such that diffusion across this scale at nucleosynthesis temperatures occurs with a timescale of **minutes**, appropriate to the entire nuclear evolution. Hence, this direct evolution scheme for the diffusion should evolve stably. In these diffusion computations we use a grid with 20 spatial zones only.

H. Kurki-Suonio (private communication, 1987) has estimated that at a temperature of $3 \times 10^{10}K$ (the starting temperature for our nucleosynthesis simulations) the typical inhomogeneity size for the transition-induced baryon inhomogeneities is about 100 meters (arising from a 1-meter scale at the quark-hadron phase transition temperature $\sim 200MeV \cong 2 \times 10^{12}K$), while Hogan (1983) estimates 10,000 meters, arising from a transition scale of 100 meters. This difference in estimates comes from factors involving quantities "of order unity". Scales on the order of one meter $\sim 10^{-4}$ horizon size at $2 \times 10^{12}K$ are too small to credibly survive to nucleosynthesis and if they did are too small a constituent of matter to have a large effect. Scales on the order of 0.1 to $1.0km$ then may, however, do so. According to Kurki Suonio (1987) the probable structure of the inhomogeneity is a large sphere ($\sim 0.01\times$ horizon size at $2 \times 10^{12}K$), of essentially homogeneous baryon abundance, surrounding a very small ($\sim 1m$) region of different (higher) baryon density. The total mass energy of the baryons involved, $\rho_{bar} \sim 10^{-9}\rho_\gamma$ at $10^{12}K$ is sufficient only to produce velocities at most $\sim 10^{-4.5}c$ after the phase transition, so larger inhomogeneities, and their larger effect on the nuclear abundances, might persist.

Our standard neutron diffusion cosmology has baryon inhomogeneities on a scale even larger than Hogan's (1983) estimate. The evolution of the baryon density inhomogeneity in this standard model (normalized to remove the overall expansion) is given in Fig. 7a (labelled Run #102). The scale is chosen so that the entire grid occupies 0.005 horizons at $T = 3 \times 10^{10}K$, corresponding to inhomogeneities of scale ~ 0.3 times horizon at $T \sim 2\times 10^{12}K$. Since our plane symmetric model has three density enhancements, we actually are dealing with lumps and voids separated by distance comparable to ~ 0.1 horizon sizes at the transition temperature, about $1km$ then, but corresponding to $\sim 100km$ at $3 \times 10^{10}K$, and $\sim 30,000km$ at 10^8K. Though these inhomogeneities are too large-scale to fit the estimates based on chromodynamics, they still lead to interesting point-by-point variations from the standard model. It can be seen from Fig. 7a that at late times, diffusion has begun to smooth the baryon density.

It should be noted again that nucleosynthesis is a complex process. In regions with low baryon density nucleosynthesis proceeds more slowly than in high density regions. Before nucleosynthesis begins, neutrons begin diffusing into the low baryon regions; they have a longer latency for nucleosynthesis there, and so these low regions develop a large neutron fraction. Nucleosynthesis proceeds rapidly in the adjacent, high baryon density regions, and neutrons actually diffuse back out of the low baryon density region they originally diffused into, before nucleosynthesis seriously gets underway in these low density regions. In this run, for instance, the neutron fraction reaches ~ 0.8 in certain low baryon density regions (Fig. 7b). Figure 7c gives the evolution of ^2H. Deuterium builds up to higher mass fraction, and leaves more residual abundance in low density regions. Figure 7d shows the ^4He abundance. This isotope shows a paradoxical peak followed by drop-off in mass fraction, in two low density regions. This behavior occurs because there is substantial burning to ^4He in these low density regions, but then a lot of late diffusion of protons into the region, which relaxes the mass **fraction** back to values more appropriate for a low density universe. The mass averaged abundances for Run #102 are: $X(^2H) = 2.7 \times 10^{-3}; X(^3He) = 3.8 \times 10^{-5}$; $X(^4He) = .250$, corresponding to $(N(^2H) + N(^3He))/N(H) = 2.0 \times 10^{-4}$. Based on standard model predictions of $^2H + ^3He$, this model overproduces ^4He by ~ 0.006. (A standard 3-neutrino model would predict over somewhat less denterium for this amount of ^4He.) Note that a relatively 'standard' result obtains from averaging $X(^4He)$ between 0.5% and 47%.

Run #101 is set up similarly to Run #102, with the single exception that the physical scale of the computational grid is taken as a factor of 10 smaller, so the entire grid covers ~ 0.0005 horizon sizes at $T = 3 \times 10^{10}K$. This corresponds to enhancements and depressions of the baryon density on scales $\sim (10^{-2}$ the horizon size at $T \sim 10^{12}K) \sim 100m$ then. Thus, this calculation is appropriate to Hogan's (1983) inhomogeneity scale estimates.

Because the scale is so much smaller, there is ample time for diffusion at late times to fill-in and smooth the baryon density abundance, as Fig. 8a shows. Figure 8b indicates that there are regions where the neutron abundance briefly exceeds that of protons. The maximum neutron fraction approaches 0.91 in certain low density regions, before the neutrons diffuse out again, and protons diffuse in, to eventually react all the neutrons. Figure 8c shows that in the lowest-density regions, the protons are substantially consumed; at late times they diffuse back to homogenize the proton distribution. As in the Run #102 case, there is substantially inhomogeneous deuterium formation, Fig. 8d, leading to a factor of four inhomogenieties in the final abundances. Also as in Run #102, the ^4He formation is dramatically inhomogeneous, leading to regions where the ^4He fraction temporarily exceeds 80%, but is brought back down to more standard (but still widely varying values) by neutron and proton diffusion (final $X(^4He) : 2\% - 44\%$); Fig. 8e. The mass averaged values for this run are: $X(^2H) = 1.8 \times 10^{-4}$; $X(^3He) = 3.8 \times 10^{-5}$, $X(^4He) = 0.244$ corresponding to: $(N(^2H) + N(^3He))/N(H) = 1.36 \times 10^{-4}$. This point lies below the homogeneous $\nu = 3$ curve in Fig. 6, in contrast to the situation for Run #102, which lies above it.

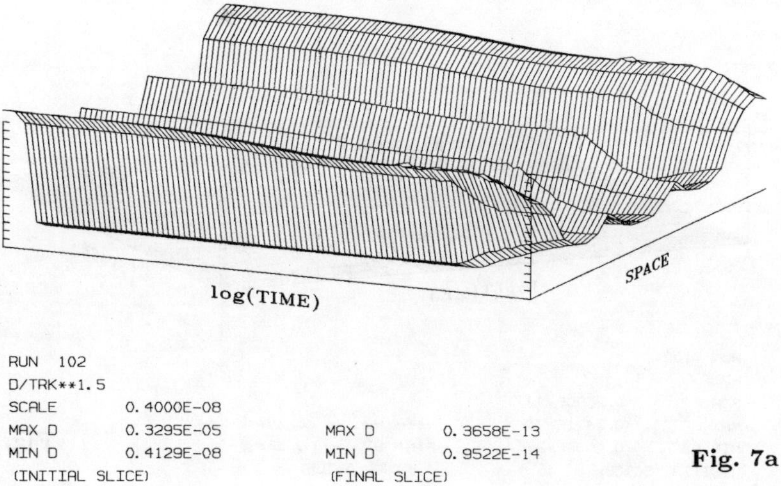

Fig. 7a

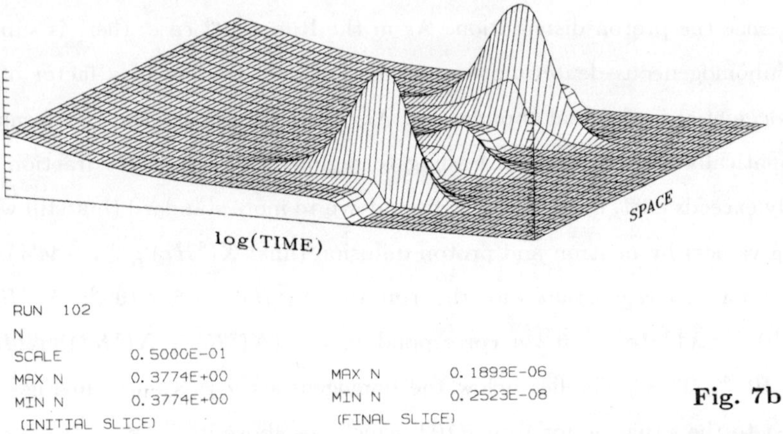
SPACE
log(TIME)
RUN 102
N
SCALE 0.5000E-01
MAX N 0.3774E+00
MIN N 0.3774E+00
(INITIAL SLICE)
MAX N 0.1893E-06
MIN N 0.2523E-08
(FINAL SLICE)

Fig. 7b

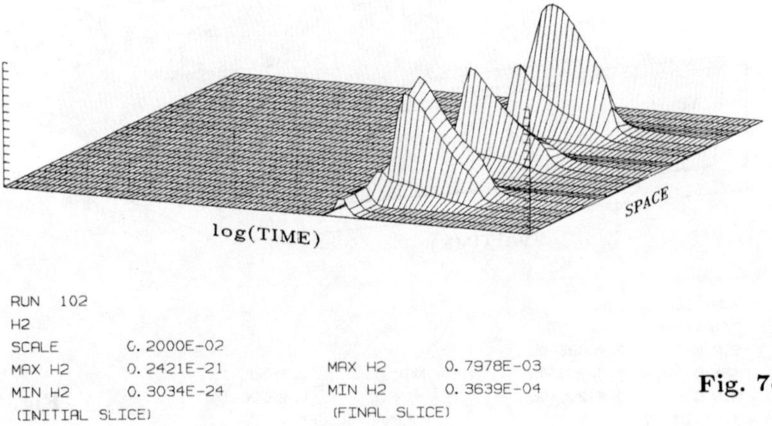
SPACE
log(TIME)
RUN 102
H2
SCALE 0.2000E-02
MAX H2 0.2421E-21
MIN H2 0.3034E-24
(INITIAL SLICE)
MAX H2 0.7978E-03
MIN H2 0.3639E-04
(FINAL SLICE)

Fig. 7c

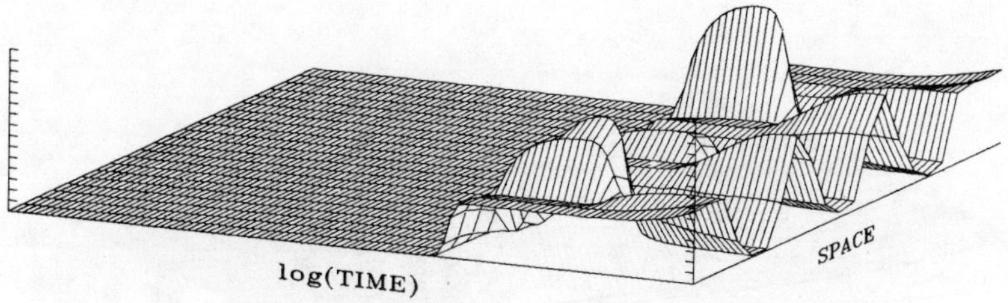

RUN 102
HE4
SCALE 0.5000E-01

MAX HE4	0.2421E-21	MAX HE4	0.4684E+00
MIN HE4	0.3034E-24	MIN HE4	0.5351E-02
(INITIAL SLICE)		(FINAL SLICE)	

Fig. 7d

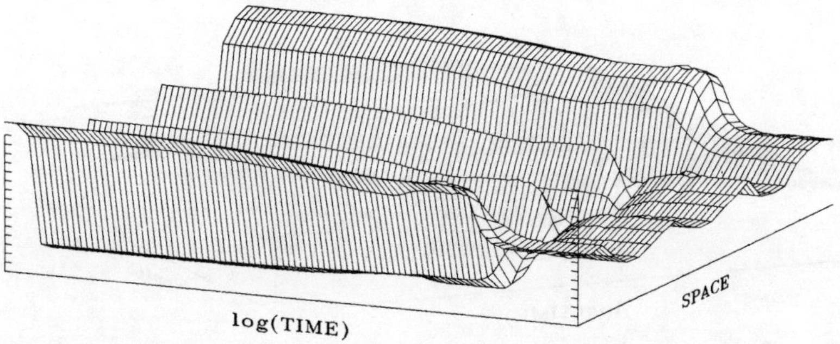

RUN 101
D/TRK**1.5
SCALE 0.4000E-08

MAX D	0.3295E-05	MAX D	0.2431E-13
MIN D	0.4129E-08	MIN D	0.1393E-13
(INITIAL SLICE)		(FINAL SLICE)	

Fig. 8a

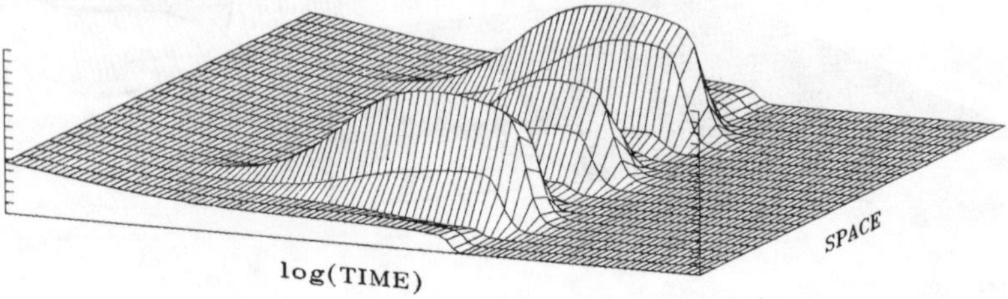

RUN 101
N
SCALE 0. 8000E-01

MAX N	0. 3774E+00	MAX N	0. 2657E-07
MIN N	0. 3774E+00	MIN N	0. 1407E-07
(INITIAL SLICE)		(FINAL SLICE)	

Fig. 8b

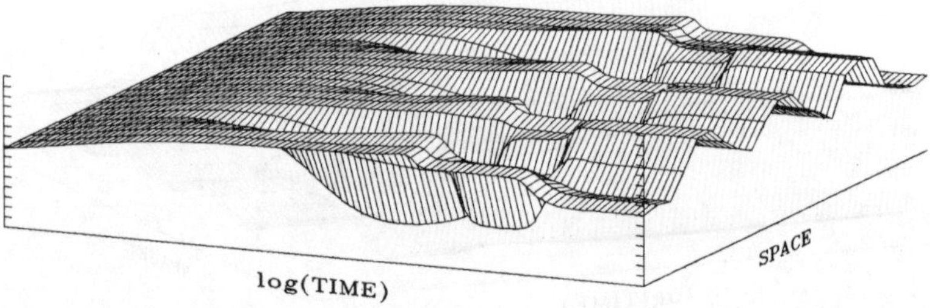

RUN 101
H1
SCALE 0. 8000E-01

MAX H1	0. 6226E+00	MAX H1	0. 9832E+00
MIN H1	0: 6226E+00	MIN H1	0. 5633E+00
(INITIAL SLICE)		(FINAL SLICE)	

Fig. 8c

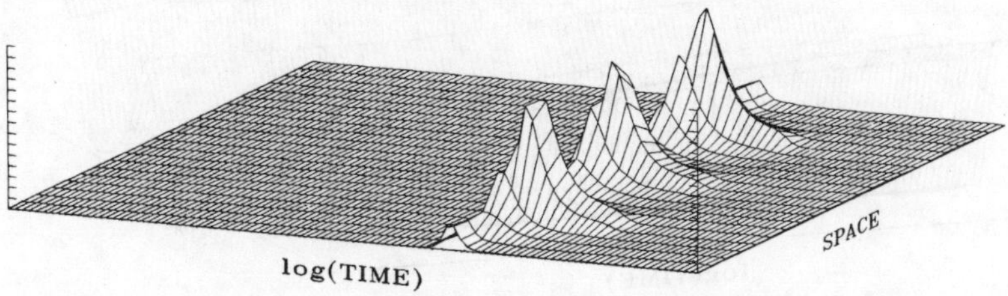

RUN 101
H2
SCALE 0.2000E-02

MAX H2	0.2421E-21	MAX H2	0.2919E-03
MIN H2	0.3034E-24	MIN H2	0.8976E-04
(INITIAL SLICE)		(FINAL SLICE)	

Fig. 8d

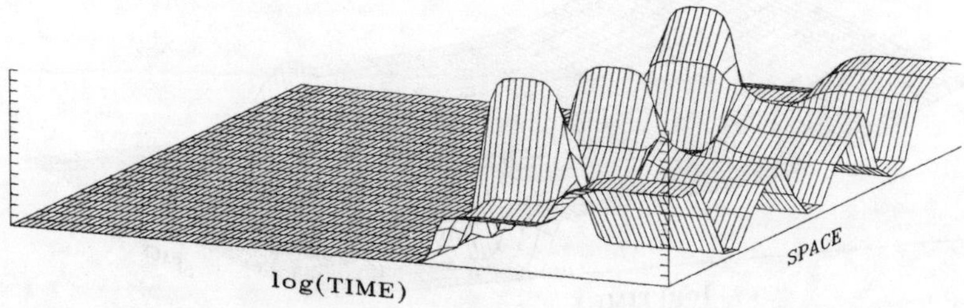

RUN 101
HE4
SCALE 0.5000E-01

MAX HE4	0.2421E-21	MAX HE4	0.4364E+00
MIN HE4	0.3034E-24	MIN HE4	0.1673E-01
(INITIAL SLICE)		(FINAL SLICE)	

Fig. 8e

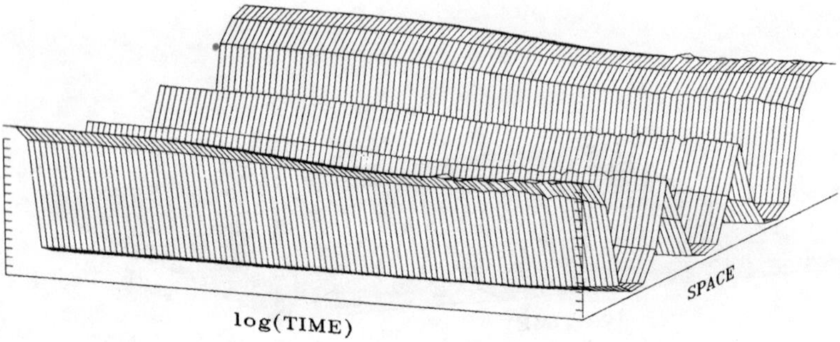

RUN 103
D/TRK**1.5
SCALE 0.4000E-08
MAX D 0.3295E-05
MIN D 0.4129E-08
(INITIAL SLICE)

MAX D 0.4733E-13
MIN D 0.1506E-14
(FINAL SLICE)

Fig. 9a

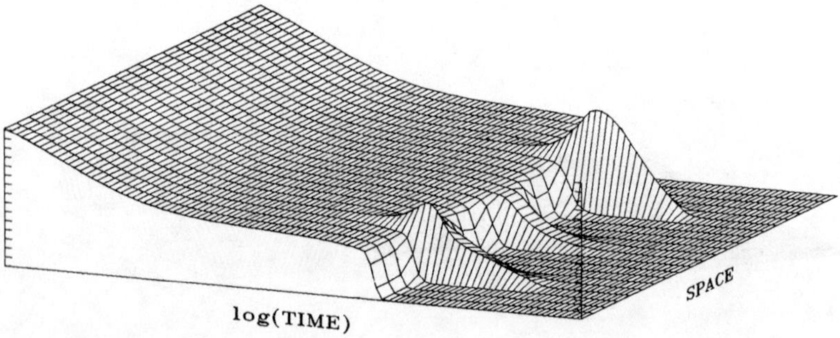

RUN 103
N
SCALE 0.2500E-01
MAX N 0.3774E+00
MIN N 0.3774E+00
(INITIAL SLICE)

MAX N 0.1186E-05
MIN N 0.8547E-09
(FINAL SLICE)

Fig. 9b

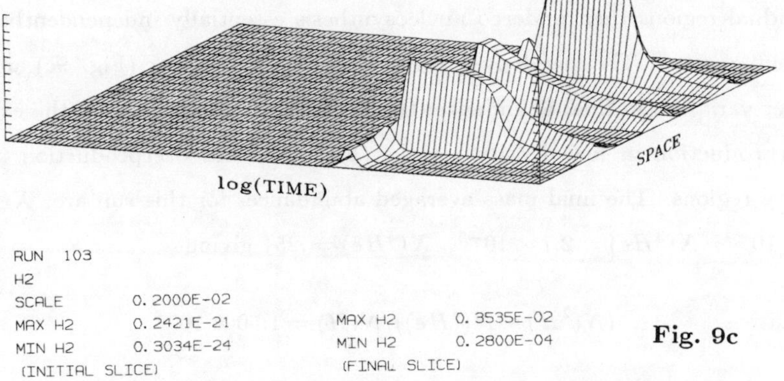
log(TIME)
SPACE
RUN 103
H2
SCALE 0.2000E-02
MAX H2 0.2421E-21
MIN H2 0.3034E-24
(INITIAL SLICE)
MAX H2 0.3535E-02
MIN H2 0.2800E-04
(FINAL SLICE)

Fig. 9c

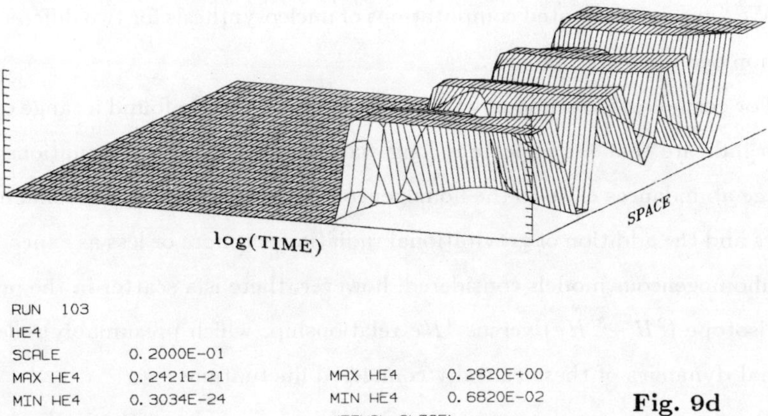
log(TIME)
SPACE
RUN 103
HE4
SCALE 0.2000E-01
MAX HE4 0.2421E-21
MIN HE4 0.3034E-24
(INITIAL SLICE)
MAX HE4 0.2820E+00
MIN HE4 0.6820E-02
(FINAL SLICE)

Fig. 9d

In Run #103, the grid size at $3 \times 10^{10} K$ is taken as 0.05 horizon sizes. Hence, the structures here are larger than the diffusion distances even at late times (Fig. 9a). There is only a slight contribution from neutron diffusion (Fig. 9b). The individual regions then undergo nucleosynthesis essentially independently in the different regions. The low density regions produce ^2H longer (Fig. 9c) and with greater variation in final abundance; the final ^4He is a reflection of the expected underproduction in low density regions (Fig. 9d) and overproduction in high density regions. The final mass averaged abundances for this run are: $X(^2H) = 2.1 \times 10^{-4}$; $X(^3He) = 2.7 \times 10^{-5}$, $X(^4He) = .251$ giving

$$(N(^2H) + N(^3He))/N(H) = 1.50 \times 10^{-4}$$

In this case, as in Run #102, the deuterium $+^3He$ abundance is somewhat above the 'standard-model' $\nu = 3$ curves. The averaged abundance points for Runs ## 101, 102, 103, are plotted in Fig. 6.

CONCLUSIONS

We have demonstrated computations of nucleosynthesis for two different classes of inhomogeneous cosmologies.

For cosmologies with long wavelength fluctuations, we found a range of effects which indicate that (as expected) long wavelength adiabatic fluctuations predict average abundances close to the homogeneous values, and that additional neutrino flavors and the addition of gravitational radiation act more or less as expected. For the inhomogeneous models considered, however, there is a scatter in the predicted light isotope ($^2H +^3 He$) versus ^4He relationship, which presumably reflects the internal dynamics of these causally connected fluctuations.

For fluctuations on a much smaller scale, we considered the isothermal fluctuations that can in principle arise from the quark-hadron phase transition. The largest scale considered, $\sim 10km$ at the transition temperature of $2 \times 10^{12} K$, showed little neutron diffusion, and essentially independent evolution at every point. The smallest scale considered $\sim 100m$ at the quark-hadron phase transition, showed substantial effects of diffusion, but the resulting averaged abundances are very comparable to 'standard' models.

Our results appear to be more consistent with those of Fuller Mathews and Alcock, (1987), and Alcock, Fuller and Mathews (1987), than with those of Applegate, Hogan and Scherrer (1987). This latter group apparently considered larger average baryon density than we, and the Fuller et al. group does. We emphasize moreover that the previous calculations **estimated** the amount of neutron diffusion in the model, and then averaged the results of independent nucleosynthesis calculations, for the resultant range of parameters, whereas we actually compute the transport.

One result from Fuller, Mathews and Alcock is that the resultant abundances reach certain fixed values for extreme values of the density contrast. We are currently investigating this phenomenon. Schramm (1987) has emphasized that the ^7Li abundance should be a good marker/test for these diffusion models. The observed abundance of ^7Li is at a minimum as a function of baryon density according to standard, homogeneous computations. Any inhomogeneous computation should result in averaging **larger** values of ^7Li. The observation in F-subdwarfs (Spite and Spite, 1982a, b) is consistent with abundance values in the minimum, so this may be a good test of these models. We, with H. Kurki-Suonio, are currently implementing ^7Li computations in our reaction chain.

ACKNOWLEDGEMENTS

We are glad to acknowledge the help and advice of Dr. H. Kurki-Suonio and of Dr. Grant Matthews. T.R. thanks Professor T. Piran for suggesting neutron diffusion computations and for discussions on this subject. This research was supported in part by National Science Foundation Grants #PHY84-04931, #PHY84-17918, #PHY84-51732, and Department of Energy Grant #DOE-W7405-ENG-48.

REFERENCES

C.R. Alcock, G.M. Fuller, and G.J. Mathews (1987) UCRL preprint 95896, Lawrence Livermore National Laboratory, Livermore, CA.

J.H. Applegate, C.J. Hogan, and R.J. Scherrer (1987) Phys. Rev. **D35** 1135-1160.

A. Boesgaard and G. Steigman (1985) Ann. Rev. Astron. Astrophys. **23** 319-378.

J.M. Centrella and J.R. Wilson (1983) Astrophys. J. **273** 428-435.

J.M. Centrella and J.R. Wilson (1984) Astrophys. J. Suppl. **54** 229-249.

J.M. Centrella, R.A. Matzner, T. Rothman and J.R. Wilson (1987) Nucl. Phys. **B266** 171-227.

M.C. Glaviano and D.J. Raymond (1981) Astrophys. J. **243** 271-278.

G.M. Fuller, G.J. Mathews and C.R. Alcock (1987) preprint Lawrence Livermore National Laboratories, Livermore, CA; also in **The Origin and Distribution of Elements**, G.J. Mathews, ed., World Scientific, Singapore.

C. Hogan (1983) Phys. Lett. **133B** 172.

K. Kajantie and H. Kurki-Suonio (1986), Phys. Rev. **D34** 1719-1739.

H. Kurki-Suonio (1985) Nucl. Phys **B255** 231.

H. Kurki-Suonio (1987) private communication.

G. Mathews (1987) private communication.

W.N. Press, B.P. Flannery, S.A. Teukolsky, W.T. Vetterling (1986) **Numerical Recipes** (Cambridge University Press, Cambridge).

D. Schramm (1987) in **The Origin and Distribution of Elements**, G.J. Mathews, ed., World Scientific, Singapore.

F. Spite and M. Spite (1982a) Astron. Astrophys. **115** 357.

M. Spite and F. Spite (1982b) Nature **297** 483.

J.M. Uson and D.T. Wilkinson (1984) Astrophys. J. Lett. **277** 11.

R.V. Wagoner (1973) Astrophys. J. **179** 343-360.

R.V. Wagoner, W.A. Fowler and F. Hoyle (1967) Astrophys. J. **148** 3-49.

J. Wilson (1979) in Sources of Gravitational Radiation, ed. L.L. Smarr (Cambridge University Press).

CHAPTER II

GALACTIC EVOLUTION

"Seas roll to waft me,
Suns to light me rise,
My footstool earth,
My canopy the skies."

Alexander Pope
Essay on Man
Epistle I, line 139

GALACTIC CHEMICAL EVOLUTION: PERSPECTIVES AND PROSPECTS

Virginia Trimble

Astronomy Program, University of Maryland, College Park MD 20742 USA

and

Department of Physics, University of California, Irvine CA 92717 USA

ABSTRACT: The first modern, quantitative models of galactic chemical and luminosity evolution appeared just 20 year ago in the PhD dissertation of the late Beatrice M. Tinsley. Such models represent a synthesis of the behavior of the 10^{11} or more stars that form over the 10^{10} year age of a galaxy like our Milky Way and are vital both for understanding how and why galaxies have the luminosities, colors, and chemical compositions we see now and for interpreting observations of distant galaxies to answer cosmological questions about the size, age, density, inhomogeneities, and geometry of the universe. A simple (one-zone, constant IMF etc.) model remains remarkably successful, though its failures (especially the G dwarf problem) have prompted many modifications. We explore here residual difficulties in the modelling of galactic evolution (e.g. the tendency to treat as adjustable parameters quantities like IMF, star formation rate, and gas in/outflows that have definite values in the real world) and some contexts where the data may be trying to tell us something.

1. HISTORY AND INTRODUCTION

We heard yesterday about processes in the early universe. What comes next is galaxies (and the things in them, ejected from them, and made up of them), whose formation is frankly not very well understood. The subsequent evolution is in slightly better condition and is generally treated as three topics, the evolution of dynamical (positions and velocities of stars, gas,...), photometric (luminosities, colors,...), and chemical properties. Clearly the three are not really separable, but the focus here will be on chemical evolution.

The history of the subject could be said to begin with the recognition of the existence of stellar populations in velocity distributions (Lindblad 1925,1926) and in HR diagrams (Baade 1944). A correlation of weak CN lines (low metallicity) with high stellar velocities was noted by Morgan, Keenan, and Kellman (1943); and our modern concepts were largely crystallized at the 1957 Vatican Conference (O'Connell 1958). The pioneering evolutionary scenario coupling dynamical and chemical evolution was that of Eggen, Lynden-Bell, and Sandage (1962).

Also in 1962, van den Bergh compared numbers of stars with different percentage contents of heavy elements to the numbers there should be if the metals had been made by (short-lived) stars formed at the same time as the ones we now see. He discovered what is now called the G dwarf problem: the local main sequence population has far fewer metal-poor stars than it would if star formation and metal formation had occurred in constant ratio over history. Schmidt (1963) identified the same problem in an analytical calculation of galactic chemical evolution and suggested as a remedy early generations consisting largely of short-lived metal-producing stars, few members of which now remain. This solution is now called variable IMF (for Initial Mass Function, the distribution function of masses for stars formed at a given time) or variable yield (= y = mass of new metals eventually ejected when a unit mass of matter goes into stars).

The first detailed numerical models of photometric and chemical evolution came from Tinsley (1968), dynamical evolution being coupled in by, for instance, Tinsley and Larson (1978). The subject has been elsewhere reviewed a countably infinite number of times. Representative examples include Trimble (1975), Tinsley (1980), Pagel (1987).

2. ANALYTICAL CALCULATIONS

Surprising insight into galactic chemical evolution can be gained from simple differential equations. These could have been written down long ago, though the standard form in fact post-dates numerical models (Tinsley 1980). Consider gas that is gradually being turned into stars which return processed gas (rapidly compared to other time scales). If u is the residual gas fraction and Z the mass-abundance of metals in the gas, then the time evolution is given by the balance between loss into stars and gain from stars as

$$\frac{d(uZ)}{dt} = Z\frac{du}{dt} - \frac{du}{dt}y(1-Z) \qquad (1)$$

where y is the yield defined above. For small Z, (1) has the exact solution $Z = y \ln(1/u)$, which, for solar neighborhood values of y (0.004-0.01) and u (0.05-0.15) gives Z correctly as 0.02 to 0.04. When Z is no longer small, incidentally, the exact solution of (1) has Z going to 1 as u goes to 0.

If we keep the time dependence in (1) and suppose that gas is fed into stars at a rate $du/dt = u^{-n}/T$, where T is a characteristic time scale, then Z(t) for the stars formed also has an exact solution (for small Z). For instance, n=2 (formation proportional, plausibly, to the square of the gas density) gives

$$Z(t) = y[1 - \exp(-t/T)] \tag{2}$$

This rises rapidly, then tapers asymptotically to Z = y, much like real abundances in the solar neighborhood.

Finally, if we consider a class of stars all of the same mass and having lifetimes longer than the age of the galaxy, we can predict what their distribution of abundances, N(Z), should be. The fractional mass in stars at any given time is just (1-u). Thus the number of stars in our class made up to time t is also proportional to [1-u(t)]. And so

$$\frac{N(t)}{N(t_o)} = \frac{1 - u(t)}{1 - u_o} \tag{3}$$

where t_o and u_o represent the present values. This can be transformed using the solution of (1) in the form $Z = -y \ln u$ and $y = Z_o/\ln(1/u_o)$ to give an expression for the fraction of stars having metal abundances equal to or less than the value Z:

$$f = \frac{N(Z)}{N(Z_o)} = \frac{1 - u_o^{(Z/Z_o)}}{1 - u_o} \tag{4}$$

Obviously, f = 1 at Z_o and, for plausible values of u (0.05-0.15) it drops away from one rather slowly as Z declines; 23% of local G-KV stars, for instance, should have Z less than 10% of the solar value. The actual fraction is more like 1%, and the discrepancy worsens at still lower Z.

This is the G dwarf problem again, motivation for many of the modifications of the simple model discussed below. Pagel (1987) has pointed out, however, that the problem is not universal. Equation (4) provides a good fit to the N(Z) observed for (a) halo globular clusters, (b) disc globular clusters, and (c) the giant stars near the galactic center seen through Baade's window. These fits are achieved with effective yields of 0.025, 0.3, and 1.8 times solar Z respectively. Analytical treatment of the evolution of galaxy luminosities and colors is also possible, particularly for the case of a single star burst, but the result is neither as simple nor as informative as the chemical version (Tinsley 1980).

3. NUMERICAL CALCULATIONS

Within living memory, galactic evolution as a subfield of astronomy essentially did not exist. Treatments other than qualitative or analytic were thought to be (a) excessively difficult (how do you add up 10^{11} stellar evolutionary tracks?!) and (b) not worth doing, in the sense that important observable properties of galaxies were not expected to change much over the time scales we could probe. The brightnesses of elliptical galaxies, for instance, come mostly from their giant stars, and it was thought that the gradually declining brightness of a typical giant (the branch being populated by stars of successively lower mass) would be nearly balanced by the gradually increasing number of giants (low mass stars being commoner than high mass ones).

The potential of the field was, however, very great. We can observe a number of different things about galaxies, which turn out to be correlated in ways that just beg for evolutionary interpretation.

The integrated properties include total luminosity, colors, mass to light ratios, integrated spectra (clearly a convolution of the masses of the stars and their compositions), isophotal diameters, residual gas fraction, and gas composition. In general, luminous galaxies are also red, massive, big, gas-poor, and metal-rich.

We also see, within single galaxies, gradients in surface brightness, colors, integrated spectra, gas fraction, and composition, as a function of distance from the center or from the galactic plane. The correlations resemble those for whole galaxies, the bright cores tending to be red, gas-poor, and metal-rich; the more diffuse outer regions the opposite.

Finally, for the solar neighborhood we can measure the average stellar composition vs. time and count the numbers of stars as a function of age and metallicity. For stars (and gas) near enough to permit high resolution spectroscopy, one can also probe variations in the relative abundances of the heavy elements as well as the gross Z(t). One would like also to know the typical dispersion of Z for stars made at a single time and place, though we have, at the moment, almost no data on this, since the stars we see here and now were formed over a wide range of times and distances from the galactic center.

3.1 The Standard or Simple Model

Tinsley (1968) solved the problem of summing stars by dividing them into 13 classes by mass and representing each class by a single evolutionary track in the HR diagram and by a single nucleosynthetic history. Then, starting at t=0 with pure H and He gas, she took 10^9 yr time steps and, at each, turned a fraction of the remaining gas into stars; plotted the old and new stars in an HR diagrm and added them up to get total luminosity, colors, and M/L; returned to the interstellar medium the nuclear reaction products and unburned H and He of the stars just made; added those up to get the new gas composition; calculated the residual gas as M(total) minus the stars; and went on to the next time step.

These models and others like them employed two free parameters, the time scale of the decline of star formation rate (which could range from a sharp single burst to essentially constant) and the shape of the initial mass function, $N(M) = N_o M^{-x}$, with x= 1-2 and an adjustable lower limit of 0.01 to 1 M_o.

Tinsley's models proved two things: galaxy evolution is calculable, and galaxy evolution is important. First it is calculable in the sense that good fits were achieved to the integrated properties of the full range of galaxy types with models having ages of order the Hubble time, a single IMF, and stellar birthrates proportional to gas mass. Rapid decline in SFR vs. time yielded models with the properties of ellipticals; slower decline made spirals and irregulars.

Second, galaxy evolution is important in the sense that total luminosities, colors, gas content, chemical composition, dominant star types, isophotal diameters, and other measurable properties have all changed since Z = 0.5 by amounts at least as large as those due to other effects you might want to look for. For instance, the evolutionary correction to the luminosities of elliptical galaxies is big enough that, if the raw data trace out $q_o = ½$ in a Hubble diagram, then the corrected data trace out $q_o = 0$ or less (Gott et al. 1974). The corrections — and the uncertainties in the corrections — are correspondingly large for other standard cosmological tests, to the point where the galactic models will have to be very good indeed before we can hope to measure the decelleration parameter or the curvature of space-time using galactic data (Brown and Tinsley 1974; Tinsley 1973, 1977).

3.2 Models with Bells and Whistles

The first-generation models did not (of course!) solve the G dwarf problem, nor (unless the lower mass limit was set very low) did they have M/L ratios as large as the ones implied by rotation curves and velocity dispersions. In addition, they had made a very wide range of simplifying assumptions. Some of these might be true (but need not be), including the constant IMF, the assumption of closed homogeneous systems, and the smoothly declining star formation rate. All of these have been relaxed in the more complex models reviewed by Trimble (1975), Tinsley (1980), Pagel (1987), and Mould (1984) and described by the last as having bells and whistles.

The modifications have, on the whole, been successful. For instance, a sharp peak in SFR reproduces properties of star burst galaxies (Larson and Tinsley 1978). Permitting dissipative gas to flow inward at the same time it is being enriched reproduces observed gradients; while allowing gas to be blown out of shallow gravitational potentials accounts for the correlation of metallicity with galactic masses (Vader 1987; Larson and Tinsley 1974; Tinsley 1974).

The G dwarf problem has been solved(?) with variable IMF (hence yields) that favor massive stars early on, by Prompt Initial Enhancement (or pre-galactic nucleosynthesis, meaning that you don't start with Z=0), with Metal-Enhanced Star Formation, with gas in- and out-flows, and, most recently (Gilmore and Wyse 1985) by adding a third, thick-disc, population component to the galaxy, which acts as a sink for halo gas and a source for thin-disc gas. Finally, large observed M/L ratios are generally attributed (Trimble 1987) to a dark matter component that does not participate in chemical evolution. It is in this sense that Pagel (1987) has remarked that the simple model, with bells and whistles, is still exceedingly successful and informative.

Others of the simplifying assumptions could not have been true and are still in the process of improvement. The instantaneous recycling approximation said that nucleosynthetic products came out as soon as the stars formed; relaxing this allows us to account for the earlier build up of (e.g.) oxygen than of iron traced out by relative abundances in old stars (Gratton and Sneden 1987). Implicit here is that one must also pull away from treating heavy elements as a unit and the yield as a single number for a given stellar population. This has also been done, with at least qualitative success (Twarog and Wheeler 1987). Tinsley's first models predated the discovery of pulsars, and so had every star leaving a 0.5 M_o remnant. This was easily fixed and tended to reduce the iron contribution by massive stars (since more got stuck in the core), which is just as well, since Type I supernovae from less massive stars are now thought to contribute most of the iron. Finally, the evolution of massive stars was (and to a considerable extent continues to be) represented by tracks for population I stars. Massive Pop II stars will, inevitably, be bluer and brighter and contribute a larger ratio of helium to heavy elements than will massive Pop I stars (Dearborn and Trimble 1978, 1980). Tinsley said 15 years ago (1972 private communication) that the lack of adequate population II evolutionary tracks was the single largest gap in her models. I suspect it may still be, with some relevance to the residual problems discussed next.

4. RESIDUAL PROBLEMS

I do not quite agree with Pagel's (1987) conclusion that the subject of galactic chemical evolution is in a confused and immature state; one can, nevertheless, identify a number of points where observation and theory disagree, or agree only by invoking dei ex computa, or have not yet really been brought into confrontation. An obvious one is the neglect of dark matter (whatever it is!) in the models. Apparently it really is not involved in chemical evolution, in that observed properties correlate best when M/L means just that in stars and gas (Garnett and

Shields 1987; Vigroux et al. 1987), but it is hard to believe that the star formation rate (for instance) is not sensitive to the confining gravitational potential.

Relative abundances of many elements and nuclides seem to be in good qualitative shape. Where metals in general are low, C tracks Fe (both being made in intermediate mass stars during asymptotic giant branch and Type I supernova phases); O, Mg, Si, and Ca are less deficient than Fe (being made in the most masssive, shortest lived stars); and the classical secondaries (whose production requires having some metals to start with) are over-deficient, including Al, Na, N, neutron-rich isotopes like $Mg^{25,26}$, and the s-process products (Gratton and Sneden 1987; François 1987; Barbuy et al. 1987; Carbon et al. 1987). Conversely, the present interstellar medium shows twice the solar ratio of C^{13} to C^{12} to go with its generally higher metal abundance (Hawkins and Jura 1987).

Quantitative agreement is less satisfactory. Twarog and Wheeler (1987) conclude that the ratios of the major nucleosynthetic products require either that the calculated yields be wrong by at least 50% or that the portion of the IMF that explodes and contributes these yields be populated very unsmoothly.

For some products, we understand fairly well the reactions needed to make them, but not the sites where those reactions occur. This is a problem for the r and p processes (discussed later in these proceedings) which make the n-rich and n-poor isotopes of elements heavier than Fe. It is something of a problem for helium, whose abundance has, observationally, gone up with that of the metals in a ratio dY/dZ of 2 to 4 from a primordial value near Y = 0.23 (Ferland 1986), while population I stars are inclined to produce a ratio of at most one. I suspect that more careful attention to evolutionary tracks and nucleosynthesis for massive Pop II stars might cure this, and perhaps also help to resolve the quantitative discrepancies identified by Twarog and Wheeler (1987).

Another widely neglected topic in the context of chemical evolution is the contribution of close binary systems. Iron and other contributions of Type I supernovae (Thielemann et al. 1986) are one aspect of binary nucleosynthesis; novae (plausible sources of Al^{26} and other rare isotopes) are another; but surely not the only ones. Even for single stars, the correlations between initial and final masses, the effects of semi-convection on the sizes of well-mixed zones, and many other aspects of stellar physics are not yet well enough known that the conversion of an IMF to yields of individual elements can be done with enough precision to separate its effects from galactic evolutionary ones.

Finally, and in the long run probably the most serious, existing models necessarily treat as adjustable parameters several quantities that must, in the real world, be deterministic. These include star formation rate as a function of time, the initial mass function, and the inflow and outflow of gas. Each must in fact depend causally upon local conditions of temperature, density, turbulence, chemistry, magnetic field strength, angular momentum distribution , and heaven knows what else, but with that many parameters one can (a) fit an elephant (according to Gamow) and (b) fulfill the conditions for Redman's theorem (that a competent theorist can explain any given set of observations using any theory) and Longair's corollary (that, in most cases, the theorist need not even be competent) to apply. These remarks should not be taken as pejorative, but only as expressing the hope that someone will come along with a new 50¢ idea for the closer coupling of chemical, photometric, and dynamical evolution, so that these parameters become calculable quantities.

5. DATA POINTERS

It is only too easy to look at a set of hard-won observations of properties of galaxies, stars, or gas and say that a valid model of galactic evolution should "predict" (in the quaint current usage of the word) them. By way of illustration, the following observations (each of which is without doubt trying to tell us something about chemical evolution) have all been culled from journals that hit the library shelves between April and July 1987.

Gradients of stellar composition (in both S and E galaxies) and of gas composition (in Ss) are unquestionably widespread and qualitatively well fit by dissipative gas flows (Thomsen and Baum 1987). But not all galaxies show the same gradients; not all elements participate equally (or in the same proportions that they do in age-dependent composition variations), and the centers of galaxies are still very poorly explored (Garnett and Shields 1987; Geisler 1987; Faundez-Abans and Maciel 1987; Pagel 1987).

Closely related is the question of when and where it is appropriate to assume homogeneity, that is, the mixing-in of nucleosynthetic products rapidly compared to the time scale of star formation. Within two regions of current rapid processing, Rosa and Mathis (1987) find that 30 Dor shows inhomogeneities due to input from (eg) Wolf-Rayet stars, while Díaz et al. (1987) find that the giant HII regions in M33 do not.

Some of them do and some of them don't also describes the observational evidence for gradual increases in gas metal abundance. QSO absorption lines enable us to probe gas in galaxies at large redshifts, and, while at least one set of lines (at Z = 2.309 in PHL 957) has C deficient by at least a factor 100 (Black et al. 1987), other sets (eg at Z = 1.345) show exactly the same abundance patterns as Milky Way gas today (Blades et al. 1987).

Some, but not all, of the interstellar gas at a given time has already been through stars. The fraction is described as the astration parameter and is an important discriminant for models with infalling primordial gas. The most sensitive probe for it is D/H, since deuterium is destroyed almost completely by passage through stars. Unfortunately, the observed interstellar value (about $2X10^{-5}$, Murthy et al. 1987) is both variable and very close to the amount we expect to have been produced in the Big Bang. As a result, one currently calculates the astration parameter and uses it to constrain the baryon density of the universe, rather than the other way round (Boesgaard and Steigman 1985).

The correlations between chemical and dynamical properties of stars and their ages (Strömgren 1987) is another case where the tail is currently wagging the dog, in that the data now serve primarily to constrain the masses of the objects that gradually heat up the stellar velocity distribution. We would like to be able to subtract off that effect and use the data to probe gas turbulence in the past.

Of the adjustable parameters that need to be converted to astrophysics, the gas inflow rate has already been noted as being constrained by astration, at least in principle. There are also, obviously, data that bear on the time dependence of the star formation rate and on the shape of the IMF.

The star formation rate in elliptical galaxies must be very much smaller now than its average over time (Vader 1987), but it did not drop instantly to zero. Interpretation of ultraviolet data suggests that either star formation with a standard IMF continued up to about 10^9 yr ago or that star formation persists now, but only intermediate and low mass stars are being made (Kjaergaard 1987; Rocca-Volmerange and Guideroni 1987). In Magellanic irregulars like NGC 4449, on the other hand, IR observations tell us that the present star formation rate is higher than the average over time (Thronson et al. 1987a). Spirals, including our own Milky Way, present the picture I find most disturbing. Looking back into the past, the ages of the oldest disc white dwarfs (Winget et al. 1987) seem to be telling us that star formation began only about 8 billion years ago. Looking forward in time, the present balance between star formation rate (from IR emissivities) and gas supply suggests that star formation must cease within the next 1-3X10^9 yr (Thronson et al. 1987b) in a large fraction of spirals, including ours. This feels a bit non-Coperican.

Finally, there are real variations in IMF, sometimes correlated with other properties of the stars and gas, though not in ways that yet enable us to "predict" what distribution of stellar masses should arise from a particular molecular cloud. Some examples include the lower main sequence IMF of globular clusters, which is flatter for the more metal-rich ones (Smith and McClure 1987) and the upper main sequences in the LMC, Milky Way, etc. which apparently have the same slopes, but upper mass cutoffs that correlate positively with degree of activity (Pierre 1987; Garmany et al. 1987). Finally, the violent activity of starburst galaxies seems to produce mostly massive stars (Belfort et al. 1987), while the relative tranquillity of cooling flows is thought to result in preferential formation of low mass stars.

There is no single coherent conclusion that can be drawn from this array, except perhaps W.L.W. Sargent's final exhortation to a NATO workshop on galactic chemistry 13 years ago, "Go home and observe!"

ACKNOWLEDGEMENTS

Owing to the pressures of time, this review was prepared entirely from published, or widely distributed preprint, literature, and the standard thank yous to contributors do not obtain. But I should like to pay tribute to the memories of Beatrice M. Tinsley, without whose pioneering efforts the subject of quantitative galactic evolution might well still not exist, and of Bengt Strömgren, whose contributions to astronomy covered a period of more than 60 years and whose last preprint pertinent to chemical evolution was distributed just a few days after his death.

REFERENCES

Baade, W. 1944. Astrophys. J. 100, 137 & 147
Barbuy, B., F. Spite & M. Spite 1987. Astron. Astrophys. 178, 199
Belfort, P. et al. 1987. Astron. Astrophys. 176, 1
van den Bergh, S. 1962. Astron. J. 67, 486
Black, J.H. et al. 1987. Astrophys. J. 314, 442
Blades, J.C. et al. 1987. IAU Symp. 119, 571
Boesgaard, A.M. & G. Steigman 1985. Ann. Rev. Astron. Astrophys. 23, 319
Brown, G.S. & B.M. Tinsley 1974. Astrophys. J. 194, 555
Carbon, D.F. et al. 1987. Publ. Astron. Soc. Pac. 99, 335
Dearborn, D.S.P. & V. Trimble 1978. IAU Symp. 80, 375
..... 1980. Nucleonika 25, 1441
Diaz, A.I. et al. 1987. Mon. Not. Royal Astron. Soc. 226, 19
Eggen, O.J., D. Lynden-Bell, & A.R. Sandage 1962. Astrophys. J. 136, 748
Faundez-Abans, X. & W.J. Maciel 1987. Astrophys. Space Sci. 129, 353
Ferland, G. 1986. Astrophys. J. 300, L67
François, P. 1987. Astron. Astrophys. 176, 294
Garmany, C.D. et al. 1987. Astron. J. 93, 1070
Garnett, D.R. & G.A. Shields 1987. Astrophys. J. 317, 82
Geisler, D. 1987. Astron. J. 94, 84
Gilmore, G. & R.F.G. Wyse 1985. Astron.J. 90, 2015
Gott, J.R., J.E. Gunn, D.N. Schramm & B.M. Tinsley 1974. ApJ 194, 543
Gratton, R.G. & C. Sneden 1987. Astron. Astrophys. 178, 179
Hawkins, I. & M. Jura 1987. Astrophys. J. 317, 926
Kjaergaard, P. 1987. Astron. Astrophys. 176, 210
Larson, R.B. & B.M. Tinsley 1974. Astrophys. J. 192, 293
..... 1978. Astrophys. J. 219, 46
Lindblad, B. 1925. Ark. f. Mat. Astr. och Fysik 19A No. 21
..... 1926. Ark. f. Mat. Astr. och Fysik 19B No. 7
Morgan, W.W., P.C. Keenan, & I. Kellman 1943. An Atlas of Stellar Spectra (Chicago: U. Chicago Press)
Mould, J. 1984. Publ. Astron. Soc. Pac. 96, 773
Murthy, J. et al. 1987. Astrophys. J. 315, 675
O'Connell, D.J.K (ed.) 1958. Ric. Astr. Speccola Vaticana 5 ("Stellar Populations")
Pagel, B.E.J. 1987. in G. Gilmore & R.F. Carswell (eds.) The Galaxy (Dordrecht: Reidel) NATO Adv. St. Inst., p. 341

Pierre, M. 1987. Astron. Astrophys. 175, 54
Rocca-Volmerange, B. & B. Guideroni 1987. Astron. Astrophys. 175, 15
Rosa, M. & J.S. Mathis 1987. Astrophys. J. 317, 163
Schmidt, M. 1963. Astrophys. J. 137, 758
Searle, L. & W.L.W. Sargent 1972. Astrophys. J. 173, 25
Smith, G.H. & R. McClure 1987. Astrophys. J. 316, 206
Strömgren, B. 1987. in G. Gilmore & R.F. Carswell (eds.) The Galaxy (Dordrecht: Reidel) NATO Adv. St. Inst., p. 229
Thielemann, F.-K., K. Nomoto & K. Yokoi 1986. AAp 158, 17
Thomsen, B. & W. Baum 1987. Astrophys. J. 315, 460
Tinsley, B.M. 1968. Astrophys. J. 155, 547
..... 1973. Astrophys. J. 178, L93
..... 1974. Astrophys. J. 192, 629
..... 1977. Astrophys. J. 210, L49 and 211, 621
..... 1980. Fund. Cosmic Physics 5, 287
Tinsley, B.M. & R.B. Larson 1974. Astrophys. J. 192, 293
..... 1978. AStrophys. J. 221, 554
Trimble, V. 1975. Rev. Mod. Phys. 47, 877
..... 1987. Ann. Rev. Astron. Astrophys. 25, 425
Twarog, B.A. & J.C. Wheeler 1987. Astrophys. J. 316, 153
Vader, J.P. 1987. Astrophys. J. 311, 128
Vigroux, L. et al. 1987. Astron. Astrophys. 172, 15
Winget, D. et al. 1987. Astrophys. J. 315, L77

THE CHEMICAL EVOLUTION OF THE GALAXY

Rosemary F.G. Wyse[♣] and Gerard Gilmore[⋈]

[♣]Astronomy Department, UC Berkeley, CA 94720
[⋈]Institute of Astronomy, Cambridge, England, CB3 0HA

ABSTRACT

The distribution of enriched material in the stars and gas of our Galaxy contains information pertaining to the chemical evolution of the Milky Way from its formation epoch to the present day, and provides general constraints on theories of galaxy formation. The separate stellar components of the Galaxy cannot readily be understood if treated in isolation, but a reasonably self-consistent model for Galactic chemical evolution may be found if one considers together the chemical properties of the extreme spheroid, thick disk and thin disk populations of the Galaxy. These three major stellar components of the Galaxy can be characterised by their distinct spatial distributions, metallicity structure, and kinematics. The vast majority of 'Population II' stars in the Galaxy belong to the thick disk, and are thus members of a metal-enriched, rotating system unlike the high-velocity subdwarfs. The thick disk provides a straightforward physical mechanism for the desired pre-enrichment to resolve the thin disk 'G-dwarf problem'. Breaks in the relations between element ratios and overall enrichment occur at the characteristic metallicities of each of these three stellar components, *i.e.*, where a change of stellar kinematics occurs, and by inference, star formation and gaseous dissipation rates also change. The scatter about the mean trends is consistent with observational error alone, and we show how detailed information about the element ratios present in spheroid stars can place constraints on the timescale on which the metal-poor system formed, and on how homogeneous the proto-Galaxy could have been at this time.

1. CHEMICAL EVOLUTION WITH A THICK DISK

1.1 The G-Dwarf Problem

The starting point for most theories of chemical evolution is the 'simple, closed box model', where one considers a cloud of gas to form stars and self-enrich under the following assumptions : There are no gas flows, either into or out of the system; the gas is homogeneous and well-mixed chemically at all times; the nucleosynthetic yield per generation of stars (and the stellar initial mass function or IMF) is time-independent; the enriched gas ejected from stars is 'instantaneously recycled' *i.e.,* stars that contribute to the chemical evolution live for an infinitesimal time; and the system starts from zero metallicity. The evolution of this model may be described by straightforward analytic equations, which have the additional advantage that the relation between metal enrichment and gas fraction is independent of the assumed star formation rate. However, as has been known for 25 years, application of this model to the Galactic thin disk fails because the predicted number of long-lived low metallicity stars greatly exceeds that observed in the solar neighborhood[1,2] – the 'G-dwarf problem'. At least one of the assumptions of the above model must be wrong, and the most appealing resolution is that the thin disk formed from gas that was not metal-free, perhaps as a result of enriched inflow[3,4], further meaning that the thin disk did not evolve as a closed system. One could also allow the stellar IMF to vary with time, allowing more massive stars at early times[5,6]. The chemical evolution of the solar neighbourhood tightly constrains the possible variations, and though not required by the observational data for our Galaxy, such a model may be viable[6]. Here we retain the more conservative assumption of an invariant IMF, in the spirit of minimising the number of parameters.

The original motivation for applying the closed box model to the Galactic disk came from the proposal[7] that the formation of the spheroid of the Galaxy, and the collapse of residual gas to a thin disk, was extremely rapid, essentially being completed on a freefall time. However, recent evidence from *in situ* observations for a stellar component of the Galaxy with kinematics and metallicity distribution intermediate between those of the extreme spheroid and thin disk[8–12] has shown that this picture of Galaxy formation and evolution is oversimplified. In any case stellar evolution in the extreme spheroid cannot produce sufficient metals to provide the pre-enrichment for the thin disk[13], even though one *can* understand the chemical structure of the spheroid by an adaptation of the 'simple model' to include outflow of the stellar ejecta[13,14]. Thus with a two-component Galaxy one can understand the spheroid if one allows for outflow of enriched material, but the sink of this gas is unknown, and one can understand the thin disk if one allows pre-enrichment by inflow prior to the bulk of star formation there, but the source of the enriched material is unknown.

This situation is rectified if one investigates[15] the metal enrichment history of a three component Galaxy with an evolutionary sequence whereby a short-lived phase of rapid star formation forms the extended, metal poor, pressure supported extreme spheroid[7], with the thick disk forming next, before a final equilibrium state is attained and the centrifugally supported thin disk forming over a long timescale once the Galaxy potential is unperturbed and steady[16,17]. The adopted model[15] starts with a cloud of zero-metal gas, and evolves by the modified 'simple model' with gas outflow at a rate proportional to the star formation rate[13], the constant of proportionality being fixed by the observations of the metallicity structure of the extreme spheroid. The gas need not actually flow away from the Galaxy, but simply be (temporarily) removed from the star-forming process by, for example, heat input due to supernovae from young, massive stars. Observational constraints set the outflow rate to be ~ 10 times the star-formation rate (see Figure 1[a]), which results in a predicted mean enrichment of the outflow to be [Fe/H] $\simeq -1.5$ dex, which is 7σ below the observed mean of the thin disk[3] – confirming the inability of spheroid stellar evolution alone to account for the thin-disk G-dwarf problem – but which is only 3σ below the thick-disk mean metallicity[9]. The process of simple model plus outflow is then started again, with initial metallicity of -1.5 dex, until the observed thick-disk metallicity distribution is obtained. The outflow here is constrained to be at a rate about equal to the star-formation rate (Figure 1 [b]), with an average enrichment in the ejecta of -0.6 dex, which is $\sim 3\sigma$ below the thin-disk mean metallicity. This latter level of pre-enrichment allows the G-dwarf metallicity distribution in the solar neighborhood to be fit[3], with an uncontrived physical mechanism for the pre-enrichment. The model here has shown that the G-dwarf problem arose in part because the older, local surveys for G-dwarfs missed the thick disk stars, due to their relatively large scale-height, $\sim$ 1kpc[8]; distant, *in situ* surveys were required in order to find the intermediate metallicity G-dwarfs in large enough numbers.

Our calculations show that the cumulative metallicity distribution for thin disk stars limits any infall of unenriched material subsequent to the prompt enrichment of the thick disk ejecta to a rate of less than 0.2 the star formation rate, assuming constant relative rates, as may be expected[15]. This is in agreement with limits on present day high velocity infall[19]. Constraints on our model from other quantities such as the age-metallicity relation and the age-velocity dispersion relation for thin-disk F-dwarfs are weak since the two recent calibrations of these relations[20,21] for the same dataset are in significant disagreement. An independent survey[22] of F-dwarfs has preliminary results in reasonable agreement with the earlier result of Twarog[20], but somewhat contradictory results[23] have also been found.

1.2 Mass Ratios

For the model discussed here, the ratio of the mass of the second zone (thick plus thin disks), to the mass of the first zone (extreme spheroid) is approximately c/α, where α is the lock-up fraction in the first zone. Adopting a value of 0.8 for α, we then predict a present day mass ratio for the extreme spheroid, to thick and thin disks together, of $\sim$ 1:15. The thick to thin disk transition as modelled here predicts a further decomposition of masses to $\sim$ 1:3:12 for the ratios extreme spheroid:thick disk:thin disk.

The global parameters of the Galaxy, which are required to estimate luminosities to compare with these predictions, are as yet ill-determined. In particular, neither the radial scale-lengths nor axial ratios of any of the three major components of the Galaxy have been measured unambiguously. The most reliable technique available at present assumes that our Galaxy is extremely similar to other well-studied Sbc disk galaxies[24]. This gives a decomposition of our Galaxy into extreme spheroid, thick disk and old thin disk[24] which implies a value of

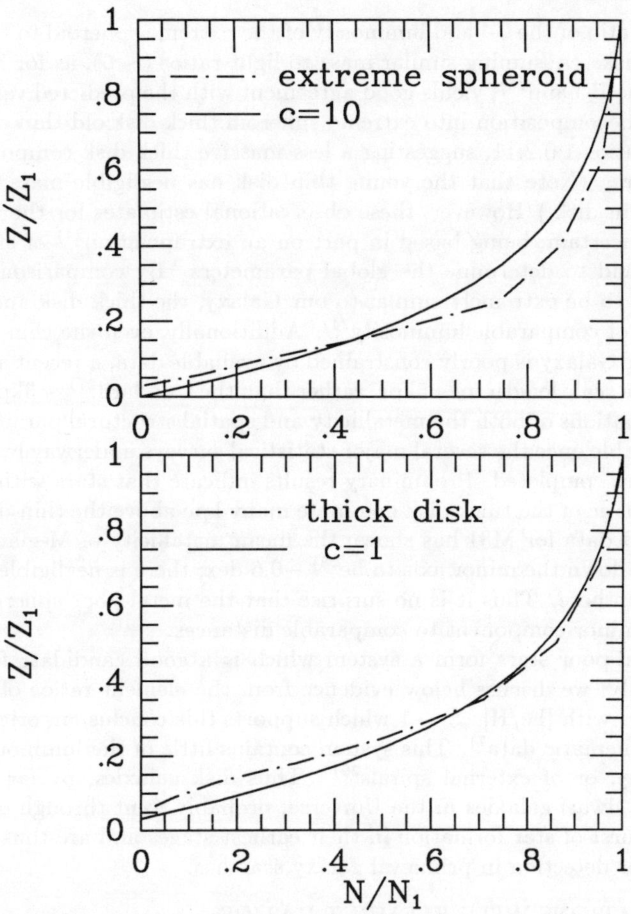

Figure 1 (a): The observed cumulative metallicity distribution of the metal-poor spheroid[10] (dashed line), *viz.* total number, N, of spheroid tracers with metallicity equal to, or less than, Z, normalized to the maximum metallicity, Z_1. The full line is the prediction of the present model with 'outflow' at a rate c = 10 times the star-formation rate.

(b): Same for the Galactic thick disk. The observed[9] log-metallicity distribution has been approximated by a truncated Gaussian ($\pm 2\sigma$), of mean [Fe/H]= -0.6, $\sigma = 0.3$. The model outflow rate is c = 1. (*cf.* ref 18 for differential metallicity distributions).

1:11 for the ratio of the J-band luminosity of the extreme spheroid to that of the sum of the disks. Assuming similar mass to light ratios (~ 6), as for SO to Sbc galaxies in the B-band[25]), yields good agreement with the predicted value above. The further decomposition into extreme spheroid:thick disk:old thin disk yields luminosity ratios 1:0.2:11, suggesting a less massive thick disk component than predicted here. (Note that the young thin disk has negligible mass compared to the old thin disk.) However, these observational estimates for the thick disk are highly uncertain, being based in part on an extrapolation[24]) of star counts in a single field to determine the global parameters. By comparison, in NGC 891, thought to be extremely similar to our Galaxy, the thick disk and extreme spheroid are of comparable luminosity[24]). Additionally, even the thin disk component in our Galaxy is poorly constrained by available data, a recent analysis[26]) increasing the scale length to $\sim$ 5kpc, rather than the standard[27]) $\sim$ 3kpc. Definitive determinations of both the metallicity and spatial structural parameters will only be available once the several major statistical surveys underway by ourselves and others are completed. Preliminary results indicate that stars with metallicity characteristic of the thick disk dominate many kpc above the thin-disk plane. Modern CCD data for M31 has shown the mean metallicity of M-giant stars in a field 12kpc down the minor axis to be[28]) -0.6 dex; there is negligible thin disk contamination here. Thus it is no surprise that the metal-poor spheroid of our Galaxy is a minor component to comparable distances.

The metal-poor stars form a system which is a good candidate for having formed rapidly; we discuss below evidence from the element ratios observed in spheroid stars with [Fe/H] $\lesssim -1$ which supports this conclusion, originally derived from kinematic data[7]). This system contains little of the luminous content of the Galaxy, or of external spirals[29]). Thus disk galaxies, by far the most common non-dwarf galaxies in the Universe, probably went through only a low luminosity burst of star formation in their earliest stages and are thus not good candidates for detection in primæval galaxy searches.

2. CORRELATIONS WITH ELEMENT RATIOS

2.1 Introduction

The chronological formation sequence of the three stellar components of the Galaxy proposed here may be tested by study of their detailed elemental and isotopic abundances. The formation of the thick and thin disks is likely to occur on timescales long compared to the lifetimes of the massive stars which explode as Type II supernovae, but of order the both the lifetimes of those stars which produce carbon and nitrogen and the evolutionary timescales for Type I supernovae. Here we will, for simplicity, consider only the classical Type II supernovae, due to exploding massive stars, and Type I supernovae, due to either the coalescence of a white-dwarf/white dwarf binary or accretion of matter onto a white dwarf from a companion during Roche Lobe overflow, thus ignoring the complications of Type Ib *etc.*

Element ratios have been measured for several samples of nearby stars, both giants and dwarfs, selected in various ways[30]. A concensus is that a significant change of slope occurs in the relationships between such element ratios as [O/Fe] or [α/Fe] (where the 'alpha' elements are those synthesised by successive capture of alpha particles, such as Mg, Si, Ca and Ti) and [Fe/H] at metallicities where there also occurs a change in the stellar kinematics, that is at [Fe/H] ~ -1 ([O/Fe], [α/Fe]) and ~ -0.4 ([α/Fe]) although this latter break is less well-established. The [O/Fe] ratio is observed to be approximately constant, independent of [Fe/H] for the most metal-poor stars, $-2.5 \lesssim$ [Fe/H] $\lesssim -1$, while [O/Fe] declines for the more metal-rich stars, [O/Fe] $\sim -1/2$ [Fe/H]; present data sets indicate an extremely small intrinsic dispersion, all scatter being consistent with observational error[22,31,32]. Assuming that [Fe/H] is a monotonically increasing function of time, this behavior can be explained if the oxygen and iron in the more metal poor stars has been produced in stars of the same lifetime, while for the more metal-rich stars, although the oxygen and iron continue to be produced together, an additional, longer timescale source now dominates the iron production. This results in [O/Fe] decreasing analogously to that of the ratio of a primary element to a secondary element. This may occur since oxygen is produced in massive stars only, while iron has a contribution from both massive and low mass stars (Type I and Type II supernovae) having an enhanced production once the much more numerous, lower mass stars contribute to its nucleosynthetic yield[33,34] An extension of this argument may be applicable to the behavior of the [α/Fe] ratio. Reliable models of this type are not yet possible as the available supernova nucleosynthesis calculations, and the fraction actually ejected, remain somewhat controversial. The canonical $25\mathcal{M}_\odot$ Type II precursor[35] produces a solar enrichment pattern, assuming the entire outer layers of the star are ejected, while the observations show a constant enhancement with respect to solar for the metal-poor stars, [O/Fe] ~ 0.5. In addition, if this argument contained the whole story over the history of star formation in the Galaxy, then it would have to be mere coincidence that the change in the predominant production mechanism of iron occurred at a metallicity, or epoch, at which the stellar kinematics change from those of a pressure-supported system, which formed stars rapidly, to those of an angular momentum-supported system. Rather, we believe that the coincidence of the value of [Fe/H] at which the Galaxy changed from a pressure-supported system to an angular momentum-supported system, with the value of [Fe/H] at which the interstellar medium became diluted by the products of long-lived stars, provides a diagnostic of the relative star-formation and dissipation rates in the proto-Galaxy.

2.2 Constraints on the Star-Formation Rate

One must explain the fact that [O/Fe] is approximately constant, at three times the solar value, for [Fe/H] $\lesssim -1$, while decreasing smoothly for [Fe/H] greater than this value, together with the fact [Fe/H] $\lesssim -1$ is only a few percent of the total stellar mass of the Galaxy, as discussed above. Clearly, the approximate constancy of the [O/Fe] ratio independently of the [Fe/H] ratio at low metallicities requires that all the stars with [Fe/H] $\lesssim -1$ formed on a timescale less than that on which a significant number of low mass (Type I) supernovae exploded. This timescale is rather difficult to estimate due to uncertainties in the mechanism of Type I supernovae and the fraction of all stars formed which are in binaries of the type that may be expected to be precursors[36]; the lowest mass and hence most numerous progenitors of CO white dwarfs have main-sequence masses and lifetimes of $\sim 5\mathcal{M}_\odot$ and $\sim 2.5 \times 10^8$ yr respectively. Thus a reasonable estimate for the characteristic time after which one expects dominance of iron from Type I supernovae is $\lesssim 10^9$ yr (but bearing in mind that some Type I systems will take a Hubble time to evolve). This general argument offers strong, direct evidence in support of rapid formation of the Extreme Population II system in the Galaxy. In principle, the actual amplitude of the oxygen over-enhancement due to the early Type II supernovae also contains information on the masses, and hence lifetimes of their progenitor stars, but this value is extremely dependent on the details of supernova model nucleosynthesis.

At metallicities above ~ -1 dex the [O/Fe] ratio shows a monotonic decline towards the solar value. The slope of this relation is ~ -0.5, implying that the amount of iron synthesised on short timescales, with oxygen, is small compared to the amount synthesised on longer timescales. This must hold even though the relevant [Fe/H] range encompasses the entire thick disk and most of the thin disk; star formation in the thick and thin disk formation stages must avoid imprinting a feature through their associated massive stars. We are currently utilising this constraint to determine the relative rates of Type I and of Type II supernovae during the epoch of formation of the disks, which in turn constrains the star formation rate during the formation of most of the stellar mass of the Galaxy.

2.3 Constraints on Inhomogeneities in the Spheroid

One of the most important constraints on the efficiency of gas motions and the mixing of the supernova ejecta, to homogenise the proto-galactic gas, is readily derivable from the observed small dispersion in the position of the breaks in element ratios as a function of [Fe/H]. The inferred lack of cosmic scatter implies little spatial variation in the star formation history of the proto-Galaxy, which is obviously of considerable importance in determining the levels of independent substructures in the gas cloud which became the Galaxy. Models[37,38] whereby the spheroid is a result of the merging of many independent 'fragments' each with its own chemical enrichment history, seek to explain the entire range of metallicity seen in halo stars as due to the statistical effects of chemical inhomogeneities, rather than as a trend with age. The 'fragments' are assumed to be disrupted by some process after some time, producing the field stars of the halo (*cf.* ref 39). Such models also predict a large spread in the element ratios observed from field star to field star if, for any one of the 'fragments' (each of which is presumed to contain many globular cluster-sized objects), star formation continued for sufficiently long that Type I supernovae became a significant source of iron. The explanation, re-iterated above, of the breaks in element ratio as a function of [Fe/H] as simply reflecting different element production sites depends on the existence of a simple correspondence between [Fe/H] and time. However, this mapping depends on the star formation rate and the stellar IMF. Assuming the IMF to be constant, at least over the stars which contribute to nucleosynthesis, 'fragments' of higher star formation rate will have attained a higher [Fe/H] prior to the fixed time at which type I supernovae dominate, so that the break in the [O/Fe]:[Fe/H] relation will occur at higher [Fe/H]. Similarly, 'fragments' of substantially lower star formation rate would lead to a break at much lower [Fe/H]. Thus samples of stars which originated in many different 'fragments' which were subsequently disrupted would not show a well-defined break at one given metallicity, contrary to the observations. The small scatter seen then requires that all 'fragments' had evolved to nearly the same gas fraction at the time when they were disrupted, which in all cases was $\lesssim 10^9$yr after significant star formation began. Such an effect is implicit in the models[13,15] whereby the metallicity distribution of the extreme spheroid is a consequence of gas being lost from the star-forming process at a rate proportional to, and greater than, the star formation rate. The required rate of gas removal is a factor ten to fifteen higher than the star formation rate, so that it is possible for each 'fragment' to evolve to completion, by losing all its gas, on a time shorter than that required by the break in the [O/Fe]:[Fe/H] relation, or $\lesssim 10^9$yr.

'Fragment' models require also that star formation began over all of that part of the proto-galaxy which became the extreme spheroid in the solar neighborhood at nearly the same time. This requirement arises since if some 'fragments' lagged behind and began their chemical evolution substantially later than most, their oxygen-rich ejecta would have enriched the gas which was to form the the thick disk population, contrary to the observations. In view of these many constraints, it seems more plausible to identify any 'fragments' with short-lived condensations which become the sites of star formation in a gaseous background, rather than as discrete structures merging to form the proto-galaxy. In general the existence of structure in the element-ratio relations is strong evidence that there was a continuous increase with time of the heavy element abundance of the Galaxy, and is difficult to reconcile with stochastic chemical evolution models.

A further constraint on models for the Galactic spheroid that appeal to merging and disruption of sub-structure is the amount of surviving structure in kinematic phase space of stars observed in the halo today. The identification of metal-poor moving groups[40] suggests that merger events such as accretion of stellar satellites (globular clusters, companion dwarf galaxies) may have contributed a significant fraction of the mass of the spheroid (*cf.* ref. 41 where it is estimated that 10^3 globular-cluster sized objects were disrupted, on the basis of Eggen's earlier observations). Studies of the stellar luminosity function both for metal poor stars in the solar neighborhood and in present globular clusters offer a severe constraint on models which appeal to disruption of systems similar to present-day Galactic globular clusters as a means to form the spheroid. McClure *et al.* have shown[42] that the shape of the luminosity function for stars with $4 < M_V < 9$ is a strong function of globular cluster metallicity. Clusters whose abundances are similar to those of the field subdwarfs have a luminosity function with a much steeper slope than do more metal-rich clusters, which in turn have a luminosity function in good agreement with that for solar abundance stars near the sun. Eggen[43] has determined the field star luminosity function for both metal rich and metal poor stars in the solar neighborhood, and shows that, over the same range in M_V as that for which cluster data exist, there is no significant dependance on metallicity. Thus, the luminosity function for metal-poor stars in globular clusters differs significantly from that for metal-poor field stars in the solar neighborhood; disruption of extant globular clusters cannot provide a significant fraction of the field spheroidal stars.

We thank NATO for a research grant (490/84) to aid our collaboration.

REFERENCES

1. van den Bergh, S., *Astr. J.*, **67**, 486-490, (1962).
2. Schmidt, M., *Astrophys. J.*, **137**, 758-769, (1963).
3. Pagel, B.E.J. and Patchett, B.E., *M. N. R. A. S.*, **172**, 13-40, (1975).
4. Ostriker, J.P. and Thuan, T.X., *Astrophys. J.*, **202**, 353-364, (1975).
5. Larson, R.B., *M. N. R. A. S.*, **218**, 409-428, (1986).
6. Wyse, R.F.G. and Silk, J., *Astrophys. J. (Lett.)*, **313**, L11-L14, (1987).
7. Eggen, O.J., Lynden-Bell, D. and Sandage, A.R., *Astrophys. J.*, **136**, 748-766, (1962).
8. Gilmore, G. and Reid, I.N., *M. N. R. A. S.*, **202**, 1025-1047, (1983).
9. Gilmore, G. and Wyse, R.F.G., *Astr. J.*, **90**, 2015-2026, (1985).
10. Zinn, R., *Astrophys. J.*, **293**, 424-444, (1985).
11. Sandage, A. R. and Fouts, G., *Astr. J.*, **93**, 74-115, (1987).
12. Norris, J., Bessel, M.S. and Pickles, A.J., *Astrophys. J. Suppl.*, **58**, 463-492, (1985).
13. Hartwick, F.D.A., *Astrophys. J.*, **209**, 418-423, (1976).
14. Fall, S.M. in *The Milky Way Galaxy,* eds H. van Woerden, R.J. Allen and W. Butler (Dordrecht, Reidel) pp 603-610, (1985).
15. Gilmore, G. and Wyse, R.F.G., *Nature*, **322**, 806-807, (1986).
16. Jones, B.J.T. and Wyse, R.F.G., *Astr. Astrophys.*, **120**, 165-180, (1983).
17. Gilmore, G., *M. N. R. A. S.*, **207**, 223-240, (1984).
18. Pagel, B.E.J., in *The Galaxy,* eds G. Gilmore and B. Carswell (Dordrecht, Reidel) pp 341-364, (1987).
19. Mirabel I.F. and Morras, R., *Astrophys. J.*, **279**, 86-92, (1984).
20. Twarog, B.A., *Astrophys. J.*, **242**, 242-259, (1980).
21. Carlberg, R.G., Dawson, P.C., Hsu, T. and VandenBerg, D.A., *Astrophys. J.*, **294**, 674-681, (1985).

22. Nissen, P.E., Edvardsson, B., Gustafsson, B., in *Production and Distribution of C,N,O Elements,* eds I.J. Danziger, F. Matteucci and K. Kjär, (Garching, ESO) pp 131-149, (1986).
23. Knude, J., Schnedler Nielsen, H. and Winther, M., *Astr. J*, **179**, 115-126, (1987).
24. Van der Kruit, P.C., *Astr. Astrophys.*, **140**, 470-475, (1984).
25. Faber, S.M. and Gallagher, J.S., *Ann. Rev. Astr. Astrophys.*, **17**, 135-187, (1979).
26. Van der Kruit, P.C., in *The Galaxy,* eds G. Gilmore and B. Carswell (Dordrecht, Reidel) pp 27-50, (1987).
27. de Vaucouleurs, G. and Pence, W., *Astron. J.*, **83**, 1163-1173, (1978).
28. Mould, J., in *Stellar Populations,* eds A. Renzini, M. Tosi and C. Norman (Cambridge, CUP) pp 9-25, (1986).
29. Wyse, R.F.G. and Gilmore, G. *Astr. J.*, submitted (1987).
30. Lambert, D. L., *J. Astrophys. Astr.*, **8**, 103-122, (1987) and refs therein.
31. Magain, P., *Astr. Astrophys*, **179**, 176-180, (1987).
32. Gustafsson, B., in *Stellar Evolution and Dynamics in the Outer Halo of the Galaxy*, ed M. Azzopardi (Garching, ESO) in press (1987).
33. Tinsley, B.M., *Astrophys. J.*, **229**, 1046-1056, (1979).
34. Matteucci, F. and Greggio, L., *Astr. Astrophys.*, **154**, 279-287, (1986).
35. Woosley, S.E and Weaver, T.A., *Ann. Rev. Astr. Ap*, **24**, 205-254, (1986).
36. Iben, I., in *Cosmogonical Processes*, eds W.D. Arnett, C.J. Hansen, J.W. Truran and S. Tsuruta, S. (Utrecht : VNU Science Press) pp 155-183, (1986).
37. Searle, L., in *The Evolution of Galaxies and Stellar Populations*, eds B.M. Tinsley and R.B. Larson (New Haven : YUP) pp 219-232, (1977).
38. Searle, L. and Zinn R., *Astrophys. J.*, **225**, 357-379, (1978).
39. Fall, S.M. and Rees, M.J., *Mon. Not. R. Astr. Soc.*, **181**, 37P, (1977).
40. Eggen, O., in *The Galaxy*, eds G. Gilmore and B. Carswell (Dordrecht, Reidel) pp 211-227, (1987).
41. Oort, J.H., in *Galactic Structure,* eds A Blaauw and M. Schmidt (Chicago, UCP) pp 455-512, (1965).
42. McClure, R.D., VandenBerg, D.A., Smith, G.H., Fahlman, G.G., Richer, H.B., Hesser, J.E., Harris, W.E., Stetson, P.B. and Bell, R.A., *Astrophys. J. (Lett.)*, **307**, L49-L53, (1986).
43. Eggen, O., *Astr. J.*, **93**, 393-420, (1987).

IRON ABUNDANCE EVOLUTION IN SPIRAL AND ELLIPTICAL GALAXIES

F. Matteucci
European Southern Observatory
Karl-Schwarzschild-Str. 2
D-8046 Garching bei München
Federal Republic of Germany

ABSTRACT

Chemical evolution models for the Galaxy in the solar neighbourhood and elliptical galaxies, which take into account the most recent developments in theories of stellar nucleosynthesis and supernova progenitors, are presented. The evolution of the abundance of iron in these systems, under the assumption that this element is mainly produced by Type I supernovae, originating from white dwarfs in binary systems, has been computed and the results compared with the observations. The general agreement between theoretical predictions and observations supports the assumptions made about the origin of iron.

INTRODUCTION

Detailed nucleosynthesis computations concerning iron production in stars have only recently become available. In fact, the idea that each outburst of a type I SN should produce a quantity of ^{56}Ni, which subsequently decays into ^{56}Co and ^{56}Fe, of the order of 0.25-0.7 $M_\odot$ in order to explain type I SN light curves (Arnett 1979; Colgate et al. 1980; Weaver et al. 1980), addressed theoretical studies towards understanding which mechanisms could produce such a huge quantity of Fe in type I SN explosions.

A good mechanism was found to be a C-O white dwarf exploding by C-deflagration in a binary system. Nomoto et al. (1984) and Woosley et al. (1986) computed the detailed nucleosynthesis occurring in a C-deflagrating C-O white dwarf, and the predicted amount of ^{56}Ni is of the order of 0.5-0.6 $M_\odot$ per event. Small quantities of the elements between C and Si were also predicted to be synthesized in these objects, in good agreement with type I SN spectra (Branch et al. 1983).

In this framework type I SNe appear to be the major contributors to iron enrichment in galaxies. In fact, the most recent nucleosynthesis results for massive stars (Woosley and Weaver 1986) indicate that the contribution to iron enrichment from these stars is about 30% of the total, the precise percentage depending on the assumed IMF.

Although many uncertainties are still present in these nucleosynthesis computations, it is now possible to compute detailed evolution of iron abundance in galaxies and try to impose some constraints on theory after comparison with well established observational data. Particularly useful to this purpose are the data on halo and disk dwarfs in the solar neighbourhood, concerning the behaviour of the abundances of some heavy elements (C, N, O, Si, Mg, S) with respect to iron (see Gustafsson 1987, for a review on the subject). It has been shown that detailed models of chemical evolution of the solar neighbourhood, taking into account the contribution to iron enrichment from type I and II SNe, in the way mentioned before, can reproduce very well the observed abundance pattern in halo and disk dwarfs (Matteucci and Greggio 1986; Matteucci 1986a, 1987), apart from some uncertainties in the results for C and N.

In this paper we present an extended version of the model of Matteucci (1987) for the solar neighbourhood, in which the evolution of ^{32}S and ^{40}Ca with respect to iron is computed, in addition to C, N, O, Si and Mg. The evolution of the iron abundance is computed also for elliptical galaxies, according to the model of Matteucci and Tornambé (1987), by means of a different chemical evolution model but with the same nucleosynthesis prescriptions as in the model for the solar neighbourhood. In particular, the average abundance of Fe and Mg in the stars of elliptical galaxies, with initial masses in the range $10^9 - 2 \cdot 10^{12}\ M_\odot$, is shown and Fe, Mg - total mass relations are predicted for these objects. Unfortunately, the observational data concerning abundances in ellipticals are based on metallicity indicators and not on real abundances, therefore any quantitative comparison with theoretical predictions is not yet possible.

Finally, some estimates are given on the amount of iron that elliptical galaxies can contribute to enrich the intergalactic medium (IGM). In fact, it is well known that X-ray observations of galaxy clusters indicate abundances of iron in the IGM comparable with the solar value (see Rothenflug and Arnaud 1985 and references therein).

NUCLEOSYNTHESIS PRESCRIPTIONS

i) Massive stars ($M > 8\ M_\odot$)

The nucleosynthesis results for C, O, Mg, Si, S and Ca in massive stars, which are supposed to end their lives as type II SNe, are taken from Woosley and Weaver (1986), where a new rate for the $^{12}C(\alpha,\gamma)^{16}O$ reaction is adopted. Concerning iron production in these stars, although many uncertainties are still present in its exact determination, we have used the following prescriptions (Woosley, private communication): stars with masses smaller than 12 $M_\odot$ do not contribute to iron enrichment at all, stars in mass range $12 \leq m/m_\odot \leq 20$ contribute at maximum 0.1 $M_\odot$ of iron, and stars with $M > 20\ M_\odot$ contribute $\sim 0.5\ M_\odot$ of iron. Uncertainties are also present in estimating the amount of mass which can be ejected by massive stars in the form of Si, S and Ca, since part of the material in the form of these elements (a few tenths of solar mass) can fall back into the collapsing core before the explosion starts (Woosley and Weaver 1986). This quantity of material falling back into the collapsing core is not known exactly and therefore should be parametrised. We call here β this quantity and try to fix the best value of it by fitting the observations. The nucleosynthesis prescriptions for ^{14}N produced in massive stars are taken from Talbot and Arnett (1973).

ii) Low and intermediate mass stars ($0.8 \leq m/m_\odot \sim 8$)

a) Type I SNe: C-deflagration in binary systems

Several configurations can be envisaged for a white dwarf exploding by C-deflagration in a binary system. The model of Whelan and Iben (1973) consists of a C-O white dwarf plus a red (super) giant companion. The white dwarf is the primary component (namely the one which is originally more massive) and its initial mass can vary between 1.5 and 8 $M_\odot$, whereas the secondary star can be as low as 0.8 and as high as 8

$M_\odot$. The lifetime of this latter gives the clock for the explosion of the primary, since this occurs when the secondary evolves off the main sequence and fills its Roche lobe, transferring material on the primary which thus reaches the Chandrasekhar mass and explodes. In computing the type I SN rate, shown in Fig. 1 for the solar neighbourhood in the framework of this model, the method suggested by Greggio and Renzini (1983) is followed. The fraction of binary systems which can give rise to type I SNe is found by the condition that the current rate in the solar neighbourhood has to reproduce the observed one. To do this we use here the new estimates of SN rates in spirals by van den Bergh et al. (1987).

Iben and Tutukov (1984, 1985) and Webbink (1984) proposed a system of two C-O white dwarfs merging after emission of gravitational wave radiation, and exploding by C-deflagration. This scenario is probably more promising than the other to explain type I SNe, as is discussed by Iben and Tutukov (1984), at least for classical type I SNe (type Ia). In fact, recently a new class of type I SNe has been envisaged, the subluminous ones or type Ib (see Panagia 1986).

In the double degenerate model the explosion times are given by the lifetime of the progenitors of the two white dwarfs, which have originally about the same mass, plus the gravitational time delay. By assuming that the progenitor masses range from 5 to 7 $M_\odot$ (Tornambé and Matteucci 1986), the explosion times range from 10^7 to $16 \cdot 10^9$ years. The resulting Type Ia SN rate is shown in Fig. 1 and refers to the solar neighbourhood. In the same figure is also plotted the expected rate of type Ib SNe, under the assumption that the progenitors of these SNe are binary systems made of a C-O white dwarf plus a non-degenerate He star merging after gravitational wave emission, as recently proposed by Iben et al. (1987) and Tornambè and Matteucci (1987). Detailed nucleosynthesis computations for this configuration are not yet available, but since the explosion mechanism should be He-detonation, it is known, in this case, from models of Nomoto (1983), that these objects should produce ~ 0.3 $M_\odot$ of iron. The progenitors of these type I SNe should be in the range 2.5-7 $M_\odot$.

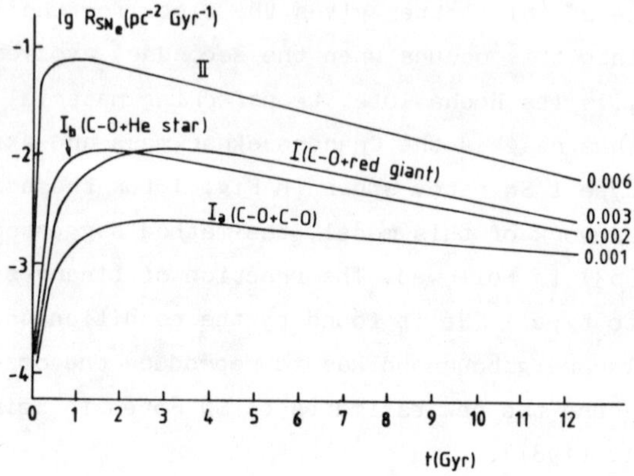

Fig. 1. The predicted SN rates as functions of time in the solar neighbourhood. The curve labelled II refers to the type II SN rate. The curve labelled I (C-O + red giant) refers to the type I SN rate computed in the framework of the Whelan and Iben (73) model. The curves labelled Ia (C-O, C-O) and Ib (C-O + He star) refer to the type Ia and b SN rates computed according to the Iben and Tutukov (84) and Iben et al. (87) models, respectively. The predicted current values of the rates are shown at the end of each curve. The observed values (van den Bergh et al. 1987) for the type II, Ia and Ib SNe, when a Hubble constant of 50 is assumed, are: 0.0055, 0.0015 and 0.002, respectively, in units of pc^{-2} Gyr^{-1}.

As one can see from Fig. 1, given the nature of the progenitors of the two kinds of type I SNe, the behaviour of the two rates is different but their current values tend to be similar, and this is in agreement with SN statistics (Panagia 1986; van den Bergh et al. 1987), which suggest that the current ratio between type Ib and type Ia is in the range 0.5-1.5. The absolute values of the current rates are also in good agreement with the observations, and in this case no constraints have been imposed "a priori" on them. From the point of view of the chemical results there is no difference whether the cumulative type I SN rate, computed in the framework of the Whelan and Iben model, or the other two (Ib and Ia) are adopted. This is due to the fact that the total number of SNe in the two cases is practically the same, as well as the current value of the total rate.

The quantity of iron per type I SN event we assume is ~ 0.45 $M_\odot$, being the average between 0.6 $M_\odot$ (type Ia) and 0.3 (type Ib).

Concerning the other elements produced by type Ia SN we take the results of Nomoto et al. (1984), their model W7.

b) Single low and intermediate mass stars

We assume the nucleosynthesis prescriptions from Renzini and Voli (1981), with $\eta = 0.33$ and $\alpha = 1.5$. These stars end their lives as white dwarfs and contribute mainly to the enrichment of He, C and N.

IRON EVOLUTION IN THE SOLAR NEIGHBOURHOOD

The chemical evolution model for the solar neighbourhood is the same as in Matteucci (1987). In summary, the model contains infall of primordial material with a time scale of 4 Gyr, a rate of star formation which is a function of the surface gas density, an initial mass function constant in space and time, the prescriptions for which are taken from Scalo (1986). This model reproduces successfully most of the basic constraints of the solar vicinity like the G-dwarf distribution with metallicity (Pagel and Patchett 1975), the current infall rate (Mirabel and Morras 1984), the current mass of gas and the current star formation rate (Miller and Scalo 1979).

The predicted absolute solar abundances of C, N, O, Mg, Si, S, Ca and Fe are well inside a factor of two with respect to the observations of Cameron (1982) (see Matteucci 1987). The iron temporal evolution in the solar neighbourhood, in particular [Fe/H] as a function of the galactic age, is shown in Fig. 2 together with the observational data of Twarog (1980), and the agreement is quite good. The assumed galactic age here is 12 Gyr, in agreement with estimates of the age of galactic globular clusters by adopting evolutionary tracks where the overabundance of O with respect to iron in metal poor stars is taken into account (Buonanno 1987). However, what matters is the qualitative behaviour of the age - metallicity relationship, and this would not change even if an age of 20 Gyr were assumed. In fact, this would only expand the period in which [Fe/H] is practically constant ($t > 5$-6 Gyr), but the time in which the rapid rise of the iron abundance takes place would always be the same, being mostly due to the lifetimes of iron producers and to the assumed infall and star formation rates.

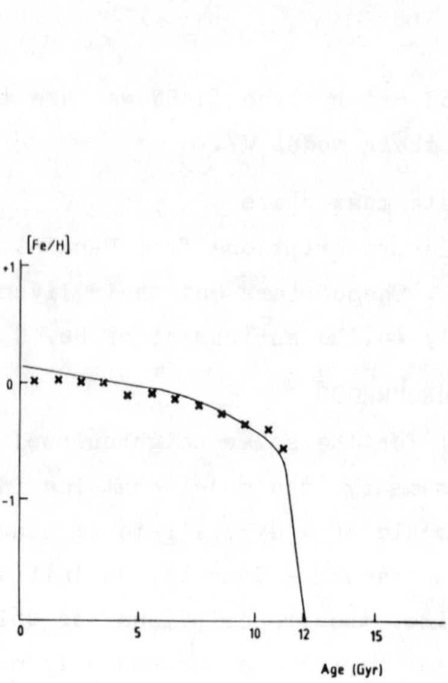

Fig. 2. The predicted age-iron abundance relation for the solar neighbourhood compared with Twarog's (80) data (crosses).

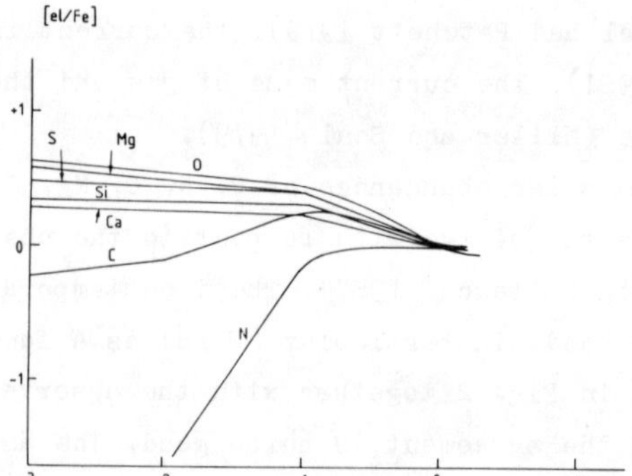

Fig. 3. The predicted behaviours of the ratios of several heavy elements to iron with respect to the iron abundance, in the solar neighbourhood.

In Fig. 3 are shown the predictions for [O/Fe], [Mg/Fe], [Si/Fe], [S/Fe], [Ca/Fe], [C/Fe] and [N/Fe] with respect to [Fe/H]. All these elements, which are mostly produced in massive stars and only in small quantities in type I SNe (see Nomoto et al. 1984), show overabundances with respect to iron at low metallicities ([Fe/H] < -1.00). The different absolute values of their overabundances are due to the different proportions in which they are produced in massive stars and in type I SNe. The absolute values of these overabundances as well as the qualitative behaviour of all these elemental ratios agree quite well with the observational data on halo and disk dwarfs. In fact, in summary, for [Fe/H] < -1.00, [O/Fe] = +0.6-0.5 dex (Sneden et al. 1979; Clegg et al. 1981; Nissen et al. 1985), [Mg/Fe] = +0.5 dex and [Si/Fe] = +0.35 dex (François 1986) [S/Fe] = +0.6 dex (François 1987) and [Ca/Fe] is not yet well established for halo dwarfs, although data from Gratton and Sneden (1987) indicate an overabundance of ~ +0.2 dex. The agreement for O, Si and Mg and probably Ca is very good, whereas the predicted overabundance of S seems to be slightly lower than the observed one. It is possible to vary this overabundance by changing the parameter β, but not too much otherwise the agreement for the other elements could be worsened. The best value of β we have found is ~ 0.6.

Observational data on N and C suggest that [C/Fe] and [N/Fe] should be solar in halo and disk dwarfs (Laird 1985). The predicted [C/Fe] is more or less solar inside an error of ±0.2 dex, marginally in agreement with the observational data.

This discrepancy could be due to the fact that the nucleosynthesis results for C produced in low and intermediate mass stars are obtained by using the old value of the rate of the $^{12}C(\alpha,\gamma)^{16}O$ reaction. Therefore, no firm conclusions on C can be drawn before new nucleosynthesis computations for low and intermediate mass stars will be available. The case of nitrogen is more complicated since this element can be both primary and secondary. If nitrogen produced in massive stars is only secondary, as is commonly believed, there is no way of finding an agreement with observations (see Fig. 3), whereas if one assumes that nitrogen in massive stars is only primary, the agreement with observations improves greatly (see Matteucci 1986b). However, more observational data on very metal poor stars as well as more up to date nucleosynthesis calculations for this element are necessary before trying to reach any conclusion.

CHEMICAL EVOLUTION OF ELLIPTICAL GALAXIES

The model for the chemical evolution of elliptical galaxies is the one of Matteucci and Tornambé (1987). We will only recall here the main assumptions of it. Galactic winds powered by SN (I and II) explosions are taken into account. No infall of external matter is considered and the rate of star formation is a function of the surface gas density. The initial mass function is assumed constant in space and time and the prescriptions for it are those of Salpeter (1955).

The average abundance of [Fe/H] and [Mg/H] at the present time in ellipticals of initial masses in the range $10^9 - 2 \cdot 10^{12}\ M_{\odot}$ is computed, by averaging of the abundances in the stars still alive at the present time (e.g. dwarfs and giants), weighted on their visual luminosities. The results obtained suggest the existence of a [Fe/H] - total galactic mass relation which is steeper than the corresponding one for [Mg/H]. In fact, $\Delta[\mathrm{Fe/H}]/\Delta \log M_G = 0.127$ and $\Delta[\mathrm{Mg/H}]/\Delta \log M_G = 0.04$. This effect is due to the different progenitors of Fe and Mg. In fact, besides the effect of galactic winds which determines the mass-metallicity relations, since more massive galaxies undergo winds later than less massive ones (Larson 1974), the different progenitors and nucleosynthesis prescriptions for different elements contribute to steepen or flatten the mass-single element abundance relations. This is a new result due to the detailed computation of single element evolution. Unfortunately, a direct comparison with observational data is not possible since the average abundances of elliptical galaxies are measured through spectrophotometric indices, and it is not well known how they relate to real abundances (see Matteucci and Tornambé 1987, for a discussion of the observational material).

Type I SN rates as functions of time are shown in Fig. 4 for a galaxy of $10^{12}\ M_{\odot}$. It can be easily seen from this figure that, in spite of the fact that the star formation rate is assumed to stop when galactic winds occur (e.g. always before 1 Gyr since galaxy formation), type I SNe continue to explode until the present time. As in the case of the solar neighbourhood, we present here the type I SN rates computed in the framework of different progenitor models, as discussed in that case.

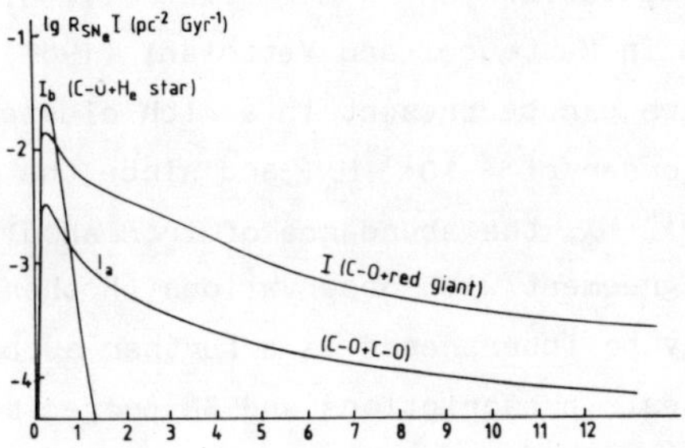

Fig. 4. The predicted type I SN rates as functions of time, for an elliptical galaxy of initial mass 10^{12} $M_{\odot}$. The labels on the various curves have the same meaning as in Fig. 1.

The predicted type Ib SN rate stops soon after star formation stops, and this is in agreement with the observations suggesting that no type Ib SNe are observed nowadays in ellipticals. On the other hand, the type Ia SNe continue to explode until the present time and the predicted current value of their rate is ~ 7.4 (-5) SNe pc^{-2} Gyr^{-1}, namely 0.04 SNu (by assuming $M/L_B \simeq 10$), and this is almost a factor of four lower than the one estimated with the other model (C-O + red giant), and a factor of 5 lower than Tammann's (1974) observational one. However, we think that this is the best estimate of the current type Ia SN rate, since by adopting the Whelan and Iben model one does not take into account the disappearance of type Ib SNe. In addition, given the many uncertainties present in the SN statistics, the agreement with the observations could be considered good even if it is inside a factor of 4. There is also the possibility that Tammann's (1974) rate for ellipticals is overestimated by a factor of 3 - 3.5, as is the case for his (1982) estimated SN rates in spirals (see van den Bergh et al. 1987).

In any case, the difference in the predicted current type I SN rates in the various models does not affect the chemical evolution of elliptical galaxies, since up to 1 Gyr the cumulative type Ia + b SN rate and the type I rate from the C-O + red giant model are almost the same, as shown in Fig. 4.

Finally, we want to mention the quantity of iron which should be expected to be found in a galaxy cluster, as a result of gas loss from ellipticals. The detailed computations and a more extensive discussion of this problem can be found in Matteucci and Vettolani (1987). The predicted amount of iron which can be present in a rich cluster like Coma turns out to be of the order of ~ 10^{11} $M_{\odot}$, and since the mass of gas measured in Coma is ~ 10^{14} $M_{\odot}$, the abundance of iron should be almost solar, in very good agreement with observations (Rothenflug and Arnaud 1985). This result may be interpreted as a further support to the assumed iron nucleosynthesis prescriptions and SN progenitors. As for the total amount of gas ejected into the IGM by ellipticals (the other galaxies would contribute negligibly to that), this turns out to be an order of magnitude lower than the observed one leading to the conclusion that most of the gas in galaxy clusters should have a primordial origin.

REFERENCES

Arnett, D.W.: 1979, Ap.J. Lett. 230, L37

Branch, D., Lacy, C.H., McCall, M.L., Sutherland, P.G., Nomoto, K., Wheeler, J.C., Wills, B.J.: 1983, Ap.J. 270, 123

Buonanno, R.: 1987, in "Stellar Evolution and Dynamics of the Outer Halo of the Galaxy", eds. M. Azzopardi, F. Matteucci and K. Kjär, E.S.O. publication, in press

Cameron, A.G.W.: 1982, in "Essays in Nuclear Astrophysics", eds. C.A. Banes, D.D. Clayton and N.D. Schramm, Cambridge University Press, p. 377

Clegg, R.E.S., Lambert, D.L., Tomkin, S.: 1981, Ap.J. 250, 262

Colgate, S.A., Petscheck, A.G., Kriese, J.T.: 1980, Ap.J. Lett 237, L81

François, P.: 1986, Astron. Astrophys. 160, 264

François, P.: 1987, in preparation

Gratton, R., Sneden, C.: 1987, Astron. Astrophys. 178, 179

Greggio, L., Renzini, A.: 1983, Astron. Astrophys. 118, 217

Gustafsson, B.: 1987, in "Stellar Evolution and Dynamics of the Outer Halo of the Galaxy", eds. M. Azzopardi, F. Matteucci and K. Kjär, E.S.O. publication, in press

Iben, I. Jr., Tutukov, A.: 1984, Astrophys. J. Suppl. 54, 335

Iben, I. Jr., Tutukov, A.: 1985, Astrophys. J. Suppl. 58, 661.

Iben, I. Jr., Nomoto, K., Tornambé, A., Tutukov, A.: 1987, Astrophys. J. 317, 717.

Laird, J.B.: 1985, Astrophys. J. 289, 556.

Larson, R.B.: 1974, Mon. Not. R. astr. Soc. 166, 585

Matteucci, F.: 1986a, Astrophys. J. Lett. 305, L81

Matteucci, F.: 1986b, Mon. Not. R. astr. Soc. 221, 911

Matteucci, F.: 1987, in "Stellar Evolution and Dynamics of the Outer Halo of the Galaxy", eds. M. Azzopardi, F. Matteucci and K. Kjär, E.S.O. publication, in press

Matteucci, F., Greggio, L.: 1986, Astron. Astrophys. 154, 279

Matteucci, F., Tornambé, A.: 1987, Astron. Astrophys. 185, 51

Matteucci, F., Vettolani, G.: 1987, Astron. Astrophys. in press

Miller, G.E., Scalo, J.M.: 1979, Astrophys. J. Suppl. 41, 513

Mirabel, I.F., Morras, R.: 1984, Astrophys. J. 279, 86

Nissen, P.E., Edvardsson, B., Gustafsson, B.: 1985, in "Production and Distribution of C,N,O Elements", eds. I.J. Danziger, F. Matteucci and K. Kjär, E.S.O. publication, p. 131

Nomoto, K.: 1983, in "Stellar Nucleosynthesis", ed. C. Chiosi and A. Renzini. Dordrecht, Reidel, p. 205

Nomoto, K., Thielemann, F.K., and Yokoi, K.: 1984, Ap.J. 286, 644

Pagel, B.E.J., Patchett, B.E.: 1975, M.N.R.A.S. 172, 13

Panagia, N.: 1986, in "High Energy Phenomena Around Collapsed Stars", ed. F. Pacini. Dordrecht, Reidel, in press.

Renzini, A., Voli, M.: 1981, Astron. Astrophys. 94, 175

Rothenflug, R., Armond, M.: 1985, Astronomy & Astrophysics 144, 431

Salpeter, E.E.: 1955, Ap.J. 121, 161

Scalo, J.H.: 1986, Fund. of Cosmic Phys. II, p. 1

Sneden, C., Lambert, D.L., Whitaker, R.W.: 1979, Ap.J. 234, 864

Talbot, R.J., Arnett, W.D.: 1973, Astrophys. J. 186, 51

Tammann, G.A.: 1974, in "Supernova and Supernova Remnants", ed. C.B. Cosmovici. Dordrecht, Reidel, p. 155

Tammann, G.A.: 1982, in "Supernovae: A Survey of Current Research", eds. M.J. Rees and R.J. Stoneham. Cambridge Univ. Press, p. 371

Tornambè, A., Matteucci, F.: 1986, M.N.R.A.S. 223, 69

Tornambè, A., Matteucci, F.: 1987, Ap.J. Lett 318, L25

Twarog, B.A.: 1980, Ap.J. 242, 242

van den Bergh, S., McClure, R.D., Evans, R.: 1987, Astrophys. J. 323, 44.

Weaver, T.A., Axelrod, T.X., Woosley, S.E.: 1980, in "Type I Supernovae" ed. J.C. Wheeler, p. 113.

Webbink, R.F.: 1984, Ap.J. 277, 355

Whelan, J.C., Iben, I. Jr.: 1973, Ap.J. 186, 1007

Woosley, S.E., Taam, R.E., Weaver, T.A.: 1986, Ap.J. 301, 601.

Woosley, S.E., Weaver, T.A.: 1986, IAU Coll. 89, eds. D. Mihalas and K.A. Winkler, p. 91

Effects of Galactic Chemical Evoluion

on Thorium/Neodymium Cosmochronoloy

G. J. Mathews

University of California
Lawrence Livermore National Laboratory
Livermore, CA 94550

and

D. N. Schramm

Department of Astronomy and Astrophysics
University of Chicago
Chicago, IL 60637

ABSTRACT

We emphasize the importance of galactic chemical evolution and the different nucleosynthetic origins for thorium and neodymium with regards to the recent observations of neighboring spectral lines from these elements in G-Dwarfs stars and the use of the correlation of the spectral line ratio with stellar age to derive an upper limit to the age of the galaxy. When effects of galactic chemical evolution and nucleosynthetic origin are included the systematic trend in the data toward constant or even increasing Th/Nd ratio with stellar age can be easily understood. A model dependent (3σ) best upper limit of $T_0 \simeq 18$ Gyr to the age of the galaxy is derived, although simpler schematic models for the evolution of the Th/Nd ratio can give (3σ) upper limits ranging from 9 to 20 Gyr.

I. INTRODUCTION

Recent observations[1] of close lying thorium and neodymium lines for a number of G-dwarf stars have led to the suggestion that the line strength ratio of Th (most likely due to unstable ^{232}Th ($\tau_{1/2}$ = 14 Gyr) to stable Nd could be used as an indicator of the age of the galaxy. Based upon the observation that the Th/Nd line ratio shows little variation as a function of stellar age, an upper limit to the age of the galaxy of 9.6 Gyr was derived which is somewhat lower than (although perhaps[2] consistent with) the age derived from other chronometers[3-7] for which $T_0 \sim 15 \pm 5$ Gyr. It should be noted that there is a lower bound from nucleochronology mean age studies of 8.7 Gyr[5] based upon assuming all production at t = 0. Thus, to have galactic ages close to this number would require the other chronometers to be produced very early. This seems unlikely since ^{244}Pu ($\tau_{1/2}$ = 8.2×10^7 yr) was present when the solar system formed, and is inconsistent with the model of constant Th and Nd production assumed in Ref. [1].

The use of the Th/Nd stellar line ratio, however, could be a significant means of determining galactic age because it would be the only nuclear cosmochronometer derived from material outside the solar system. Also since observations are made on stars of different galactic ages it may be possible to better understand the galactic chemical evolution parameters for this chronometer. In this letter, however, we wish to highlight two difficulties associated with utilizing this chronometric ratio which can significantly affect the derived upper limit to the galactic age and possibly eliminate the above mentioned inconsistencies.

The first and most important difficulty is due to the different stellar mechanisms[1,8] contributing to the abundances of these elements. Although ^{232}Th is surely only produced in an explosive rapid-neutron-capture environment (the r-process), solar-system neodymium isotopes are have about equal abundances[9,10] of material from both the r-process and quiescent slow neutron capture (the s-process). Since sand r-process elements have different galactic evolution histories[11], the relative Th and Nd production rates should have changed during the lifetime of the galaxy. Indeed, there are by now numerous observations[12-15] indicating enhanced r-process to s-process abundances in metal poor stars. The implication of these data are that the Th/Nd production ratio could have been significantly higher in the past.

A second important feature to consider is the destruction term due to the loss of Nd and Th from the interstellar medium (ISM) by star formation and subsequent trapping of material into stellar remnants. The equation for the rate of change of the abundance of an element in the ISM used in Ref. [1] absorbs this term into a generalized production function. It is, however, important[16,17] to explicitly include a linear term describing the net loss of material from the ISM. As we shall see, the explicit inclusion of this term can lead to a smaller estimated galactic age.

In what follows we utilize a model which includes both the explicit destruction due to star formation, and the different production histories for ^{232}Th and Nd. In the context of this model we find that it is imperative to include both of these effects. Doing so improves the fit to the data and allows for substantially different galactic ages.

II. GALACTIC CHEMICAL EVOLUTION

Including the additional terms into the dfferential equation for the rate of change of species, i, we write;

$$dN_i/dt = P^s{}_i p(t)^s + P^r{}_i p(t)^r + E_i(t) - B(t)N_i - \lambda_i N_i \ , \qquad (1)$$

where $P^j{}_i p(t)^j$ are source terms for r- and s-process material, $E_i(t)$ is the contribution from the ejection from stars which have not destroyed or added to their initial Th or Nd abundance, B is the fractional rate of loss of mass from the ISM by star formation, and λ_i is the radioactive decay rate. In Eq. (1) we neglect any contribution from infalling extragalactic s-process or r-process material, or radial mixing in the disk.

All of the quantities in Eq. (1) can be defined in terms of standard separable galactic chemical evolution models[18], based upon an initial mass function, $\phi(m)$, and a stellar birthrate function, f(t). The fractional birthrate, B, is;

$$B = \int dm\, m\, \phi(m) f(t)/M_G(t) \quad , \qquad (2)$$

where $M_G(t)$ is the mass of the interstellar medium. Similarly, the source terms in Eq. (1) become;

$$P^j{}_i p(t)^j = \int dm(m-m_r)\phi(m) f(t-\tau(m)) X^j{}_i(m) \quad , \qquad (3)$$

where m_r is the mass of the stellar remnant for a star of initial mass, m, which dies after an age $\tau(m)$, and $X^j{}_i$ is the mass fraction of species, i, in the ejecta. For the ejection, $E_i(t)$, from non r- or s-process stars, the expression is the same as Eq. (3) but with the $X^j{}_i$ given by the star's initial mass fraction for Th or Nd. Therefore, in what follows we absorb $E_i(t)$ into the source terms, $P^j{}_i p(t)^j$.

III. r-PROCESS AND s-PROCESS PRODUCTION FUNCTIONS

Although the details of the mechanism for s-process nucleosynthesis are still being debated[10,19-22], it is now widely accepted that the s-process occurs during the thermal pulse phase of asymptotic giant branch stars[23]. This restricts the range of stellar masses to those in which shell helium burning occurs above an electron-degenerate carbon-oxygen core. This places an upper limit[24] on the initial mass of such stars of about 8-9 $M_\odot$ depending on the composition. For the lower initial mass limit, stars as small as 0.8-1.0 $M_\odot$ eventually enter the thermally pulsing AGB phase. The enrichment of s-process material in the outer envelope, however, requires core masses greater than 0.6 $M_\odot$. Allowing for mass loss from the main sequence, this places a lower limit of about $1M_\odot$. For the purposes of this study we therefore consider s-process elements to occur in the mass range of 1.0-9.0 $M_\odot$. The mass fraction of produced Nd will be primarily a function of the neutron exposure and the initial Fe abundance[8]. For the purposes of this study, however, we will assume two possibilities: 1) a constant s-process mass fraction as a function of stellar mass; and 2) s-process production proportional to the Fe mass fraction, where the Fe abundance is estimated by assuming a constant type-I supernova rate and instantaneous recycling.[18] In a future study we will consider the effect of the time dependence of the metallicity and the variation of neutron exposure with stellar mass in more detail. As in the work of Clayton[25] we find that the dependence of the s-process on preexisting Fe seed can increase the inferred galactic ages by delaying the appearance of s-process neodymium, and thus increasing the Th/Nd ratio for old stars. This effect is further enhanced by the fact that the s-process probably occurs in longer-lived lower-mass stars than the r-process.

The site for the r-process is much less well understood[3,6,26] than for the s-process. Numerous sites have been proposed such as ejection of neutronized core material from a supernova, shock-induced explosive helium burning, or neutron-star collisions. Most proposed sites are associated in one way or another with massive star ($M \geq 9\ M_{\odot}$) evolution, and the most viable sites at the present time involve the build up of r-process material from a neutronized gas, independent of the presence of preexisting "seed" material. Indeed, observationally it appears that the r-process elements are primary[27]. Thus, we assume a constant r-process mass fraction for massive stars independent of initial composition.

IV. MODEL CALCULATIONS

The general solution to equation (1) for the observed Th/Nd ratio at time T_0 on a star which formed at time, T, after the birth of the galaxy is;

$$Th/Nd = \exp\{-\lambda T_0\}\int_0^T P^r_{Th} p(t)^r \exp\{\int_0^t (\lambda+B)d\tau\}dt$$

$$\times \left[\int_0^T P^r_{Nd} p(t)^r \exp\{\int_0^t B d\tau\}dt + \int_0^T P^s_{Nd} p(t)^s \exp\{\int_0^t B d\tau\}dt\right]^{-1} . \qquad (4)$$

We consider this solution in three simple models.

Model a) no explicit loss term; constant equal production functions

Consider first the limit used in Ref. [1], which for our purposes corresponds to $B = 0$, and $X^s{}_{Nd} = 0$. Then for a constant production rate, the ratio of Th/Nd for a star formed at time, T, becomes;

$$Th/Nd = (P_{Th}/P_{Nd})\exp(-\lambda T_0)[\exp\{\lambda T\}-1]/(\lambda T) \ , \qquad (5)$$

where, $P_{Th}/P_{Nd} \sim 1\text{-}2$, was adjusted for each calculation to optimally fit the data[1] which are normalized to a mean of unity.

This ratio is plotted in figure 1 as a function of stellar age for various choices for T_0. These are compared with the line ratios given in Ref. [1]. To the data of Table 3 in that paper we have added a point (presumably HR7869) with an age of 12 ± 3 Gyr which appears in Fig. 3 of that paper. In addition, the sun was always taken to have been born at T_0 - 4.6 Gyr. The ages of the other stars in the sample were scaled by a factor of $T_0/20$.

The calculation of χ^2 for fits to these data included the uncertainties in the line ratio and stellar age by weighting the square of the resuduals with a variance[29] of $\sigma^2{}_i = \sigma^2{}_{yi} + (dy/dx)\sigma^2{}_{xi}$, where y is the Th/Nd ratio, x is the stellar age, and the derivative is evaluated from the model. There is a rather broad minimum in χ^2 for this model at an age of 5.5 Gyr. The 95% confidence (2σ) upper limit for a χ^2 distribution would be at an age of 15 Gyr, and the plotted 99.7% (3σ) upper limit is 20 Gyr. The reason our 95% confidence limit is so much larger than that quoted in Ref.

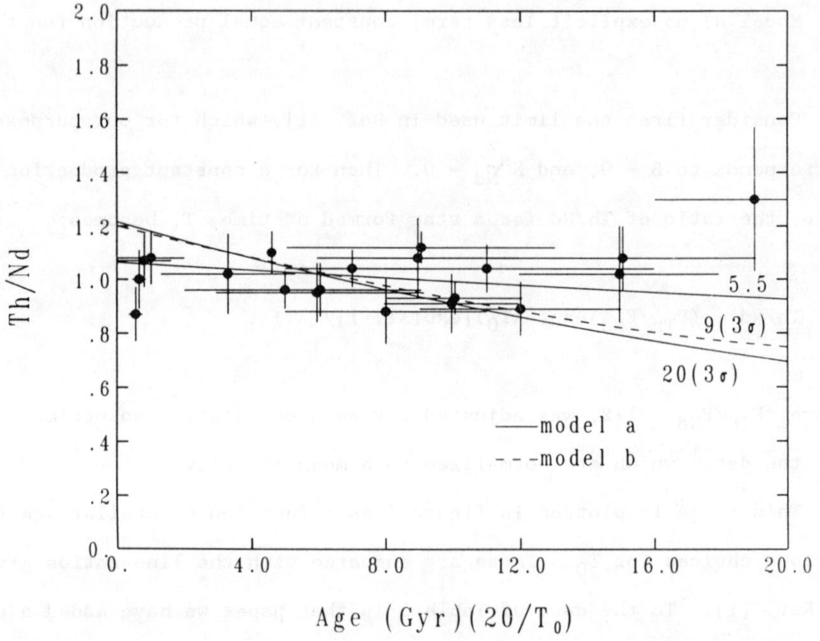

Figure 1. Measured Th/Nd line ratios[1] as a function of apparent stellar age compared with calculated ratios. The solid lines are from a model similar to that used in Ref. [1]. The curve for T_0 = 5.5 Gyr corresponds to the optimum galactic age in this model. The curve for 20 Gyr corresponds to the 3σ upper limit to this galactic age. The dashed line shows the 3σ upper limit from a model which explicitly includes the loss of Th and Nd from the ISM as new stars and stellar remnants form. This term increases the dependence of the Th/Nd ratio on stellar age and decreases the upper limit to the galactic age to 9 Gyr.

[1] can be traced[30] to the neglect of higher order terms in the expansion of Eq. (5) which was used in the analysis of Ref. [1].

Model b) loss term included

Consider next the effect of adding the linear loss term in Eq. (1) but still keeping the production rates for Th and Nd constant and equal. For a constant loss term, the analytic solution for Th/Nd becomes;

$$\mathrm{Th/Nd} = (P_{Th}/P_{Nd})\exp\{-\lambda(T_0-T)\}[B/(\lambda+B)]$$
$$\times\ [1-\exp\{-(\lambda+B)T\}]\ /\ [1-\exp\{-BT\}]\quad . \qquad (6)$$

The fractional birthrate, B, in the solar neighborhood at the present time is in the range[30,31], $0.2 < B < 2.5\ \mathrm{Gyr}^{-1}$. Explicitly including the effect of the ejection of nondestroyed Th or Nd in the instantaneous recycling approximation would reduce this rate by a factor of 0.6 to 0.9[30]. This value for B is to be contrasted with the much smaller ^{232}Th decay rate ($\lambda_{Th} = 0.049\ \mathrm{Gyr}^{-1}$). Thus, the loss due to star formation should completely dominate the loss from ^{232}Th decay. The increased removal of Th and Nd from the ISM will mean that the Th/Nd ratio will depend more sensitively on the production just before incorporation into a star. Therefore, the Th/Nd ratio will show more dramatic dependence on stellar age. For a constant $B = 1\ \mathrm{Gyr}^{-1}$, there was no minimum in χ^2 for any galactic age, however all ages less than 0.8 Gyr yield a reduced $\chi^2 \leq 1$. The dashed curve for T_0 = 9 Gyr on Fig. 1 corresponds to a 99.7% confidence (3σ) upper limit to exceed a reduced χ^2 of 1. We note that this result is largely independent of the value for B as long as $B \gg \lambda$.

Model c) different production rates and loss terms included

Finally, we consider the different production histories for the s- and r-process material. To estimate these we must explicitly compute the quantities in Eq. (4) from an initial mass function, $\phi(m)$, and birthrate function, f(t). We find little dependence of the Th/Nd ratio on the stellar birthrate for reasonable variations[31.32] from exponentially decreasing to constant functions. Therefore, for illustration, we use a constant star formation rate, $f(t) = 1/T_0$. For $\phi(m)$ we use the Miller and Scalo[31] three segment power-law initial mass function (for constant f(t) and $T_0 = 12$). With f(t) constant, the different nucleosynthetic histories are only reflected in the time delay before low-mass stars eject s-process material into the ISM and the timescale for the growth of s-process seed material. For the stellar ages we take,[32] $\tau(m) = 11.7/m^2 + .001$ Gyr, and for the remnant masses we take;[34]
$m_r = m$ for $m \le 0.7$; $m_r = 0.17m + 0.55$ for $0.7 \le m \le 5$;
and $m_r = 1.4$ for $m \ge 5$.

With these input data, Eq. (4) can be numerically solved. We have studied the effects on Eq. (4) of varying rates of infalling extragalactic material, and reasonable variations in the time dependence of M_G. These features, although important for galactic evoltion, seem to have little effect on the calculated Th/Nd ratio. Therefore, for simplicity, assume an infall rate such that M_G remains constant at a present value[30] of $7.5 \pm 1.5\ M_\odot\ pc^{-2}$. The relative neodymium production mass fractions, $X^s{}_{Nd}$ and $X^r{}_{Nd}$, are fixed by the solar abundances, i.e. the s-process and r-process contributions are equal for a star born at $T = T_0 - 4.6$.

The results for different T_0 are given in Figs. 2 and 3. The curves in Fig. 2 are for models in which the Fe seed abundance is constant. The curves in Fig. 3 are for models which include the time dependence of the Fe seed abundance, and therefore represent our best model. In this model values of T_0 from 10 to 15 Gyr give equivalently good fits and a reduced $\chi^2 \le 1$. The (2σ) and (3σ) upper limits are at 16 and 18 Gyr respectively (10 and 13 Gyr for a constant s-process seed as in Fig. 2). Note that with the different production rates included there now appears to be a natural explanation for the apparently high Th/Nd ratio in HR3018 as well as the nearly constant ratio for stars born later. It would be extremely useful to have more Th/Nd measurements for stars with large ages and low metallicity. It appears from this analysis that HR3018 may have formed within a few billion years of the first heavy-element nucleosynthesis. Also note that the predicted increased Th/Nd ratio early on allows for much larger galactic ages than when the s-process contribution to Nd is neglected (Model b).

From these analyses we conclude that simple models of galactic chemical evolution can fit the Th/Nd line ratios with ages as large as 20 Gyr, although our most detailed analysis favors an upper limit of 18 Gyr. This limit, however, could further increase in models with a bimodel initial mass function.[35] Therefore we conclude that the Th/Nd line ratios do not necessarily require short (< 10 Gyr) galactic ages, but they do give us interesting information on the early history of the galaxy.

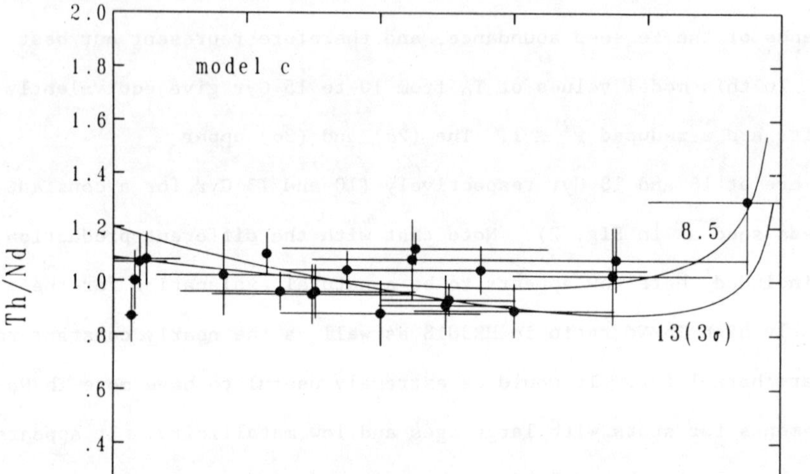

Figure 2) Th/Nd line ratios compared with calculated ratios as a function of stellar age. In this calculation a constant rate of star formation and constant s-process seed abundance was assumed but the s- and r-process contributions were allowed to take place in stars of different mass. The effect of these different histories is to improve the fit to the observed ratios. The curve for T_0 = 8.5 Gyr corresponds to the maximum galactic age which yields a reduced $\chi^2 \leq 1$. The curve for 13 Gyr corresponds to the 3σ upper limit (99.7% confidence) to exceed a reduced χ^2 of 1.

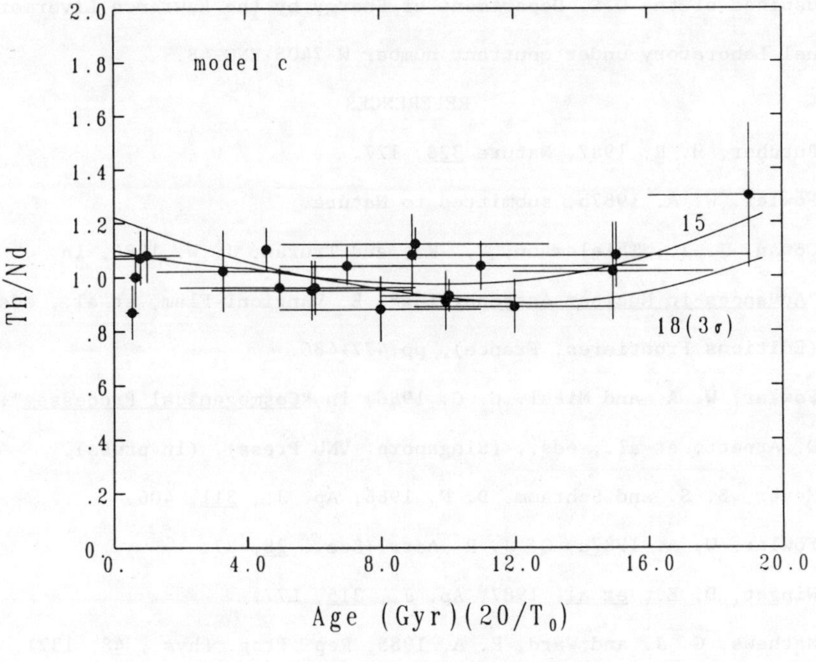

Figure 3) Th/Nd line ratios compared with calculated ratios as a function of stellar age. In this calculation a constant rate of star formation was assumed and the time dependence of the s-process seed abundance was included. These calculated curves represent our best model. The curve for $T_0 = 15$ Gyr corresponds to the maximum galactic age which yields a reduced $\chi^2 \leq 1$. The curve for 18 Gyr corresponds to the 3σ upper limit (99.7% confidence) to exceed a reduced χ^2 of 1.

V. ACKNOWLEDGEMENT

Informative discussion with. Prof. William A. Fowler, Dr. B. E. J. Pagel, and Dr. R. F. G. Wyse is gratefully acknowledged as well as useful correspondence from H. R. Butcher and D. D. Clayton. Work performed under the auspices of the U.S. Department of Energy by the Lawrence Livermore National Laboratory under contract number W-7405-ENG-48.

REFERENCES

1] Butcher, H. R. 1987, Nature 328, 127.

2] Fowler, W. A. 1987b, submitted to Nature.

3] Cowan, J. J., Thielemann, F. -K., and Truran, J. W. 1986, in "Advances in Nuclear Astrophysics", E. Vangioni-Flam, et al., eds., (Editions Frontieres, France), pp.477-486.

4) Fowler, W. A. and Miesl, C. C. 1986, in "Cosmogenical Processes", W. D. Arnett, et al., eds., (Singapore: VNU Press), (in press).

5] Meyer, B. S. and Schramm, D. N. 1986, Ap. J., 311, 406.

6] Fowler, W. A. 1987a, Q. J. R. Astr. Soc., 28, 87.

7] Winget, D. E., et al. 1987, Ap. J., 315, L77.

8] Mathews, G. J. and Ward, R. A. 1985, Rep. Prog. Phys., 48, 1371.

9] Mathews, G. J. and Kappeler, F. 1984, Ap. J., 286, 810.

10] Howard, W. M., Mathews, G. J., Takahashi, K., and Ward, R. A. 1986, Ap. J., 309, 633.

11] Schramm, D. N. and Tinsley, B. M. 1974, Ap. J., 193, 151.

12] Spite, M. and Spite, F. 1978, Astron. Astrophys., 67, 23.

13] Sneden, C. and Parthasarathy, M. 1983, Ap. J., 267, 757.

14] Luck, R. E. and Bond, H. E. 1985, Ap. J. 292, 559.

15] Sneden, C. and Pilachowski, C. A. 1985, Ap. J. Lett., 288, L55.

16] Schramm, D. N. and Wasserburg, G. J. 1970, Ap. J., 163, 57.

17] Schramm, D. N. 1974, Ann. Rev. Astr. Astrophys., 12, 383.

18] Audouze, J. and Tinsley, B. M. 1976, Ann. Rev. Astron. Astrophys., 14, 43.

19] Iben, I. and Truran, J. W. 1978, Ap. J., 220, 980.

20] Cosner, K. R., Iben, I., and Truran, J. W., 1980, Ap J. Lett., 238, L91.

21] Mathews, G. J., Ward, R. A., Takahashi, K., and Howard, W. M. 1986, in "Nucleosynthesis and Its Implications on Nuclear and Particle Physics", Audouze and Mathieu, eds., (Nato ASI Series, Vol. 163), (D. Reidel, Dordrecht) pp. 277-284.

22] Malaney, R. A. and Boothroyd, A. I. 1987, Ap J. (in press).

23] Iben, I., Jr., 1977, Ap. J., 217, 788.

24] Iben, I. and Renzini, A. 1983, Ann. Rev. Astron. Astrophys., 21, 271.

25] Clayton, D. D. 1987, Nature, 329, 397.

26] Norman, E. B. and Schramm, D. N. 1979, Ap. J., 228, 881.

27] Truran, J. W. 1981, Asrton. Astrophys., 97, 391.

28] Bevington, P. R. 1969, "Data Reduction and Analysis for the Physical Sciences" (McGraw-Hill, New York).

29] Pagel, B. E. J. 1987, Priv. Comm.

30] Tinsley, B. M. 1977, Ap. J, 216, 548.

31] Miller, G. E. and Scalo, J. M. 1979, Ap. J. Suppl., 41, 513.

32] Twarog, B. A. 1980, Ap. J., 242, 242.

33] Talbot, R. J. and Arnett, W. D. 1971, Ap. J., 170, 409.

34] Paczynski, B. and Ziolkowski, J. 1968, Acta Astronomica, 18, 255.

35] Larson, R. B. 1986, Mon. Not. R. Astr. Soc., 218, 409.

Pregalactic Nucleosynthesis and Population III Stars

B.J.CARR

School of Mathematical Sciences
Queen Mary College
Mile End Road
London E1 4NS

ABSTRACT

Population III stars have been invoked in two contexts: to produce an initial enrichment and to generate the dark matter in galactic halos. Those of the first kind definitely exist but may not warrant a special name. Those of the second kind definitely warrant a special name but may not exist. There is no necessity for either kind to be pregalactic, although recent claims to have detected a far-IR background may provide evidence for some pregalactic star formation.

1. INTRODUCTION

There is considerable confusion in the literature because the term "Population III" has been used to describe three distinct types of stars: (1) the first ones which form out of the pristine gas left over after cosmological nucleosynthesis; (2) the ones which generate the first metals; and (3) the ones which have been hypothesized to provide the dark matter in galactic halos. These usages of the term are based on temporal, nucleosynthetic, and spatial criteria, respectively, and are not necessarily equivalent. Attempts to allay the confusion by introducing further classifications, such as "Population 0" and "Population IV", have merely compounded the problem.

The confusion is minimized if the dark matter in galactic halos is provided by some sort of elementary particle, so that stars of type (3) do not exist. In this case, the first stars are necessarily the ones that generate the first metals, so definitions (1) and (2) are equivalent. However, there remains the issue of whether the stars warrant a separate name. This only makes sense if they are qualitatively different from ordinary "Population II" stars. For example, it would not be justified if the first metal-producing stars were merely the ones at the high mass end of the Population II mass spectrum. For in this case they would generate the first metals simply because they evolved fastest. The introduction of a new term would only be warranted if the first metal-producing stars formed at a distinct epoch or if the IMF of the first stars was bimodal (i.e. with a distinct population of high and low mass stars forming in different locations). As discussed in Section (3), this relates to the observational issue of whether or not there is a lower cut-off in the metallicity distribution of Population II stars.

In this context, it is important to decide whether the appellation "Population III" is to be assigned to the first stars or their parent clouds. If one calls the clouds Population III, then all the stars which they spawn must also be called Population III. But if the stars formed in two stages or had a bimodal mass spectrum, this would be inappropriate since one would want to differentiate between them. For this reason, I would advocate using the term "Population III" for the stars rather than the parent clouds, although this convention is not adopted universally.

A more serious semantic issue arises when one contemplates the existence of stars of type (3). In this case, as discussed in Section (4), the dark matter in galactic halos must be either jupiters or black holes. An ambiguity now arises depending upon whether these dark objects form before, after, or contemporaneously with the stars which make the first metals. All three cases are possible. The first situation would arise if the most of the baryons go into a first generation of stars whose IMF is such that they produce no metals at all. The second situation would arise if most of the baryons go into dark stars only after some metal-producing stars have formed. The third situation would arise if the baryons go into a first generation of stars which have an IMF such that most become dark and a few provide an initial enrichment. In any case, it is important to distinguish between metal-producing Population III and dark-matter producing Population III, although they could in principle be the same.

2. STELLAR NUCLEOSYNTHESIS

Since the emphasis of this talk is on nucleosynthesis by Population III stars, I will begin by summarizing the basic features of stellar nucleosynthesis. Only stars larger than 0.8 $M_{\odot}$ will have completed their main-sequence evolution by the present epoch. Those between 0.8 and 10 $M_{\odot}$, which have a main-sequence time between 10^{8} and 10^{10}y, either evolve to white dwarfs or undergo carbon deflagration [1]. They are the major source of carbon, nitrogen and s-process elements [2]; those in binaries could also produce iron peak nuclei via type I supernovae [3]. Stars in the mass range 10-100 $M_{\odot}$, which have a main-sequence time between 10^{7} and 10^{8}y, evolve to type II supernovae and are the major source of oxygen, neon to calcium, and the r-process elements [4]. They may also leave neutron star remnants or, if their mass exceeds some critical value M_{BH}, black holes. The value of M_{BH} is unknown but it might be around 50 $M_{\odot}$ [5].

These properties are already sufficient to explain many of the abundance characteristics of Population I and II stars. For example, field halo stars with $Z < 0.1\ Z_{\odot}$ have enhancements in the ratios of O, Mg, Si and Ca to Fe by a factor of 3 relative to the Sun and also depletions in the s-process elements Sr and Ba [6]. This is naturally explained by the fact that these elements are preferentially produced by the sort of massive stars which would complete their evolution on the timescale (10^{8}y) associated with the formation of the Galactic halo. Elements produced by stars below 10 $M_{\odot}$ would not appear until the Galactic disk formed. This also explain why the poorest disk stars have a metallicity of about 0.1 $Z_{\odot}$.

Stars more massive than 100 $M_\odot$, which are termed "Very Massive Objects" or VMOs, are radiation-dominated and therefore unstable to pulsations. These pulsations would lead to considerable mass loss but are unlikely to be completely disruptive, so there is no reason in principle why such stars should not exist. VMOs are certainly rare at the present epoch but they could have been more abundant in the past [7]. They would all have a main-sequence time of around 10^7y and they would all experience the electron-positron instability during their oxygen-burning phase [8]. This has two possible consequences: those below some critical mass M_{crit} would explode, generating a lot of oxygen and oxygen-burning products; those larger than M_{crit} would collapse directly to black holes, although they might return helium via mass loss before doing so. The value of M_{crit} is about 200 $M_\odot$ if there is no rotation [9] but it could be as large as $2x10^4 M_\odot$ if rotation is maximal [10].

Stars larger than $10^5 M_\odot$ are termed "Supermassive Objects" or SMOs. If they were metal-free, they would collapse directly to black holes before any nuclear burning [11] and therefore have no nucleosynthetic consequences. However, they could explode in some mass range above $10^5 M_\odot$ if they had non-zero metallicity [12]. It must be stressed that the existence of VMOs and SMOs is entirely speculative and they are primarily invoked for the purpose of making dark matter. VMOs could theoretically be responsible for some of the abundance anomalies in metal poor stars, such as the possible existence of primary nitrogen [13], but less massive stars might also explain these.

3. METAL-PRODUCING POPULATION III STARS

Since metals are generated exclusively by stars, one might expect stars smaller than 0.8 $M_\odot$ (whose lifetime exceeds the age of the Universe) to display arbitrarily low metallicity. [In fact, this is not strictly true since the first stars in a cloud might complete their evolution and contaminate the rest of the cloud before the smaller stars have had time to complete their Kelvin-Helmholtz phase. However, for a reasonable IMF, the associated minimum metallicity for low mass stars would be very tiny.] On the other hand, Bond [14] has claimed that there is a metallicity cut-off Z_{min} of order 10^{-5} below which no stars are found. If this were true, and if it could not be attributed to self-enrichment within the first clouds, it would suggest that the first stars had an IMF with a lower cut-off M_{min} in the range 0.8 $M_{\odot}$ to M_{BH}. For only then could they produce a minimum enrichment Z_{min} without surviving until the present epoch [15].

Since the stars observed today have M_{min} around 0.1 $M_{\odot}$, this would imply that the first stars had a different IMF from ordinary Population II stars. Various reasons have been suggested for this. For example, the fragment mass could be smaller before metals formed because cooling would be less efficient [16]. There is also observational evidence that the IMF may become shallower as Z decreases [17] and this would at least reduce the fraction of low mass stars. But why should Z_{min} have the value 10^{-5} claimed by Bond? One possibility is that the Universe will be reionized once the number density of UV photons generated by the stars becomes comparable to the number density of atoms [18]. This happens when the fraction of the gas processed through the stars reaches about 10^{-5} (the precise value depending upon the IMF) which, for a metal yield of 0.1, corresponds to an enrichment of about 10^{-6}. If star formation is temporarily halted at reionization (eg. due to the Compton drag of the microwave background), one might naturally get the value of Z_{min} required. A second possibility is that $Z \simeq 10^{-5}$ may correspond to the threshold at which metallic cooling becomes important enough to change the characteristic star mass [16].

Cayrel [19] has proposed a third scheme, invoking self-enrichment of the first clouds, to explain Z_{min}. The idea is that each cloud will first form a compact cluster of O stars at its centre. These stars will explode and produce mass loss; a shock will then form where the outflowing material hits the inflowing material and the stars forming in this shock should be smaller than the first ones because the density is high and metals are now present. One thus generates a bimodal IMF. Another feature of the scenario is that a large fraction of the stars and gas will be blasted away by the supernovae in the core and this naturally produces an initial metallicity of order Z_{min}. Cayrel identifies globular clusters with the stars which remain bound, field Population II objects with the ejected stars, and Population I objects with the stars which form out of the ejected gas. The model also predicts a halo to disk mass ratio of order 0.1.

[Incidentally, the title of Cayrel's paper, "And if Population III were Population II?", illustrates the problem with semantics. In this particular scenario, it would surely be sensible to call the first stars "Population III" rather than "Population II" since they are a distinct generation.]

All of these models aim to explain a lower cut-off in the metallicity distribution of Population II stars. Unfortunately, the evidence for such a cut-off is now in dispute. Beers *et al.* [20] claim that the Z distribution for Population II stars extends well below 10^{-5} and there exists one object with $Z=6\times10^{-7}$ [21]. In any case, the number of low Z objects is not necessarily incompatible with the assumption that the IMF has always been the same [22], so it is by no means certain that the first stars were qualitatively different from the usual Population II stars. They could just have been part of a continuous IMF, in which case the introduction of the term "Population III" , at least in the sense of criterion (2), would be unnecessary.

4. DARK-MATTER-PRODUCING POPULATION III STARS

Cosmological nucleosynthesis arguments [23] imply that the baryon density parameter satisfies $0.014h^{-2} < \Omega_b < 0.035h^{-2}$, where h is the Hubble parameter H_o in units of 100 km/s/Mpc, whereas the visible density parameter is $\Omega_{vis} \simeq 0.01$. If $H_o=100$, one could have $\Omega_{vis}=\Omega_b$ but in this case the Universe would be younger than some globular clusters unless one invokes a cosmological constant [24]. On the other hand, if $H_o=50$, a large fraction of the baryons must be dark and the density of dark baryons could at least be sufficient to provide the dark matter in galactic halos. The dark matter in clusters could also be baryonic, although the background dark matter (required to make up the critical density if one believes inflation) would have to be some kind of elementary particle. The usual prejudice that the halo dark matter is also an elementary particle just seems to stem from an expectation that the number of types of dark matter should be as small as possible. However, dark matter could take as many different forms as visible matter and there is really no reason why baryons should not turn dark as efficiently as they turn visible.

If the dark matter in halos is baryonic, background light and nucleosynthetic constraints require that it must be either jupiters or the black hole remnants of VMOs [7]. This is the basis of the suggestion that most of the baryons in the Universe have been processed through a generation of dark-matter-producing stars. Such stars would definitely warrant the name "Population III" since they could not result from a standard IMF. However, as discussed in Section (5), they would not necessarily have to be pregalactic.

Let us first focus on the jupiter option. This is particularly topical because there may already be evidence that baryons can be converted into jupiters with high efficiency in the cooling flows observed in cores of rich clusters [25]. It is not fully understood *why* cooling flows make jupiters but the high pressure means that the Jeans mass (assumed to provide an upper limit to the star mass) is much less than it is in the interstellar medium. Could the Population III stars which make jupiters also produce an initial enrichment? Although jupiters themselves do not undergo any nucleosynthesis, one could always arrange to have an IMF which produces a small number of metal-producing stars. For example, let us assume that the number of stars in the mass range (M,M+dM) is

$$n(M)dM \propto M^{-\alpha}dM \qquad (M_{min}<M<M_{max}). \qquad (1)$$

Then most of the mass is in the smallest stars providing $\alpha>2$ but one will produce some enrichment providing M_{max} exceeds 1 $M_{\odot}$. Indeed one can generate any required initial enrichment Z by choosing the lower cut-off of the IMF at

$$M_{min} = (Z_{ej}f_*/Z)^{1/(2-\alpha)}M_\odot \qquad (2)$$

where Z_{ej} is the yield of the individual star. However, to explain the particular value of Z advocated by Bond [9] would require an implausibly small value for M_{min} unless α were very large. Indeed, one would generally *overproduce* metals unless M_{max} were itself below 1 $M_\odot$, in which case there would be no nucleosynthesis at all.

Let us now consider the black hole option. If galactic halos are made of black holes, their precursors were probably VMOs in the mass range above $M_C \simeq 200\ M_\odot$. Such stars would not themselves produce heavy elements but one could always have an IMF extending down to a mass M_{min} below M_C. In this case, one expects to produce a small enrichment providing $\alpha<2$, so that most of the mass is in the largest stars. To produce an initial metallicity Z, one would need the IMF to have an upper cut-off at

$$M_{max} = M_C(Z_{ej}f_*/Z)^{1/(2-\alpha)} \qquad (3)$$

In this situation, the metal-poorest stars would be expected to have the high O/Fe ratio associated with exploding VMOs. However, as in the jupiter case, one would tend to make too many metals unless M_{max} were very large. The most natural assumption is therefore that M_{min} itself exceeds M_C, in which case there would be no enrichment.

We conclude that metal-producing and dark-matter-producing Population III stars are unlikely to be part of a single IMF. However, although VMOs larger than M_C do not produce heavy elements, they could generate helium. This is because they undergo considerable mass-ejection during their hydrogen-burning phase and the material lost would be progressively enriched in helium. If the mass loss fraction is Φ_L and the initial helium abundance is Y_i, one can show [26] that the helium yield per star is

$$\Delta Y = (1-Y_i/2)\Phi_L^2 \text{ for } \Phi_L<(1-Y_i)/(2-Y_i) \qquad (4)$$

The limit on Φ_L just requires that the boundary of the star shrink more slowly than the convective core. The maximum helium yield, which occurs when Φ_L has the limiting value, is

$$(\Delta Y)_{max} = (1-Y_i)^2/2(2-Y_i) \ . \qquad (5)$$

It might seem unlikely that Φ_L would have this value. However, it may be achieved necessarily due to envelope ejection during hydrogen-shell burning [26]. In this case, the helium yield $(\Delta Y)_{max}$ is inevitable. If Y_i=0.23, corresponding to the usual primordial helium abundance, $(\Delta Y)_{max}$=0.17, which means that helium would be grossly overproduced if most of the baryons in the Universe were processed through such stars. However, if Y_i=0, we get $(\Delta Y)_{max}$=0.25. In this case, a mean helium abundance Y is generated when the fraction of the Universe processed through VMOs is

$$F = \int_0^Y \frac{2(2-Y)}{(1-Y)^2} \exp\left[\frac{2Y}{Y-1}\right] dY \simeq 4Y \text{ for } Y \ll 1. \qquad (6)$$

One naturally gets Y=0.25 if F≃1, so one has an alternative explanation for the primordial helium abundance. More precisely, $0.22<Y<0.24$ for $0.90<F<0.95$.

In order to suppress cosmological helium production, one would need to invoke lepton degeneracy or a cold early Universe [27]. Naturally enough, most cosmologists would prefer cosmological production of helium to VMO production, especially since, in the latter case, one still has to account for the abundances of the other light elements (such as deuterium and lithium). However, it is important to appreciate that non-standard options are not completely eliminated. It is in any case an amusing coincidence that VMOs produce roughly the helium abundance required, at least if one believes in envelope ejection.

4. WERE POPULATION III STARS PREGALACTIC?

It is usually assumed that the first stars formed protogalactically, that is to say, when galaxies first settled out from the cosmic expansion. However, in most cosmological scenarios, one would expect the first objects to form pregalactically and be smaller than galaxies. For example, in the *hierarchical clustering* scenario the first bound objects have a mass around $10^6 M_\odot$ and bind at a redshift in the range 30–100; these clouds then cluster gravitationally to make galaxies and clusters of galaxies [28]. A currently popular version of this model is the "Cold Dark Matter" scenario [29], in which the density of the Universe is dominated by some cold elementary particle (such as the photino or axion). In the *pancake* scenario [30], the first objects to form are pancakes of cluster or supercluster scale and they do so at a redshift in the range 3–10. They first fragment into 10^6–$10^9 M_\odot$ clumps and so these clumps have to cluster before galaxies can form. A particularly interesting version of this theory [31] applies if the Universe is dominated by some hot elementary particle (such as the neutrino). In the *explosion* scenario [32,33], the first objects to form are explosive seeds (stars or clusters of stars). These generate shocks which sweep up vast shells of gas; when the shells overlap, most of the gas gets compressed into thin sheets [34]. The sheets fragment either directly into galaxies or into lower mass objects, depending on the cooling mechanism.

In all these scenarios, one can have an appreciable fraction of the Universe going into subgalactic-scale clouds before galaxies themselves form. The question then arises: what happens to these clouds? If they are too small or too large, one expects them to be disrupted by collisions with other clouds because the cooling time is too long for them to collapse within a Hubble time. However, there is a mass range 10^6–$10^{11} M_\odot$ in which clouds would cool and collapse fast enough to survive. In this case they could face various possible fates. Ashman & Carr [35] argue that they could turn into jupiters via pregalactic cooling flows; these would be analagous to the cluster flows observed at the present epoch but on a smaller scale. Other people argue that they could fragment into VMOs [7] or into a mixture of small and large stars [36]. They might even collapse directly to supermassive black holes.

All these proposals could lead to a large fraction of the baryons in the Universe becoming dark. However, it is important to appreciate that dark-matter-producing Population III stars, even if they exist, would not necessarily have to be pregalactic. It is possible that the first clouds do not produce any stars at all but just become Lyman-α clouds [37]. In this case, the formation of the first stars could be postponed until galaxy formation and the Population III objects might even be postgalactic. Nevertheless, there is at least the theoretical possibility that the Population III stars were pregalactic.

What would constitute evidence for pregalactic star formation? It is hard to envisage such evidence in the jupiter scenario but the situation is more hopeful in the VMO case because VMOs would generate a lot of background light. Each VMO has a luminosity $L \simeq 10^{40}(M/10^2 M_\odot)$ erg/s and a surface temperature of about 10^5K. In the absence of dust absorption, one would expect a background density $\Omega_R = 0.004\Omega_*(1+z_*)^{-1}$ peaking at a redshifted wavelength $\lambda = 0.04(1+z_*)\mu$; here Ω_R is measured in units of the critical density, Ω_* is the density of the stars in the same units, and z_* the redshift at which they burn. Comparison with the upper limits on the background radiation density in the near-IR to UV bands [38] shows that VMOs with $\Omega_* \simeq 0.1$ would have to burn at a redshift exceeding 30; otherwise Ω_R would be too large. This would imply that they were indeed pregalactic. However, if dust were present, the radiation could be reprocessed into the far-IR, where the limit on the background density is much weaker. Even the dust in galaxies could serve this purpose, providing galaxies cover the sky [39]. For a dust density comparable to that associated with galaxies, one would only require that the VMOs burn at redshifts exceeding 4 [38], which does not necessarily force one to a pregalactic epoch.

If the light from VMOs is reprocessed by dust, it would have roughly the same density as the original starlight but simple analytical considerations show that the spectrum should peak at a wavelength [40]

$$\lambda_{peak} = 700 \left[\frac{\Omega_R h^2}{10^{-6}}\right]^{-1/5} \left[\frac{r_d}{0.1\mu}\right]^{1/5} \left[\frac{1+z}{10}\right]^{1/5} \mu \qquad (7)$$

where r_d is the grain size and z is the redshift of light production or grain production (whichever is smaller). For comparison, the microwave background has a density $\Omega_R \simeq 2\times10^{-5}h^{-2}$ and peaks at 1400μ. The total background spectrum should thus have three parts: the microwave background component, the far-IR dust component, and the attenuated starlight component.

There have, in fact, been been two claimed detections of a far-IR background. Rowan-Robinson [41] used IRAS data to infer a background with $\Omega_R \simeq 10^{-5}h^{-2}$ at 100μ, although this now seems rather dubious. More recently, Matsumoto *et al.* [42] have claimed to detect a background with $\Omega_R \simeq 5\times10^{-6}h^{-2}$, peaking at 700$\mu$, as a result of a rocket experiment. If this radiation is from grains, they would need to have a temperature of 3.5 K and the radiation would probably have to be generated pregalactically [40]. Thus this result, if confirmed, could provide direct evidence for a generation of pregalactic stars. In this context, it should be noted that some people have proposed that the microwave background is itself grain-thermalized starlight [43]. This is possible in principle but the grains would have to form at very high redshift ($z>200$) and be very elongated in order to thermalize at long wavelengths [44].

6. CONCLUSIONS

In concluding, let us revert to the semantic problem on which we focussed in the Introduction. We have seen that one can usefully distinguish between Population III stars of the metal-producing and dark-matter-producing type. Those of the first kind must exist but they only warrant a special name if there is a lower cut-off in the metallicity distribution of Population II stars. It is not clear that this is the case. Stars of the second kind may not exist but, if they do, they certainly warrant a separate name. In principle, both kinds of Population III stars could be the same, in the sense that they derive from a single mass spectrum, but that would require the IMF to be finely tuned.

ACKNOWLEDGMENT

I thank Bernard Pagel and Rosie Wyse for interesting remarks which resulted in the written version of this talk being somewhat different than the spoken version!

REFERENCES

1. Renzini,A., in "Stellar Nucleosynthesis", eds. C.Chiosi and A.Renzini, Reidel, Dordrecht (1984).
2. Truran,J.W., Ann.Rev.Nucl.Part.Phys. 34, 53 (1984).
3. Iben,I., Quart.J.R.astr.Soc. 26, 1 (1985).
4. Arnett,W.D., Astrophys.J. 219, 1008 (1978).
5. Schild,H. and Maeder,A., Astr.Astrophys. 143, L7 (1985).
6. Spite,M. and Spite,F., Ann.Rev.Astr.Astrophys. 23, 225 (1985).
7. Carr,B.J., Bond,J.R. and Arnett,W.D., Astrophys.J. 277, 445 (1984).
8. Fowler,W.A. and Hoyle,F., Astrophys.J.Supp. 9, 201 (1964).
9. Bond,J.R.,Arnett, W.D. and Carr,B.J., Astrophys.J. 280, 825 (1984).
10. Glatzel,W., El Eid,M.F. and Fricke,K.J., Astr.Astrophys. 149, 419 (1985).
11. Fowler,W.A., Astrophys.J. 144, 180 (1966).
12. Fricke,K.J., Astrophys.J. 183, 941 (1973).
13. Pagel,B.E.J. and Edmunds,M.G., Ann.Rev.Astr.Astrophys. 19, 77 (1981).
14. Bond,H.E., Astrophys.J. 248, 606 (1981).
15. Truran,J. and Cameron,A.G.W., Astrophys.Space Sci. 14, 179 (1971).
16. Silk,J., Astrophys.J. 211, 638 (1977).
17. Terlevich,R.J., in "Star Forming Dwarf Galaxies", eds. D.Kunth,T.X.Thuan and J.Tran Thanh Van, Editions Frontieres, Paris, p 395 (1985).
18. Hartquist,T.W. and Cameron,A.G.W., Astrophys.Space Sci. 48, 145 (1977).
19. Cayrel,R., Astr.Astrophys. 168, 81 (1987).
20. Beers,T.C., Preston,G.W. and Shectman,S.A., Astron.J. 90, 2089 (1985).

21. Bessel,M.S. and Norris,J., Astrophys.J. 285, 622 (1984).
22. Pagel,B.E.J., in "The Galaxy", eds. G.Gilmore and B.Carswell, Reidel, Dordrecht, p 341 (1987).
23. Yang,J.,Turner,M.S.,Steigman,G.,Schramm,D.N. and Olive,K.A., Astrophys.J. 281, 493 (1984).
24. Tayler,R.J., Quart.J.R.astr.Soc. 27, 367 (1986).
25. Fabian,A.C.,Nulsen,P.E.J. and Canizares,C.R., Nature 310, 733 (1984).
26. Bond,J.R.,Carr,B.J. and Arnett,W.D., Nature 304, 514 (1983).
27. Kaufman,M., Astrophys.J. 160, 459 (1970).
28. Peebles,P.J.E. and Dicke,R.H., Astrophys.J. 154, 891 (1968).
29. Blumenthal,G.R.,Faber,S.M.,Primack,J.R. and Rees,M.J., Nature 311, 517 (1984).
30. Zeldovich,Ya.B., Astr.Astrophys. 5, 84 (1970).
31. Bond,J.R.,Centrella,J.,Szalay,A.S. and Wilson,J.R., Mon.Not.R.astr.Soc. 210, 515 (1984).
32. Ostriker,J.P. and Cowie,L.L., Astrophys.J. 243, L127 (1981).
33. Ikeuchi,S., Pub.astr.Soc.Japan 33 211 (1981).
34. Carr,B.J. and Ikeuchi,S., Mon.Not.R.astr.Soc. 213, 497 (1985).
35. Ashman,K.M. and Carr,B.J., preprint (1987).
36. Kashlinsky,A. and Rees,M.J., Mon.Not.R.astr.Soc. 205, 955 (1983).
37. Rees,M.J., Mon.Not.R.astr.Soc. 218, 25P (1986).
38. McDowell, J.C., Mon.Not.R.astr.Soc. 223, 763 (1986).
39. Ostriker,J.P. and Heisler,J., Astrophys.J. 278, 1 (1984).
40. Bond,J.R., Carr,B.J. and Hogan,C., Astrophys.J. 306, 428 (1986).
41. Rowan-Robinson,M., Mon.Not.R.astr.Soc. 219, 737 (1986).
42. Matsumoto,T.,Hayakawa,S.,Matsuo,H.,Murakami,H.,Sato,S.,Lange,A.E. and Richards,P.L., preprint (1987).
43. Bond,J.R.,Carr,B.J.,Hogan,C.J., preprint (1987).
44. Rees,M.J., Nature 275, 35 (1978).
45. Hawkins,I. and Wright,E.L., preprint (1987).

CHAPTER III

ABUNDANCES

"God's chase
Round vase
What say?
What play?
Don't know
Nice, though."

Desmond Skirrow
Ode on a Grecian Urn Summarized

The Chemical Composition of Stars in Globular Clusters: A Glimpse of the Contribution by Population III?

George Wallerstein
Astronomy Department FM-20
University of Washington
Seattle, WA 98195
USA

I. INTRODUCTION

Early work on the chemical composition of stars in globular clusters was oriented toward the solution of problems of galactic structure. This was stimulated by Baade's suggestion of the two types of stellar populations and the extension of Baade's ideas at the 1958 Vatican Conference on Stellar Populations. Hence, most of the work on the composition of globular cluster stars was concentrated on determining their overall metallicity and its dependence on such parameters as the distance of the clusters from the Galactic Center, the distance from the Galactic plane and the kinematics of the clusters.

During that early work there was no doubt that the authors wondered – sometimes in print – as to the nature of the origin of the traces, i.e., 1-10% of the solar abundances, of the elements from sodium to barium. As improved data has become available, first with photographic echelle spectrographs and now with CCD detectors, it has become possible to derive sufficiently accurate abundances of key elements to make interesting inferences regarding the origin of the elements seen on the surfaces of globular cluster stars.

Unfortunately the observed abundances are not necessarily the abundances with which the observed star was born. Nuclear reactions in the stars' own interiors followed by convective mixing has modified the surface composition. This has been demonstrated in a striking way by Sneden, Pilachowski and Vanden Berg[1] who found $^{12}C/^{13}C \simeq 4$ in metal-poor halo stars that are very similar to red giants in globular clusters. Such mixing should effect only the CNO elements and isotopes. For an excellent review of low resolution surveys of CH and CN in globular clusters see Graeme Smith's[2] review.

In this report I want to emphasize work at high spectral resolution from which it is possible to obtain accurate abundances of C, N, O, $^{12}C/^{13}C$, and various other elements. I will present the data and the derivation of abundance information first and then comment on possible interpretations rather than endeavor to fit the data into a general scheme.

II. AVAILABLE QUANTITATIVE ANALYSES

A – CNO in M13

This cluster has long been known to show a metal deficiency of a factor 40. It is a classic cluster of intermediate metallicity, but with a surprisingly blue horizontal branch. From coude CCD spectra[3] we have found a spread in the oxygen abundances. This has been confirmed from spectra of higher resolution and signal-to-noise by Hatzes[4]. A copy of his Fig. 2 is shown below:

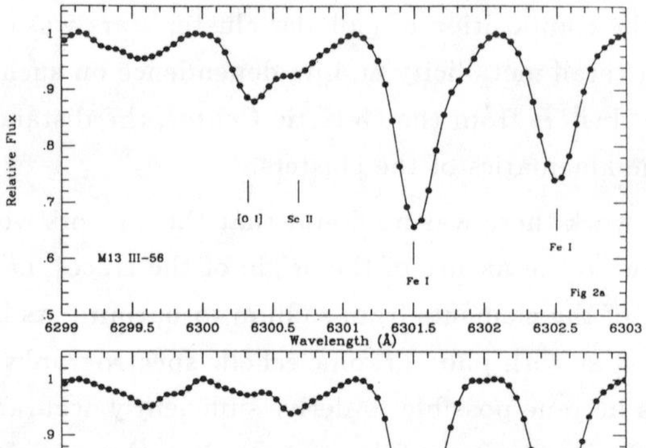

FIG. 2–The spectral region around 6300 Å for III-56 (2a) and II-67 (2b) after removal of all atmospheric features.

Recent Palomar observations indicate that two additional stars in M13 have weak oxygen lines; but due to blending with ScII and atmospheric H_2O a full reduction and analysis will be necessary to confirm that conclusion.

Is the oxygen deficiency due to complete CNO cycling and mixing? Observations of CN show this not to be so[5]. From Palomar CCD spectra and a model atmosphere analysis we can use the CN molecular lines to derive a line in the log N/H, log C/H plane. We show this below:

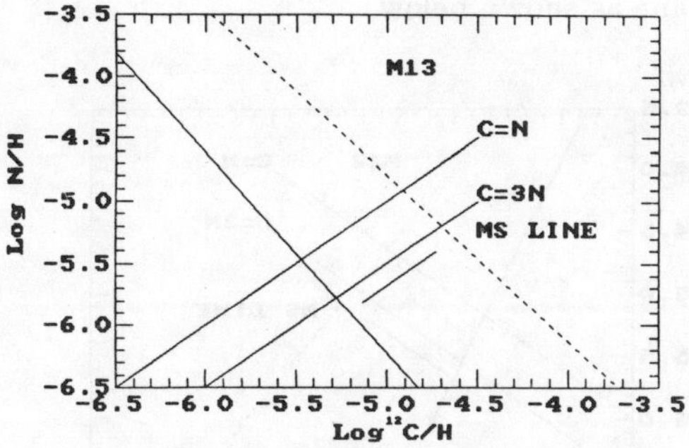

FIG. 1. Relationship between the carbon and nitrogen abundances as derived from CN lines in the two stars in M13. The dashed line is the CN locus for III-63 and the continuous line is the CN locus for II-67.

The dashed line shows that the locus of possible values of N/H and C/H for the star with high oxygen is substantially above the locus for the oxygen-deficient star, II-67. The absence of enhanced nitrogen has also been established by Langer *et al.* [6] who analyzed the NH bands in II-67. Transforming into English this says that the sum of C + N +O is much larger – at least a factor 5 – in star III-63 as compared to II-67. This cannot be due to CN cycling but must reflect the original composition. It is very strange that one star in M13 should be deficient in these light elements especially since they are produced by different nuclear processes: helium burning for ^{12}C and ^{16}O and CNO cycling for nitrogen. One has to suspect red giants, in which helium burning can be combined with CNO cycling by an appropriate mixing scheme, as the source of most carbon and nitrogen during late stages of stellar evolution. Perhaps appreciable oxygen is also produced this way rather than in massive supernovae.

B – CNO in M22

This nearby globular cluster shows a metal deficiency of about a factor 50 and large variations from star-to-star in the strengths of the CN and CH bands[7].

Those authors report a significant spread in the calcium abundances and Pilachowski *et al.* [8] report a likely (but not certain) spread in other metallic abundances. A quantitative study of two stars with different band strengths by Wallerstein *et al.* [5] using the same methods as used for M13 yielded lines in the log N/H, log C/H plane as shown below

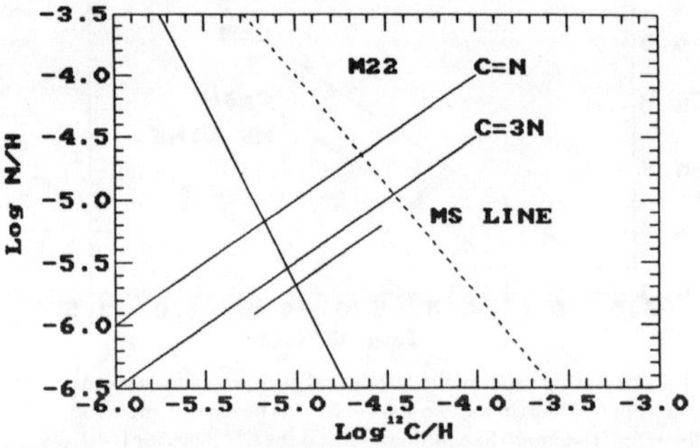

FIG. 4. Relationship between the carbon and nitrogen abundances as derived from CN lines in M22, III-3, and M22, IV-102. The dashed line is the CN locus for III-3, and the continuous line is for IV-102.

Combining the data shown above with an O/H difference of a factor of 5[8] allows us to conclude that similar to M13 all three elements are deficient in star IV-102 as compared to III-3 by roughly a factor 5. For the strong CN star we found a $^{12}C/^{13}C$ ratio near the CN cycle equilibrium value of 4. Such a low $^{12}C/^{13}C$ ratio has also been seen in metal-poor halo stars by Sneden *et al.* [1] It implies much deeper mixing than predicted by any models. Even our neighbor Arcturus which is metal-poor by only a factor 4 shows $^{12}C/^{13}C$ of only 7.2 and shows neither the nitrogen excess nor the carbon depletion expected for the depth of mixing implied by the $^{12}C/^{13}C$ ratio[9]. In my opinion we do not understand the evolution of these stars, at all.

C – Elements Uneffected by CN Cycling

While we have suggested above that some of the CNO seen in globular clusters came from the atmospheres of evolved red giants of substantial mass, there can be little doubt that the bulk of the elements from sodium through the iron peak were made in supernovae. With the certain exception of ω Cen and the possible exception of M22 the abundances of these elements appear to be the same from star-to-star within a given clusters. However, the pattern is not solar and hence, shows a possible difference between the supernovae that produced the solar heavy elements and those (or even a single supernova for a very metal-poor cluster) that produced the heavy elements now seen in globular cluster stars.

First let's look at oxygen. Most globular cluster stars show a smaller deficiency in oxygen than iron. Values of [O/Fe] are usually around 0.3 to 0.7 dex with the reality of the spread being questionable. For metal-poor field stars with a wide range of [Fe/H] values Barbuy[10] finds [O/Fe] = +0.35 with a very small spread. Her data were obtained with a CCD and with S/N on the order of 200. Although oxygen is usually attributed to supernovae of high mass (witness the fast moving knots in Cas A) perhaps much of it is contributed by helium burning and mass-loss in red giants as we discussed for M13 above.

For heavier elements it becomes more likely that they were produced by supernovae, but we should not forget that ^{23}Na can be produced by successive proton captures by ^{20}Ne or even by a single proton capture by ^{22}Ne, and ^{27}Al can be produced by a single similar sequence on ^{24}Mg or a single capture by ^{26}Mg.

The behavior of sodium and aluminum in globular cluster stars is rather uncertain but intriguing. Begley and Cottrell[11] have pointed to technical difficulties in the analysis of these elements but the anomalies may exceed the uncertainties. For the cluster M4 the [Na/Fe] excess was found to be 0.5 dex while the [Al/Fe] excess was evaluated at 0.8 dex.[5] The latter is substantially more than the uncertainties in the analysis indicated by Begley and Cottrell.[11] We are probably seeing a real enhancement of aluminum in the M4 stars. Enhancements of light, odd elements like Na and Al are not seen in metal-poor main sequence stars.

Anomalies are more often seen in the ratio-to-iron of the light, even elements, such as Mg and Si whose nuclei consist of integral numbers of α-particles. Positive values of [α-elements/Fe] were first seen in subdwarfs by Aller and Greenstein[12] and first recognized as a general property of metal-poor dwarfs by Wallerstein.[13] This effect is clearly seen in recent work by Gratton *et al.* [14] on five globular clusters with [Fe/H] ranging from −.8 to −1.5. They found [α/Fe] = +0.30 very similar to field stars for which Gratton and Sneden[15] found α/Fe = +0.25. Their determination is probably most accurate for silicon which is in excess by 0.37 and 0.36 dex respectively for globular cluster stars and halo stars. The evidence clearly favors a difference between α-element and iron-peak element production in the supernovae that contributed to metal-poor stars vis-a-vis stars of solar composition.

IV. RECENT RESULTS FOR $^{12}C/^{13}C$ IN GLOBULAR CLUSTER STARS

About six weeks ago I obtained echelle CCD observations at Cerro Tololo for OI, CN, and $^{12}C/^{13}C$ in 4 clusters, M4, M22, ω Cen and 47 Tuc. A "quick look" data reduction without a formal analysis allows some very interesting conclusions to be reached regarding $^{12}C/^{13}C$.

Lets start with M22, which shows a wide range of CN and CH band strengths with a good correlation of CN and CH.[7] The ^{12}CN lines are strong in the 4 previously recognized strong CN stars. All of them show such a strong ^{13}CN feature at λ8005 Å that the $^{12}C/^{13}C$ ratio must be near the equilibrium value of 4. The marginal CH star, III-106, in M22 also shows strong ^{13}C. The same holds for two stars in 47 Tuc and 4 stars in M4. The weak CN stars in M22 do not show strong enough bands to evaluate the $^{12}C/^{13}C$ ratio. Apparently deep mixing, as originally seen by Sneden *et al.* [1] in the field, is pervasive among the red giant tip stars in globular clusters as well.

Turning to Omega Cen we see the most interesting results. As the only globular with a wide range of metal abundance combined with the largest of any globular mass, ω Cen provides an opportunity to study stellar evolution and local enrichment as does no other accessible stellar system. My work on ω Cen is in cooperation with Verne Smith and I will report only on $^{12}C/^{13}C$ at this time. The three most metal poor stars show such weak CN that no isotope ratios are possible. A star of intermediate metallicity, ROA 287, shows substantial ^{13}C, probably near the equilibrium $^{12}C/^{13}C$ ratio. One star that may be a barium star has strong CN and strong ^{13}CN, very likely near the equilibrium value. Another object is a carbon star with very strong ^{13}CN. Finally we have CN spectra for two stars reported by Lloyd Evans[16] to show ZrO bands and hence, to be stars of type S. The degree of heavy element enhancement is being determined by Verne Smith. Both stars show strong ^{12}CN and weak ^{13}CN. In fact, they are the only stars of this survey to show strong CN and weak ^{13}C. My guess from looking at the spectra is that $^{12}C/^{13}C$ is around 10 in one star and perhaps 20 in the other.

This has significant implications for nuclear astrophysics.

In the temporary absence of an analysis to derive the total carbon abundance we can suggest two obvious hypotheses. Either the ^{12}C has been enhanced by $3\,^4\mathrm{He} \rightarrow ^{12}\mathrm{C}$ or the ^{13}C has been depleted by $^{13}\mathrm{C}\,(\alpha, \mathrm{n})\,^{16}\mathrm{O}$.

Of course both could well have occurred simultaneously under similar physical conditions, viz $T_6 \simeq 100\,\mathrm{K}$. Before proceeding we should note that these are not highly evolved stars. Both have $M_v \simeq -1.0$ which is much less luminous than expected for an AGB star. For this reason Lloyd Evans[16] suggested that these stars show enhanced heavy elements as evidence of their original composition rather than their recent evolution. Our high $^{12}\mathrm{C}/^{13}\mathrm{C}$ ratio separates them from the other red giants in the globular clusters and causes me to suspect that they actually are evolved stars. The evolved Galactic S Star o^1 Ori shows $M_v = -1.2$ (from its K-line width) and a high Tc abundance[17] so it is surely a recently evolved star with shell flashes etc. We are in need of a scheme to create S stars with a luminosity well below that of the canonical AGB.

REFERENCES

1. Sneden, C., Pilachowski, C. A., and VandenBerg, D. A. 1986, *Ap. J.*, **311**, 826.
2. Smith, G. H. 1987, *Pub.A.S.P.*, **99**, 67.
3. Leep, E. M., Wallerstein, G. and Oke, J. B. 1986, *A. J.*, **91**, 1117.
4. Hatzes, A. P. 1987, *Pub.A.S.P.*, **99**, 369.
5. Wallerstein, G., Leep, E. M., and Oke, J. B. 1987, *A. J.*, **93**, 1137.
6. Langer, G. E., Friel, E., Kraft, R. P. and Suntzeff, N. B. 1987, *Pub.A.S.P.*, **99**, 15.
7. Norris, J. and Freeman, K. C. 1983, *Ap. J.*, **266**, 130; **273**, 837 (E).
8. Pilachowski, C., Wallerstein, G., Leep, E. M. and Peterson, R.. C. 1982, *Ap. J.*, **263**, 187.
9. Lambert, D. L. and Ries, L. 1981, *Ap. J.*, **248**, 228.
10. Barbuy, B. 1987, *Astron. and Astrophys.* (in press).
11. Begley, M. J. and Cottrell, P. L. 1987, *M.N.R.A.S.*, **224**, 633.
12. Aller, L. H. and Greenstein, J. L. 1960, *Ap. J. Suppl.*, **5**, 139.

13. Wallerstein, G. 1962, *Ap. J. Suppl.*, **6**, 407.

14. Gratton, R. G., Quarta, M. L., Ortolani, S. 1986, *Astron. and Astrophys.*, **169**, 208.

15. Gratton, R. G. and Sneden, C. 1987, *Astron. and Astrophys.*, **178**, 179.

16. Lloyd Evans, T. 1983, *M.N.R.A.S.*, **204**, 975.

17. Smith, V. V. and Lambert, D. L. 1985, *Ap. J.*, **294**, 326.

Probing the Possibility of a $^{12}C/^{13}C$ Abundance Gradient from Observations of Interstellar CH^+

Isabel Hawkins
IGPP, Lawrence Livermore National Laboratory

ABSTRACT

I have performed high signal-to-noise ($S/N = 300 - 800$) observations of interstellar CH^+ at Lick Observatory and at CTIO of the reddened, early-type stars *HD 183143*, *HD 24432*, and *HD 157038* in an effort to probe the existence of a $^{12}C/^{13}C$ abundance gradient in our Galaxy.

Previous very high quality optical observations of interstellar CH^+ toward the stars ς Oph, ξ Per, P Cyg, 20 Tau, and 23 Tau, located within a few hundred parsecs of the Sun, have yielded a precise weighted mean $^{12}C/^{13}C$ isotope ratio of 43 $\pm$ 4 (1σ) (Hawkins, Jura, and Meyer 1985; Hawkins and Jura 1987). Since our derived local ISM $^{12}C/^{13}C$ ratio is a factor of two smaller than the value of 89 in the solar system, we have concluded that the ISM has evolved chemically since the formation of the Sun, in accordance with the predictions put forth by the majority of Galactic chemical evolution models. The isotope ratios derived toward the above five stars and four lines of sight in the local ISM are uniform within 12%. The similarity among these carbon isotope ratios determined in diffuse clouds possessing different physical conditions precludes the possibility that the CH^+ molecule is suffering from isotope selective effects in these regions. These results have yielded concrete and reliable evidence in favor of a locally homogeneous ISM.

The precise $^{12}C/^{13}C$ ratios derivable from high quality CH^+ observations provide the unique opportunity to study the homogeneity of the ISM in a large scale and the history of nucleosynthesis in our Galaxy. Since CH^+ seems to be the most sensitive probe of $^{12}C/^{13}C$ in the diffuse ISM, observations toward more distant stars located up to 2 kpc from the Sun allow us to study the possibility of a Galactic $^{12}C/^{13}C$ abundance gradient. The preliminary results I have derived from recently completed observations toward the stars *HD 183143*, *HD 24432*, and *HD 157038* are consistent with the local ISM $^{12}C/^{13}C$ value of 43 $\pm$ 4.

INTRODUCTION

The accurate determination of element abundances throughout the plane of the Milky Way, in particular ratios of primary-to-secondary elements, serves as a sensitive probe of the chemical history of our Galaxy, and helps to discriminate among different theories of Galactic chemical evolution. Most of these models (Güsten and Mezger 1982) predict the existence of a temporal gradient in the abundance of certain elements. In particular, primary-to-secondary element abundance ratios, like $^{12}C/^{13}C$, are expected to decrease with time. In the models, the interstellar material out of which stars form is assumed to be locally homogeneous and well mixed. The most direct way to assess the degree of homogeneity and the amount of chemical evolution of the ISM, is by an accurate determination of element abundances toward different lines of sight. The measured abundances can then be compared with solar system values, which are representative of the ISM abundances of 4.6 billion years ago. Recent very accurate results have shown that the local ISM $^{12}C/^{13}C$ ratio within 1 kpc of the Sun is homogeneous within 12%, and has decreased by a factor of two since the formation of the Sun (Hawkins, Meyer, and Jura 1985; Hawkins and Jura 1987). These results stem from optical observations of the interstellar CH^+ molecule in nearby diffuse clouds. We believe that the measured $^{12}CH^+/^{13}CH^+$ ratios are representative of $^{12}C/^{13}C$ in the gas because of the close similarity among the values toward five stars and four lines of sight. This finding coupled with theories of CH^+ formation provide evidence that, unlike CO and other species, CH^+ does not suffer from chemical fractionation or other isotope-selective effects in these regions. Because this molecule seems to be a sensitive and unambiguous probe of $^{12}C/^{13}C$ in diffuse clouds, observations toward more distant, reddened stars up to 2 kpc from the Sun can assess the existence of a carbon isotope abundance gradient on a larger scale.

Oxygen and nitrogen negative radial abundance gradients have been observed in the plane of our Galaxy (Shaver *et al.* 1983). In addition, similar trends are seen in the N/O abundance ratio (Lester *et al.* 1986) and possibly in $^{12}C/^{13}C$ (Henkel, Wilson and Bieging 1982; Henkel, Güsten, and Gardner 1985). Gradients in the abundance ratio of closely related isotopic species are easier to assess because the isotopes are most likely not affected by "selective" depletion onto dust grains or molecules. Models of Galactic chemical evolution which relax the instantaneous recycling approximation and include bimodal star formation (variable element yields) and infall of halo gas have successfully explained the observed gradients (Güsten and Mezger 1982). However, the evidence for a $^{12}C/^{13}C$ abundance gradient comes from measurements of isotopic formaldehyde absorption toward strong continuum sources. These determinations suffer from uncertainties due to saturation and fractionation effects which are very difficult to take into account. Despite the smaller distances which can be sampled with optical observations of CH^+, this molecule is a more sensitive probe of the true ^{12}C and ^{13}C abundances in the observed regions since, unlike H_2CO, the line formation process is not strongly coupled to the physical conditions of individual clouds.

OBSERVATIONS

The advent of sensitive solid state detectors with a very stable, linear response, has made possible the acquisition of high resolution (< 0.1 Å), high signal-to-noise (> 500) data toward stars as faint as B$\approx$ 8, and the detection of features as weak as 0.3 mÅ. Observations with an 1872-element Reticon detector, coupled with the coudé spectrograph of the 3.0 m telescope at Lick Observatory, have yielded 5 σ detections of the weak $^{13}CH^+$ feature toward five bright ($B < 5$) stars located within a few hundred parsecs of the Sun. These spectra, with S/N ratios of 1000 – 2500, have allowed us to make a very accurate determination of the local ISM $^{12}C/^{13}C$ isotope ratio. The measured values toward five stars and four lines of sight have a weighted mean value of 43 $\pm$ 4 (1σ), and are uniform within 12%. Because the CH^+ molecule appears to be the most sensitive probe of the carbon isotope abundance ratio in diffuse clouds, we can safely state that the molecular isotope ratios we have measured are representative of $^{12}C/^{13}C$ in the gas, and that the ISM is locally homogeneous at the 12% level. These results have been described in detail elsewhere (Hawkins, Jura, and Meyer 1985; Hawkins and Jura 1987).

In an effort to assess the possibility of a $^{12}C/^{13}C$ abundance gradient in the plane of our Galaxy within 2 kpc of the Sun, I have extended the local CH^+ observations by acquiring data toward more distant, moderately reddened stars in the Northern and Southern hemispheres. I have analyzed observations of the (0,0) CH^+ band at 4232 Å toward the stars *HD 183143* and *HD 24432* which were obtained at Lick Observatory in the Fall of 1986 using the 3.0 m coudé spectrograph and an 800 × 800 TI CCD. I also analyzed (1,0) band 3957 Å data toward *HD 157038*, which I obtained at CTIO in June of 1986 and 1987, with the 4.0 m echelle spectrograph and a 512 × 354 RCA CCD. Both spectrograph setups yielded a resolution of 0.1 Å. In these CCD reductions I implemented the technique of adding several wavelength-aligned stellar scans taken at slightly different grating tilts, a procedure which damps out the low-level, pixel-to-pixel systematic effects unremovable by flat field division.

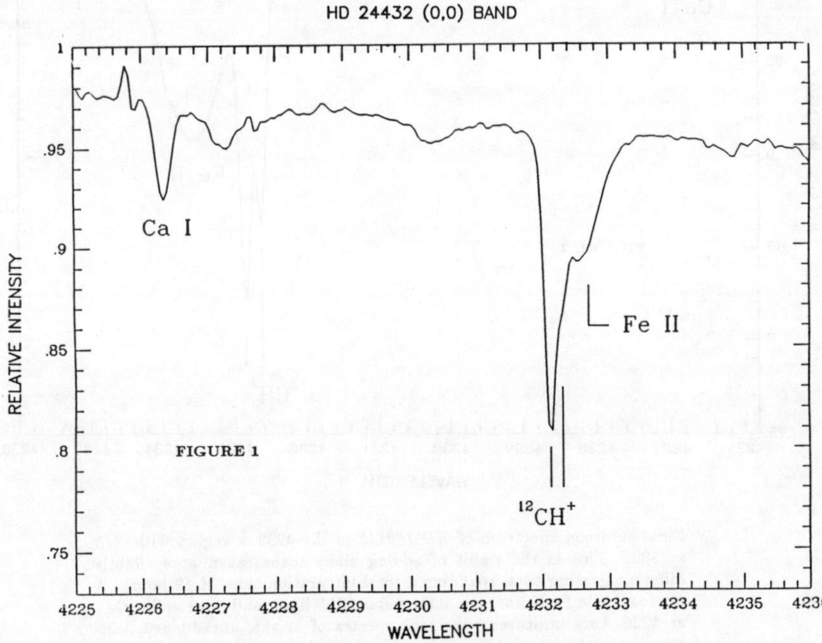

FIGURE 1

Final summed spectrum of *HD 24432* in the 4232 Å region with $S/N \approx 300$. This is the result of adding several scans taken at 4 slightly different grating tilts, spanning 10 hours of integration.

The final spectra, shown in Figures 1 through 4, were obtained by adding CCD columns which were first individually weighted by their S/N ratios. In this way, the quality of the final scans was enhanced, achieving S/N ratios of 300 – 800. Data acquisition in the (0,0) as well as the (1,0) bands is essential for this project for several reasons. Observing these two molecular bands can yield two independent measurements of the $^{12}C/^{13}C$ ratio. Equivalent width measurements of CH^+ in the (0,0), (1,0), and also the (2,0) bands which arise from the ground vibrational level can be used in a curve-of-growth analysis to assess the amount of saturation suffered by the stronger $^{12}CH^+$ lines. Additionally, the identification of different velocity CH^+ cloud components which could hinder the measurement of the weak $^{13}CH^+$ feature is imperative. I am in the process of analyzing recently completed (1,0) 3957 Å band observations toward *HD 183143*, and (2,0) band 3745 Å observations toward *HD 157038*, which will complement the results presented here.

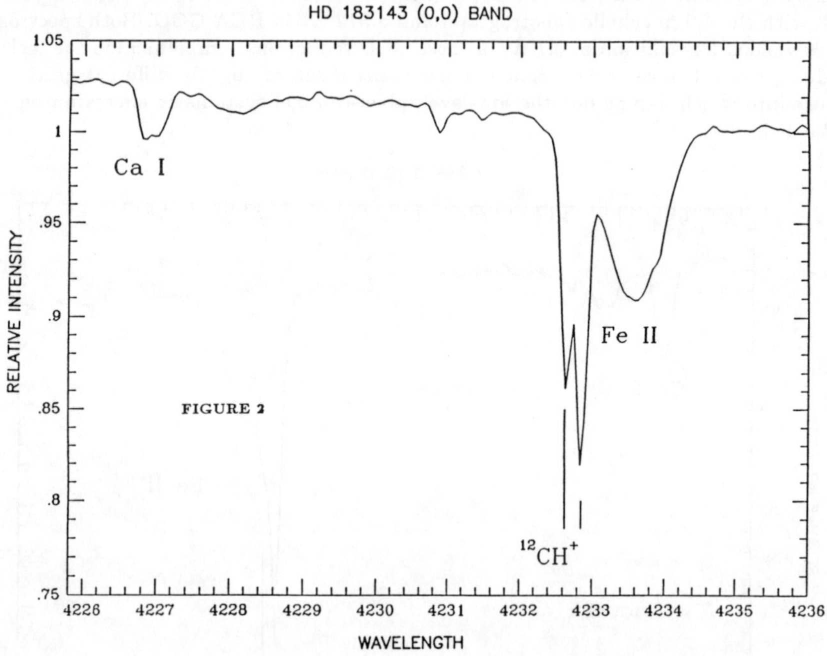

Final summed spectrum of *HD 183143* in the 4232 Å region with $S/N \approx 500$. This is the result of adding many scans taken at 4 slightly different grating tilts, spanning a total integration time of 12 hours. A photospheric Fe II line, two interstellar CH^+ lines, and interstellar Ca I at 4226 Åare indicated. We took spectra of bright, unreddened, fast-rotating, early-type stars to be able to iden:ify weak atmospheric or instrumental features.

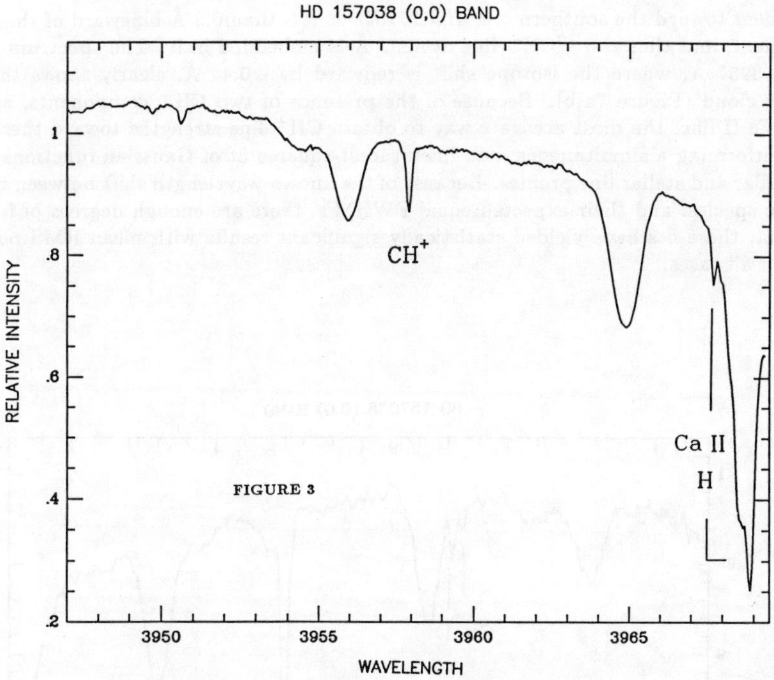

Final summed spectrum of *HD 157038* in the 3957 Å region with $S/N \approx 800$. This is the result of adding many scans taken at 4 slightly different echelle grating tilts, spanning 10 hours of integration time. Photospheric lines of N II and He I bracket the interstellar CH^+ feature, and interstellar Ca II H absorbs against photospheric Hϵ.

RESULTS

Even though previous photographic plate observations toward the stars *HD 183143*, *HD 24432*, and *HD 157038* (Cohen 1973, 1975) indicated the presence of a single CH^+ component, the spectra I obtained using more sensitive detectors show the presence of two interstellar CH^+ clouds toward each star. Because of this complication, the satellite (0,0) band $^{13}CH^+$ line (isotope shift −0.26 Å) of the redward component is blended with the absorption profile of the higher velocity line. In the case of *HD 24432* and *HD 183143* as seen from Figures 5 and 6, the separation between the components and their strength are large enough so that, in principle, it is still possible to derive a $^{12}C/^{13}C$ ratio from the blueward CH^+ component. There is a further complication at 4232 Å toward these two stars due to a stellar Fe II line at 4233 Å. As seen from Figure 7, the high velocity component toward the southern star *HD 157038* is less than 0.3 Å blueward of the strong component, and thus the $^{13}CH^+$ line at 4232 Å is embedded in it. The spectrum of this star at 3957 Å, where the isotope shift is redward by +0.44 Å, clearly shows the high velocity cloud (Figure 7a,b). Because of the presence of two CH^+ components, and the stellar Fe II line, the most accurate way to obtain CH^+ line strengths toward these stars is by performing a simultaneous, non-linear, least-squares fit of Gaussian functions to the interstellar and stellar line profiles. Because of the known wavelength shift between the two isotopic species, and their expected equal FWHM's, there are enough degrees of freedom such that these fits have yielded statistically significant results with mean RMS residuals $< 3\%$ in all cases.

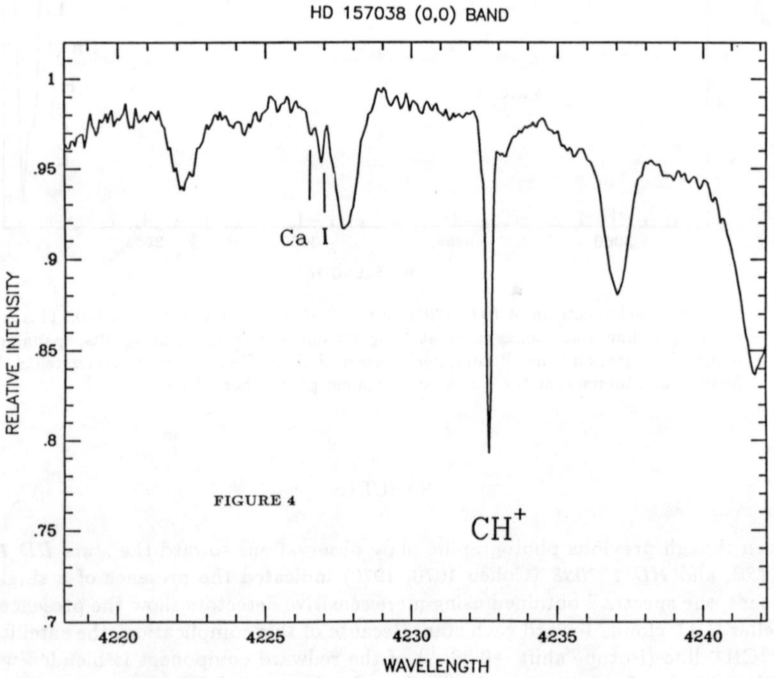

FIGURE 4

Final summed spectrum of *HD 157038* in the 4232 Å region with $S/N \approx 500$. This is the result of adding many scans taken at 2 slightly different echelle grating tilts, spanning 5 hours of integration time. Two components of interstellar Ca I are seen.

The absence of stellar line blends in the (1,0) band region, coupled with a redward isotope shift, has the potential for making the measurement of the $^{13}CH^+$ line strengths toward *HD 157038* more straightforward. However, the weak isotopic species in this region appears to be contaminated by several telluric features which most likely arise from atmospheric water vapor. Despite the high statistical quality of these data, the above complications make the measurement of the weak $^{13}CH^+$ line strengths toward these stars a difficult task. Figures 6, 7 and 8 show evidence of the $^{13}CH^+$ lines toward *HD 183143* and *HD 157038*. Because of the lower S/N ratio of the *HD 24432* summed scan, the weak isotopic line was not detected (Figure 5), and thus only an upper limit on the strength of $^{13}CH^+$ was obtainable. Table 1 lists the the properties of the observed stars, and their respective $^{12}C/^{13}C$ ratios, which were obtained from the ratio of equivalent widths derived from the Gaussian fits to the data.

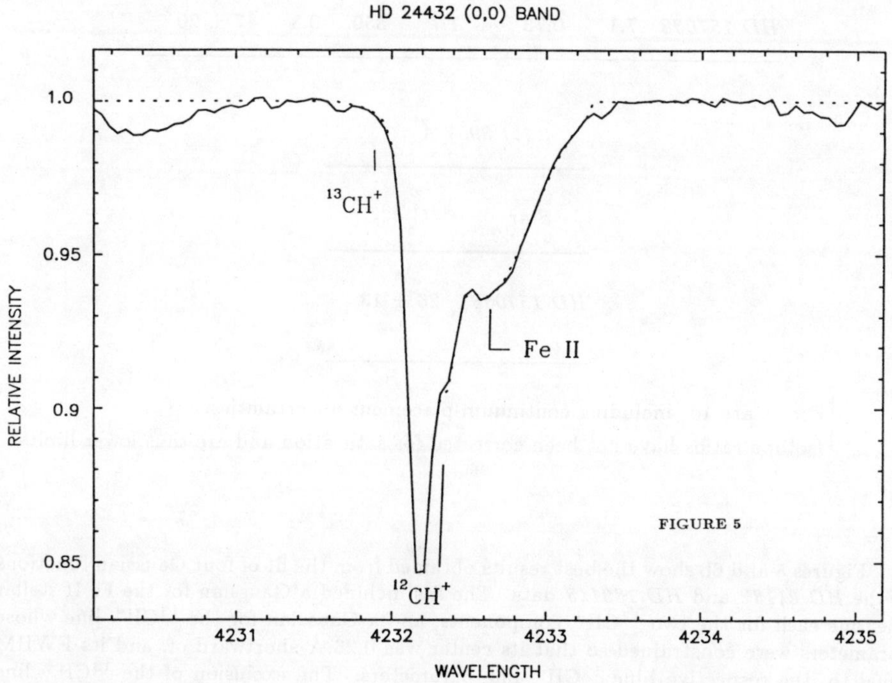

FIGURE 5

HD 24432 in the 4232 Å region. The dotted line is the best fit of four Gaussian functions to the data with a mean RMS residual < 1% (see text).

TABLE 1

(a) 4232 Å†

Star	*B*	*E(B–V)*	*D(kpc)*	l_{II}	b_{II}	$^{12}C/^{13}C$‡
HD 24432	8.1	0.76	0.7	151	−4	> 27
HD 183143	7.4	1.3	1.2	56	+6	20 ± 10
HD 157038	7.1	0.73	1.7	350	-0.8	47 ± 20

(b) 3957 Å†

Star	$^{12}C/^{13}C$‡
HD 157038	26 ± 13

†Errors are 1σ, including continuum-placement uncertainties.

‡Isotope ratios have not been corrected for saturation and are thus lower limits.

Figures 5 and 6b show the best results obtained from the fit of four Gaussian functions to the *HD 24432* and *HD 183143* data. The fits included a Gaussian for the Fe II stellar line, one each for the two $^{12}CH^+$ components, and a Gaussian for the $^{13}CH^+$ line whose parameters were constrained so that its center was 0.26 Å shortward of, and its FWHM equal to, the respective blue $^{12}CH^+$ line parameters. The exclusion of the $^{13}CH^+$ line from the fit to *HD 24432* yielded an equally good fit, thus only an upper limit to the strength of this line was obtained. However, as seen from Figure 6a, the exclusion of the weak isotopic line yields a significantly worse fit to the data of *HD 183143*.

Figure 7a shows the fit of a Gaussian function to the strong $^{12}CH^+$ component, and another one displaced by the expected isotope shift (+0.44 Å) and having the same FWHM, which was fit to the weak isotopic component. It is evident that, since the $^{12}CH^+$ line is fit well by a Gaussian function and the $^{13}CH^+$ line should have the same profile, this weak satellite component is contaminated by blends of telluric features. This is supported by the identification of several weak lines at the same wavelength in the scan of a telluric standard star, α Eri. These features show up clearly in the high S/N scan of this bright, fast-rotating, early-type star.

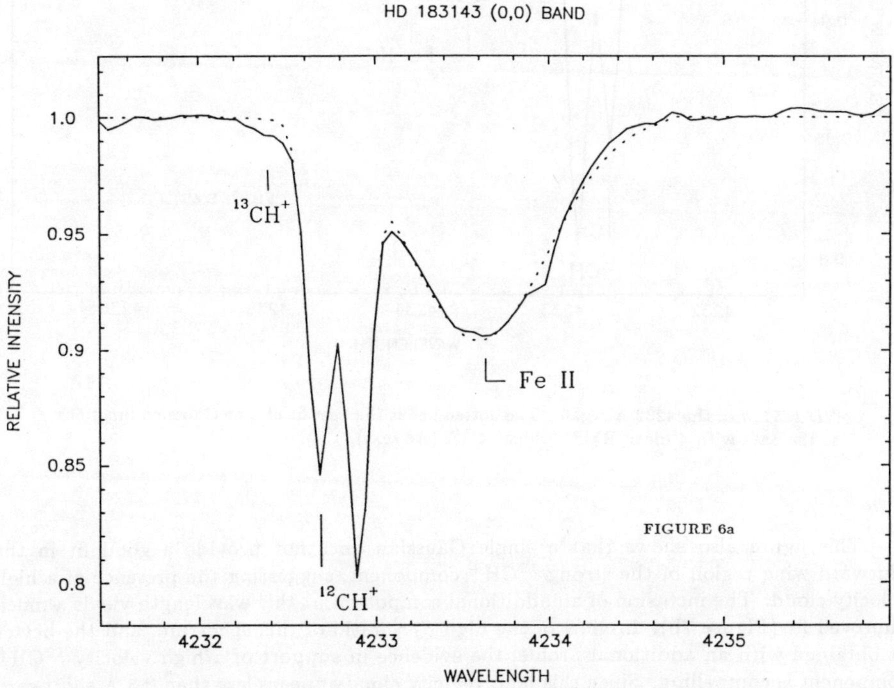

HD 183143 in the 4232 Å region. The dotted line is the best fit of three Gaussian functions to the data with a mean RMS residual < 1% (see text).

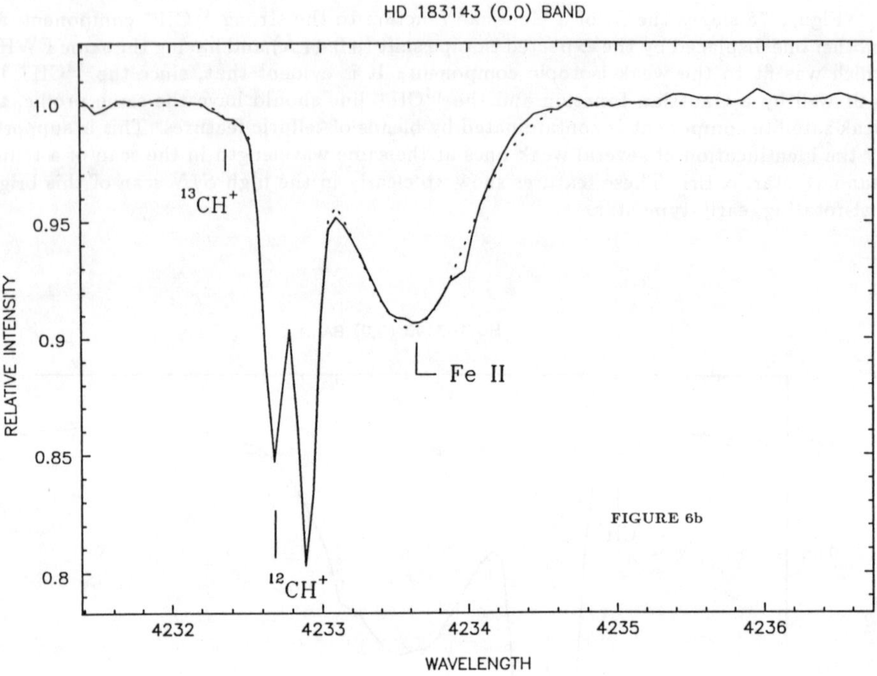

HD 183143 in the 4232 Å region. The dotted line is the best fit of four Gaussian functions to the data with a mean RMS residual < 1% (see text).

This figure also shows that a single Gaussian does not provide a good fit in the blueward wing region of the strong $^{12}CH^+$ component, suggesting the presence of a high velocity cloud. The inclusion of an additional component at this wavelength yields a much improved fit (Figure 7b). In view of the high S/N ratio of this spectrum, and the better fit obtained with an additional profile, the evidence in support of a high velocity $^{12}CH^+$ component is compelling. Since this high velocity cloud appears less than 0.3 Å shortward of the principal component, the $^{13}CH^+$ line at 4232 Å is blended with it. Thus, to be able to extract an equivalent width of this weak line in the (0, 0) band region, I fit three Gaussians simultaneously to the 4232 Å scan. One to fit the principal $^{12}CH^+$ component, one with the same FWHM and displaced by -0.26 Å to fit the $^{13}CH^+$ line, and one to fit the high velocity $^{12}CH^+$ cloud with the same wavelength shift as in the 3957 Å region and whose strength was constrained by the ratio of the vibrational oscillator strengths $f_{1,0}/f_{0,0} = 0.57$ (Elander, Oddershede, and Beebe 1977). The final fit is shown in Figure 8.

The 1σ errors quoted in Table 1 include continuum-placement uncertainties in addition to the error induced by statistical fluctuations. The values for $^{12}C/^{13}C$ listed in the table are preliminary, and have not been corrected for the effects of saturation. These values are thus only lower limits. Even though observed CH^+ Gaussian line width parameters ($b = \Delta V_{FWHM}/1.665$) are considerably greater than those measured for other molecules (Hobbs 1973, Hawkins and Jura 1987), it is likely that the $^{12}CH^+$ lines are significantly saturated toward these reddened, early-type stars. It is difficult to ascertain the amount of saturation of the $^{12}CH^+$ lines in stars which exhibit multiple absorption components. Observations in the (1,0) and (2,0) bands toward two of the stars, which I am in the process of analyzing, will help determine the degree of saturation of these lines.

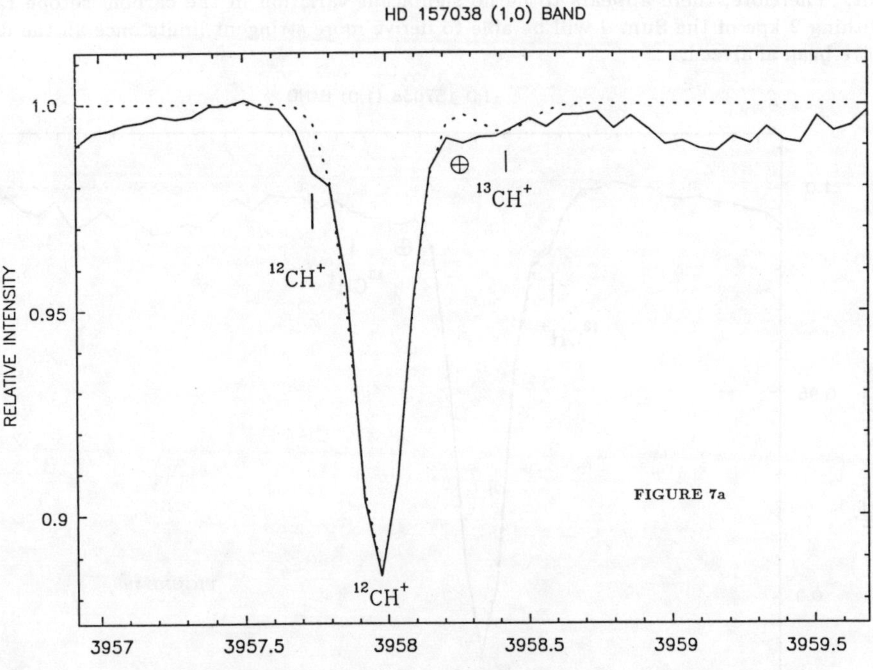

HD 157038 in the 3957 Å region. The dotted line is the two-Gaussian function fit with a mean RMS residual of 4% arising mainly from the discrepancy between the fit and the data points immediately redward of the principal $^{12}CH^+$ component. Telluric features are indicated (⊕) (see text).

DISCUSSION

Even though the CH^+ molecule suffers no significant fractionation in diffuse clouds and is thus a faithful representative of the carbon abundance in the ISM, the complications that arise due to the presence of multiple velocity components makes the determination of $^{12}C/^{13}C$ ratios toward distant, reddened stars quite difficult. The technique of extracting line strengths by the simultaneous fitting of multiple Gaussian profiles yields statistically significant results because of the theoretical constraints one can impose on some of the parameters. However, the uncertainties inherent in this procedure are difficult to quantify. The $^{12}C/^{13}C$ ratios quoted in Table 1 are lower limits, and saturation corrections will increase their final value, thus their consistency with our accurate local ISM $^{12}C/^{13}C$ value of 43 ± 4 (Hawkins, Jura, and Meyer 1985; Hawkins and Jura 1987) cannot be ruled out. Therefore, there appears to be no significant variation in the carbon isotope ratio withing 2 kpc of the Sun. I will be able to derive more stringent limits once all the data have been analyzed.

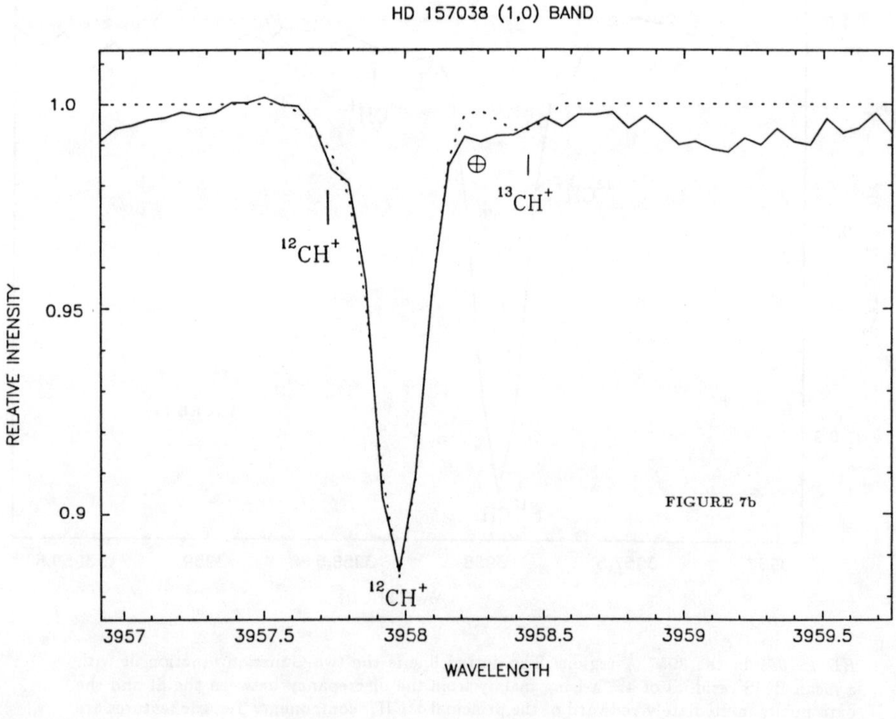

HD 157038 in the 3957 Å region. The dotted line is the three-Gaussian function fit with a smaller RMS residual of 3% (see text).

Despite having acquired the most sensitive data towards *HD 24432*, *HD 183143*, and *HD 157038* to date, results of the quality we have achieved previously from observations of nearby, brighter stars will be obtainable only when higher resolution spectrographs coupled with more blue-sensitive detectors become available. It is important to keep in mind that, in spite of the complications I have encountered in this study, the CH^+ molecule is still the best one to use in the determination of $^{12}C/^{13}C$ ratios because it is not strongly affected by any known isotope-selective mechanism, and because its larger intrinsic line width implies small saturation corrections.

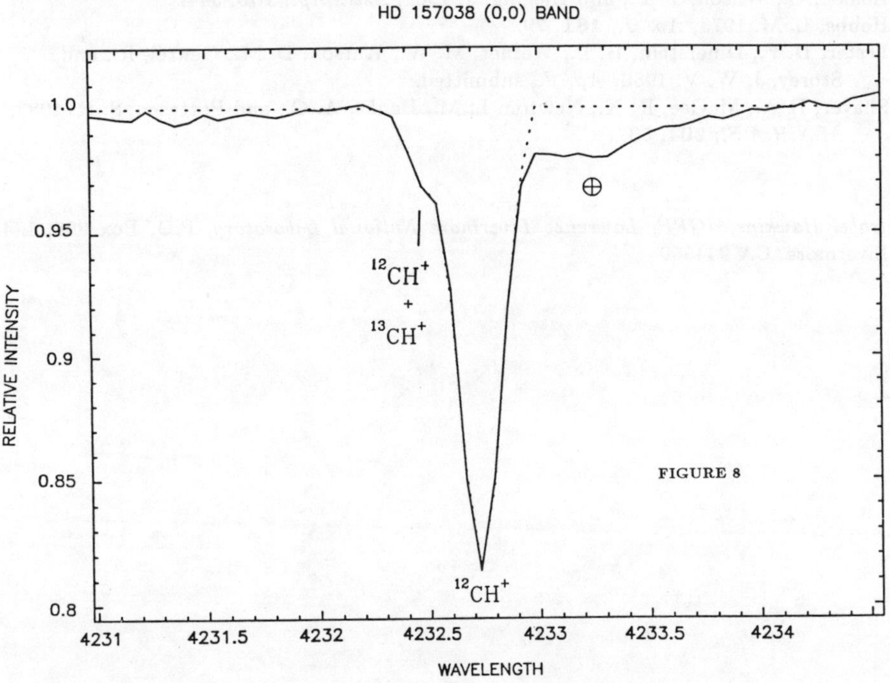

HD 157038 in the 4232 Å region. The dotted line is the best fit of 3 Gaussian functions to the data with a mean RMS residual $< 1\%$. Telluric features are indicated ($\oplus$) (see text).

REFERENCES

Cohen, J. 1973, *Ap. J.*, **186**, 149.
Cohen, J. 1975, *Ap. J.*, **197**, 117.
Elander, N., Oddershede, J., and Beebe, N. H. F. 1977, *Ap. J.*, **216**, 165.
Güsten, R., and Mezger, P. G. 1982, *Vistas Astr.*, **26**, 159.
Hawkins, I., Jura, M., and Meyer, D. M. 1985, *Ap. J. (Letters)*, **294**, L131.
Hawkins, I., and Jura, M. 1987, *Ap. J.*, **317**, 926.
Henkel, C., Güsten, R., and Gardner, F. F. 1985, *Astr. Ap.*, **143**, 148.
Henkel, C., Wilson, T. L., and Bieging, J. 1982, *Astr. Ap.*, **109**, 344.
Hobbs, L. M. 1973, *Ap. J.*, **181**, 79.
Lester, D. F., Dinerstein, H. L., Werner, M. W., Watson, D. M., Genzel, R., and Storey, J. W. V. 1986, *Ap. J.*, submitted.
Shaver, P. A., McGee, R. X., Newton, L. M., Danks, A. C., and Pottasch, S. R. 1983, *M.N.R.A.S.*, **204**, 53.

Isabel Hawkins, IGPP, Lawrence Livermore National Laboratory, P.O. Box 808, L-413, Livermore, CA 944550

THE ORIGIN AND DISTRIBUTION OF THE ELEMENTS*

B E J Pagel
Royal Greenwich Observatory
Herstmonceux Castle, Hailsham, Sussex, BN27 1RP, UK

Abstract

An acceptable framework for understanding the origin of the elements is provided by combining the ideas on nucleosynthesis in stars, put forward by Burbidge, Burbidge, Fowler and Hoyle and by Cameron thirty years ago, with current models of the first few minutes in the history of the Universe after the Big Bang. The so-called Standard Big Bang Nucleosynthesis model is in excellent agreement with primordial abundances of the light elements (D, ^{3}He, ^{4}He, ^{7}Li) deduced by extrapolating observational data provided that the half-life of the neutron is not above 10.2 minutes, and computer models of stellar evolution can now account for relative abundances of several different species formed in similar nucleosynthetic environments, although there are serious problems such as mixing, mass loss and the evolution of close binary stars.

Total abundances in the local part of our Galaxy, and large-scale trends in space and time (eg abundance gradients and differences between galaxies), involve questions related to star formation, the initial mass function and material flows into and out of galaxies about which very little is known for certain at present, and many different hypotheses seem to be able to account for the same data. In addition, there is evidence that abundances in some systems are affected by local pollution.

* Paper delivered to American Chemical Society Symposium on Origin and Distribution of the Elements, New Orleans, Aug 31 - Sept 4 1987.
G J Mathews (ed), Singapore : World Scientific Publishing Company

1. Introduction

It is now 30 years since the publication of the classic papers by B^2FH [1] and Cameron [2] on nucleosynthesis in stars; and 20 odd years since that of realistic calculations of nucleosynthesis in the Big Bang [3,4]. Supplemented by improved data on nuclear and particle physics and by computer calculations of stellar evolution, [5] these ideas still provide the basis for our current understanding of the origin of the elements and of relative abundances of species formed in similar nucleosynthetic environments that are found in the Solar System [6,7] and the local part of our Galaxy. [8]

Understanding the overall abundances and their distribution is a still more complex problem, however, involving models of galactic chemical evolution (GCE) and their various ingredients: star formation rates, initial mass functions (IMF) and hypotheses about mixing and gas flows, all involving great uncertainties in addition to those arising from stellar evolution itself, and in this field I sometimes wonder whether there has really been much progress since the introduction of the instantaneous simple model by Schmidt. [9] On the other hand, the data base on abundances in stars and nebulae has been vastly improved since then [10-17] and the results can be used in conjunction with rather simple models of GCE (applied with due caution) to make certain inferences as to which kinds of stars do what in the way of nucleosynthesis. In this paper I make some comments on how the situation looks to me now with respect to both Big Bang and stellar nucleosynthesis and some implications for GCE.

2. Big Bang Nucleosynthesis and the Half Life of the Neutron

Theoretical and observational aspects of the Standard Big Bang nucleosynthesis (SBBN) model have been very thoroughly discussed by Schramm, Steigman and their colleagues [18-20] who have emphasised the very good agreement between SBBN predictions and primordial abundances of D, ^3He, ^4He and ^7Li deduced from astronomical data over 9 orders of magnitude and used it to place limits on the cosmological baryon density and the number of light neutrino species. Other people have questioned the perfection of the agreement, however, on the grounds that the best estimate we can make on primordial helium, $Y_p \simeq 0.235$[21-23], is below the minimum value predicted from limits on primordial D or D + ^3He in the context of fairly simple GCE models.[23,24] Basically there are two ways out of this dilemma that have been considered: (a) that SBBN could be wrong owing to isothermal density fluctuations [25] or for some other reason; and (b) that there could have been more destruction of both D and ^3He than is allowed by simple GCE assumptions.[26,27]

However, there now seems to be quite an easy way out resulting from recent measurements of the half life of the neutron. The theoretically predicted mass fraction of helium is given by [19]

$$Y_p = 0.261 - 0.018\,[\log 10^5(\mathrm{D}+{}^3\mathrm{He})/\mathrm{H}] + 0.014\,(N_\nu - 3) + 0.014\,(\tau_{\frac{1}{2}} - 10.6) \quad (1)$$

$$> 0.240 + 0.014\,(\tau_{\frac{1}{2}} - 10.4) \quad (2)$$

if $(\mathrm{D} + {}^3\mathrm{He})/\mathrm{H} < 10^{-4}$ and $N_\nu = 3$ [18]), where $\tau_{\frac{1}{2}}$ is the neutron half life in minutes. Yang et al. [18] took a lower limit to $\tau_{\frac{1}{2}}$ of 10.4 minutes on the basis of laboratory measurements available up to 1984, but the validity of this lower limit has been questioned [28] and the questioning receives support from more recent experimental data (Fig 1). Clearly 10.2 min is a more safe lower limit than 10.4 min, reducing the lower limit on Y_p to 0.237 which fits the astrophysical value within uncertainties. The experimental situation is still in need of clarification: if the lower values of $\tau_{\frac{1}{2}}$ are confirmed, then this will be another triumph for SBBN and also a confirmation of reasonably simple GCE models for the destruction of D and ^3He.

3. Big-Bang Nucleosynthesis and Primordial Lithium

Recent, improved calculations of the primordial ^7Li abundance within the framework of SBBN [20] give a range of cosmological density

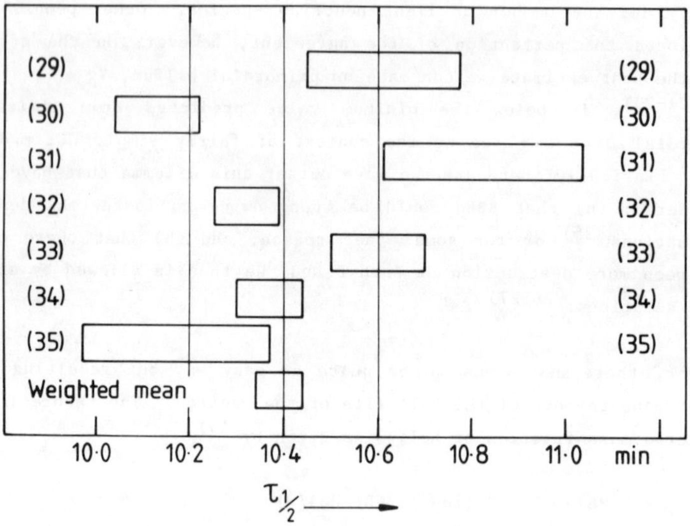

Fig 1 - Selected laboratory measurements of the neutron half life between 1972 and 1987 [29-35] with ±1σ errors and reference numbers.

parameters that is in excellent agreement with that derived from D + ^{3}He provided that one takes the observational value from extreme subdwarfs of Population II, $Li/H \simeq 10^{-10}$, rather than the Population I value, $Li/H \simeq 10^{-9}$. [19,23] In this case, however, there are difficulties in understanding how the primordial lithium in subdwarf atmospheres with effective temperatures near that of the Sun has survived all this time without the convective destruction that has operated in a shorter time in the Sun itself. Consequently the possibility that the Population I ^{7}Li abundance is largely primordial has not been fully ruled out, [19] although there is no shortage of likely stellar production mechanisms. [36]

If (as seems more likely) ^{7}Li in the local interstellar medium is largely due to stellar nucleosynthesis, then we should expect its abundance in the interstellar media of other galaxies to follow that of other metals, after making due allowance for ionisation and locking on to dust grains, whereas if it is due to the Big Bang its abundance should remain the same or go down with increasing degrees of astration like deuterium. The appearance of supernova 1987A in the Large Magellanic Cloud has provided an opportunity to put these alternatives to the test using spectra of very high resolution. From a rather marginal detection of LiI at the LMC radial velocity, Vidal-Madjar _et al_ [37] estimate that the ratio of column densities LiI/KI is about half of what it is in the local Galactic interstellar gas, which sounds favourable to stellar production, but any firm conclusion will have to await more thorough evaluations of the data.

4. Local Galactic Abundances

The standard source for abundances in the solar neighbourhood is the Solar System distribution [6,7] based on spectroscopy of the photosphere, analytical measurements of carbonaceous chondrites and some interpolations based on theoretical arguments. The chief remaining uncertainties relate to inert gases, especially helium for which solar models suggest $Y = 0.25$ to 0.30 in fair agreement with the Orion nebula [23] making $Z = 0.02$ and $X = 0.68$ to 0.73, given $Z/X = 0.025$. [6]. The heavy-element fraction Z is made up predominantly of O, C, Ne, N, Mg, Si and Fe. Nearby stars, diffuse clouds and emission nebulae like Orion have broadly similar similar compositions, as far as we can tell, apart from some features which include the following [8]:-

1. Variations in stellar metallicity up to twice solar with most observable elements varying more or less in lockstep except for oxygen and other elements with α-particle nuclei which vary less, especially oxygen [12,13]. Thus "metallicity", as measured by the easily observed parameter [Fe/H] for the cooler stars, is not a good indicator of the heavy-element mass fraction Z unless corrected for the so-called oxygen anomaly. [38] ([] indicate the logarithm of a stellar quantity relative to its value in the Sun.) Population II stars have $[Fe/H] < -1$ and are vary rare in the solar neighbourhood; when $[Fe/H] < -1.5$, they show overdeficiencies (relative to iron) of elements formed predominantly (in the Solar System at least) by the s-process, [14]) whereas carbon

and r-process elements vary more or less in lockstep with iron.[14,17,39] N/Fe shows no clear systematic trend with Fe/H, but substantial scatter at the lowest metallicities.[17]

2. The Orion Nebula is deficient in oxygen, compared to the Sun, by about a factor of 2 [40] with smaller (or perhaps non-existent) deficiencies in carbon and nitrogen. There are indications from mm wave and optical data that $^{13}C/^{12}C$ is enhanced over the Solar System value by a factor of 2 in Orion [41] and in nearby diffuse clouds.[42]

How well are these local abundances explained by theory? In single-star evolution the outcome mainly depends on the initial mass of the star. The smallest stars below a mass (in solar units) $m_1 = 0.9$, say, have such long lifetimes that they do little except lock up diffuse interstellar material; but precisely because of this their contribution to the IMF is an important parameter in fixing the lock-up fraction and hence the nucleosynthetic yield [8] and is poorly known because of ambiguities associated with the "missing mass".[43] This in itself leads to an uncertainty of a factor of 2 in the yield.

Intermediate mass stars with masses up to $m_2 \simeq 9$ form degenerate CO cores and undergo thermal pulses in late stages of evolution with double shell sources leading to production of C, N and s-process elements which are mixed out to the surface and ejected in winds and planetary nebulae leaving a white-dwarf remnant.[44] The biggest stars, above m_2, eject large amounts of mass while still near the main sequence, in some cases exposing and ejecting hydrogen and helium burning products from the core to form Wolf-Rayet and related stars.[45] In the meantime, deeper inside the core, carbon, oxygen and silicon burning take place leading to the canonical onion-skin structure until iron-group elements are synthesised at the centre and it collapses to nuclear densities leading in some or maybe all cases to a supernova outburst in which explosive nucleosynthesis modifies some abundances, the products are ejected and a neutron star is left behind. Above $m_3 \simeq 100$ other mechanisms cause various instabilities which may or may not be relevant to the distribution of the elements. Massive supernovae of Types II and Ib are held responsible for most of the production of O, Ne, Si and neighbouring elements and some iron, but the bulk of the iron in the Solar System probably comes from Type Ia supernovae [46]) which are interacting binaries with a white dwarf component.

Calculations of relative abundances synthesised in massive stars with m around 25 and Z = .02, making various assumptions about the $^{12}C(\alpha, \gamma)$ ^{16}O rate, give fair agreement (to factors of 2 either way) with relative Solar System abundances of species from O to Ni, but with glaring discrepancies in some cases.[47,48]) The absolute abundances depend on the IMF assumed and the GCE model. In earlier work, [49-51]) simple GCE models were combined with a flexible parametrisation of the IMF and the solar abundances were fitted with a total yield somewhat below 1 per cent. More recent authors [52-54]) have taken the IMF as given and were then forced to resort to various devices (like inflow and an upper limit to the masses of stars which explode) to keep the yield down. For reasons given above, I prefer the first approach. The ratios of C and N to oxygen depend on the detailed shape of the IMF, the mixing length and the value of m_2, so that by playing around with these parameters it is not too difficult to fit them.

Other significant features of the solar neighbourhood are the age-metallicity relation (if any) and the statistical distribution of abundances (or "G-dwarf problem"). The simple GCE model (closed, mixed and with constant yields of primary elements, ie those synthesised from hydrogen and helium in the original make-up of the progenitor) predicts metallicity increasing with time, a broad distribution of metallicities, constant ratios of primary elements (except helium) in proportion to their yields and a decrease in the abundance ratio of secondary elements (ie those whose synthesis depends on primary elements in the original make-up of the progenitor) to primary elements as the metallicity decreases; the two last predictions, however, need to be corrected in the light of possible departures from the instantaneous recycling approximation and of the likelihood that mixing processes giving rise to secondary elements like the s-process nuclei and some proportion of nitrogen may be more vigorous at lower metallicities where hydrogen burning (at least) takes place at higher temperatures.

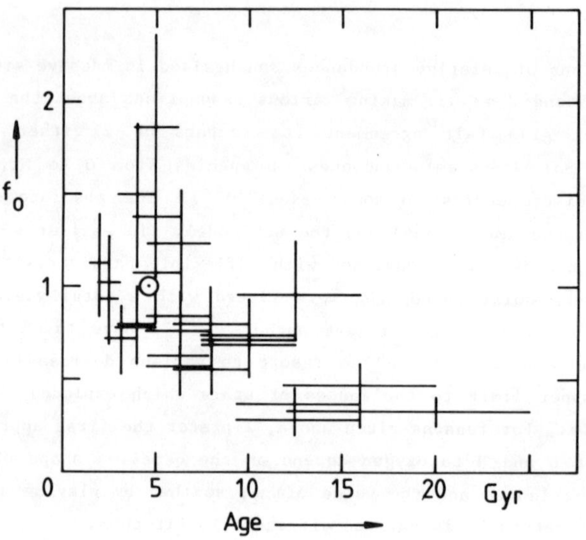

Fig 2. Relation between stellar age and oxygen abundance, in solar units.[57] Errors of ± 0.1 dex have been assumed in both coordinates.

The age-metallicity relation is still poorly known, presumably because of systematic errors in the determination of both parameters from intermediate-band photometry.[55,56] We shall have much better estimates of stellar ages once we have parallaxes from the forthcoming HIPPARCOS satellite mission. Because oxygen is a better candidate than iron for treatment in the instantaneous recycling approximation, the relation between age and oxygen abundance is of particular interest (Fig 2). The data are compatible with a linear or possibly a rising trend.

Statistical metallicity distributions in the solar neighbourhood and elsewhere[8]) show that the notorious shortage of low-abundance stars is a feature of the solar neighbourhood which is absent from the other samples studied and therefore needs to be explained on the basis of local effects in the formation of the disk,[58,59]) rather than by initial enrichment from massive objects or much higher yields at earlier times for which there is no discernible reason why they should have happened here and not elsewhere. One still needs to study the oxygen distribution rather than the metallicity distribution, because of the instantaneous recycling problem, but the qualitative results are unlikely to be very different.

Trends in relative abundances of different elements are partly understandable within the canonical theoretical framework. The O/Fe "anomaly" can be explained by differences in evolutionary times[46]) and there seem to be odd-even effects in ratios like Na/Mg and Al/Mg[60]) consistent with the odd isotopes being largely secondary products from massive supernovae.[61,62]) Nitrogen and s-process elements do not show the secondary: primary behaviour relative to carbon and iron that one would expect on the basis of simple models[63]) and it seems that a source of primary nitrogen in low-metallicity massive stars is needed.[64]) One candidate is a "clean" 500 solar mass star with zero heavy elements[65]), but it is by no means clear to me that the appropriate mixing processes could not have taken place in a clean star with a more ordinary mass.

5. Large-scale Trends in Abundances

The main characteristics of abundances in galaxies are (i) an increase in average heavy-element abundances in the bright inner regions with the total mass of the system;[10,63]) and (ii) existence of radial abundance gradients in the disks and bulges of spirals and in at least some ellipticals.[10,66-70]) The gradients in spiral disks are closely related to the surface density of mass deduced from rotation curves[71,72]) and application of the simple GCE model suggests an increase in the (effective) yield towards regions with higher abundance.[72,73] Three different sorts of explanation suggest themselves for radial gradients: (i) variations in the true yield, associated with the IMF;[73,74] (ii) inflow of material towards the disk and centre;[75,76] and (iii) reduction of the effective yield in low-abundance regions by outflow of processed gas.[77] The third explanation has been applied with very considerable success to explain the mass-metallicity relation for galaxies as a result of galactic winds,[78,79] covering a range of nearly 3 orders of magnitude between the smallest dwarf spheroidals and the nuclear regions of giant spirals and ellipticals.

Large-scale trends in relative abundances of some of the more common elements can be deduced from studies of emission lines in HII regions[10] and supernova remnants.[80] Stars make substantial quantities of helium for each unit of oxygen that they make: $dY/dZ \simeq 4$ [21-23] and this is quite a strong constraint on stellar evolutionary models.[45,81] Nitrogen keeps a more or less constant ratio to carbon at abundance levels up to that of Orion,[82] but far more data are available for the relation between N/O and O/H. This shows a richly complex pattern, not very different from the stellar one, [67] and having a dispersion that is at least partly real (see next section). At first glance the pattern suggests primary nitrogen ($N/O \simeq$ const) at low abundances and secondary nitrogen ($N/O \propto O/H$) at higher abundances, but if nitrogen were replaced by iron a similar pattern would be seen in the data for stars! Furthermore, some spiral galaxies show little gradient in N/O, [70] although it has to be said that in our own Galaxy there are disturbing discrepancies between N/O ratios deduced from optical and infra-red data [83] and the dust still has to settle. Ingenious GCE mechanisms that have been devised to reconcile the data with secondary nitrogen production [84,85] may play some role, but they certainly fail to cover such extreme cases as I Zw 18 [82] so that the H II region data confirm the deduction from stars that primary nitrogen production dominates at the lower abundance levels. [86]

Other elements observed in H II regions are neon, sulphur and argon, which naively at least we would expect to vary in lockstep with oxygen. Neon indeed seems to do this, [87] but the data for sulphur are more ambiguous. In irregular galaxies with rather low abundances, S/O appears to be more or less solar, [88,89] but in spirals the gradient in S/H appears to be significantly flatter than in O/H and *a fortiori* N/H, [90,69,66] and the effect does not seem able to be explained away by technical arguments about the analysis. [91] The large ratios of N^+/S^+ detected in the nuclei of very luminous spirals [92] could thus be telling us something significant about abundances, although this is not yet clear.

6. Small-scale Trends in Abundances

Wolf-Rayet stars, planetary nebulae, young supernova remnants and the atmospheres of carbon stars and other giants show abundance anomalies due to their internal stellar evolution with products that have not undergone large-scale mixing through the interstellar medium. These are examples of what might be termed "local environmental pollution", an effect that falls outside the normal smoothed-out approximation used in simple GCE models. Other examples of such pollution or limited-mixing effects are provided by the isotope anomalies in meteorites [93,94] and possibly some effects in globular clusters. [95,8]

A topic that I should like to discuss in more detail is the pollution of significant portions of giant H II regions by supernova ejecta and especially winds from massive stars, which can in principle cause them to have abundances that are unrepresentative of the underlying galaxy. A microcosm of this kind of effect is displayed by the ring nebulae surrounding certain WR stars in our own Galaxy, in which several tens of solar masses of what is presumably swept-up interstellar material, extending over 15 pc or so, have large overabundances of helium and nitrogen from the wind of the central star.[96,97] A smaller region of the 30 Doradus nebula, extending over about 1 pc, has overabundances of He, O and related elements (excluding nitrogen), believed to result from pollution by the wind from a massive star that has reached (or passed) the WC stage. [98]

Processes of a similar kind could be operating on a larger scale in giant extragalactic H II regions which often show WR features in the underlying continuous spectrum from hot, young stars. They would be particularly noticeable in the blue compact dwarf galaxies whose light is dominated by a single H II complex (also known as H II galaxies), in which the underlying heavy-element abundance is very low and local enhancements correspondingly easier to notice. Star formation in these systems takes place in bursts involving something like 10^7 solar masses at a time, and the long intervals between bursts enable primary nucleosynthesis products like carbon and nitrogen, formed in part by the evolution of intermediate-mass stars, to "catch up" with the oxygen (formed in massive stars) even when the total oxygen abundance is very low.[86)] It has been suggested that the blue compact galaxy I Zw 18, which has the lowest abundances of heavy elements known in any H II galaxy, is now experiencing only its first or second burst and that the abundances in the visible part (ie the H II region) are enhanced by supernova remnants from its own ionising star cluster, the underlying galaxy being much more deficient still.[99)] There are several difficulties with this hypothesis: one is the time scale, because supernovae are expected to come out in large numbers only after the ionisation by stars near the main sequence has faded away, and another is the relatively high abundance of carbon and nitrogen,[82)] if these need to catch up. Furthermore, I Zw 18 appears to have too little oxygen abundance for its mass, rather than too much[63)] although the mass is very uncertain and difficult even to define in view of the existence of good evidence for large quantities of dark matter in objects of this class.[100, 101)]

A more promising candidate for explanation as a result of pollution effects is the dispersion in N/O in irregular and H II galaxies, which is now well established, and associated variations in the helium abundance (for given O/H) which are rather less well established because they are very small.[22,102)] The ratio of excess helium to excess nitrogen (above a constant N/O ratio of about 0.033) is consistent with an admixture of one or two per cent of material from WN star winds with secondary production of nitrogen, as opposed to the primary nitrogen that accounts for the constant N/O ratio in the underlying systems. In one such galaxy, NGC 5253, one can actually see an enhanced abundance of nitrogen in the brightest (and presumably youngest) H II knot, with normal N/O in the remaining nebulosities[102,103)] while the oxygen abundance is constant throughout,[102)] and the same thing seems to be happening in the more remote H II galaxy POX 4.[104)]

7. Conclusion

Even from such an incomplete review as the foregoing, it is clear that the distribution of the elements has resulted from many different sorts of process acting at the same time, each of which is only partially understood if at all. Among these the Standard Big Bang Nucleosynthesis model seems to be in relatively good shape subject to the physicists getting their act together and agreeing on the neutron half life to better than 2 per cent, and this in turn suggests reasonably simple initial conditions for the chemical evolution of galaxies. Beyond that point, however, our understanding of stellar evolution is defective at many points and that of the large-scale processes governing star formation and the overall evolution of galaxies is very rudimentary indeed. One hopes that further information on abundances in different places will eventually help to throw light on these basic astronomical questions as well as on nucleosynthesis as such.

I thank W Priester for enlightening correspondence and the Director and staff of the Instituto de Fisica, Universidade Federal do Rio Grande do Sul, Porto Alegre, Brazil, for their hospitality during the preparation of this review.

References

1. Burbidge, E.M., Burbidge, G.R., Fowler, W.A., and Hoyle, F. (B^2FH) 1957, Rev. Mod. Phys., **29**, 547.

2. Cameron, A.G.W. 1957, Chalk River Report, CRL 41.

3. Peebles, P.J.E. 1966, Astrophys. J., **146**, 542.

4. Wagoner, R.V., Fowler, W.A., and Hoyle, F. 1967, Astrophys. J., **148**, 3.

5. Chiosi, C., and Renzini, A. (eds) 1984, Stellar Nucleosynthesis, Dordrecht:Reidel.

6. Cameron, A.G.W. 1982, in Essays in Nuclear Astrophysics C.A. Barnes, D.D. Clayton and D.N. Schramm (eds), Cambridge University Press, p.23.

7. Anders, E., and Ebihara, M. 1982, Geochim. Cosmochim. Acta, **46**, 2363.

8. Pagel, B.E.J. 1987a, in The Galaxy, G. Gilmore and R.F. Carswell (eds), Dordrecht: Reidel, p. 341.

9. Schmidt, M. 1963, Astrophys. J., **137**, 758.

10. Pagel, B.E.J., and Edmunds, M.G. 1981, Ann. Rev. Astr. Astrophys., **19**, 77.

11. Freeman, K.C., and Norris, J. 1981, Ann. Rev. Astr. Astrophys., **19**, 319.

12. Clegg, R., Lambert, D.L., and Tomkin, J. 1981, Astrophys. J., 250, 262.

13. Luck, R.E., and Bond, H.E. 1985 Astrophys. J., **292**, 559.

14. Spite, M., and Spite, F. 1985, Ann. Rev. Astr. Astrophys., 23, 225.

15. Zinn, R. 1985, Astrophys. J., **293**, 424.

16. Norman, C.A., Renzini, A., and Tosi M. (eds) 1986 Stellar Populations Cambridge University Press.

17. Carbon, D.F., Barbuy, B., Kraft, R.P., Friel, E.D., and Suntzeff, N.B. 1987, Pub. Astr. Soc. Pacific, **99**, 335.

18. Yang, J., Turner, M.S., Steigman, G., Schramm, D.N., and Olive, K.A. 1984, Astrophys. J., **281**, 493.

19. Boesgaard, A.M. and Steigman, G. 1985, Ann. Rev. Astr. Astrophys., **23**, 319.

20. Kawano, L., Schramm, D., and Steigman, G. 1987, Preprint.

21. Peimbert, M. 1985, in Star Forming Dwarf Galaxies D. Kunth, T.X. Thuan and J.T.T. Van (eds) Paris: Ed. Frontieres, p. 403.

22. Pagel, B.E.J., Terlevich, R.J., and Melnick, J. 1986, Pub. Astr. Soc. Pacific, **98**, 1005.

23. Pagel, B.E.J., 1987b, in First ESO-CERN School on Astro-Particle Physics, A. de Rujula, D. Nanopoulos and P.A. Shaver (eds) Singapore: World Scientific Pub. Co.

24. Vidal-Madjar, A., and Gry, C. 1984, Astr. Astrophys., **138**, 285.

25. See contributions by Fuller, Mathews and Alcock, and by Malaney and Fowler, this conference. Also Alcock, C., Fuller, G. M., and Mathews, G. J. 1987, Astrophys. J., **320**, in press.

26. Delbourgo-Salvador, P., Gry, C., Malinie, G., and Audouze, J. 1985, Astr. Astrophys., **150**, 53.

27. Delbourgo-Salvador, P., Audouze, J., and Vidal-Madjar, A. 1987, Astr. Astrophys., **174**, 365.

28. Ellis, J., Enqvist, K., Nanopoulos, D.V., and Sarkar, S. 1986, Phys. Lett. B, **167B**, 457.

29. Christensen, C.J., Nielsen, A., Bahnsen, A., Brown, W.K., and Rustad, B.M. 1972, Phys. Rev. D., **5**, 1628.

30. Bondarenko, L.N., Kurguzov, V.V., Prokofiev, Yu.A., Rogov, E.V., and Spivak, P.E. 1978, JETP Lett., **28**, 303.

31. Byrne, J., Morse, J., Smith, K.F., Shaikh, F., Green, K., et al. 1980, Phys. Lett B, **92**, 274.

32. Frank, A.I. 1982, Sov. Phys. (Uspekhii), **25**, 280.

33. Byrne, J. 1984, Nature (London), **310**, 212.

34. Bopp, P., Dubbers, D., Klemt, E., Last, J., and Schultze, H. 1984, J. Phys (Paris), **45**, C3.

35. Measurements by K.J. Kügler, W. Paul and D. Dubbers quoted by Blome, H.J., and Priester, W. 1984, Naturwiss., **71**, 456, and private communication from W. Priester (1987).

36. Audouze, J., Boulade, O., Malinie, G., and Poilane, Y. 1983, Astr. Astrophys., **127**, 164.

37. Vidal-Madjar, A., Andreani, P., Cristiani, S., Ferlet, R., Lanz, T., and Vladilo, G. 1987, Astr. Astrophys., **177**, L17.

38. Strömgren, B. 1987, in The Galaxy G. Gilmore and R.F. Carswell (eds), Dordrecht: Reidel, p. 229.

39. Butcher, H.E. 1975, Astrophys. J., **199**, 710.

40. Peimbert, M. 1986, in IAU Symp. No. 115: Star Forming Regions, M. Peimbert and J. Jugaku (eds), Dordrecht: Reidel.

41. Johanesson, L., et al, 1984, Astr. Astrophys., **130**, 227.

42. Hawkins, I., and Jura, M. 1987, Astrophys. J., **317**, 926.

43. Bahcall, J.N. 1984, Astrophys. J., **287**, 926. But cf. Gilmore, G., and Wyse, R.F.G. 1987, in The Galaxy, G. Gilmore and R.F. Carswell (eds), Dordrecht:Reidel, p. 247.

44. Renzini, A. 1984, in Stellar Nucleosynthesis C Chiosi and A. Renzini (eds), Dordrecht: Reidel, p. 99.

45. Maeder, A. 1984, in Stellar Nucleosynthesis, C. Chiosi and A. Renzini (eds), Dordrecht: Reidel, p. 115.

46. Matteucci, F., and Greggio, L. 1986, Astr. Astrophys., **154**, 279.

47. Thielemann, F.K., and Arnett, W.D. 1985, Astrophys. J., **295**, 604.

48. Woosley, S.E., and Weaver, T.A. 1986, in Nucleosynthesis and its Implications on Nuclear and Particle Physics, J. Andouze and N Mathieu (eds), Dordrecht: Reidel, p. 145.

49. Arnett, W.D. 1978, Astrophys. J., **219**, 1008.

50. Alloin, D., Collin-Souffrin, S., Joly, M., and Vigroux, L. 1979, Astr. Astrophys., **78**, 200.

51. Lequeux, J., Peimbert, M., Rayo, J.F., Serrano, A., and Torres-Peimbert, S. 1979, Astr. Astrophys., **80**, 155.

52. Twarog, B., and Wheeler, J.C. 1982, Astrophys. J., **261**, 636.

53. Díaz, A.I., and Tosi, M. 1984, Mon. Not. R. astr. Soc., **208**, 365.

54. Tosi, M., and Díaz, A.I. 1985, Mon. Not. R. astr. Soc., **217**, 571.

55. Twarog, B. 1980, Astrophys. J., **242**, 242.

56. Carlberg, R.G., Dawson, P.C., Hsu, T., and Vandenberg, D.A. 1985, Astrophys. J., **294**, 674.

57. Nissen, P.E., Edvardsson, B., and Gustafsson, B. 1985, in Production and Distribution of the CNO Elements, I.J. Danziger, F Matteucci and K. Kjär (eds), ESO, Garching, p.131.

58. Larson, R.B., 1976, Mon. Not. R. astr. Soc., **218**, 409.

59. Lynden-Bell, D. 1975, Vistas in Astron. 19, 299.

60. Tomkin, J., Lambert, D.L. and Balachandran, S. 1985, Astrophys. J., **290**, 289.

61. Arnett, W.D. 1971, Astrophys. J., **166**, 153.

62. Woosley, S.E., and Weaver, T.A., 1982a, in Essays in Nuclear Astrophysics, C.A. Barnes, D.D. Clayton and D.N. Schramm (eds), Cambridge University Press, p.377.

63. Pagel, B.E.J. 1986, in Cosmogonical Processes, W.D. Arnett, C. Hansen, J. Truran and S. Tsuruta (eds), Utrecht: VNU Science Press, p.66.

64. Matteucci, F. 1986, Pub. Astr. Soc. Pacific, **98**, 973.

65. Woosley. S.E., and Weaver, T.A. 1982b in Supernovae: a Survey of Current Reseach M.J. Rees and R.J. Stoneham (eds), Dordrecht:Reidel, p.79.

66. Shaver, P.A., McGee, R.X., Danks, A.C., and Pottasch, S.R. 1983, Mon. Not. R. astr. Soc., **204**, 53.

67. Pagel, B.E.J. 1985, in Production and Distribution of the CNO Elements, I.J. Danziger, F. Matteucci and K. Kjär (eds) ESO, Garching, p.155.

68. Efstathiou, G., and Gorgas, J. 1985, Mon. Not. R. astr. Soc., 215, 37P.

69. Evans, I.N. 1986, Astrophys. J., **309**, 544.

70. Garnett, D., and Shields, G. 1987, Astrophys. J., **317**, 82.

71. McCall, M.L. 1982, Thesis, University of Texas, Austin.

72. Edmunds, M.G., and Pagel, B.E.J. 1984, Mon. Not. R. astr. Soc., **211**, 507.

73. Peimbert, M., and Serrano, A. 1982, Mon. Not. R. astr. Soc., **198**, 563.

74. Güsten, R., and Mezger, P.G. 1982, Vistas in Astr., **26**, 159.

75. Larson, R.B. 1974a, Mon. Not. R. astr. Soc., **166**, 585.

76. Lacey, C.G., and Fall, S.M. 1985, Astrophys. J., **290**, 154.

77. Hartwick, F.D.A. 1976, Astrophys. J., **209**, 418.

78. Larson, R.B. 1974b, Mon. Not. R. astr. Soc., **169**, 229.

79. Arimoto, M., and Yoshii, Y. 1987, Astr. Astrophys., **173**, 23.

80. Blair, W.P., and Kirshner, R.P. 1985, Astrophys. J., **289**, 582.

81. Serrano, A. 1986, Pub. Astr. Soc. Pacific, **98**, 1066.

82. Dufour, R.J. 1986, Pub. Astr. Soc. Pacific, **98**, 1025.

83. Lester, D.F., Dinerstein, H.L., Werner, M.W., Watson, D.M., Genzel, R., and Storey, J.W.V. 1987, Astrophys. J., **320**, in press.

84. Serrano, A., and Peimbert, M. 1983, Rev. Mex. Astr. Astrofis., **8**, 117.

85. White, S.D.M., and Audouze, J. 1983, Mon. Not. R. astr. Soc., **203**, 603.

86. Matteucci, F., and Tosi, M. 1985, Mon. Not. R. astr. Soc., **217**, 391.

87. Vigroux, L., Stasinska, G., and Comte, G. 1987, Astr. Astrophys., **172**, 15.

88. Pagel, B.E.J. 1978, Mon. Not. R. astr. Soc., **183**, 1P.

89. Dennefeld, M., and Stasinska, G. 1983, Astr. Astrophys., **118**, 234.

90. Talent, D.L., and Dufour, R.J. 1979, Astrophys. J., **233**, 888.

91. Vilchez, J.M. 1987, Thesis, Universidad de la Laguna, Tenerife.

92. Rubin, V.C., Ford, W.K., Jnr., and Whitmore, B.C. 1984, Astrophys. J. Lett., **281**, L21.

93. Wasserburg, G.J., and Papanastassiou, D.A. 1982, in Essays in Nuclear Astrophysics, C.A. Barnes, D.D. Clayton and D.N. Schramm (eds), Cambridge University Press, p.77.

94. Clayton, D.D. 1986, in Cosmogonical Processes, W.D. Arnett, C. Hansen, J. Truran and S. Tsuruta (eds), Utrecht: VNU Science Press, p.101.

95. Kraft, R.P. 1985, in Production and Distribution of the CNO Elements, I.J. Danziger, F. Matteucci and K. Kjär (eds), ESO, Garching, p.21.

96. Kwitter, K. 1984, Astrophys. J., **287**, 840.

97. Rosa, M. 1987. in IAU Symposium no. 122: Circumstellar Matter Dordrecht : Reidel.

98. Rosa, M., and Mathis, J.S. 1987, Astrophys. J., **317**, 163.

99. Kunth, D., and Sargent, W.L.W. 1985, Astrophys. J., **300**, 496.

100. Aaronson, M., and Olszewski, E. W. 1987, in IAU Symposium No. 130: Evolution of Large Scale Structure in the Universe.

101. Brinks, E., and Klein, U. 1987, submitted.

102. Pagel, B.E.J. 1987c, in Starbursts and Galaxy Evolution, T. Montmerle and J.T.T. Van (eds), Paris: Ed. Frontières, in press.

103. Walsh, J., and Roy, J.-R. 1987, Astrophys. J. Lett., **319**, L57.

104. Kunth, D., and Sargent, W.L.W. 1983, Astrophys. J., **272**, 81.

NUCLEAR REACTIONS DURING COSMIC-RAY TRANSPORT

V.E. VIOLA, L.W. WOO and K. KWIATKOWSKI
Department of Chemistry and IUCF
Indiana University
Bloomington, IN 47405 USA

ABSTRACT

Nuclear reactions which occur during the passage of galactic cosmic rays through the interstellar medium are important to our understanding of a number of astrophysical phenomena. Foremost among these are the nucleosynthesis of the elements lithium, beryllium and boron and modifications of the primary source composition of the galactic cosmic rays. In this paper cross-section measurements for nuclear reactions relevant to these problems are discussed. New data are presented for the A = 6-15 isobars produced in the 100-MeV p + ^{16}O reaction. These results demonstrate the need for the inclusion of mean-field effects in nuclear reaction models for nucleon-induced reactions on light target nuclei.

I. INTRODUCTION

Nuclear reactions induced by galactic cosmic rays as they traverse the interstellar medium play an important role in modifying the elemental and isotopic composition of the universe. Such collisions are believed to be primarily responsible for creation of the elements lithium, beryllium and boron.[1] At the same time they contribute minor abundance modifications to many of the heavier elements. Reactions in the ISM also alter the isotopic distribution which characterizes the primary GCR flux, thereby complicating attempts to understand the origin of cosmic rays.[2] Of more practical concern, these reactions also lead to problems associated with space travel, e.g. radiation damage effects in biological systems[3] and microelectronic devices.[4]

The major contribution of GCR-ISM interactions to nucleosynthesis is in the production of the elements Li, Be and B. In Fig. 1 the relative abundances of the elements in the GCR flux are compared with those in the ISM. Two obvious features of this comparison are: (1) the greatly enhanced values for LiBeB and (2) the smoothing of odd-even irregularities in the GCR flux relative to the ISM. As Austin[5] has pointed out, the GCR-ISM mechanism is an inevitable one for LiBeB nucleonsynthesis and our extensive knowledge of the properties of both components make this a particularly attractive model for confronting astrophysical theory with experimental measurements.

In order to derive the primary composition of the galactic cosmic ray flux at its source, it is essential to unfold the effects of secondary GCR-ISM interactions. Recent satellite and balloon measurements[6-8] have studied the isotopic ratios of elements as heavy as magnesium in the GCR flux. These observations have demonstrated enhanced abundances for the neutron-rich isotopes of these elements, which may provide important constraints on models which seek to explain cosmic ray origins and the nucleosynthetic processes which may be associated with their acceleration. Many viable theories exist in which the primary GCR and ISM compositions are the same; others involve dynamic phases of stellar evolution in which the production of neutron-excess isotopes plays an important role. Reduction of the model-space for explaining these phenomena depends critically on the ability to correct the transported GCR spectrum to its primary distribution - an undertaking which relies heavily on our understanding of nuclear reaction probabilities.

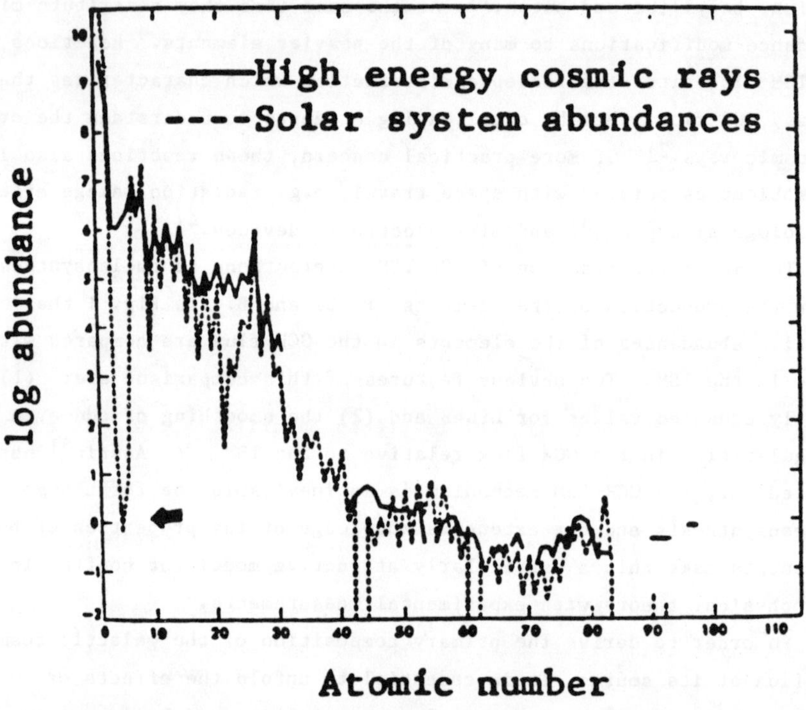

Fig. 1. Comparison of instellar medium abundances (dotted line) with abundances in the galactic-cosmic-ray flux (solid line).

Ideally, to solve the above problems, one would desire cross sections for all possible GCR-ISM reactions from threshold to the upper energies of the cosmic ray spectrum. Fortunately , this huge reaction/energy matrix can be reduced immensely by virtue of several factors. First, from examination of Fig. 1, one notes that on the basis of the exponential dependence of the elemental abundances, only a few reactions will dominate the production spectrum of $Z > 3$ elements in GCR-ISM interactions. These are:

$$p + {}^{12}C,\ {}^{14}N \text{ and } {}^{16}O \text{ and}$$

$$ {}^{4}He + {}^{4}He,\ {}^{12}C,\ {}^{14}N \text{ and } {}^{16}O. \qquad (1)$$

Next in importance are targets of ^{20}Ne, ^{24}Mg and ^{28}Si. Further, nuclear reaction cross sections at intermediate energies do not vary strongly from one target to another;[9,10] hence, once a number of critical reactions are well-defined, reliable estimates for other reactions can be made. Finally, above about 100-200 MeV/nucleon, the cross sections for proton and alpha-particle reactions with light target nuclei ($A < 30$) show very little energy dependence, which, coupled with the decreasing GCR flux as a function of energy, removes the necessity to perform experiments over a large energy region.

Thus, the experimental objective for the purposes of investigating GCR-ISM-related phenomena is as follows. Detailed reaction data are being measured for the critical reactions described above. In addition, the $p + {}^{28}Si$ and $^{4}He + {}^{28}Si$ reactions are being studied as representative of reactions in the $20 < A < 40$ region. Measurements include mass, charge, energy and angular distributions for all reaction products. With such data in hand, nuclear reaction models - which at present account for these reactions poorly - will be investigated in order to provide a reliable basis for calculation of unmeasured reactions required to perform rigorous tests of GCR-ISM phenomena.

In this paper we first review the experimental techniques involved in these measurements. Next, we examine the question of LiBeB nucleosynthesis via the GCR-ISM mechanism.[1] Finally, we discuss reactions of importance to modifications of the GCR flux during transport and point out the implications of the results on nuclear reaction models.

II. EXPERIMENTAL MEASUREMENTS

In order to test models of nuclide formation in GCR-ISM interactions one must solve the production equation:

$$\frac{dN_L}{dt} = \sum_i N_i \int_{Eo} \sum_j^{\infty} \phi_j(E)\sigma_{ijL}(E)\ S_{ijL}(E)\ dE. \qquad (2)$$

Here, N_i is the abundance of each target nucleus in the ISM, $\phi_j(E)$ is the flux of a given GCR projectile, σ_{ijL} is the cross section for formation of nuclide L from the reaction i + j, and $S_{ijL}(E)$ contains the physical parameters of the model - e.g. stopping powers, medium thickness, etc. Both normal (i + j) and inverse (j + i) reactions must be considered. One then compares the formation rate with experimental values, or conversely, uses the data to unfold the flux function $\phi_j(E)$.

Several factors conspire to complicate the determination of reaction parameters for light-ion-induced reaction on light target nuclei, discussed in detail in Ref. 5. One of these problems is that the major part of the yield feeds stable isotopes; thus, radiochemical cross-section measurements are limited to a few species, although chemical separations coupled with conventional mass spectrometry has also proven useful for some cross section determinations[11] for stable nuclei. Another major difficulty is the very low kinetic energies which characterize much of the heavy recoil yield. This difficulty can be surmounted by the use of very thin targets and time-of-flight techniques which minimize degradation of the fragment energy spectrum.[10,12,13] Such techniques have been used to obtain all of the data reported here. An alternative approach is the use of inverse reactions (heavy projectile incident on light target).[14]

The experimental configuration employed in our studies is shown in Fig. 2. The basic concept involves the use of a pair of detector telescopes. The first of these (A arm) serves to define complete energy spectra and angular distributions for all isobars which make up the total reaction cross section, $d^2\sigma(A)/d\Omega dE$. This system uses a channel-plate fast-timing detector[15] (or a beam rf pickoff signal) as a time-of-flight (TOF) start signal and a thin, totally depleted silicon detector as a TOF stop and energy (or energy-loss) detector to identify

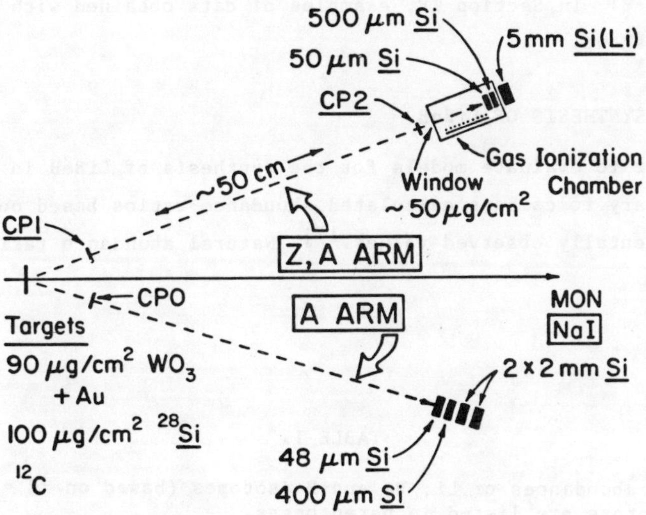

Fig. 2. Schematic diagram of experimental apparatus for nuclidic yield measurements.

all fragments with A > 6. Additional silicon or NaI detectors in the telescope identify all nuclides that penetrate the initial elements. Cross sections for all fragments with E/A > 0.05 MeV can be determined with this technique.

The A arm is supplemented by a second telescope (Z,A arm) which yields partial nuclide identification data $d^2\sigma(Z,A)/d\Omega dE$. This system utilizes two channel-plate devices for TOF purposes, followed by a low-pressure (~15 torr) gas-ionization ΔE detector and silicon detectors for Z identification via (dE/dx)-E techniques. Nuclides with E/A > 0.2 MeV can be identified with this arm; however, for targets as heavy as silicon, this system may fail to detect as much as 50 percent of the total yield for near-target recoils. In order to derive total nuclide (Z,A) cross sections, it is assumed that the charge distribution for a given isobar is independent of fragment energy. Hence, the A arm data, which detects the reaction cross section, can be corrected accordingly. For lighter systems (e.g. p + ^{16}O) this procedure has been shown to work rather well.[16] In Section IV, examples of data obtained with these systems are shown.

III. NUCLEOSYNTHESIS OF LiBeB

In order to evaluate models for the synthesis of LiBeB in nature, it is necessary to compare calculated abundance ratios based on Eq.(2) with experimentally observed values.[17] Natural abundance ratios (based

TABLE I.

Observed abundances of Li, Be and B isotopes (based on $^{1}H = 10^{12}$); Error factors are listed in parentheses.

	Meyer[18]	Austin[5]	Cameron[19]	Anders[20]
$^{6}Li/H$	76(2)	70(2)	180	160
$^{7}Li/H$	900(2)	900(2)	2100	2000
$^{9}Be/H$	14(1.6)	14(1.6)	45	29
$^{10}B/H$	30(2)	30(2)	70	180
$^{11}B/H$	120(2)	120(2)	270	710
$^{7}Li/^{6}Li$	12.6 ± 0.2	12.6 ± 0.2	12.6 ± 0.2	12.6 ± 0.2
$^{11}B/^{10}B$	4.05 ± 0.10	4.05 ± 0.10	4.05 ± 0.10	4.05 ± 0.10
$^{6}Li/^{9}Be$	5.7(3.2)	5.0(3.2)	3.7	5.7
$(^{10}B + ^{11}B)/^{9}Be$	14(3.2)	11(3.2)	7.5	31

on $^1H = 10^{12}$) are summarized in Table I for various recent compilations,[5,18-20] along with error factors in parentheses and the important isotopic ratios $^7Li/^6Li$ and $^{11}B/^{10}B$. It is noted that for LiBeB the abundance tables of Reeves[18] and Austin[5] are significantly lower than those of Cameron[19] and Anders and Ebihara[20], although both sets are consistent within experimental uncertainties. The primary differences arise from the predominant weight given to photospheric abundances in the former case, whereas in the latter compilations primary weight is given to meteoritic values.

Since we are concerned with light element production in global astrophysical environments, we take N_j to be interstellar medium abundances $_L$ [Cameron 1982]. In this work the cross section systematics, $\sigma_{ij}(E)$, of Ref. 9 have been incorporated into the calculations for the above reactions. These include all p and ^4He cross sections for both normal and inverse reactions with ^4He, ^{12}C, ^{13}C, ^{14}N and ^{16}O from threshold to the high-energy cross section saturation region. The range-energy relations of Northcliffe and Schilling[21] have been used as a basis for the determination of stopping powers. For production by galactic cosmic rays[1,22,23] we use spectra of the form

$$\phi_i(E) = \alpha_i(E + E_o)^{-2.6} \quad (3)$$

with $E_o = M_oc^2$ and the set of α_i normalized to the observed galactic cosmic ray proton flux[24] and $\phi_p = 12.5\ cm^{-2}\ sec^{-1}\ GeV^{-1}\ nucleon^{-1}$.

The energy loss term, $S_L(E)$, takes a form corresponding to the leaky-box model of GCR propagation.[25] For the purposes of this work we choose a value of 5 g/cm^2 which adequately reproduces the light element-to-CNO ratio observed in hgh energy cosmic rays.[22,23] Finally, in order to compare with absolute abundances, the rate equation (2) is multiplied by an appropriate time scale for nucleosynthesis; we have used $\Delta t \cong 10^{10}$ years for the galactic-cosmic-ray calculations. Implicit in these calculations is the assumption that cosmic-ray nucleosynthesis has been approximately constant as a function of time, after the effects of galactic chemical evolution and infall have been considered.

In Table II the results of calculations based on the galactic cosmic ray nucleosynthesis model are presented. Based on the analyses of GCR and ISM abundance errors and uncertainties in the nuclear reaction cross sections and the function $S_L(E)$, we estimate the errors in these calculations are about 30% for all isotopes.

As in earlier such comparisons[22,23], the present GCR nucleosynthesis calculations, which now include complete experimental cross sections, are found to reproduce successfully the observed abundances of ^{6}Li, ^{9}Be, ^{10}B and ^{11}B, within the quoted limits of error. It is further reassuring that the new calculations reproduce the absolute ^{9}Be abundance quite well, since this isotope is subject to the smallest uncertainties, both in terms of experimental and calculated values. However, ^{7}Li remains underproduced by nearly an order of magnitude and the $^{11}B/^{10}B$ ratio is also unsatisfactory. As a possible remedy for the noted discrepancies, it is reasonable to consider a low-energy component of the cosmic ray flux which generates the needed ^{11}B via the $^{14}N(p,\alpha)^{11}C(\beta^+)^{11}$ reaction and some ^{7}Li via $^4He(\alpha,p)^7Li$, while leaving the ^{6}Li, ^{9}Be, and ^{10}B abundances relatively undisturbed. In Table II the results of such a low-energy component for three different spectral indicies (γ = 3,5,7) of a power-law spectrum $E^{-\gamma}$ are combined with the GCR component and normalized so that $^{11}B/^{10}B = 4.05$. Thus our GCR flux function becomes:

$$\phi_i(E) = \alpha_i(E + E_o)^{-2.6} + f_i E^{-\gamma}, \quad (4)$$

where f_i is the normalization factor for the low-energy component. We have taken the flux ratios of the various low energy cosmic ray nuclei to be the same as their high energy counterparts. Since ^{4}He + ^{4}He reactions dominate the low-energy production of Li, the He/H ratio is particularly relevant to these studies, but not so important when viewed in comparison with observational uncertainties in light element abundances. Note also that such low-energy fluxes must not over-ionize the ISM at low energies.

TABLE II

Calculated abundances of LiBeB isotopes based on (1) a galactic cosmic ray model (GCR) and (2) a GCR model which includes a low energy flare component (GCR + $fE^{-\gamma}$), normalized to $^{11}B/^{10}B = 4.05$. Abundances are based on $^{1}H = 10^{12}$.

	GCR	GCR + E^{-3} (f = 1.05)	GCR + E^{-5} (f = 0.131)	GCR + E^{-7} (f = 0.028)
$^{6}Li/H$	69	380	120	83
$^{7}Li/H$	99	1000	350	180
$^{9}Be/H$	15	31	17	15
$^{10}B/H$	71	170	91	82
$^{11}B/H$	180	680	370	330
$^{7}Li/^{6}Li$	1.4	2.63	2.92	2.17
$^{11}B/^{10}B$	2.5	≡ 4.05	≡ 4.05	≡ 4.05

TABLE III

Calculated abundance of LiBeB isotopes based on production in a transported flare spectrum, $E^{-\gamma}$, where E is in energy/nucleon. All values are normalized to $^{9}Be \equiv 1$. Experimental values (expt) are based on Austin[5]

	expt	$\gamma = 3$	$\gamma = 5$	$\gamma = 7$
$^{6}Li/^{9}Be$	5.7	20	26	34
$^{7}Li^{9}Be$	64	57	130	180
^{9}Be (≡1)	1	1	1	1
$^{10}B/^{9}Be$	2.9	6.1	10	26
$^{11}B/^{9}Be$	11	32	99	360
$^{7}Li/^{6}Li$	12.6	2.9	4.8	5.2
$^{11}B/^{10}B$	4.05	5.2	9.9	14

From Table II, we see that a high-energy GCR model supplemented with a softer low-energy component with spectral index $\gamma \cong 7$ can reproduce the $^{11}B/^{10}B$ isotopic abundance ratio without overproducing ^{6}Li, ^{9}Be, ^{10}B, and ^{11}B. An additional virtue of this spectrum is the rather low total energy requirements placed on the accelerating environment in order to produce the additional ^{11}B. We conclude that GCR nucleosynthesis with a judiciously chosen low-energy component is a reasonable model for the production of ^{6}Li, ^{9}Be, ^{10}B, and ^{11}B observed in the galaxy. However, ^{7}Li cannot be produced in adequate amounts by this mechanism and another source is therefore required to account for the natural abundance of ^{7}Li.

Other plausible mechanisms involving energetic nuclear reactions may also account for LiBeB synthesis. Two cases which are representative of such mechanisms are synthesis in flare spectra[26] and in supernovae shock waves.[27] In Table III the calculated abundance ratios for flare source spectra are presented as a function of the power law index γ. The present results argue rather decidedly against a flare mechanism as a significant source of LiBeB synthesis. In all cases the calculated abundance ratios exceed the experimental values, with the discrepancies being particularly severe for large values of γ. Compared to previous estimates, a large overproduction of ^{11}B in the present calculations is attributed to computational improvements involving the use of very small energy steps in calculating the production integrals in the low-energy resonance regions of the excitation functions. This procedure particularly enhances the $^{14}N(p,\alpha)^{11}C(B^{+})^{11}B$ reaction, which has the lowest threshold energy of any of the principal reactions involved in LiBeB synthesis, as well as a very rich resonance structure. The best agreement is found for low values of γ, which, it should be noted, begin to approximate the GCR flux of Eq. (3). However, low values of γ imply an accelerating mechanism which utilizes a significant fraction of the total gravitational energy of the source. Under any circumstances the $^{7}Li/^{6}Li$ and $^{11}B/^{10}B$ ratios are unsatisfactory at all values of γ.

In summary, while the flare mechanism may be important in accounting for small anomalies in isotopic ratios (e.g. $^{11}B/^{10}B$), the present analysis appears to rule out this mechanism as a primary source of LiBeB synthesis.

Results of the shock wave abundance calculations are summarized in Table IV. The strong variations in these ratios demand that the energy of the shockwave be very finely tuned in order to reproduce the observed abundances, a constraint which would appear to limit the usefulness of this mechanism as a global model. Furthermore, even allowing for a source spectral shape defined precisely enough to satisfy the observed ratios, a severe limitation is imposed on this mechanism due to the tendency of such a shock wave to disperse by energy-loss when passing through any given medium. Thus, although LiBeB abundance ratios may serve as a useful monitor of shock wave processes (if such observations could be made), these calculations place very severe constraints on this mechanism as a global model for LiBeB synthesis. As in the case with flare spectra, this mechanism may be useful for second-order tuning of the isotope ratios, it is unlikely that it can account for the bulk of the LiBeB abundances.

In summary, the present results for the GCR + ISM calculation are able to account for the absolute abundances of ^{6}Li, ^{9}Be, ^{10}B and ^{11}B satisfactorily, reinforcing previous conclusions.[22,23] The only free parameter in this calculation is the time over which nucleosynthesis has occurred, which is taken to be 10^{10} years. On the other hand, the more-sensitive $^{7}Li/^{6}Li$ and $^{11}B/^{10}B$ isotope ratios are found to be well outside the experimental limits of error. The $^{11}B/^{10}B$ ratio can be explained by the existence of a low-energy component of the GCR transported flux of the form $\phi(E) \alpha E^{-\gamma}$ where $\gamma \approx 5$-7. Such a spectrum contributes the additional ^{11}B and a small amount of ^{7}Li, but does not change the abundances of the other isotopes. In addition, this flux places only minor energy demands on the accelerating source. The inability of the GCR + ISM mechanism to account for the $^{7}Li/^{6}Li$ ratio has long been recognized and some other source must be identified to account for the ^{7}Li in Nature; e.g. the Big Bang and/or the ^{4}He + ^{3}He reaction in certain phases of stellar burning.

TABLE IV

Calculated abundance of Li, Be, and B isotopes based on monoenergetic flux spectrum (in total energy E and E/A, MeV/nucleon). Nuclidic abundances are in arbitrary units.

E(MeV)	6	7	9	10	11	7/6	11/10	Li/Be	B/Be
12	2.5	.09	0	14.7	187.6	.04	12.8		
13	3.8	0.69	0	12.1	268	.18	22.2		
14	4.6	6.9	9.4	8.4	274	1.5	32.6	6.5	706
15	5.7	30	1.9	11.6	192	5.3	16.6	18.8	107
16	6.1	52	2.7	38.5	203	8.5	5.3	21.5	89
17	6.1	71.2	3.7	68.1	175	11.7	2.6	20.9	68
18	6.3	84.3	4.3	86.6	160	13.4	1.9	21.1	57
19	6.2	96.0	4.5	104	174	15.5	1.7	22.7	62
20	6.4	102	4.7	110	251	15.9	2.3	23.1	77
25	41.2	65.6	7.3	158	815	1.6	5.2	14.6	133
E/A									
7.5	8.5	.7	.16	42.9	589	.08	13.9	57.5	4006
8.0	13.2	1	0.5	45.6	470	.08	10.3	26.6	1031
8.5	18.8	1.8	1.1	45.5	230	.10	5.1	18.7	251
9.0	21.2	116	1.7	43.6	220	5.5	5.1	80.6	155
9.5	23.0	2165	3	42.5	363	94	8.5	729	35
10.0	26.4	30,000	4.6	47.4	330	1136	7.0	6527	82
10.5	35.1	13,000	7.6	67.8	478	370	7.1	1715	72
11.0	50.0	19,000	11.5	94.6	483	380	5.1	1657	50
11.5	2051	17,000	17.4	113.5	548	8.3	4.8	1095	38
12.0	2772	11,220	24.0	129.1	530	4.1	4.1	804	28
15.0	5675	10,600	40.0	178.2	527	1.9	3.0	407	17.6
20.0	5698	10,680	51.0	269.1	513	1.9	1.9	321	15.3
expt.	-	-	-	-		12.6	4.05	69	14

IV. MODIFICATIONS OF THE GCR FLUX

Measurements of the isotopic composition of heavy galactic cosmic rays[2] have demonstrated clear differences with those observed in the interstellar medium. For example cosmic ray ratios for ^{22}Ne:^{21}Ne:^{20}Ne are found to be 0.67:0.25:1 compared to 0.12:0.003:1 in the ISM; for Mg one obtains GCR ratios for ^{26}Mg:^{25}Mg:^{24}Mg of 0.30:0.28:1 and 0.14:0.13:1 in the ISM. In order to relate these values to the composition of the primary GCR source, corrections must be made for nuclear reactions which occur during transport through the ISM. As discussed earlier, these corrections are strongly dependent on cross sections for nuclear reactions between heavy cosmic rays and the H and He of the ISM.

While extensive data exist for the ^{4}He + ^{4}He , p + ^{12}C and ^{4}He + ^{12}C reactions, the situation for heavier targets is considerably more fragmentary. Using the techniques described in Section II, we have recently studied the p + ^{16}O, p + ^{27}Al, p + ^{28}Si and ^{4}He + ^{28}Si reactions between 50 and 200 MeV, an energy region of primary relevance to GCR + ISM studies. In this section, we first describe results for the p + ^{27}Al rection,[10] which are representative of reactions with ^{28}Si as well. These reactions should serve as important indicators of nuclidic modifications experienced by heavy cosmic rays during transport through the ISM. Comparison of these data with theoretical calculations provides an important barometer of the reliability of these models for more general use in treating GCR source modifications. In addition, very recent data on the 100-MeV + ^{16}O reaction are presented.

In Fig. 3 isobaric cross sections for the 180 MeV p + ^{27}Al system are shown. It is observed that the yields are spread broadly over the entire range of possible products, with a large contribution from relatively light (A < 20) fragments. Peak yields are found to be in the A = 20-22 region, hence ^{28}Si spallation serves perimarily to enrich Ne isotopes. Under the assumption that nucleon removal is correlated with energy deposition, the results imply that relatively large amounts of energy must be transferred to the struck nucleus in these collisions. Also shown in Fig. 3 are yield predictions based on the intranuclear-cascade-plus-evaporation code,[28] preequilibrium-hybrid-

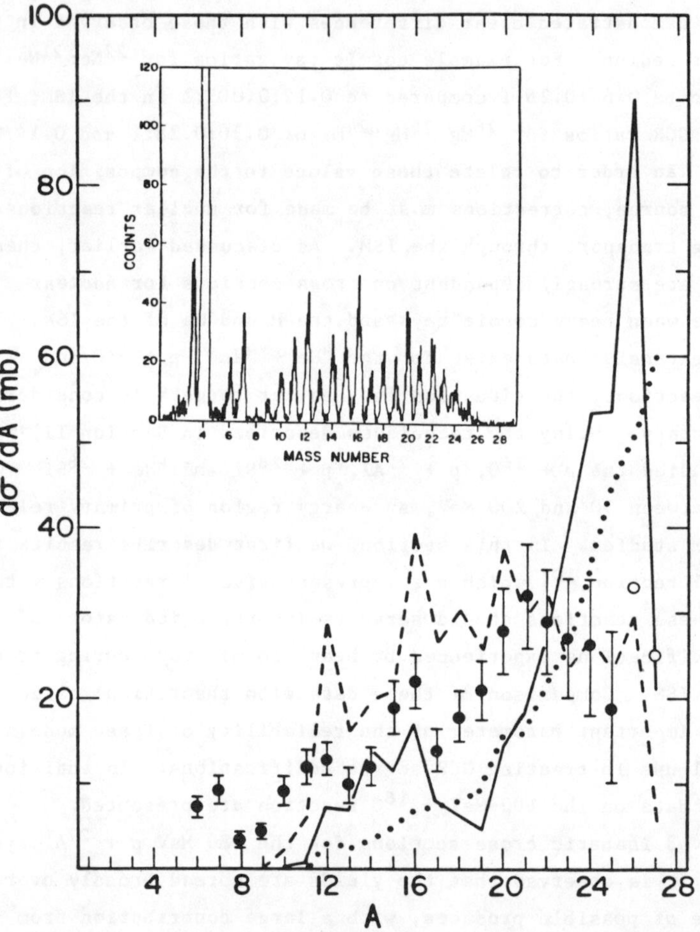

Fig. 3. Isobaric cross sections for fragments from the 180-MeV proton + ^{27}Al reaction. Inset: On-line mass spectrum for fragments with E > 10 MeV emitted at 20 deg. Symbols are as follows: closed circles, these data; open circles, data from Ref. 33; solid line, intranuclear-cascade- and -evaporation code of Ref. 28; dashed line preequilibrium exciton model of Ref. 29; and dotted line, semiempirical calculation of Ref. 30.

model calculations,[29] and the semiempirical estimates of Silberberg and Tsao.[30] The semiempirical values clearly underestimate the probability for high-energy-deposition events leading to light-fragment production. This shortcoming may significantly influence current calculations of cosmic-ray-related phenomena,[3,4,7] which depend upon these calculated cross sections.

The intranuclear cascade calculations also overpredict the yields of products near the target; this is due to the relatively low excitation energies of residual heavy nuclei formed in simple quasifree scattering processes. The experimental results suggest a much broader distribution of energy deposition in which the initial collision step is more effective in transferring linear momentum and excitation energy than predicted by the model. Specifically, the cascade code overpredicts the probability for few-nucleon removal processes (A = 23-27) by about 30%, while failing to account for this same amount of cross section for lighter fragments (A = 6-22). Thus, it would appear from these comparisons that, in addition to the basic nucleon-nucleon aspects of these collisions, as represented by the cascade code, some additional mechanism for energy deposition must also be present.

On the other hand, agreement with the preequilibrium-hybrid-plus-evaporation predictions[28] is relatively good, suggesting that the energy sink associated with particle-hole excitations in this model is accounting for energy deposition in the initial stages of the nucleon-nucleon cascade more successfully. In this respect we note that, whereas for heavier targets such as Ni the cascade and preequilibrium exciton models predict rather similar product mass distributions, for ^{27}Al the predictions are distinctly different. This behavior is explained in terms of the diffuse nature of the nucleon distribution in ^{27}Al relative to heavier targets, which leads to a significantly lower probability for multiple collision steps in the cascade in ^{27}Al. Since multiple collisions also serve to enhance energy deposition, signatures of the initial collision step are masked in heavier target nuclei.

Representative angular distributions are shown in Fig. 4. Here one observes a trend from forward-peaked yields for the lightest fragments to two-component distributions for fragments a few nucleons removed from the target. The two components presumably arise from (1) high-linear-momentum-transfer collisions which give rise to a forward component and (2) peripheral processes such as (p,p') and (p,pn) in which the light ions are strongly forward peaked, leading to excited recoiling heavy fragments which are focussed in the laboratory system near 50-70 deg. The strong forward-peaking of the angular distribution again suggests that the initial nucleon-nucleon collision step in these reactions involves a significant amount of energy deposition in the struck nucleus. On comparison with predictions of the cascade model, qualitative agreement between experiment and theory is found for the backward-angle component ($\theta > 50$ deg) of the data, whereas a marked divergence is evident at forward angles. The observation implies that, although the cascade model may describe peripheral, quasifree processes relatively well, it exhibits a deficiency in accounting for the strength of high-linear-momentum-transfer events leading to forward-peaked recoil nuclei.

In Fig. 5, representative energy spectra for A=7, 16, 22 fragments at angles of 20, 40, and 70 deg are shown. The spectra are characterized by a broad peak at low energies which demonstrates the low energies that are detected in our counters. Higher energy fragments are characterized by an exponential decrease, extending up to relatively high energies. The slopes of the exponential tails become systematically steeper as a function of both increasing fragment mass and angle. The cascade calculations yield energy spectra with spectral shapes that describe the data relatively well. However, because of the high yields of light fragments and the forward peaking of the angular distributions, the angle-integrated energy spectra for the data contain a much larger fraction of high-energy fragments than predicted by the cascade model.

In summary, these results demonstrate that for light-ion interactions with light target nuclei ($A < 40$), cascade and semiempirical cross sections tend to overpredict quasi-free processes.

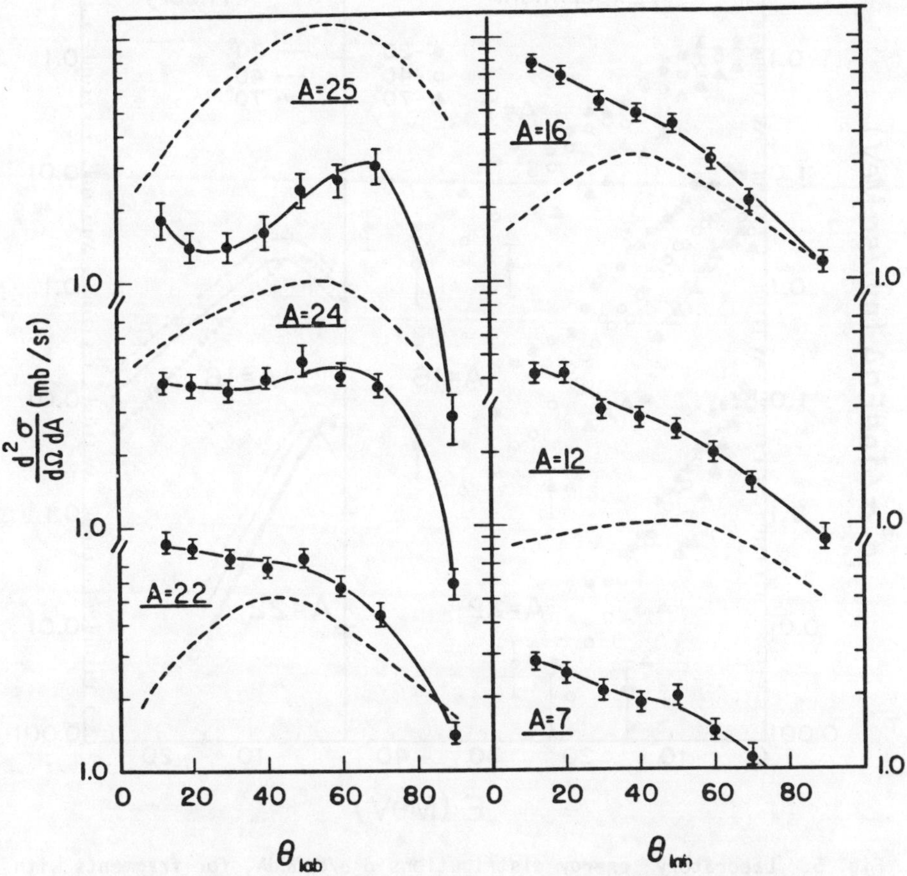

Fig. 4. Fragment angular distributions for the masses indicated, $d^2\sigma/d\Omega dA$, from the 180-MeV p + ^{27}Al reaction. Dashed lines are predictions of the intranuclear-cascade-plus-evaporation code; solid lines represent a smooth average of the data (closed circles). The calculations predict no significant yield for A = 7.

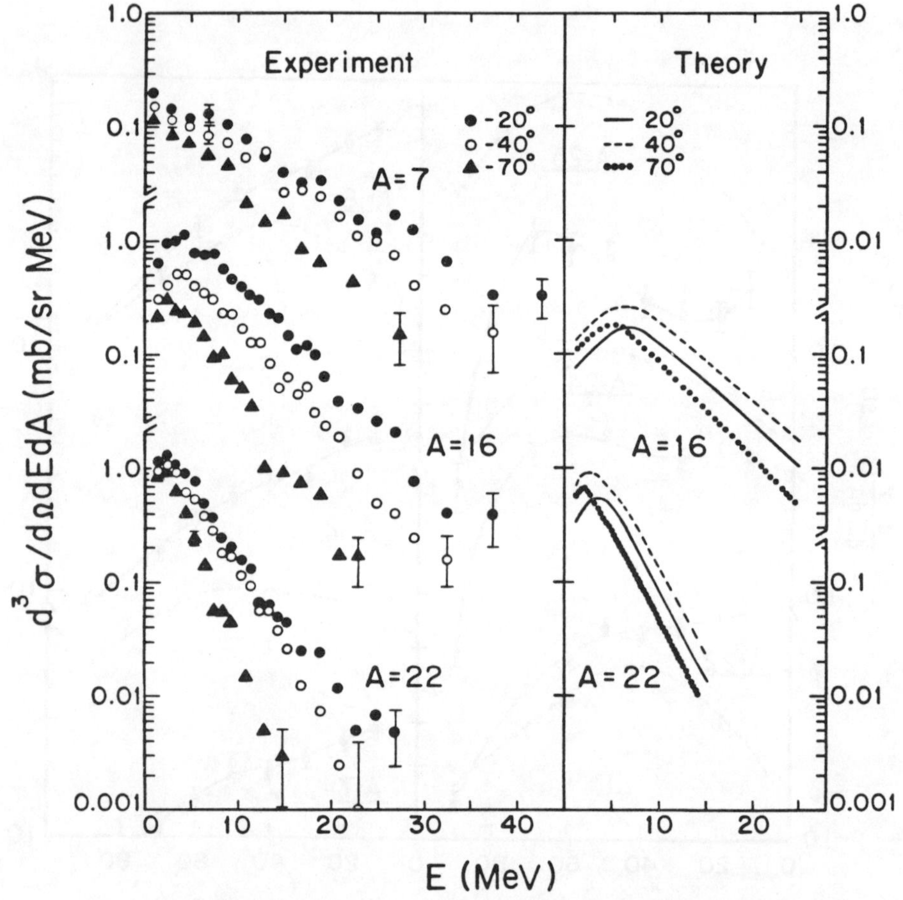

Fig. 5. Laboratory energy distributions $d^3\sigma/d\Omega dEdA$, for fragments with A = 7, 16, and 22 in the 180-MeV p + ^{27}Al reaction. Points represent data obtained at laboratory angles of 20 (closed circles), 40 (open circles), and 70 (triangles) deg in the laboratory system. Lines are predictions of the intranuclear-cascade-plus-evaporation code.

While pre-equilibrium models reproduce the cross sections more satisfactorily, such models do not yield angular distributions and energy spectra. It is hoped that Boltzmann-Uehling-Uhlenbeck calculations,[31] which contain both mean-field and nucleon-nucleon effects, will eventually provide a reliable framework for calculating these reaction probabilities.

Finally, in Table V, we present cross sections for the yields of A = 6-15 isobars in the p + ^{16}O reaction at 100 MeV. These data are in good agreement with higher energy studies at the LBL Bevalac using the inverse reaction technique.[32] However, the A = 15 cross section of 59 nb is substantially smaller than that expected in the basis of radiochemical studies. We are presently investigating possible sources of this discrepancy. The agreement of our cross sections with previously measured values for lower A, the quality of our low-energy spectra for A = 15, and the agreement of our total cross section (σ_R = 320 nb) with other data serve to support our A = 15 measurement.

TABLE V

Isobaric yields from the 100 MeV p + ^{16}O reaction

A	6	7	8	9	10	11	12	13	14	15
σ(mb)	15	20	5	6	19	46	58	44	58	59

V. SUMMARY

In conclusion, reactions between galactic rays and the interstellar medium have a significant impact on the origin and distribution of the elements. The synthesis of ^{6}Li, ^{9}Be, ^{10}B and ^{11}B can be accounted for well by such a model; if a low-energy component exists in the GCR flux, then the model is remarkably successful. On the other hand, ^{7}Li is underproduced by an order of magnitude in GCR and ISM interactions and must have another source.

Reaction studies of the p + ^{27}Al and p + ^{28}Si systems demonstrate that current nuclear reaction models encounter significant deficiencies in accounting for the cross sections of heavy cosmic rays with the ISM. Most successful among these models are pre-equilibrium calculations, which better account for the greater energy dissipation and reduced quasi-free scattering observed in these reactions. In addition, new data for A = 6-15 isobars formed in the 100-MeV p + ^{16}O reaction, which exhibit a nuclidic yield distribution analogous to the heavier targets.

The authors wish to acknowledge their many collaborators in various phases of this work: these include G.J. Mathews at Lawrence Livermore National Laboratory, S.H. Zhou, T.P. Walker, W.G. Wilson, D. Bonser, M. Fatyga, D.E. Fields, T.E. Ward, and R.W. Viola at Indiana University; and H. Breuer and A.C. Mignerey at the University of Maryland. This work was supported by the U.S. Department of Energy, IBM Corporation and the National Science Foundation.

REFERENCES

1. Reeves, H., Fowler, W.A., and Hoyle, F., Nature 226, 727 (1970).
2. Wiedenbeck, M.E. and Greiner, D.E. Phys. Rev. Lett. 46, 682 (1981).
3. Jackson, D.F., Nucl. Phys. A354, 237 (1981).
4. McNulty, P.J., Physics Today 36, 9 (1983).
5. Austin, Sam M., Prog. Part. and Nucl. Phys. I, 1 (1981).
6. Waddington, C.J., these proceedings and references therein.
7. Mewaldt, R.A. et al., Ap.J. Lett. 235, L95 (1980).
8. Wiedenbeck, M.E. and Greiner, D.E., Ap.J. 247, L119 (1981).
9. Read, S.M. and Viola, V.E., At. and Nucl. Data Tables 31, 359 (1984).
10. Kwiatkowski, K. et al., Phys. Rev. Lett. 21, 1648 (1983).

11. Raisbeck, G.M., Lestringuez, J. and Yiou, F., Phys. Rev. C6, 685 (1972); Phys. Lett. B57, 186 (1975).

12. Davids, C.N., Laumer, H.W. and Austin, Sam M., Phys. Rev. C1, 270 (1970).

13. Jacobs, W.W. et al., Phys. Rev. C9, 2134 (1974).

14. Lindstrom, P. et al., Lawrence Berkeley Laboratory Report 3650 (1975); Greiner, D.E. et al., Phys. Rev. Lett. 35, 152 (1975).

15. Kwiatkowski, K., et al., Nucl. Instr. Meth. 225, 65 (1984).

16. Moyle, R.A., et al., Phys. Rev. C 19, 631 (1978).

17. Walker, T.P., et al, Ap.J. 229, 745 (1985).

18. Reeves, H. and Meyer, J.-P., Ap. J. 226, 613 (1978).

19. Cameron, A.G.W., Essays in Astrophys. (Editors: C. Barnes, D. Clayton, and D. Schramm) 35 (1982).

20. Anders, E. and Ebihara, M., Geochim and Cosmochim Acta 46, 2363 (1982).

21. Northcliffe, L. and Schilling, R.F. At. and Nucl. Data Tables, 7A, 233 (1970).

22. Mitler, H.E., Astrophys. Space Sci. 17, 186 (1972).

23. Meneguzzi, M. et al., Astron. Astrophys. 15, 337 (1971).

24. Gloeckler, G. and Jokipii, J.R., Ap. J. 148, 141 (1967).

25. Cesarsky, C.J., Ann. Rev. Astron., Ap. 18, 289 (1980).

26. Fowler, W.A. et al., Geophys. J. Royal Astron. Soc. 6, 148 (1962).

27. Colgate, S.A., Ap. J. 181, L53 (1973).

28. Mathews, G.J. et al., Phys. Rev. C 25, 2181 (1982).

29. Blann, M. Ann. Rev. Nucl. Sci. 25, 123 (1975).

30. Silberberg, R., and Tsao, C.H., Ap. J. 25, 315 (1973).

31. Kruse, H. et al., Phys. Rev. Lett. 54, 284 (1985).

32. Mitchell, J. and Wefel, J.P., private communication.

33. Artun, O., Phys. Rev. Lett. 35, 773 (1975).

THE ABUNDANCES OF THE ELEMENTS AS OBSERVED IN THE COSMIC RADIATION

C. Jake Waddington
School of Physics and Astronomy, University of Minnesota
Minneapolis, Mn. 55455, USA

ABSTRACT

The cosmic radiation consists of energetic nuclei of all the elements in the periodic table, ranging from hydrogen to uranium. The relative abundances are roughly similar to those deduced for other samples of undifferentiated cosmic matter, but there are significant differences. Some of these are introduced while the particles travel through interstellar matter from the source of the cosmic rays to us, while others appear to reflect the influence of a selection process occuring during the acceleration to cosmic ray energies. However, even when these effects are taken into account, there still appear to be some significant differences between the abundances in the source of the cosmic rays and those typical of the solar system or normal main sequence stars. Since the matter in the cosmic rays represents the only sample of extra-solar matter that we can observe directly the observation of the relative elemental abundances provides unique information. This talk will discuss some of the recent measurements of these abundances and describe how they can be interpretated.

1. INTRODUCTION

Space is filled with a thin gas of extremely energetic particles moving with velocities close to that of light. These relativistic particles are mostly atomic nuclei which have been fully stripped of their electrons and they make up the "cosmic radiation" that is the subject of this talk. This gas has a very low density, with only about one particle per thousand cubic meters, but an energy density in the galaxy that at 1 eV/cm^3 is comparable to that of the visible star light, or of the magnetic fields, or of the turbulent gas motions. As a result the cosmic ray particles play a vital role in the dynamics of the galaxy and their existence has to be taken into account in any discussion of the properties of galaxies.

Here, however, we are mainly concerned with the nature of the particles and what they can tell us about the sources from which they come. The fact that these particles are travelling with velocities close to that of light implies that they can have come from very far away. In practice the charged nature of the particles means that they are deflected by the magnetic fields that pervade space and as a consequence most of them can hardly have travelled inter-galactic distances. Instead we believe that they are mostly confined within galaxies and that those we observe originated within our own galaxy, travelling on average for a few tens of millions of years through the interstellar medium before being detected or escaping from the galaxy. Those few particles with the very highest energies which cannot be significantly deflected by the magnetic fields may be of extra-galactic origin. Nevertheless, the cosmic ray particles clearly have come from much more distant regions than any other sample of matter which we can hope to examine directly. The elemental and isotopic composition of these particles allows us to study matter that has had a very different history from the matter that we see around us in the solar system and its neighborhood. Unfortunately there are both experimental and interpretative difficulties in observing and understanding the abundances of these particles, and we still cannot fully describe the source regions, although observations over the past few years on the heavier nuclei have greatly improved our understanding of the probable nature of the source.

2. Experimental Observations

Experimentally there are a number of problems that have to be solved in order to measure the intensities of the different types of particle in the cosmic rays. Firstly, in spite of the great energies of these particles, they are unable to penetrate the atmosphere of the earth without undergoing nuclear interactions which destroy them. Consequently observations must be made in space or at least near the top of the atmosphere. Secondly, the particles are detected one at a time, which means that the charge, Z, and, if possible, the mass, A, have to be measured on individual nuclei. Thirdly, the intensity of these particles is low, and in order to observe significant numbers in a reasonable time the detectors have to be made large, but without sacrificing the ability to resolve Z and A. Finally, there is the added complication that the abundances are energy dependent, and hence it is necessary to determine them as a function of the energy, even though the intensities fall off rapidly with increasing energy. Because of these difficulties we still can only resolve the individual elements over a rather small range of energies, and not over the entire periodic table. Over the energy range of 10^7 to 10^{20} eV covered by cosmic ray particles, it is only possible to resolve charges of $Z \leq 60$ for $E \leq 10^{12}$ eV. In general these limitations are because of the lack of intensity at higher energies and higher charges.

It is reasonable to expect, and proves to be generally true, that the composition of the cosmic rays should be quite similar to that of the solar system, at least at moderate (0.1 to 10 GeV per nucleon) energies. As a consequence one can expect to observe the same large intensity variations with atomic number that are characteristic of the solar system abundances. Fig. 1 shows the elemental abundances derived by Anders and Ebihara[1], plotted on a strongly logarithmic scale. It can be seen that hydrogen and helium are overwhelmingly predominant, carbon to nickel have roughly similar abundances, and the heavier elements are much rarer. The same is true in the cosmic rays, where hydrogen is 90% and helium is 9% of the particles, and all the heavier nuclei make up only some 1%. The elements heavier than iron-nickel make up less than 0.001% of the particles. Individual observations therefore tend to be generally restricted to just one of these groups of elements and to a moderate range of energies. At higher energies the numbers of particles are much smaller, due to the rapidly falling energy spectra, and little is known about the details of the relative abundances, apart from the fact that they are probably significantly different from those at lower energies. Since in this paper we are mainly concerned with the abundances that we can deduce at the source, the energy dependent variations in the abundances are only of interest in so far as they allow us to understand the effects of propagation through the interstellar medium on the source abundances. Nevertheless, care has to be taken when comparing different measurements of abundances to take account of the energies at which these measurements are made

2.1 Hydrogen and Helium

Hydrogen, and to a lesser extent helium, are the dominant elements in the cosmic radiation, they are numerically by far the most abundant and they carry most, although not all, of the energy density. At low energy, less than a few times 10^9 eV, the stable isotopes can be resolved and the relative abundances studied, but at higher energies only the elemental abundances can be measured, while at the highest energies, above 1 PeV (10^{15} eV) the particles are so rare that they cannot be observed directly and even element resolution is not possible. Figs. 2a[2] and 2b[3] show some of the values of the abundance ratios $^2H/^4He$ and $^3He/^4He$ that have been reported at the relatively low energies where mass resolution has been achieved. These abundances reflect the influence of the propagation processes and do not represent those at the source.

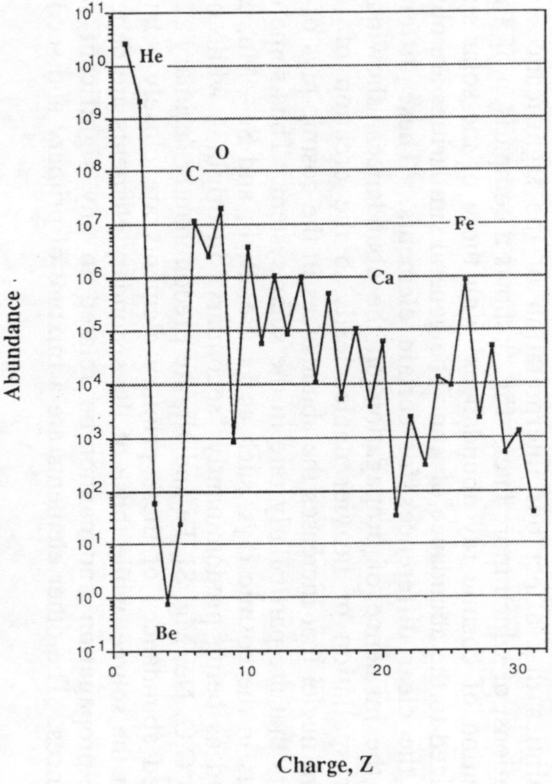

Fig. 1 The elemental abundances of solar system material, as determined by Anders and Ebihara[1].

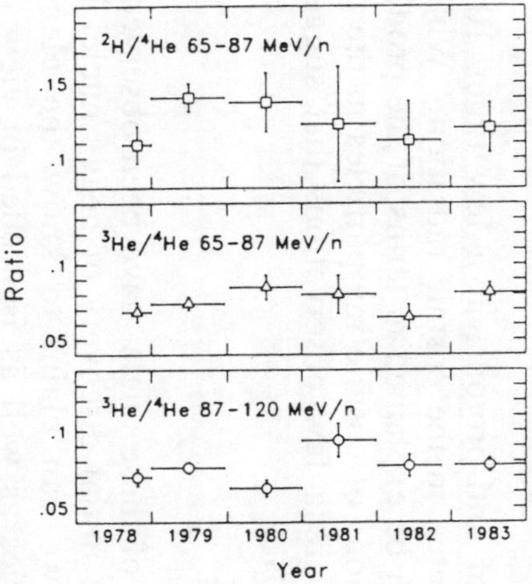

Fig. 2a Isotopic ratios for hydrogen and helium as a function of time[2].

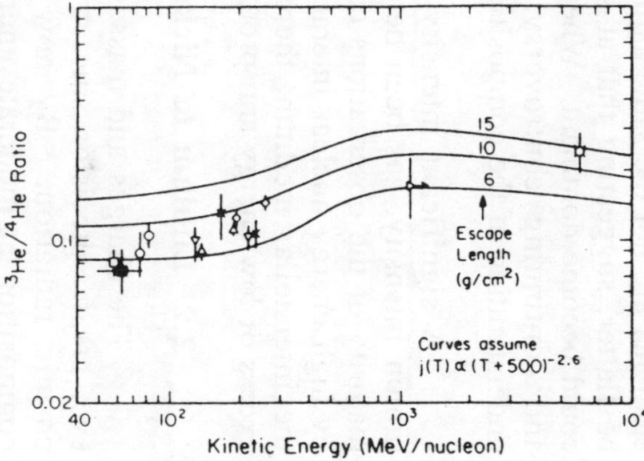

Fig. 2b Ratio of ^3He to ^4He as a function of energy[3].

At higher energies, although mass resolution is not generally possible we can still readily resolve the elements. It is found that the relative abundance of Hydrogen and Helium is approximately constant at least up to 1 TeV, with energy spectra that are power laws in kinetic energy having a slope of $\approx$ - 2.6. The energy spectra of the heavier elements are not the same as those of Hydrogen and Helium, but seem to be flatter, suggesting that at still higher energies these heavy elements could become dominant. Whether in fact they do is the subject of a long and continuing controversy, which, however, does not affect our understanding of the composition at the source.

A significant intensity of anti-protons, at a few times 10^{-4} the proton intensity, has been detected in the cosmic radiation. While the majority of the observations can be explained in terms of the production by high energy nuclear interactions of cosmic ray particles in the gas of the interstellar medium, there are a few observations that suggest an excess of low energy anti-protons.

2.2 Lithium to Nickel

The charges and masses of these nuclei have been observed with ever greater precision since the initial discovery of heavy nuclei in the cosmic radiation. By now we can claim to know the elemental composition at moderate energies as well as is needed in view of the uncertainties that affect our interpretation of the significance of the observations, of which more later. Fig. 3 shows a comparison of a recent compilation of cosmic ray abundances[4] with those of the solar system, normalized to the abundance of iron The general similarities are obvious, as are the clear differences for certain elements. These differences reflect the influence of propagation on the abundances, showing how nuclear spallation of heavier nuclei leads to the creation of lighter fragment nuclei that increases the abundances in the cosmic rays of those elements that are particularly rare in the solar system. Thus some of the elements in the cosmic rays, such as Li, Be, B; F; and Sc - Mn, can be regarded as being predominantly "secondary" in origin, , while others, such as C,O, Ne, Mg, Si, Fe and Ni, are predominantly "primary", . The observed abundances of these primary elements are closely related to those at the source, while those of the secondary elements are dependent on the propagation and cannot be related to any significant source abundances. The other elements are a mixture of primary and secondary

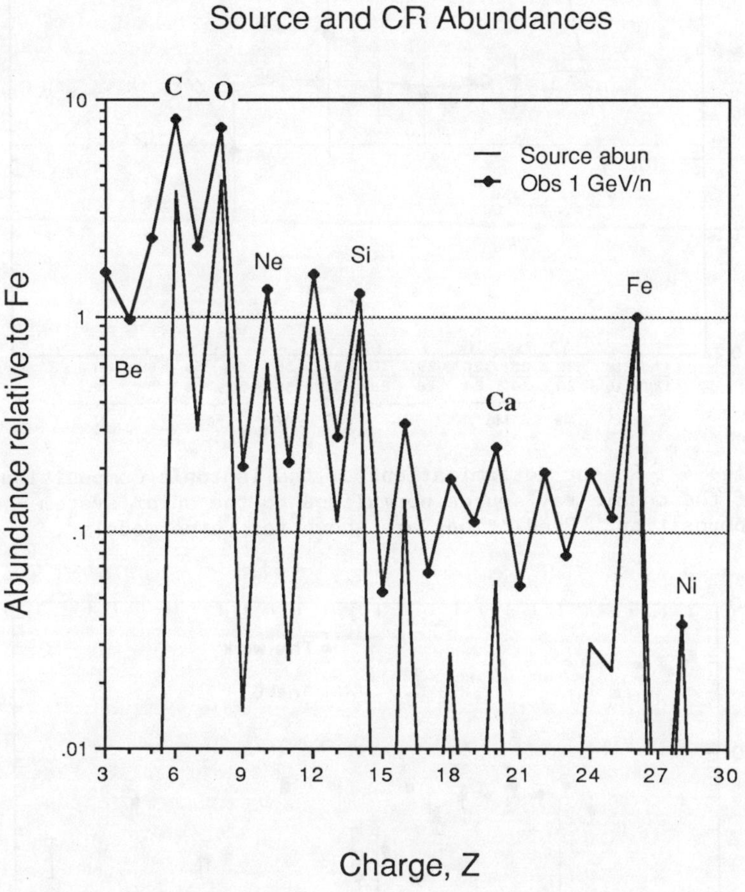

Fig. 3 A comparison of cosmic ray abundances[4] with those in the solar system, normalized to the abundance of iron, $_{26}Fe$.

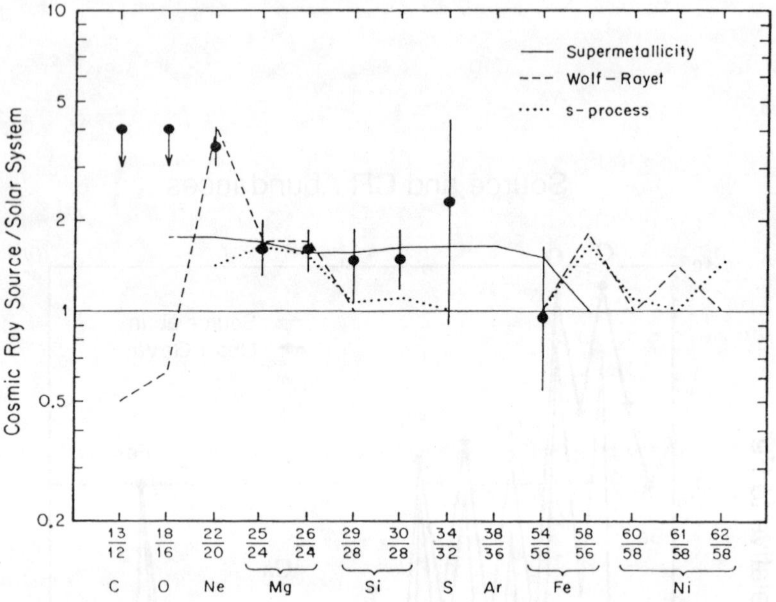

Fig. 4 Recent determinations of the isotopic composition of the cosmic ray source normalized to the solar system composition. Predictions are shown for three models[5].

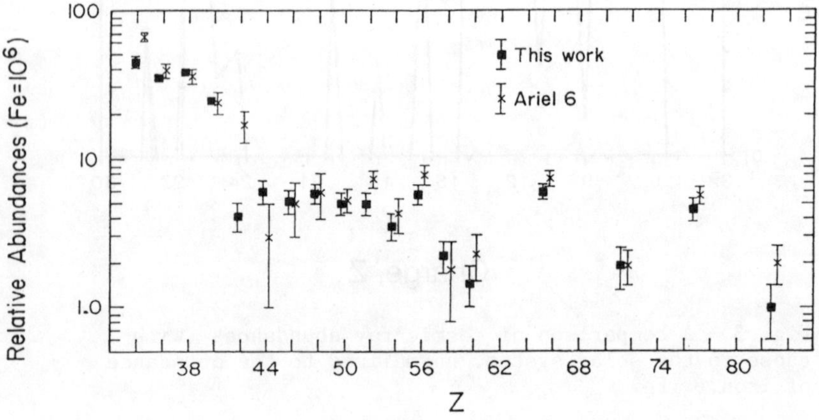

Fig. 6 Comparison between the HEAO-3 and Ariel 6 abundances.

components and any deduced source abundances are very sensitively dependent on the details of the corrections for secondary production.

At moderate energies (≈ 1 GeV/nucleon) a number of the isotopes of the more abundant elements can be resolved. Some of these show clear evidence for an excess of the neutron rich isotopes compared with those observed in the solar system. Fig. 4 shows[5] a recent compilation which illustrates the large excess of ^{22}Ne and the smaller, but still significant excesses of ^{25}Mg, ^{26}Mg, ^{29}Si and ^{30}Si. Various models have proposed to explain these excesses, but none has been totally satisfactory; also there is still some uncertainty regarding the validity and coherence of the observational data, which makes it premature to draw too strong conclusions. This is a subject that urgently needs more good observations, particularly in the iron region.

2.3 Ultra Heavy Nuclei

These UH-nuclei are so rare that their very presence in the cosmic radiation was not confirmed for twenty years after the discovery of the presence of heavy nuclei, and observations of significant weight were not made before the mid 1970's, using detectors flown on balloons or exposed on SpaceLab. Only in 1979 were two satellite detectors flown to look specifically at these nuclei, HEAO - 3 and Ariel-6, and even these lacked the charge resolution or collecting power to do more than establish the broad features of the composition. For example the total number of individual actinide nuclei observed by these two detectors over several years of exposure, during which some 2.10^7 iron nuclei were detected, is between 3 and 5, a number which is clearly inadequate for much more than a proof of existence. Nevertheless, these UH- elements are particularly interesting when studying the cosmic ray source since they should have abundances that would be particularly sensitive to the conditions present in the source. In particular the great majority of these nuclei are believed to be synthesized in the conceptually simple process of neutron addition, either rapidly or slowly, via the r- or s-processes. Solar system material is a mixture of r- and s- process nuclides, so the question that confronts the cosmic ray abundances is whether they contain a similar mixture, or whether they show an enhancement due to one of other of these process, reflecting the very different histories of these samples of matter.

Fig. 5 Composition of heavy cosmic ray nuclei as measured by the HEAO-3 detector for various charge ranges.

a) For 31 ≦ Z ≦ 44;

b) For 40 ≦ Z ≦ 62;

c) For 60 ≦ Z ≦ 94

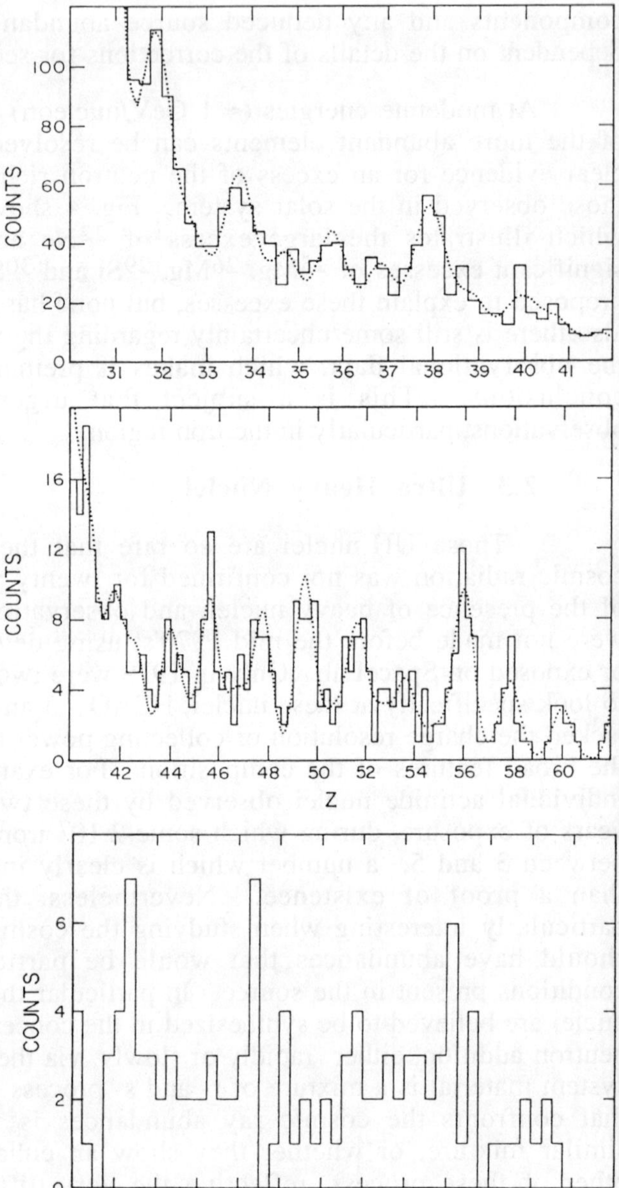

The results from the HEAO-3 detector show the best charge resolution obtained thus far. Fig. 5 shows[6] the composition observed by HEAO-3 in various parts of the charge range covered. It can be seen that as the charges increase, and the intensities decrease, the resolution and the statistics deteriorate. The generally good agreement with the Ariel-6 results[7], which have lower resolution but similar statistical weight, is illustrated in figure 6, where it should be noted the abundances are plotted relative to 10^6 iron nuclei, illustrating just how rare are these UH-nuclei.

3. DISCUSSION

The interpretative difficulties in understanding the abundances come from the problems of relating the observed abundances to those that must have originally been present in the source region. As the particles propagate through the interstellar medium the abundances are changed by the nuclear interactions that occur. The primary nuclei emitted from the source may interact with the gas and dust in the interstellar medium and either be destroyed, or more probably, fragmented into lighter secondary nuclei, thus changing the observed abundances. Any estimate of the abundances at the source must involve the application of a correction for the effects of propagation being applied to the observed abundances. A modern propagation model[8] applied to a charge distribution which at the source is similar to that of the solar system will reproduce the observed cosmic ray spectrum rather well (see Fig.3) All the "secondary" elements and the relative rare odd charged elements are predicted to have enhanced abundances in the cosmic rays. It is these secondary elements that provide the most sensitive tests of the models of propagation and have led to the development of the models that we now generally accept.

In the simplest form of these models the cosmic ray nuclei are assumed to be injected from the sources into the interstellar medium and then diffuse through the galaxy as a consequence of resonant scattering by hydromagnetic waves and interactions with the local magnetic fields[9]. The nuclei then face loss by two competing mechanisms, nuclear interaction and leakage from the galaxy. This model, the "leaky box" model, appears to fit the data rather well, and although many more complicated variants have been tried, none appear to be markedly better at describing the observations, although one will find considerable discussion on "double" leaky boxes; truncated path length distributions

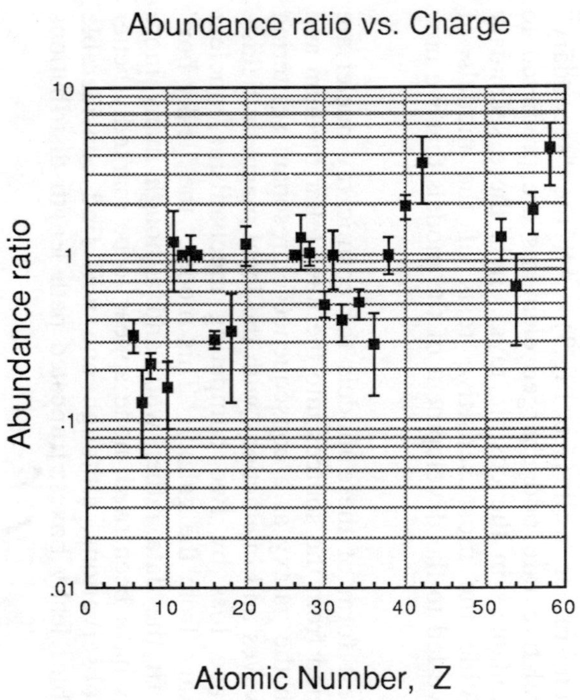

Fig. 7 Ratio of cosmic ray elemental abundances propagated back to the source to those in the solar system as a function of the atomic number.

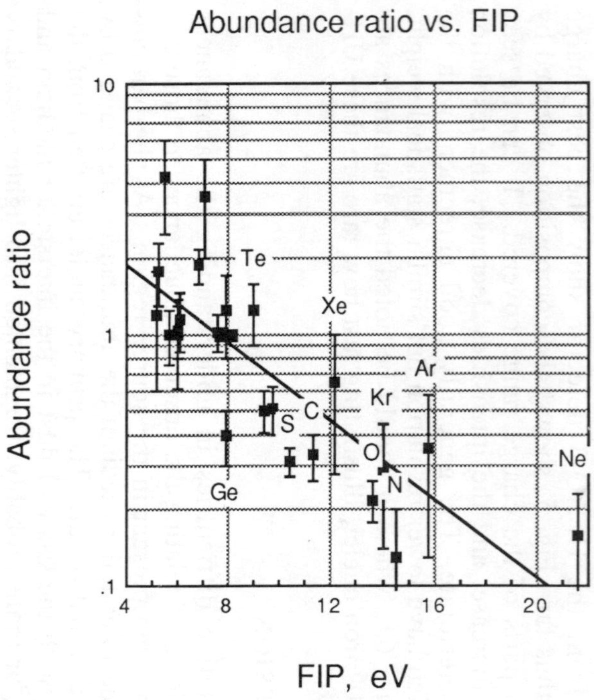

Fig. 8 Ratio of cosmic ray elemental abundances propagated back to the source of those in the solar system as a function of the first ionization potential of each element.

Fig. 9 Agreement between the deduced abundances of the UH nuclei from the HEAO-3 observations and those calculated from the solar system abundances and a step FIP dependence.

Fig. 10 Abundance ratio of lead-like nuclei to platinum-like nuclei as observed by HEAO and Ariel compared with the predictions of various models.

Fig. 11 The ratio of actinides to lead plus platinum nuclei compared with solar system, s- and r-process abundances.

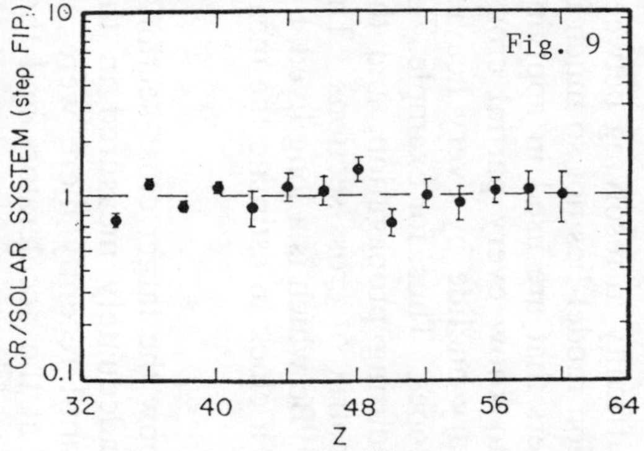

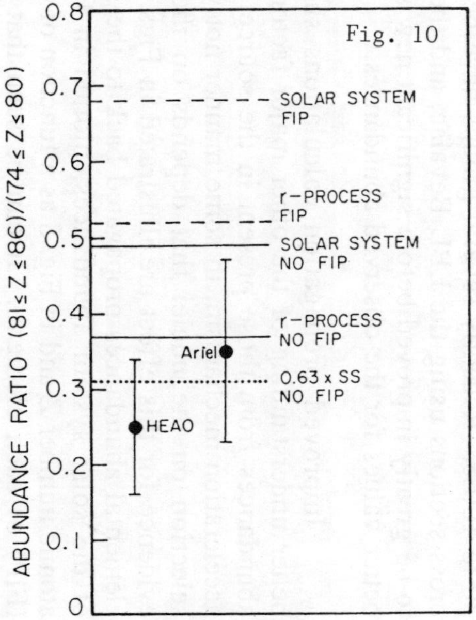

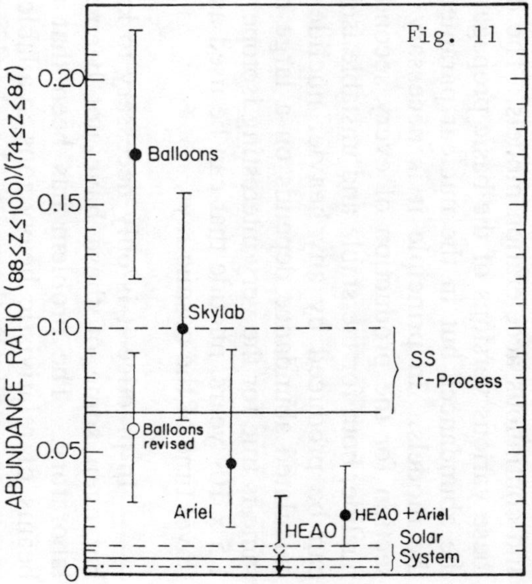

and continuous acceleration models. The difficulty in resolving between these various versions of the basic propagation model lies not so much in the abundances but in the nuclear parameters that are used in applying the models. In principle it is necessary to know every partial cross-section for the production of every secondary nuclide by every heavier nuclide, both for the stable and unstable isotopes. Thus, for example, ^{7}Li can be produced by any heavier nuclide during propagation, and the calculated abundance depends on a large matrix of cross-sections. The same is true for the very interesting isotope ^{10}Be which is a long lived, t = 1.5 x 10^6 years, nuclide that can be used as a clock to estimate the mean travel time of the cosmic rays.

In practice it is only necessary to know the larger cross-sections, but even some of these have not been adequately measured in the laboratory. The problem has been that until recently there were no beams of relativistic heavy ions available at the accelerators and that measurements from the proton bombardment of heavy targets have been restricted to determining the yields of a few radio active fragments. As a consequence most propagation calculations have been dependent on the use of semi-empirical relations that combine some nuclear theory and the few measured cross-sections to fit multiparametric expressions to calculate the unknown cross-sections. These expressions have been generally found to agree with newer measurements only to within some 30 to 100%. Recently several groups have begun to measure the needed cross-sections using the LBL Bevalac, and the situation can be expected to be greatly improved before significant new space missions can produce better values for the observed abundances.

Improved propagation calculations should allow us to reach a better understanding of the other major factor that distorts the observed abundances from those present in the source. It appears likely that the acceleration mechanism, in some manner not well understood, imposes a selection on the nuclei that depends on the atomic properties. The evidence for this effect are illustrated in Figs. 7 and 8, which shows the elemental abundances propagated back to the source expressed as a ratio of the solar system abundances, plotted in Fig 7 as a function of the atomic number Z and in Fig. 8 as a function of the first ionization potential (FIP) of each element.. It can be seen that although these abundance ratios show a wide spread in values the FIP parameter appears to organize them in a rather clear manner, suggesting that the initial source abundances could have been closely similar to those in the solar system and that the differences are all due to a selection process during injection and acceleration. Various other atomic parameters have been suggested that give equally, or better, fits to the data, such as the ionisation cross-section[10]. The comparatively large uncertainties on the abundances of the UH-nuclei suggest that the above conclusion might not be valid for these elements, but a detailed study does not confirm this for any element lighter than $_{60}$Nd. Fig. 9 illustrates the good agreement between the deduced abundances of the UH nuclei and those calculated from the solar system abundances and a so-called "step FIP" dependence.

Only in the lead-platinum region is there strong evidence for a significant deviation from a solar system like source. Although it is not possible to resolve the individual elements in this charge region, both the two satellite detectors claim to be able to separate the mostly s-process lead-like nuclei from those in the almost entirely r-process osmium - platinum group. Fig. 10 shows that the abundance ratio of these two groups of elements is significantly less than that predicted for a FIP adjusted solar system composition. Interestingly, these observations have emphasized the fact that the lead abundance in the solar system is not well known. If the photospheric abundances are used[11], then the cosmic ray data is in good agreement with a solar system abundance, if it is <u>not</u> corrected for any FIP dependence, or with an appreciable enhancement of r-process material adjusted for a FIP dependence. This conclusion is neither contradicted nor confirmed by the few actinide nuclei observed. Fig. 11 shows the abundance ratio of the actinides to the lead plus platinum nuclei as reported by HEAO and Ariel and show that although these results are inconsistent with freshly synthesised r-process material they are not able to distinguish between a solar system or an enriched r-process source. These results are summarised in Fig. 12, which shows the observed abundances compared with those predicted for source abundances that are solar system like, pure s-process and pure r-process, respectively. Only for the heaviest nuclei do the predictions for a solar system like source differ from the observations.

4. CONCLUSIONS

We can conclude that the cosmic radiation nuclei do not originate directly from supernova, in spite of the spectacular energies involved, but that instead the most probable mechanism is one where the energy of the supernova is only later converted into cosmic ray energy. Several such mechanisms have been suggested. The most popular at the present time

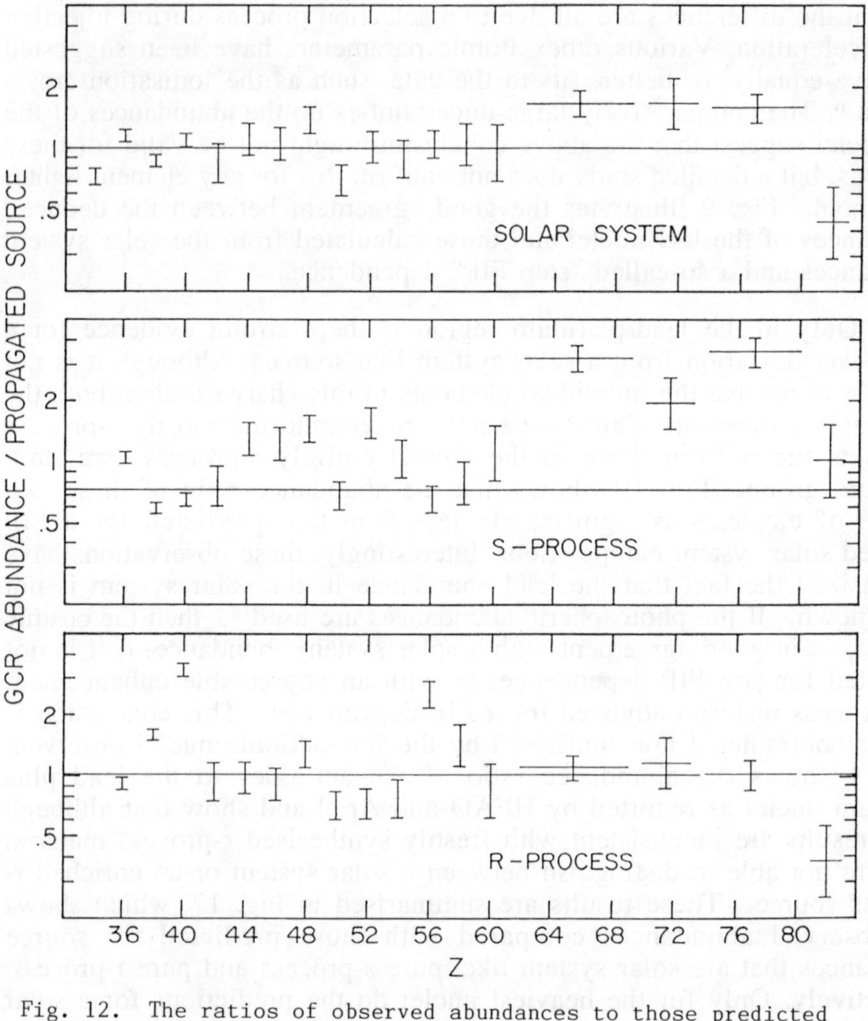

Fig. 12. The ratios of observed abundances to those predicted for source material that is (a) solar system-like, (b) pure s-process, (c) pure r-process.

is acceleration of the ambient matter near a supernova by shock waves generated during the explosion. The matter accelerated is thus not affected by the nucleosynthesis that occurs in the interior of the supernova, but the energy is obtained from the explosion. An alternative mechanism would have the outer layers of matter in the supernova being accelerated, either by shock waves again, or by a pulsar formed during the supernova collapse. Here again the matter accelerated is not affected by the nucleosynthesis occurring deep within the supernova and the composition could well be quite similar to that in the solar system, although, being younger, enriched in ejecta from earlier supernova. Only when it becomes possible to resolve the isotopes of the lighter nuclei and all the elements of the heavier nuclei will we be able to be more definitive about the true nature of the source and the acceleration process.

Acknowledgements: This work was partially supported by NASA under Grants NAG 8-500 and NGR 24-005-050. I am indebted to my colleagues on the HEAO collaboration for providing some of the diagrams used.

---

1 Anders,E and Ebihara, M., 1982, Geochimica et Cosmochimica Acta **46**, 2363

2 Evenson,P., Kroeger, R., Meyer,P. and Muller D., 1985, ICRC Vol.2 p 60

3 Mewaldt, R.A., 1985, ICRC Vol 2 p.64

4 Dwyer, R. and Meyer, P., 1985, Ap J. **294**, 441

5 Mewaldt, R.A., 1987, 13th Texas Sypmo.on Rel.Astrophys.Ed. M.P.Ulmer, p. 573

6 Stone,E.C.,Waddington,C.J.,Binns,W.R.,Garrard,T.L.,Gibner,P.S.,Israel,M.H.,Kertzman,M.P.,Klarman,J. and Newport,B.J.,1987, ICRC, **Vol 1**,366

7 Fowler,P.H.,Walker,R.N.F.,Masheder,M.R.W.,Worley,A.,and Gay,A.M. 1987 Ap.J., **314**, 739

8 Protheroe,R.J., Ormes, J.F. and Comstock, G.M., 1981, Ap.J. **247**, 362

9 Cesarsky,C.J. and Ormes, J.F., 1987, Essays in Space Science, NASA Conf.Pub. 2464, p 191

10 Arnaud,M.,Rothenflug,R and Casse, M., 1987 ICRC, **Vol 2**, 146

11 Grevesse, N and Meyer, J.P. 1985, ICRC, **3**, 5

COSMIC RAYS: MATERIAL FROM CORONAE OF ORDINARY STARS AND FROM He-BURNING ZONES

Jean-Paul MEYER
Service d'Astrophysique
Centre d'Etudes Nucléaires de Saclay
France

ABSTRACT

The Galactic Cosmic Ray (GCR) composition seems to tell us that the bulk of the GCR nuclei were first extracted out of the coronae of ordinary F to M stars. They were most likely first accelerated to MeV energies by stellar flares, and later on boosted up to GeV energies by SNR (and possibly massive star wind) shocks. A small fraction of them (~ 2%) should originate in He-burning material, plausibly in Wolf-Rayet stars. All these views might however have to be questioned if the recent indications of a deficiency of N in GCR sources (discussed in the Appendix) are confirmed. There is also evidence for an excess of Pt-peak (but not of lower Z) r-nuclei in GCR's.

1. Introduction

The main questions I will try to answer are: from where can the Galactic Cosmic Ray (GCR) nuclei have been extracted ? What does their composition tell us on this point ? And, conversely, what can the GCR composition tell us on other fields of astrophysics, such as the outer atmosphere of later-type stars and Wolf-Rayet star physics ? Note that the extraction sites of the GCR nuclei need not be the same as their acceleration sites.

On this line, a key step is to clarify which branches of physics (nuclear, atomic) are relevant in interpreting the various features of the Galactic Cosmic Ray Source (GCRS) composition. The discussion will be largely centered on sixteen trace heavy elements between C and Zn, which constitute a rich and self-consistent source of information to be deciphered. The dominant elements, H and He, which are conspicuously deficient in GCRS relative to all heavier nuclei, will not be discussed here. There are strong indications that the H and He abundances are controlled by other processes than those governing the relative abundances of heavy elements [e.g. 42, 43].

This short note is largely an updated report, at the time of the Moscow Cosmic Ray Conference, on the comprehensive studies [41, 42, 43], to which the reader is referred for more details, justifications and references. Note that the preliminary evidence we had two years ago for an excess of s-nuclei in GCRS [43] is not confirmed by further analysis of the HEAO-C3 data [54]; by contrast, there is probably an excess of Pt-peak r-nuclei (§ 6.2). In addition, there is serious new evidence that N could be very low in GCRS (Appendix) ; if this deficiency of N is confirmed, the otherwise consistent interpretation of the GCRS composition in terms of atomic selection effects in stellar atmospheres ([42]; § 4) may have to be questioned (§ 6.3).

2. Discussion of the Galactic Cosmic Ray Source (GCRS) elemental composition up to Z = 30

I am going to discuss the GCR composition, not as directly observed near Earth, but at their sources (i.e. after correction for rigidity dependent escape from the galaxy, spallation reactions and energy loss between sources and Earth).

Fig. 1 shows the deviations of the GCRS elemental composition relative to standard Local Galactic (LG, ≈ solar photospheric ; see [41, 42] and [44] at this conference) composition, up to Z = 30. (The GCRS composition is less well known beyond Z = 30, see § 6.2). I am now going to ask a few simple questions about what the GCRS composition anomalies displayed in fig. 1 tell us.

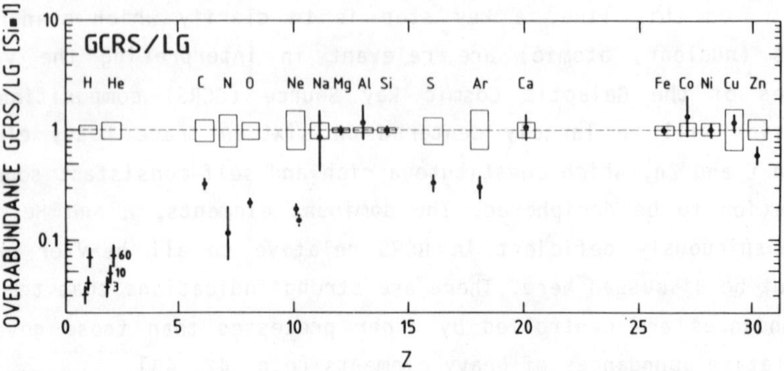

Fig. 1 :
Overabundance of elements up to Z = 30 in Galactic Cosmic Ray Source (GCRS) relative to Local Galactic (LG ; i.e. essentially solar photospheric) composition, versus Z [42,43]. Normalized to Si. Boxes : error on LG composition ; the LG estimates are still those of [41,42] ; the picture would not be significantly changed with the new estimates [44]. The relative abundances of heavy elements seem independent of energy over the well investigated range between ~ 0.5 and 20 GeV/n. The H and He source abundances are given at 3, 10 and 60 GeV/n [43]. The "anomalous" ^{22}Ne has been subtracted in the lower point plotted for Ne, which thus refers specifically to ^{20}Ne (§ 6.1).

2.1. Is the GCRS elemental abundance pattern a signature of specific nucleosynthetic processes ? Do the bulk of the GCR nuclei originate in sites of recent nucleosynthesis ?

Fig. 2 shows that there is no correlation whatsoever between the GCRS composition anomalies and specific cycles of nucleosynthesis [42,2]: the bulk of the material cannot originate in H-burning material (He/H roughly normal), and in particular not in CNO cycle material (N/C,O not enhanced) ; nor in He- burning material (no huge excesses of C,O,^{22}Ne ; no absence of N ; see however § 6.1 and 6.3) ; nor specifically in C-burning or O-burning or Si-burning or e-process material (Na/Mg/Al/Si/Ca/Fe/Ca/Ni/Cu normal ; while among C-burning products Ne/Na,Mg,Al is low ; and among O- and Si-burning products S,Ar/Si,Ca is low [45, 55, 68]).

Accordingly, there are at present no hints suggesting that the GCRS composition might resemble the bulk yield of type II SN

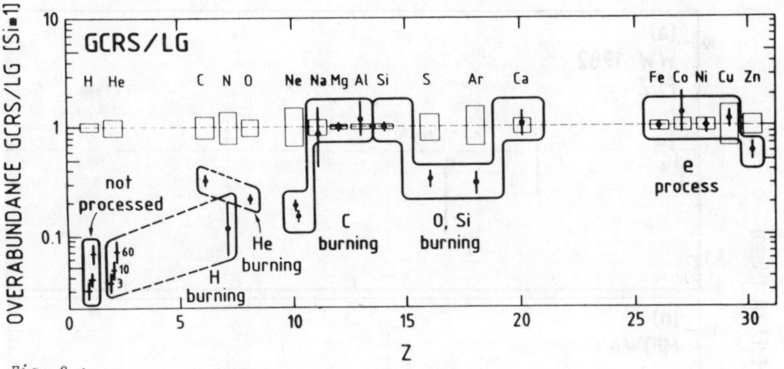

Fig. 2 :
Same as fig. 1. Elements presumably synthesized within the same nucleosynthesis cycle are grouped together.

ejecta. In fig. 3c I have plotted the GCRS/SN ejecta abundance ratios vs. Z, based on the recent SN calculations by [68] for a 25 M_0 SN (a typical mass for the SN that do eject heavy elements ; lower mass SN, ~ 9 to 12 M_0, are thought to eject almost no heavies [69]). Clearly, the two compositions are widely different. Although the model accounts for part of the (energy dependent !) deficiency of H in GCRS, highly reliably known GCRS ratios such as, e.g., C/O, O/Mg, Ne/Mg, Si/Mg, S/Mg differ by factors of 2.5 to 9 from the SN predictions.

Now, the calculations of the composition of the SN ejecta is, of course, subject to numerous uncertainties, among which the estimate of the $^{12}C(\alpha,\gamma)^{16}O$ reaction rate plays a major role. One of the main improvements in the recent SN calculations by [68] over the earlier one by [67] lies in their adopting of a 3-fold higher $^{12}C(\alpha,\gamma)^{16}O$ rate, following the recent reevaluation of this rate from the earlier canonical "FCZ" value. Our knowledge of this rate remains, however, quite inaccurate. I have, therefore, also plotted, in fig. 3a, the GCRS/SN ratios based on the earlier calculations of [67] using the, lower, "FCZ" rate. In this older SN calculation, O and O-burning products (Si,S,Ar,Ca) were consistently lower with respect to C and C-burning products (Ne,Na,Mg,Al). (The corresponding points for the GCRS/SN

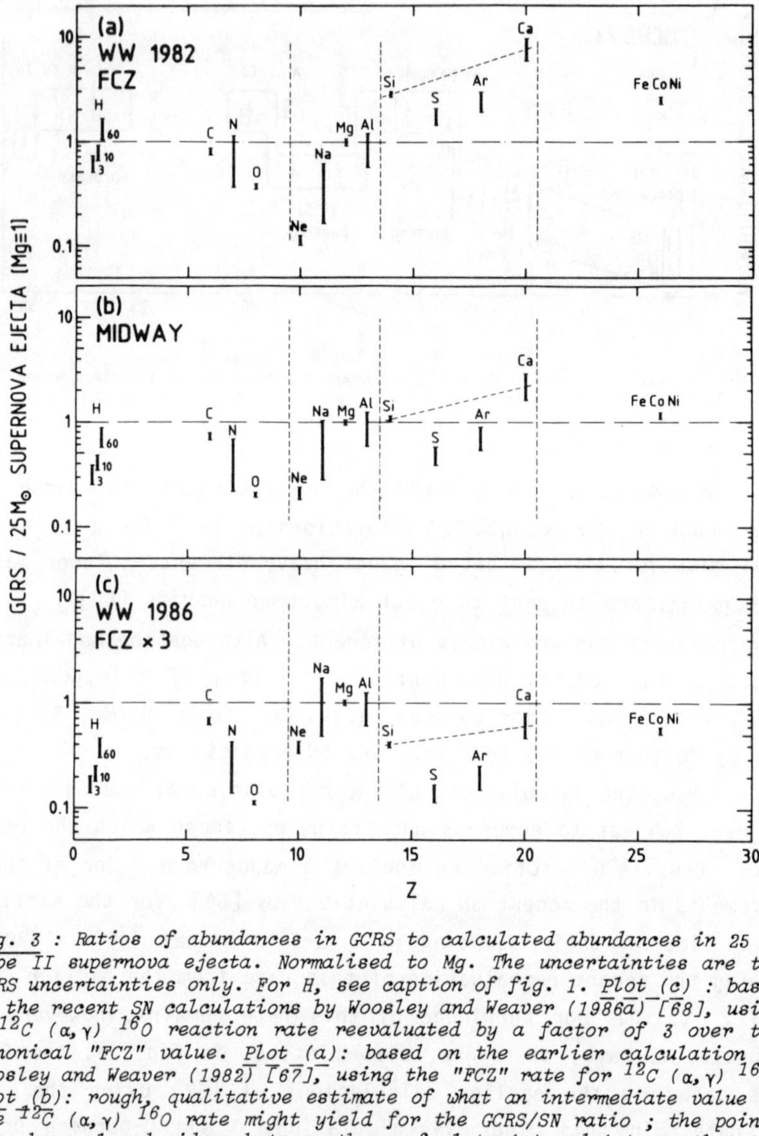

Fig. 3 : Ratios of abundances in GCRS to calculated abundances in 25 $M_{\odot}$ type II supernova ejecta. Normalised to Mg. The uncertainties are the GCRS uncertainties only. For H, see caption of fig. 1. Plot (c) : based on the recent SN calculations by Woosley and Weaver (1986a) [68], using a ^{12}C (α,γ) ^{16}O reaction rate reevaluated by a factor of 3 over the canonical "FCZ" value. Plot (a): based on the earlier calculation by Woosley and Weaver (1982) [67], using the "FCZ" rate for ^{12}C (α,γ) ^{16}O. Plot (b): rough, qualitative estimate of what an intermediate value of the ^{12}C (α,γ) ^{16}O rate might yield for the GCRS/SN ratio ; the points have been placed midway between those of plots (c) and (a), so that the GCRS Mg/Si/FeCoNi and Ne/O (which are equal to the LG ratios, fig. 1) are roughly accounted for by the SN model (see also discussion in [68]). Vertical dotted lines isolate the products of C-burning (Ne, Na, Mg, Si) and those of O- and Si-burning (Si, S, Ar, Ca). A dotted line from the Si to the Ca point stresses the GCRS underabundance of S and Ar relative to Si and Ca as predicted by all quasi-equilibrium hydrostatic as well as explosive nucleosynthesis models [45,55,68].

ratio are accordingly higher in plot (a)!). It is clear from fig. 3a that the GCRS composition agrees even worse with that yielded by the 25 M_0 SN calculation using the "FCZ" rate for $^{12}C(\alpha,\gamma)^{16}O$.

When comparing fig. 3 c and a, one may note that some differences between GCRS and calculated SN abundances, such as Si,S,Ar,Ca/Mg or Ne/O, have changed sign. Since the uncertainty on the $^{12}C(\alpha,\gamma)^{16}O$ rate remains large, one may meaningfully wonder whether the GCRS composition would not resemble that calculated for SN ejecta, if the $^{12}C(\alpha,\gamma)^{16}O$ rate turned out to be actually intermediate. It was already noted in [68] that the relative abundances of O and O-burning products (Si,S,Ar,Ca) to C-burning products (Ne,Na,Mg,Al) in LG matter would be readily accounted for with an intermediate value of the $^{12}C(\alpha,\gamma)^{16}O$ rate. To get a very rough, qualitative idea of the situation regarding GCRS's, I have plotted in fig. 3b GCRS/SN abundance ratios just midway between those of fig. 3 a and c, which account roughly for the GCRS (= LG, see fig. 1) Mg/Si/FeCoNi and Ne/O ratios. This is probably the best case one can make with existing SN models to fit the GCRS abundances. Does the plot in fig. 3b show a reasonable similarity between GCRS and estimated SN ejecta abondances ? Definitely not ! Whatever the value of the $^{12}C(\alpha,\gamma)^{16}O$ rate, one cannot account for the lower GCRS O,Ne/Mg,Si,Fe ratios (and more specifically for the lower Ne/Mg ratio among the products of C-burning), and S,Ar/Si,Ca ratios among the O- and Si-burning products (cf. [45]; see also [55]). The C,N/O ratios are also always higher in GCRS than in 25 M_0 SN ejecta ; this is due to the high yield of O in massive SN nuclei synthesis ; it is actually believed that most of the galactic C and N are synthesized in lower mass stars [68, 69].

For a further, excellent, broad discussion of GCRS abundances versus SN yields, see [2].

I conclude that, as regards nucleosynthetic history, the major fraction of GCR nuclei originate in ordinary, "solar mix"-type, material (see however § 6.1).

2.2. Is the GCRS elemental abundance pattern correlated with First Ionization Potential (FIP) ?

The three sharp jumps between Ne and Na, Si and S, and Ar and Ca in fig. 1 are characteristic of an organization of the GCRS composition anomalies in terms of FIP [13,42]. Fig. 4d shows the data of fig. 1 plotted against FIP instead of Z. Clearly, FIP organises the data, except for H and He : heavy elements with FIP $\geqslant 9$ eV ("high-FIP") are comparatively ~ 6 times less abundant than elements with FIP $\leqslant 9$ eV ("low-FIP").

This splitting around FIP = 9 eV tells us that the genesis of GCR's has, at some stage, something to do with an ion-neutral selection in a plasma at $\sim$ 5000 to 8000 K (possibly under a radiation field at a comparable temperature), in which "high-FIP" elements are largely neutral while "low-FIP" elements are predominantly ionized.

(This picture would, however, have to be questioned if the recent indications that the N abundance actually lies near the lower bound of its error bar in fig. 4d were confirmed (Appendix ; § 6.3)).

2.3. Can the bulk of the GCR nuclei originate in the ISM ?

(i) The ISM gas phase

A striking feature of fig. 1 is that the more condensable metals, which are highly depleted in most of the ISM gas phase (locked in grains), are not at all depleted in GCRS. So we need a GCR source medium in which the refractory elements are not at all condensed in grains [14,45,42]. This requirement absolutely excludes the ordinary cold ($\leqslant$ 100 K) and warm (10^4 K) gas phases of the ISM as source media for GCR's. High velocity clouds, in which grains are destroyed, are not a likely source site for GCR's. Grains present in the gas also get destroyed in SNR shock waves, but these are preceded by a hot ionizing precursor, so that a correlation of abundances with FIP would again not be plausible. As regards the hot (10^6 K) ISM phase, X-ray data indica-

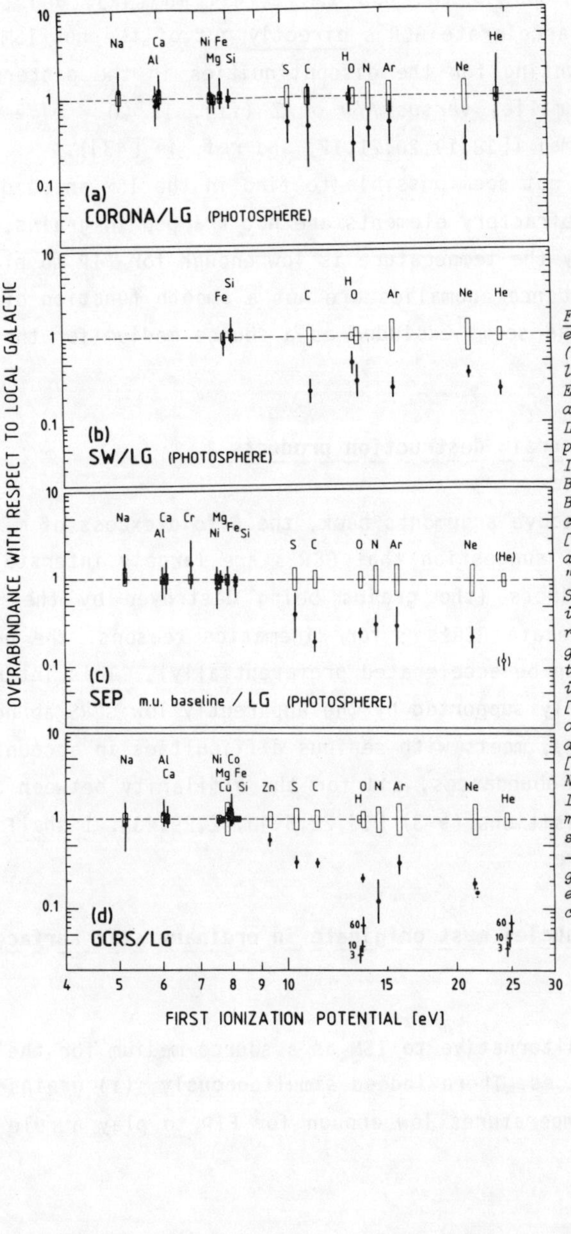

Fig. 4 : Overabundance of elements up to Z = 30 in, (a) Solar Corona, (b) Solar Wind (SW), (c) Solar Energetic Particles (SEP), and (d) GCRS relative to LG (or photospheric) composition, versus First Ionization Potential (FIP). Based mainly on [42]. Boxes: errors on the LG composition ([42], not [44]). SW plot : updated after [27]. SEP plot : "mass unbiased baseline" SEP composition, defined in [41] ; the recent average values by [10] would give a very similar picture ; the point for He is not really significant [41]. GCRS plot : based on fig. 1; H and He given at 3, 10, 60 GeV/n, from [43]; the lower point for Ne refers to ^{20}Ne only. In addition, recent gamma-ray line data on the solar upper chromosphere -lower Transition Region give a similar picture, except for Ne [48]. See caption of fig. 1.

te that grains may be sometimes evaporated there. But, once again, why a correlation of GCRS abundances with FIP if cosmic rays originate from such a hot medium ? (And X-ray observations indeed exclude a 6-fold excess of metals over CNO in the hot ISM ; ref. in [42], and [1]). Models which tried to accelerate GCR's directly out of the hot ISM gas indeed failed in accounting for the discontinuities in the pattern of the GCRS abundance anomalies versus A/Q or Z (fig. 1) (Q = effective charge in the hot plasma) ([18,19,20,21,12] and ref. in [43]).

It thus does not seem possible to find in the ISM an adequate environment in which refractory elements are not trapped in grains, and in which simultaneously the temperature is low enough for FIP to play a role (so that the abundance anomalies are not a smooth function of Z). The entire ISM gas phase seems excluded as a source medium for the GCR nuclei [45,42].

(ii) GCR's as ISM grain destruction products ?

Turning the above arguments back, the 6-fold excess of metals in GCRS has led to the suggestion that GCR's are largely interstellar grain destruction products (the grains being destroyed by the very shock waves that accelerate GCR's ; for kinematics reasons, the grain destruction products can be accelerated preferentially). This interesting hypothesis, possibly supported by the apparently low GCRS abundances of Ge and Pb (§ 6.2), meets with serious difficulties in accounting for the GCRS noble gas abundances, and for the similarity between GCRS and solar coronal compositions (§ 3) [16,22,5,40,42,29,43]. I shall not consider it further here.

2.4. Conclusion: GCR nuclei must originate in ordinary star surface material

The obvious alternative to ISM as a source medium for the GCR nuclei is stellar surfaces. There indeed simultaneously, (i) grains are destroyed, and (ii) temperatures low enough for FIP to play a role are found.

So, by itself, the GCR elemental composition tells us that GCR's must have been extracted from unevolved star surface material [45,42].

3. GCR's and composition of the solar outer atmosphere

We now take another line of analysis, and compare the GCRS composition anomalies with those observed in the Solar Corona and Transition Region (TR) (from 2.10^6 down to 70000 K), Solar Wind (SW ; ~ 1 keV/n) and Solar Energetic Particles (SEP ; ~ 1 MeV/n). This comparison is made in fig. 4, which shows that all four compositions are alike, showing very similar biases with respect to FIP [42] (regarding the problem of N, see Appendix and § 6.3). The heavy element compositions of SW and SEP most probably just reflect that of the Corona, their source medium.

So, the similarity in heavy element composition between GCRS and Solar Corona, SW, SEP is a strong hint confirming our above conclusion (§ 2.4) that GCR's were first extracted out of ordinary star surface material. In addition it tells us, more precisely, that the GCR material originates, one way or another, not in the Photospheres but in the Coronae of these stars.

(Again, this conclusion may have to be questioned if the N source abundance is low (Appendix ; § 6.3)).

4. A plausible scenario for the genesis of the bulk of the GCR nuclei

A few remarks are in order at this point of our argument. (i), all the clues discussed here are derived from the relative abundances of GCR heavy nuclei ; of course, we assume implicitly that at least a large fraction of the (comparatively deficient) H and He are accelerated together with the heavies ; (a fraction of the protons might originate in "thick" cosmic ray sources out of which no heavy nucleus can escape, e.g. [34]). (ii), it is observed that the structure of the solar atmosphere (Chromosphere, TR, Corona) is very generally present in all F to M (at least main sequence) stars, with chromospheric temperatures essentially independent of spectral type within this range (earlier type stars are too hot for neutral Hydrogen to survive and therefore to maintain a cool chromosphere) (ref. in [42]). (iii), the energies available in ordinary later-type star flaring are by far insufficient to account for the GCR acceleration to energies ≥ GeV in the galaxy (even taking in consideration very active young stars and binary dMe stars ([47], and ref. in [42,46]). So, an energetic external agent is necessary to boost up particles to GeV energies, and the obvious candidates are SNR and possibly massive star wind shocks (ref. in [42,4,61]). (iv), if this main acceleration process to GeV energies was acting directly on later-type star thermal coronal material, I do not see how it could avoid accelerating simultaneously much more ISM material, whose composition does not fit the observations ([42] ; § 2.3). This problem no longer exists if mainly pre-accelerated MeV seed particles with frozen-in coronal composition, similar to SEP's, are efficiently accelerated to GeV energies ; this hypothesis is quite plausible ([20,4] ; and ref. in [42,46]). With these remarks in mind, the following scenario emerges for the genesis of the GCR nuclei [42]:

first, for the same reason as on the Sun (next § 5), the coronae of ordinary F to M stars have commonly a FIP-biased composition ; second, like on the Sun, some of these coronal ions get pre-accelerated as MeV Stellar Energetic Particles, without large change of heavy element composition on the average ; third, some of these MeV-particles get later on boosted up to the GeV energies of GCR's upon encounter with strong shock waves of SNR and possibly massive star winds, again without major change of composition (at least among heavy nuclei).

In this scenario the shaping of the GCRS heavy element composition is totally distinct from the acceleration stage of the particles.

(If the GCRS abundance of N turned out to be low, this otherwise consistent scenario would have to be questioned (Appendix, § 6.3)).

5. The clue to understanding the bulk GCRS elemental composition : why is the composition of the Solar Corona FIP-biased ?

At this point, the problem of understanding the GCRS composition has been transferred to that of understanding the cause of the (FIP-related) anomaly of the composition of the Solar Corona.

Now, how can the solar corona, a medium at 10^6 K, have its composition biased according to FIP ? The Corona is fed from the Chromosphere, which does have the right temperature (~ 6000 K ; note that photoionization is also important) for "low-FIP" elements to be essentially ionized and "high-FIP" elements to be largely neutral [59,26].

So, the data tell us that, for some reason, neutral heavies rise from solar Chromosphere into Corona ~ 4 times less efficiently than ionized heavies. How can such an ion-neutral separation take place ?

We require a process that acts on the bulk for the matter upflow from Chromosphere into Corona. We know that spicules are a major (plausibly the major) "elevator" from Chromosphere to Corona [e.g., 3,66]. Spicules are short lived (~ 5 min) narrow upward jets (radius ~ 500 km) of chromospheric material, rising ~ 5000 km into the corona (most of the uplifted material later on falls back, either fast in the spicule channel itself, or gently in the surrounding space ; on this way back, all heavy elements are ionized, so that FIP cannot play a role at this stage). So, one way or another, the material carried up by spicules will have to be FIP-biased.

Two scenarios are currently being investigated along this line, both based on the same principle: neutral heavies diffuse away from the medium out of which the corona is being fed, while ions stay because they are unable to diffuse across the magnetic field embedded in this medium.

The first scenario, investigated by [57], is based on the gravitational settling of neutral heavies in the quiet chromospheric plateau at ~ 6000 K. On the other hand, the ionized heavies cannot settle down across the mainly horizontal chromospheric magnetic field.

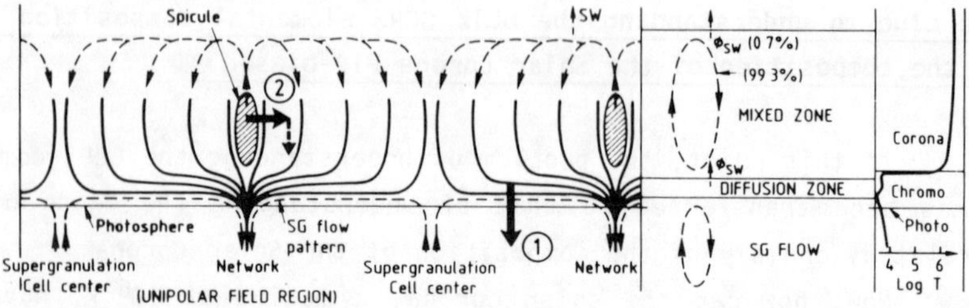

Fig. 5 :

Illustration of the two scenarios currently proposed to account for the deficiency of "high-FIP" elements in the Solar Corona. Dashed : matter flows ; full : magnetic field lines. Model 1 : gravitational settlement of neutral heavies in the chromospheric plateau [57] ; the time scales are marginal. Model 2 : diffusion of neutrals across spicules [26,27] ; the time scales pose very serious problems. See text § 5.

It is this material, impoverished in "high-FIP" heavy elements, that is from time to time fed into the Corona, largely via spicules (fig. 5(1)). This scenario is marginal as regards time scales. It requires that the upper chromospheric layers remain uncontaminated by lower altitude material over the diffusion time scale, which is of the order of ~ 2 to 3 days, while the characteristic time for the steady renewal of the supergranulation network pattern is ~ 1 day. The way the upper chromospheric material is fed into spicules is also not obvious.

In the second scenario, developed by [26,27], the selection operates dynamically during the transport process into corona itself: neutrals diffuse through the radial density gradient across the spicules, thus leaving the "elevator". Again, ions cannot diffuse across the magnetic field directed along the spicule (fig. 5(2)). In this second senario, the time scale problem is even much more serious. With realistic figures concerning spicules, the diffusion time scales are also of the order of a day, while the typical spicule lifetime is 5 minutes !

The hydrogen abundance in the corona, which is not perfectly settled observationally [58,42,27], will also be an important additional clue in the selecting of a scenario for the heavy element fractionation.

6. Other anomalies in GCRS composition, not related to FIP, and possibly related to nucleosynthesis

So much for the bulk GCRS elemental composition up to Z = 30, characterized by a FIP bias, and most probably related to later-type star chromospheric-coronal physics. Are there other anomalies, not related to FIP, in the GCRS composition ?

6.1. The ^{22}Ne, ^{25,26}Mg, C, O excess

Large isotopic anomalies in the GCRS composition, if well established, are almost certainly the signature of specific nucleosynthetic processes associated with the birth of GCR's. (No large change on isotopic ratios is expected from galactic evolution since solar birth).

The detection of isotopes in GCR's is by far less advanced than that of elements [65,43]. It is, however, now well established that the ratio ^{22}Ne/^{20}Ne is in excess by a factor of ~ 3.2 in GCRS ; a slight excess of ^{25,26}Mg/^{24}Mg is also suggested, by a factor of ~ 1.0 to 1.7 ; a ^{29,30}Si excess has also been reported [65], but it is not confirmed by the more recent analysis [43].

Let us for a moment turn back to the similarity between the GCRS and the SEP, SW and Solar Coronal compositions (fig. 4). To check this similarity more precisely, let us plot in fig. 6 the GCRS/SEP ratio vs. FIP. As expected, the bias with FIP has, to first order, cancelled (although "high-FIP" elements appear slightly more depleted in GCRS than in SEP's, by a factor of ~ 6 instead of ~ 4, as evidenced by the dashed line drawn to guide the eye in fig. 6). But the striking point is that, as compared to SEP's, C is definitely in excess in GCRS relative to O and ^{20}Ne (C/O ≈ 0.80 in GCRS, as compared to C/O ≈ 0.40 to 0.55 in SEP and SW, as well as in the Photosphere) ; O is probably also slightly in excess relative to ^{20}Ne and other "high-FIP" elements (excess factors : C/^{20}Ne ≈ 2.3 to 3, O/^{20}Ne ≈ 1.5).

I regard this simultaneous ^{22}Ne and C excesses by comparable factors as the signature of the presence of He-burning material in GCRS [39,42]. This may seem conflicting with the above evidence that the

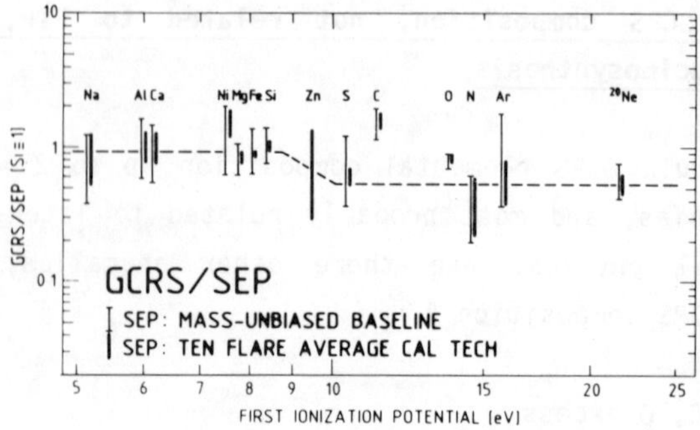

Fig. 6 : GCRS/SEP abundance ratios for Z < 30, versus FIP [after 43]. Normalized to Si. Two estimates of the SEP abundances are used : the "mass-unbiased baseline" of [41], and the 10-flare average of [10]. The line is drawn only to guide the eye (see text § 6.1). Among the three heavy element compositions probably representative of that of the solar corona, that of SEP's has been chosen for comparison with GCR's because it is the best determined (fig. 4).

main features of the GCRS elemental abundance pattern implies a "solar-mix" type material as regards nucleosynthetic history (§ 4). There is, however, no contradiction if a "processed component", highly enriched in specific species, is highly diluted in a "main component" made of FIP-biased "solar-mix" material (weak enrichments, and a fortiori depletions of specific species in the "processed component" are then swamped in the final mixture).

At the onset of He-burning in massive stars, the entire original CNO is turned into ^{22}Ne, resulting in a ^{22}Ne excess by factor of ≈ 120, a very model-independent figure, which persists through most of the He-burning phase. Meanwhile, C steadily builds up, before being turned into O by the end of the He-burning phase. During this phase, on the average ~ 25% of the original (H + He) is thus in the form of C. This leads to a time-averaged C excess factor of ~ 65, again a rather model independent figure, even when the uncertainty on the $^{12}C(\alpha,\gamma)^{16}O$ reaction rate is taken into account [51]. If this He-burning material is diluted in "solar-mix" material, it turns out that the same dilution factor of 120/(3.2-1) ≈ 65/(2.3-1) ≈ 50 simultaneously accounts for both the ^{22}Ne and the C excess factors of ~ 3.2 and ~ 2.3 in GCRS [39,42].

As a site for the origin of this He-burning material, Wolf-Rayet (WR) stars have been proposed, since in the WC-WO phases the star is peeled off to the point that He-burning products appear at the surface, and are even expelled in the form of strong winds [15,37,9]. WR star models including the entire sequence WN, WC, WO have been recently developed [49,50,51]. If their supply of material to cosmic rays is on the average constant over time, they lead to an associated excess of N that is too weak to be observable after dilution (a factor of ~1.2; [42,43,49,51]) and to coupled small excesses of O and ^{25,26}Mg (factors of 1.4 and 1.5; [50,51,43]).

As regards the predicted ^{25,26}Mg excess, one should however be careful. In the main component in which the WR component is being diluted, Mg, a "low-FIP" element, is comparatively ~ 6 times as abundant as "high-FIP" Ne, C, O (fig. 4 d). Unless the WR component is affected by the same FIP-bias, which seems highly implausible since WR stars are too hot to maintain a chromosphere, the ^{25,26}Mg output predicted by WR star evolution calculations is ~ 6 times more diluted than the associated ^{22}Ne, C, O yields [42,43]. The predicted excess of ^{25,26}Mg (and associated weak s-process low-FIP species; [50,43]) in GCRS then becomes minute (fig. 7). This effect, pointed out in [42,43], has been overlooked in all the earlier work.

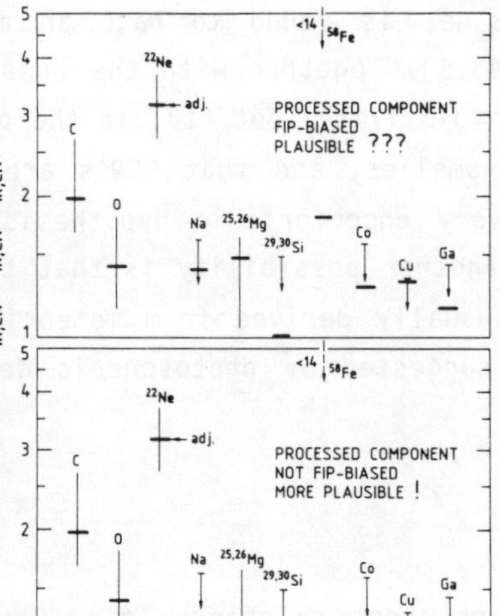

Fig. 7 : GCRS excesses of He-burning and weak s-process species, relative to the main GCRS component, after [43]. Vertical error bars, identical in both graphs : GCRS excesses derived from observations. Thick horizontal marks : predictions based on [50]'s model of He-burning and weak s-process in WC-WO stars, diluted by a factor of 50 for ^{20}Ne (adjusted as to fit the "observed" GCRS ^{22}Ne/^{20}Ne excess). Top graph : the processed WC-WO component is assumed to be FIP-biased like the main component, so that all species are diluted by the same factor ; this assumption has been made in most existing studies, but is not very plausible. Bottom graph : the processed component is not FIP-biased, so that its low FIP elements are ~ 6 times as diluted as its high-FIP elements ; this hypothesis is much more plausible [42,43].

6.2. GCRS composition for Z > 30 ("Ultra-Heavy" nuclei, UH)

The GCRS abundance of the major elements up to Z = 30, discussed above, is well established. Beyond Z = 30, we enter a "frontier" field since the fluxes of detectable nuclei become minute and charge resolution is not so easy to achieve up to Z = 90 ! The field has, however, very significantly progressed in the last few years, thanks to two spacecraft experiments, HEAO-C3 and Ariel VI [e.g. 6,7,8,54,24, and ref. in 43].

a. The range Z = 31 to 60

Up to Z = 60, the most recent results [54,24] are fully consistent with a "solar-mix" source composition distorted by the FIP-bias seen for Z ⩽ 30 (fig. 8). The data are inconsistent with a strong predominance of, either s-, or r-process. (This contrasts with the HEAO-C3 preliminary results presented orally at the 1985 La Jolla Cosmic Ray Conference, which I had strongly stressed in my Rapporteur talk at this Conference [43]).

One element however, $_{32}$Ge, is found to have an abundance twice lower than expected [11,6,43,8]. Together with the possibly low Pb abundance, it may suggest that volatility, not FIP, is the parameter governing the GCRS composition anomalies, and that GCR's are largely grain destruction products, a very uncomfortable hypothesis however (§ 2.3 ; [16,22,5,40,42,29,43]). Another possibility is that the Solar System Ge and/or Pb abundances, usually derived from meteorites, are overestimated. This is possibly suggested by photospheric determinations [29,43,44].

b. Abundances for Z > 60

Above Z = 60, the picture seems to change (fig. 8): Pt-peak elements (as well as the secondary nuclei from Z = 62 to 73 formed by their spallation) seem in excess by a factor of ~ 1.6 relative to regular FIP-biased "solar mix" source composition [54,24]. If confirmed by detailed analysis, this excess will probably imply the admixture

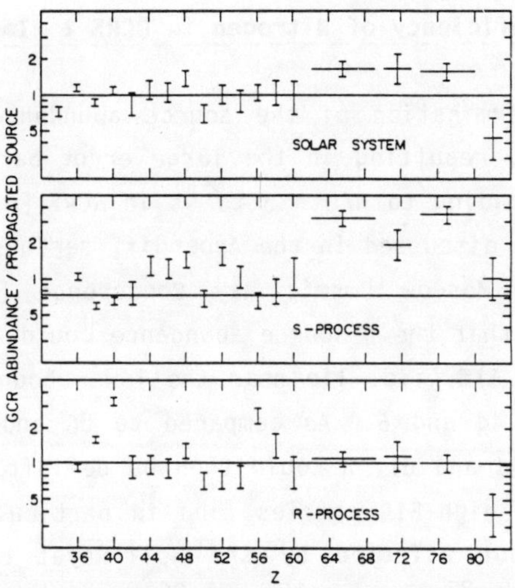

Fig. 8 : Ratio of the GCR abundances measured by HEAO-C3 for Z > 34 to calculated abundance near Earth, for three hypothesis regarding source composition : standard FIP-fractionated solar-mix (LG) material, pure s- and pure r-process material (from [54]).

of a minor GCR component highly enriched in highest-Z r-process products, highly diluted in "solar-mix" material (similar to the ^{22}Ne-C-^{25,26}Mg-O-rich component ; § 6.1).

Relative to Pt-peak elements, Pb is certainly low. Relative to all lighter nuclei, a normal Pb abundance is found by [24], while the, better resolved, data of [54] imply a deficiency by a factor of ~ 2 (fig. 8). Such a deficiency of Pb on nucleosynthetic grounds would require that the bulk of GCR's originate in sites of current nucleosynthesis (§ 6.1), which would conflict with the very strong evidence that most GCR nuclei are primarily "solar-mix" material (§ 4). As just mentioned in connection with the Ge problem, other possible explanations for a low Pb abundance could be that GCR's are primarily grain destruction products (§ 2.3 ; [16,22,5,40,42,29,43]), or that the adopted Solar System Pb abundance is too high [29,43].

No compelling excess of actinide nuclei is presently being reported [24,54].

6.3. A specific deficiency of Nitrogen in GCRS ? Implications.

The determination of the source abundance of N has been a problem for years, resulting in the large error bars for N in fig. 1, 4d and 6, corresponding to N/O ≈ 3 to 9% in GCRS [e.g., 38, 42, 43 and ref. therein]. As discussed in the Appendix, serious new evidence just presented at the Moscow Cosmic Ray Conference [64; see also 33] strongly suggests that the N source abundance could well be actually as low as N/O ≈ (3 ± 1)%, i.e. lie near the lower bound of the error bar for N in fig. 1, 4d and 6. As compared to LG and especially to SEP abundances (fig. 4d and 6), N could then be definitely deficient relative to all other high-FIP species (and in particular with respect to ^{20}Ne, which is not affected by the Wolf-Rayet component plausibly responsible for the ^{22}Ne, ^{12}C, ^{16}O, ^{25,26}Mg excess ; § 6.1 [51]). The condition that N be not deficient relative to ^{20}Ne in GCRS, as compared to SEP's, is equivalent to the condition that N/O ⩾ 6% in GCRS.

The matter is of extreme importance. An excess of specific nuclei may indeed always be accounted for by the presence, in GCR's, of a minor component highly enriched in these nuclei and highly diluted in a dominant component made of FIP-biased "solar mix" material (the abundances of other nuclei then remain unaffected by the presence of the minor component; § 6.1). But the depletion of a single, isolated heavy species, if really proven, would imply that the bulk of GCR's originate in a medium specifically depleted in that species. Such an evidence would be sufficient to question the relevance of the entire apparent correlation with FIP and of the similarity with SEP and Solar Coronal compositions. So, a new scenario, in which FIP may play a role, or not, would have to be found to explain the GCRS abundance of all elements.

Considering specifically the nuclei N, ^{22}Ne, ^{25,26}Mg in GCR's, [64] have made the intriguing remark that if indeed N/O ≈ 3%, the number of "missing" ^{14}N nuclei, ~ 6% of O (relative to general SEP-coronal abundances ; fig. 6), is about equal to the number of excess ^{22}Ne and ^{25,26}Mg nuclei (§ 6.1). Since in He-burning ^{22}Ne and ^{25,26}Mg are made by destruction of N through α-captures (§ 6.1) , it is very tantalizing to suggest that about 2/3 of the GCR's originate in a "N-free-^{22}Ne-rich" medium having undergone He-burning [64].

In the above argument, the amount of "missing" N is defined using the LG ("solar mix") abundance of N as a reference baseline (relative to some other high-FIP species, say ^{20}Ne, so as to avoid the atomic selection effects). Now, if the "N-free-^{22}Ne-rich" material processed by He-burning has earlier gone through the CNO-cycle, then its N abundance <u>at the start</u> of the He-burning is ~ 17 times its LG value [42]. The deficiency of N with respect to its LG abundance is then not relevant to the problem. And one expects ~ 17 times as many ^{22}Ne, ^{25,26}Mg nuclei as in [64]'s reasoning (for the same fraction, 2/3, of "N-free-^{22}Ne-rich" material in GCRS, still required to account for the N deficiency).

So, the numerical similarity noted by [64] can be of significance only if ~ 2/3 of the GCR's originate in a gas in which ^{14}N has been turned into ^{22}Ne by He-burning, <u>without having earlier gone though CNO-cycle</u>.

Does such a medium exist ? It is far from obvious. Let me just note that in lower mass stars in which quiescent H-burning takes place via the pp-cycle, He is later burnt at low temperatures so that ^{18}O, not ^{22}Ne, is formed [56]. For a further discussion, see the excellent review by [2].

[Let me stress the essential difference between the scenario proposed by [64], and the now classical He-burning Wolf-Rayet scenario discussed in § 6.1 (which explains the ^{2}Ne... excess, but not a N deficiency). Both are based on the same nuclear reactions. The Wolf-Rayet scenario starts from CNO-cycled material with N enhanced by a factor of 17, thus leading to a ^{22}Ne enhancement factor of 120 requiring a dilution by a factor of 50, which allows the abundances of most other elements not to be affected by this processed component. By contrast, the scenario of [64], by starting with the much lower LG N abundance, yields a much smaller quantity of ^{22}Ne allowing a dilution factor of 1.5 only. The final GCRS composition of all elements will then largely reflect that of the processed component].

7. Summary

- The major features of the GCRS elemental composition do not show any hint for a special nucleosynthetic history. In this respect, the bulk of GCR's really look like "solar-mix" material ! By contrast, the GCRS composition is strikingly similar to those of Solar Corona, SW and SEP. They all differ from LG or photospheric composition by the same bias as a function of FIP. Note also that the refractory elements, highly depleted in most of the ISM gas phase, are not at all depleted in GCRS.

- For these and other strong reasons, the bulk (98%) of the GCR heavy nuclei are most likely Stellar Energetic Particles (~ MeV), energized by surface activity out of the Coronal gas of solar-like F to M stars, that got later on reaccelerated to GCR energies ($\geqslant$ GeV) by strong shock waves of SNR and possibly massive star winds.

- In a way, GCR's thus give us a strong hint that the composition anomalies found in the solar corona are a very common feature in stellar coronae. This feature is not astonishing in view of the strong similarities between the temperature profiles of chromosphere, transition region and corona on the Sun and on F to M main sequence stars.

- The problem of understanding the composition of GCR's is thus transferred to that of understanding that of the Solar Corona. On the Sun, an ion-neutral segregation must take place in the cool chromospheric material, before or during the rise of matter from chromosphere to Corona. Two such scenarios are at present being developed, in which heavy neutrals diffuse away from the material to be fed into the Corona, while ions do not, since they are blocked by the magnetic field.

- There is a serious possibility that N be deficient in GCRS. If this were confirmed, all the above statements would have to be questioned.

- The ^{22}Ne, C, ^{25,26}Mg, O excesses in GCRS strongly suggest that about 2% of the GCR heavy nuclei originate in He-burning material, plausibly in WC-WO Wolf-Rayet stars. [In assessing the expected GCRS

^{25,26}Mg excess from nucleosynthesis models, FIP-dependent differential effects in the dilution of the synthesized ^{25,26}Mg in "solar-mix" material must be considered.]

- Present-day observations of GCR nuclei with Z = 30 to 92 are by and large consistent with the above conclusions. There is an evidence for a specific excess of Pt-peak r-nuclei. Ge and Pb are possibly deficient.

- The low H and He abundances relative to heavies in GCR's pose specific problems. These can perhaps be better tackled when we have more accurate information on their energy spectra.

APPENDIX

THE B-^{15}N CONTRADICTION, DISTRIBUTED REACCELERATION, AND THE GCR SOURCE ABUNDANCE OF NITROGEN

The source abundance of N is not easy to determine, because a major fraction of the N we observe is "secondary" (i.e. produced by spallation reactions during the interstellar propagation of GCR's). So, the N source abundance is obtained by subtraction of two comparable quantities.

For several years now, there has been an unresolved contradiction between the low source abundances N/O ≈ 3% derived from the low energy isotopic ^{14}N observations, and the higher ones, N/O ≈ 6 to 8% derived from the high energy, mainly elemental, measurements [e.g., 28, 38,62,42,43 and ref. therein,23,33,64].

At low energy (≤ 500 MeV/n), one cannot hope to reliably determine the ^{14}N source abundance as long as the "B-^{15}N" contradiction", which tells us that something is wrong in our basic hypothesis or propagation calculation, has not been cleared up. What is it all about ? B and ^{15}N are two (presumably) purely secondary species, whose flux ratio is directly predicted by the propagation models, and is quite inensitive to the details of this model. Now, at low energy, the observed ^{15}N/B ratio differs significantly from the predicted one [e.g.,30,32,43,33,64]. This implies that, either <u>(i)</u> the cosmic-ray observations are grossly incorrect, which does not seem plausible ; or <u>(ii)</u> ^{15}N is enhanced by a factor of ~ 100 in GCRS's (as compared to 3.2 for ^{22}Ne), which seems very difficult to accept in the absence of other large anomalies ; or <u>(iii)</u> the cross-sections used in the propagation calculation are inaccurate ; or <u>(iv)</u> the nuclei of the ^{22}Ne-C-$^{25,26}Mg$-O-rich component presumably originating in Wolf-Rayet (WR) stars (§ 6.1) traverse on the average much less interstellar matter than the other nuclei between their source and the Earth [43] ; B is indeed largely made from C, ~ 60% of which is believed to originate in WR stars, while ^{15}N is largely made from O, only ~ 33% of which originate in WR's ; it thus turns out that ~ 40% of the B, but only ~25% of the ^{15}N are secondaries of "WR C and O nuclei" ; if the "WR nuclei" traverse very little matter, B will be comparatively deficient relative to ^{15}N and heavier secondary species ; this hypothesis is entirely ad hoc, and very difficult to prove or disprove ; or (v) I suggested [43] that one should investigate whether the hypothesis of a distributed reacceleration of GCR's during their propagation [e.g., 52,53,17, 60] could account for the observed low energy $^{15}N/B$ ratio ; this could be the case if the cross sections for formation of ^{15}N had much higher low energy peaks than those for the formation of B, in an adequate low energy range ; a success could lead to a strong substantiation of the reacceleration hypothesis.

Along this line (v), [35,36] have recently concluded that the B-^{15}N contradiction is indeed removed with the hypothesis of distributed reacceleration. I think this conclusion is erroneous, and that distributed reacceleration definitely cannot be the solution of the puzzle. First, contrary to the authors' estimate, the effective cross section for the formation of B is not at all flat at low energy ; more than 92% of the low energy B is formed from nearby CNO parents [43], and the cross sections for formation of B from both C and O have strong peaks around 50 MeV/n [e.g.,25]. So, I don't see how the effective cross section for the formation of B could help peaking at low energy in much the same way as that for ^{15}N does. Further, with the small mean reacceleration factors of ~ 2 now preferred in this reacceleration model [36], cross section peaks located below 100 MeV/n [30,32,25] cannot influence the observed fluxes of B and ^{15}N up to 450 MeV/n ; and even more so when the ⩾ 200 MeV/n interplanetary deceleration is taken into account. The model shown to fit the $^{15}N/O$ ratio in fig. 2 of [35] does not at all fit simultaneously the observed low energy B/C ratio when solar modulation is taken into account (cf. fig. 7a of [60]). This conclusion, that distributed reacceleration cannot significantly modify the predicted $^{15}N/B$ ratio, has just been confirmed by a full calculation by [31].

On the other hand, the possibility (iii) that the cross sections used in the propagation calculations were inaccurate has been explored by [64]. They have developed a new semi-empirical estimate of unmeasured cross sections based on their recent cross section measurements [63], which leads to changes in the predicted formation yields for B and ^{15}N. These changes are due mainly to modifications in the estimated cross sections for production of these species (i) from much heavier parents, based on several new measurements, and (ii) from nearby neutron-rich parents such as ^{13}C, ^{15}N, ^{22}Ne, based on an extrapolation of a comparison of cross sections measured for the ^{40}Ar and ^{40}Ca isobars.

With these new cross section estimates, the discrepancy between observed and predicted low energy $^{15}N/B$ ratios is reduced from $\sim$ 27% down to $\sim$ 14% [64]. This is a great step towards a solution of the puzzle, although the remaining discrepancy is still significant.

With the new cross sections and a source ratio $^{14}N/O$ = 3%, [64] still fit perfectly the observed elemental N/O ratio at all energies, and much better the low energy isotopic $^{14}N/O$ ratio (especially the accurate new point by [33]) ; there are, however, difficulties with the high energy $^{14}N/O$ points, and in particular with the recent, "best" HEAO-C2 point by [23] which may, however, still be affected by systematic errors.

Clearly, the new cross section estimates, associated with a lower N/O source ratio $\sim$ 3%, have led to an improved consistency of the picture. Just like with the earlier estimates and N/O $\sim$ 6 to 8% at the sources, the very well observed elemental B/C and N/O ratios are simultaneously perfectly fitted at all energies. The low energy fits of the observed isotopic $^{15}N/O$ and $^{14}N/O$ ratios are much improved, although that for $^{15}N/O$ is not yet satisfactory. The fit of the high energy $^{14}N/O$ ratio has however got poorer, but here the data are less reliable. In spite of these encouraging hints, more work is probably necessary to qualify the new cross section systematics.

The implications of a source N/O ratio as low as $\sim$ (3+1)% are briefly discussed in the text, § 6.3.

8. References ("ICRC" stands for International Cosmic Ray Conference)

1. Arnaud, M., and Rothenflug, R. 1986, Adv. Space Res. 6, N°2, 119.
2. Arnould, M. 1984, Adv. Space Res. 4, N°2-3, 45.
3. Athay, R.G. 1976, The Solar Chromosphere and Corona : Quiet Sun (Dordrecht Reidel).
4. Axford, W.I. 1981, 17th ICRC, Paris, 12, 155.
5. Bibring, J.P., and Cesarsky, C.J. 1981, 17th ICRC, Paris, 2, 289.
6. Binns, W.R., et al. 1984, Adv. Space Res. 4, N°2-3, 25.
7. Binns, W.R., et al. 1985, Ap. J. 297, 111.
8. Binns, W.R., et al. 1987, in "Essays in Space Science, NASA-CP 2464, p.173.
9. Blake, J.B., and Dearborn, D.S.P. 1984, Adv. Space Res. 4, N°2-3, 89.
10. Breneman, H.H., and Stone, E.C. 1986, Ap. J. Letters, 299, L57.
11. Byrnak, B., et al. 1983, 18th ICRC, Bangalore, 2, 29.
12. Cassé, M. 1983, in "Composition and Origin of Cosmic Rays", M.M. Shapiro ed. (Reidel), p.193.
13. Cassé, M., and Goret, P. 1978, Ap. J. 221, 703.
14. Cassé, M., Goret, P., and Cesarsky, C.J. 1975, 14th ICRC, Munich, 2, 646.
15. Cassé, M., and Paul, J.A. 1982, Ap. J. 258, 860.
16. Cesarsky, C.J., and Bibring, J.P. 1980, IAU Symp. N°94, Origin of Cosmic Rays, G. Setti, G. Spada, A.W. Wolfdendale ed., p.361.
17. Cowsik, R. 1986, Astr. Ap. 155, 344.
18. Eichler, D. 1979, Ap. J. 229, 419.
19. Eichler, D., and Hainebach, K. 1981, Phys. Rev. Letters 47, 1560.
20. Ellison, D.C. 1981, Ph. D Thesis.
21. Ellison, D.C., Jones, F.C., and Eichler, D. 1981, J. Geophys. 50, 110.
22. Epstein, R.I. 1980, MNRAS 193, 723.
23. Ferrando, P., et al. 1987, Astr. Ap., in press.
24. Fowler, P.H., et al., 1987, Ap. J. 314, 739.
25. Garcia-Munoz, M., et al. 1987, Ap. J. Suppl. 64, 269.
26. Geiss, J., and Bochsler, P. 1985, in "Isotopic Ratios in the Solar System", Paris (CNES), June 1984 (Cepedues Editions), p.213.
27. Geiss, J., and Bochsler, P. 1986, in "The Sun and the Heliosphere in three dimensions", R.G. Marsden ed. (Reidel), p.173.
28. Goret, P., et al. 1981, 17th ICRC, Paris, 9, 122.
29. Grevesse, N., and Meyer, J.P. 1985, 19th ICRC, La Jolla, 3, 5.
30. Guzik, T.G. 1981, Ap. J. 244, 695.
31. Guzik, T.G., Wefel, J.P., and Beatty, J.J. 1987, 20th ICRC, Moscow, 2, 226.
32. Guzik, T.G., et al. 1986, 19th ICRC, La Jolla, 2, 80.
33. Krombel, K.E., and Wiedenbeck, M.E. 1987, 20th ICRC, Moscow, 1, 360.
34. Lagage, P.O., and Cesarsky, C.J. 1985, Astr. Ap. 147, 127.
35. Letaw, J.R. 1987, Ap. J. Letters 317, L69.
36. Letaw, J.R. et al. 1987, 20th ICRC, Moscow, 2, 222.
37. Maeder, A. 1983, Astr. Ap. 120, 130.
38. Mewaldt, R.A. 1981, 17th ICRC, Paris, 13, 49.

39. Meyer, J.P. 1981a, 17th ICRC, Paris, 2, 265.
40. Meyer, J.P. 1981b, 17th ICRC, Paris, 2, 281.
41. Meyer, J.P. 1985a, Ap. J. Suppl. 57, 151.
42. Meyer, J.P. 1985b, Ap. J. Suppl. 57, 173.
43. Meyer, J.P. 1985c, 19th ICRC, La Jolla, 9, 141.
44. Meyer, J.P. 1987, this conference.
45. Meyer, J.P., Cassé, M., and Reeves, H. 1979, 16th ICRC, Kyoto, 12, 108.
46. Montmerle, T. 1984, Adv. Space Res. 4, N°2-3, 357.
47. Mullan, D.J. 1979, Ap.J. 234, 588.
48. Murphy, R.J., Ramaty, R., Forrest, D.J., and Kozlovsky, B. 1985, 19th ICRC, La Jolla, 4, 249 and 253.
49. Prantzos, N. 1984, Adv. Space Res. 4, N°2-3, 109.
50. Prantzos, N., Arnould, M., Arcoragi, J.P., and Cassé, M. 1985, 19th ICRC, La Jolla, 3, 167.
51. Prantzos, N., Doom, C., Arnould, M., and de Loore, C. 1986, Ap. J. 304, 695.
52. Silberberg, R., Tsao, C.H., Letaw, J.R., and Shapiro, M.M. 1983, Phys. Rev. Letters 51, 1217.
53. Simon, M., Heinrich, W., Mathis, K.D. 1986, Ap. J. 300, 32.
54. Stone, E.C., et al. 1987, 20th ICRC, Moscow, 1, 366.
55. Thielemann, F.K., and Arnett, W.D. 1985, Ap. J. 295, 604.
56. Thielemann, F.K., and Arnould, M., 1979, in "Les Elements et leurs Isotopes dans l'Univers", 22nd Liège Internat. Astrophys. Symp. (U. of Liège Press), p.59.
57. Vauclair, S., and Meyer, J.P. 1985, 19th ICRC, La Jolla, 4, 232.
58. Veck, N.J., and Parkinson, J.H. 1981, MNRAS 197, 41.
59. Vernazza, J.E., Avrett, E.H., and Loeser, R. 1981, Ap. J. Suppl. 45, 635.
60. Wandel, A., et al. 1987, Ap. J. 316, 676.
61. Webb, G.M., Forman, M.A., and Axford, W.T. 1985, Ap. J. 298, 684.
62. Webber, W.R. 1983, 18th ICRC, Bangalore, 9, 151.
63. Webber, W.R. 1987, 20th ICRC, Moscow, 2, 463.
64. Webber, W.R., and Gupta, M. 1987, 20th ICRC, Moscow, 2, 129.
65. Wiedenbeck, M.E. 1984, Adv. Space Res. 4, N°2-3, 15.
66. Withbroe, G.L., and Noyes, R.W. 1977, Ann. Rev. Astr. Ap. 15, 363.
67. Woosley, S.E., and Weaver, T.A. 1982, in "Essays in Nuclear Astrophysics", C.A. Barnes, D.D. Clayton, and D.N. Schramm ed. (Cambridge U. Press), p.377.
68. Woosley, S.E., and Weaver, T.A. 1986a, in "Nucleosynthesis and Its Implications on Nuclear and Particle Physics, J. Audouze, and N. Mathieu eds. (Reidel), p.145.
69. Woosley, S.E., and Weaver, T.A. 1986b, Ann. Rev. Astr. Ap. 24, 205.

EVERYTHING YOU ALWAYS WANTED TO ASK ABOUT LOCAL GALACTIC ABUNDANCES, BUT WERE AFRAID TO KNOW

Jean-Paul MEYER
Service d'Astrophysique
Centre d'Etudes Nucléaires de Saclay
France

ABSTRACT

I first give an overall view, intended for non-specialists, of the problems involved in determining a Local Galactic (LG) standard of abundances. I briefly discuss the recently proposed "SEP-derived photospheric" abundances. I finally give an updated table of LG abundances.

1. Introduction

In this short overview on abundance determinations, intended for non-specialists, I am going to mention a number of elementary facts which are so obvious to people in the field that they are seldom expressed. I also briefly discuss the recently proposed "SEP derived photospheric" abundances. In addition, I present an updated table of Local Galactic (LG) abundances up to Zn, which is essentially a critical merging of my earlier studies [23,18,19,20,21] and of the excellent recent photospheric and meteoritic tables by [13,14,2] (I have not recently reviewed the literature regarding abundance determinations outside the Solar System).

Note that, for most elements, the isotopic ratios are much easier to determine (at least in the Solar System ! § 3.1) than the elemental abundances I am going to discuss.

2. General : Solar System and non-Solar System data

What we want is some typical composition of our galactic neighbourhood, valid for the last few billion years (during which the evolution of abundances is believed to have been quite slow for most species).

Fig. 1 gives an overall view of the sources of information we have available for the abundances of the various elements. They should be split into two groups :

(i) data from the Solar System ; there, the idea is to try to find media whose compositions are as similar as possible to that of the protosolar nebula. We will actually find two adequate media, discussed in § 3.1 : the solar photosphere (but not the corona ! § 3.2), and the C1 meteorites. These data pertain to the interstellar medium at the time of solar birth, 4.6 billion years ago.

(ii) Data on nearby HI gas, HII regions and stars give us a more general (though usually less accurate) information on our galactic neighbourhood. Data on HI and HII media refer to the present-day interstellar gas, while stellar data span the entire galactic lifetime.

In actual reality, the data from the Solar System are by far dominant (available for many more elements, and usually much more accurate), except for noble gases (fig. 1 ; § 4).

This, of course, poses the problem of the representativity of the composition of the protosolar nebula as regards our galactic neighbourhood. First, we have the systematic spatial and temporal abundance gradients (determined from HII region and stellar data [e.g., 27,30]) ; these can be rather well handled, if one sticks to our ~ kpc neighbourhood and to epochs between solar birth and to-day, especially when one considers the relative abundances of heavy species, which generally change much more slowly than abundances relative to H [e.g., 9,12] ; (however, a ratio such as $^{13}C/^{12}C$ might have doubled since solar birth!) Second, we have the random fluctuations of abundances between stars born at the same location and time [30]. The Sun being just one ordinary star, this problem may be quite serious. There is conflicting evidence as to whether the Sun is "average" or particularly rich in heavy elements [e.g., 30,24].

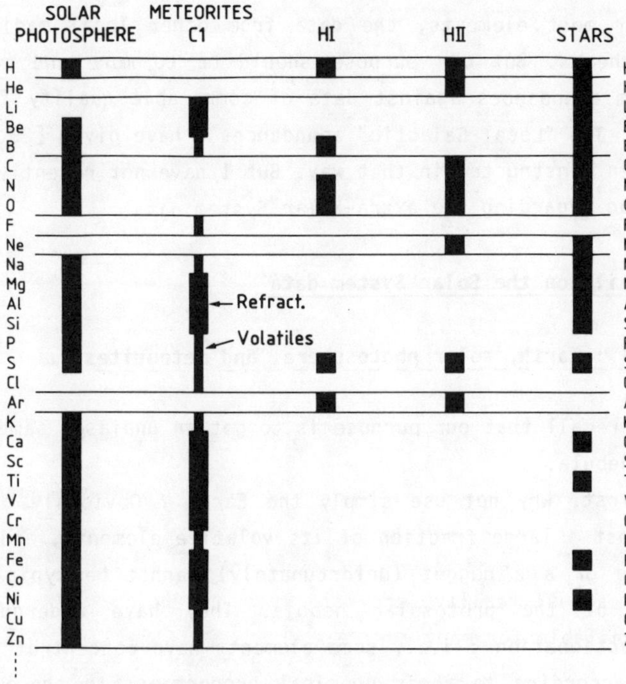

Fig. 1 : Available sources of information for the abundances of the various elements. Solar Photosphere and C1 meteorites refer specifically to the composition of the protosolar nebula; HI gas, HII regions and stellar data give us a broader (though usually less accurate) view of the abundances in our galactic neighbourhood. They are used, both to check the significance of the Solar System data, and to investigate the spatial gradient (mainly HII) and time evolution (stars) of the abundances. They are also our prime source of information for noble gases, for which we have almost no information from the Solar System (there are data from Solar Corona, Solar Wind and Solar Energetic Particles ; but these are not so easy to exploit because of the abundance fractionation between Photosphere and Corona (§ 3.2) ; the best case is possibly Ar, whose FIP is quite close from that of O). The abundances of He and Ne are quite reliably determined from HII and hot star data, and that of Ar from HI and HII data. The HI and HII gas data do not give us information on the abundances of metals in the interstellar medium, since these are almost always largely locked in grains. For meteorites, I have indicated which elements are more refractory or volatile ; the abundances of refractories are more certain than those of volatiles (§ 4).

These considerations show the need for abundance compilations covering simultaneously Solar System and extra-Solar System sets of data. Of course, the information from the Solar System still remains dominant for most elements, the data from other local media serving mainly as checks. But our purpose should be to more and more check Solar System abundances against data of comparable quality from other local media. The "Local Galactic" abundances I have given [23,18,19,20, 21] have been constructed in that way. But I have not recently reviewed the situation regarding the extra-Solar System data.

3. More details on the Solar System data

3.1. General : Earth, solar photosphere, and meteorites

I recall that our purpose is to get an unbiased sample of the protosolar nebula.

First, why not use simply the Earth ? Obviously, the Earth crust has lost a large fraction of its volatile elements, and a chunck of limestone or a a nugget (unfortunately) cannot be typical of the composition of the protosolar nebula. They have undergone heavy chemical fractionation : i.e., some elements have concentrated, others have gone, according to their chemical properties, to the volatility (condensation temperature) of the chemicals the various elements tend to form. So, Earth cannot be used to determine elemental abundances relevant to the protosolar nebula.

By contrast, this is not true for the isotope ratios. For isotopes, the risk of a fragmentation is much lower, at least in large bodies. The isotope ratios adopted in tables such as [8,2] are therefore usually the terrestrial ones, which are very accurately determined. Exceptions are H and noble gases, which pose very specific problems [e.g., 25].

Back to elemental abundances in the Solar System, we have actually two main sources of information :

(i) direct spectroscopic observation of the solar photosphere, which is generally believed to have kept the composition of the protosolar gas (except for D, ^{3}He and Li for which nuclear reactions at the bottom of the surface convective zone have modified the original abundances). Note that the composition of the Solar corona is not directly relevant ; it might, however, be used indirectly (§ 3.2).

(ii) Chemical analysis of C1 carbonaceous chondrite meteorites (hereafter C1's).

Both are supposed to give us the same composition of the protosolar nebula. One may wonder why one appeals to meteorites, if one can directly observe the solar photosphere, a certainly more undisputable unbiased sample of the protosolar nebula. The reason is that, due mainly to our limited knowledge of atomic physics and of the structure of the solar atmosphere, the photospheric abundance determinations of many elements have long been lacking or unreliable. In meteorites, on the other hand, the abundance measurement in itself can often be excellent (and this is why all earlier abundance standards are based primarily on meteorites [29,6,7,8,2]). But we do not know what we are measuring !

In other words, why believe that the C1 abundances have anything to do with those of the protosolar nebula ? Among meteorites, C1's are the only ones in which there is no obvious fractionation, except for H, C, N, O and noble gases (in particular C2's, sometimes proposed as a standard, are made of 50% "pebbles" which have lost all their volatiles !!! [1,18,19,11,2]) ; but this does not imply that C1's are totally unfractionated ! The real arguments are only a posteriori. They are twofold. First, when comparison is possible between C1 abundances and reliable photospheric determinations, the values generally agree very well (especially for refractory elements ; for volatiles, there might commonly be deviations by ≤ 20% [2,13], and possibly sometimes more [15] ; there is also a very serious problem for Fe, see § 4). So, one feels justified to assume that the C1 abundances are generally significant also for elements for which we lack reliable photospheric data. Second, the smoothness of the C1 abundances as a function of Z for elements beyond Fe (and especially the smooth $N_s \sigma_n$ pattern derived from C1 data for pure s-species [e.g., 16]) for elements belonging to widely different volatility classes [e.g., 28,1,18] cannot be pure chance ("Suess curves") : it must mean that C1 abundances do largely reflect truly nuclear properties.

From C1's, one cannot get any information on abundances of H and C, N, O, simply because these elements do not form condensable solids between themselves (except for some graphite), and are comparatively so abundant that only a small fraction of them can combine with other elements into condensable solids (such as silicates, Fe-oxydes..). Noble gases are obviously not either retained in meteorites, nor do they have observable lines in the (comparatively cool) solar photosphere.

The lack of data on H abundance in meteorites poses the problem of the normalization of the meteoritic and photospheric abundances (the latter being always determined relative to H). This is achieved by a best average matching of the abundances obtained in both media for about 10 elements whose relative abundances are accurately measured and found in good agreement in both media. This normalisation is good to whithin ~ 5%.

The situation regarding the photospheric abundance determinations has improved a lot in recent years, and is still improving rapidly, thanks to many high quality measurements of atomic oscillator strengths [13,14]. In turn, these laboratory data have allowed an extreme refinement of the solar atmosphere model, which is now checked with high redundancy : it is able to reproduce the relative intensities of a huge number of observed atomic and molecular lines formed in widely different physical conditions. Every year, the abundance of a few additional elements is quite accurately determined. The striking point is that, in most cases, the new photospheric determination agrees very well with the meteoritic value, especially for refractory elements (a very noticeable exception is Fe ; § 4).

3.2. Solar Corona, and tentative "SEP derived photospheric" abundances

As discussed in another paper at this conference [22, fig. 4a,b,c] ; it seems now clear that the composition of the Solar corona differs from that of the photosphere [21,10]. Some fractionation, apparently related to First Ionization Potential (FIP), must take place in the rise of matter between photosphere and corona. So, observations of the composition of the corona, or of populations originating in the corona such as Solar Wind (SW) and Solar Energetic Particles (SEP), cannot be directly used to help determine the LG standard. However, [4] have recently proposed to indirectly determine, first, "SEP-derived coronal" and, second, "SEP derived photospheric" abundances, from their very accurate average SEP composition data. The idea is very interesting, but not without danger: it requires, (i) an accurate estimate of the average magnetic rigidity dependent bias of SEP abundances with respect to corona (which is actually determined by reference to the spectroscopic photospheric abundances of selected elements, and affected by the puzzle regarding the Fe abundance discussed in § 4), and (ii) a precise assumption regarding the shape of the functional dependence of the FIP distortion of coronal abundances with respect to photospheric ones. The latter point is very touchy, especially for elements with FIP ≈ 8 to 12 eV, and has in particular led the authors to propose a "SEP derived photospheric" abundance of C which is in flat contradiction with the quite accurate (± 25%) direct photospheric determinations (Table 1). So, I feel deriving photospheric abundances from SEP data is, by and large, still a bit premature : we would like to have a more precise and firmly established background regarding the fractionation processes which have to be corrected for.

But for specific, adequately chosen elements, I feel the SEP data may already allow a reliable refinement of the photospheric abundances. In particular, the C/O and N/O ratios are observed to be extremely stable in SEP's (no rigidity correction required ; cf. same A/Q values) [e.g., 4, 20, 17] ; the FIP related depletion is almost certainly the same for N and O, and not more pronounced for C than for O [22, fig. 4c]. This leads to a "SEP derived photospheric" abundance of N in perfect agreement with, but more accurate than the direct photospheric determination, and to an upper limit to the C abundance in the lower half of the error bar of the direct photospheric determination (see values in § 4 and table 1).

When applying such procedures, one should also keep in mind the risk of roundabout arguments. In particular, the procedure should never be applied to obtain a "SEP derived photospheric" Ne abundance ; the FIP of Ne is much higher than that of all other heavy elements [22, fig. 4c] ; it is therefore only the observed depletion of SEP Ne with respect to its accepted LG value that tells us that, above ≈ 11 eV, the FIP bias is constant up to as far as 22 eV ; this FIP bias cannot, in turn, tell us the LG (or photospheric) abundance of Ne ! And, in a more straightforward way, one should later be careful when comparing SEP abundances to "photospheric" abundances that were derived from these same SEP's !

Note also that the "mass unbiased baseline" SEP composition concept [20] is another, essentially equivalent, way of deriving a coronal composition from the SEP data.

4. An updated table of Local Galactic (LG) abundances

Table 1 gives my updated LG abundance evaluation for H, He and elements from C to Zn. It is based on my earlier studies covering both Solar System and HI, HII gas and stellar data [23,18,19,20,21] and, to a very large extent, on the meteoritic and photospheric values recently summarized by [2,13,14]. As usual, the important point is not so much the abundance values themselves (these converge well now, and are very similar from one table to another) as the realistic estimate of their uncertainties, given for all elements. These have decreased a lot, as compared to those in [20,21], which were based on earlier work. I stress, however, that the uncertainties are essentially those on the Solar System abundances (except for noble gases) ; the, maybe dominant, uncertainties associated with a possible deviation of the Solar System abundances from a typical local average are not considered (cf. § 2).

I now comment on various elements in Table 1.

H : The uncertainty given for H is related to that on the normalisation of the meteoritic and photospheric scales (§ 3.1).

Table 1

LOCAL GALACTIC ABUNDANCES *

(1987)

H	2.79×10^6	(1.05)
He	0.28×10^6	(1.15)
C	1360.	(1.25)**
N	270.	(1.40)**
O	2320.	(1.05)
F	0.084	(1.25)
Ne	330.	(1.50)
Na	5.8	(1.05)
Mg	107.	(1.04)
Al	8.5	(1.05)
Si	≡ 100.	(1.04)
P	1.04	(1.10)
S	45.	(1.15)
Cl	0.52	(1.15)
Ar	10.7	(1.50)
K	0.38	(1.08)
Ca	6.3	(1.04)
Sc	0.0034	(1.10)
Ti	0.27	(1.15)
V	0.0295	(1.05)
Cr	1.31	(1.06)
Mn	0.87	(1.20)
Fe	110.	(1.27)
Co	0.225	(1.07)
Ni	4.93	(1.06)
Cu	0.049	(1.19)
Zn	0.111	(1.20)

* Error factors ("within a factor of ...") are in parenthesis. Normalized to Si ≡ 100.

** "SEP-derived photospheric" values (§ 3.2) would yield C ⩽ 1050(1.20), i.e., strictly C ⩽ 1260, and N = 280(1.15).

He : The value for He, whose abundance is very uniform in the universe, is derived from a very broad set of evidences including HII region, hot star observations and best agreement with models of solar interior [e.g., 26,32,31,3,5]. (Do not use possibly fractionated solar wind or coronal values ; the latter are, in addition, extremely poorly measured [21]).

CNO : For CNO, we have no data from meteorites (§ 3.1). The accuracy of the photospheric value for O, determined with high redundancy from all sorts of atomic and molecular lines [13,14], is excellent. Larger error bars have still to be assigned to C and especially N. For the C/O and N/O ratios, one may want to try to derive values from Solar Energetic Particle data (§ 3.2, [4]). For N the procedure is safe and one indeed gets N ≈ 280 (1.15), in excellent agreement with the photospheric determination. For C, the procedure is more touchy (§ 3.2), and [4] get "SEP derived" C abundance twice lower than the photospheric value. I think the procedure can give only an upper limit for "SEP-derived photospheric" C, which, based on all existing data, lies around 1050(1.20), i.e., C ⩽ 1260.

Ne,Ar : Ne and Ar are based on HII region, HI gas and stellar data [18, 19, updated in 20]. These abundances are reasonably well determined. The constancy of the Ne^{++}/O^{++} ratio observed in a very large number of hotter HII regions is particularly impressive [18]. The Ar abundance predicted by the O and Si-burning models [8] agrees well with the observed value.

Na,P,S,Cl,K,Mn,Cu,Zn,and Ge,Pb (not in the table) : Quite generally, the abundances of volatile elements vary much more among the various types of chondrites than those of refractories. So the C1 values as well have more chances to be biased. For many volatile elements (S,Mn, Cu,Zn, but not Na,K), there is a hint that the C1 value are slightly higher (~ 15%) than the photospheric values [2,13]. Recall also that existing data, however meager as regards photosphere, suggest that volatile Ge and Pb could be ~ 1.6 times as abundant in C1's as in the Photosphere (Ge: Photo.: 72(1.38), C1: 118(1.10); Pb: Photo.: 1.97(1.12), C1: 3.15(1.08) on the scale Si ≡ 10^6) [15]. For the above elements, I have taken into account both error bars, not only their narrower range of overlap. For Cl and P, we have no reliable photospheric value to compare the C1 value with.

Fe,Ti : We have a very serious (and possibly ominous) problem with Fe. The photospheric value, based on a refined analysis of many neutral FeI lines with highly accurate atomic data, is higher by a factor of 1.45 ± 0.11 than the equally well defined C1 value [13,14] ! This is all the more a puzzle since the siderophile elements Cr, Co, Ni, Pd, which condensate together with Fe in meteorites, definitely do not show the same trend and are found equally abundant in C1's and in the photosphere [13]. By contrast, there could be another discrepancy for Ti, a refractory, not siderophile element ! The reality of this difference for Fe will be confirmed, or not, when we have accurate atomic data to reliably analyze the ionized Fe II photospheric lines. In Table 1, I have adopted a broad error bar for Fe, encompassing the photospheric and the C1 values.

8. References ("ICRC" stands for International Cosmic Ray Conference)

1. Anders, E., 1971, GCA, 35, 516.
2. Anders, E., and Ebihara, M., 1982, GCA, 46, 2363.
3. Boesgaard, A.M., and Steigman, G., 1985, Ann. Rev. Astr. Ap. 23, 319.
4. Breneman, H.H., and Stone, E.C. 1986, Ap. J. Letters, 299, L57.
5. Cahen, S., Doom, C., and Cassé, M., 1987, submitted to Ap. J.
6. Cameron, A.G.W., 1968, in Origin and Distribution of the Elements, L.H. Ahrens ed. (Pergamon), p. 125.
7. Cameron, A.G.W., 1973, Space Sci. Rev., 15, 121.
8. Cameron, A.G.W., 1982, in "Essays in Nuclear Astrophysics", C.A. Barnes, D.D. Clayton, and D.N. Schramm ed. (Cambridge U. Press), p. 23.
9. Clegg, R.E., Lambert, D.L., and Tomkin, J., 1981, Ap. J. 250, 262.
10. Cook, W.R., Stone, E.C., and Vogt, R.E. 1984, Ap. J. 279, 827.
11. Ebihara, M., Wolf, R., and Anders, E., 1982, GCA, 46, 1849.
12. Gratton, R.G., and Sneden, C., 1987, Astr. Ap. 178, 179.
13. Grevesse, N., 1984a, Phys. Scripta T8, 49.
14. Grevesse, N., 1984b, in Frontiers of Astronomy and Astrophysics, ed. R. Pallavicini (FLorence: Italian Astron. Soc.), p. 71.
15. Grevesse, N., and Meyer, J.P., 1985, 19th ICRC, La Jolla, 3, 5.
16. Howard, W.M., Matthews, G.J., Takahashi, K., and Ward, R.A., 1986, Ap. J. 309, 633.
17. McGuire, R.E., Von Rosenvinge, T.T., and McDonald, F.B., 1986, Ap.J., 301, 938.
18. Meyer, J.P., 1979a, in "Les Eléments et leurs Isotopes dans l'Univers", 22nd Liège Internat. Astrophys. Symp. (U. of Liège Press), p. 153,465,477,489.
19. Meyer, J.P., 1979b, 16th ICRC, Kyoto, 2, 115.
20. Meyer, J.P., 1985a, Ap. J. Suppl. 57, 151.
21. Meyer, J.P., 1985b, Ap. J. Suppl. 57, 173.
22. Meyer, J.P., 1987, this conference.
23. Meyer, J.P., and Reeves, H. 1977, 15th ICRC, Plovdiv, 2, 137.
24. Peimbert, M., 1987, in : Star Forming Regions, IAU Symp. 115, M. Peimbert and J. Jugaku ed. (Reidel), p. 111.
25. Podosek, F.A., 1978, Ann. Rev. Astr. Ap. 16, 293.
26. Shaver, P.A., 1983, ESO Workshop on Primordial He, p. 299.
27. Shaver, P.A., et al., 1983, MNRAS 204, 53.
28. Suess, H.E., 1947, Z. Naturforsch. 22, 311.
29. Suess, H.E., and Urey, H.C., 1956, Rev. Mod. Phys. 28, 53.
30. Twarog, B.A., 1980, Ap. J. 242, 242.
31. Wolf, S.C., and Heasley, J.N., 1985, Ap. J. 292, 589.
32. Yang, J., et al., 1984, Ap. J. 281, 493.

SOLAR-SYSTEM ABUNDANCES OF THE ELEMENTS

Edward Anders

Enrico Fermi Institute and Dept. of Chemistry,
University of Chicago, Chicago, IL 60637, USA

It is generally accepted that Cl chondrites contain most elements in unfractionated, solar proportions, and thus are a good source of solar-system abundances of the elements. Table 1 lists two of the most recent compilations, by Cameron (1982) and Anders and Ebihara (1982; hereafter AE). The AE values are based on more recent data (literature coverage through mid-1982 rather than 1976), and more stringent data selection, but the differences between the two tables are usually less than 20%, indicating convergence of the analytical data.

A few comments on Table 1. The σ values in column 4 are standard deviations of the *population*; these can be converted to σ_{mean} by dividing them by the square root of the number of analyses from column 5. Parenthesized values in column 5 mean that the abundances was estimated indirectly from data for other meteorite classes; zeros indicate interpolated or theoretical estimates, whereas dashes mean that the abundance was based on astronomical data. Blanks are shown for the rare earths as these elements were estimated by a more involved procedure, using high-quality data from several other chondrite classes in addition to Cl's. The "Orgueil" column gives abundances *by mass*, (or by volume for noble gases) for this, particularly well-studied, Cl chondrite. These values generally are not quite consistent with the atomic abundances of column 4, being based on one rather than all Cl chondrites. (Moreover, they are low for elements that are incompletely condensed in Cl chondrites: H, C, N, O, and noble gases). If mass abundances consistent with atomic abundances are needed, they can be calculated with the formula given in the third footnote.

Table 1. Abundances of the Elements (Atoms/10^6Si)

Element	Cameron (1982)	Anders and Ebihara 1982	σ(%)	# Anal.	Orgueil[+‡]	# Anal.
1 H	2.66×10^{10}	2.72×10^{10}	5	-	2.02 *%*	2
2 He	1.8×10^{9}	2.18×10^{9}	12	-	56 *nL/g*	
3 Li	60	59.7	2.7	3	1.59 *ppm*	2
4 Be	1.2	0.78	7.5	(3)	26.7 *ppb*	0
5 B	9	24	30	(34)	1.25 *ppm*	6
6 C	1.11×10^{7}	1.21×10^{7}	30	-	3.45 *%*	7
7 N	2.31×10^{6}	2.48×10^{6}	40	-	3180 *ppm*	4
8 O	1.84×10^{7}	2.01×10^{7}	20	-	46.4 *%*	4
9 F	780	843	15	7	58.2 *ppm*	5
10 Ne	2.6×10^{6}	3.76×10^{6}	30	-	203 *pL/g*	
11 Na	6.0×10^{4}	5.70×10^{4}	7.3	17	4830 *ppm*	12
12 Mg	1.06×10^{6}	1.075×10^{6}*	3.6	14	9.55 *%*	10
13 Al	8.5×10^{4}	8.49×10^{4}	4.9	18	8620 *ppm*	12
14 Si	1.00×10^{6}	1.00×10^{6}	4.2	9	10.67 *%*	5
15 P	6500	1.04×10^{4}	10	4	1180 *ppm*	3
16 S	5.0×10^{5}	5.15×10^{5}*	13	5	5.25 *%*	2
17 Cl	4740	5240	15	10	698 *ppm*	8
18 Ar	1.06×10^{5}	1.04×10^{5}	20	-	751 *pL/g*	7
19 K	3500	3770	7.9	27	569 *ppm*	18
20 Ca	6.25×10^{4}	6.11×10^{4}	7.1	15	9020 *ppm*	12
21 Sc	31	33.8	9.0	15	5.76 *ppm*	10
22 Ti	2400	2400	5.0	7	436 *ppm*	7
23 V	254	295	4.8	8	56.7 *ppm*	6
24 Cr	1.27×10^{4}	1.34×10^{4}	8.0	12	2650 *ppm*	8
25 Mn	9300	9510	9.7	16	1960 *ppm*	12
26 Fe	9.0×10^{5}	9.00×10^{5}*	2.7	19	18.51 *%*	14
27 Co	2200	2250	6.9	16	509 *ppm*	10
28 Ni	4.78×10^{4}	4.93×10^{4}	5.7	25	1.10 *%*	14
29 Cu	540	514	12	6	112 *ppm*	3
30 Zn	1260	1260	4.2	21	308 *ppm*	11
31 Ga	38	37.8	6.9	14	10.1 *ppm*	10
32 Ge	117	118	9.4	28	32.2 *ppm*	20
33 As	6.2	6.79	11	14	1.91 *ppm*	10
34 Se	67	62.1	6.4	18	18.2 *ppm*	11
35 Br	9.2	11.8	19	(18)	3.56 *ppm*	10
36 Kr	41.3	45.3	11	-	8.7 *pL/g*	7
37 Rb	6.1	7.09	7.0	17	2.30 *ppm*	11
38 Sr	22.9	23.8	8.3	15	7.91 *ppm*	12
39 Y	4.8	4.64	8.2	3	1.50 *ppm*	2
40 Zr	12	10.7	13	7	3.69 *ppm*	7
41 Nb	0.9	0.71	14	(9)	250 *ppb*	0
42 Mo	4.0	2.52	5.4	2	920 *ppb*	2
44 Ru	1.9	1.86	5.4	9	714 *ppb*	5
45 Rh	0.40	0.344	8	(7)	134 *ppb*	0
46 Pd	1.3	1.39	7.7	18	557 *ppb*	12
47 Ag	0.46	0.529	9.5	11	220 *ppb*	8
48 Cd	1.55	1.59	6.5	24	673 *ppb*	17
49 In	0.19	0.184	6.4	24	77.8 *ppb*	16
50 Sn	3.7	3.82	9.4	11	1680 *ppb*	9
51 Sb	0.31	0.352	18	34	155 *ppb*	22

Element	Cameron (1982)	Anders and Ebihara 1982	σ(%)	# Anal.	Orgueil†‡	# Anal.
52 Te	6.5	4.91	12	10	2280 *ppb*	7
53 I	1.27	0.90	21	(11)	430 *ppb*	0
54 Xe	5.84	4.35	13	0	8.6 *pL/g*	6
55 Cs	0.39	0.372	6	18	186 *ppb*	9
56 Ba	4.8	4.36	4.5	7	2270 *ppb*	7
57 La	0.37	0.448	5.1	10	236 *ppb*	7
58 Ce	1.2	1.16			619 *ppb*	8
59 Pr	0.18	0.174			90 *ppb*	2
60 Nd	0.79	0.836			462 *ppb*	9
62 Sm	0.24	0.261			142 *ppb*	8
63 Eu	0.094	0.0972			54.3 *ppb*	15
64 Gd	0.42	0.331			196 *ppb*	5
65 Tb	0.076	0.0589			35.3 *ppb*	4
66 Dy	0.37	0.398			242 *ppb*	4
67 Ho	0.092	0.0875			54 *ppb*	
68 Er	0.23	0.253			160 *ppb*	4
69 Tm	0.035	0.0386			22 *ppb*	1
70 Yb	0.20	0.243			166 *ppb*	12
71 Lu	0.035	0.0369			24.3 *ppb*	10
72 Hf	0.17	0.176	6.6	3	119 *ppb*	3
73 Ta	0.020	0.0226	10	1	17 *ppb*	1
74 W	0.30	0.137	7.3	1	89 *ppb*	1
75 Re	0.051	0.0507	12	20	36.9 *ppb*	13
76 Os	0.69	0.717	9.8	18	504 *ppb*	12
77 Ir	0.72	0.660	6.3	35	473 *ppb*	26
78 Pt	1.41	1.37	10	9	953 *ppb*	6
79 Au	0.21	0.186	15	39	145 *ppb*	25
80 Hg	0.21	0.52	50	0	390 *ppb*	1
81 Tl	0.19	0.184	9.4	18	143 *ppb*	12
82 Pb	2.6	3.15	7.8	3	2430 *ppb*	1
83 Bi	0.14	0.144	8.2	13	111 *ppb*	7
90 Th	0.045	0.0335	5.7	9	28.6 *ppb*	1
92 U	0.027	0.0090	8.4	16	8.1 *ppb*	7

*Average of mean values for individual meteorites. For the remaining elements, a straight average of all acceptable analyses was used.

†Abundances of light elements in Orgueil are from the following sources: H *(B4)*; C *(B5, G12, W3)*; N *(G12)*; O *(W3)*. Abundances of noble gases (in nL/g or pL/g at STP) are not for the elements but for the principal isotopes He^4, Ne^{20}, Ar^{36}, Kr^{84}, and Xe^{132} *(J2, M13)*. As He and Ne in bulk Orgueil are derived mainly from the solar wind, the values for these two are based on ratios for NaOH-etched Orgueil silicates: $Ne^{20}/Ar^{36} = 0.27$ and $He^4/Ne^{20} = 277$.

‡Abundances in Orgueil are adequately close to the Cl chondrite mean, except for REE, where the values in Table 6 are preferable. For maximum accuracy, a Cl chondrite mean may be calculated from the atomic abundances in column 3, using the relation $C = 3.788 \times 10^{-3} HA$, where C = weight concentration (ppm), H = atomic abundance, and A = atomic weight.

An update of the AE table (Anders and Grevesse, 1988), confirms that the data for Cl chondrites are converging. There are small changes for most elements, but with two exceptions (one of them interpolated) they are within one standard deviation of the AE table. The indirect values have fared well, being almost exactly confirmed in those (two) cases where they have been checked by direct measurements on Cl chondrites.

One must continue to ask, of course, whether Cl chondrites are truly unfractionated relative to the Sun (except for HCNO and noble gases). AE listed 7 cases — Ga, Ge, Nb, Ag, Lu, W, and Os — where meteoritic and solar abundances differed by more than twice the nominal error of the solar data. Most of these discrepancies have been eliminated by better solar data or *f*-values, but Ge has remained, now also joined by Pb (Grevesse and Meyer, 1985). Though past experience has tended to support the meteoritic data, all such discrepancies must be approached with an open mind.

Acknowledgements

This work was supported in part by NASA Grant NAG 9-52.

REFERENCES

Anders, E. and Ebihara, M. (1982) Solar-system abundances of the elements. *Geochim. Cosmochim. Acta* **46**, 2362–2380.

Anders, E. and Grevesse, N. (1988) Abundances of the elements: meteoritic and solar. Manuscript in preparation.

Cameron, A. G. W. (1982) Elementary and nuclidic abundances in the solar system. In *Essays in Nuclear Astrophysics* (eds. C. A. Barnes, D. N. Schramm, and D. D. Clayton), Cambridge Univ. Press.

Grevesse, N. and Meyer, J.-P. (1985) Germanium and lead: significant differences between meteoritic and photospheric abundances? *19th Intern. Cosmic Ray Conf.* 3, 5, LaJolla, CA.

UBIQUITOUS METEORITIC ISOTOPIC ANOMALIES IN Ca, Ti, AND Cr

S. NIEMEYER
Lawrence Livermore National Laboratory
P.O. Box 808, L-232
Livermore, CA 94550

Isotopic anomalies in meteorites provide a means of tracing the evolution of the solar system prior to the assembling of the meteorites. In this review talk, I consider the elements Ca, Ti, and Cr where isotopic anomalies are commonly present in refractory inclusions in Allende; the only other element to have such an ubiquitous anomaly is oxygen. Data for these 3 elements in Allende inclusions are shown in Figures 1-3 for Ti [1-4], Ca [5,6], and Cr [7-9] respectively. Every inclusion shows an excess of Ti-50 relative to the isotopic composition of terrestrial Ti (epsilon units are deviations in parts per ten thousand from the terrestrial ratio). Refractory inclusions from other CV meteorites show similar effects. Resolving the effect in Ca is more difficult due to the large mass fractionation corrections that are required. But repeated analyses enabled Jungck et al. [6] to resolve excesses of Ca-48 for the majority of their inclusions. The data, however, do not show a quantitative correlation between the magnitudes of the Ca and Ti isotopic anomalies. This suggests that Ti exhibits a more refractory behavior than Ca during thermal processing. More recently, similar isotopic anomalies have been resolved for Cr. For all three elements the isotopic patterns are consistent with the relative enrichment of the heaviest isotope, but this assignment for Cr is rather arbitrary because it has only 3 non-radiogenic isotopes. The typical excesses are 4, 9, & 6 epsilon units for Ca, Ti, & Cr respectively. Initial efforts to resolve Ni anomalies were unsuccessful, but Birck et al. [10] recently identified small 1-2 epsilon excesses of Ni-64.

Further Ti isotopic work has demonstrated that these anomalies are not limited to these refractory inclusions. Figure 4 shows data for whole-rock samples of meteorites from all 4 classes of carbonaceous chondrites [11,12]. The stippled bars are the mean for each class; Ti-50 excesses are clearly present in all 4 classes. This is true even for the average of 3 CI meteorites, where refractory inclusions are not observed. The fact that the Ti anomalies are so similar for all these meteorites, even though the proportion of inclusions varies considerably among them, suggests that constituents other than the inclusions carry anomalous Ti. More direct evidence is apparent in Figure 5 for my chondrule data taken in the past year [13]. Four out of eight Allende chondrules show well-resolved Ti anomalies. Moreover, at least 2 isotopically distinct reservoirs are required in which one is enriched in Ti-50 relative to the earth, whereas the other is depleted.

The first evidence for a Ti anomaly in a non-carbonaceous chondrite is for the chondrules from Chainpur, which are also shown in Figure 5. All results lie below the normal, and #1 & #4 give well-resolved deficits of Ti-50. This is in contrast to the carbonaceous chondrites chondrules which generally show excesses of Ti-50. This relationship is intriguing since the oxygen isotopes gives excesses of O-16 for the Allende chondrules, whereas ordinary chondrite chondrules are depleted in O-16. This suggests a possible coherence between the oxygen and Ti isotopic anomalies.

Experiments to measure Ti isotopes in matrix from carbonaceous chondrites are difficult to interpret because of the difficulty in purifying matrix separates. My data for matrix-rich separates from Allende and 2 CM meteorites [14] suggest, however, that matrix also carries Ti-50 excesses on the order of 1-2 epsilon units. This along with the whole-rock CI meteorite data indicates that matrix in carbonaceous chondrites typically carries this isotopic signature. Thus, generally all the 3 major constituents of these most primitive meteorites are enriched in Ti-50 relative to the earth and ordinary chondrites.

Why do these anomalies exist when so many elements show no isotopic variations? The first issue to consider is the source of these particular isotopes. Strong arguments can be made against a solar system source itself [2], e.g. T Tauri irradiation; nucleosynthetic models indicate that most likely these nearly mono-isotopic contributions to the various Fe-peak elements were synthesized under nuclear equilibrium (or near equilibrium) conditions in a neutron-rich environment in massive stars. Judging between specific models, e.g. Si-burning vs. full nuclear equilibrium, is hampered by the uncertainties introduced by: 1) the chemical fractionations which probably occurred, as evidenced by the lack of quantitative correlation between the Ca and Ti anomalies in inclusions, and 2) deciding whether the history of the exotic component warrants comparing the nucleosynthetic model predictions to the relative number of exotic atoms for each isotope, as opposed to the more typical comparison to the epsilon values. Next one needs to consider the history of these exotic atoms from the time they are expelled from the star to their eventual incorporation into the meteorites. Little is known about the evolution of dust grains in the interstellar medium, but a plausible view is that these exotic atoms were condensed into a refractory phase soon after leaving the star, and that this carrier was more resistant to interstellar destruction than most carriers of Ti, Ca, and Cr isotopes. Then one needs to consider the distribution and response of the various types of dust grains within the solar nebula as it evolves to eventually form our solar system.

Even though our understanding of the connection between star and meteorite is quite tenuous, the isotopic anomalies also provide important insights into the relationships between the various meteorites and their constituents. The precursor assemblages of the various carbonaceous chondrites must be fairly similar, at least in regard to the relative proportion of the exotic carrier grains, because of the overall similarity in the magnitudes of the Ti-50 enrichments. But there was also a large-scale inhomogeneity in the solar nebula because the more abundant matter represented by the earth and ordinary chondrites have distinctly different isotopic signatures. Even within the broadly homogeneous region where carbonaceous chondrites formed, a fine-scale heterogeneity is evident, witnessed by the isotopic diversity in the Allende chondrules. The similarity of the Ti anomalies for chondrules and matrix is consistent with the precursor dustballs for the chondrules being closely related to the matrix found in the carbonaceous meteorites. The Ti anomalies for chondrules are viewed as features which are inherited from the precursor dustballs, rather than being generated by the transient heating event itself on otherwise homogeneous precursor dust. Likewise, the anomalies in the refractory inclusions are attributed to the precursor assemblages rather than the final heating event which formed these objects. The larger anomalies for the inclusions are a consequence of the complex processing of the dust which resulted in a concentration of refractory elements, and this could naturally account for the relatively higher proportion of the refractory carrier of these particular isotopes. But consideration of the full data base indicates that factors other than simply the refractory nature of the carrier phase played a role in producing the observed distribution of these isotopic anomalies in meteorites. Obviously we have only begun to answer the question of why these elements show isotopic anomalies in these elements, and a fuller understanding will be closely tied to unravelling the complicated evolution of the dust phase in the early solar nebula.

This work was performed under the auspices of the U.S. Dept of Energy by LLNL under No. W-7405-Eng-48.

References

[1] Niemeyer S. and Lugmair G.W., Ubiquitous isotopic anomalies in Ti from normal Allende inclusions, Earth Planet. Sci. Lett. 53, 211-225 (1981).

[2] Niemeyer S. and Lugmair G.W., Titanium isotopic anomalies in meteorites, Geochim. Cosmochim. Acta 48, 1401-1416 (1984).

[3] Niederer F.R., Papanastassiou D.A., and Wasserburg G.J., The isotopic composition of titanium in the Allende and Leoville meteorites, Geochim. Cosmochim. Acta 45, 1017-1031 (1981).

[4] Niederer F.R., Papanastassiou D.A., and Wasserburg G.J., Absolute isotopic abundances of Ti in meteorites, Geochim. Cosmochim. Acta 49, 835-852 (1985).

[5] Niederer F.R. and Papanastassiou D.A., Ca isotopes in refractory inclusions, Geochim. Cosmochim. Acta 48, 1279-1293 (1984).

[6] Jungck M.H.A., Shimamura T., and Lugmair G.W., Ca isotope variations in Allende, Geochim. Cosmochim. Acta 48, 2651-2658, (1984).

[7] Birck J.-L. and Allegre C.J., Chromium isotopic anomalies in Allende refractory inclusions, Geophys. Res. Lett. 11, 943-946 (1984).

[8] Birck J.-L. and Allegre C.J., Evidence for the presence of 53-Mn in the early solar system, Geophy. Res. Lett. 12, 745-748 (1985).

[9] Papanastassiou D.A., Chromium isotopic anomalies in the Allende meteorite, Ap. J. 308, L27-L30 (1986).

[10] Birck J.L., Prombo C.A., and Lugmair G.W., Ni and Cr isotopes in Allende inclusions, in Lunar and Planetary Science XVIII, 79-80, (1987).

[11] Niemeyer S., Systematics of Ti isotopes in carbonaceous chondrite whole-rock samples, Geophys. Res. Lett. 12, 733-736 (1985).

[12] Niemeyer S., Ti isotopes in Allende and Chainpur chondrules and in the Kaidun breccia, in Lunar and Planetary Science XVIII, 718-719 (1987).

[13] Niemeyer S., Titanium isotopic anomalies in chondrules from carbonaceous chondrites, submitted to Geochim. Cosmochim. Acta (1987).

[14] Niemeyer S., Ti isotopic anomalies in Orgueil and constituents of CM and CV meteorites (abstract), Meteoritics 21, 470-472 (1986).

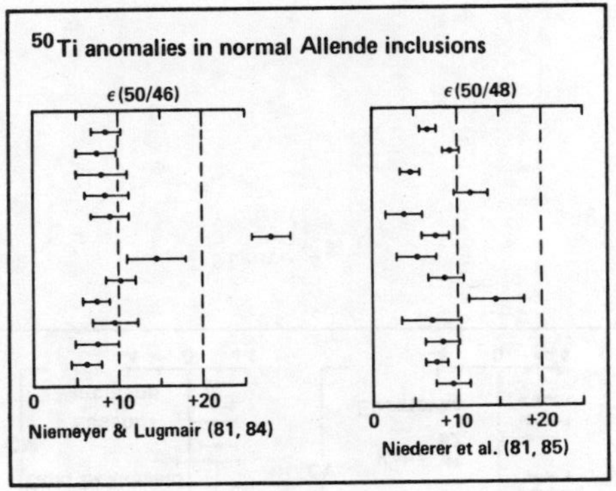

Figure 1.

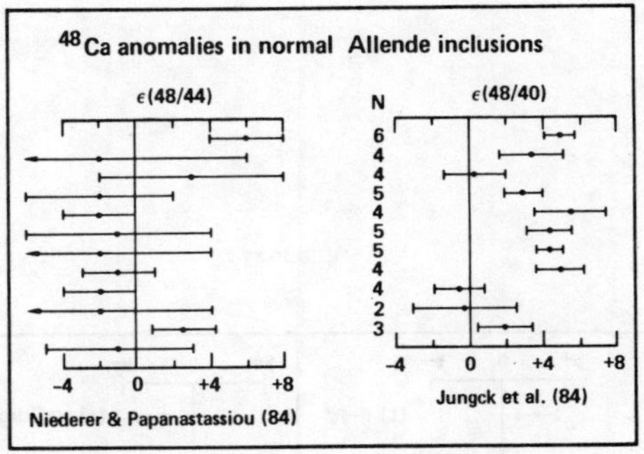

Figure 2.

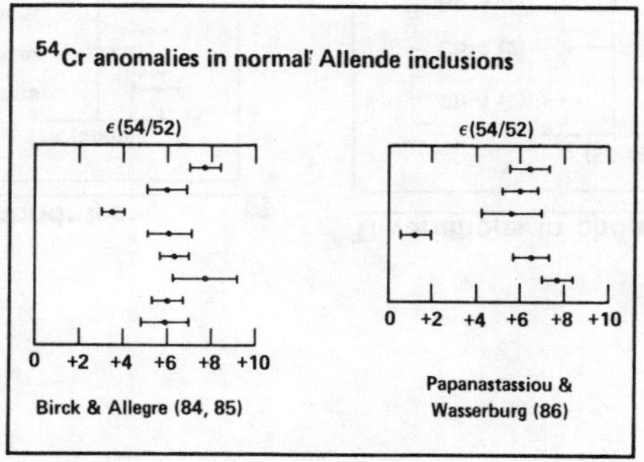

Figure 3.

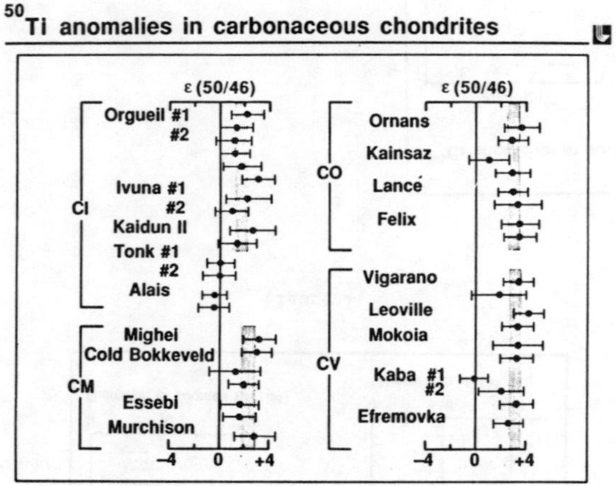

Figure 4.

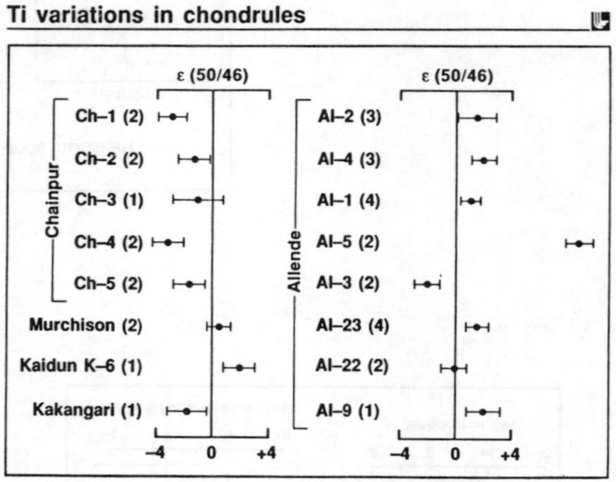

Figure 5.

NEUTRON-RICH EQUILIBRIUM NUCLEOSYNTHESIS AND METEORITIC ABUNDANCE ANOMALIES

D. Hartmann

Board of Studies in Astronomy and Astrophysics
University of California at Santa Cruz, CA 95064

Abstract

Experimental data on correlated ^{48}Ca, ^{50}Ti, and ^{54}Cr abundance anomalies in meteoritic inclusions lend support to the idea that the natural synthesis of these rare isotopes occurs within neutron-rich stellar zones that have achieved nuclear statistical equilibrium (NSE). We have studied matter in the neutron-rich environment near the mass-cut of a proto neutron star that is shock heated to NSE and subsequently ejected in the supernova event. The correlated abundance anomalies can be explained if a typical supernova event contributes only a small amount of such neutron-rich matter. We discuss the sensitivity of the neutron-rich equilibrium model to nuclear binding properties.

Introduction

The observation of isotopic anomalies in meteorites demonstrates that the pre-solar nebula was not well-mixed and of chemically and isotopically uniform composition as it was believed two decades ago (Suess 1965). In fact, it is the preservation of isotopic heterogeneities in primitive meteoritic samples that offers a valuable opportunity to study the nucleosynthetic history of solar system material (Lee 1987).

The group of meteorites that is believed to represent the primordial abundances of the solar nebula are the carbonaceous chondrites. In several of them "large" scale (linear dimensions ~1-10 mm) Ca-Al-rich inclusions (CAI) have been found. Owing to their high transition temperatures they are expected to be among the first elements to condense out of the cooling pre-solar nebula (for an alternative to the hot solar nebula model see *e.g.* Trivedi 1987). The data seem to indicate that the CAI's constitute the oldest solid material of the solar system.

In particular, the Allende meteorite has provided a rich source of isotopically anomalous material. The observed anomalies are the direct experimental proof that not a single but several different nucleosynthetic processes are needed to explain the origin and the distribution of the elements.

Correlated Abundance Anomalies

Anomalous abundances of neutron-rich isotopes (^{48}Ca, ^{50}Ti, and ^{54}Cr) found at the low mass wing of the iron peak have a significant impact on the nucleosynthesis theory (see Lee 1987 for a recent review). The favored interpretation for them invokes the neutron-rich e-process (nnse) described briefly in the following section (see also Hartmann *et al.* 1985 and references therein). The nnse-process is thought to operate in extremely neutron-rich sites situated near the neutron star in exploding massive stars. The origin of the less neutron-rich isotopes in the Ca-Ti-Cr range is due to a variety of nucleosynthetic processes (see Table 12 in Woosley, 1986) located farther out in the same star.

Non-linear, or nuclear, effects have been detected in a large number of FUN and non-FUN inclusions. It appears that the size of the anomalies increase in the following order: bulk meteorites, non-FUN CAI, FUN CAI, and hibonite grains, in other words, the largest anomalies are found in smallest samples (Lee 1987). Perhaps it was this trend that prevented an earlier recognition of the heterogeneous nature of pre-solar gas.

The Ca data reveal substantial effects in ^{48}Ca without any indication of non-linear effects in ^{46}Ca (Lee, Papanastassiou, and Wasserburg 1978; Lee, Russel, and Wasserburg 1979; Niederer and Papanastassiou 1984; Jungck *et al.* 1984; Zinner *et al.* 1986; Prombo and Lugmair 1986). Ti isotopic anomalies are quite wide spread among meteoritic samples (Niederer *et al.* 1980, 1981, 1985; Niemeyer and Lugmair 1981, 1984; Niemeyer 1985; Hutcheon *et al.* 1983; Fahey *et al.* 1985, 1986; Ireland *et al.* 1985; Hinton *et al.* 1987). The largest nuclear effects appear at ^{50}Ti and all lighter Ti isotopes appear to have a normal distribution. Recently large Cr anomalies, in particular at ^{54}Cr, have been measured in FUN and non-FUN inclusions (Birck and Allegre 1984; Papanastassiou 1986). For more details on the Ca, Ti, and Cr data see the contribution of S. Niemeyer in this volume.

There exists a rough correlation between the observed anomalies of ^{48}Ca and ^{50}Ti in the sense that their signs are almost always the same. The situation for ^{54}Cr and ^{50}Ti is similar. The data indicate that the ^{50}Ti and ^{54}Cr effects for normal Ca-Al-rich inclusions are comparable and both are somewhat larger than the ^{48}Ca effect (Lee 1987).

The observed correlations together with the absence of ^{46}Ca anomalies are the salient features that must be explained by the nucleosynthetic model. Several theoretical models have been put foreward during recent years (Cameron 1979; Sandler *et al.* 1982; Harris 1983; Ziegert *et al.* 1985; Hillebrandt 1986; Hillebrandt *et al.* 1986; Hartmann, Woosley, and El Eid 1985). Comparison of the most recent data with the predictions made by the above models leaves only the multi-zone neutron-rich equilibrium (Hartmann *et al.* 1985) as a valid candidate to explain the Ca-Ti-Cr anomalies (Lee 1987). Because of the apparent success of this model, it appears useful to consider the intrinsic accuracy of e-process calculations in neutron-rich matter. Under equilibrium conditions the abundances do not depend on reaction rates but are sensitive to the nuclear binding energies.

Neutron-rich Equilibrium Abundances

The compositions discussed here are those of material which has been heated to some maximum temperature T_{max} greater than 5×10^9K at reasonably high density ρ and subsequently ejected on a hydrodynamic time scale

$$\tau_{dyn} \sim 446\ \rho^{-1/2}\ \ s. \tag{1}$$

At this high temperature the time scale for any initial composition to approach equilibrium is small compared to the dynamic time. The expanding matter cools adiabatically and eventually drops out of equilibrium as the temperature falls below the critical value required to maintain strong flows from species to species ($T\sim3\times10^9$K); hence the evolution has to be followed by using a nuclear reaction network. The composition at "freezeout" is reasonably represented by the equations of nuclear statistical equilibrium (Delano and Cameron 1971; Hartmann, Woosley, and El Eid 1985) but still undergoes substantial changes by β decay. To study the effect of high degrees of neutron enrichment it is useful to approximate the frozen composition by an equilibrium distribution, taking advantage of the simplicity of the method.

All components of the system, nuclei, electrons, and free nucleons, are assumed to be in thermal equilibrium at a given temperature T (β=1/kT), *i.e.*, all strong and electromagnetic reactions (not the weak ones) occur at rates balanced by their inverses. In that case the composition of the system is specified by T, ρ, and the mean electron molecular weight Y_e. Starting from the basic assumption that nuclei can be treated as an ideal, non-relativistic, non-degenerate gas, the mass fraction of the nucleus (Z,A) is given by

$$X_{Z,A} = \frac{A}{N_A\,\rho}\,\omega_{Z,A}(\beta)\left(\frac{2\pi m^0_{Z,A}}{h^2\,\beta}\right)^{3/2} exp\left(\beta(\mu_{Z,A}+B^0_{Z,A})+\delta\eta_c\right) \quad , \tag{2}$$

where $\omega_{Z,A}(\beta)$ is the nuclear partition function, $B^0_{Z,A}$ is the ground state binding energy, and $\mu_{Z,A}$ is the nuclear chemical potential without rest mass. In statistical equilibrium, the nuclear chemical potential of free and bound nucleons are equal, so that

$$\mu_{Z,A} = (A-Z)\mu_n + Z\mu_p \; . \tag{3}$$

In addition one must solve the equations of mass conservation

$$X_n + X_p + \sum_{Z,A} X_{Z,A} = 1 \tag{4}$$

and charge neutrality

$$X_p + \sum_{Z,A}\frac{Z}{A}\,X_{Z,A} = Y_e = \frac{n_e}{N_A\,\rho} = \frac{1}{2}(1-\eta) \; , \tag{5}$$

where η is the neutron excess. The free nucleons are treated as non-relativistic Fermi gases so that their densities are given by

$$\rho_{n,p} = \frac{1}{2\pi^2}\left(\frac{2m^*_{n,p}}{\hbar^2\,\beta}\right)^{3/2} F_{1/2}(\eta_{n,p}) \; , \tag{6}$$

where $F_{1/2}(\eta)$ is a well known Fermi integral. The chemical potential of free nucleons is given by

$$\mu_{n,p} = \eta_{n,p}/\beta + V_{n,p}(\rho_n,\rho_p) + V_{Coul}\delta_{(n,p),p} \; . \tag{7}$$

The potential $V_{n,p}$ and the nuclear effective mass $m^*_{n,p}$ are due to strong interactions which become important at very high densities and can be calculated utilizing Skyrme type forces, but are neglected here. On the other hand, the multi-component plasma at temperatures and densities under consideration is strongly Coulomb coupled, *i.e.*, the ratio of the average Coulomb energy per particle to the thermal energy is not small compared to unity. The corrections to the equilibrium abundances introduced by the particle-particle correlations mediated by the Coulomb interactions are contained in V_{Coul} and $\delta\eta_c$ and will be discussed elsewhere. We now turn our attention to a discussion of the sensitivity of the neutron-rich equilibrium model to uncertainties in the nuclear binding properties.

Sensitivity to Nuclear Binding Energies

It is apparent from the equations presented above, that there are several quantities that could cause substantial shifts in the final abundances: nuclear partition functions at finite temperature, nuclear binding energies, nuclear force parametrizations, Coulomb shifts in a strongly coupled plasma, freezeout correction factors determined from a complex nuclear reaction network, and some more. Here we consider only the effect of uncertainties in nuclear mass laws.

As described in Hartmann *et al.* (1985) the binding energies of experimentally unexcessable nuclear masses are determined by an extrapolation scheme that relies on a nuclear mass law. The errors associated with this procedure increase with the distance from the valley of β stability. Figure 2a shows the differences between the binding energies used in our 1985 study (von Groote, Hilf, and Takahashi, 1976) and those from the recent mass evaluation data base (Wapstra and Audi 1985). The theoretical evaluation is based on the so-called macroscopic-microscopic model in which the smooth behavior is represented by the droplet mass model and the fluctuating part by a shell correction term. As is apparent from the figure typical differences are of order 1 MeV and there is a strong systematic trend with increasing neutron excess. In Figure 2b we compare the experimental data with recent calculations of ground-state masses for 4678 nuclei utilizing the finite-range droplet model for the macroscopic part and a folded Yukawa single particle potential for the microscopic contribution (Möller *et al.* 1987; for more details on this mass formula see the contribution of M. Howard in this volume). It is apparent that though the overall dispersion is substantially larger the new mass law has the advantage of exhibiting no systematic trends with increasing neutron excess.

We have calculated neutron-rich equilibrium mass fractions with binding energies obtained separately from the two mass laws. Final abundances are calculated after β decay to the stable isotopes and all mass fractions are normalized to solar values (Cameron 1982). Concentrating on the abundances of those Ca-Ti-Cr isotopes that exhibit strong meteoritic anomalies we compare in Figure 1 the results for both sets of binding energies. We find that mass fraction ratios do not change by more than about one order of magnitude. This is, however, an upper limit. When experimental masses are used in conjunction with a mass law based extrapolation scheme (Hartmann *et al.* 1985), substantially smaller uncertainties result. Still, with the availability of more and improved meteoritic data it becomes important to use a mass law that is free of systematic effects and accurate to better than 1 MeV.

We would like to thank P. Möller, W. D. Myers, W. J. Swiatecki, and J. Treiner for kindly providing their recent nuclear mass formula prior to publication and for helpful discussions. It is a pleasure to thank S. Woosley, R. Hoffman, and G. Mathews for their assistance and many clarifying discussions. It is also a pleasure to acknowledge stimulating discussions with D. Arnett, M. F. El Eid, M. Howard, G. Mathews, P. Möller, S. Niemeyer, G. Stringfellow, and K. Takahashi during the Symposium. This research has been supported by the National Science Foundation under grant AST-84-18185, by the Cal Space Institute under grant CS-56-84, and the Institute of Geophysics and Planetary Physics at Lawrence Livermore National Laboratory under grant IGPP-LLNL-88-27.

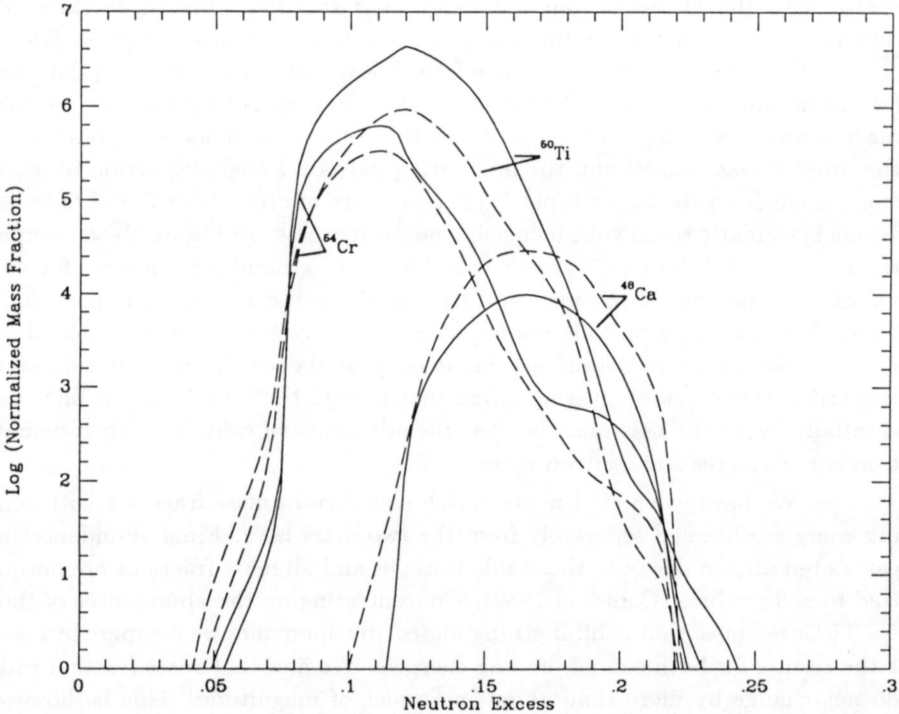

Fig. 1 — NSE mass fractions (after β-decay) normalized to solar abundances for ^{48}Ca, ^{50}Ti, and ^{54}Cr as a function of neutron excess. Shown are calculations for binding energies from von Groote, Hilf, and Takahashi (1976) (dashed lines) and from Möller *et al.* (1987) (full lines).

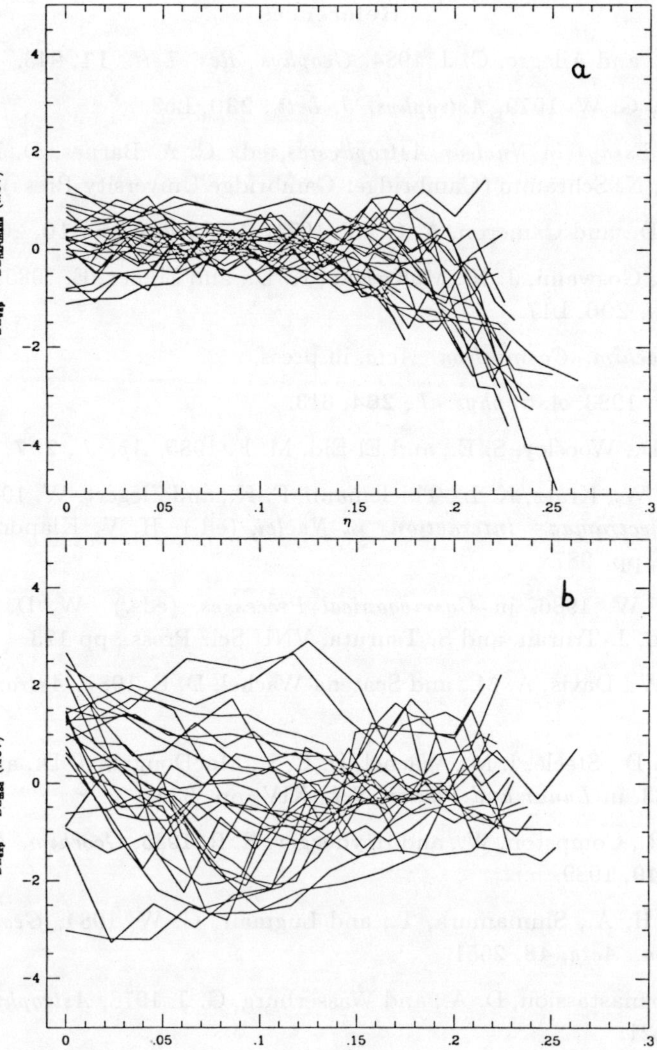

Fig. 2 — Comparison of experimental binding energies (Wapstra and Audi 1985) with mass model predictions taken from a) von Groote, Hilf, and Takahashi (1976) (see Hartmann, Woosley, and El Eid 1985), and b) Möller, Myers, Swiatecki, and Treiner (1987) as a function of neutron excess η. The lines connect isotopes of constant Z (14 to 42).

References

Birck, J. -L., and Allegre, C. J. 1984, *Geophys. Res. Lett.*, **11**, 943.

Cameron, A. G. W. 1979, *Astrophys. J. Lett.*, **230**, L53.

— 1982, in *Essays in Nuclear Astrophysics*, ed. C. A. Barnes, D. D. Clayton, and D. N. Schramm (Cambridge: Cambridge University Press), p. 23.

Delano, M. D., and Cameron, A. G. W. 1971, *Ap. Spac. Sci.*, **10**, 203.

Fahey, A. J., Goswami, J. N., McKeegan, K. D., and Zinner, E. 1985, *Astrophys. J. Lett.*, **296**, L17.

— 1986, *Geochim. Cosmochim. Acta*, in press.

Harris, M. J. 1983, *Astrophys. J.*, **264**, 613.

Hartmann, D., Woosley, S. E., and El Eid, M. F. 1985, *Ap. J.*, **297**, 837.

Hillebrandt, W., Kratz, K.-L., Thielemann, F.-K., and Ziegert, W. 1986, in *Weak and Electromag. Interactions in Nuclei*, (ed.) H. V. Klapdor, Springer Verlag, pp. 987.

Hillebrandt, W. 1986, in *Cosmogonical Processes*, (eds.) W. D. Arnett, C. Hansen, J. Truran, and S. Tsuruta, VNU Sci. Press., pp 123

Hinton, R. W., Davis, A. M., and Scatena-Wachel, D. E. 1987, *Astrophys. J.* , in press.

Hutcheon, I. D., Steele, I. M., Wachel, D. E. S., MacDougall, J. D., and Phinney, D. 1983, in *Lunar and Planet. Sci.*, XIV, pp. 339.

Ireland, T. R., Compston, W., and Heydegger, H. R. 1985, *Geochim. Cosmochim. Acta*, **49**, 1989.

Jungck, M. H. A., Shimamura, T., and Lugmair, G. W. 1984, *Geochim. Cosmochim. Acta*, **48**, 2651.

Lee, T., Papanastassiou, D. A., and Wasserburg, G. J. 1978, *Astrophys. J. Lett.*, **220**, L21.

Lee, T., Russel, W. A., and Wasserburg, G. J. 1979, *Astrophys. J. Lett.*, **228**, L93.

Lee, T. 1987, in *Meteorites and the Early Solar System*, (ed.) John F. Kerridge, Univ. of Arizona Press, Tucson.

Möller, P., Myers, W. D., Swiatecki, W. J., and Treiner, J. 1987, *Atomic Data Nucl. Data Tables*, submitted.

Niederer, F. R., Papanastassiou, D. A., and Wasserburg, G. J. 1981, *Geochim. Cosmochim. Acta*, **45**, 1017.

— 1985, *Geochim. Cosmochim. Acta*, **49**, 835.

— 1980, *Astrophys. J. Lett.*, **240**, L73.

Niederer, F. R., and Papanastassiou, D. A. 1984, *Geochim. Cosmochim. Acta*, **48**, 1279.

Niemeyer, S., and Lugmair, G. W. 1981, *Earth Planet. Sci. Lett.*, **53**, 211.

— 1984, *Geochim. Cosmochim. Acta*, **48**, 1401.

Niemeyer, S. 1985, *Geophys. Res. Lett.*, **12**, 733.

Papanastassiou, D. A. 1986, *Astrophys. J. Lett.*, **308**, L27.

Prombo, C., and Lugmair, G. W. 1986, *Lunar Planet. Sci.*, **17**, 685.

Sandler, D. G., Koonin, S. E., and Fowler, W. A. 1982, *Astrophys. J.*, **259**, 908.

Suess, H. 1965, Ann. Rev. Astron. Astrophys., **3**, 217.

Trivedi, B. M. P. 1987, *Astrophys. J.*, **320**, 430.

Von Groote, H., Hilf, E. R., and Takahashi, K. 1976, *Atomic Data Nucl. Data Tables*, **17**, 418.

Ziegert, W., Wiescher, M., Kratz, K. L., Moeller, P., Krumlinde, J., Thielemann, F. K., and Hillebrandt, W. 1985, *Phys. Rev. Lett.*, **55**, 1935.

Zinner, E. K., Fahey, A. J., Goswami, J. N., Ireland, T. R., and McKeegan, K. D. 1986, *Astrophys. J. Lett.*, **311**, L103.

Wapstra, A. H., and Audi, G. 1985, *Nucl. Phys.*, **A432**, 1.

Woosley, S. E. 1986, in *16th Adv. Course of the Swiss Soc. Astr. Astrophys.*, Saas Fee, 1986, pp. 1-195.

CHAPTER IV

STELLAR EVOLUTION AND SUPERNOVAE

"The brightest flashes in the world of thought are incomplete until they have proved to have their counterparts in the world of fact."

John Tyndall
Fragments of Science Vol. II
Scientific Materialism

Main Sequence Mass Loss

by

Wendee M. Brunish
Los Alamos National Laboratory
P.O. Box 1663, MS F665
Los Alamos NM 87545

Joyce A. Guzik, Lee Anne Willson and George Bowen
Department of Physics
Iowa State University
Ames Iowa 50011

It has been hypothesized that variable stars may experience mass loss, driven, at least in part, by oscillations [1]. The class of stars we are discussing here are the δ Scuti variables. These are variable stars with masses between about 1.2 and 2.25 $M_\odot$, lying on or very near the main sequence. According to this theory, high rotation rates enhance the rate of mass loss, so main sequence stars born in this mass range would have a range of mass loss rates, depending on their initial rotation velocity and the amplitude of the oscillations. The stars would evolve rapidly down the main sequence until (at about 1.25 $M_\odot$) a surface convection zone began to form. The presence of this convective region would slow the rotation, perhaps allowing magnetic braking to occur, and thus sharply reduce the mass loss rate.

We have constructed models and evolved them to reproduce the sun's mass radius and luminosity, at approximately the same time as the standard age [2]. Our best fit was derived with models with an initial mass of 2.0 M_Θ, and an initial helium abundance of Y= 0.20 to 0.22. Initial mass loss rates were on the order of 10^{-8} M_Θ/yr, decreasing exponentially with an e-folding time of about 2×10^8 years.

The basic structure of our models was similar to the standard model. However, in our models the core is somewhat hotter and denser and therefore more of the hydrogen has been exhausted. This results in a total lifetime for our models of about 10 percent less than for the standard model. Because of the core conditions, the neutrino flux is also slightly higher than in the standard model. The depth of the convection zones in our models in somewhat deeper than in the standard models (approximately 0.25 R_Θ compared to 0.20 R_Θ). This structure should result in improved agreement over the standard model of the solar oscillation spectrum with observations. A detailed study of solar oscillations and calculations of the p-mode spectrum for our models are planned [3].

Another consequence of the deeper convection zone is that the temperature at the base of the zone is hot enough so that all of the ^{7}Li in the surface layers will be destroyed. Therefore the observed lithium in the solar atmosphere cannot be primordial and must have been created by some other mechanism. We believe that it is energetically feasible for solar flare and coronal spallation to account for the observed abundances of ^{7}Li and ^{9}Be in the sun and main sequence A-F stars, and work is now underway on the details of the process and the resulting abundances. [4].

The luminosity history of these models is obviously quite different from the standard models. Instead of gradually increasing over billions of years, the luminosity starts out much higher and drops rapidly to about the current solar luminosity. This will obviously have considerable effects on the early solar system [5]. The "faint early sun paradox", where many solar system bodies including the earth, suggest a more luminous sun in the past, is solved. These models also provide an explanation for the gravitational locking of Venus, since the gravitational tidal forces would have been much larger in the early epochs of the solar system. The large core found in the planet Mercury can also be explained by the presence of the much higher solar flux in the past, boiling off a large percentage of the planet's mantle. This provides a good constraint on the 2 $M_\odot$ initial mass, because if the sun were initially much more massive, the entire planet would have been boiled away, and a significantly less luminous early sun would not have stripped off as much of Mercury's mantle.

This work has many interesting implications not only for the solar system but for the galaxy as well. Recent determinations of the IMF have suggested that their may be a "bump" at about 1 $M_\odot$ [6]. Observations of clusters and binary stars reveal a paucity of stars between 1.2 and 2.3 $M_\odot$, coinciding with the δ Scuti instability region, and a definite gap at 2.0 to 2.3 $M_\odot$ (see [7] and references therein). These observations are nicely explained if some stars formed with masses of 2 to 2.5 $M_\odot$ evolve down the main sequence until formation of a convective surface layer effectively halts the mass loss.

Another consequence of this behavior is that the ages of the oldest clusters in the galaxy cannot be accurately determined from the cluster turnoff mass. An intrinsic uncertainty is introduced if stars evolve down the main sequence since stars with the same mass do not necessarily have the same age. The universe may well be younger than the age determinations based on the oldest clusters lead us to believe.

This work was supported by the United States Department of Energy.

References

Willson, L.A., Bowen, G., and Struck-Marcell, C., **Comm. Ap.** 12, 17 (1987).

Guzik, J.A., Willson, L.A. and Brunish, W.M., **Ap. J.** 319, 957-965 (1987).

Guzik, J.A. 1987, private communication.

Willson, L.A. 1987, private communication.

Bowen, G. 1987, preprint.

Scalo, J.M., **Fund. Cosmic Phys.** 11, 3-278 (1986).

Willson, L.A., Proceedings Trieste Workshop on Pulsation and Mass Loss, eds. Stalio and Willson, Reidel Publishing, in press (1987).

Core Helium Flash, Convection, and Surface Abundance Anomalies

RICHARD K. WALLACE
Los Alamos National Laboratory
X-DO, MS B218
Los Alamos, NM 87545
USA

ABSTRACT

Two core helium flash calculations of varying intensity are presented. The intensity is governed by the degree of degeneracy where the explosion occurs. Residue nuclear species of the core helium flash are mixed into the outer hydrogen shell. The peak temperatures for these flashes were 9 and 10 x 10^8 K.

The primary material mixed into the hydrogen shell is usually ^{12}C, unless the peak temperature is near 10^9K. Other major products (^{20}Ne, ^{24}Mg, ^{28}Si, and ^{32}S) result from the hot alpha captures that occur during the flash. The order of their abundance depends on the peak temperature, with the higher atomic weight species dominating for higher temperature flashes. Other products that are generated are ^{27}Al, ^{31}P, ^{35}Cl, ^{36}Ar, and some ^{14}N.

The intermediate flashes produce the most mixing, with the lower temperature flashes not producing as much processed material and the highest temperature flashes generating sufficient envelope expansion that the hydrogen shell is pushed so far away that it is more difficult to penetrate.

Comparisons of these calculations with several stellar observations are encouraging, but abundance anomolies cannot be unambiguously attributed to the core helium flash mechanism.

INTRODUCTION

The core helium flash occurs at the tip of the giant branch in stars of about 2 $M_{\odot}$ and less. The event is produced by the ignition of helium by the 3α process in a stellar core where the density is so high that the electrons are degenerate. The location of the flash is at some distance from the stellar center, owing to cooling of the center by neutrino losses [1,2].

As nuclear energy generation builds up due to local heating, neither radiation nor electron conduction can transport the energy generated very quickly. This localized heating leads to superadiabatic gradients and convection is initiated. Therefore, convection becomes the primary energy-loss mechanism from the flash location before the degeneracy is lifted, and it is the competition between the thermonuclear energy generation and the energy losses due to convective energy transport that determine the time scale of the temperature (and hence energy) growth.

Cole and Deupree [3,4] have calculated the core flash by solving the time-dependent conservation laws in two spatial dimensions with standard two-dimensional finite difference techniques. With this approach, two dimensional convective flow patterns are automatically generated by the unbalanced buoyancy forces in the convectively unstable region produced by the thermonuclear energy generation. The time-dependent interaction between the convective flow (and hence the convective energy transport) and the thermonuclear energy generation is governed by the conservation laws, rather than by some arbitrary prescription. Recent calculations by Deupree [5] have improved the previous calculations by increasing the two-dimensional spacial resolution, extending the outer radial boundary further from the location of the thermonuclear runaway, and extending the calculation to a three-dimensional grid.

The qualitative results of the two- and three-dimensional calculations are similar, although the quantitative results differ slightly. Small nonradial temperature fluctuations are produced by the convective flow, and are quickly and strongly amplified by the highly temperature sensitive nuclear energy generation rates. As long as convection can transport energy inward into the previously neutrino cooled inner core, the thermonuclear runaway can be contained. Once the inner core becomes sufficiently hot that this downward energy transport is no longer efficient, one the these localized regions of high temperature and high energy generation produces a short timescale hydrodynamic explosion. Deupree [6] has shown that this explosion can be treated as an off-center point source explosion three-dimensionally (or on the polar axis two-dimensionally).

The peak temperature reached during the explosion is governed by how degenerate the exploding material is when the explosion begins. The range of flash models includes those that produce very mild explosions, similar to those found in traditional stellar evolution calculations, to violent explosions which disrupt the entire core. The relative distribution of stars among these states is probably related to their core rotation parameters. We can only parameterize our calculations by the relative degeneracy of the core at the onset of the flash.

We have extended Deupree's calculations to include the hydrogen shell so that we could calculate any interaction between the core flash nuclear residue and the hydrogen shell [7]. Each of our calculations begins as an off-center point source explosion along the polar axis in a two-dimensional finite difference grid. This was shown to well mimic the full three-dimensional results that are obtained by following a full calculation from a traditional stellar evolution structure. The calculational details and a larger range of flash models are described elsewhere [8].

HYDRODYNAMIC EVOLUTION OF THE CORE AND THE HYDROGEN SHELL

Figure 1 shows a contour plot of the carbon abundance, which outlines the bubble location, at various times for case 2, along with the location of the hydrogen shell interface.

Figure 2 shows a similar contour plot for a more violent case. At later times, the hydrogen shell location at $\theta = 180°$ also begins to rise. This is caused by the convergence at this angle of the acoustic disturbance produced by the thermonuclear explosion. The convergence causes radial outward expansion along the density gradient, resulting in the ultimate expansion of the material above the convergence point, including the hydrogen shell. This expansion of the hydrogen shell results in much less mixing than might otherwise be the case.

ABUNDANCES MIXED INTO THE HYDROGEN SHELL

The transport and mixing of individual isotopes required to correctly calculate the energy generation rate are included in the hydrodynamic simulations. These are the α chain nuclei from ^{4}He to ^{28}Si. An additional 250 isotopes were followed by postprocessing with the nuclear reaction network of Wallace and Woosley [9].

The amount of material mixed into the hydrogen envelope is presented in Figure 3, where the initial relative abundances in the envelope are assumed to be solar and the bubble material is assumed to mix throughout the envelope. The envelope abundance enhancement is defined as the ratio of the species mass mixed into the envelope to the species mass initially present in the envelope. This is plotted against the product of the metal abundance and the envelope mass (both scaled to solar units). Carbon is the element most enhanced in all but the most extreme cases. The elements which show the next greatest enhancements are those in the direct α capture sequence. ^{20}Ne is most prominent for fairly low temperature flashes, with ^{24}Mg and then ^{28}Si and ^{32}S becoming more prominent as the flash temperature increases. ^{14}N can also be enhanced if the carbon is sufficiently hot when it penetrates the hydrogen shell that proton captures on the ^{12}C are initiated on a relatively short time scale. Finally come the elements ^{27}Al, ^{31}P, ^{35}Cl, and ^{36}Ar. However, for these to be produced in any observable amount, there must be sizable enhancements of at least an appropriate member of the α capture nuclei mentioned above. The strongest flashes can also enhance helium by up to 50%, through the mixing of core material into the hydrogen shell.

There are several notable absences in the enhancements. ^{16}O is effectively destroyed as soon as it is produced in all of these flashes. There is no s-processing in any of the flashes, as the time scales are far too short for the achieved temperatures. Only a trace of ^{23}Na is produced, and an extremely metal poor envelope would be required for any enhancement to be noticeable. Aluminum enhancement only occurs in the presence of an even larger magnesium enhancement.

Violent flashes would be rare, since they span such a limited degeneracy range. Several very weak hydrodynamic events may occur during the core flash in an individual star; however, one violent flash sufficiently alters the core structure that it prevents any further violent events as a part of the core helium flash. Thus, while it may be possible to mix a suitable amount of carbon (and perhaps oxygen and neon) above the hydrogen shell from a succession of weak flashes, higher Z elements require the one violent flash to occur.

COMPARISON WITH OBSERVATIONS

Some globular clusters, such as ω Cen and M22, show complex abundance anomalies; however, many of the observed anomalies cannot be accounted for with this core helium flash model. Perhaps red giant stars are not the correct objects to examine for the more violent events, since we are not sure where the evolutionary H-R diagram track will go following the event. There are other objects that show anomalies similar to those predicted by our models.

HD 46703 shows an order of magnitude enhancement of carbon, nitrogen, oxygen and sulfur [10]. These are predicted by a flash similar to our case 2, except for oxygen. The oxygen could come from a second mixture of oxygen-rich material occurring at a later evolutionary phase, or some previous mixing of oxygen by weak core helium flashes. Unfortunately, this is probably a post-AGB star so it is conceivable that the α capture excesses could have come from helium shell flashes.

Heber [11] has observed BD +10°2179 to be overabundant in carbon and magnesium; normal in nitrogen, aluminum, silicon, phosphorous, and sulfur; and underabundant in oxygen and elements of higher Z than sulfur. These conditions could be explained by a core helium flash somewhat weaker than case 1 with an initial envelope metal abundance between one-third and one-tenth solar. This could produce pronounced excesses of carbon and magnesium and sufficient enhancements of the elements between magnesium and sulfur to make them appear normal. Nitrogen could appear solar because of the enhancement produced by previous hydrogen burning, while the other elements would appear underabundant because they have not been enhanced by the flash.

Wallerstein (private communication) has suggested that the "R Stars", carbon stars with no s processing, might be explained by this mechanism. He also suggested the sub-giant CH stars as possible candidates for such enhancements.

REFERENCES

1. Thomas, H.-C., Zs. Ap., 67, 420 (1967).
2. Thomas, H.-C., Ap. Space Sci., 6, 400 (1970).
3. Cole, P.W., and Deupree, R.G., Ap. J., 239, 284 (1980).
4. Cole, P.W., and Deupree, R.G., Ap. J., 247, 607 (1981).
5. Deupree, R.G., Ap. J., 282, 274 (1984).
6. Deupree, R.G., Ap. J., 287, 268 (1984).
7. Deupree, R.G., Ap. J., 303, 649 (1986).
8. Deupree, R.G., and Wallace, R.K., Ap. J., 317, 724 (1987).
9. Wallace, R.K, and Woosley, S.E., Ap. J. Suppl., 45, 389 (1981).
10. Bond, H.E., and Luck, R.E., Ap. J., 312, 203 (1987).
11. Heber, U., Astr. Ap., 118, 39 (1983).

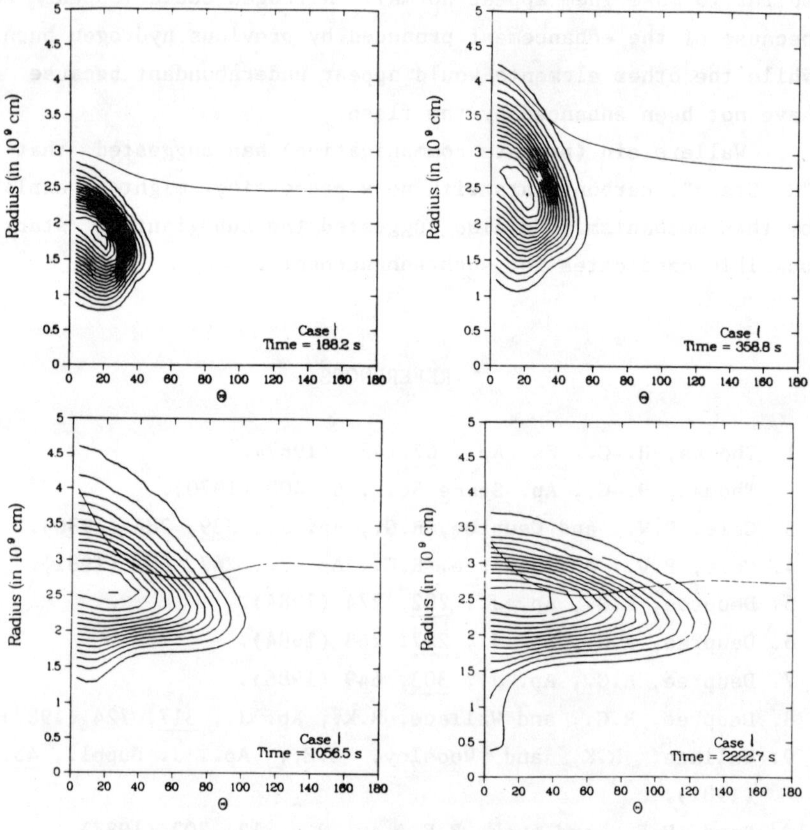

Fig. 1. Plots of the carbon abundance contours and the H shell location at various times. The carbon contour interval is 0.015.

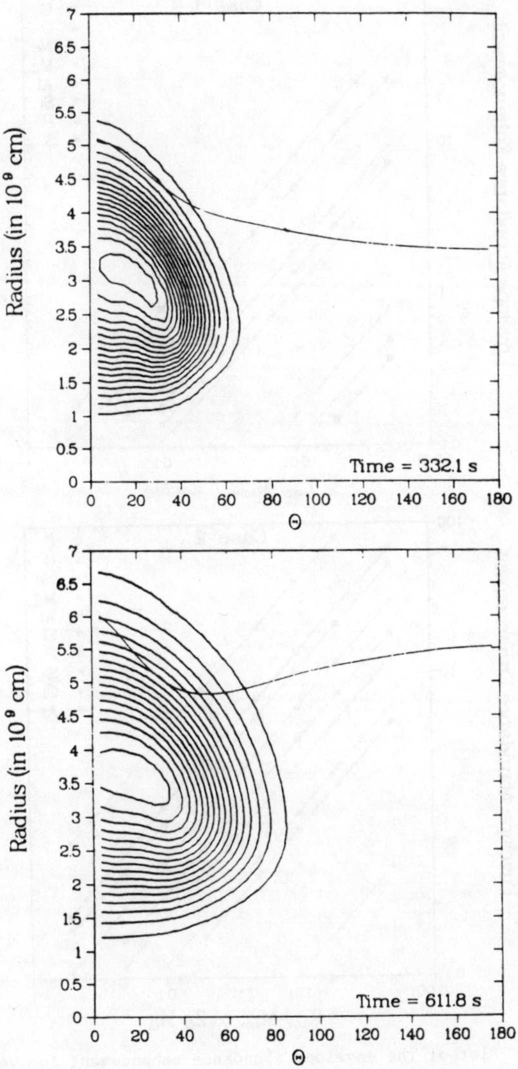

Fig. 2. The carbon abundance contours and the hydrogen shell location at various times for a more violent flash.

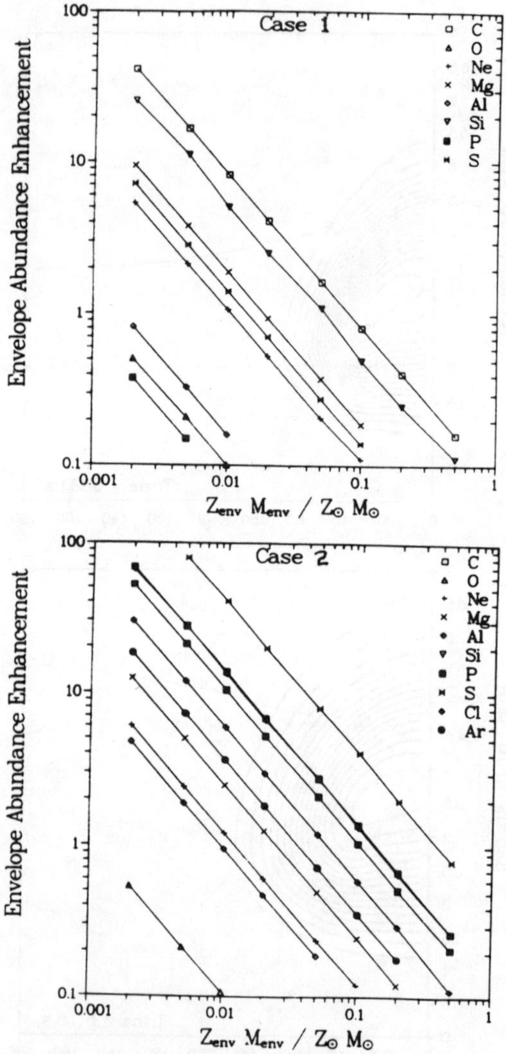

Fig. 3. Plot of the envelope abundance enhancement for various elements as a function of the product of the initial abuncances and envelope mass for case 1 and case 2.

AGB - STAR EVOLUTION

S. A. BECKER

Los Alamos National Laboratory
University of California
Los Alamos, NM 87544
USA

ABSTRACT (U)

Asymptotic giant branch stars are red supergiant stars of low-to-intermediate mass. This class of stars is of particular interest because many of these stars can have nuclear processed material brought up repeatedly from the deep interior to the surface where it can be observed. A review of recent theoretical and observational work on stars undergoing the asymptotic giant branch phase is presented.

I. INTRODUCTION

The term asymptotic giant branch (AGB) originally referred to globular cluster stars located on a sequence in the H-R diagram which appeared to merge asymptotically with the giant branch at high luminosity. With time, the meaning of the term has been generalized to now refer to the evolution of any red giant star that possesses a degenerate carbon-oxygen core. AGB stars are characterized by one of two evolutionary states. Early AGB stars are dominated by a powerful He-burning shell while the H-burning shell is either inactive or very weak. Mature AGB stars have two active shell sources closely spaced in mass with the H-burning shell being dominant most of the time. The H-burning shell is normally active at about a hundredth of the luminosity of the H-burning shell; however, the

He-burning shell is unstable and it will periodically undergo a thermal runaway known as a thermal pulse or He-shell flash. A thermal pulse will extinguish the H-burning shell, and it could lead to some of the products of He-burning being dredged up to the surface of the star. Some important parameters of AGB stars are:

LUMINOSITY	$-7.3 < M_{bol} < -3$
	$3.0 < \log L/L_\odot < 4.8$
TEMPERATURE	$3.4 < \log Te < 3.6$
	$2500K < Te < 4000K$
RADIUS	$200 < R/R_\odot < 1000$
CORE MASS	$0.5 < M_{co}/M_\odot < 1.4$

Single stars with an initial mass in the range from slightly less than $1M_\odot$ to somewhere in the range of 5 to 9 $M_\odot$ will evolve into the AGB phase. More massive stars will avoid the AGB phase by igniting carbon in their cores before they becomes electron degenerate while low mass stars below the limit will not have had sufficient time during the age of the universe to evolve to the AGB. For binaries the situation is more complex and the question can either component evolve into the AGB phase will depend on the initial seperation and masses of the two stars.

Excellent detailed reviews of the pre-AGB and AGB evolutionary phases of single stars can be found in references 1-6. As a result, this information will not be repeated here, instead a brief summary of work completed since these reviews were published is presented. As a practical working point this paper will focus on developments from 1983 onward.

II. NEW THEORETICAL DEVELOPMENTS

A. Pre-AGB Evolution

One of the more interesting recent developments concerning tne modeling of stellar evolution is the application of connective overshoot and semiconvection to tne H-burning (references 7-9) and He-burning (references 7-10) convective core. Both convective overshoot and semiconvection provide more fuel to the convective core compared to cases where these effects are not included, and as one would expect, the lifetimes of various pre-AGB phases are changed. In addition during a given evolutionary phase a stellar model will attain a greater luminosity and will evolve onto the AGB with a larger degenerate carbon-oxygen core. In a way convective overshoot and semiconvection cause a stellar model to act as if it were more massive than it really is. For example, M_{up} tne maximum mass a star can have that can evolve onto the AGB is about $9M_{\odot}$ in the standard evolutionary calculation while the limit may drop to as small as $5M_{\odot}$ when convective overshoot and semiconvection are included.[7,10]

While the application of convective overshoot and semiconvection to stellar evolution models shows promise with regards to low and intermediate mass stars, there is controversy as to when to apply these effects and the degree to which they are present (see especially reference 11). For example reference 7 applies convective overshoot to both the H-burning and He-burning convective core phases while reference 10 applies convective overshoot and semiconvection to only the He-burning convective core. Both approaches produce similiar models that evolve onto the AGB, but the previous pre-AGB evolutionary behavior is quite different especially with regards to the lifetimes of the various phases. Comparisons with observations may help to resolve some of the controversary.

Another fascinating development is the effect of a binary companion on the evolution of a star as discussed in references 5 and 12. The effects of mass transfer and system mass loss can drastically alter the evolutionary behavior from that expected of a single star so that the AGB phase may not be attained or it may be achieved by accretion after the star has become a white dwarf. Such accretion could permit the core of an AGB star to grow past the Chandrasekhar limit of $1.4M_{\odot}$ and therby result in a carbon deflagration supernova.

B. AGB Evolution

New models of AGB stars have been evolved especially for stars having an initial mass $\leq 3\ M_{\odot}$[13-19] The third dredge-up phase is now found to occur at lower core masses and total stellar masses than previously thought[4,6,16,18] which is more consistent with the lower lumonosity limit of the observed distribution of carbon stars. For dredge-up to occur, it is necessary to have $= {}^{1}/H_p$ be > 1 in the calculations. In addition for models of low Z and small envelope mass, the effect of carbon recombination on the opacity during the power down phase of the He-burning shell can result in the formation of a semiconnective region which can mix together carbon and hydrogen.[4,6,19,15] When the hydrogen-burning shell reforms this mixed region is converted into ^{13}C which is later absorbed into the helium-burning convective shell during the next thermal pulse. This mechanism provides a neutron source for S-processing via the $^{13}C(\alpha,n)^{16}O$ reaction which can operate at much cooler temperature than the $^{22}Ne(\alpha,n)^{25}Mg$ reaction. It appears that the maximum temperature attained by the helium-burning shell for stars with $M_{co} < 0.8M_{\odot}$ is too cool to make effective use of the $^{22}Ne(\alpha,n)^{25}Mg$ reaction.

Other developments include a new look at the core-mass luminosity relation for low mass stars,[13, 16] and the evolution of a $3M_{\odot}$ model from the main sequence to the AGB to the white dwarf stage,[19] and the role of post AGB helium-shell flashes causing a "born again" AGB phase.[6] Private conversations with researchers reveal much work in progress which indicate this topic is certainly a fruitful one for future research.

C. Nucleosynthesis and the S-Process

As discussed in reference 4 the details of the dredge-up process and the effect of multiple thermal pulses allowing some material to have repeated neutron exposures have in principle the right ingredients to explain the solar system distribution of the elements. However, attempts to achieve this explaination have run into problems primarily because of the high temperature ($T > 3 \times 10^8$ K) needed to drive the $^{22}Ne(\alpha, n)^{25}Mg$ reaction. Models which can achieve these high temperatures suffer from a dearth of observational counterparts, while models which have observational counterparts (such as the carbon stars) are unable to burn much of the ^{22}Ne. For example reference 20 finds that a helium-burning shell temperature of 310×10^6 K would be required to explain the solar system distribution of the elements if it were to come from stars having core masses of 0.6 to 0.8 $M_\odot$. This is at least 10×10^6 K hotter than any model in that core range has been able to achieve. Since the reaction rate for the $^{22}Ne(\alpha, n)^{25}Mg$ is uncertain, an increase in the rate would help to lower the needed burning temperature to perhaps the temperature range achieved by current models, however, this remains to be seen. References 21-26 represent a sample of current work on various aspects of the S-process in intermediate mass stars.

While the $^{13}C(\alpha, n)^{16}O$ reaction provides an alternate source of neutrons, the mechanism for production of ^{13}C is only likely to work for low mass stars with small envelope masses. In addition the size of the neutron absorption cross section for the relevant nuclei around ^{13}C will not provide the proper regulation of the nuetron flux (unlike the case for ^{22}Ne) so that the S-processes elements produced will be unlikely to have a solar system distribution. Despite these theoretical problems nature is clearly telling us that the S-process is taking place because of the presence of ^{99}Tc in the atmospheres of some AGB stars. Clearly the problem has yet to be solved.

III. NEW OBSERVATIONAL DEVELOPMENTS

Observational studies of AGB stars have shown that carbon stars have a luminosity distribution of $-4 > M_{bol} > -6$ [27] which raises the question where are the brighter AGB stars predicted by theory? Some studies of AGB stars in clusters and field regions of the Magellanic Clouds have shown a lack or scarcity of AGB stars more luminous than $M_{bol} = -6$ [9,27,28,29], however, references 30-34 have found that AGB stars do exist up to the theoretical limit of $M_{bol} = -7.3$. These bright AGB stars are not seen as carbon stars, but rather as long period variables and they appear to be about a factor of 10 less abundant than standard theory predicts. The lower than expected number of bright AGB stars has been interpreted as indicating the lifetime of a star on this part of the AGB is relatively short on the order of 10^5 yr with mass loss being evoked as the cause. Steady mass loss rates on the AGB are observed to be large and there exists the possibility that the entire envelope can be ejected. Indeed, reference 35 calculates that for stars $< 3M_{\odot}$ with cores $> 0.86\ M_{\odot}$ at peak surface luminosity following a helium-shell flash, radiation pressure within the envelope was sufficient to eject it. Other factors which may have some bearing on the lower than expected numbers of bright AGB stars are the effects on convective overshoot, variations in the stellar birthrate function, and hot envelope burning of the dredge-up carbon. Studies of stars leaving the AGB are given in references 36 and 37.

Finally, detailed observations are being made of both the elemental and isotopic abundance in AGB stars. Examples of such studies can be found in references 38-41. Besides providing qualitative agreement with some of the predictions of the stellar evolution models these detailed observations also provide interesting points of disagreement which will require further theoretical study.

IV. SUMMARY

For AGB stars there is a fair agreement between theory and observation, but many details need to be worked out. AGB stars are observed to exist and they provide direct evidence of the third dredge-up phase bringing up ^{12}C and S-processed elements to the surfaces of stars where by mass loss these products are released into the interstellar medium for further recycling into the next generation of stars. Whether the integrated effect of AGB stars of all masses can reproduce the observed solar system distribution of elements remains to be determined. In any case AGB stars should be an important contributor of ^{4}He, ^{12}C, and some S-processed elements to the Galaxy.

* This work was performed under the auspices of the U. S. Department of Energy by the Los Alamos National Laboratory under contract number W-7405-ENG-36.

References

1) Iben, I. Jr., "Post Main Sequence Evolution of Single Stars," Ann. Rev. Astr. Ap. 12, pp. 215-256 (1974).

2) Iben, I. Jr., and A. Renzini, "Asymptotic Giant Branch Evolution and Beyond," Ann. Rev. Astr. Ap. 21, pp. 271-342 (1983).

3) Iben, I. Jr., and A. Renzini, "Single Star Evolution I. Massive Stars and Early Evolution of Low and Intermediate Mass Stars," Physics Reports 105. pp. 329-406 (1984).

4) Iben, I. Jr., "Nucleosynthesis in Low and Intermediate Mass Stars on the Asymptotic Giant Branch," in Nucleosynthesis, Challenges and New Developments ed. W. A. Arnett and J. W. Truran (Chicago: University of Chicago Press) pp. 272-291 (1985).

5) Iben, I. Jr., "The Life and Times of an Intermediate Mass Star - In Isolation/In A Close Binary," Q Jl. R. Astr. Soc. 26, pp. 1-39 (1985).

6) Iben, I. Jr., "Helium Shell Flashes and Their Consequences," in Late Stages of Stellar Evolution, ed. S. Kwok and S. R. Pottasch (Dordrecht: Reidel) pp. 175-195 (1987).

7) Bertelli, G., Bressan, A. G., and Chiosi, C., "Evolution of Intermediate Mass Stars: the Role of Convective Overshooting and Stellar Wind," Astron, Astrophys. 150. pp. 33-52 (1985).

8) Bertelli, G., Bressan, A., Chiosi, C., and Angerer, K., "Evolutionary Models of Low and Intermediate Mass Stars with Convective Overshooting," Astron. Astrophys. Suppl. 66, pp. 191-234 (1986).

9) Chiosi, C., Bertelli, G., Bressan, A., and Nasi, E., "Magellanic Cloud Star Clusters: the Problems of Age Determination, Metalticity - Age Relationship and AGB Star Luminosity Function," Astron. Astrophys. 165, pp. 84-94 (1986).

10) Castellani, V., Chieffi, A., Pulone, L. and Tornambe, A. "Convective Cores and Carbon Ignition in Intermediate-Mass Stars," AP. J. 294, pp. L31-L34 (1985).

11) Langer, N., "Non-local Treatment of Convection and Overshooting from Stellar Convective Cores, " Astron. Astrophys. 164, pp. 45-50 (1986).

References (continued)

12) Iben, I. Jr. and Tutukov, A. V., "Evolution Scenarios for Intermediate-Mass Stars in Close Binaries," Ap. J., 313. pp. 727-742 (1987).

13) Boothoyd, A. I. and Sackmann, I. J., "The Core Mass-Luminosity Relation for Low Mass Stars," Ap. J., submitted, (1987).

14) Hollowell, D. E., "Dredge-up and Nucleosynthesis in Low-Mass AGB Stars," Bull Am. Astr. Soc. 18, p. 963 (1986).

15) Hollowell, D. E., "Mixing at the Carbon-Hydrogen Discontinuity During the Dredge-up Phase in Low-Mass AGB Stars," in Late States of Stellar Evolution, ed. S. Kwok and S. R. Pottasch (Dordrecht: Reidell) pp. 239-240 (1987).

16) Lattanzio, J. C., "The AGB Evolution of 1.0-3.0 $M_\odot$ Stars as a Function of Mass and Composition," Ap. J. 311, pp. 708-730 (1986).

17) Lattanzio, J. C., "Asymptotic Giant Branch Evolution in the Magellanic Clouds," in Late States of Stellar Evolution, ed. S. Kwok and S. R. Pottasch (Dordrecht: Reidel) pp. 235-238 (1987).

18) Lattanzio, J. C., "Formation of a 1.5 $M_\odot$ Carbon Star with $M_{bol} = -4.4$," Ap. J., 313, L15-L18 (1987).

19) Mazzitelli, I. and D'Antona, Jr., "Evolution from the Main Sequence to the White Dwarf Stage for a 3 $M_\odot$ Star," Ap. J., 308, pp. 706-720 (1986).

20) Malaney, R. A. and Boothoyd, A. I., "On the Orgin of the Solar System S-Process Abundances," Ap. J., submitted to letters, (1987).

21) Beer, H., "S-Process Nucleosynthesis Below A=90," in Nucleosynthesis and Its Implications on Nuclear and Particle Physics, ed. J. Audouze and N. Mathieu (Dordrecht: Reidel) pp. 263-270 (1986).

22) Howard, W. M., Mathews, G. J., Takahashi, K. and Ward R. A., "Parametric Study of Pulsed Nuetron Source Models for the S-Process," Ap. J., 309, pp. 663-652 (1986).

23) Käppeler, F., "S-Process Nucleosynthesis - Stellar Aspects and the Classical Model," in Nucleosynthesis and Its Implications on Nuclear and Particle Physics, ed., J. Audouze and N. Mathieu (Dordrecht: Reidel) pp. 253-262 (1986).

References (continued)

24) Mathews, G. J., Ward, R. A., Takahashi, K. and Howard, W. M., "Stellar S-Process Diagnostics," in Nucleosyntheses and Its Implications on Nuclear and Particle Physics, ed., J. Audouze and N. Mathieu (Dordrecht: Reidel) pp. 277-284 (1986).

25) Mathews, G. J. Takahashi, K. Ward, R. A., and Howard, W. M., "Stellar Technctium and Niobium Abundances as a Measure of the Lifetime of AGB Stars in the Third Deredge-up Phase," Ap. J., 301, pp. 410-414 (1986).

26) Takahashi, K., Mathews, G. J., Ward, R. A., and Becker, S. A., "Production and Survival of ^{99}Tc in He-Shell Recurrent Thermal Pulses," in Nucleosynthesis and Its Implications on Nuclear and Particle Physics, ed., J. Audouze and N. Mathieu (Dordrecnt: Reidel) pp. 285-292 (1986).

27) Mould, J. and Aaronson, M., "The Formation and Evolution of Carbon Stars," Ap. J. 303, pp. 10-18 (1986).

28) Mould, J. and Reid, N. "AGB Evolution in Shapley Constellation III," in Late Stages of Stellar Evolution, ed. S. Kwok and S. R. Pottasch (Dordrecht: Reidel) pp. 209-212 (1987).

29) Renzini, A., Bernazzani, M., Buonanno, R., and Corsi, C. E., "Asymptotic Grant Branch Stars in the Large Magellanic Cloud: Results from a Deep BV Survey and Theoretical Implications," Ap. J., 294, L7-L11 (1985).

30) Wood, P. R., Bessell, M. S., and Fox, M. W., "Long Period Variables in the Magellanic Clouds: Supergiants, AGB Stars, Supernova Precursors, Planetary Nebula Precursors, and Enrichment of the Interstellar Medium," Ap. J., 272, pp. 99-115 (1983).

31) Wood, P. R. Bessell, M. S., and Paltoglou, G., "A Sample of Long-Period Variables in the Bar of the LMC and Evidence for a Recent Burst of Star Formation," Ap. J., 290, pp. 477-486 (1985).

32) Wood, P. R., Bessell, M. S., and Whiteoak, J. B., "Detection of an OH/IR Star in the Large Magellanic Cloud," Ap. J. 306, L77-L80 (1986).

33) Wood, P. R., "The Magellanic Clouds as a Testing Ground for Theories of the Late Stages of Stellar Evolution," in Late Stages of Stellar Evolution, ed., S. Kwok and S. R. Pottasch (Dordrecht: Reidel) pp. 197-208 (1987).

References (continued)

34) Mould, J. and Reid, N., "The Evolution of Asymptotic Giant Branch Stars in the Magellanic Clouds III, The Problem of Intermediate Mass Stars," Ap, J. 321, pp. 156-161 (1987).

35) Wood, P. R. and Faulkner, D. J., "Hydrostatic Evolutionary Sequences for the Nuclei of Planetary Nebulae," Ap. J., 307, pp. 659-674 (1986).

36) Baird, S. R. and Flower, P. J., "Superluminous Giant Candidates in NGL 1966," A. J. 93, pp. 851-855 (1987).

37) Jura, M., "RV TAURI Stars as Post-Asymptotic Giant Branch Objects," Ap. J. 309, pp. 732-736 (1986).

38) Dominy, J. F. and Wallerstein, G., "Quantitive Technetium Abundance in the Long Period Variables X Cygui and σ Ceti," Ap. J. 310, pp. 371-377 (1986).

39) Harris, W. M., Lambert, D. L., Hinkle, K. H., Gustafsson, G. and Eriksson, K., "Oxygen Isotopic Abundances in Evolved Stars III. 26 Carbon Stars," Ap. J. 316, pp. 294-304 (1987).

40) Smith, V. V., and Lambert, D. L., "The Chemical Composition of Red Giants II. Helium burning and the S-Process in the MS and S Stars," Ap. J., 311, pp. 843-863 (1986).

41) Smith, V. V., "Nucleosynthesis in the MS and S Stars," in Late Stages of Stellar Evolution, ed., S. Kwok and S. R. Pottasch (Dordrecht: Reidel) pp. 241-244 (1987).

AN UPDATE ON THE CARBON STAR MYSTERY

J. C. Lattanzio[1]
Canadian Institute for Theoretical Astrophysics
University of Toronto
60 St. George Street
Toronto, Ontario
M5S 1A1 CANADA

ABSTRACT

A summary of our knowledge of the formation of Carbon stars is given. The emphasis is on the previous disagreement between synthetic and observed Carbon star distributions, and recent developments which improve the agreement.

I shall divide this paper into three sections, each dealing with one of the key elements—"Carbon star", "mystery" and "update". I'll address each of these in turn, firstly by giving some details about Carbon stars. Then I'll outline the mystery, which may be conveniently divided into two questions, as given by Icko Iben's (1981) paper "The Carbon Star Mystery: Why Do the Low Mass Ones Become Such and Where Have the High Mass Ones Gone?" Finally I discuss recent developments which may mean that the mystery is solved.

Carbon stars may seem to be a small detail in a meeting which addresses the elemental abundances of the entire universe. But consider. Most, if not all, stars between about 0.6 and 5 $M_{\odot}$ pass through a carbon star phase. The shape of the initial mass function ensures that this is a very large number of stars. During this phase they possess an envelope enriched in carbon together with a high mass loss rate. Hence this carbon is injected back into the interstellar medium. Carbon stars may well be a very important source of carbon in the universe. In particular, stellar models with low metallicity ($Z \lesssim 0.01$) enter this phase more readily than stars of near solar metallicity. Qualitative confirmation of this trend is provided by the observed fact that the ratio of the number of Carbon stars to

[1] Present Address: I.G.P.P., Lawrence Livermore National Laboratory, L-413, Box 808, Livermore, California, 94550, U.S.A.

M stars increases in populations with lower Z. Thus, in early stellar populations in particular, Carbon stars may provide a significant contribution to the overall carbon content of the universe.

I. WHAT IS A CARBON STAR ?

Any star which shows a higher number density of C^{12} than O^{16} in its envelope is called a Carbon star. This, of course, is the opposite to the situation in the majority of stars. Carbon stars occur during the latter stages of evolution. They are red-giants which will soon exhaust their envelope, through nuclear burning at its base and mass loss from its edge. They end their life as white dwarfs.

Note that there are low luminosity Carbon stars, typically found near the base of the giant branch. I will not discuss these. They do not form from the mechanism to be described, but possibly from mass transfer in binaries or from atypical evolution during the core helium flash.

When does a star become a Carbon star? Figure 1 shows the various stages of stellar evolution as a function of mass. For example, a star with $M < M_0$ will never ignite hydrogen (or helium) and will end its life as a brown dwarf. For masses between M_1 and M_2 the star burns hydrogen on the main sequence and then ignites its helium supply in the core helium flash at the tip of the first giant branch. It then burns helium on the horizontal branch, but has insufficient mass

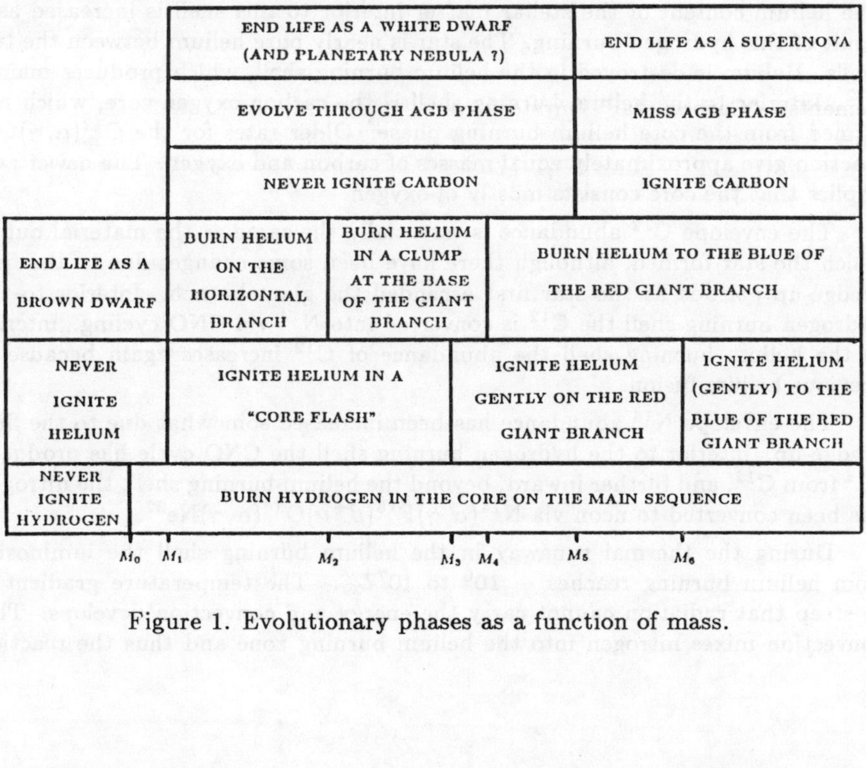

Figure 1. Evolutionary phases as a function of mass.

to ignite carbon. It evolves in the HR diagram along the second giant branch which approaches the first giant branch at high luminosities, and is hence called the Asymptotic Giant Branch (hereafter AGB). During this stage there is a recurring thermal instability in the star which may lead to the formation of a Carbon star. Thus a potential Carbon star must pass through the AGB phase. That is, it must ignite its core helium supply but not its core carbon supply. This means that it must have a mass between the two critical masses, M_1 and M_5. Note that these (and all other) critical masses depend on the composition of the star.

How does a star become a Carbon star? During its ascent of the AGB the star consists of an inert core composed mainly of carbon and oxygen. This is surrounded by a helium-burning shell. A little further outward there is a hydrogen burning shell, and beyond that the envelope, which is largely convective. As the star evolves up the AGB, the two nuclear burning shells move closer together in mass and this proximity, combined with the partial degeneracy of the helium burning shell, leads to a recurring thermal instability in this shell. It is these so-called "thermal pulses" or "shell flashes" that lead to the formation of Carbon star characteristics.

Consider the abundance distribution in the star prior to the first pulse. The hydrogen in the envelope is converted to helium in the hydrogen burning shell. The helium content of the stellar matter interior to this shell is increased as a result of this hydrogen burning. The star is nearly pure helium between the two shells. Helium is destroyed in the helium burning shell, which produces mainly C^{12}. Interior to the helium burning shell is the carbon-oxygen core, which has formed from the core helium burning phase. Older rates for the $C^{12}(\alpha,\gamma)O^{16}$ reaction give approximately equal masses of carbon and oxygen. The newer rate implies that the core consists mostly of oxygen.

The envelope C^{12} abundance is essentially the same as the material out of which the star formed, although there have been some changes due to the first dredge-up episode as the star first ascended the giant branch. Interior to the hydrogen burning shell the C^{12} is converted into N^{14} via CNO cycling. Interior to the helium burning shell the abundance of C^{12} increases again because of previous helium fusion.

The envelope N^{14} abundance has been increased somewhat due to the first dredge-up. Interior to the hydrogen burning shell the CNO cycle has produced N^{14} from C^{12}, and further inward, beyond the helium burning shell, the nitrogen has been converted to neon via $N^{14}(\alpha,\gamma)F^{18}(\beta^+\nu)O^{18}(\alpha,\gamma)Ne^{22}$.

During the thermal runaway in the helium burning shell the luminosity from helium burning reaches $\sim 10^6$ to $10^7 L_\odot$. The temperature gradient is so steep that radiation cannot carry the energy and convection develops. This convection mixes nitrogen into the helium burning zone and thus the reaction

$Ne^{22}(\alpha,n)Mg^{25}$ occurs. This releases neutrons which may react with iron to produce s-process elements.

The energy liberated by the runaway expands the star. The hydrogen burning shell is propelled outward to regions of low temperature and density, and is extinguished. The energy generation in the helium burning shell drops to normal values. During the next few thousand years the expanded envelope becomes convective to greater and greater depths. After a few pulses this convection reaches down beyond the hydrogen-helium discontinuity, mixing helium and nitrogen to the surface. The strength of the pulses increases with each succeeding pulse, and hence the maximum depth of the interpulse convective envelope also increases. Eventually convection reaches down to depths where the intershell convection has deposited freshly synthesized carbon. Carbon is then mixed to the surface.

After a sufficient number of these dredge-up episodes, the number density of carbon exceeds that of oxygen and the star is a Carbon star.

There is a linear relation between luminosity and the mass of the hydrogen exhausted core. Stellar behaviour on the AGB is largely determined by the core mass and hence by the surface luminosity. It is thus possible, in principle, to determine a minimum luminosity, as a function of stellar mass and composition, beyond which carbon is dredged to the surface. Likewise, one may calculate the amount of carbon added to the envelope with each pulse (it is not constant) and hence determine a minimum luminosity for the appearance of Carbon stars.

II. THE MYSTERY

And hence the mystery. One can take standard stellar birthrates, an initial mass function, and parametrized AGB evolution (including dredge-up) and construct synthetic AGB stellar populations. This was done by Renzini and Voli (1981) and Iben (1981). When compared with the observed distribution of Carbon stars and M stars, one notices two things: 1) theory gives too many high luminosity Carbon stars, and 2) too few low luminosity Carbon stars. Hence the title of Iben's paper—"The Carbon Star Mystery: Why Do the Low Mass Ones Become Such and Where Have the High Mass Ones Gone?"

Note also that theory predicts that thermally pulsing M stars should outnumber Carbon stars by about a factor of four. In fact it is the Carbon stars which outnumber the M stars by about a factor of 2. Carbon stars are the rule, not the exception. The conclusion reached was that thermal pulses begin and end at lower luminosities, which means lower core masses, than predicted by theory.

III. WHY UPDATE ?

Firstly we look at why the theoretical calculations predict too many high luminosity Carbon stars. Recall that we need $M < M_5$, which is called M_{UP} in the literature. If we can reduce M_{UP} then we stop high mass (*i.e.* high luminosity) stars from reaching the AGB, and they then cannot become Carbon stars. It was suggested by Renzini *et al.* (1985) the perhaps our estimates of M_{UP} were too high.

Previous estimates of M_{UP} were made without including the effects of semiconvection, convective overshooting, or the slightly controversial core breathing pulses (which will be discussed below). Becker and Iben (1979, 1980) found $M_{UP} \simeq 9M_\odot$. Castellani *et al.* (1985a,b) included semiconvection and core breathing pulses and find $M_{UP} \simeq 5$ or $6M_\odot$. Chiosi *et al.* (1987) included convective overshooting (in which case their models never develop semiconvective zones or show core breathing pulses) and find $M_{UP} \simeq 4$ or $6M_\odot$. If correct, this will remove high mass, high luminosity stars from the AGB. Stars with masses greater than M_{UP} ignite their core carbon supply under non-degenerate conditions and avoid the AGB phase. Less massive stars, however, will join the AGB. They may still rise above the luminosity at which a star of mass M_{UP} would join the AGB. Thus it may still be necessary to evoke higher mass loss rates for these stars, in order to remove them from the AGB before they become high luminosity carbon stars. But with most of the massive stars no longer experiencing the AGB phase this looks like a promising solution to the question "where have all the high mass Carbon stars gone?"

So why do theoretical models fail to make low mass Carbon stars? The synthetic distributions discussed earlier were based on model calculations for masses greater than 3 $M_\odot$ and less than 1 $M_\odot$. Results in this mass range, which constitute the majority of Carbon stars in the observed sample, were found by interpolation between these results. So I have constructed models in the mass range 1 to $3M_\odot$ for many different abundances (Lattanzio 1986, 1987a; hereafter Papers 1 and 2, respectively). I included semiconvection and approximated convective overshooting in a very crude way. At the edge of a convective core I estimate the ratio of the radiative to adiabatic temperature gradients at the first radiative mass shell beyond the convective core by linearly extrapolating from the values at the two outer most convective points in the convective core. If this ratio is greater than unity (in the usual Schwarzschild criterion) then I include this shell into the convective region, even though the ratio is less than one when calculated formally. On the next iteration for this model, this point now has an abundance as would have been appropriate if it had been in the convective core. That is, if matter had overshot the edge of the convective region and mixed with this shell. Whether this shell remains convective or not depends on the current value of the ratio of the gradients, calculated with the new abundances and thermodynamic variables. This procedure is repeated at the edge of the core at each iteration, thus allowing for rapid growth of the convective zone if that is what is required physically. The motivation for this procedure is the fact that the initial configuration gives a non-zero buoyancy at the convective core edge which would lead to overshooting, and produce a configuration like that which results from this method. This is a simple, albeit crude, formulation of semiconvection and convective overshooting, but it gives excellent results and is motivated by physics (more details may be found in Paper 1).

But there is still another problem. Many studies have shown that when the central helium abundance gets small (less than about 0.1 by mass), the convective core repeatedly grows and shrinks in so-called "breathing pulses" (*e.g.* Castellani *et al.* 1971a,b; Demarque and Mengel 1972; Gingold 1976; Castellani *et al.* 1985a,b). This is believed to be a physical, rather than numerical, instability (Sweigart and Demarque, 1973). A small growth in the convective core will mix in helium from the helium-rich surrounding layers. If the central helium abundance is small, then this may constitute a significant percentage increase in the central helium abundance. Because energy generation at this stage varies as the helium abundance to the third power, this extra helium results in a large perturbation in the energy liberation. This in turn generates a growth in the convective core, and we have a runaway. Almost instantaneously the central helium abundance jumps to a value where a small increase no longer causes a large change in the energy generation. Castellani and his co-workers find that there are typically three of these pulses prior to core helium exhaustion. This is in good agreement with my results, which used a very different method of calculating semiconvection.

I will not discuss the other details of the interior evolution, because this is now well understood. The important point to be made is that the details of helium burning with convection and/or semiconvection, possibly including overshooting, are not well understood. Chiosi and co-workers allow convective overshooting and never find semiconvection, nor do they find core breathing pulses. Yet their models exhaust core helium with a structure similar to those which do find these two phenomena. The models are telling us something. I think they are saying "Listen here, mate. You may not know how to calculate convective overshooting properly, but the way you are doing it means that I don't need to become semiconvective. In which case I don't need to display those vulgar breathing pulses either. But if you are not going to let me overshoot at all, well I'll go semiconvective just to spite you! And I'll develop breathing pulses which you can't calculate properly. So there! In either case, I demand that my hydrogen and helium exhausted cores be much larger than you used to think. And I'll make that happen by which ever route you leave open to me."

Now to the main results of these calculations. Paper 1 showed the core mass at the first pulse as a function of stellar mass for various abundances. Previously it had been assumed that for $M \lesssim 2M_{\odot}$ the core mass was essentially independent of total mass. While this is seen to be roughly true for $Z \gtrsim 0.01$, the mass dependence for $Z = 0.001$ is exponential. There is also a significant dependence on the helium content, Y. As Y decreases from 0.3 to 0.2 the core mass at the first pulse increases by $\simeq 0.05M_{\odot}$. The calculations must be compared with the extensive observations of Carbon stars in the Magellanic Clouds. Further calculations were made of stellar models in this metallicity range (Paper 2) which showed that the mass of the hydrogen exhausted core at the first pulse increases roughly linearly with total mass.

Figure 2 shows the luminosity at the first pulse versus mass for my models of various composition. The solid line is from Iben and Renzini (1983), and represents previous estimates for this luminosity. Note that Carbon stars hav $M_{bol} \lesssim -4.0$. If the solid line indicated the beginning of thermally pulsing evolution, which is a pre-requisite for the formation of Carbon stars, then it would be impossible to match the observed luminosities of the Carbon stars. But the new models clearly begin their thermally pulsing evolution well below the luminosity appropriate for Carbon stars. Indeed, many of my models begin pulsing up to 1 magnitude lower in luminosity than previously thought.

The next question is how do these stars evolve? I have continued the evolution of all of the 1.5 $M_{\odot}$ models through at least 10 pulses. For the models with

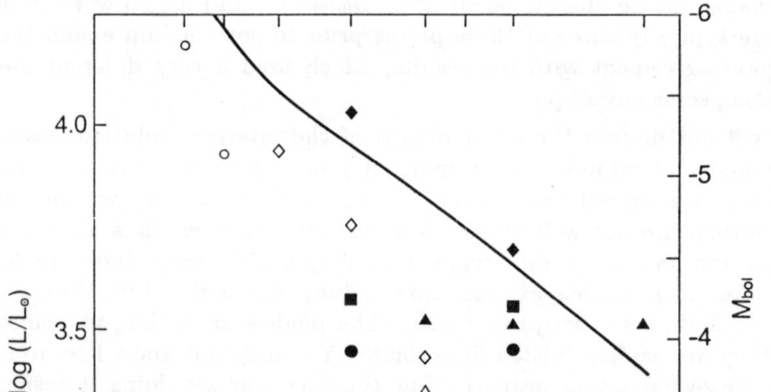

Figure 2. Luminosity at the first pulse (from Papers 1 and 2) for various masses and abundances. The solid line is from Iben and Renzini (1983) for $Z = 0.02$ and $Y = 0.28$.

$Z = 0.003$ and 0.006 I found carbon to be dredged to the surface in three of the four I was studying (Lattanzio 1987b, hereafter Paper 3). One of these became a Carbon star at a luminosity well below previous estimates of the minimum possible for Carbon stars. Indeed, note that my Carbon star shows $M_{bol} = -4.4$, approximately the luminosity previously thought to be the minimum for thermal pulses. (I should say that when I stopped the calculations, two of the other models were dredging carbon to their surfaces and one had not yet begun. There is no reason to believe that, given further calculations, these other stars will not become Carbon stars also.)

Bessel, Wood and Lloyd Evans (1983) conclude that Carbon stars first appear at $M_{bol} \simeq -4.3$ to -4.6. Similarly, Mould, DaCosta and Wieland (library) conclude that the critical luminosity for the appearance of Carbon stars is $M_{bol} = -4.3 \pm 0.3$. The agreement with the Carbon star model of Paper 3 is excellent.

Yet a small problem may remain. Comparison of the effective temperature of my Carbon star model with those of the Bar West field of the Large Magellanic Cloud (Richer, 1981) shows that my model is quite hotter than is observed. One could reduce the temperature by decreasing the mixing-length parameter $\alpha = \ell/H_p$. But note that carbon is not dredged to the surface for $\alpha = 1.0$ (Paper 1), although it is for $\alpha = 1.5$, which was used in the model of Paper 3 (see also Wood 1981). To agree with the observations would require $\alpha = 1.2$ or 1.3. It is unknown whether carbon dredge-up will occur for this value of α. In any case, the opacity used in calculating the models neglected molecules and the extra carbon in the stellar envelope, both of which would act to cool the star; hence the problem may not be serious.

CONCLUSION

In conclusion: Carbon stars form due to thermal pulses which occur during the Asymptotic Giant Branch phase of the evolution of stars. Previous calculations, which ignore convective overshooting, semiconvection and core breathing pulses are unable to match the observed Carbon star distributions. They predict too many high luminosity Carbon stars and too few of low luminosity. Including convective overshooting or semiconvection (which entails core breathing pulses) reduces M_{UP} to about 5 or 6 $M_{\odot}$. This possibly resolves the discrepancy at high luminosities. Explicit calculations with semiconvection and core breathing pulses in models of low mass stars provide agreement with the observations at low luminosities.

Sincere thanks to Bernard Carr for presenting the oral version of this paper at the meeting in New Orleans.

REFERENCES

Becker, S. A., and Iben, I., Jr., 1979, *Astrophys. J.*, **232**, 831.

Becker, S. A., and Iben, I., Jr., 1980, *Astrophys. J.*, **237**, 111.

Bessell, M. S., Wood, P. R., and Lloyd Evans, T., 1983, *Mon. Not. R. astr. Soc.*, **202**, 59.

Castellani, V., Giannone, P., and Renzini, A., 1971a, *Ap. Space Science*, **10**, 340.

Castellani, V., Chieffi, A., Pulone, L., and Tornambè, A., 1985a, *Astrophys. J. Lett.*, **294**, L31.

Castellani, V., Chieffi, A., Pulone, L., and Tornambè, A., 1985b, *Astrophys. J.*, **296**, 204.

Castellani, V., Giannone, P., and Renzini, A., 1971b, *Ap. Space Science*, **10**, 355.

Chiosi, C., Bertelli, G., Bressan, A., 1987 in *Late Stages of Stellar Evolution*, ed. S. Kwok and S. R. Pottasch (Dordrecht: Reidel), p213.

Demarque, P., and Mengel, J. G., 1972, *Astrophys. J.*, **171**, 583.

Gingold, R. A., 1976, *Astrophys. J.*, **204**, 116.

Iben, I., Jr., 1981, *Astrophys. J.*, **246**, 278.

Iben, I., Jr., and Renzini, A., 1983, *Ann. Rev. Astr. Astrophys.*,**21**,271.

Lattanzio, J. C., 1986, *Astrophys. J.*, **311**, 708.

Lattanzio, J. C., 1987a in *Late Stages of Stellar Evolution*, ed. S. Kwok and S. R. Pottasch (Dordrecht: Reidel), p235.

Lattanzio, J. C., 1987b, *Astrophys. J. Lett.*, **313**, L15.

Mould, J. R., Da Costa, G. S., and Wieland, F. P., 1986, *Astrophys. J.*, **309**, 39.

Renzini, A., Bernazanni, M., Buonanno, R., and Corsi, C. E., 1985, *Astrophys. J. Lett.*, **294**, L7.

Renzini, A., and Voli, M., 1981, *Astron. Astrophys.*, **94**, 175.

Richer, H. B., 1981, *Astrophys. J.*, **243**, 744.

Sweigart, A. V., and Demarque, P., 1973, in *Variable Stars in Globular Clusters and Related Systems*, ed. J. D. Fernie (Dordrecht: Reidel), p221.

Wood, P. R., 1981 in *Physical Processes in Red Giants* ed. I. Iben and A. Renzini (Dordrecht: Reidel),p135.

INTERPRETATION OF SPECTRA OF SUPERNOVA REMNANTS AND SUPERNOVAE

I.J. Danziger
European Southern Observatory
Karl-Schwarzschild-Str. 2
D-8046 Garching bei München
Federal Republic of Germany

Supernova remnants can be used to study different types of chemistry in an astrophysical setting. Old supernova remnants can be and have been used to trace abundances in the interstellar medium of our galaxy and nearby galaxies such as the LMC, M31 and M33[1,2,3]. In these old remnants the visibly excited gas is predominantly the result of swept-up interstellar material hardly any longer contaminated by the anomalous abundances ejected by the evolved exploded star.

These same remnants, and younger ones, can be used as laboratories for studying atomic and molecular chemical processes apart from just the details of chemical abundances. The comparison of observations with models of propagating shocks has become an industry in which the profits are gradually dwindling due to the inability of observers to provide sensitive discriminatory criteria and of theoreticians to model the probably complex physical situations that in reality exist. Molecular chemistry in this context is in its infancy, at least observationally, because of the paucity of data, IC443 being the only SNR until recently in which a molecular species (H_2) has been observed.

While it is intended to make some comments in these areas most of the talk will concentrate on young supernova remnants because they are the objects where one might hope to see in a direct way the results of nuclear chemistry in the progenitor star or as a result of the SN explosion itself. Young SNR will be either those resulting from historical SN, or those in which filaments have been identified to have velocity spreads exceeding several thousand km/sec.

Until relatively recently (mid 1970's) a review of this kind would be a review of individual SNR one at a time because each had its own special idiosyncracies - a somewhat disconcerting fact given that they were then generally accepted to originate from only 2 types of SN, types I and II (pace F. Zwicky). While those individual characteristics still remain many of them are conceivably better understood, and certainly groupings of young SNR with pronounced common characteristics have become apparent. (Just at the same period, and possibly accelerated by the occurrence of SN 1987a, when the need for a diversity of SN types is gaining acceptance.)

ABUNDANCES IN YOUNG SNR

What follows will be a discussion of some of the successes and some of the problems that exist. The following table lists all the young SNR that are relevant to a discussion of nuclear chemistry. In general X-ray results have not been discussed because the uncertainties due to non-equilibrium modelling are too daunting.

The Crab Nebula continues to occupy a unique position in the pantheon of SNR, even though LMC 0540-69.3 has been proclaimed to be very similar (because of the presence of a 50 ms pulsar and a non-thermal X-ray spectrum). In recent years the Crab has been associated with a Type II SN (hydrogen lines in the nebula) coming from an 8-10 $M_{\odot}$ star (abundance pattern). It has been known for some considerable time[4] that helium is overabundant and for a shorter time[5,6] that this abundance varies in a systematic way over the nebula, i.e. the outer shell has a normal He abundance while the inner regions tend to be high. No other abnormal abundances appear to be present. This itself is difficult to understand because overproduction of He in the CNO cycle requires an overproduction and resulting overabundance of nitrogen. There are of course scenarios in which N^{14} is converted into

Abundance Effects in SNR

SNR	Historical	RV km/sec	Optical Abundance Peculiarities
Crab	✓	1200	He variable N normal
Cas A	(✓)	5700	
HVF		Large PM	N, O, S, Ar, Ca
Q			N, He
Puppis A			
HVF		300	O var.
LVF		1500	N, O var.
LMC 132D		2200	C, O, Ne
SMC 0102-72.3		2000	C, O, Ne, Mg
G292.0+1.8		1600	O, Ne
LMC 0540-69.3		1500	O
NGC 4449		3500	O, Ne, S
Kepler	✓	300	N
3C 58	✓		N
SN 1006	✓	200 Large PM	Fe ?

Ne^{22} as a result of He burning in a shell of a 9 $M_{\odot}$ star[7]. The presence of a 9 $M_{\odot}$ star at a distance of 200 pc above the galactic is itself an unusual but not impossible feature.

The next group of SNR's have the common characteristic that they contain high velocity oxygen-rich filaments. It is this characteristic that might lead one to believe that they all result from the explosion of a star in the restricted mass range 15-25 $M_{\odot}$.

In the table are also presented a list of other elements that are suggested observationally to be overabundant in the oxygen-rich filaments. These vary somewhat from remnant to remnant possibly in part because some remnants are better observed than others but probably also as a result of real differences in the progenitor star. There has also been the suggestion in at least 3 cases that a ring structure describes the morphology and motion of the high velocity filaments.

It is worth discussing in detail recent results for 4 of these objects.

The recent discovery of high velocity nitrogen-enriched filaments in Cas A [8] has caused a doubt whether the progenitor could be a WO star and it is suggested that it is a WN8 star having a mass $M \sim 60\ M_\odot$, even though the abundance pattern is similar to that expected from an exploding 15-25 $M_\odot$ star. The nitrogen enriched low velocity filaments have been postulated to originate in the material that has been driven off by a stellar wind prior to the explosion. This would be qualitatively consistent with such a scenario.

The first successful IUE observation of the spectrum of SMC 0102-72.3 [9] (see Fig. 1) allows us to conclude that O, C, Ne, Mg are all overabundant in this remnant. This combined with the lack of a detectable helium overabundance suggests that modelled exploding stars of 15-25 $M_\odot$ [10] may well be the culprits (see Fig. 2). A recent IUE observation (August 17) of N132D suggests a similar pattern of overabundance. The absence of an overabundance of nitrogen represents a problem. It is worth noting here that silicon ought also to be overabundant. The precise determination of abundance ratios Ne/O for example may lead in the future to a better determined mass of the progenitor star.

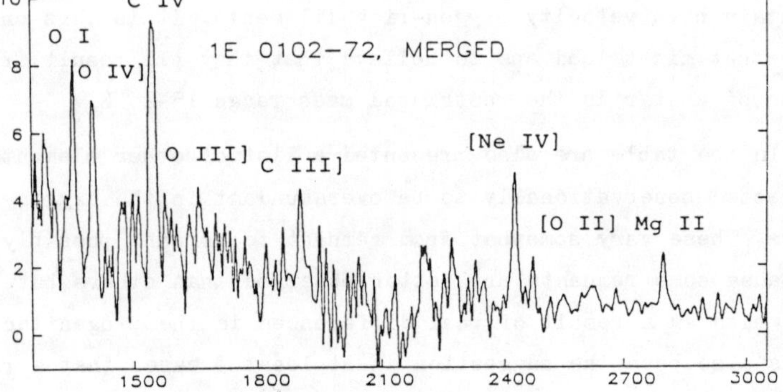

Fig. 1. UV spectrum of SMC IE 0102-7219

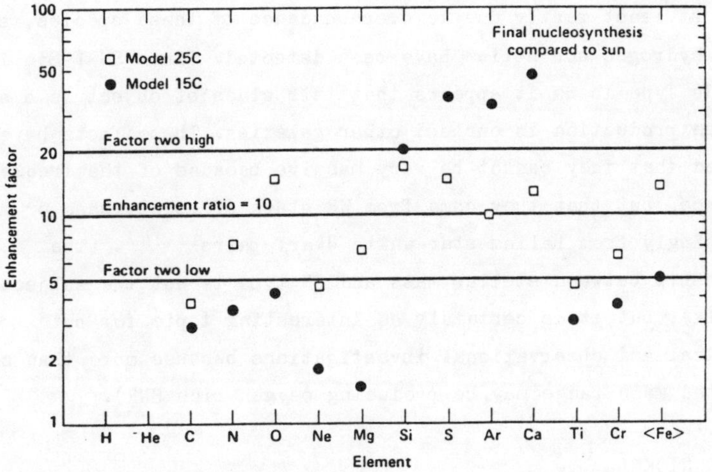

Fig. 2. Comparison of bulk elemental nucleosynthesis in 15 and 25 $M_\odot$ supernovae to solar abundances.

The discovery of high velocity oxygen filaments in Puppis A [11] suggests that the SNR may be younger than otherwise thought. The huge overabundance of nitrogen in the low velocity filaments also suggests that the progenitor could have been similar to that of Cas A i.e. a WN star of some considerable mass. This seems to be one of the few objects where the X-ray spectroscopy supports conclusions from optical wavelengths at least concerning oxygen.

Kepler and 3C 58 both pose the problem that although they result from historical SN, high velocity filaments have not yet been detected. The presently known small overabundances of nitrogen in the low velocity filaments are postulated to result from the gas ejected by a stellar wind prior to the explosion and now excited by the passage of the shock wave.

Apart from the inferences from models concerning progenitors for these oxygen-rich SNR's, is there a more direct way of linking these remnants to supernovae? SN 1983n and SN 1985f may provide that link[12] because their late time spectra (8 months for SN 1983n) show greatly enhanced lines of oxygen, calcium, sodium and magnesium which must be at least partly due to overabundance of these species, since neither hydrogen nor helium have been detected. Since SN 1983n is a bona fide Type Ib SN it appears that this class of object is a source of oxygen production in our and other galaxies. These facts have suggested that they cannot be very massive because of the frequency of occurrence, but that they come from WR stars of modest mass or more interestingly from helium star-white dwarf pairs[13,14]. (The relationship between stellar mass and SN type is not the subject of this talk), but it is certainly an interesting topic for both theoretical and observational investigations because more than one restricted mass range may be producing osygen-rich SNR).

IRON AND NICKEL

Because of the strong theoretical arguments concerning the mechanism of Type I SN and their light curves, there has been much emphasis on the fact that several tenths of a solar mass of iron are produced in Type I SN. Clearly iron must be produced somewhere contemporaneously (albeit under extreme conditions) because we see its temporal and spatial variation amongst galactic stars. What is the observational evidence? Firstly the light curves of Type I SN after maximum light require a mechanism by which stored energy inside the remnant can be released to produce the characteristic light curve. The radioactive decay of Ni^{56} into Co^{56} and thence to Fe^{56} seems to be the only viable mechanism and leads to the requirement of production of significant masses of Fe mentioned above.

There has been a claim[15] that broad absorption lines of FeII multiplets are seen in the spectrum of a star situated behind the remnant of SN 1006. The presence of the absorption features seems to have been confirmed by subsequent spectra[16]. While this strengthens the case for FeII in SN 1006 the interpretation may still be considered to be less than certain.

Iron in SN will be discussed later.

Because of the nickel-cobalt-iron decay mechanism, there has been some interest in whether reports of nickel lines in remnants such as the Crab implying overabundances of this element have validity and therefore significance. This problem has been dealt with[17] and it has been concluded that reported overabundances may result from uncertain atomic parameters. A high S/N spectrum of the old SNR N49 has been used as a partial test of this. Accurate determinations of the strengths of the [NiII]7413/[NiII]7380 ratio have been made. ([NiII]7380 is blended with [FeII]7388 which contributes ~ 20 % to the blend). This ratio (0.056) when compared with the theoretical ratios of transition strengths[18] does indeed suggest that for reasonable values of the conditions in the N49 filaments the ratios could be in error by at least a factor 2 in the direction that when corrected would decrease the computed value of the abundance of nickel.

ABUNDANCES IN SN

Although most workers in this field would argue that abundances are not solar in Type I SN there has not been a strong consensus on what the real abundances are except for the almost complete absence of hydrogen. This is particularly true concerning the early absorption spectra and applies to early Type II spectra as well.

At later phases (~ 250 days) in Type I spectra a strong broad emission line spectrum has developed and the strongest optical features in SN 1972e have been interpreted[19,20] as being due to lines of [FeII] and [FeIII] suggesting a large overabundance of iron in SN 1972e. Kirshner and Oke suggest that 10^{-2} $M_\odot$ of Fe are seen while Meyerott deduces that 0.5 $M_\odot$ of Fe are produced by radioactive decay, this energy being necessary to produce the light curve and spectra at 245 days.

While the interpretation of the spectra as the blended emission of [FeII] and [FeIII] is attractive, it is worth re-examining some of these questions with currently available higher resolution spectra of other SN. Spectra of the Type I SN 1986g at 264 and 303 days show a superficial similarity to that of SN 1972e at 245 days. Nevertheless the strong features between 4200 and 5500 Å in SN 1986g correspond to only the [FeIII] features proposed by Meyerott while the features at just > 7000 Å ascribed to [FeII] are present. It requires much more work to ascertain whether this difference results from a difference in excitation and ionization balance or whether some new fundamental identifications are required. Our spectra show many as yet unidentified emission features. In addition at these late phases strong broad features at 5876 and 6562 Å are evident with possible self-absorption features at 264 days but not at 303 days. It will be important to know whether these features due to HeI and Hα result from the original envelope of expelled gas or whether they are swept up material accelerated to large velocities. The self absorption must occur reasonably close to the continuum forming region.

The other piece of observational evidence for the production of iron in SN resulted from the observation of the strong IR line of [FeII] 1.644 μ [21] in the spectrum of Type I SN 1983n one year after maximum light. Since 1983n subsequently was recognised as a Type Ib and was oxygen rich (see discussion above) the results of the modelling to arrive at 0.3 $M_\odot$ of iron need to be revised. In any case the interpretation is irrelevant to the problem of iron production in low mass Type Ia SN, but not necessarily irrelevant to iron production in the galaxy. Alternatively however there has been a suggestion[23,24] that because strong SiI transitions occur at almost the same wavelength one might be seeing silicon in approximately normal abundances in SN 1983n. This can and will be checked by trying to observe other lines of Fe and Si in the IR spectrum. It should be noted however that [FeII] 1.64 μ has been detected in older remnants though this does not imply an overabundance in these objects.

The emission line spectra of Type II supernovae have proved a little more tractable concerning abundances of a few lighter elements. In the UV spectrum of SN 1979c, emission lines of He, C, N, Si have been identified[24] with the conclusion that N/C ~ 8 (30× cosmic) and O/C ⩽ 4. This enhancement of N has led to the conclusion that the progenitor was a red supergiant which had undergone mass loss with the emission occurring in the outer layers of the SN ejecta. From what we now know of SN 1987a these results do not preclude the possibility that the progenitor was a blue supergiant.

The UV spectrum of SN 1987a obtained in June 1987 i.e. ~ 100 days after outburst has recently been analysed[25]. It is found that N/C ~ 60× normal and N/O ~ 6× normal, not so different from the result for SN 1979c. However because the lines are narrow it is suggested that they originate in circumstellar material resulting from mass loss. It is probably not irrelevant that analyses[26] of blue supergiant stars in the LMC with estimated masses ~ 30 $M_{\odot}$ give significantly enhanced abundances of helium and nitrogen.

ATOMIC AND MOLECULAR CHEMISTRY IN SNR

An attempt has been made recently[27] to use IR spectra to study conditions in old SNR. Lines of [FeII] have been detected in all observed remnants and the line ratio [FeII]1.60/1.64 has given electron densities higher by a factor 2 than those obtained from conventional optical lines. This is not surprising if we are observing cooler denser regions behind the shock. The [FeII](1.64)/Hβ ratio is remarkably uniform ~ 0.2 amongst the remnants. The simplest interpretation of this is that the Fe/H abundance ratio is constant in all remnants as is the depletion factor resulting from grain formation. We find this depletion factor to be $\gtrsim 0.9$ for Fe, thus suggesting that shock heating is not very effective in destroying grains, contrary to expectation[28].

Another interesting result has been the detection of the H_2 (1-0 S(1)) line of molecular hydrogen (2.12 μ) in RCW 103 where its strength is ~ 1/3 × [FeII]1.644. Its spatial distribution is more extended than the optical filaments and a partial map of RCW 103 in this line shows that it peaks ~ 20 arcsecs (0.3 psc) further from the centre than [FeII] and Hα. Shock excitation is the obvious explanation for the emission although UV photon excitation from the SNR may be effective. In the future IR line ratios can be used as diagnostics for the two mechanisms.

REFERENCES

1) Danziger, I.J. and Leibowitz, E.M., Mon. Not. R. astr. Soc. **216**, 365 (1985).
2) Blair, W.P. and Kirshner, R.P., Astrophys. J. **289**, 582 (1985).
3) Blair, W.P., Kirshner, R.P. and Chevalier, R.A., Astrophys. J. **254**, 50 (1982).
4) Woltjer, L., Bull. Astr. Inst. Netherlands **14**, 39 (1958).
5) Clark, D.H., Murdin, P., Wood, R., Gilmozzi, R., Danziger, J. and Furr, A.W., Mon. Not. R. astr. Soc. **204**, 415 (1983).
6) Uomoto, A.K., MacAlpine, G.M. and Henry, R.B.C., in The Crab Nebula and Related Supernova Remnants, ed. M.C. Kafatos and R.B.C. Henry (Cambridge University Press), p. 9 (1985).
7) Nomoto, K., in The Crab Nebula and Related Supernova Remnants, ed. M.C. Kafatos and R.B.C. Henry (Cambridge University Press), p. 97 (1985).
8) Fesen, R.A., Becker, R.H. and Blair, W.P., Astrophys. J. **313**, 378 (1987).
9) Blair, W.P., Raymond, J.C., Danziger, J. and Matteucci, F., Astrophys. J., submitted (1987).
10) Woosley, S.E. and Weaver, T.A., in Nucleosynthesis and its Implications on Nuclear and Particle Physics, ed. J. Audouze and N. Mathieu (D. Reidel Publishing Company), p. 145 (1986).
11) Winkler, P.F. and Kirshner, R.P., Astrophys. J. **299**, 981 (1985).
12) Gaskell, C.M., Cappellaro, E., Dinerstein, H.L., Garnett, D.R., Harkness, R.P. and Wheeler, J.C., Astrophys. J. **306**, L77 (1986).
13) Iben, I., Jr., Nomoto, K., Tornambè, A., Tutukov, A.V., Astrophys. J. **317**, 717 (1987).
14) Tornambè, A., Matteucci, F., Astrophys. J. **318**, L25 (1987).
15) Wu, C.-C., Leventhal, M., Sarazin, C.L. and Gull, T.R., Astrophys. J. Lett. **269**, L5 (1983).
16) Fesen, R.A., Wu, C.-C. and Leventhal, M., private communication (1987).
17) Henry, R.B.C., Pub. Astron. Soc. Pacific **98**, 1044 (1986).
18) Nussbaumer, H. and Storey, P.J., Astron. Astrophys. **110**, 295 (1982).
19) Kirshner, R.P. and Oke, J.B., Astrophys. J. **200**, 574 (1975).
20) Meyerott, R.E., Astrophys. J. **239**, 257 (1980).
21) Graham, J.R., Meikle, W.P.S., Allen, D.A., Longmore, A.J. and Williams, P.M., Mon. Not. R. astr. Soc. **218**, 93 (1986).
22) Oliva, E., Astron. Astrophys., in press (1987).
23) Fransson, C., private communication (1987).
24) Fransson, C., Benvenuti, P., Gordon, C., Hempe, K., Palumbo, G.G.C., Panagia, N., Reimers, D. and Wamsteker, W., Astron. Astrophys. **132**, 1 (1984).
25) Cassatella, A., ESO Workshop on SN 1987a, in press (1987).
26) Kudritzki, R.P., ESO Workshop on SN 1987a, in press (1987).
27) Moorwood, A.F.M., Oliva, E. and Danziger, I.J., IAU Colloquium No. 101, Penticton, Canada, in press (1987).
28) Seab, G. and Shull, J.M., Astrophys. J. **275**, 652 (1983).

ON THE TIMING OF THE NEUTRINO AND PHOTON BURSTS FROM SUPERNOVA 1987A

W. David Arnett
Enrico Fermi Institute and
Departments of Astronomy and Physics
5640 S. Ellis Avenue
University of Chicago
Chicago IL 60637, USA

Abstract. The timing of the neutrino burst and the early behavior of the optical light from Supernova 1987a provide a unique challenge for the theory of Type II supernovae. Using realistic initial models for the presupernova and quantitative calculations of the explosion, it is concluded that the early light curve data require a time of explosion close to that of the escape of antineutrinos detected by the Kamiokande-II and the IMB experiments, and inconsistent with the time of the Mont Blanc event.

1. History

The early history of the observations of Supernova 1987a may be summarized as follows. On Feb. 23.124 UT a neutrino signal was detected at the Mont Blanc Neutrino Observatory (Castagnoli 1987), and subsequently identified with Supernova 1987a. On Feb. 23.316 UT a neutrino (antineutrino) burst was observed with the Kamiokande-II detector (Koshiba 1987), and "no events were recorded around Feb. 23.12 UT, the time of the reported Mont Blanc detection". The nucleon decay detector of the Irvine-Michigan-Brookhaven (IMB) collaboration (Svoboda 1987) observed a signal coincident with that of the Kamiokande-II collaboration. R. H. McNaught (1987a) reports that "A. Jones (Nelson, New Zealand), who easily discovered the supernova visually on Feb. 24.37 UT with a 30 x 78 finder telescope, did not notice the object with the same instrument on Feb. 23.39 UT, suggesting that it was then fainter than mag 7.5". The official discovery by Ian Shelton (1987), University of Toronto Las Capanas Station, was made on Feb. 24.23 UT from an exposure begun Feb. 24.06 UT, and independently by O. Duhalde (also at Las Capanas) around Feb. 24.2. R. H. McNaught (1987a) reports that the object appeared on films from the previous night: Feb. 23.443 at 6.0 and Feb. 23.445 at 6.2 magnitude; and that films by Garradd confirm that the field was identical down to 14.5 mag on Jan. 24 and Feb. 22. Wampler (private communication) reports revised values from the McNaught films of $V = 6.4 \pm 0.1$ at 23.444 and 6.11 ± 0.1 at 23.621. By Feb. 24.8 the International Ultraviolet Explorer (Sonneborn 1987) and other major instruments were observing the supernova, as they have since. McNaught (1987b) noted that the star Sanduleak -69° 202 was a possible progenitor. The timing of these events is shown in Table 1.

TABLE 1. EARLY HISTORY OF SUPERNOVA 1987A

Date (Feb. UT)	Event	Interval (t - Feb. 23.316)
23.124	Mont Blanc event	-0.192 d = -276.5
23.316	Kamiokande-II event	0.
23.316	IMB event	0.
23.39	A. Jones upper limit on optical	0.074 d = 106.5 m
23.444	V = 6.4 (on film,	0.127 d = 182.8 m
23.621	V = 6.11 McNaught)	0.129 d = 185.8 m
24.23	V = 5.0 (Shelton)	0.914 d
24.37	V = 5.1 (Jones)	1.054 d
24.46	V = 4.81 (Moreno and Walker)	1.144
24.455	V = 4.8	1.139 d
24.472	4.8	1.156 d
24.635	4.4	1.319 d
24.679	4.5	1.363 d
24.717	4.4	1.401 d
	(McNaught and Garradd)	
24.813	V = 4.514 (IUE)	1.497 d

These data offer an exceptional opportunity to quantitatively explore a well developed astrophysical theory: that of Type II supernovae. Indeed, the reaction of several authors (e. g. Harkness, Wheeler, Sutherland, and Swartz 1987, and Ostriker 1987) has been that moderate to severe modifications to this theory are required by the observations. I shall argue to the contrary: the theory fares spectacularly well.

Preliminary analysis (Sonneborn and Kirshner 1987) indicated that star 1 (Sanduleak -69° 202) did not explode. Despite this conclusion several theorists (Woosley, Pinto, Martin and Weaver 1987; Arnett 1987a; Schaeffer, Calle, Mochkovitch and Cahen 1987) obstinately insisted that the presupernova was something like star 1. It was explicitly shown that (1) a star of 15 $M_\odot$ and LMC abundances could have core collapse as a B supergiant (Woosley, et al. 1987; Arnett 1987a), and that (2) for an explosion energy of 2×10^{51} ergs such a star would give a light curve (Arnett 1987a; Shigeyama, Nomoto, Hashimoto, and Sugimoto 1987) similar to that observed for the first 20 days of 1987a. News of the nonsurvival of star 1 (Kirshner, Nassiopoulos, Sonneborn and Crenshaw 1987) was greeted with some enthusiasm in these quarters; for the first time we have direct information on the progenitor of a Type II supernova, and the theoretical indications were confirmed.

From an analysis of both UV and optical line spectra of 1987a shortly after outburst, Lucy (1987) finds they are accurately matched by an expanding spherical shell with a power law dependence of gas density on radius. In particular he finds that $5 \leq n \leq 7$ with $n = 6.3$ giving good fits to the line profiles on Feb. 26 UT. Extrapolating these structures back in time, Wampler, Truran, Lucy, Hoeflich, and Hillebrandt (1987) find it extremely difficult to understand how the neutrino signal detected by both Kamiokande-II and IMB experiments can correspond to the core collapse event. On the contrary, they feel their results to be consistent with core collapse occurring about 5^h earlier (near the time of the Mont Blanc signal). This conclusion is an important one; and has been used by Hillebrandt, Hoeflich, Kafka, Muller, Schmidt, and Truran (1987) to argue that the Mont Blanc event corresponds to the formation of a neutron star, and the Kamiokande-II/IMB event to its subsequent collapse to a black hole.

There are three arguments against this interesting hypothesis. First is the experimental conflict between Mont Blanc and Kamiokande-II (because of a higher threshold, IMB might not have seen the Mont Blanc signal). Second, Schaeffer, Declais, and Jullian (1987) have shown that this picture requires a neutron star binding energy which they feel are impossible to supply with presently available equations of state. The third involves a quantitative analysis of the emergence of the supernova shock and subseqent behavior; and will be given below.

2. Numerical Results

Figure 1 shows the behavior of the supernova and a theoretical model (Arnett 1987b) for a time of 2.3 days, around the time of discovery.

The supernova time and the model time are synchronized by assuming that core collapse coincided with the Kamiokande-II/IMB time.

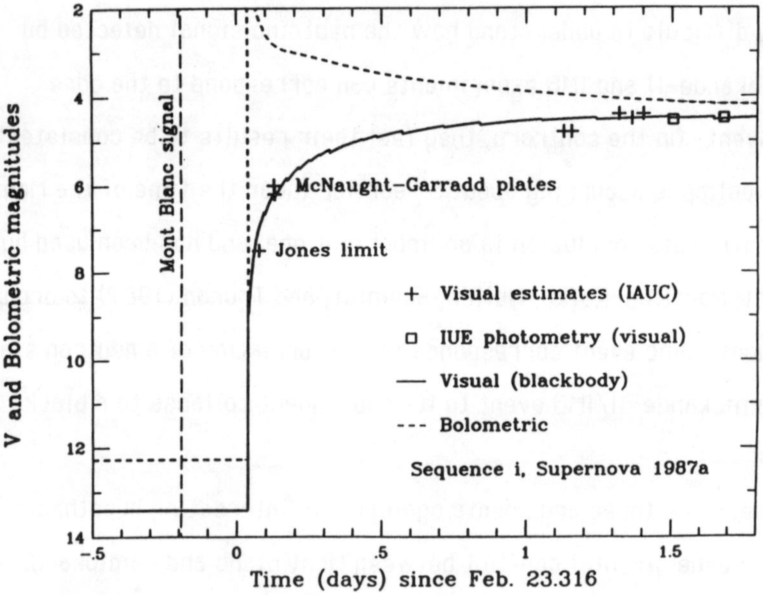

Figure 1. Supernova 1987a: Visual and bolometric magnitudes as a function of time, for the first 2.3 days around the event. The theoretical and observational time axes were synchronized by taking the time of core collapse to coincide with the detection of antineutrinos by the Kamiokande-II and the IMB experiments. After a short period of apparent inactivity the presupernova suddenly brightens by 4.5 magnitudes in the visual in one minute, followed by a more gradual increase to a plateau level. The (unobserved) bolometric luminosity much exceeds the visual at these early times. The time of the Mont Blanc signal is shown for comparison. The theoretical curves correspond to "sequence i" (see text).

The initial star had a radius of $2X10^{12}$ cm, a mass of 15 $M_{\odot}$, and an explosion energy of $2x10^{51}$ ergs (the "i sequence"). It was rather like Sanduleak -69° 202; although further "fine tuning" could be done. Notice that the theoretical V magnitude goes through both the upper limit due to Jones and the early McNaught point, as well as smoothly into the later photoelectric data. The later McNaught point lies below the curve. It is unclear if this is an essential disagreement. The effects of expansion opacities (lines smeared by large doppler shifts) and special relativity ($v/c > 0.1$) are probably important at just this epoch, and may modify the detailed shape of this part of the visual light curve, but were not included in the numerical calculation. The time for the bolometric luminosity (small dashed curve) to rise from twice its initial (presupernova) value to its maximum (off the figure) is one *minute*. This is roughly the light travel time across the presupernova radius. The visual luminosity changes by 4.5 magnitudes in this time interval. Between core collapse and this rapid brightening, the shock takes about 3,000 seconds to travel from the core to the stellar surface. The time of the Mont Blanc signal is shown by a vertical (large dashed) line.

The upper limit due to Jones is crucial in determining the curvature of the line denoting visual luminosity. In fact the value of 7.5 magnitudes used here (and taken from the IAU circular) may be too relaxed; Wampler et al. (1987) suggest that mag. 8 or even 9 might be more realistic. Clearly the supernova must have brightened very rapidly (as did the theoretical model).

This may be seen more clearly in Figure 2, which has an expanded time scale covering 0.6 days around the neutrino signals. The optical discovery date is offscale to the right. The theoretical curve ("sequence i") rises too fast, exceeding the Jones upper limit. This sequence is a counter example to the constraint of Wampler et al. (1987) in that its rise is too fast, not too slow. The time interval between core collapse and the

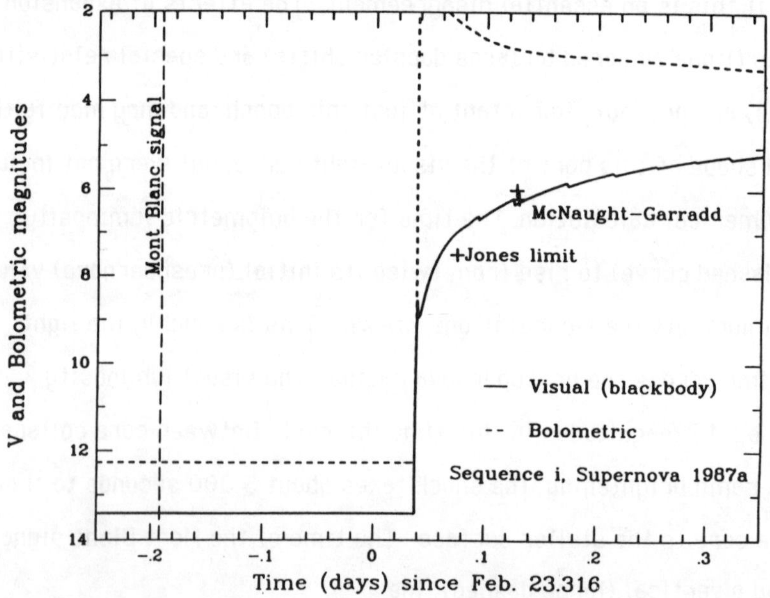

Figure 2. Supernova 1987a: Visual and bolometric magnitudes as a function of time, for the first 0.6 days around the event. The theoretical curves correspond to "sequence i" (see text) which has a presupernova radius of 2×10^{12} cm. On this time scale the slightly too rapid rise of this model sequence is clear; the theoretical curve exceeds the observational upper limit due to A. Jones. The weak sawtooth pattern in the theoretical curve is due to numerical effects in treating the recombination of ionized matter in individual zones and their resulting transparency.

rapid optical rise is a measure of the initial radius divided by the average shock velocity (Arnett 1987a; Shigeyama et al. 1987).

The luminosity approaches a value which is proportional to both the mean velocity squared and the initial radius (an opacity dependence is

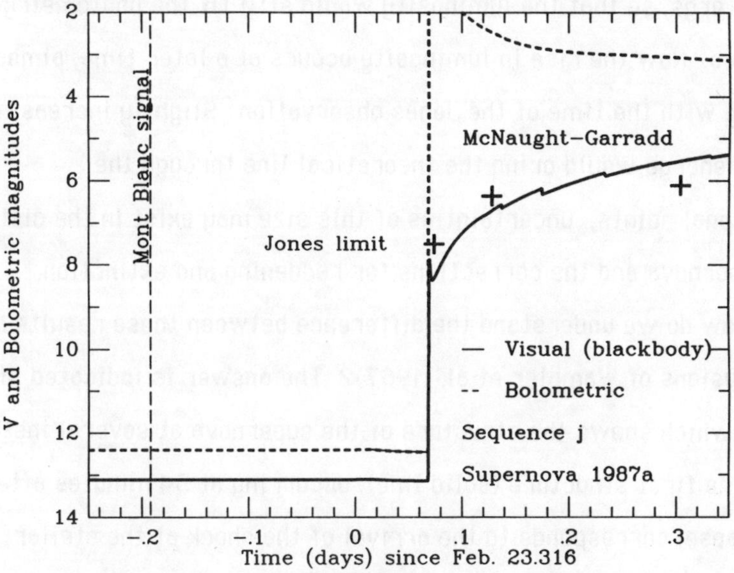

Figure 3. Supernova 1987a: Visual and bolometric magnitudes as a function of time, for the first 0.6 days around the event. The theoretical curves correspond to "sequence j" (see text) which has a presupernova radius of 3×10^{12} cm. This radius is close to that obtained from the evolutionary calculation, or alternatively, inferred from the star Sanduleak -69° 202 which disappeared. This theoretical curve is consistent with the observational upper limit due to A. Jones. Any attempt to shift the time scales so that core collapse would coincide with the Mont Blanc signal would imply that the supernova was well above this upper limit.

scattering). The Jones limit sets a minimum initial radius for the supernova, and one consistent with Sanduleak -69° 202 as the progenitor.

In Figure 3 is shown the behavior of a supernova of initial radius of 3×10^{12} cm, a value close to that found in the stellar evolutionary calculations (Arnett 1987a). The explosion energy was chosen to be 1.33×10^{51} ergs, so that the luminosity would still fit the photometry at later times. Now the rise in luminosity occurs at a later time, almost coinciding with the time of the Jones observation. Slightly increasing the explosion energy would bring the theoretical line through the observational points; uncertainties of this size may exist in the distance to the supernova and the corrections for reddening and extinction.

How do we understand the difference between these results and the conclusions of Wampler et al. (1987)? The answer is indicated in Figure 4, which shows the structure of the supernova at several instants in time. The first structure (solid line), occurring at 54 minutes after core collapse, corresponds to the arrival of the shock at the stellar surface. This steep dependence of density upon radius is characteristic of the surface layers of a hydrostatic object (Chandrasekhar 1939). Formally, the logarithmic derivative,

$$n = d\ln(\rho)/d\ln(r),$$

becomes infinite as the density ρ approaches zero at the surface. With a stellar structure code using finite difference equations, this is approximated by a large but finite number. With more accurate zoning the value of n increases. For calculations of initial models used here, $n > 30$.

As the shock propogates into this steep density gradient, it strengthens, giving rise to greater velocities. This results in a nonhomologous

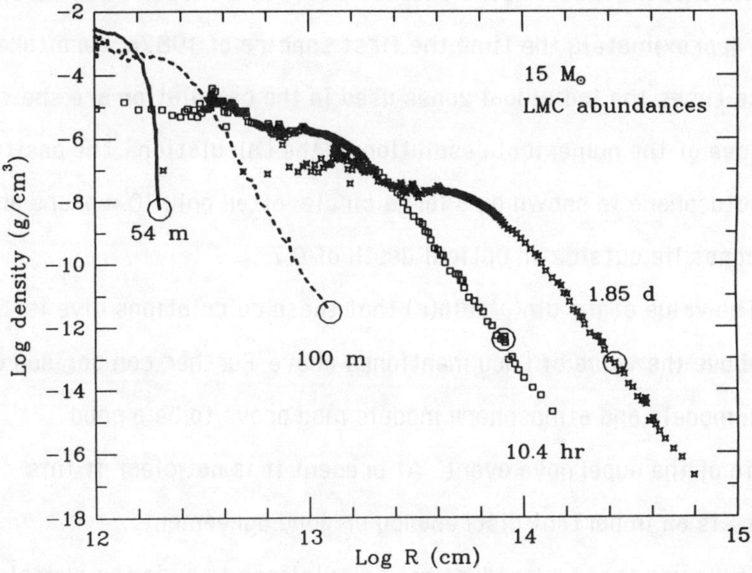

Figure 4. Density versus radius for the supernova (sequence i) at various instants in time. The solid line denotes the structure at 54 minutes after core collapse, when the shock arrives at the stellar surface. The shape has not yet lost the steep gradient in density near the surface which is characteristic of hydrostatic equilibrium. At 100 minutes the structure is given by the dashed line. Already the higher velocities produced by shock acceleration in the density gradient are changing the shape to that of a power law. At 10.4 hours the structure is fairly well relaxed, and coasting. Each square plotted represents an individual zone, giving a sense of the numerical resolution. The large circles are centered on the zone with an optical depth of 0.7 (the photosphere) at the respective times. At 1.85 days, a time characteristic of the earliest spectra taken, 18 zones lie above the photosphere; this corresponds to a mass of 0.0015 $M_\odot$.

expansion which is the cause of the power-law structure seen at later times. After 100 minutes (dashed line), the power-law is approaching an asymptotic shape. At 10.4 hours and 1.85 days, this shape is realized; the latter is approximately the time the first spectra of 1987a were taken. For these times the individual zones used in the calculation are shown to give a sense of the numerical resolution of the calculation. The position of the photosphere is shown by a large circle; after only 10.4 hours about a dozen zones lie outside an optical depth of 0.7.

The value of $n = d\ln(\rho)/d\ln(r)$ that these calculations give is 8 to 9, a bit above the value of Lucy mentioned above. Further comparison of explosion models and atmosphere models may prove to be a good diagnostic of the supernova event. At present it is not clear if this difference is an important discrepancy or good agreement.

If we use these hydrodynamic calculations to redo the Wampler et al. (1987) argument, the discrepancy with the Kamiokande-II/IMB time and core collapse disappears. In fact, we can make a more positive statement: the early light curve data *requires* an explosion time, and therefore we presume a core collapse time, which is close to that of the Kamiokande-II/IMB events. Although we expect the theory and the interpretation of the observational data to improve further, a strong statement can be made now. The early data on the supernove defines its initial radius and its shock velocity. This fixes the time from explosion in the core to outburst at the surface, the absolute luminosity, the effective temperature and the doppler shifts of spectral lines. The models given above represent these quantities fairly well; some fine details need improvement. Making the explosion time coincide with the Mont Blanc event would require drastic changes in these quantities. For example, consider the following rough estimates, which have been tested by comparison of different numerical models. The luminosity at early times scales as $L \sim v^2 R$, where R is the initial radius and $v^2 \sim E_{sn}/M$ is the kinetic energy per unit mass. The effective temperature T_e is related to L and R by $L \sim R^2 T_e{}^4$. The time between the start of the explosion and the emergence of the shock at the stellar surface is $\tau \sim R/v$. If we use the Mont Blanc time instead of the Kamiokande/IMB time, and the Jones upper limit on the luminosity, $\tau \to 3.65\,\tau$, a factor $f = 3.65$. Thus, from the relations above we find

$$v \to v f^{-1/3} = 0.65\, v,$$

$$R \to R f^{2/3} = 2.4\, R, \text{ and}$$

$$T_e \to f^{-1/3} T_e = 0.65\, T_e.$$

This implies velocities and effective temperatures two thirds as large as those observed and radii more that twice as large as those for Sanduleak 202. It should be stressed that the constraint comes from observations of effective temperature and expansion velocity, and not just luminosity alone. The Kamiokande/IMB time agrees well with the optical observations of the supernova; the scales of effective temperature and velocity agree with those observed to about 10% or so. This would imply a range of time for the explosion of Feb. 23.316±0.024 which excludes the

Mont Blanc time of Feb. 23.124.

However, if we arbtrarily exclude the Jones limit, the constraint is not so tight. The variation in the brightness of the supernova on the first two McNaught plates is not large, and therefore does not constrain the time of rapid rise in luminosity well. Data on the change in effective temperature or spectral features comes from the optical discovery epoch, a day later. I estimate the uncertainty to be ±0.2 days, giving a range of Feb. 23.316±0.2 which is marginally consistent with Mont Blanc time as well as that of Kamiokande/IMB. The presupernova characteristics remain the same, only the time of explosion may be shifted. Thus, in addition to the experimental conflict between the Mont Blanc and the Kamiokande results, there is also a conflict between the Mont Blanc results and the Jones limit.

Acknowledgements

Special thanks are due the many authors who provided versions of their work prior to publication, and to the NSF Astronomy Instrumentation program and Floating Point Systems, Inc., who helped provide the significant computing power needed to complete these calculations before the supernova did! Particular thanks are due Joe Wampler for sharing his analysis of the early data. The author apologizes for any oversight due to finite mailing times, which are becoming obvious for research on SN 1987a.

References

Arnett, W. D. **Astrophys. J.**, in press (1987a).

Arnett, W. D., preprint (1987b).

Castagnoli, C. International Astronomical Union Circular (IAUC) 4223 (1987).

Chandrasekhar, S. **An Introduction to the Study of Stellar Structure** (Chicago: University of Chicago Press), (1937); see Chapter VIII.

Harkness, R. P., Wheeler, J. C., Sutherland, P. G., and Swartz, D., preprint (1987).

Hillebrandt, W., Hoeflich, P., Kafka, P., Muller, E., Schmidt, H. U., and Truran, J. W., preprint (1987).

Kirshner, R. P., Nassiopoulos, G. E., Sonneborn, G., and Crenshaw, D. M., preprint (1987).

Koshiba, K. IAUC 4338 (1987).

Lucy, L., preprint (1987).

McNaught, R. H. IAUC 4316 (1987a).

McNaught, R. H. IAUC 4317 (1987b).

Ostriker, J. P., preprint (1987).

Schaeffer, R., Casse, M., Mochkovitch, R. and Cahen, S., preprint (1987).

Schaeffer, R., Declais, Y., and Jullian, S., preprint (1987).

Shelton, I. IAUC 4317 (1987).

Shigeyama, T., Nomoto, K., Hashimoto, M., and Sugimoto, D., preprint(1987).

Sonneborn, G. IAUC 4317 (1987).

Sonneborn, G. and Kirshner, R. IAUC 4333 (1987).

Svoboda, R. IAUC 4340 (1987).

Wampler, E. J., Truran, J. W., Lucy, L. B., Hoeflich, P., and Hillebrandt, W., preprint (1987).

Woosley, S. E., Pinto, P. A., Martin, P. G., and Weaver, T. A. **Astrophys. J.**, in press (1987).

Nucleosynthesis and Supernovae

Produced by Late Time Neutrino Heating

Ronald W. Mayle and James R. Wilson
Lawrence Livermore National Laboratory
B Division L-35
Livermore, CA 94550
USA

ABSTRACT

During a supernova explosion a small amount (a few hundreths of a solar mass) of material composed of neutron rich nuclei heavier than iron group elements may be expelled into the interstellar medium. The composition of this neutron rich material is not determined by the classical r-process. In this paper we discuss the physical conditions in a supernova that produce neutron rich material, and the environment in which the final heavy element composition is determined.

I. INTRODUCTION

It has long been known that the source of heavy element nucleosynthesis from carbon to iron group elements is likely found in exploding massive stars [1],[2],[3]. However, even prior to explosion, nuclear reactions have transformed the core region of these stars from hydrogen, through successive burning stages, to heavier elements, and supplied the energy needed to keep the star from collapsing under the influence of gravity. Stars in the mass range $10< M <70\ M_\odot$ burn nuclear fuel up to iron group nuclei, and begin to collapse due to the photodisintegration instability and the electron capture instability [4]. Stars in the range of $8< M <10\ M_\odot$ collapse due to the electron capture instability prior to the ignition of oxygen burning. Electron capture on the iron group nuclei is sufficiently rapid to overcome the rise in thermal energy produced by the subsequent oxygen detonation, and the core continues to collapse [4],[5],[6]. For all massive stars ($8< M <70\ M_{\odot}$), the collapse halts when the central densities reach and exceed nuclear matter densities; a shock wave forms near the surface of the rebounding inner core, at about the $.55\ M_{\odot}$ point. Initially the shock wave rapidly proceeds outward, but stalls at a radius of about 300 km due to energy losses from disassociation of heavy nuclei and neutrino cooling. At this time the neutrinosphere is located about 60 km from the center of the star, and is slowly moving inward with time.

The stalled shock wave can be revived by late time neutrino heating of the post shocked material, as soon as the shock encounters matter that, after passing through the shock, has a temperature and density low enough to reduce the local rate of neutrino cooling by neutrino production. A source of heating for this material comes from interaction of the free neutrons and protons with the hot core electron neutrinos and electron antineutrinos streaming from the neutrinosphere. Thermal pressure is increased by this net gain of internal energy and the shock wave begins to proceed outward into the hydrogen-helium envelope of the star.

In calculations we have carried out [4], the heavy elements from carbon to iron group nuclei synthesized during the star's lifetime do not escape in a supernova explosion, but become part of the neutron star (or black hole) that forms. The heavy elements from carbon to iron group nuclei expelled into the interstellar medium are produced as the shock wave passes outward into the hydrogen-helium envelope, shock heating material to temperatures where nuclear reactions are fast compared to dynamical timescales.

Figure 1 shows the density structure of a massive star just before collapse begins, and after a proto-neutron star has been born (on one second timescales for late time neutrino heating supernovae). A hot bubble region, just outside the surface of the proto-neutron star, forms as the material is heated by interaction of the hot core neutrinos to large values of entropy (i.e. s about 100). This bubble remains hot as it expands in volume due to neutrino heating by emission absorption reactions with free baryons as well as neutrino scattering reactions with electron-positron pairs found in the bubble region. The shock is moving down the density gradient and will reach the surface of the star on timescales of hours or days (depending on the extent of the hydrogen-helium envelope).

So far, we have only discussed heavy element production by explosive burning. In the remaining part of the introduction, nucleosynthesis in supernovae that produce elements heavier than the iron group will be discussed. In section II of this paper we will present quantitative results from a numerical calculation that illustrates an extreme case of the nucleosynthesis we now consider.

Some of the material decomposed by the shock will be ejected, along with the newly formed carbon through iron group elements produced by the revived shock. Early on, after the shock has stalled, material falling through the shock will move down close to the protoneutron star. This material may come close enough to the protoneutron star so that its electron fraction is reduced to a value as low as .2; later on, it is ejected. As the material heats and expands, the neutron fraction begins to decrease in response to the decreasing electron chemical potential. If the expansion is rapid enough, the weak interaction rates for positron capture on neutrons and neutron decay can decrease to the point that the neutron fraction freezes at a value greater than .5 (i.e. about .6). The rapid expansion will also inhibit the recombination of the free baryons to helium and the helium to heavier elements. In massive stars (i.e. $10 < M < 70\ M_\odot$) exploding by late time neutrino heating, the expansion of the neutron rich material is slow enough that the neutron fraction can decrease to values near .5 before freezing out. The recombination of the free baryons to heavier elements is also more complete. The expansion rate is partially determined by the amount of matter overlying the bubble region. Stars in the range $10 < M < 70\ M_\odot$ have large hydrogen-helium envelopes that slow down the bubble expansion. Stars in the mass range

$8 < M < 9\ M_\odot$ have more tenuous envelopes, and the bubble expansion can be much more rapid. In the case of rapid expansion, the final composition of this neutron rich material needs to be determined using a complex nuclear reaction network. It is not the classical r-process, because at temperatures sufficiently low for (n,γ) reactions to proceed to nuclei having short beta decay lifetimes, the free neutron abundance is very small. It is not quite the results of unmodified nuclear statistical equilibrium [7] since the final abundance of ^{4}He is large. We now turn to a description of the results of a numerical calculation illustrating the rapid bubble expansion scenario.

II. SUPERNOVAE FROM COLLAPSE OF OXYGEN, MAGNESIUM, NEON CORES

A general description of the evolution of a collapsing oxygen, magnesium, neon core can be found in Mayle and Wilson [8]. The core studied in [8] was the result of evolving a $2.2\ M_\odot$ helium core (equivalent to an $8\ M_\odot$ star after hydrogen burning, with the hydrogen envelope removed) by Nomoto [6]. We have already described the collapse and explosion seen in a core of this type in the introduction. Here we are going to concentrate on discussing the quantitative results of the neutron rich nucleosynthesis seen in [8].

Figure 2 shows a matter trajectory of material that will eventually be ejected. First the material is in quasi-free-fall, then is decomposed to free baryons by the shock wave, heated by neutrino interactions after passing close to the surface of the proto-neutron star, and finally is ejected. In Figure 3 we display a number of physical quanities for the matter element whose trajectory is seen in Figure 2. The proton fraction is seen to fall from .5 (oxygen) to a value of .23 as the material nears the surface of the proto-neutron star remnant. The proton fraction then begins to increase as the material is heated by neutrino interactions and begins to accelerate away from the remnant. Also shown is the neutrino fraction (i.e. the number of electron neutrinos per baryon) present in the environment of the matter element. The difference between the electron neutrino and electron antineutrino number (multiplied by 100) is also shown, and decreases with time as the neutrinos streaming from the neutrinosphere begin to be produced dominately by thermal processes. The fraction of nucleons in helium is seen to increase at about the time the proton fraction has frozen. Note that this figure also graphs the equilibrium value of the proton fraction that would be seen if the matter element were in chemical equilibrium as well as thermal equilibrium. The density is also shown in Figure 3, discontinuous at the time the material encounters the shock wave, increasing as the material moves close to the neutron star remnant, and finally decreasing as the material accelerates outward. The two timescales of interest in this nucleosynthesis are the weak interaction timescale for the increase of the proton fraction, and the timescale for the triple alpha reaction to occur for the recombination of the helium. Both of these timescales are shown in Figure 4, along with the proton fraction and the helium mass fraction displayed in Figure 3.

The elemental composition of this ejected matter parcel (proton fraction of .4) is very neutron rich. Stanford Woosley has studied the nucleosynthesis from several of the neutron rich mass zones ejected from the collapsed oxygen, magnesium, neon core studied by us. He concluded that his reaction network [9], which includes isotopes up to and including ^{87}Kr, was inadequate to calculate the nucleosynthesis for the material ejected with proton fraction less than .487 . Only one of the matter trajectories studied by Woosley, with proton fraction equal to .487, was accurately determined in Woosley's opinion. The final composition was chiefly ^{58}Ni (65.3% by mass), ^{4}He (11.1%), ^{56}Ni (10.1%), ^{57}Ni (3.5%), and ^{62}Zn (8.1%). Later, these latter three species would of course decay to their stable isobars, ^{56,57}Fe and ^{62}Ni. The other neutron rich material, with proton fraction less than .487, that was ejected displayed flows that would have left the network from its upper end (atomic number equal to 87) had adequate nuclear data for heavy, neutron rich nuclei been available and included.

III. CONCLUSIONS

The novelty of this neutron rich nucleosynthesis is to be emphasized, as well as its extremely non-solar character. It is not the classical r-process, because the temperatures sufficiently low for (n,γ) reactions to proceed to nuclei having short beta decay lifetimes, the free neutron abundance is very small. It is also not quite the results of unmodified nuclear statistical equilibrium [7] since the final abundance of ^{4}He is large. It certainly warrents further study.

We would also like to thank Stanford Woosley for useful conversations we have had with him on the subject of this paper and for sharing his results on the neutron rich nucleosynthesis discussed here.

REFERENCES

1. Burbidge, E.M., Burbidge, G.R., Fowler, W.A. and Hoyle, F., Rev. Mod. Phys. 29, 547 (1957)
2. Arnett, W.D. and Schramm, D.N., Ap. J. Lettr. 184, 147 (1973)
3. Weaver, T.A. and Woosley, S.E., Ann. N.Y. Acad. Sci. 336, 335 (1980)
4. Wilson, J.R., Mayle, R.W., Woosley, S.E. and Weaver, T.A., Ann. N.Y. Acad. Sci. 470, 267 (1986)
5. Bruenn, S.W., Ap. J. Suppl. 24, 283 (1972)
6. Nomoto, K., Ap. J. 277, 791 (1984)
7. Hartmann, D., Woosley, S.E. and El Eid, M.F., Ap. J. 297, 837 (1985)
8. Mayle, R.W. and Wilson, J.R. ,"Supernovae from Collapse of Oxygen, Magnesium, Neon Cores", submitted to Ap. J. (1986)
9. Woosley, S.E., "Nucleosynthesis and Chemical Evolution" ; Saas-Fee Lecture Notes, eds. B. Hanch and A. Maeder (Swiss. Acad. Sci: Geneva) (1986)

FIGURE CAPTIONS

Fig. 1-Density versus radius for two times in the collapse of the core of a massive star. Time equal to zero is just before the core becomes unstable. Time equal to one second is after the core has bounced and a shock wave has entered the hydrogen-helium of the star. A proto-neutron star remnant is left at the center of the star surrounded by a hot bubble of radiation and electron-positron pairs (which dominate the energy density), and free baryons (which dominate the matter density). This figure is more appropiate for massive stars in the mass range $10 < M < 70\ M_{\odot}$.

Fig. 2-Matter parcel trajectory and shock wave position versus time for the oxygen, magnesium, neon core. The initially free-falling oxygen is shock decomposed to free baryons before passing close to the proto-neutron star remnant (not shown). The material is heated by neutrino reactions before rapidly expanding outwards. The rapid expansion results in material that is neutron rich.

Fig. 3-Various physical quanties associated with the matter element trajectory shown in Figure 2. The fractions with respect to the number of baryons of non-baryonic matter is denoted by capital Y with a subscript (i.e., neutrinos). The proton fraction is denoted by Y_p, the fraction of nucleons in helium is given by X_{He}, and the equilibrium fraction of protons is given by $(Y_p)_{eq}$. Note that the value of X_{He} given after .14 seconds is not correct since the timescale for the triple alpha reaction to occur becomes on the order of the dynamical timescale, and the helium fraction was calculated assuming nuclear statistical equilibrium . The density of the material is also shown in this figure.

Fig. 4-The two timescales of interest to the neutron rich nucleosynthesis problem are the timescale for the weak interaction to increase the proton fraction of the ejected material, and the timescale for the recombination of helium to heavier elements (estimated by the triple alpha reaction timescale). Both these are shown in this figure as well as the proton fraction and the fraction of nucleons in helium shown in figure 3.

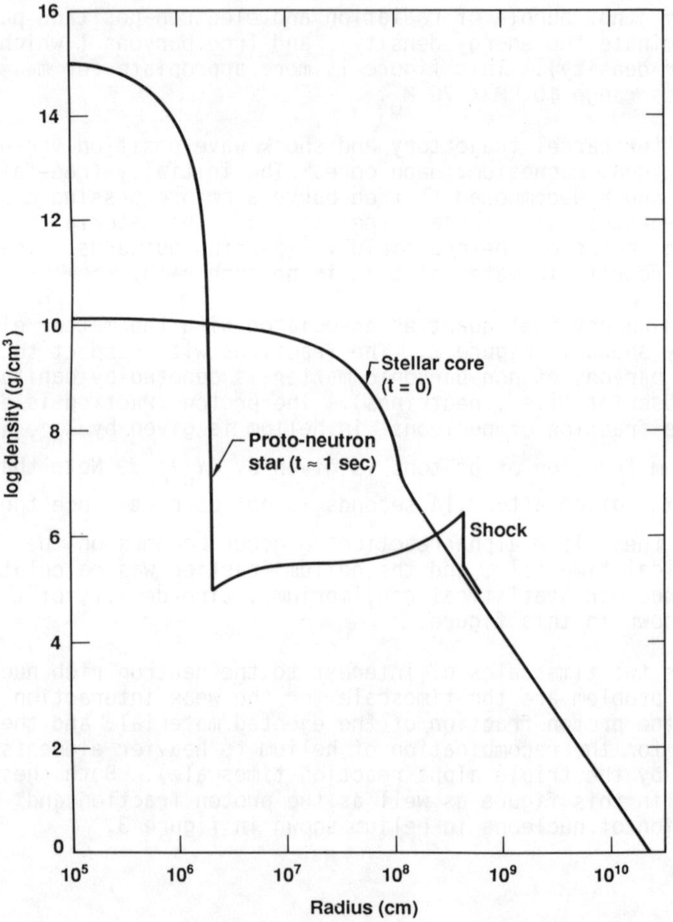

16
14
12
10
8
6
4
2
0
log density (g/cm³)
Stellar core (t = 0)
Proto-neutron star (t ≈ 1 sec)
Shock
10⁵
10⁶
10⁷
10⁸
10⁹
10¹⁰
Radius (cm)

Figure 1

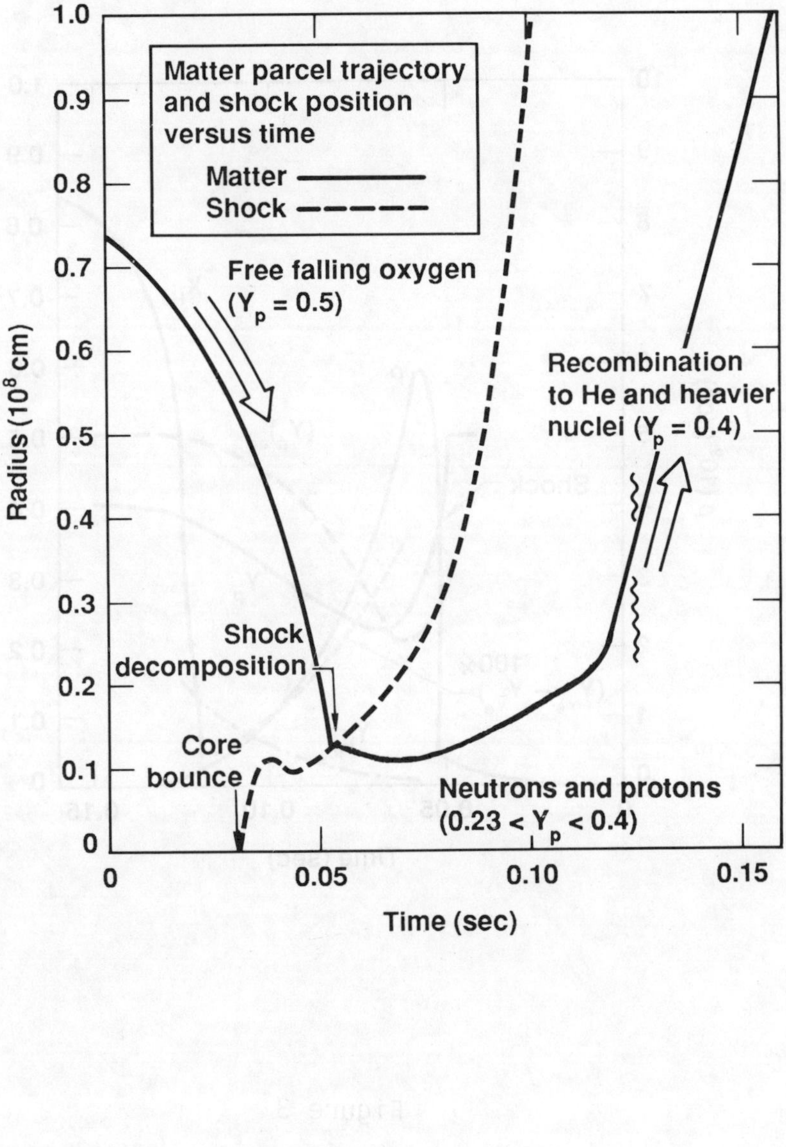

Matter parcel trajectory and shock position versus time
Matter
Shock
Free falling oxygen (Y_p = 0.5)
Recombination to He and heavier nuclei (Y_p = 0.4)
Shock decomposition
Core bounce
Neutrons and protons (0.23 < Y_p < 0.4)
Radius (10^8 cm)
Time (sec)
1.0
0.9
0.8
0.7
0.6
0.5
0.4
0.3
0.2
0.1
0
0.05
0.10
0.15

Figure 2

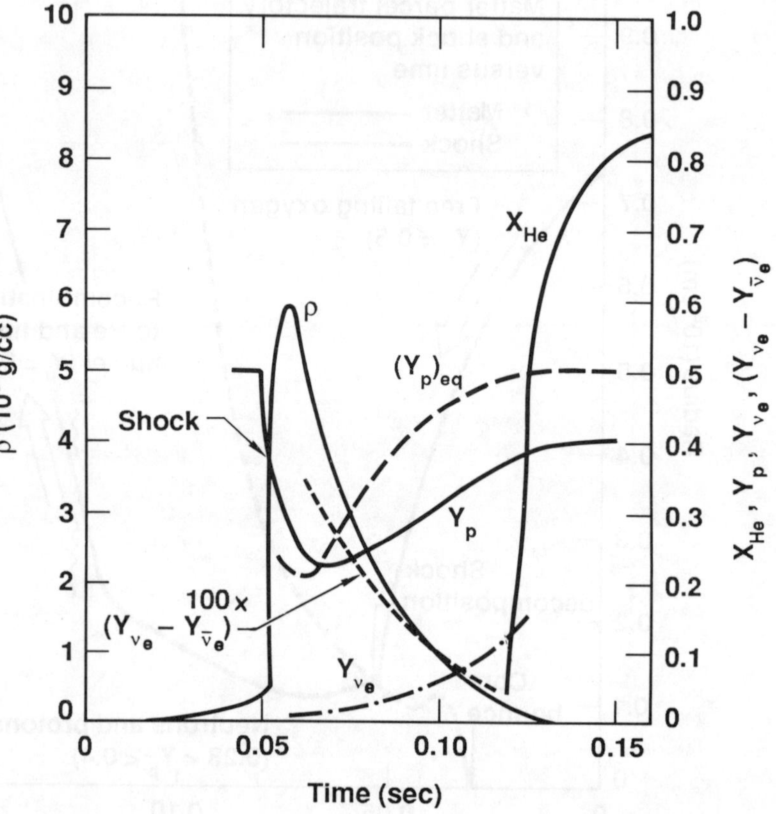

10
9
8
7
6
5
4
3
2
1
0
1.0
0.9
0.8
0.7
0.6
0.5
0.4
0.3
0.2
0.1
0
0
0.05
0.10
0.15
ρ (10^9 g/cc)
Time (sec)
X_{He}, Y_p, Y_{ν_e}, $(Y_{\nu_e} - Y_{\bar{\nu}_e})$
ρ
X_{He}
$(Y_p)_{eq}$
Shock
Y_p
100 x
$(Y_{\nu_e} - Y_{\bar{\nu}_e})$
Y_{ν_e}

Figure 3

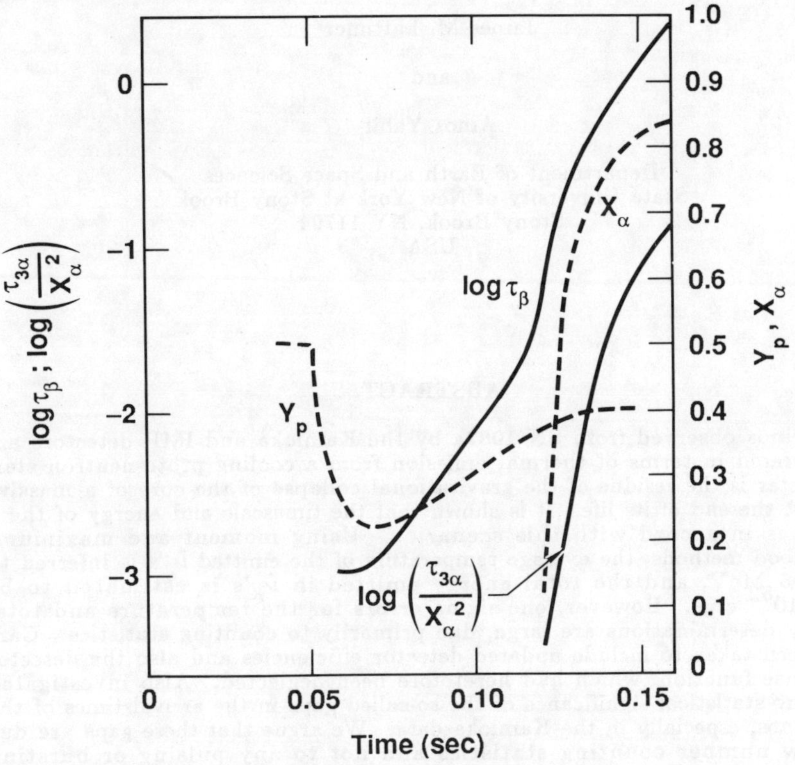
log τ_β ; log (τ_3α / X_α^2)
0
−1
−2
−3
1.0
0.9
0.8
0.7
0.6
0.5
0.4
0.3
0.2
0.1
0
Y_p, X_α
X_α
log τ_β
Y_p
log (τ_3α / X_α^2)
0
0.05
0.10
0.15
Time (sec)

Figure 4

Interpretation of Neutrinos from SN 1987a[1]

James M. Lattimer

and

Amos Yahil

Department of Earth and Space Sciences
State University of New York at Stony Brook
Stony Brook, NY 11794
USA

ABSTRACT

Neutrinos observed from SN 1987a by the Kamioka and IMB detectors are interpreted in terms of thermal emission from a cooling proto-neutron star. This star is the residue of the gravitational collapse of the core of a massive star at the end of its life. It is shown that the timescale and energy of the ν burst is in accord with this scenario. Using moment and maximum-likelihood methods, the average temperature of the emitted $\bar{\nu}_e$'s is inferred to be 3–5 MeV, and the total energy emitted in $\bar{\nu}_e$'s is estimated to be $3\text{–}5\times10^{52}$ ergs. However, one sigma errors for the temperature and total energy determinations are large, due primarily to counting statistics. Care has been taken to include updated detector efficiencies and also the detector response function, which had heretofore been neglected. Also investigated was the statistical significance of the so-called gaps in the arrival times of the neutrinos, especially in the Kamioka data. We argue that these gaps are due to low number counting statistics and not to any pulsing or bursting behavior of the source as has been suggested in the literature.

1. INTRODUCTION

It has long been a tenet of stellar astrophysics that neutrino, not photon, emission dominates the last phases of the evolution of massive stars[1)]. The lengths of the final burning stages are shortened considerably,

---

1. Invited talk at the New Orleans ACS Symposium on the "Origin and Distribution of the Elements", 30 Aug - 4 Sept 1987

the entropy of the core is reduced, and the core structure converges to a degenerate configuration with nearly the Chandrasekhar mass. The final stage is the formation of an iron ($M > 11 M_\odot$) or O-Ne-Mg ($8 M_\odot < M < 11 M_\odot$) core, which soon becomes unstable and implodes due to the continued neutrino emission during the final quasistatic burning stages. The collapse and subsequent supernova phases are accompanied by the most energetic neutrino burst of all as a neutron star, or black hole, is formed.

This prediction was dramatically confirmed on 23 Feb 1987 by the simultaneous observation of a neutrino bursts by the Kamioka[2] and IMB[3] collaborations, about 4 hours before the optical burst was first observed from the supernova SN1987a in the Large Magellanic Cloud. The 8 IMB neutrino events, of average energy 34 MeV, spread over 5.6 s, and the 11 Kamioka events, of average energy 17 MeV, spread over 12.5 s, are in substantial agreement with the standard predictions of supernova theory. The average energies and relative counts are consistent with a (redshifted) emission temperature of 4–5 MeV. The total counts imply that roughly $2\text{–}3\times 10^{53}$ ergs was emitted in neutrinos of all types. This is the same as the binding energy of a 1–1.5 $M_\odot$ neutron star. Finally, the multisecond timescales indicate that the neutrinos did indeed diffuse out of a cooling neutron star. In other words, the standard theory was verified to an amazing degree.

In the rest of this talk, we give a short summary of the predictions of the standard supernova model as they apply to neutrino emission. We compare semi-realistic neutron star cooling models to the observations. Next we perform statistical analyses of the neutrino data to derive mean emission temperatures and total emitted energies. We point out the importance of including the detector response in these analyses. Monte Carlo simulations are performed to investigate the statistical significance of the gaps observed in the neutrino data.

2. THE STANDARD MODEL

The history of supernova theory is a long and often turbulent story. The importance of neutrinos in collapsing stars has been recognized for over 20 years. We cannot give an historically accurate rendition in the alotted space, so for a more detailed treatment, the reader is referred to some of the more influential papers[4]. In the final stage of the life of a massive star, a degenerate iron or O-Ne-Mg core with nearly the Chandrasekhar mass forms. Continued shell burning on the periphery of the core slowly adds to the core's mass. Simultaneously, neutrinos produced from neutronizing electron capture reactions immediately escape; the subsequent pressure reduction causes a slow contraction of the core. Soon, the core mass exceeds the effective Chandrasekhar mass. When this occurs, the core collapses dynamically, and pulls away from the rest of the star. The flow separates into an inner subsonic core and an outer, infalling, supersonic mantle[5].

Compression leads to further electron capture and neutrino emission. However, when the central density reaches 10^{12} g cm^{-3}, the opacity of the neutrinos rises sufficiently to trap them within the core on dynamical timescales[6]. Then neutronization ceases, well before nuclear densities (3×10^{14} g cm^{-3}) are reached. A chemical beta equilibrium between the baryons and leptons is established, with the net number of leptons per baryon between 0.35 and 0.4, far above the canonical value of 0.05 for a cold, catalyzed neutron star[7]. A Fermi sea of degenerate ν_e's is added to the sea of degenerate e^-'s; typical Fermi energies in the range 200–300 MeV develop. Meanwhile, the nuclei, which until a decade ago were thought to photodisintegrate during the collapse, remain bound due to energy storage in nuclear excited states, which keeps temperatures low during the collapse[8]. As a result, the equation of state is dominated by the relativistic lepton component, and the collapse cannot be reversed until supernuclear densities are reached.

When the inner core finally does rebound, its kinetic energy is converted into a shock wave at its outer boundary, which subsequently drives into the still infalling mantle. Within milliseconds, the inner core is left behind as a hydrostatic proto-neutron star[9]. Depending on details of the initial core size, the amount of neutronization that occurred during the collapse, and the equation of state around and above nuclear density, it is thought that one of the following three situations develop. If the initial iron core, the amount of neutronization, and the incompressibility of nuclear matter are all relatively small, the shock is able to overcome the debilitating effects of nuclear dissociation and electron neutrino emission and a prompt Type II supernova occurs[10]. If this prompt mechanism fails, the shock may be revived, due to neutrino heating, after a pause of 0.1–1 second[11]. If this delayed mechanism also fails, there seems to be no choice but for the infalling outer layers eventually to overwhelm the shock; a black hole forms, and ν emission ceases.

In either successful case, the protoneutron star is in hydrostatic equilibrium. It is hot, lepton-rich, and only marginally bound at first. Neutrino losses during the collapse amount to about 10^{51} ergs, and those accompanying the shock's breakout through the neutrinosphere, from where ν's can again freely escape, add up to less than 10^{52} ergs[12]. Most of the energy and leptons that must be radiated to form a neutron star, whose binding energy is $2–4\times10^{53}$ ergs, is yet to be released.

The neutrino emission from a protoneutron star is in two phases[13]. The initial emission is dominated by the cooling and neutronization of the outer shocked core, and by matter accreted through the shock and falling onto the hydrostatic remnant. This phase lasts no more than half a second and blends into a long term diffusion from the inner core. The ν_e's in the core have energies in excess of 200 MeV. As they diffuse to the surface, their degeneracy energy is converted into thermal energy[14], i.e. ν-$\bar{\nu}$ pairs of all types. These eventually diffuse to and are emitted from the neutrinosphere, but with energies in the range 10–20 MeV. The emission timescale is thus the diffusion timescale and is on the order of seconds[14,15]. It is set by the opacity of dense matter, and is uncertain by at least a factor of three. The standard model thus predicts that neutrinos of all species should carry away the binding energy in roughly equal amounts. It is expected that $\bar{\nu}_e$'s, because of their large absorption cross section on protons, will dominate the signal in water Cerenkov detectors.

There were few detailed models[13] of neutron star cooling that existed before the supernova was observed, but they are in basic agreement with the observations from Kamioka and IMB. These models neglected accretion through the shock and set $\bar{\nu}_e$ opacities equal to ν_e ones. Because this overestimates the $\bar{\nu}_e$ opacities, it underestimates their effective temperatures; they should have been emitted from deeper, hotter, regions. Recent calculations[16] in which more attention was paid to details of neutrino transport suggest a 30–50% increase in neutrino temperature is appropriate. Figure 1 compares the experimental and predicted accumulations of signal for the Kamioka and IMB detectors. The nonconvecting model of ref. 13 is used, but the $\bar{\nu}_e$ temperature is scaled up by 50%, to an average value of about 4 MeV. The total emitted $\bar{\nu}_e$ energy, about 3×10^{52} ergs, is unaltered. It should be noted that although the zero times of the experiments and the model have all been equated they might well be different by a few tenths of a second.

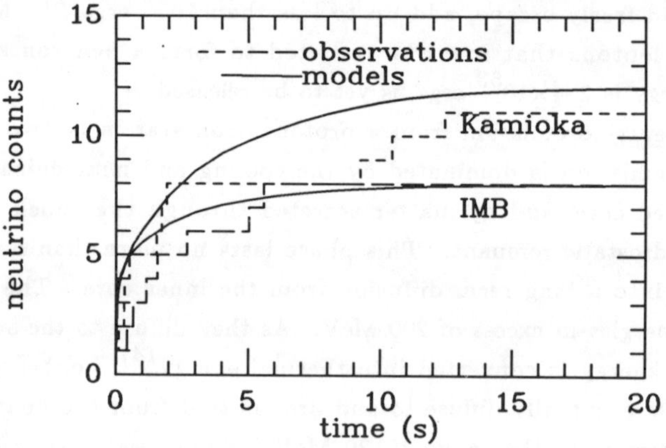

Figure 1. Model vs. observed signal accumulations. See ref. 13 and text for details of the models.

3. ENERGY AND TEMPERATURE ESTIMATES

There are two general methods we will employ in order to derive estimates of the temperature and total energy of the observed neutrinos. In both methods, some assumptions about the spectrum and time development of the signal must be made. For simplicity, we assume that the time-averaged spectrum is Fermi-Dirac, characterized by an average temperature and zero chemical potential. We have shown elsewhere[17)] that these assumptions are not too drastic. The data are too meagre to allow more spectral parameters to be fitted, anyway. The first method, the so-called moment method, only uses the average neutrino energy and the total number of neutrinos. The maximum-likelihood method, on the other hand, employs the individual energies and therefore should give more reliable results. In fact, both approaches give nearly identical results.

Before proceeding, some comments on the observations are in order. The published data consist of detection times, estimated electron energies and electron angles with respect to the Large Magellanic Cloud. It is not possible, except in a statistical sense, to determine which events are due to ν scattering and which are due to $\bar{\nu}_e$ absorption on protons. The distribution of angles suggests that all but at most one or two of the detections were absorptions. Since theoretical expectations are that less than 5% of the events were due to scattering, we assume here that none were. Our conclusions are not sensitive to this assumption. Moreover, we take the incident neutrino energy to be the detected energy plus 1.3 MeV, the neutron-proton mass difference, regardless of the observed angle. This is a good approximation. Finally, we take the detector masses, thresholds and efficiences from refs. 2 and 3, and as updated at this meeting (in a parallel session). The IMB detector is characterized by a threshold energy of about 20 MeV, as compared to 7 MeV for the Kamioka detector. Thus the former will only sample the high energy tail of the spectrum and will provide estimates of somewhat lower reliability. Nevertheless, since the characteristics of these detectors are so different, a separate determination of the $\bar{\nu}_e$ temperature can be made.

The average energy of the neutrinos at the source is

$$<\epsilon>_s = \int_0^\infty \epsilon^3 f d\epsilon / \int_0^\infty \epsilon^2 f d\epsilon = [F_3(0)/F_2(0)]\ T \simeq 3.15\ T \tag{1}$$

where $f = [1+\exp(\epsilon/T)]^{-1}$ is the Fermi distribution, T is the effective temperature as viewed at the Earth, and F_i is the standard Fermi integral. The average energy of the detected neutrinos must involve corrections for the neutrino absorption cross-section ($\propto [\epsilon\text{-}1.3]^2$), the threshold energy H, and the detector efficiency $W(\epsilon)$:

$$\Sigma\epsilon_i/N = <\epsilon>_d = \int_H^\infty \epsilon^3[\epsilon\text{-}1.3]^2 fW d\epsilon / \int_H^\infty \epsilon^2[\epsilon\text{-}1.3]^2 fW d\epsilon = [G_5/G_4]T. \tag{2}$$

We have defined the modified, truncated, Fermi integral, G_i, which is a function of H and T. The expected rate of accumulation of detected neutrinos, N, depends on the $\bar{\nu}_e$ luminosity L, the absolute value of the cross-section, the detector mass, M (in kilotons), and the distance D to the supernova:

$$\dot{N} = 0.0127\ (L/10^{51}\text{ergs})(T/\text{MeV})(50\text{kpc}/D)^2(M/1\text{ kt})G_4/F_3(0). \quad (3)$$

The total energy emitted in $\bar{\nu}_e$'s can then be estimated, if the temperature is assumed to be constant, by integrating eq. (3):

$$E_{\bar{\nu}} = 78.5\ (D/50\text{kpc})^2(1\text{ kt}/M)\ [F_3(0)G_5/G_4^{\ 2}]\ N/<\epsilon>_d. \quad (4)$$

In the moment method, equation (2) is implicitly solved for the temperature T, while in the maximum-likelihood method, the best fitting Fermi-Dirac spectrum is determined from the individual neutrino energies. Some results for these approaches are presented in the Table, under the heading of "no convolution". Notable are the close agreement of these two methods with each other. Equally notable are the approximately one sigma discrepancies of the temperatures and energies deduced from the two data sets. This should not come as much of a surprise, given the different detector characteristics and the simple assumption of a Fermi-Dirac spectrum.

TABLE

Derived temperatures (MeV) and total energies (10^{51}ergs)

	$T_{Kamioka}$	T_{IMB}	$E_{Kamioka}$	E_{IMB}
No convolution				
Moment	2.81±0.6	4.62±0.75	61(-30/+50)	29(-16/+26)
Max-Likelihood	2.85±0.4	4.60±0.8	59(-28/+34)	27(-10/+26)
" combined	3.8±0.4		43(-18/+20)	
Convolution				
Moment	2.75±0.5	4.0±0.7	69(-38/+55)	57(-28/+100)
Max-Likelihood	2.80±0.4	4.2±1.1	63(-31/+40)	45(-34/+120)
" combined	3.7±0.4		48(-18/+21)	
Relative counts	5.0(+1.25/-1.0)		26(-16/+19)	

In the maximum-likelihood approach, it is also possible to infer a temperature from the combined experiments. The error bars on these combined estimates are relatively smaller than the individual determinations. These results are also displayed in the Table.

Kolb, Stebbins and Turner[18] have pointed out that the observed energies are subject to a systematic bias known as detector response. Because the combination of cross-section, efficiency and incoming spectrum is a steep function of the energy, the error in the determination of the detected neutrino energy is not evenly distributed among lower and higher energies around this value. In fact, it is likely that the true energy lies closer to the median energy of the distribution. According to the papers announcing the neutrino observations[2,3], the response function can be approximated by a Gaussian. We proceed to include the detector response into our calculations.

Denote the integrand of the denominator of equation (2) by $I(\epsilon)$. If the detector response function is Gaussian, then

$$R(\epsilon\text{-}x) = (2\pi\Delta^2)^{-1/2}\ \exp[-(\epsilon\text{-}x)^2/2\Delta^2]. \tag{5}$$

Here Δ is the error at the observed energy ϵ, and x is the true neutrino energy. In the limit when Δ goes to zero, R is just a delta function. The integrals G_4 and G_5 represent the zeroth and first moments over the observed neutrino distribution. Therefore the modified G integrals, $\tilde{G}$, may be written

$$\tilde{G}_{4,5} = \int_H^\infty \int_0^\infty [1,x]\ I(x)R(\epsilon\text{-}x)\ dx\ d\epsilon \tag{6}$$

where the 1 in the integrand refers to $\tilde{G}_4$, the x to $\tilde{G}_5$. If the standard approximation is used of treating Δ as a function of x instead of ϵ, the integral over ϵ may be performed first, and we find

$$\tilde{G}_4 = \int_0^\infty I(x)\ (1/2)[1\text{-erf}((H\text{-}x)/\sqrt{2}\Delta)]\ dx \tag{7}$$

and

$$\tilde{G}_5 = \int_0^\infty I(x)\ [(x/2)[1\text{-erf}((H\text{-}x)/\sqrt{2}\Delta)] + \Delta^2\ R(H\text{-}x)]\ dx, \tag{8}$$

where erf is the error function. Note from these equations that $\tilde{G}_4 < G_4$, in general, since $|\text{erf}|<1$ and $\int_0^H I(x)dx$ is small. But $\tilde{G}_5 > G_5$ because of the additional term proportional to Δ^2. Therefore, including detector response will result in a decrease of the estimated temperature and an increase in the estimated total emitted energy. Our results are displayed in the Table, under the heading "convolution". Because the IMB ν's lie in the tail of the Fermi distribution, the effect of detector response is much greater for the IMB detector. Therefore, the IMB and Kamioka temperatures and, in particular, the total energies are pushed closer together.

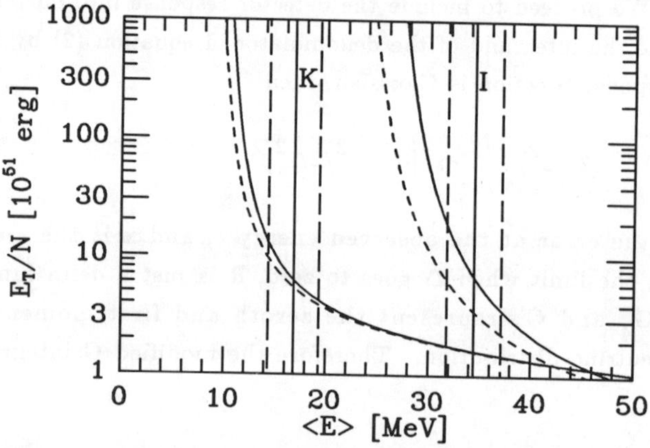

Figure 2. Inferred total $\bar{\nu}_e$ energy per detected neutrino in units of 10^{51} ergs (foes) as a function of the average detection energy. K refers to Kamioka, I to IMB. Solid curves include detector response (see text); dashed curves do not. Solid vertical lines are the observed values and dashed lines are one sigma deviations.

The effects of detector response are more easily seen in Figure 2, which displays the total energy per detected neutrino as a function of the average energy. Eq. (4) demonstrates that this quantity, $E_{\bar{\nu}}/N$, is independent of N. Although this figure is drawn from the moment method analysis, it is clear that the maximum-likelihood case will be very similiar. Results with and without detector response are shown for each detector. Clearly, the high threshold energy of the IMB detector makes it depend more sensitively on detector response effects. In addition, note that the one sigma error estimates lead to a rather large uncertainty in the determination of the total energy whether or not detector response is included.

We now point out that we have not yet used an important observation, namely the relative numbers of detected neutrinos. This ratio is expected to be an important indicator of temperature or average source energy. Forcing the total energies and temperatures inferred by the two experiments to be equal, respectively, gives the condition that the quantity G_4M/N must also be. Figure 3 shows the temperature as inferred from this condition for the two cases in which the response function is and is not included. Results from this calculation are included in the Table. The derived temperature, 5 MeV, is in excess of those determined individually from the two detectors, although the one sigma error bars overlap those determined from the combined maximum-likelihood results. The relatively low energy inferred, although in contrast to that found in the maximum-likelihood and moment methods, is in good agreement with that determined from the model calculations of neutron star cooling reported in the previous section.

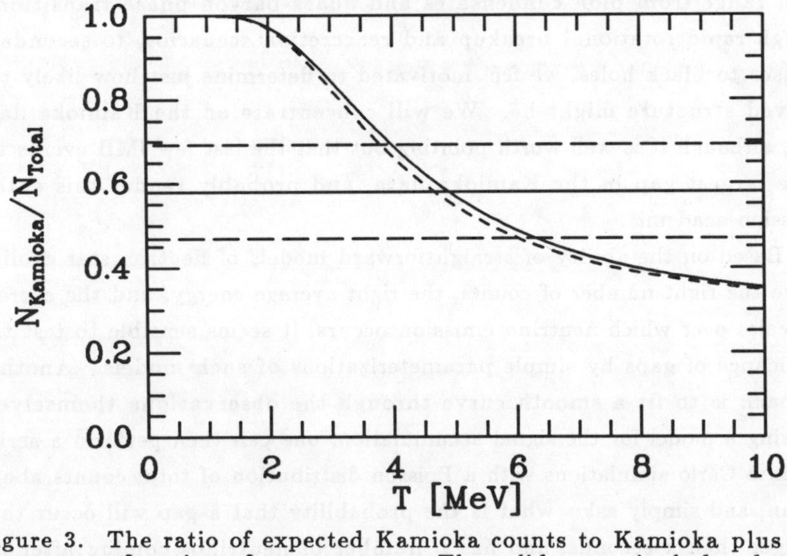

Figure 3. The ratio of expected Kamioka counts to Kamioka plus IMB counts as a function of temperature. The solid curve includes detector response, the dashed curve does not. The solid horizontal line is the observed ratio and the dashed lines represent one sigma errors.

We conclude this section by noting that the inferred total $\bar{\nu}_e$ energy can be used to estimate the mass of the neutron star that might have formed.

According to detailed models of neutrino emission, each of the six known neutrino species carries a nearly equal share of the total energy. Thus, the total binding energy of the neutron star is about 6 times E_{ν}, or $2\text{–}3\times10^{53}$ ergs. Making reference to recent equations of state[19] this translates into neutron star gravitational masses of 1.2–1.5 $M_{\odot}$. It will thus be difficult to determine from the energetics alone which explosion mechanism occurred in the case of SN 1987a.

4. ANALYSIS OF ARRIVAL TIMES

Many authors have commented on the bunched structure that the Kamioka data show in the arrival times. It has prompted some to conclude the distribution is inconsistent with that predicted by a simple cooling model, namely an exponential or power-law signal decay, and instead to propose a variety of explanations for this behavior. Without detailing the list of ideas, which range from pion condensates and quark-baryon phase transitions, through rapid rotational breakup and reaccretion scenarios, to secondary collapses to black holes, we feel motivated to determine just how likely the observed structure might be. We will concentrate on the Kamioka data alone, although it is well worth pointing out that the last few IMB events fill in the largest gap in the Kamioka data, and probably render this entire discussion academic.

Based on the ability of straightforward models of neutron star cooling to give the right number of counts, the right average energy, and the correct timescales over which neutrino emission occurs, it seems sensible to test the significance of gaps by simple parameterizations of such models. Another approach is to fit a smooth curve through the observations themselves. Choosing a model for the signal accumulation, one can then perform a series of Monte Carlo simulations with a Poisson distribution of total counts about a mean, and simply ask: what is the probability that a gap will occur that is longer than x seconds, and has n number of neutrinos coming after it? We used two model functions: first, a smooth curve fit to the data (it was not required to pass exactly through the data points), and second, a double exponential decay model that fits both the data and models of neutron star cooling. For the latter, the signal accumulation rate is

$$\dot{N} \propto \exp(-t/t_1) + (t_1/t_2)\exp(-t/t_2) \tag{9}$$

where $t_1 \simeq 0.5$ s and $t_2 \simeq 6$ s. The adopted form of eq. (9) has the property that both timescales contribute equally to the integrated signal. The two timescales correspond to the two phases of ν emission discussed in Section 2.

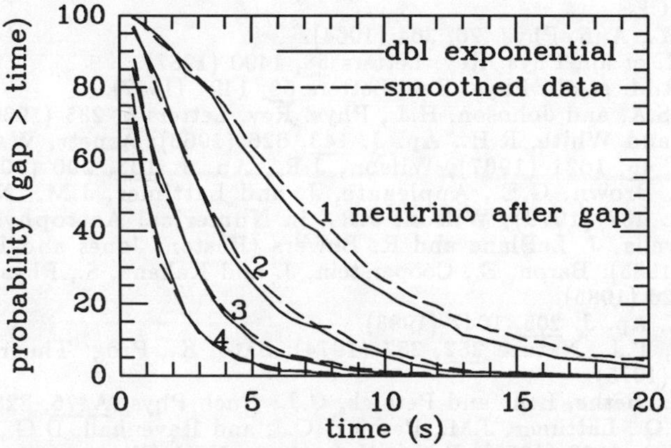

Figure 4. Probabilities that gaps with duration larger than t seconds and with n trailing neutrino events occur in randomly generated data sets which have a Poisson mean of 11 events. The parent signal accumulation function follows eq. (9) for the solid curves, and a smoothed fit to the Kamioka observations for the dashed curves.

Results are shown in Figure 4. We interpret them as follows: the situation of having a gap of 7 s, with 3 neutrinos following the gap, occurs about 5% of the time. One way to understand the magnitude of this number is to approximate the counting rate at late times (3 s<t<10 s) as a constant value $\simeq 3/10$ counts per second. Then the odds of having no counts in the first seven seconds of a ten second period is $(0.7)^7 \simeq 0.085$. The presence of such a gap is thus improbable, but not infinitessimally so. Before we decide that this gap is even marginally significant, however, we have to enquire whether a gap of the same duration, but having only two trailing neutrinos, or whether a gap of, say, 5 s duration with 3 trailing neutrinos, would have elicited the same response (i.e., number of preprints). We think they would have, and the higher probability of the latter situations would make us think very hard before proposing that the observed gap is significant. The fact is that small number statistics are important and gaps are not uncommon in such data sets.

We appreciate conversations with E. Beier, G.E. Brown, A. Burrows, J. Cooperstein, A. Mann and S. Kahana. This research was supported in part by USDOE Grant DE-FG02-87ER40317.A000.

REFERENCES

1)Chiu, H.Y., Ann. Phys. 26, 364 (1964).
2)Hirata, K. et al., Phys. Rev. Letters 58, 1490 (1987).
3)Bionta, R.M. et al., Phys. Rev. Letters 58, 1494 (1987).
4)Colgate, S.A. and Johnson, H.J., Phys. Rev. Letters 5, 235 (1960); Colgate, S.A. and White, R.H., Ap. J. 143, 626 (1966); Arnett, W.D., Can. J. Phys. 45, 1621 (1967); Wilson, J.R., Ap. J. 163, 290 (1971); Bethe, H.A., Brown, G.E., Applegate, J. and Lattimer, J.M., Nucl. Phys., A324, 487 (1979); Wilson, J.R., in Numerical Astrophysics, ed. J. Centrella, J. LeBlanc and R. Bowers (Boston: Jones and Bartlett), p. 422 (1985); Baron, E., Cooperstein, J. and Kahana, S., Phys. Rev. Lett. 55, 126 (1985).
5)Yahil, A., Ap. J. 265, 1047 (1983).
6)Mazurek, T.J., Nature 252, 287 (1974); Sato, K., Prog. Theor. Phys. 54, 1325 (1975).
7)Baym, G., Bethe, H.A. and Pethick, C.J., Nucl. Phys. A175, 225 (1971).
8)Lamb, D.Q., Lattimer, J.M., Pethick, C.J. and Ravenhall, D.G., Phys. Rev. Lett. 41, 1623 (1979); Bethe, H.A. et al., Nucl. Phys. A324, 487 (1979).
9)Brown, G.E., Bethe, H.A. and Baym, G., Nucl. Phys. A375, 481 (1982).
10)Lattimer, J.M. and Burrows, A., Ap. J. 270, 735 (1983).
11)Wilson, J.R., in Numerical Astrophysics, ed. J. Centrella, J. LeBlanc and R. Bowers (Boston: Jones and Bartlett), p. 422 (1985).
12)Burrows, A. and Mazurek, T.J., Ap. J. 259, 330 (1982).
13)Burrows, A. and Lattimer, J.M., Ap. J. 307, 178 (1986).
14)Burrows, A., Mazurek, T.J. and Lattimer, J.M., Ap. J. 251, 325 (1981).
15)Sawyer, R.F. and Soni, A., Ap. J. 230, 859 (1979).
16)Bruenn, S.A., Phys. Rev. Lett., in press (1987); Mayle, R. and Wilson, J.R., preprint (1987).
17)Lattimer, J.M. and Yahil, A., preprint (1987).
18)Kolb, E.W., Stebbins, C. and Turner, M.S., submitted to Phys. Rev. D. (1987).
19)Lattimer, J.M., Nucl. Phys. A, in press (1987); Sato, K. and Suzuki, H., preprint UTAP-52 (1987).

CORE BOUNCE SUPERNOVAE

J. Cooperstein
Physics Department
Brookhaven National Laboratory
Upton NY 11973

ABSTRACT

The gravitational collapse mechanism for Type II supernovae is considered, concentrating on the direct implosion - core bounce - hydrodynamic explosion picture. We examine the influence of the stiffness of the dense matter equation of state and discuss how the shock wave is formed. Its chances of success are determined by the equation of state, general relativistic effects, neutrino transport, and the size of presupernova iron core.

Theorists have long worked on unravelling the supernova mechanism. When Supernova 1987a first burst upon the scene in February, we were in a typical state of agreement about various main features, a typical state of disagreement over numerous specifics, and a typical state of confusion about what were the most pressing questions demanding resolution and how to go about it. We were comforted by the unlikelihood of being put to the test of detailed observations about the most crucial few seconds of the explosion. After all, there hasn't been a naked eye supernova explosion in the Milky Way since the seventeenth century. Progress was sought in numerical simulations performed by a number of groups, but despite improvements in many of the physical inputs, no clear consensus had emerged about the specific mechanisms for Type II supernovae.

But then came Supernova 1987a and suddenly supernovae made the front page of Time magazine. And we had to learn to work in the more pressured mode of confronting incoming data. Always one of the most difficult areas in astrophysics due to its complexity and range, the supernova problem has now taken on greater urgency.

Today I am talking about core bounce supernovae, those associated with the catastrophic gravitational collapse of massive stars with $M \gtrsim 10\ M_{\odot}$, Type II supernovae. Supernova 1987a was surely of this type; the neutrinos detected on earth have proven this unambiguously. Type I supernovae are not expected to produce measureable neutrinos; in type II they are paid for by the binding energy of the emerging neutron star, or if it accretes too much mass, the resulting black hole. While Supernova 1987a is not a prototypical Type II, the complications are generally believed to arise from the progenitor's envelope. The explosive events deep in the iron core of the star don't really care about the envelope, like whether the star is blue or red, for example. The fate of the explosion is already sealed before the mantle knows it has happened, although the overlying layers eventually determine the optical signature of the supernova. Those investigators who evolve the progenitor to the point of instability have been given some difficult new homework problems. But those of us in the business of exploding the stars they deliver have generally found it important to consider possible variations in the progenitors and work out what they imply for the future evolution into explosion.

There exist, at present, two basic methods theoretically of producing Type II supernovae. First, there is the "direct" explosion scenario. Here the hydrodynamical shock wave unleashed by the rebound of the collapsing iron core proceeds to the outer edge of the iron core and then propagates through the overlying elemental shells, without pause or significant slowing down. The work I shall describe has been done in conjunction with E. Baron and S. Kahana, and more details are published in references [1-3], (hereafter BCK). As I shall explain, our group has argued that this should work for stars with iron core masses $M \lesssim 1.4\ M_{\odot}$. The most recent presupernova calculations indicate such cores should be found in stars of up to $M \sim 20\ M_{\odot}$. The second method was invented by Jim Wilson, and is the "delayed" mechanism[4,5]. Here the shock wave lacks sufficient energy to produce a direct explosion and stalls, entering into an accretion mode. But about a second after bounce, neutrinos emerging from the nascent neutron star manage to deposit about 0.1% of their energy in material near the shockfront. This gives new life to the explosion, and after this pause, it continues to propagate outward and produce a supernova. If the direct explosion succeeds, the delayed one becomes irrelevant, and thus we believe it probably is not required to explain supernovae except for stars with $M \gtrsim 20\ M_{\odot}$, since no other method has yet been found to generate supernovae from them.

Our group has concentrated on the direct mechanism, so I will mostly restrict my discussion to the prompt scenario. The long time calculations are quite difficult to perform reliably; the longer time scales mean approximations developed for the shorter direct explosion time scales can break down, and various implicit-oriented hydrodynamical and neutrino transport schemes need to be used to make the calculations computationally feasible. We are presently pursuing this path and should report results later this year.

I've made this sound simple, but what I've said is controversial, and there have been a host of claims that the direct model does not work. Indeed, until recently, despite almost two decades of calculations, shock waves failed to produce direct supernovae when what was believed to be the most reasonable physics was employed in simulations. But three new developments in the past several years have broken through the impasse. These are:

1. the use of softer equations of state for dense neutron-rich matter;
2. the inclusion of general relativity in the simulations;
3. the lowering in size of the initial iron core mass.

The first two points have quite similar effects. They cause the star to collapse to a higher density before the shock wave forms and thereby make more energy available to it from gravitation. The third point enables a shock wave with less energy to produce a successful explosion. We will return to the discussion of how the higher densities extract energy for the explosion from gravitation, and first examine what a successful supernova demands and how nature goes about satisfying these requirements.

First let's take a brief look at the infall stage. The iron core of the star beomes unstable due to a combination of photodissociation reactions and electron captures[6]. Once the central density reaches $\rho \sim 10^{11}\mathrm{gm/cm^3}$ the electron captures accelerate, and the electron neutrinos produced in the reaction $\rho + e^- \rightarrow n + \nu_e$ at first easily escape the star. The pressure is reduced due to the loss of leptonic pressure and implosion hastens, reaching an appreciable fraction of free fall velocity. By $\rho_{trap} \sim 10^{12}\mathrm{gm/cm^3}$ neutrino mean free paths become too short for the neutrinos to diffuse out of the collapsing core within the shrinking dynamical timescale, which is now a few milliseconds. Further electron captures continue, but neutrinos equilibrate with the matter until absorption balances emission, and the neutrinos actually become degenerate.

There are two quantities established during infall which are critical for determining the success or failure of the shock wave. These are the total lepton number after trapping, Y_L, and the entropy S[7,8]. Of these

$$Y_L = Y_e + Y_\nu \tag{1}$$

where Y_e and Y_ν are the number of electrons and neutrinos per nucleon, is the most critical. This is because of how the shock forms. As we shall see, this occurs not at the center of the star but at a point further out in mass. In fact, this mass point, M_{form}, depends sensitively on Y_L, and roughly $M_{form} \propto Y_L^2$. An overestimate of the energy the shock requires to reach the edge of the iron core can be obtained by noting that the energy required to dissociate complex nuclei into free neutrons and protons through shock heating is*

$$10\ \mathrm{MeV/nucleon}\ \sim 2\ \mathrm{foes}/0.1\mathrm{M}_\odot\ . \qquad (2)$$

Since the total energy released in the explosion itself, E_{expl}, can be as high as 3 foes, each tenth of a solar mass dissociated could cost almost as much as the explosion itself. The total mass which must be dissociated must satisfy

$$M_{shock} = M_{core} - M_{form} \lesssim E_{expl}/20\ \mathrm{foes} \qquad (3)$$

with M in units of $\mathrm{M}_\odot$, for success to follow. Thus, if we use typical numbers like $M_{core} = 1.35\ \mathrm{M}_\odot$, $M_{form} = 0.7\ \mathrm{M}_\odot$, we'd need 13 foes of energy in the shock for it to succeed. Equation (3) is actually an overestimate[8] because if the shock wave is successful, its rapid expansion lowers the temperature at the front to ~ 1 MeV before the entire iron core has been traversed. Below $T \sim 1$ MeV, dissociation goes only to α-particles, which costs only ~ 3 MeV/nucleon, since the α-particles are still bound by ~ 7 MeV/nucleon.

It is rather obvious that explosions will be encouraged either by increasing M_{form} or decreasing M_{core}. As noted, M_{form} depends sensitively on the neutrino losses during the infall of the star, and we will return to its determination. M_{core} depends completely upon the presupernova evolution of the star. Over the past few years it has steadily decreased in calculations, and this has aided shock propagation.

To understand the decisive role of the dense matter equation of state, we must look at how the shock forms. The collapse of the star continues unabated (in fact it accelerates) until at the center the density is greater than nuclear matter saturation density[9],

$$\rho_0\,(Y_e) \simeq \rho_0\,(1/2)\left(1 - \frac{3}{4}\,(1-2Y_e)^2\right)\ , \qquad (4)$$

where the symmetric saturation density $\rho_0\,(1/2) = 0.16\ \mathrm{fm}^3 = 2.7\times 10^{14}\mathrm{gm/cm}^3$. In the dense core the matter is neutron-rich, so $Y_e \sim 1/3$, which gives $\rho_0\,(1/3) \sim 0.145\ \mathrm{fm}^{-3}$. It is extremely important that this density is lower than the symmetric nuclear matter value, as it is $\rho_0\,(1/3)$, which is the relevant density where the center begins to scream against further compression. In fact, the matter merges into a uniform medium at $\rho \sim 2/3\rho_0$, even though the nucleons contribute negatively

* Our unit of energy is "foes", which is ten to the fifty one ergs.

to the pressure there, since $\rho < \rho_0$. However, this is where presuure begins to increase rapidly, and signals of resistance to collapse begin to propagate.

As the density continues to increase, acoustical waves increase in amplitude, and as more material is brought over ρ_0, further signals are generated. These propagate out towards the "sonic point", the region where the infall velocity becomes supersonic. Being sound waves, they cannot propagate further in space and eventually will crest at the sonic point, add together, and begin to impart enough momentum reversal to the infalling matter to give birth to the discontinuities in velocity pressure and density which define the developing shock wave. The sonic point roughly becomes M_{form}, and its position in mass depends quadratically upon Y_L.

If the equation of state is "soft", the amplitude of the acoustic waves is lessened, and the core will move to higher densities before they have become strong enough to combine into a shock wave. In fact, if the equation of state is too soft, they never become strong enough, and the star collapses directly into a black hole. When the equation of state is "hard", the shock wave is generated earlier in time. It begins moving outward before it has become as strong as it could have been if it had had more time to gather up energy as it does when the equation of state is softer.

We can see the operation of these considerations quantitatively by inspection of Table 1, which presents a compilation of the BCK numerical simulations. The quantity $U_{max} = \rho_{c,max}/\rho_0\,(1/3)$ is the central density compression factor at the instant of "maximum scrunch." We see that in the successful explosions $U_{max} \sim 4.1$. The numerical simulations used the BCK form for the nuclear matter equation of state,

$$P = \frac{K_0\,(Y_e)\,\rho_0\,(Y_e)}{9\gamma}\,\left(U^\gamma - 1\right)\;;\quad U = \rho\Big/\rho_0\,(Y_e)\;, \tag{5}$$

where γ is the high density adiabatic index, and

$$K_0\,(Y_e) \equiv 9\frac{dP}{d\rho}\bigg|_{\rho=\rho_0(Y_e)} \tag{6}$$

is the nuclear bulk compression modulus. The equation of state can be softened either by decreasing K_0 or γ. The energy in the shock wave when the calculation is terminated (when the shock wave exits the iron core) is denoted as E_{shock}. This somewhat less than E_{expl}, the explosion energy, obtained from E_{shock} by subtracting about 0.5 foe for further binding of material past the iron core, and

Table 1

Models 40 and 41 are based on Woosley and Weaver's $12M_\odot$ model, and 42-45 on their $15M_\odot$ model. Models 59, 61, 62, and 63 use Hashimoto and Nomoto's $13M_\odot$ star. The bulk-symmetry coefficient W_s is given in MeV. The effective trapping density, ρ_{trap}, is given in units of 10^{12} gm/cm^3. This is obtained by adjustment of the parameters of the leakage scheme to simulate a simple transport cutoff density. All calculations employ full general relativity, and use the BCK phenomenological equation of state for high density. Y_L is the final trapped lepton fraction during the collapse phase. U_{max} is the maximum central density achieved just prior to bounce and is given in units of its ratio to $\rho_0\,(0.33) = 2.4 \times 10^{14}$ gm/cm^3 appropriate for neutron-rich matter. The explosion energy E_{expl} is obtained from E_{shock} by accounting for the binding energy of the mantle and envelope and the additional input from explosive nuclear burning. E_{loss} is the total neutrino loss when the calculation was stopped (roughly 30 msec after bounce). These energies are given in units of 10^{51} ergs (foes).

Model	Initial Star	$K_0\,(1/3)$	γ	W_s	ρ_{trap}	Y_L	U_{max}	E_{shock}	E_{loss}
40	WW-12	140	2	29.3	0.4	0.390	12.0	3.2	2.2
41	"	"	3	"	"	"	3.1	0.8	3.3
42	WW-15	"	"	"	"	"	"	-	2.5
43	"	"	2.5	"	"	"	4.1	1.7	3.4
44	"	120	3	"	"	"	3.3	-	3.0
45	"	90	"	"	"	"	4.0	0.8	3.2
59	HN-13	140	2.5	"	1.0	0.365	4.1	-	4.6
61	"	"	"	"	0.4	0.390	"	1.9	4.9
62	"	"	"	36.0	1.0	0.385	"	2.1	2.1
63	"	"	"	34.0	"	0.375	"	1.4	4.6

adding about 1 foe for the additional energy released by nuclear burning of the oxygen shell.

In these calculations BCK preferred to use $K_0\,(1/3) = 140$ MeV, which corresponds to a symmetric value $K_0\,(1/2) \sim 180$ MeV. Many authors have failed to discern this isospin-dependence and have claimed our results to be at variance with measurements of the giant monopole resonance, which give $K_0\,(1/2) \sim 210 \pm 30$ MeV. This, however, is in our range. Moreover, Eq. (5) has been misapplied to calculate neutron star properties, and it is not designed for this as it is meant only for matter with a small neutron excess. Furthermore, we should note that the decrease in ρ_0 with Y_e, which we have also included, is harmful to the shock, since it induces its formation at an earlier time. For the smaller iron cores being produced recently a higher value of K_0 could be lived with.

The softer equation of state and higher U_{max} are helpful because they lead to a more coherent transfer of energy to the shockwave from the gravitational potential. They cause the entire nuclear matter region to come to rest as a unit, and then rebound strongly; when the equation of state is hard the center begins to move out while the outer edges of the nuclear matter region are still falling inward. The result is a shattering of the acoustical signals, a destructive interference. In the soft case, a resonance-like condition develops in which the "don't squish me" signals from the center accumulate at the sonic point and add coherently just at the time when the shock wave has developed to the point where it is getting ready to move outward; the child on the swing is being pushed at the proper time. The result is a greater push.

All of the calculations reported here employ full general relativity. In Newtonian calculations U_{max} is reduced to $1.7 \sim 2.1$ and the shock wave thereby weakened, because general relativistic gravitaiton is stronger and thus things tend to behave as if the equation of state were softer. Of course, the general relativistic hydrodynamics is highly non-linear and introduces many other changes such as a decrease in M_{form} which is quite harmful to the shock. But, overall, its effect is to strengthen the shock wave in the case where the shock wave has some chance of success.

Two of the parameters listed in the table, W_s, the bulk symmetry coefficient, and ρ_{trap} (in units of 10^{12}gm/cm^3) help determine Y_L, which is also listed. W_s helps control how much electron capture goes on in the early stages of the collapse, a low value of W_s, making it easier for protons to boil off of nuclei and find electrons to capture, and hence reduce Y_L. ρ_{trap} is the effective trapping density above which neutrinos no longer escape from the star. A high value of ρ_{trap} thus induces a lower Y_L. In our earlier calculations (models 40-45) we made two largely cancelling errors. We tuned our neutrino transport scheme to give a trapping density too low ($\rho_{12} \sim 0.4$), and we have now adjusted this upwards ($\rho_{12} \sim 1$), to match more detailed published calculations[10] and new unpublished detailed transport calculations of our own. On the other hand, the value of W_s used was too low, and did not fit properly the nuclear symmetry energies in the region of the atomic mass table of interest.*

* To be more precise, it is $\hat{\mu} = \mu_n - \mu_p$ which must be compared to the mass tables, and this involves contributions both from bulk and surface energies. See ref. [6] for further clarification.

Three different initial models are used in the table: the first two are the Woosley-Weaver 12 and 15 $M_\odot$ cores[11], which have quite similar iron cores with $M_{core} = 1.35\ M_\odot$. The third model, the Nomoto-Hashimoto 13 $M_\odot$ star[12] with $M_{core} = 1.18\ M_\odot$, is far easier to explode. The reduction in M_{core} has come about for several reasons, the most important being the proper inclusion of Coulomb lattice connections to the equation of state. These reduce M_{core} by 0.1 to 0.15 $M_\odot$, and have recently also been included in the newest Woosley-Weaver calculation.

It seems clear from the table that there seems to be no trouble in accomplishing direct explosions for the new reduced cores. Only in the worst case (low W_s, high ρ_{trap}) of model 59 or the $\gamma = 3$ stiff equations of state of models 42-44 do failures ensue. (Remember the M_{core} of these models is too high in that the calculations did not include the Coulomb corrections.)

For the above reasons, we feel quite confident that prompt explosions are viable as long as $M_{core} \lesssim 1.35\ M_\odot$. One is not yet sure what total mass progenitor this correlates to because the new series of pre-supernova calculations has not yet been completed. But the indications are that this critical iron core mass will correspond to a stellar mass, $M \sim 20\ M_\odot$. For stars with $M \gtrsim 20\ M_\odot$, the direct explosion probably will not work.

We are still calculating the mass of the remnant neutron star, but our preliminary results for the Hashimoto-Nomoto star (calculations 59-63) indicate a compact remnant with the mass of the iron core itself. Once the correction for gravitational binding is applied this leaves a gravitational mass for the neutron star of only $\sim 1.1\ M_\odot$, quite small. If this general feature continues through a series of configurations and calculations, that the mass of the neutron star is approximately that of the iron core, it will be hard to produce any large neutron stars via direct explosions. However, the measured neutron star masses (generally greater than this value) are all for components of binary systems, and so there need be no contradiction with low neutron star masses from at least some direct supernovae.

ACKNOWLEDGEMENTS

The calculations reported here were done in collaboration with E. Baron and S. Kahana, and were performed on the CRAY computer at the National Magnetic Fusion Computer Center under the auspices of the U. S. Department of Energy. This work has been supported by the U. S. Department of Energy under contract no. DE-AC02-76CH00016.

REFERENCES

1. E. Baron, J. Cooperstein and S. Kahana, Phys. Rev. Lett. 55 (1985)126.
2. E. Baron, J. Cooperstein and S. Kahana, Nucl. Phys. A440 (1985)744.
3. E. Baron, H. A. Bethe, G. E. Brown, J. Cooperstein and S. Kahana, Phys. Rev. Lett. 59 (1987)7360.
4. J. R. Wilson, in *Numerical Astrophysics* eds. J. Centrella, J. LeBlanc and R. Bowers (Boston: Jones & Bartlett 1985).
5. H. A. Bethe and J. R. Wilson, Ap. J. 295 (1985)14.
6. H. A. Bethe, G. E. Brown, J. Applegate and J. Lattimer, Nucl. Phys. A334 (1979)487.
7. A. Burrows and J. Lattimer, Ap. J. 270 (1983)735.
8. J. Cooperstein, H. A. Bethe and G. E. Brown, Nucl. Phys. A429 (1985)527.
9. J. Cooperstein, Nucl. Phys. A438 (1985)722.
10. S. Bruenn, Ap. J. Suppl. 58 (1985)771.
11. S. E. Woosley and T. A. Weaver, Bull. Am. Astron. Soc. 16 (1984)971.
12. K. Nomoto and M. Hashimoto, to appear in Physics Reports (1987).

TABLE 1

Models 40 and 41 are based on Woosley and Weaver's 12 $M_\odot$ model, and 42-45 on their 15 $M_\odot$ model[12]. Models 59, 61, 62, and 63 use Nomoto and Hashimoto's 13 $M_\odot$ star[11]. The bulk-symmetry coefficient W_s is given in MeV. The effective trapping density, ρ_{trap}, is given in units of 10^{12} gm/cm^3. This is obtained by adjustment of the parameters of the leakage scheme to simulate a simple transport cutoff density. All calculations employ full general relativity, and use the BCK phenomenological equation of state for high density. Y_L is the final trapped lepton fraction during the collapse phase. U_{max} is the maximum central density achieved just prior to bounce and is given in units of its ratio to $\rho_0\,(0.33) = 2.4 \times 10^{14}$ gm/cm^3 appropriate for neutron-rich matter. The explosion energy E_{expl} is obtained from E_{shock} by accounting for the binding energy of the mantle and envelope and the additional input from explosive nuclear burning. E_{loss} is the total neutrino loss when the calculations were stopped (roughly 30 msec after bounce). These energies are given in units of 10^{51} ergs (foes).

Model	Initial Star	$K_0\,(1/3)$	γ	W_s	ρ_{trap}	Y_L	U_{max}	E_{shock}	E_{loss}
40	WW-12	140	2	29.3	0.4	0.390	12.0	3.2	2.2
41	"	"	3	"	"	"	3.1	0.8	3.3
42	WW-15	"	"	"	"	"	"	-	2.5
43	"	"	2.5	"	"	"	4.1	1.7	3.4
44	"	120	3	"	"	"	3.3	-	3.0
45	"	90	"	"	"	"	4.0	0.8	3.2
59	HN-13	140	2.5	"	1.0	0.365	4.1	-	4.6
61	"	"	"	"	0.4	0.390	"	1.9	4.9
62	"	"	"	36.0	1.0	0.385	"	2.1	2.1
63	"	"	"	34.0	"	0.375	"	1.4	4.6

DYNAMICS AND NUCLEOSYNTHESIS IN ROTATING PAIR-INSTABILITY SUPERNOVAE

Guy S. Stringfellow and S. E. Woosley
Lick Observatory, Board of Studies in Astronomy and Astrophysics
University of California, Santa Cruz, California 95064
and
Institute of Geophysics and Planetary Physics
Lawrence Livermore National Laboratory

ABSTRACT

The effects of angular momentum on the dynamical collapse and final state of very massive stars are presented. Spherically symmetric stars with helium core masses larger than about 110 $M_\odot$ collapse directly to black holes without producing (supernova) explosions or ejecting iron peak nuclei. However, the inclusion of a reasonably small amount of angular momentum J at the beginning of the collapse produces enough support so that energy deposition due to explosive oxygen burning halts and reverses the infall, resulting in a very energetic explosion ($\gtrsim 10^{53}$ erg) which leaves no remnant behind; complete disruption. Specifically, helium cores of 200, 300, and 500 $M_\odot$ explode for minimum *specific* angular momenta $J/M \gtrsim 5\times10^{17}$, 2×10^{18}, and 4×10^{18} $cm^2\ s^{-1}$, respectively. Below these values, the cores collapse to black holes. Since the lower values of J drive the collapse deeper, higher temperatures are achieved and significant quantities of "Fe" produced. For example, the smallest J value that yielded an explosion in the 200 $M_\odot$ core produced about 90 $M_\odot$ of "Fe". Furthermore, the composition profile is asymmetrical, a consequence of mixing by vortices established during the time when the collapse was reversed, while the explosion itself is essentially spherically symmetric. For the *failed* explosions which collapse completely to black holes, matter passing through the event horizon produces large amounts of gravitational radiation resulting from severe asymmetries induced by rotation. Cosmological implications resulting from the existence of such very massive stars are briefly considered, along with constraints imposed on their number density.

INTRODUCTION

Interest in very massive stars (VMS's) has risen dramatically over the last decade, primarily due to their potential impact on fundamental cosmological problems. For example, based on their spherical evolution (to be described below) black hole remnants derived from VMS's have been advocated to explain the dark matter problem[1–3] without producing an over enrichment of heavy elements through explosions. These remnants, either alone or in conjunction with their stellar progenitor, could also account for the microwave[4,5], IR and X-ray[6] background radiation.

Furthermore, explosions (isolated or in unison) produced by VMS's, generating energies $\gtrsim 10^{53}$ erg each, may result in the formation of galaxies[7–10], be important in the dynamics of the interstellar medium[11–13], and in either maintaining an ionized Universe or reionizing it at a later epoch[14,15]. An initial enrichment of heavy elements has also been attributed to VMS's[16], along with a nonstandard helium production[17,18]. These investigations are based on the spherical evolution of VMS's, the results of which could be (are) drastically altered when rotational effects are considered. Thorough discussions concerning the *spherical* evolution of VMS's and their cosmological consequences are given by Bond, et al.[19] and Carr, et al.[2,3], respectively.

It is interesting to note that there are in fact VMS's currently observed in the local Universe. The best known Galactic example is η-Car. This star is believed to have a present mass of about 180 $M_\odot$[20,21], being roughly 2/3 through its hydrogen burning lifetime. Correcting for the mass lost since its formation by comparing with the evolutionary tracks of Maeder[22], η-Car had a zero-age main sequence mass of $\sim$ 220 $M_\odot$; indeed a VMS. There are other examples of similar, but perhaps less massive stars in the LMC, Var 83 in M33, and a class of stars known as S-Dor variables. The most exotic case is that of SN1961v, which 'exploded' in the outskirts of the face on galaxy NGC1058[23]. The precursor star was *observed* to have an apparent magnitude $m_{pg} \approx 18$ for at least 37 years **prior** to outburst. Given the duration and magnitude of energy output, SN1961v was (is) a VMS, regardless of the nature of the outburst[24,25] or of the distance scale adopted.

VMS's are known to experience high mass loss rates. During quasi-static hydrogen burning on the main-sequence, it is estimated that about half of the star's initial mass will be lost, either continuously in the form of a wind ($\dot{M} > 10^{-5}$ $M_\odot$/yr), by large episodic outbursts (e.g., η-Car: Genderen and Thé[20]; SN1961v: Penrod, et al.[24]), or a combination thereof. Stars which, after hydrogen burning, develop helium cores with masses M_α in the range $35\ M_\odot \lesssim M_\alpha \lesssim few \times 10^4\ M_\odot$ encounter the pair-instability; at temperatures $\gtrsim 10^9$ °K, wherein energy is consumed by the creation of the (rest) mass of $e^\pm$-pairs[26]. The process escalates with increasing temperature under nondegenerate conditions, at least up to the point when the electron gas becomes relativistic, and so a profuse amount of pairs can be created. In these radiation pressure dominated stars, thermal pressure support is robbed at the expense of the pairs, resulting in the adiabatic index Γ falling below 4/3 when effects like rotation are neglected. Hence, the star becomes globally unstable to hydrodynamical collapse. The pair-instability occurs just prior to and during explosive oxygen burning. The mass of the carbon-oxygen core is $\sim \pm 10\%$ of M_α, depending on whether mass loss (+) is taken into account or not (-)[19,31]; the CO-core mass depends on the convective history of the star during helium burning.

The spherical evolution of VMS's at constant mass has been computed in recent years by Bond, Arnett, and Carr[19], Woosley and Weaver[27], and, with mass loss included, by Ober, El Eid, and Fricke[30–32]. These investigations, along with earlier studies[33–35], find that the low-mass helium cores ($\sim 35 \lesssim M_\alpha/M_\odot \lesssim 50$) generate explosions with mass ejection (some even rendering nuclear generated pulsational explosions[51]) coupled with the formation of massive black hole remnants. However, when M_α is above about 55 $M_\odot$ explosive oxygen burning results in the complete disruption of the core; no remnant is left. While copious amounts of intermediate mass elements (Si thru Ca) are synthesized and ejected in these cases, *no iron peak elements are produced* ! This remains so until M_α exceeds $\sim$ 110 $M_\odot$, beyond which the (spherically symmetric) cores fail to explode, instead collapsing completely to black holes.

Woosley and Weaver[28,29] have recently recalculated the spherical collapse of nonrotating helium cores in the range $35 \lesssim M/M_\odot \lesssim 130$ for the new enhancement ($\sim$ factor 3) in the $^{12}C(\alpha,\gamma)^{16}O$ rate. They find that for those cores which explode (see below) significant "Fe" production can occur; e.g., $\sim$0.6 $M_\odot$ and 32 $M_\odot$ for $M_\alpha = 75$ and 110 $M_\odot$, respectively. This is in marked contrast with earlier results[27,30,33–35] which, for the same core masses and the range explored, found < 0.5 $M_\odot$ of "Fe" produced, if any at all. The underlying differences are still under investigation but due at least in part to a change in the structure of the star because of the enhanced rate; essentially no carbon survives early on in the evolution. This leads to an effective mass scaling; lower mass cores calculated with the enhanced rate have nucleosynthetic yields similar to higher mass cores calculated years earlier with the lower rate. Also, as the explosion/collapse boundary is approached in mass, the collapse prior to complete explosion becomes deeper and more "Fe" is produced (see results section below). It is curious that previous calculations close to the collapse boundary[30] did not produce any "Fe"; less than a few tenths of a solar mass. Nevertheless, our 2D calculations remain unique in the sense that iron-peak elements were produced and ejected whereas the 1D results collapsed completely to black holes *for identical cores*!

We shall address here the dynamical evolution and fate of *rotating* VMS's pertaining to the production of Fe-peak elements and their potential of forming black hole remnants, thereby contributing to the dark matter problem. We investigate whether a *reasonable* amount of angular momentum present within the central regions of the core at the onset of dynamical collapse will alter the evolutionary outcome of those cores which collapsed completely to black holes in the spherical calculations ($M_\alpha \gtrsim 110$ $M_\odot$). We find that it does, both qualitatively and quantitatively. The results presented here and by Stringfellow, et al.[36] supersede our earlier results[37,38], including those reviewed by Woosley, et al.[39].

PHYSICS AND CODE

In this study, we have utilized an existing 2-dimensional explicit hydrodynamic computer code, written in the Eulerian frame, whose original application was directed toward understanding the formation of rotating protostars[40]. As will be shown below, VMS's radii undergo a reduction of only about a magnitude or so from their value attained during formation of their He-core, ensuring greater numerical accuracy in such physical quantities as the angular momentum J. Fortunately, VMS's also are quite 'fluffy'; i.e., their interiors achieve high temperatures, $\geq 10^9$ °K, but with densities less than 10^8 g cm^{-3}. Hence, even complete degeneracy is avoided and additional exotic and rather ill understood physics, such as the properties of matter at nuclear densities, neutrino trapping and associated energy transport, and so on need not be considered. This rather simplifies the numerical computations of their dynamical collapse. Additional physics has been added to study the problem of supernovae explosions under these desirable conditions.

Even though relatively unimportant, radiative transfer is calculated implicitly. Tables of Rosseland mean opacities were obtained from the Kepler supernovae code of Weaver, et al.[41]. The dominant contribution to the opacity arises from electron scattering, corrected for both degeneracy of the electrons and the relativistic decrease in their cross section. The lepton component of the equation of state is also included (in tabular form[41]) in the calculation of the pressure and energy density, obtained for arbitrary degeneracy and relativicity. The nuclei and radiation components remain ideal gases and are explicitly calculated.

We have developed a 9-element reaction network[28,36] to follow the *energy generation rate* (not necessarily the nucleosynthesis). The elements followed are ^{12}C, ^{16}O, ^{20}Ne, ^{24}Mg, "^{28}Si" (an average intermediate mass nucleus representing Si - Ca), "^{56}Ni" (representing the Fe-peak group), along with α particles and free nucleons. These are primarily linked through (α, γ) reactions and their inverse rates. This network has been extensively tested, the energy generation rate comparing favorably with much larger networks and also gives, to first order, the resulting nucleosynthesis. Advection of all elemental species in our network, whether due to collapse, explosion, or rotational mixing, are individually followed. Our initial helium core models (see below) were computed with the older (smaller) ^{12}C$(\alpha, \gamma)^{16}$O rate[42], and for sake of consistency we also adopt this rate. Changes in this rate do not greatly affect the energy generation or nucleosynthesis in our models since they are practically void of any carbon when we begin their dynamical evolution, but the entropy structure of the presupernova star may differ somewhat. Photodisintegration of and electron-positron capture[43] processes by the nuclei provides the major energy sink after endoergic Si-burning; energy is lost through ν-emission and stored in the free particles. We include $e^{\pm}$ capture onto *free nucleons* only and the associated $\nu - \bar{\nu}$ emission. Rates for the $\nu - \bar{\nu}$ emission arising from plasma processes were obtained from Beaudet, et al.[44].

INITIAL CONDITIONS

Spherically symmetric non-rotating helium cores of 200, 300, and 500 $M_\odot$ were evolved (with no angular momentum) through explosive oxygen and all subsequent burning phases. All of these models collapsed completely to black holes with no explosion or ejection of matter. In principal, the core is expected to undergo some mass loss from its surface in the form of (He- or O-)winds, but the cores studied here were evolved at constant mass and for the unrevised $^{12}C(\alpha,\gamma)^{16}O$ rate[42]. In each of these models there comes a time after helium burning but *prior* to explosive oxygen burning when they begin to dynamically collapse; defined here to be the point when most of the core achieves infall velocities of $\sim 100-1000$ km s^{-1}. This typically corresponds to central densities and temperatures of 10^4-10^5 g cm^{-3} and 1-2$\times10^9$ °K, respectively. The density, temperature, velocity, and composition profiles for these 1-D models are then mapped spherically into our 2-D grid (all matter with $\rho \geq 1$ g cm^{-3}), after which rotation is applied. The code is written in cylindrical coordinates, R being the equatorial axis and Z the polar axis with azimuthal symmetry. A grid size of 60$\times$60 zones in R and Z has been used. We followed the 2-D collapse of each of these models in the absence of rotation and obtained essentially identical results (collapse to black holes) as in the 1-D calculations.

The initial angular momentum distribution was chosen such that the ratio of centrifugal force to the axial component of the gravitational force, C/G, was constant. This distribution satisfies the Solberg-Høiland stability criterion[45]: *i*) the angular momentum per unit mass, $j(m_i)$, increases with distance from the *rotation* axis R, and *ii*) j increases from pole to equator along a surface of constant entropy. The Goldreich-Schubert-Fricke stability criterion[45], however, is violated: the angular velocity is *not* constant on cylindrical surfaces. Rectification of the thermal imbalance is expected to occur via meridional circulation currents, provided there are no significant (physical) viscous effects or mean molecular weight gradients to inhibit bulk mass motion. Smoothing takes place over a Kelvin-Helmholtz timescale though, and since our cores are initially in a state of dynamical collapse, there is insufficient time for smoothing to occur and the GSF instability can be disregarded. The particular angular momentum distribution the core will have attained when our calculations begin depends on the details of the prior evolutionary history of the star and is quite uncertain.

The initial C/G values are listed in the first column of Table 1 for each core mass, while J/M, the *global* specific angular momentum, is given in column 2. Based on observed surface rotational velocities for main-sequence stars with masses between

$\sim 2 - 40\ M_\odot$, Kraft[46] finds that $J/M \propto M^{0.57}$, assuming the stars rotate as rigid bodies; the higher mass stars have more angular momentum, on average, than do their lower mass counterparts. This relation yields values comparable to those listed in Table 1, even if we assume half the mass to be lost during the formation of M_α. Bodenheimer and Ostriker[47] find that *differential* rotation can increase the deduced main-sequence value of J/M by a factor of 10, perhaps more. Furthermore, the collapse (or contraction) will amplify $j(m_i)$ [$= \Omega_i R_i^2$] in the central regions. Thus, if our stars are differentially rotating when on the main sequence and conserve angular momentum during their evolution, they could have substantially higher values of J/M than those given in Table 1.

TABLE 1

PROPERTIES OF MODELS AT MAXIMUM COLLAPSE

C/G	J/M (10^{18} cm^2 s^{-1})	ρ_c (10^6 g cm^{-3})	T_c (10^9 °K)	T/W
		200 $M_\odot$		
0.03	1.50	1.98	4.53	0.026
0.02	1.30	2.77	5.00	0.019
0.01	0.916	4.58	5.72	0.011
0.008	0.820	5.31	5.93	0.009
0.005	0.619	7.72	6.44	0.006
		300 $M_\odot$		
0.05	2.60	2.15	4.90	0.044
0.03	2.02	9.49	7.26	0.038
		500 $M_\odot$		
0.06	4.66	1.79	4.90	0.064
0.04	3.83	6.19	6.83	0.058

The real uncertainty in all this lies with how much angular momentum the star retains during its formation and evolution, along with the manner in which it is (re)distributed within the star. The convective history of the star is probably the single most important process concerning the *distribution* of angular momentum, $j(m_i)$, with time. VMS's are nearly completely convective during their main-sequence phase and retain significant convective regions beyond hydrogen burning. This acts to redistribute the angular momentum within the core, the details still being quite uncertain. Furthermore, viscous shear effects in radiative regions can result in angular momentum transfer internally. At the present time, it is impossible to say anything definitive with regard to these queries. The initial angular momentum *distribution* must be treated as another independent parameter to be explored.

RESULTS

All of the models listed in Table 1 exploded as a result of *rotation coupled with explosive oxygen burning.* Such a model was suggested earlier by Fowler and Hoyle[26], while Bodenheimer and Woosley[48] have explored a similar interaction for very low mass cores involved with traditional Type II supernovae (mantle ejection, not core explosion). Thermonuclear energy derived from oxygen burning alone is insufficient to explosively disrupt such massive cores. Instead, additional pressure support from rotation is required. In these radiation dominated cores,

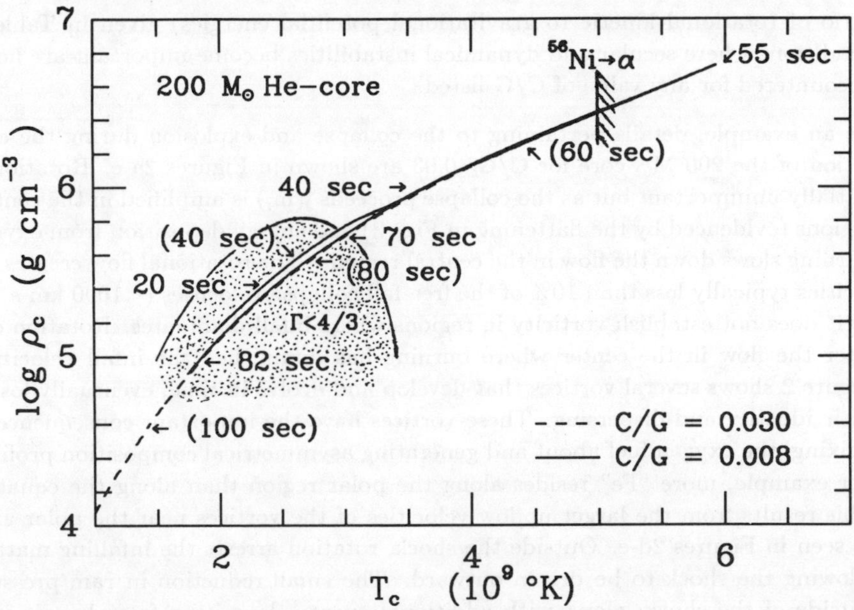

Figure 1: Evolution of central density and temperature for the 200 M$_\odot$ core, with two values of C/G. Both calculations begin with the same initial conditions at a time t=0 well within the pair-instability region where $\Gamma < 4/3$. Cumulative times are shown along the track; those in parentheses are for the C/G=0.030 case. The C/G=0.008 case evolves to higher ρ and T conditions quicker, well into the region of photodisintegration prior to reaching maximum collapse, 55 sec, and exploding. Maximum collapse occurs at (60 sec) for the C/G=0.030 case. The collapse and explosion are very nearly adiabatic.

the adiabatic index Γ is so close, but $< 4/3$ that the slight extra pressure support from rotation raises Γ sufficiently, enabling the infalling matter to be reversed and driven outward in an explosion. However, rotation never becomes important enough to significantly change the physical structure of the star. This is evident in Figure 1, in which the evolution of the central density (ρ_c) and temperature (T_c) for two values of C/G (0.008 and 0.030) are shown for the 200 $M_\odot$ core. Indicated is the pair-instability region and the approximate conditions under which about half of the "Fe" has photodisintegrated into ^{4}He. The two cases apparently evolve with nearly identical entropy structure, since both follow the same $\rho_c - T_c$ path; although high, ν-losses are not very important in altering the structure of the core in either case. The C/G=0.008 case collapses faster and deeper, reaching higher values of ρ_c and T_c before the collapse is reversed. Properties of the cores at maximum collapse are also given in Table 1. From the values of T/W (the ratio of rotational kinetic to gravitational potential energies) given in Table 1, conditions where secular and dynamical instabilities become important are never encountered for any value of C/G listed.

As an example, details pertaining to the collapse and explosion during the evolution of the 200 $M_\odot$ core for C/G=0.03 are shown in Figures 2a-e. Rotation is initially unimportant but as the collapse proceeds $j(m_i)$ is amplified in the central regions (evidenced by the flattening in Fig. 2) and energy deposition from oxygen burning slows down the flow in the central regions. The rotational flow creates velocities typically less than 10% of the free-fall collapse velocities (<1000 km s^{-1}), so it does not establish vorticity in regions where infall dominates. Rotation can alter the flow in the center where burning has diminished the infall velocities. Figure 2 shows several vortices that develop and evolve in time, eventually losing their identity and dispersing. These vortices have the important consequence of 'mixing' the oxygen fuel about and generating asymmetrical composition profiles. For example, more "Fe" resides along the polar region than along the equator. This results from the larger upflow velocities of the vortices near the polar axes as seen in Figures 2d-e. Outside the shock, rotation arrests the infalling matter, allowing the shock to be driven outward. The small reduction in ram pressure outside of the shock, along with additional energy deposition from burning allows the flow to be reversed resulting in an explosion. Final expansion velocities $\geq 10,000$ km s^{-1} and energies $\gtrsim 10^{53}$ erg are achieved. As in the spherical cases for $M_\alpha \lesssim 110\ M_\odot$, the core is completely disrupted! Nothing remains.

For each core mass there exists a minimum J/M below which complete collapse to a black hole ensues. Figure 3 summarizes our results, illustrating this dependence. The dashed curve is an estimated boundary between explosion/collapse, whereas the fate of the points lying above and below this curve for each core mass have been determined from our calculations. Lower J/M cases which ex-

Figure 2: Infall velocities (**v**) and density contours of the inner 25% of the core radius for the 200 $M_\odot$ C/G = 0.030 collapse case. The maximum velocity is indicated horizontally in the upper right corner of each plot, with the magnitude given above the plots along with its coordinates. Variations in the density contour levels are also given above each plot, specified by the multiplicative factor $\delta\rho$. Each velocity vector corresponds to an actual grid point, with the coordinate (2,2) being (R=0,Z=0). Cumulative time from the start of our collapse calculation is given at the bottom of each plot. a) Response of core to initial rotation; spherical infall and density contours. b) Slight flattening of ρ-contours; Z/R$\sim$0.90 near 10^{10} cm. The flow in the central region is becoming distorted, yet the ρ-contours remain closely spherical there. c) *Vorticity* (one complete vortex) is established in the center. A shock, thick at the axes and thin along the diagonal, is also beginning to form at R$\sim$Z$\sim$ 8×10^9 cm. d) The flow has become quite complex, with at least three separate vortices having developed which retain their identity and evolve in time. The shock is well established and growing in radius. e) Essentially all vortices have been dispersed and a bifurcated flow is seen. The region inside the shock shows nearly spherically symmetric contours, while just outside the shock rotation is still quite important. The shock propagates spherically and rotation soon becomes completely unimportant again.

$\rho_{max}(2,2) = 1.68\times10^5$; $\delta\rho = 1.261$

$v_{max}(24,24) = 1,490$ km s^{-1}

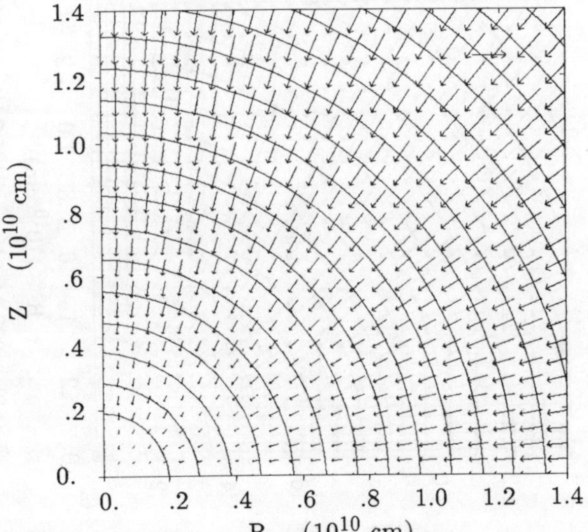

Fig. 2a: 200 $M_\odot$ He-core C/G = 0.030

Time = 6.83 sec.

$\rho_{\max}(2,2) = 1.08 \times 10^6 \quad ; \quad \delta\rho = 1.538$
$v_{\max}(23,31) = 5,602 \text{ km s}^{-1}$

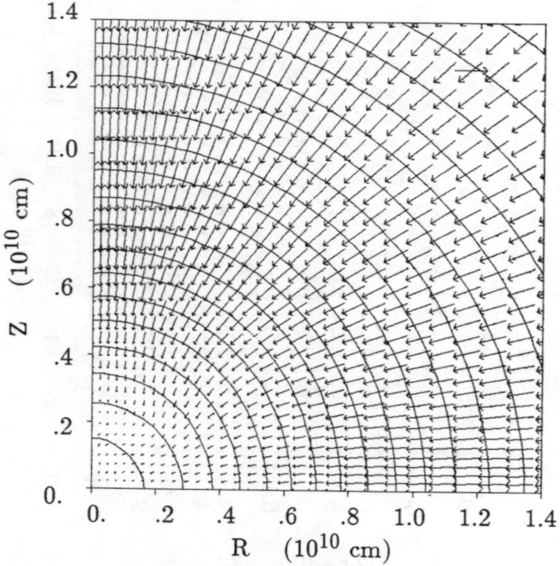

Fig. 2b: 200 $M_\odot$ He-core C/G = 0.030
Time = 49.89 sec.

$\rho_{\max}(2,2) = 1.91 \times 10^6 \quad ; \quad \delta\rho = 1.646$
$v_{\max}(3,33) = 8,764 \text{ km s}^{-1}$

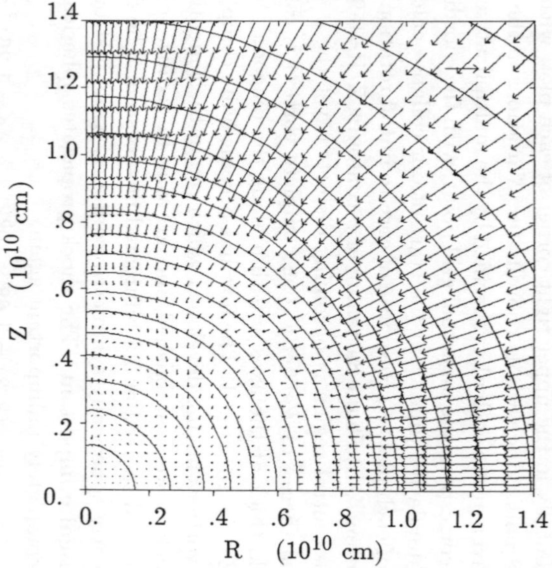

Fig. 2c: 200 $M_\odot$ He-core C/G = 0.030
Time = 59.71 sec.

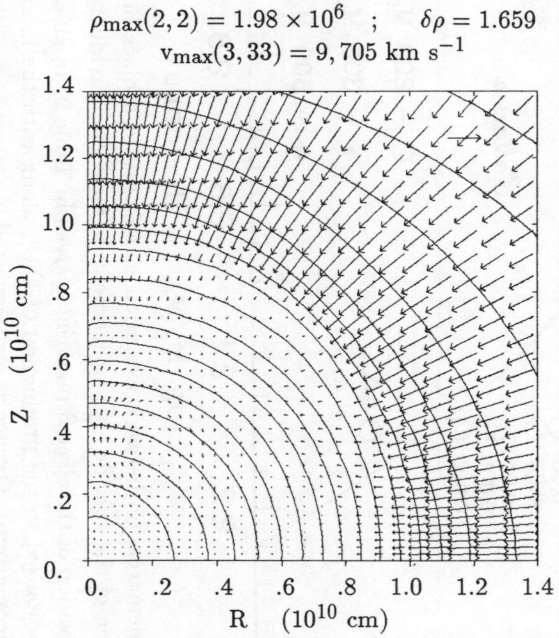

Fig. 2d: 200 $M_\odot$ He-core C/G = 0.030
Time = 61.84 sec.

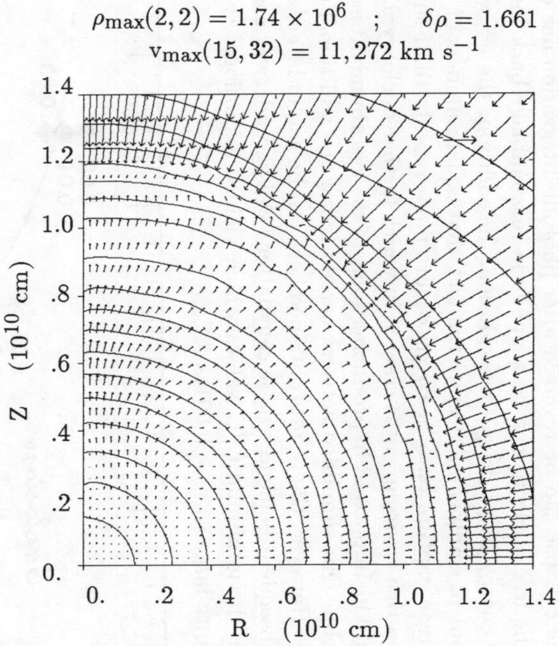

Fig. 2e: 200 $M_\odot$ He-core C/G = 0.030
Time = 66.13 sec.

ploded close to the transition boundary (explosion/collapse) experience considerable "Fe"-photodisintegration into free nucleons. This energy is potentially recoverable. If the collapse can be reversed *after* photodisintegration has occurred, the energy in the free nucleons can be tapped via reassemblage back into nickel. The additional energy helps drive the explosion. In addition, the $e^{\pm}$ weak capture rates are out of equilbrium, despite the fact that equal numbers of pairs are produced; positron capture onto neutrons occurs more rapidly than the inverse rate[43]. A proton rich environment is created in which further nucleosynthesis can occur. This could be important in producing some of the less abundant isotopes in the Fe-peak group. Our simplified network cannot follow the additional reactions that may ensue. However, the results obtained here can be run in larger networks to follow the subsequent behavior; these calculations will be underway soon. Note though that for the minimum C/G values which yielded explosions, the proton abundance is quite large ($\eta \approx -0.001$).

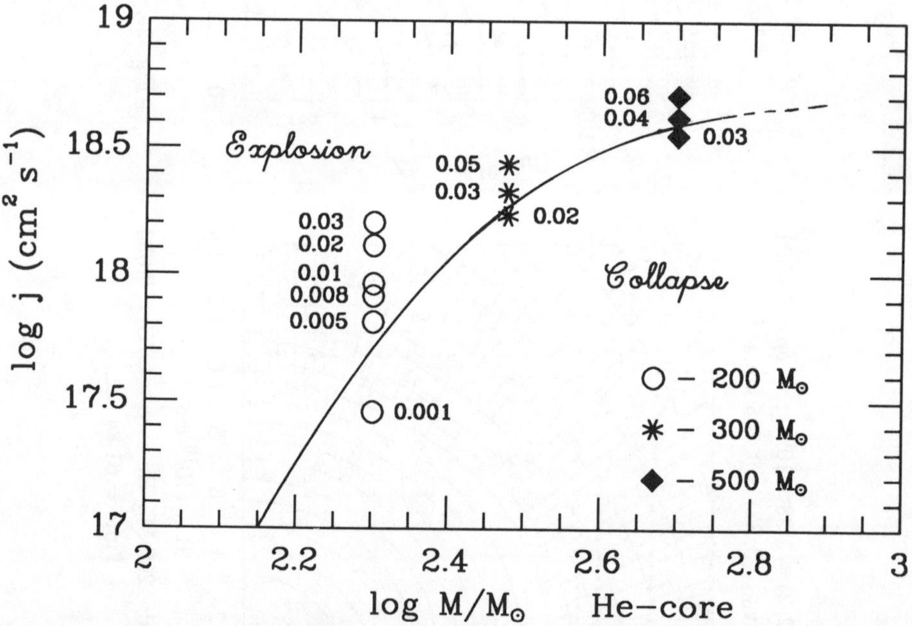

Figure 3: The demarcation between explosion or collapse to a black hole (solid line) is a function of the specific angular momentum for each individual mass.

The bulk composition of the ejected material is given in Table 2. In all cases, the inclusion of rotation led to the production of "Fe", along with the intermediate mass nuclei (silicon group). Oxygen is overproduced in comparison with carbon, although mass lost from the star *and* core during the evolution must be taken into account when discussing overall yields. This is beyond the scope of this presentation. Carbon presumably comes from stars of lower mass anyway.

TABLE 2

NUCLEOSYNTHETIC YIELD[a] FROM EXPLOSION

M_α [b] :	200 $M_\odot$			300 $M_\odot$		500 $M_\odot$	
C/G :	0.03	0.008	0.005	0.05	0.03	0.06	0.04
^{12}C ...	3.4	2.0	1.8	4.8	2.9	7.0	4.0
^{16}O ...	73.	45.	40.	93.	51.	157.	92.
^{20}Ne ...	7.0	4.0	3.5	8.6	4.3	14.	8.5
^{24}Mg ...	17.4	10.5	9.5	22.	12.	35.	22.
^{28}Si ...	59.	49.	46.	83.	55.	137.	87.
^{56}Ni ...	35.3	84.	93.	83.	164.	145.	273.
^{4}He ...	4.7	6.3	7.0	6.4	11.8	6.0	15.
p ...	1.9(-9)	9.6(-4)	2.1(-3)	1.4(-6)	2.6(-2)	3.7(-6)	2.9(-2)
n ...	1.6(-9)	5.6(-4)	8.4(-4)	1.2(-6)	2.0(-3)	2.8(-6)	3.2(-3)
ϵ_p [c] ...	3.0(-10)	4.0(-4)	1.3(-3)	2.0(-7)	2.4(-2)	9.0(-7)	2.6(-2)

[a]Numbers listed for each element are in solar masses.
[b]Initial He-core mass for the model.
[c]$\epsilon_p \equiv M_p - M_n$, the difference in total mass contained in free nucleons, gives the excess proton abundance in solar masses.

Glatzel, et al.[49] (GEF) have simulated the collapse of rotating cores through the use of Maclaurin spheroids, which imply uniform rotation for a homogeneous (one-zone) model. This corresponds to an n=0 polytrope, but these can be corrected for polytropes of n>0. In accord with our results, they also obtain explosions, given 'enough' J, for masses which in 1D would have collapsed to black holes. In as far as these approximations of the core hold, their results require much higher values of J; e.g., from their Table 2, if one takes a value $j = 1.73 \times 10^{18}$ cm^2 s^{-1} for a 102 $M_\odot$ exploding core with a polytropic index n=0 and then correcting both the mass and J/M for an n=3 polytrope (multiplying M by 3.36 and j by 5.86) yields $j \sim 1 \times 10^{19}$ cm^2 s^{-1}. This is to be compared to an interpolated value of j for a similar core mass (about 343 $M_\odot$) from our Figure 3; $j \sim 2.0 \times 10^{18}$ cm^2 s^{-1}. Their value of j is about a factor of 5 higher than ours (note that Figure 4 of GEF[50] does not agree with our analysis here; they give only a factor of 1.9). GEF's calculations give an "Fe" abundance $\leq 0.5 M_\odot$, consistent with the spherical explosions but markedly different from our 2D results. Furthermore, we turned off ν-losses and recomputed the collapse for the cases which did not explode; these still form black holes, whereas GEF found much lower values of j for black hole formation. Their curve for the explosion/collapse boundary was shifted down considerably to lower j's for a given mass in Figure 3.

DISCUSSION

Previous studies have concluded that, in the absence of rotation, the contribution to the dark matter in the form of black hole remnants arising from VMS's could be cosmologically significant. In fact, the only other plausible contribution from black holes comes from the primordial formation of tiny ($<< 1\ M_\odot$) black hole remnants[15,16]. However, we have shown that angular momentum considerations cannot be ignored. In order for black hole remnants to be significant contributors to the dark matter problem in light of our results, either *i*) very efficient angular momentum transport has occurred during the evolutionary phases prior to the formation of the helium core, or *ii*) the distribution of the angular momentum *at* the time of helium core formation is quite different than assumed here, with most of the momentum residing in the outer layers of the star. Both cases advocate the presence of little angular momentum deep within the core during the dynamical collapse.

All of our explosions produce large quantities of "Fe". Those with the lowest angular momentum can even be somewhat embarrassing in the amount of iron synthesized. This places rather severe restrictions on the number density of VMS's existing over the lifetime of the Universe, especially at early times. We shall not go into a detailed analysis here, but a few simple arguments can be presented. First, envision the case where many VMS's form at an early time in the expansion of the Universe ($z > 10$). Provided they explode, VMS's could then generate the initial enrichment of heavy metals which seems to be observed in the most metal poor subdwarfs and globular clusters. This corresponds to a minimum metallicity $Z_{min} \leq 10^{-4}$ throughout the Galaxy. If we take the mass of the Galaxy, M_G, to be $2\times10^{11}\ M_\odot$, then the total mass of "Fe" in the early Galaxy, mostly bound in stars, is given by

$$M_{Fe} \approx M_G Z_{min} Z_{Fe} \leq 2\times10^6\ M_\odot$$

where Z_{Fe}, ~ 0.1, is the mass fraction of "Fe" contained in Z_{min}. Our results can be utilized by assuming an average VMS core mass for the yield of "Fe"; $M_\alpha = 200\ M_\odot$ with a mass fraction $X_{Fe} \sim 0.15$, which minimizes the iron production by VMS's (see Table 2). Hence,

$$M_{Fe} \approx \langle X_{Fe} M_\alpha \rangle N_{VMS}\ f_J$$

with N_{VMS} giving the total number of VMS formed and f_J (set here to unity) being the fraction of rotating VMS's with enough angular momentum to explode.

Combining the two expressions gives

$$N_{VMS} \approx \frac{M_G \, Z_{min} \, Z_{Fe}}{\langle X_{Fe} \, M_\alpha \rangle \, f_J} \lesssim 7 \times 10^4 \quad ,$$

which is not a particularly large number compared to the $\sim 10^8$ Type II supernovae expected in Galactic history. If the progenitor main-sequence mass was about 400 $M_\odot$, this amounts to only 3×10^7 $M_\odot$ going into VMS's; much less than the mass required to solve the dark matter problem. However, the total explosion energy liberated from all these going off roughly in unison (VMS's have Eddington limited lifetimes of $\sim 3 \times 10^6$ yrs) is $\gtrsim 10^{58}$ erg, which could be quite important indeed in galaxy formation and shell structure, as discussed in the introduction. Given the same conditions as above, $\lesssim 10^7$ VMS's could provide the observed Pop I abundance of iron and intermediate mass elements, neglecting the contribution from Type I and II supernovae. This also does not admit the possible loss of any ejected VMS matter from the Galaxy, despite the fact that the explosion velocities are much greater than the escape velocity of the Galaxy. Constraints can be placed though on when this enrichment could have occurred. Could VMS's provide an enrichment of "Fe" at late times? Observations of the age-metallicity relation for field stars implies that the iron abundance cannot have more than doubled in the interstellar medium during the last 5 billion years. Taking the amount of material in the interstellar medium to be about 5% of the total baryonic mass in the Galaxy, and assuming a mass fraction of iron of 0.001, yields an iron mass of $\sim 10^7$ $M_\odot$. This corresponds to $N_{VMS} \lesssim 3 \times 10^5$ or an explosion rate of one every 15,000 years. One Type II supernova occurs about every 100 years in our Galaxy. Comparing the rates of VMS's with Type II's, about 1 out of every 150 Type II supernova events could be an explosion of a VMS. There are ~630 known supernovae which have gone off in the last century and about half these are Type II's. This implies that 1-2 VMS explosions could have occurred during the last century, or might be observed in the near future. Since many of the supernovae detected have very little observational data (light curve or spectra), some of these may in fact be the explosion of VMS's. This is entirely consistent with (lack of) observations of VMS explosions.

In conclusion, Figure 4 shows what one would expect to observe if a rotating VMS were to explode. Plotted is the light curve for a 200 $M_\odot$ helium core with C/G = 0.008 which produced 84 $M_\odot$ of ^{56}Ni. This extremely brilliant display would be easily seen even in quite distant galaxies. The initial rise and peaks at early times are due to oxygen and helium recombination, but the bulk of the main display including the broad hump peaking at about 100 days is from nickel and cobalt decay. The long tail is due to ^{56}Co decay. At peak, the VMS is more than 10 times brighter than the brightest Type Ia supernova and remains so for over a year. Such a display apppears inconsistent with SN1961v, as does the expansion velocities, etc., and we believe, contrary to Utrobin[25], that 1961v *was not* an explosion of a star having a mass more than a few hundred solar masses.

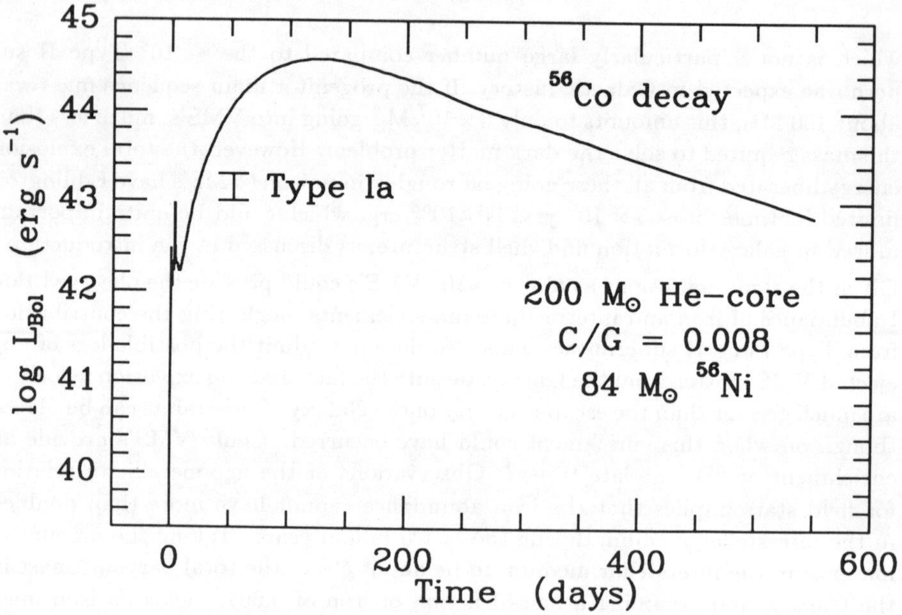

Figure 4: Bolometric light curve for one of our exploding VMS's. The luminosities of Type Ia supernovae are shown for comparison. A γ-ray opacity of 0.07 cm^2 g^{-1} has been used.

Acknowledgments

Our appreciation is extended to our colleague and co-worker Peter Bodenheimer for support and informative conversations regarding multidimensional calculations. We would like to acknowledge the support and hospitality received from Claire Max, Head of the Institute of Geophysics and Planetary Physics at Lawrence Livermore National Laboratory. We acknowledge useful discussions with George Fuller during the course of this investigation and stimulating conversations with David Arnett, Bernard Carr, George Fuller, Dieter Hartmann, and David Schramm during this meeting. We also thank Lisa Ensman for assistance with obtaining the light curve for our exploding VMS. We thank the organizer of this symposium, Grant Mathews, for putting together such an exciting and enjoyable symposium and to the participants for many friendly discussions. GSS thanks Michelle House and Dieter Hartmann for encouragement and comments on the manuscript. This research has been supported in part by grants from the IGPP at

LLNL (#85-14), the National Science Foundation (AST81-08509,AST84-18185), and NASA (NCA2-66).

References

1 - White, S.D.M., and Rees, M.J. 1978 M.N.R.A.S. 183, 341.

2 - Carr, B.J., Bond, J.R., and Arnett, W.D. 1984 Ap. J. 277, 445.

3 - Carr, B.J. 1987 This proceeding.

4 - Rowan-Robinson, M., Negroponte, J., and Silk, J. 1979 Nature 281, 635.

5 - Carr, B.J. 1981 M.N.R.A.S. 195, 669.

6 - Carr, B.J., McDowell, J., and Sato, H. 1983 Nature 306, 666.

7 - Ostriker, J.P., and Cowie, L.L. 1981 Ap. J. Letters 243, L127.

8 - Ikeuchi, S. 1981 Publ. Astr. Soc. Japan 33, 211.

9 - Carr, B.J., and Rees, M.J. 1984 M.N.R.A.S. 206, 801.

10 - Carr, B.J., and Ikeuchi, S. 1985 M.N.R.A.S. 213, 497.

11 - Heiles, C. 1979 Ap. J. 229, 553.

12 - Meaburn, J. 1980 M.N.R.A.S. 192, 365.

13 - Williams, R.E., and Christiansen, W.A. 1985 Ap. J. 291, 80.

14 - Doroshkevich, A.G., Zel'dovich, Ya.B., and Novikov, I.D. 1967 Soviet Astr. - AJ 11, 231.

15 - Hartquist, T.W., and Cameron, A.G.W. 1977 Ap. Space Sci. 48, 145.

16 - Truran, J.W., and Cameron, A.G.W. 1971 Ap. Space Sci. 14, 179.

17 - Hoyle, F., and Fowler, W.A. 1963 Nature 197, 533.

18 - Talbot, R.J., and Arnett, W.D. 1971 Nature 229, 250.

19 - Bond, J.R., Arnett, W.D., and Carr, B.J. 1984 Ap. J. 280, 825.

20 - van Genderen, A.M., and Thé, P.S. 1984 Space Sci. Review 39, 317.

21 - Davidson, K., Dufour, R.J., Walborn, N.R., and Gull, T.R. 1986 Ap. J. 305, 867.

22 - Maeder, A. 1980 Astr. Ap. 92, 101.

23 - Zwicky, F. 1964 Ap. J. 139, 514.

24 - Penrod, G.D., Goodrich, R.W., Stringfellow, G.S., and Fileppenko, A.V. 1987 In preparation for Ap. J.

25 - Utrobin, V.P. 1984 Ap. Space Sci. 98, 115.

26 - Fowler, W.A., and Hoyle, F. 1964 Ap. J. Suppl. 9, 201.

27 - Woosley, S.E., and Weaver, T.A. 1982 In *Supernovae: A Survey of Current Research*, ed. Rees, M.J. and Stoneham, R.J. Dordrecht: Reidel. pg. 79

28 - Woosley, S.E. 1986 in Nucleosynthesis and Chemical Evolution, Saas-Fee Lecture Notes. p 1.

29 - Woosley, S.E., and Weaver, T.A. 1987 In preparation for Ap. J.
30 - Ober, W.W., El Eid, M.F., and Fricke, K.J. 1983 Astr. Ap. 119, 61.
31 - El Eid, M.F., Fricke, K.J., and Ober, W.W. 1983 Astr. Ap. 119, 54.
32 - El Eid, M.F. 1987 This proceedings.
33 - Barkat, Z., Rakavy, G., and Sack, N. 1967 Phys. Rev. Letters 18, 379.
34 - Fraley, G.S. 1968 Ap. Space Sci. 2, 96.
35 - Arnett, W.D. 1972 Ap. J. 176, 681.
36 - Stringfellow, G.S., Woosley, S.E., and Bodenheimer, P. 1987 In preparation for Ap. J.
37 - Stringfellow, G.S., and Woosley, S.E. 1983 Bull. Am. Astr. Soc. 15, 955.
38 - Stringfellow, G.S., Woosley, S.E., and Bodenheimer, P. 1985 Bull. Am. Phys. Soc. 31 No.4, p 779.
39 - Woosley, S.E., Axelrod, T.A., and Weaver, T.A. 1984 In *Stellar Nucleosynthesis*, ed. C. Chiosi and A. Renzini p 263. Dordrecht: Reidel.
40 - Black, D.C., and Bodenheimer, P. 1975 Ap.J. 199, 619.
41 - Weaver, T.A., Zimmerman, G.B., and Woosley, S.E. 1978 Ap.J. 225,1021.
42 - Fowler, W. A., Caughlan, G. R., Zimmerman, B. A. 1975, Ann. Rev. Astr. Ap. 13, 69.
43 - Fuller, G.M., Fowler, W.A., and Newman, M.J. 1985 Ap.J. 293, 1.
44 - Beaudet, G.V., Petrosian, V., and Salpeter, E.E. 1967 Ap.J. 150, 979.
45 - Tassoul, J-L. 1978 Theory of Rotating Stars. Princeton University Press.
46 - Kraft, R.P. 1970 in Spectroscopic Astrophysics, ed. G. Herbig (University of California Press) pg 385.
47 - Bodenheimer, P., and Ostriker, J.P. 1970 Ap.J. 161, 1101. Fig. 4.
48 - Bodenheimer, P., and Woosley, S.E. 1983 Ap. J. 269, 281.
49 - Glatzel, W., El Eid, M.F., and Fricke, K.J. 1985 Astr. Ap. 149, 413.
50 - Glatzel, W., El Eid, M.F., and Fricke, K.J. 1986 in Nucleosynthesis and Its Implications on Nuclear and Particle Physics, ed. J. Audouze and N. Mathieu, pg 167.
51 - Woosley, S.E., and Weaver, T.A. 1986 in Radiation Hydrodynamics in Stars and Compact Objects, ed. D. Mihalas and K.-H.A. Winkler, pg 91.

EXPLOSION AND NUCLEOSYNTHESIS IN VERY MASSIVE STARS

M.F.El Eid[1], and N.Prantzos[2]

[1] Universitäts-Sternwarte Göttingen, FRG

[2] Institut d'Astrophysique, Paris, France

Abstract

To illustrate a possible evolutionary scenario for very massive stars ($M > 60\,M_\odot$), we briefly describe the evolution with mass loss of a $100\,M_\odot$ Pop I star through the hydrogen and helium burning phases. This star may ends core helium burning as a $45\,M_\odot$ Wolf-Rayet (W-R) star of type WN. Following further evolution of this WN star, we find that it encounters the electron- positron pair creation instability prior to oxygen ignition, and is likely to be disrupted by the ensuing explosive oxygen burning. The resulting nucleosynthetic yield from such explosion is compared with observed abundances in the oxygen-rich supernova remnant Cas A. It is found that several features of the model calculations are compatible with observations. Thus, it appears that exploding massive WN stars can be envisaged as progintors for at least some oxygen-rich supernova remnants of Cas A type.

Introduction

There are several reasons for studying the evolution of very massive stars (hereafter, called VMS) of masses in excess of $\sim 60\,M_\odot$. Such stars have been proposed, (Woosley and Weaver, 1982; El Eid et al. 1983; Ober et al. 1983; Bond et al. 1984; Carr et al. 1984) as possible candidates for the so called Pop III (or zero metal) stars.The evolution and nucleosynthesis of these objects have been studied in the first three references for the aim to estimate their contribution to an early chemical enrichment of the Galaxy. An excellent discussion of the physics of these stars, and their cosmological consequences if they have been formed as first stars has been presented in the last two references.

There is growing observational evidence that the most luminous stars in the Milky Way, and in several nearby galaxies are the most massive ones. Famous examples are η Car, P Cyg in the Milky Way, S Doradus in the Large Magellanic Cloud, and many so called Hubble- Sandage variables in M31 and M33. A composite Hertzsprung–Russell (H-R) diagram for the most luminous stars (Humphreys and Davidson, 1984) show that stars with luminosities in excess of $10^6 L_\odot$ do not exhibit effective temperatures less than $\sim 1.5\ 10^4\ K$. This seems to indicates that some kind of instability occurs when a VMS evolves redwards across the upper part of the H-R diagram which may lead to a partial or complete ejection of its hydrogen-rich envelope. In this way a star becomes a Wolf-Rayet (W-R) star having a hot surface which is helium-rich and either nitrogen-rich (spectral type WN) or carbon/oxygen-rich (spectral type WC/WO). However, mass loss may not be the only process leading to the formation of W-R stars. Convective overshooting may also have an important effect.

Many uncertainties are still involved in the theoretical description of the combined effect of mass loss and convection during the quasistatic evolution of massive stars. In particular, the mechanism of mass loss is not well known. Surface radiation pressure certainly plays a key role, but not in all evolutionary phases. As suggested by De Jager (1984) turbulent pressure near the surface may also cause heavy mass loss when a convective envelope is established during the post main sequence evolution. Selfconsistent calculations including these effects are not simple and still not available.

In the context of the evolutionary scenario described above (see also Maeder, 1984; Langer, 1987), the final evolution of massive W-R stars is interesting in many respects. Firstly, it seems that some of the oxygen-rich supernova remnants (SNR) of Cas A type may have resulted from the explosion of such stars (Begelmann and Sarazin, 1986; Fesen et al. 1987). Secondly, their contribution to nucleosynthesis like the s-process (Prantzos et al. 1987) and the p-process (see Rayet et al., this proceeding) cannot be estimated without knowing their final fate.

Recent computation (Langer and El Eid, 1986; El Eid and Langer, 1986; Woosley, 1986) have emphasized the role of the electron-positron ($e^{\pm}$) pair creation instability in determining the final fate of a massive W-R star. Due to this instability oxygen burning proceeds explosively and may lead to supernova explosion, usually called pair creation supernova. Depending sensitively upon the mass of the star at helium exhaustion, the outcome of such explosion may lead to a total disruption of the star or to a violent pulsation accompanied by some mass ejection according to Woosley (1986).

In the present contribution the evolutionary scenario scetched above is illustrated in the case of a $100\,M_{\odot}$ Pop I star. In sect. 2 we summarize the main characteristics of its early evolution through the hydrogen and helium burning phases, and point out some involved uncertainties. This star may terminate its core He burning as a WN star of $45\,M_{\odot}$ as described in sect. 2 (see Langer and El Eid, 1986 for more details). In sect. 3 we investigate the advanced burning phases of this WN star including its supernovae stage. The nucleosynthetic yield which may result from such explosions is presented in sect. 4 and is compared with the observed abundances in the supernova remnant Cas A.

2. Early evolution of a 100 $M_{\odot}$ Pop I star

In a recent work, Langer and El Eid (1986) have investigated the hydrogen and helium burning phases of this star assuming Pop I initial composition and including mass loss as well as convective overshooting in a parameterized way, since no consistent theory exists so far for both effects. Comparative study of the model calculations with observations strongly favoured an amount of overshooting for VMS much smaller than one pressure scale hight (H_p). The recent paper of Maeder and Meynet (1987) is in support of this view. These authors have adopted an overshooting distance of $0.25H_p$. In contrast, the computations of

Prantzos et al. (1986) based on the Roxburgh criterion for convection (Roxburgh, 1978) suggest a very large overshooting distance. As indicated in Table 1, the uncertainty in the treatment of convective overshoot has a strong impact on the mass of the W-R star.

Table 1. Comparison of the remaining mass M_α at the end of He burning for a star of initially 100 $M_\odot$ obtained by several authors. The remaining star may be identified with a W-R star. d_{over}/H_p is the overshooting distance in terms of the pressure scale hight H_p . M_O is the mass of the oxygen core.

Authors	1	2		3	4
$M_\alpha/M_\odot$	50	45	61	65	42
$M_O/M_\odot$	42	38	52	54	36
d_{over}/H_p	0.25	0.	1.90	Roxburgh Crit.	0. (probably)
W-R Type	WC	WN	WC	WC	WN

1: Maeder and Meynet (1987), 2: Langer and El Eid (1986)

3: Prantzos et al. (1986), 4: Woosley and Weaver (1986)

Table 1 shows that the Roxburgh criterion leads to the largest mass of the W-R star, and may correspond to an overshooting distance of $\sim 1.9H_p$. Whether that criterion is suitable for determining the degree of overshooting or not seems to be questionable according to Baker and Kuhfuss (1986).

We adopt the point of view that a Pop I star of initially $100\,M_\odot$ is likely to end its core He burning as a W-R star having a mass of $45\,M_\odot$ if overshooting can be neglected, or $50\,M_\odot$ with a moderate amount of overshooting. In the following we focus on the advanced and final evolution of the $45\,M_\odot$ star. The main characteristcs of this model at the end of He burning are summarized by El Eid and Langer (1986), and it serves as initial model for the present computations.

3. Advanced and final evolution of a 45 $M_\odot$ WN star

3.1 Numerical method and input physics

To follow further evolution of the model presented above, we have used an implicit hydrodynamical code that solves iteratively the difference equations of stellar structure in a Lagrangian grid. Spherical symmetry has been assumed, and the energy transport was due to radiative diffusion. Convection has been turned off, since it seems not to be effective in the present calculations as described below. The early work of Fraley (1968) suggests that convection has no important effect on the evolution through the pair creation supernova.

The equation of state has been calculated explicitly for a mixture of ions, black body radiation, and $e^{\pm}$- pairs in thermodynamic equilibrium in which the ions are treated as a Boltzmann gas. The effect of the $e^{\pm}$- pairs is taken into account by assuming that they can be created by the radiation field via the equilibrium process

$$2\gamma \rightleftharpoons e^- + e^+. \tag{1}$$

The chemical potentials of the e^- and e^+ are then related to each other by

$$\mu_+ = -\mu_- - 2\, m_e\, c^2. \tag{2}$$

For a given temperature T and density ρ, we insert these chemical potentials into the Fermi-Intergrals, and calculate the concentrations of the e^- and e^+ imposing charge neutrality (see Cox and Giuli, 1968 for details).

The thermodynamic conditions encountered in the pair unstable stars require a correction of the Thomson scattering opacity, which should include the effect of inelastic compton scattering in the presence of $e^{\pm}$- pairs, and the degeneracy of the electrons. The correction without degeneracy has been introduced by fitting the results of Sampson (1959) by the following expression:

$$\kappa(T_9) = \kappa_{TH}\ (1.001 + 2.394\ T_9 - 0.347\ T_9^2)^{-1}, \tag{3}$$

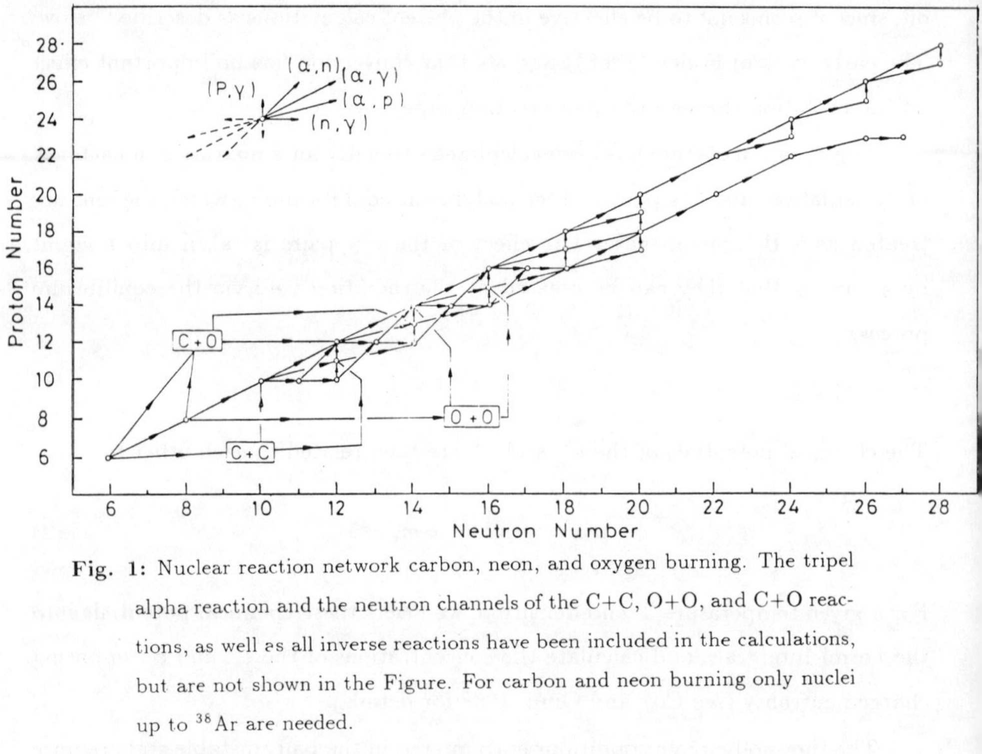

Fig. 1: Nuclear reaction network carbon, neon, and oxygen burning. The tripel alpha reaction and the neutron channels of the C+C, O+O, and C+O reactions, as well as all inverse reactions have been included in the calculations, but are not shown in the Figure. For carbon and neon burning only nuclei up to ^{38}Ar are needed.

where κ_{TH} is the usual Thomson scattering opacity, and T_9 is the temperature in units of $10^9 K$. The above expression should only be used up to $T_9 = 9.0$. At higher temperatures a better fitting formula which include the degeneracy of electrons has been derived by Buchler and Yueh (1976). The corrected opacity is then:

$$\kappa(T,\eta) = \sigma_0 \; n \; G(T,\eta)^{-1}/\rho, \tag{4}$$

where $\sigma_0 = 6.652 \; 10^{-25} \; cm^2$ is the Thomson cross section, $n = n_+ + n_-$ denotes the combined number density of electrons and positrons, and ρ is the mass density. The function G has been taken from Buchler and Yueh (1975), and will not be given here.

Neutrino energy loss rates have been calculated according to the new compilations of Munakata et al. (1985) in the framework of the Weinberg- Salam theory.

An important ingredient in the present calculations is the nuclear reaction network, since it determines the energy generation rate as well as the nucleosynthesis. As displayed in Fig. 1 the nuclei between ^{12}C and ^{56}Ni have been linked by (p,γ), (α,p), (α,γ), (n,γ), and (α,n) reactions and their inverse reactions. In addition, the triple- alpha reaction, the α and p channels of the ^{12}C+^{12}C, ^{12}C+^{16}O, and ^{16}O+^{16}O reactions have been included. The reaction rates are according to Gaughlan et al. (1985) or earlier versions up to Si. For heavier nuclei new theoretical predictions according to Thielemann et al. (1985) (see also Rayet et al., this proceeding) have been utilized.

With the numerical code and input physics described above, we have followed the evolution of the 45 $M_\odot$ WN star through the carbon, neon, and oxygen burning phases. The results are presented in the following section.

3.2 Results

For calculating the carbon and neon burning phases the nuclear reaction network shown in Fig.1 has been utilized up tp ^{38}Ar, which is sufficient for this

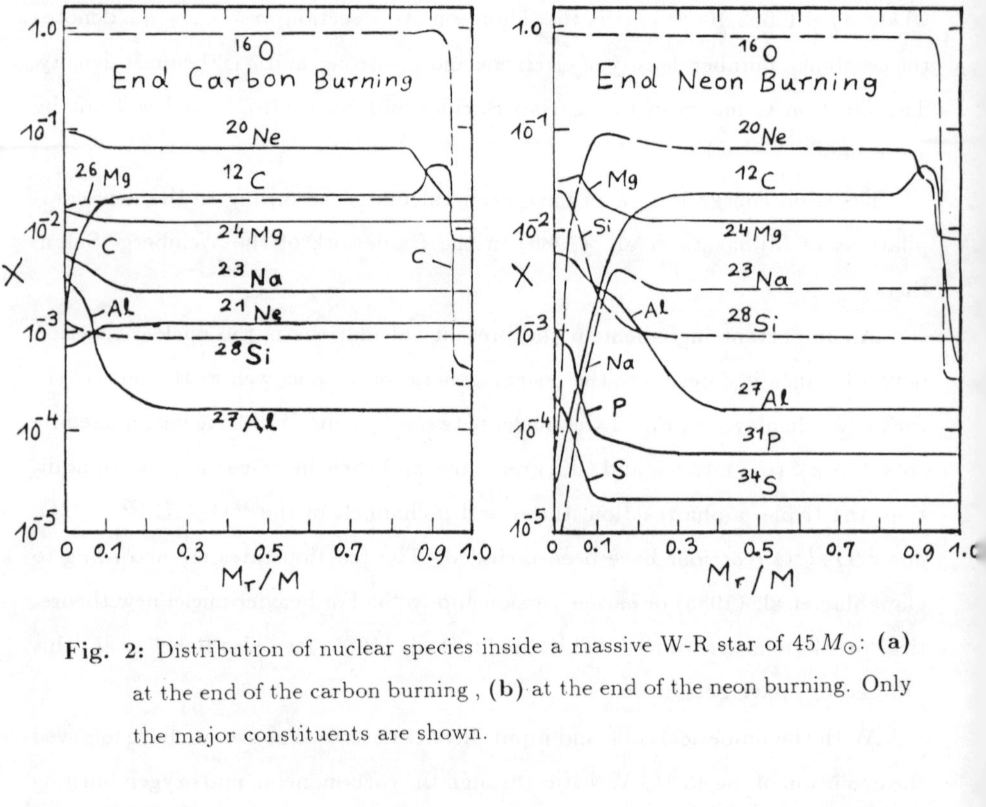

Fig. 2: Distribution of nuclear species inside a massive W-R star of 45 $M_\odot$: **(a)** at the end of the carbon burning , **(b)** at the end of the neon burning. Only the major constituents are shown.

purpose. The initial abundances for the elements C, Ne, and O are those given by EL Eid and Langer (1986). For the other nuclear species we have adopted the average values calculated by Prantzos (1986) (see also Prantzos et al. 1986) to account for the nuclear transmutations during He burning. The evolution towards carbon ignition is mediated by a Kelvin- Helmholtz contraction phase which lasts $\sim 10^3$ yrs, and is accelerated by enhanced neutrino energy losses. Central carbon burning occurred at $T_9 \sim 8.0$, $\rho \sim 2.5\ 10^4 g\ cm^{-3}$, and a dimensionless entropy $S_c \sim 7.0$. This burning phase is characterized by a negative effective (nuclear minus neutrino) energy generation rates (see Fig. 3a). Hence, no convective core develops under these conditions. Fig. 2a shows that carbon burning has occurred within 20% of the mass.The nuclear species comprising about 98% in mass fraction are ^{16}O, ^{20}Ne, ^{24,25,25}Mg. ^{27}Al and ^{28}Si have been enhanced by a factor of ~20 and ~14 respectively compared with the initial values. We note that the effect of $e^{\pm}$- pairs was not present during this phase.

Central neon burning commenced near to $T_c = 1.46\ 10^9 K$, $\rho_c = 2.35\ 10^5 g\ cm^{-3}$, and $S_c = 3.77$. As seen in Fig. 3a, the global effective energy generation rate during this phase was also negative which again inhibits the formation of a convective core. Fig. 2b displays the composition profiles at the end of Ne burning, where $T_c = 2.03\ 10^9 K$, $\rho_c = 6.81\ 10^5 g\ cm^{-3}$, and $S_c = 3.42$. The major nuclear constituents are ^{16}O, ^{24,25,26}Mg ^{27}Al, ^{28,29,30}Si. ^{31}P and ^{34}S have been enhanced. At this time it has been found that η_e was close to -1.0 at the center supporting our view that the opacity should be calculated according to Eq.(4) rather than (3). The opacity had a central value of 0.044 $cm^2\ g^{-1}$ and 0.17 $cm^2\ g^{-1}$ at the surface of the star. The distribution of the opacity inside the star plays a key role in determining the final outcome of the supernova calculations (see recent calculations by El Eid , 1986).

Further evolution of our model has been calculated using the full network as in Fig. 1. As indicated in Figs. 3a and 3b the sharp increase in the total kinetic energy prior to oxygen ignition marks the onset of dynamical instability due to

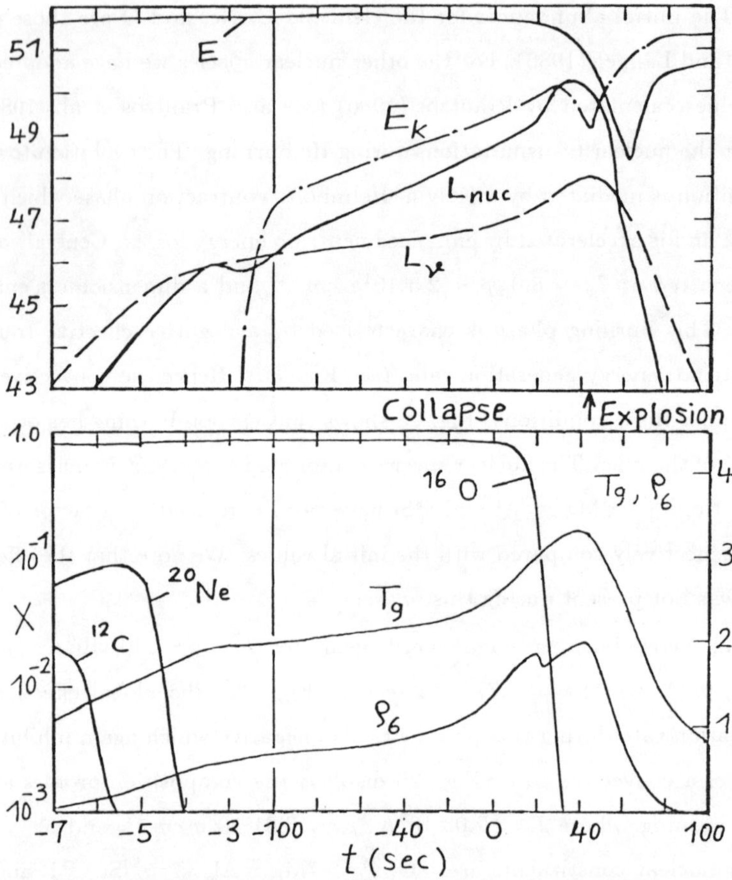

Fig. 3: Energetics and central conditions for the final evolution of the 45 $M_\odot$ WN star as a function of time t. t=0 is chosen arbitrary at oxygen ignition. Note that the time scale is logarithmic up t= -100 sec. Top Fig., 3a, show the logarithmic integrated values of the kin. energy, E_k, the energy generation rate, L_{nuc}, the rate of neutrino energy loss, L_ν, and the energy $E = |E_{pot} + E_{th} + E_k|$, where E_{pot}, E_{th} are the potential and thermal energies respectively; all quantities in cgs units. Bottom Fig., 3b, displays the fraction by mass at center for C, Ne,and O (left scale), and the central temperature (in 10^9 K) and density (in 10^6 gcm^{-3}).

conspicuous $e^{\pm}$- pair creation. The ensuing collapse rises the temperatures and densities such that oxygen ignites at the center at $T_9 = 2.51K$, $\rho = 1.32\ 10^6\ g\ cm^{-3}$ (see Fig. 3b). Since oxygen burning is firstly restricted to the central region of the star, the collapse was halted there at $T_c = 2.93\ 10^9\ K$, and $\rho = 1.84\ 10^6\ g\ cm^-3$ (first maximum in the density curve in Fig. 3b). However, only the inner layers comprising $\sim 1\,M_\odot$ attain positive velocities, while the outer layers were still falling in. This phase lasted until oxygen exhaustion at the center. Afterwards, the star collapsed again achieving a peak central temperature of $3.37\ 10^9\ K$ and density $2.05\ 10^6\ g\ cm^-3$ (second maximum in Fig. 3b). At this time pair instabilty comprised $\sim 70\%$ of the star's mass between 6 and 37 $M_\odot$, and $\sim 2.5\,M_\odot$ of oxygen have been consumed, which was now sufficient to reverse the whole collapse phase smoothly without shock formation. We note that no indication for the formation of a convective core has been found during explosive oxygen burning. The entropy had always a positive gradient away from the center, and the radiative temperature gradient remained below the adiabatic one in the region where oxygen burning has occurred.

The explosion phase has been so far followed until $T_c = 8.0\ 10^8 K$ and $\rho_c = 2.3\ 10^4\ g\ cm^{-3}$. Up to this stage the kinetic energy was $2.60\ 10^{51} erg$, and the surface velocity $\sim 6900\ \ km/s$. It is expected that the kinetic energy will increase further to $\sim 3\ 10^{51}\ erg$, since in total about $3.3\,M_\odot$ of oxygen have been burnt. The present computations clearly indicate that energy release from explosive oxygen burning is enough to disrupt the star. We note, however, that further evolution of the present model should include radiation transport in order to give a definite conclusion about its final fate of total disruption. These calculations are in progress, and will also enable us to calculate the light curve of such explosion.

The present results differ from the preceding ones (cf. El Eid and Langer, 1986) in two respects. Firstly, we have extended the nuclear reaction network including 39 nuclei instead of 13 nuclei previously. The net effect is an increase up to $\sim 25\%$ in the energy generation rate during oxygen burning, which favors a stronger explosion. Secondly, the correction of the opacities (see sect. 3.1) decline their values throughout the star which enhances the energy transport and explains why the peak temperature is higher than previously.

Comparing our results with those summarized by Woosley (1987), we find that the present model resembles his model for the $55\,M_\odot$ core rather than the $45\,M_\odot$ core. Though the reason for this discrepancy is unclear, we think that the treatment of explosive oxygen burning seems to be the major difference between the tow calculations.

4. Nucleosynthetic yield in comparison with Cas A

The nucleosynthesis which may result from the explosion described above is displayed in Fig. 4, where the composition profiles of several nuclear species (including their isotopes, see Fig. 1) are plotted against the mass. The typical oxygen burning products can be clearly identified in the central region of the star. We note that only a tiny amount of $\sim$ 4.8 10^{-6} $M_\odot$ of ^{56}Ni is synthesized in this model. The abundances in Fig. 4 have bee normalized to oxygen for the aim to compare them with those observed in the SNR Cas A. This comparison is motivated by the recent discovery of fast moving nitrogen-rich knots in Cas A by Fesen et al. (1987), and the arguments gathered in that work which stongly suggest a W-R star of type WNL as a progenitor for this object. These stars rank among the most massive W-R stars (Schild and Maeder, 1984; Langer, 1987). An exploding massive W-R star would also satisfy the remnant x-ray mass estimated to be larger than $15\,M_\odot$ (Fabian et al. 1980; Markret et al. 1983).

The spectral observations of the fast moving knots (FMK) in Cas A (e.g. Chevalier and Kirshner, 1979) suggest a comparison of the nucleosynthesis calculations with individual layers in the stellar model. These observations clearly indicate that oxygen burning had occurred in the gas, and show that the typical oxygen burning products (S, Ar, and Ca) are related to each other. However, the fact that no correlation between the abundances of the FMK and their velocities is observed is very difficult to accommodate in any theoretical model. Despite of these problems, it is useful to see what kind of predictions we can make on the basis of our probably simple model of shperically symmetric explosion without mixing during the expansion.

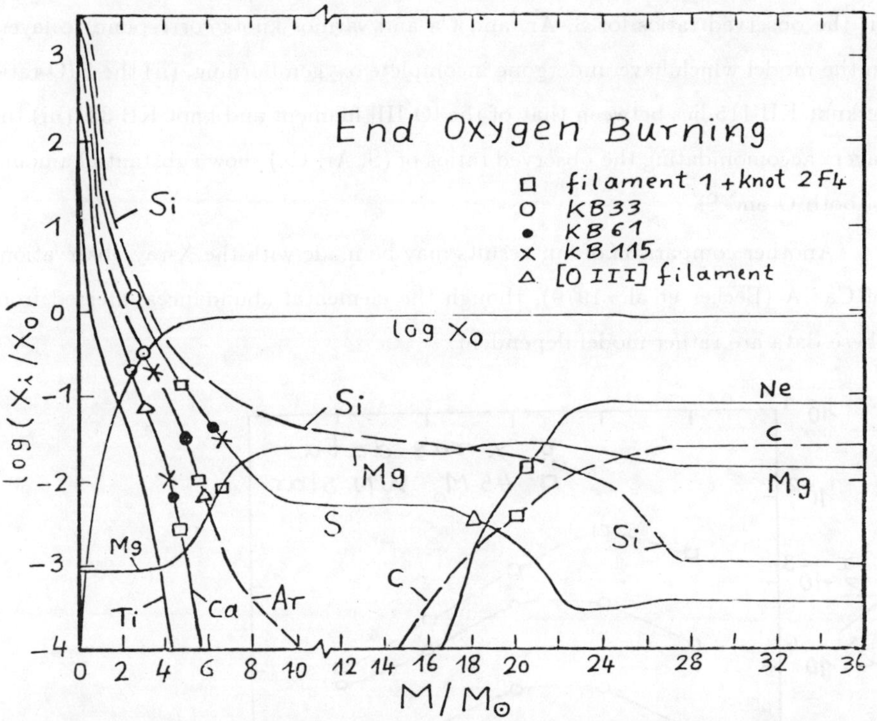

Fig. 4: Comparison of the observed abundance ratios relative to oxygen of the fast moving knots of Cas A with the calculated ones for the exploding 45 $M_\odot$ WN star (only 36 $M_\odot$ are shown). The observed values taken from Chevalier and Kirshner (1979) are plotted on the calculated profiles (see text). Note that the oxygen profile (log X_O) is shown in addition.

In Fig. 4, the abundances in of several FMK of Cas A as found by chevalier and Kirshner (1979) are superimposed on the calculated profiles. The upper limits for C and Ne are found at the same position in the outer layers where these elements are still abundant. The limit for Mg is encountered deeper in the star where the observed ratios for S, Ar, and Ca are found to be the natural products of explosive oxygen burning.

Several features of our results are compatible with the optical observations: (i) the observed ratios for S, Ar, and Ca and variuos knots correspond to layers in the model which have undergone incomplete oxygen burning, (ii) the S/O ratio of knot KB 115 lies between that of the [O III] filament and knot KB 33, (iii) the layers accomomdating the observed ratios of (S, Ar, Ca) show substantial amount of both O and Si.

Another comparison of our results may be made with the X-ray observations of Cas A (Becker et al. 1979), though the elemental abundances inferred from these data are rather model dependent.

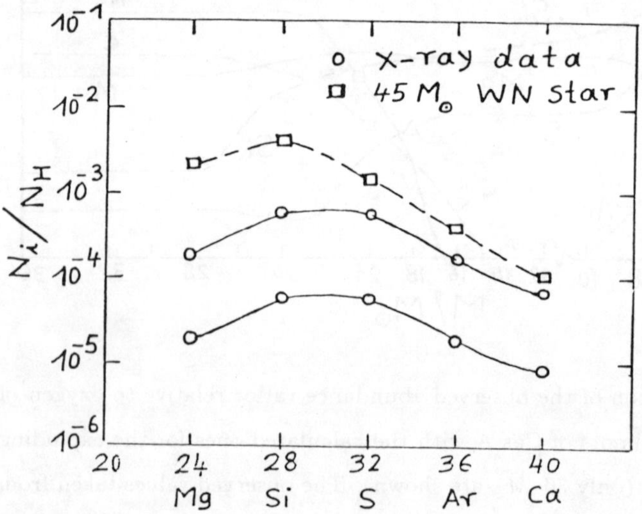

Fig. 5: Comparison of elemental abundances inferred from X-ray observations of Cas A (Becker et al. 1979) with the calculated ones. N_i/N_H are ratios by number relative to hydrogen (see text). The lines connecting the data are ment as a guide for the eyes, and have no physical meaning.

In Fig. 5 the ratios by number relative to hydrogen are shown for the indicated elements. The normalization in our case is based on 27.6 $M_\odot$ of hydrogen which has been ejected via stationary mass loss during the evolution of the 100 $M_\odot$ star to the W-R stage (cf. Langer and El Eid, 1986). This is of course subject to modification depending on the exact amount of mass loss. Taken into account the uncertainty (faktor of 10 is suggested by Becker et al. 1979) in the X-ray data, our calculated values shown in Fig. 5 are in reasonable overall agreement with the observed ones. Only Mg and Si appears to be overproduced in our model.

In summary, tough we cannot claim to have found a model for the complicated object Cas A, there are several features in the present model of an exploding WN star of 45 $M_\odot$ which are compatible with available observational data of this object. It appears that the explosion of massive W-R stars via the pair creation supernovae can be envisaged as an explanation for at least some of the oxygen-rich supernova remnants of type Cas A. It also seems worthwhile to to extend the present calculations in two directions. Firstly, radiation transport should be included in order to calculate the light curve resulting from such explosions. Secondly, to investigate the final evolution of less massive W-R stars which are expected to exhibit violent pulsational behavior after they become pair unstable (Woosley, 1987).

Acknowledgement: M.F. EL Eid is grateful to the Univ. of Göttingen for the guest professorship during the summer term 1987 and the winter term 1987/1988, and thanks the DFG for the travelling support to New Orleans. N. Prantzos thanks the CNRS for supporting his trip to New orleans. The numerical calculations have been done with the Univac 1100 of the GWDG at Göttingen and the CRAY X-MP/24 of the Konrad-Zuse Zentrum at Berlin.

References

Baker, N.H., and Kuhfuss, R. 1986, Preprint, MPI für Physik und Astrophysik, no **256**

Becker, R.H., Holt, S.S., Smith, B.W., White, N.E., Boldt, E.A., Mushotzky, R.F., and Serlemitsos, P.J. 1979, Astrophys. J. **234**, L73

Begelman, M.C., and Sarazin, C.L. 1986, Astrophys. J. **302**, L59

Bond, J.R., Arnett, W.D., and Carr, B.J. 1984, Astrophys. J. **280**, 825

Buchler, J.R., and Yueh, W.R. 1976, Astrophys. J. **210**, 440

Carr, B.J., Bond, J.R., and Arnett, W.D. 1984, Astrophys. J. **277**, 445

Caughlan, G.R., Fowler, W.A., Harris, M.J., and Zimmerman, B.A. 1985, Atomic Data Nucl. Data Tables **32**, 197

Chevalier, R.A., and Kirshner, R.P. 1979, Astrophys. J. **233**, 154

Cox, J.P., and Giuli, R.T. 1968, Principal of Stellar Evolution, Cordon and Breach

De Jager, C. 1984, Astron. Astrophys. **138**, 252

EL Eid, M.F., Fricke, K.J., and Ober, W. 1983, Astron. Astrophys. **119**, 54

El Eid, M.F., and Langer, N. 1986, Astron. Astrophys. **167**, 274

El Eid, M.F., 1986, in Advanced in Nuclear Astropysics, eds. E. Vangioni-Flam et al.(edition Frontieres), Paris, p 283

Fabian, A.C., Willingale, R., Pye, J.P., Murray, S.S., and Fabbiano, G. 1980, M.N.R.A.S. **193**, 175

Fesen, R.A., Becker, R.H., and Blair, W.P. 1987, Astrophys. J. **313**, 378

Fraley, G.S. 1968, Astrophys. Sp. Sci. **2**, 96

Humphreys, R.M., and Davidson, K. 1984, Science **23**, 243

Langer, N., El Eid, M.F. 1986, Astron. Astrophys. **167**, 265

Langer, N. 1987, Astron. Astrophys. (Letters) **171**, L1

Maeder, A., Meynet, G. 1987, Astron. Astrophys. **182**, 243

Munakata, H., Kohyama, Y., Itoh, N. 1985, Astrophys. J. **296**, 197

Ober, W., El Eid, M.F., and Fricke, K.J. 1983, Astron. Astrophys. **119**, 61

Prantzos, N. 1986, Ph.D. Thesis, Universite De Paris VII, France

Prantzos, N., Arnould, M., Arcoragi, J–P. 1987, Astrophys. J. **315**,209

Prantzos, N., Doom, C., Arnould, M., DE Loore, C. 1986, Astrophys. J. **304**, 695

Roxburgh, I.W. 1978, Astron. Astrophys. **65**, 281

Sampson, D.H. 1959, Astrophys. J. **129**, 734

Thielemann, F.-K., Arnould, M., and Truran, J.W. 1986, in Advances in Nuclear Astrophysics, eds. E. Vongioni-Flam et al. (edition Frontieres), Paris, P 525

Woosley, S.E., and Weaver, T.A. 1982, in Supernovae: A Survey of Current Research, eds. M.J. Rees, and R.J. Stroneham (D. Reidel), p 79

Woosley, S.E., and Weaver, T.A. 1986 IAU Coll. **89**, eds. D. Mihalas, and K.H. Winkler

Woosley, S.E. 1986, in Nucleosynthesis and Chemical Evolution, eds. Hauck et al. (Observatoire de Geneve), p 1

CHAPTER V

THE s–PROCESS
THE r–PROCESS
THE p–PROCESS

"Nature never deceives us; it is always we who deceive ourselves."

Jean Jaques Rousseau
Emile, Ou De L'Education
Book III

s-PROCESS AND SOLAR ABUNDANCES

H. Beer
Kernforschungszentrum Karlsruhe
Institut für Kernphysik III
P.O.B. 3640, D-7500 Karlsruhe 1
Federal Republic of Germany

ABSTRACT

The origin and distribution of the heavy elements (A>56) are attributed to a large extent to s-process nucleosynthesis. Structures from the s-process are clearly visible in the distribution of solar abundances. It is shown that the s-process abundances are a superposition of different s-process components which are distinguished both by the character and strength of the neutron exposure. An exponential exposure distribution with pulsed irradiations and single flux exposures for the beginning and termination of the s-process path yielded the best reproduction of the s-only isotopes. An important prerequisite for the analysis in these terms is a reliable input data basis, especially of the neutron capture rates of all the nuclei on the synthesis path. Special cases are discussed in connection with their interrelation with the calculations. Solar abundances of heavy elements are classified and decomposed into its s- and r-process contributions. s-process branchings are examined because of their crucial role in elucidating the nature of the irradiations.

1. INTRODUCTION

Since the establishment and verification of a uniform distribution of the elements (latest compilation [1]) over solar and possibly cosmic dimensions it was always a challenge to understand and reproduce more and more of the details of its structures. The original concept already presented by Burbidge et al. [2] is up to this day still valid fundamentally. Progress is, therefore, mainly due to the tremendous improvement in the input data basis, forcing us to a continuously increasing refinement of the original picture.

Transmutations of the elements are caused by nuclear reactions and radioactive decay. Therefore, the nuclear properties of the elements are the key of their understanding. In this work we are concerned with elements heavier than ^{56}Fe brought into existence mainly by successive neutron additions and radioactive decay of unstable products. Essentially it is sufficient to subdivide these neutron capture processes into two fundamental classes, a neutron capture process on a rapid time scale (r-process) and a slow neutron capture process (s-process) (To explain rare proton rich isotopes a neutron independent process called p-process is needed, too). The progress in the field of the s-process and the implications on the solar abundances which are primarily a mixture of s- and r-process are the main topic of this article. We will present the s-process concept used, the data basis and then the special impact the application of the s-process model has on the solar abundances. Model and empirical data excert a mutual influence and most of the progress is due to nuclear input data for the s-process calculations.

2. CHARACTERISTICS OF THE s-PROCESS

The s-process nucleosynthesis in its mathematical formulation is represented by a system of linear differential equations [3,4]. A small percentage of the abundances of the iron peak nuclei from explosive oxygen and silicon burning [5] acts as seed material for the heavy elements up to ^{209}Bi. The abundance change dN(A)/dt of an isotope A on the synthesis path is given by the equation

$$dN(A)/dt = \lambda(A-1)N(A-1) - \lambda(A)N(A) \quad (1)$$

Modifications of this equation occur at a radioactive nucleus A' where the neutron capture rate λ(A') is in competition with the β-decay rate $\lambda_-(A')$.

$$dN(A')/dt = \lambda(A'-1)N(A'-1) - [\lambda_-(A') + \lambda(A')]N(A') \quad (2)$$

and the stable isobar A" of A' obtained by β-decay from A' is given by

$$dN(A'')/dt = \lambda_-(A')N(A') - \lambda(A'')N(A'') \quad (3)$$

The occurring neutron capture rate, λ, is defined in the usual way as the product of neutron density, n, times the capture cross section, σ, averaged over a Maxwell-Boltzmann distribution of velocities v:

$$\lambda = n\langle\sigma v\rangle = nv_T\langle\sigma v\rangle/v_T \quad (4)$$

$\langle\sigma v\rangle/v_T$: Maxwellian averaged capture cross section

This Maxwellian averaged capture cross section, in the following simply designated as σ, was defined by introducing the thermal velocity v_T. The equations (2) and (3) describe a branching of the synthesis path and it is practice to introduce a branching factor f defined as

$$f = \lambda_{-}(T,n_e)/[\lambda_{-}(T,n_e) + \lambda(n)] \quad (5)$$

f is always a function of the neutron density n via λ and can be additionally dependent on temperature, T, (and electron density, n_e) if the half-life of the radioisotope is sensitive on temperature (and electron density).

In general, the solutions of the equations of type (1), (2) and (3) are functions of the time dependent neutron density, temperature and electron density of the s-process environment. Therfore, the search for a solution of the set of differential equations of type (1), (2) or (3) which describes the empirical abundance data satisfactorily yields then the special conditions under which the s-process nucleosynthesis may have occurred.

3. THE EMPIRICAL DATA BASIS

Basic requirement for an s-process analysis is the precise knowledge of the capture and beta decay rates λ, λ_{-}, respectively, of all the isotopes on the synthesis path from ^{56}Fe to ^{209}Bi. The essential part of the capture rates are the Maxwellian averaged capture cross sections. The beta decay rates of those species are needed which give rise to a branching of the s-process synthesis path.

It is most favourable that the s-process takes place in the stability valley of nuclides. Therfore, the required nuclear properties of the isotopes can be either measured directly or derived with confidence from the known properties of the neighbouring isotopes. The compilation of the capture cross sections [6] shows that most of the experimental contributions come from ORELA at Oak Ridge and the Van de Graaff accelerator at Karlsruhe. The relevant beta decay rates as functions of temperature and electron density have been studied in great detail [7].

The s-process analysis is greatly simplified through the existence of s-only isotopes which are shielded from r-process contributions by stable neutron richer isobars. For these nuclei the empirical distribution of Maxwellian averaged capture cross section times abundance $\sigma N(A)$ can be specified because the solar abundance is identical with the s-process abundance. This distribution is supplemented by s-process dominant isotopes where an r-process correction can be carried out sufficiently accurate. Minor additional corrections of the s-only nuclei are due to a

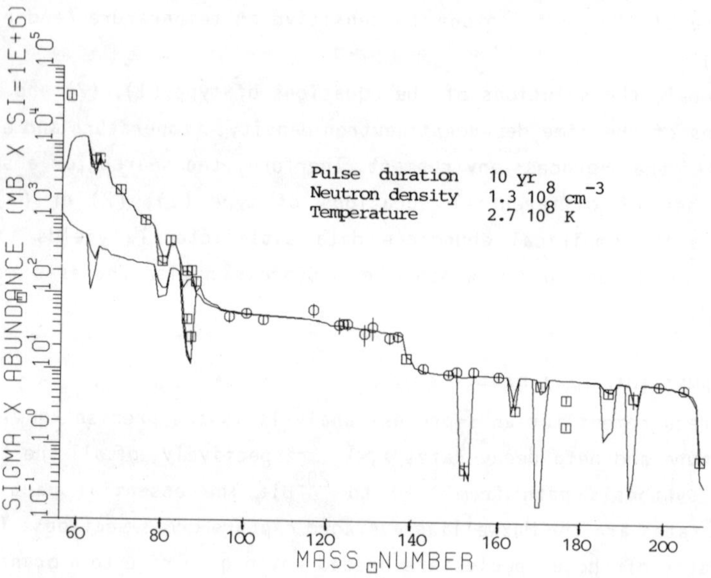

Fig.1 The product capture cross section and s-process abundance as a function of mass number. The symbols are empirical data. Significant branchings were idendified due to low empirical values on one of the branches. Note that the ^{86}Kr and ^{87}Rb data located in the ^{85}Kr branching are not corrected for r-process contributions. Therefore, the theoretical calculation should remain below the empirical values to avoid overproduction. Below A=90 and above A=205 the weak and the strong components are superimposed as s-process with single flux exposures. The course of the main component in these regions is plotted, too. No fit of ^{176}Hf and ^{176}Lu which require special treatment as cosmic clock or stellar thermometer is shown.

p-process contribution estimated from neighboring p-only isotopes and neutron capture on thermally populated excited states leading to an effective capture cross section different from ground state capture.

The empirical $\sigma N(A)$ data points are plotted in Fig.1. They form a smooth over large parts rather flat distribution with steps at the magic neutron numbers 50, 82, and 126. Other irregularities can be interpreted as s-only or preferentially s-process isotopes located in a branching of the synthesis path. Therefore, they are partially bypassed in the nucleosynthesis and lie distinctly below the distribution of the other data points. This occurs for ^{80}Kr, ^{86}Kr, ^{87}Rb, ^{152}Gd, ^{164}Er, ^{170}Yb, ^{176}Hf, ^{176}Lu, ^{186}Os, and ^{192}Pt. Not shown are ^{113}In, ^{115}Sn and $^{180}Ta^m$ which also might be partially s-process isotopes and ^{187}Os which contains a large radiogenic component from the ^{187}Re decay.

4. s-PROCESS CALCULATIONS

4.1. s-Process With Exponential Exposure Distribution

Already the early calculations showed that the gross feature of empirical s-process abundances (Fig.1) cannot be reproduced by single neutron exposures τ [3] ($\tau=\int\Phi(t)dt$, Φ Neutron flux), but a distribution of different exposures is required over which the single exposure solutions have to be integrated. In the search for the best exposure distribution $\rho(\tau)$ an exponetial distribution $\rho(\tau)\sim\exp(-\tau/\tau_o)$ [8] turned out to be superior to other suggestions. This kind of s-process is then characterized by two free parameters. The average exposure τ_o and the required iron seed abundance N(56) given by the normalization condition $N(56)=\int\rho(\tau)d\tau$.

This model of an s-process with one exponential exposure distribution was extended to a distribution consisting of a sum of three exponentials with different values for τ_o and the iron seed abundances to reproduce besides the abundances of the bulk of the elements between A=90 and 200 also the abundances at the beginning of the s-process path and at the termination [9,10]. These additional components are called weak and strong, respectively, because the average exposure is much smaller and larger than that of the main component.

4.2 Single Flux s-Process

The weak and strong s-process components can alternatively be formed as a single exposure s-process. The solutions of the differential equations as a function of τ [3,4] are taken without integrating over a neutron exposure distribution. This single exposure s-process is characterized again by two free parameters, the iron seed abundance and the exposure τ. Single exposure components for the weak and the strong components supplementing the main component can meet the constraints imposed by the empirical data much more flexible than the application of an s-process constructed with the sum of three exposure distributions.

4.3. Pulsed s-Process

The idea that s-process nucleosynthesis may have occurred as a series of separate neutron bursts instead of a continuous irradiation is very attractive. This pulsed s-process can provide a natural explanation for the required exponential exposure distribution of the main component through the introduction of a mixing mechanism effective in the interpulse periods replacing part of the exposed seed by fresh seed material [12,13]. Pulse duration and average exposure τ_o are then interrelated with the fraction of material, r, remaining after each pulse for further exposure [$r=\exp(-\Delta\tau/\tau_o)$, $\Delta\tau$ exposure per pulse].

If we model the main s-process nucleosynthesis as a pulsed process changes are supposed to occur at branchings [14]. The radioactive isotope which gives rise to a branching is decaying during the neutron burst free interval and has to be rebuilt in the following burst. A criterion for the sensitivity of a branching to the pulse conditions is the ratio of pulse duration Δt and the total lifetime $\tau_{tot}= [\lambda+\lambda\_]^{-1}$ of the branch point isotope [11]. If this ratio is exceedingly larger than unity the succession of neutron bursts works as one long continuous neutron irradiation. For sufficiently long pulses the pulsed s-process approaches this continuous s-process asymptotically.

4.4 The σN(A)-Curve

The fit of the empirical σN(A) values is a two step calculation. The main component is fitted in the mass region A=90-200 and then the computer calculation is extended below A=90 and above A=205 using the derived astrophysical parameters (Table 1). This main component fit forms the basis for an adjustment of the weak component below A=90 and the strong component above A=205. Weak and strong component are superimposed on the main component and have to exert a significant influence only over the mass ranges A=56-90 and A>205, respectively, to avoid overproductions of the isotopes already accounted for by the main s-process. Single exposure components for weak and strong component are well adapted to this role. The weak component avoids overproductions of ^{58}Fe and the Zr isotopes and allows for a perfect fit of the s-only nuclei ^{70}Ge and ^{76}Se. The strong component accounts for a significant part of the solar ^{208}Pb abundance otherwise left unexplained. The adjusted parameters for the fraction of solar iron abundance and exposure are given in Table 1 for the three components. Parameters deduced from exponential exposure fits for the weak and strong components are indicated for comparison.

Table 1. Parameters for the various s-process components

s-process component	Fraction of iron seed[%]		Exposure[mb^{-1}]	
	SF	EE	SF	EE
Main	-	0.048	-	τ_o=0.3
Weak	0.32	1.6	τ=0.2	τ_o=0.06
Strong	0.9 10^{-4}	1.2 10^{-4}	$\tau \geq 2.5$	$\tau_o \geq 6$

SF single flux s-process

EE s-process with exponential exposure distribution

Neutron density and temperature are adjusted via the s-only isotopes in the branchings at ^{151}Sm, ^{163}Dy, ^{170}Tm, ^{185}W, and ^{192}Ir. Neutron density and temperature of the weak component can be determined from the ^{79}Se and ^{85}Kr branchings after the contribution of the main component has been taken into account. First results showed that main and weak component require significantly different values for neutron density and temperature [15]. But after the temperature dependence of the ^{79}Se half life could be based on a measured log(ft)-value for the crucial excited state at 96 keV these parameters turned out to be nearly the same for both components [16,17]. For the strong component there is no branching to measure these quantities.

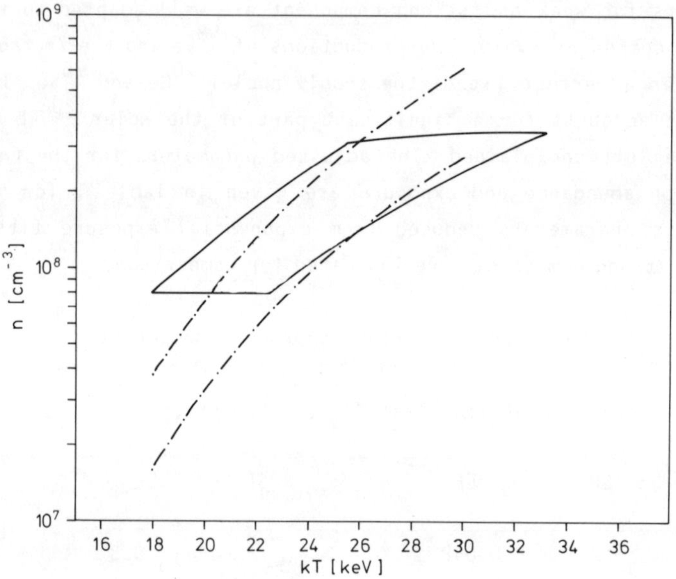

Fig.2 The s-process neutron density n and temperature kT derived from the various branchings. The solid line circumscribes the allowed range of values calculated from the ^{151}Sm, ^{170}Tm and ^{185}W branchings. The dashed dotted lines designate the range of temperatures and neutron densities from ^{176}Lu treated as a stellar thermometer.

The predictions for s-process neutron density and temperature of the main component derived from branchings are further complicated as their dependence on the pulse conditions (pulse width, interpulse interval) has to be considered, too, although the additional parameters may allow for a better adjustment to the empirical data. The investigations showed that a solution is required which is for the majority of branchings close to the asymptotic solution. These branchings are, therefore, insensitive to the pulse conditions. The valid range of neutron density n and temperature kT [18] is displayed in Fig.2, and pulse widths between 3 to 50 yr were estimated.

An overall s-process calculation from ^{56}Fe to ^{209}Bi, where the main component is modeled as a pulsed s-process and the weak and the strong components as single flux components is shown in Fig.1. Neutron density, temperature and electron density are adjusted to $n=1.3\ 10^{8}\ cm^{-3}$, $T=2.7\ 10^{8}K$ and $n_e=2.0\ 10^{27}\ cm^{-3}$, respectively, for all components. The pulses of the main s-process are assumed rectangular with a width of 10 yr. The calculated curve is adequate to the quality of the empirical data, therefore, an extension to time dependent n-, T- and n_e-values was not considered. An attempt [19] to apply the thermal pulses of current stellar models [13] did not result in a good fit of the solar σN(A) curve.

5. SPECIAL ISSUES OF THE s-PROCESS ANALYSIS

5.1. The Solar Abundances of the Nobel Gases and Mercury

For the determination of the elemental solar abundances C1 type carbonaceous meteorites represent the primary source of information [1]. This is true for most of the heavy elements. Exceptions are the nobel gases and the highly volatile mercury which shows a large variability in meteorite samples. Also a determination of these abundances from the optical spectra of the sun fails. In this situation reliable elemental abundances can be deduced from a σN(A)-calculation via the s-process isotopes ^{82}Kr, $^{128,130}Xe$ and ^{198}Hg [15,20,21,22] as for these cases the s-process abundance is identical with the solar abundance. In general, the reliability of measured solar abundances from meteorites can be checked using s-process systematics. There are a number of cases, where the inconsistencies between σN(A) systematics and the measurements lead to a revision of the empirical data [9,23].

5.2. Classification of Elements and Isotopes

From the calculated σN(A)-curve the s-process abundances can be deduced by dividing by their capture cross sections. These data subtracted from the solar abundances leads to the distribution of the r-process residuals (Fig.3) which are an important prerequisite for the theoretical r-process calculations. From the isotopic s- and r-process distributions the respective elemental distributions can be formed and compared to the abundances of certain stars which still exhibit the traces of past nucleosynthetic events [24].

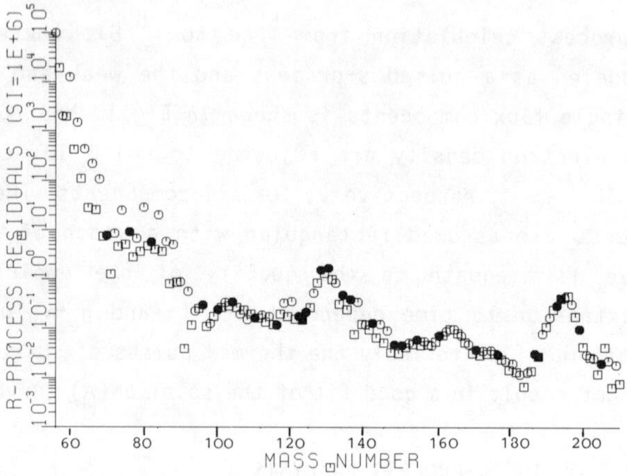

Fig.3 The r-process residuals obtained by subtracting the s-process abundances from the solar abundances are plotted. Full black points are r-only isotopes. No more oversubtractions in the critical mass range 80-100 were found.

At the termination of the s-process path the analysis of the s-process in terms of main and strong component leads to a consistent description of all the various lead components, including radiogenic lead from the decays of ^{232}Th and ^{235,238}U and r-process contributions from the decays of the short lived transbismuth elements [22].

The close inspection of the excited state properties of the isotopes ^{113}Cd, ^{115}In, and ^{180}Hf revealed that there are possibilities (isomeric states with significant beta decay branches) to synthesize the rare isotopes ^{113}In, ^{115}Sn, and ^{180}Ta^{m} via s- and r-process [25,26,27,28]. These nuclei were before a kind of mystery because they could not be classified easily to one of the traditional processes.

Another field where s-process nucleosynthesis, especially the weak component, helps classifying are rare neutron rich isotopes of elements between neon and iron. These nuclei are produced significantly by neutron capture on highly abundant progenitor isotopes which act as seed material.

A dominant s-process origin was found especially for ^{36}S, ^{40}Ar and ^{40}K [29].

5.3. Chronology

Solar abundances are the ashes of the nuclear burning from stars passed away long ago. But the time history is conserved through long lived radioactive isotopes the abundances of which continually decrease like the sand in the sand-glass. Therfore, if one succeeds in an assessment of the abundance created in the nucleosynthesis the time elapsed since then can be calculated and this age of the synthesis is a measure of the age of the Galaxy. These so called cosmic clocks are represented by either isotopes classified as r- or s-process nuclei. For both types s-process nucleosynthesis offers ways to determine the original abundance.

5.3.1 ^{187}Re with a half life of 5 10^{10} yr is an r-process chronometer. The amount of radiogenetic decay to ^{187}Os, an s-process isotope, is determined by subtracting the s-process contribution from the solar abundance $N_\odot$ [30,31,32].

$$N_{rad}(^{187}Re) = N_\odot(^{187}Os) - N(^{187}Os) \qquad (6)$$

$N(^{187}Os)$ is obtained via s-only ^{186}Os and the local approximation.

$$\sigma(^{186}Os)N(^{186}Os) \simeq \sigma(^{187}Os)N(^{187}Os) \qquad (7)$$

Unfortunately this relatively simple method is hampered by a number of significant corrections, e.g. the excited state capture of ^{187}Os [33], the temperature sensitivity of the ^{187}Re half life [34], the s-process branching at ^{185}W [35].

Other r-process chronometers which may be treated in a similar way are ^{87}Rb [36] and ^{235}U [22].

5.3.2 ^{176}Lu (3.6 10^{10} yr) is an s-process isotope shielded from r-process contributions by ^{176}Yb. But the direct calculation of the original s-process abundance is complicated because of an isomeric state at 127 keV through which a significant part of the s-process flow is immediately directed to ^{176}Hf without reaching the long lived ground state. The percentage f of this flow can be measured by means of the activation technique [37,38,18]. On the other hand Beer [39] showed that the branching factor f as well as the age can be determined independently from an activation measurement via the s-process properties itself.

$$(1-f)/f = [\sigma N^*(^{176}Lu)]/[\sigma N^*(^{176}Hf)] \quad (8)$$

$$N^*(^{176}Lu) + N^*(^{176}Hf) = N_\odot(^{176}Lu) + N_\odot(^{176}Hf) \quad (9)$$

$$\sigma N^*(^{176}Lu) + \sigma N^*(^{176}Hf) \simeq \sigma N_\odot(^{160}Dy) \quad (10)$$

The asterisk indicates the original abundances at the time of the s-process.

In the analysis it turned out that f from the s-process calculations (0.5±0.1) is smaller than the directly measured activation value (0.64±0.04). In addition the age is larger than 25 Gyr [18], therefore, hardly compatible with astronomical determinations of the age of the univers. This discrepancy suggests that after termination of the s-process the ^{176}Lu half life was temporary thermally enhanced [38,18]. The difference in the determinations of the branching factor f indicates that also during the s-process thermal effects are present [18] which have changed the populations of ground and isomeric states. It is also likely that these thermal effects have even caused thermal equilibrium between the ground and isomeric states, because the calculated temperature and neutron densities under this assumption are in good agreement with the respective values derived from other branchings (Fig. 2) [18]. In order to solve this question of thermal effects on the ^{176}Lu ground and isomeric states through internal electromagnetic transitions the level scheme of ^{176}Lu which was incomplete is presently studied [40].

Other s-process chronometers which were investigated are ^{40}K [29] and ^{205}Pb [41].

5.4 Special s-Process Branchings

5.4.1 <u>^{163}Dy.</u> Under stellar s-process conditions ^{163}Dy, a stable isotope on the synthesis path, becomes radioactive so that competition between beta decay and neutron capture allows for an s-process synthesis of ^{164}Er via neutron capture of ^{163}Ho and subsequent beta decay of ^{164}Ho to ^{164}Er. The surprising fact that ^{163}Dy in stellar interiors might be radioactive is due to the small Q value between ^{163}Ho and ^{163}Dy of only 2.3 keV which makes possible bound state beta decay [34] and facilitates excited state beta decay. Besides temperature there is also a strong dependency on electron density of the ^{163}Dy, ^{163}Ho and ^{164}Ho half lives [7]. Therefore, the reproduction of the s-process abundance of ^{164}Er (solar minus p-process abundance) allows for a determination of the electron density. For $n=(0.8\text{-}1.8)\ 10^8\ cm^{-3}$ and kT=18-28 keV, n_e was found to be $(0.8\text{-}4)\ 10^{27} cm^{-3}$ [42].

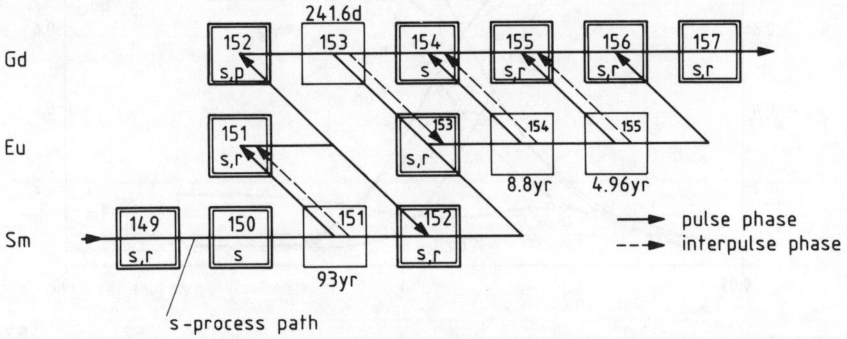

Fig.4 The synthesis of Sm, Eu, and Gd isotopes during the pulse (solid line) and in the interpulse phase (dashed lines), where the ^{151}Sm, ^{153}Gd, and ^{154}Eu isotopes which are built up disappear each time. As ^{150}Sm, ^{152,154}Gd are purely s-process in origin the real course of the synthesis for solar system s-process material can be checked. The designated terrestrial half lives can change drastically at the s-process site.

5.4.2 ^{151}Sm. A lower bound for the burst width of a pulsed s-process can be derived from the ^{151}Sm branching (Fig.4). Neutron capture and beta decay rate of this isotope are so that already a very small [~6%] percentage of the s-process flow is sufficient to reproduce the abundance of ^{152}Gd [43]. This sets a lower limit for the allowed pulse width of the neutron bursts. The analysis shows that for shorter and shorter pulse durations the amount of ^{152}Gd produced is strongly increasing if the pulse width becomes comparable to and smaller than the total half life of ^{151}Sm. A decrease in pulse width compensated by a corresponding increase in neutron density or a decrease in temperature leads to conflicts with other branchings. Therfore, to avoid overproduction of ^{152}Gd the neutron bursts should not be smaller than 3 yr (Fig.5) [11,44].

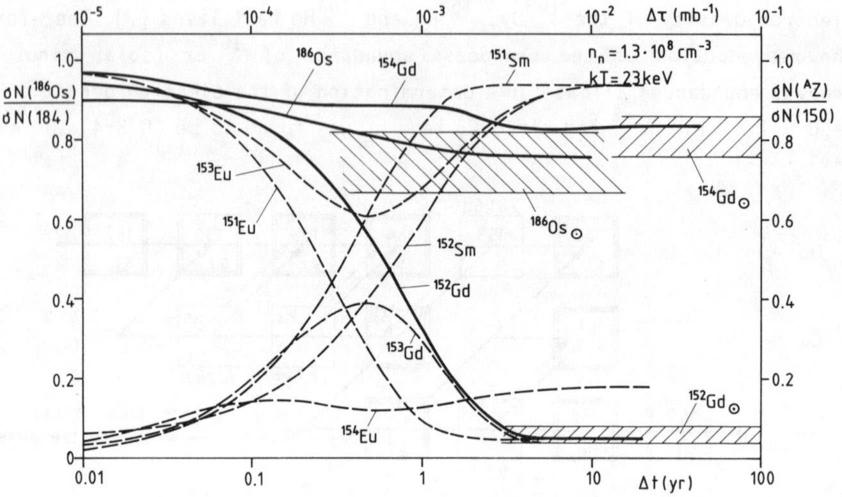

Fig.5 The isotopes located in the branchings at ^{151}Sm, ^{152}Eu , ^{153}Gd, and ^{154}Eu. The σN values normalized to ^{150}Sm are shown as a function of the pulse width and the exposure per pulse, respectively. A calculation of the ^{186}Os isotope located in the ^{185}W branching is also shown. The isotopes for which empirical values indicated by hatched areas can be given, are drawn as solid lines. Others are plotted as dashed lines. The branchings were analyzed with a neutron density of n_n=1.3 10^8 cm^{-3} and a temperature kT=23 keV. The pulse width is constrained by ^{152}Gd to values larger than 3 yr.

5.4.3 ^{85}Kr. The ^{85}Kr branching can possibly set an upper limit for the pulse width. Although a significant part of the s-process flow is directed immediately to ^{85}Rb and ^{86,87}Sr via an isomeric state [45,46,47,48] the percentage of the flow to ^{86}Kr and ^{87}Rb classifies these isotopes as almost exclusively s-process nuclei. This is, however, in contradiction to the fact that the abundances of ^{86}Kr and ^{87}Rb must contain also sizeable r-process contributions. Fig.6 demonstrates that a pulse width less than about 50 yr greatly helps avoid these complications. The pulses effect the branching to ^{85}Rb from the ^{85}Kr ground state to be strenghtened.

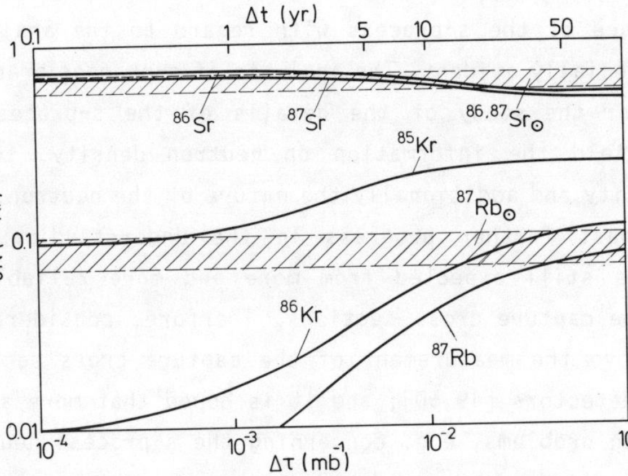

Fig.6 The analysis of the isotopes located in the ^{85}Kr branching. The weak component contribution and the ^{85}Kr isomeric population by neutron capture are taken into account. Neutron density and temperature for both components are the same as for the ^{151}Sm analysis. The shaded areas are empirical values for ^{86,87}Sr and ^{87}Rb, respectively. Above Δt=40yr the whole solar ^{87}Rb abundance would be reproduced by the s-process. The best agreement of the empirical ^{86,87}Sr abundances with the respective theoretical curves lies between Δt=3-20yr.

6. CONCLUSIONS

Solar s-process abundances can be represented as a mixture of three components. At the beginning and possibly at the termination of the s-process path single flux components have to be superimposed to a pulsed s-process (Δt≃3-50 yr) which describes the bulk of nuclei between A≈90 and 200.

This concept is very successful in predicting particular features of the solar abundances. It allows for a reliable estimate of solar abundances of the nobel gases and mercury, it yield a quantitative classification of all the heavy isotopes with respect to s- and r-process origin, inclusive the rare species ^{113}In, ^{115}Sn and ^{180}Ta^{m} and some rare light isotopes between neon and iron.

The importance of the s-process with regard to the analysis of cosmic clocks is especially evident. The analysis of s-process branchings opens a rich field for the study of the details of the s-process itself. The branchings yield the information on neutron density, temperature and electron density and additionally the nature of the neutron irradiations.

As in the past further progress in the understanding of the solar abundances is still expected from more and more reliable input data, especially the capture cross sections. Therfore, considerable effort is made to improve the measurement of the capture cross sections with new methods and detectors [49,50], and it is hoped that more and more of the remaining open problems, e.g. concerning the s-process neutron source or the types of stars which produced the solar s-process abundances are solved.

REFERENCES

1. Anders, E., Ebihara, M., Geochim. et Cosmochim. Acta 46,2363(1982)
2. Burbidge, E.M., Burbidge, G.R., Fowler, W.A., Hoyle, F., Rev. Mod. Phys. 29,547(1957)
3. Clayton, D.D., Fowler, W.A., Hull, T.E., Zimmerman, B.A., Ann. of Physics 12,331(1961)
4. Ward, R.A., Newman, M.J., Clayton, D.D., Ap. J. Suppl. 31,33(1976)
5. Woosley, S.A., Arnett, W.D., Clayton, D.D., Ap. J. Suppl. 26,231(1973)
6. Bao, Z.Y., Käppeler, F., Atomic and Nucl. Data Tables 36,411(1987)
7. Takahashi, K., Yokoi, K., Atomic and Nucl. Data Tables 36,375(1987)
8. Seeger, P.A., Fowler, W.A., Clayton, D.D., Ap. J. Suppl. 97,121(1965)
9. Käppeler, F., Beer, H., Wisshak, K., Clayton, D.D., Ward, R.A., Ap. J. 257,821(1982)
10. Beer, H., 5th Moriond Astrophysics Meeting, Nucleosynthesis and its Implications on Nuclear and Particle Physics, Les Arcs, 1985, eds. J. Audouze, N. Mathieu (D. Reidel Pub. Company 1986)p. 263
11. Beer, H., 2nd IAP Rencontre on Nucl. Astrophysics, Paris, 1986, in Advances in Nucl. Astrophysics, eds. E. Vangioni-Flam, J. Audouze, M. Casse, J.P. Chieze, J. Tranthanhvan (EDITIONS FRONTIERES)p. 375
12. Ulrich, R.K., Explosive Nucleosynthesis, eds. D.N. Schramm, W.D. Arnett (Austin University of Texas Press 1973)p. 139
13. Iben, I.Jr., Ap. J. 196,525(1975)
14. Ward,R.A., Newman, M.J., Ap. J. 219,195(1978)
15. Walter, G., Beer, H., Käppeler, F., Reffo, G., Fabbri, F., Astron. & Astrophys. 167,186(1986)
16. Klay, N., Käppeler, F., Int. Conf. on Weak and Electromagnetic Interactions in Nuclei (Spinger-Verlag Berlin, Heidelberg 1986) p. 977
17. Käppeler, F., MPI Workshop on Nuclear Astrophys., Ringberg 1987
18. Beer, H., Walter, G., Macklin, R.L., Patchett, P.J., Phys. Rev.

C30,464(1984)
19. Howard, W.M., Mathews, G.J., Takahashi, K., Ward, R.A., Ap. J. 309,633(1986)
20. Beer, H., Käppeler, F., Reffo, G., Venturini, G., Astrophys. Space Sci. 97,95(1983)
21. Walter, G., Beer, H., Astron. & Astrophys. 123,279(1983)
22. Beer, H., Macklin, R.L., Phys. Rev. C32,738(1985)
23. Beer, H., Käppeler, F., Wisshak, K., Astron. & Astrophys. 105,270(1982)
24. Beer, H., Walter, G., Proc. Colloquium on Cold Stars with Excess of Heavy Elements, Astrophs. & Space Sci. Lib. 114 (D. Reidel 1985)p. 373
25. Ward, R.A., Beer, H., Astron. & Astrophys. 103,189(1981)
26. Beer, H., Ward, R.A., Nature 291,308(1981)
27. Beer, H., Macklin, R.L., Phys. Rev. C26,1404(1982)
28. Yokoi, K., Takahashi, K., Nature 305,198(1983)
29. Beer, H., Penzhorn R.-D., Astron. & Astrophys. 174,323(1987)
30. Clayton, D.D., Ap. J. 139,637(1964)
31. Winters, R.R., Macklin, R.L., Phys. Rev. C25,208(1982)
32. Browne, C., Berman, B.L., Phys. Rev. C23,1434(1981)
33. Hershberger, R.L., Macklin, R.L., Balakrishnan, M., Hill, N.W., McEllistrem, M.T., Phys. Rev. C28,2249(1983)
34. Takahashi, K., Yokoi, K., Nucl. Phys. A404,578(1983)
35. Arnould, M., Takahashi, K., Yokoi, K., Astron. & Astrophys. 137,51(1984)
36. Beer, H., Walter, G., Astrophys. Space Sci. 100,243(1984)
37. Beer, H., Käppeler, F., Phys. Rev. C21,534(1980)
38. Beer, H., Käppeler, F., Wisshak, K., Ward, R.A., Ap. J. Suppl. 46,295(1981)
39. Beer, H., Ap. J. 262,739(1982)
40. Klay, N., Beer, H., Börner, H., Käppeler, F., Schatz, G., Krusche, B., Spring Meeting, Groningen 1987 (Verhandlungen DPG)p. D-3.5
41. Yokoi, K., Takahashi, K., Arnould, M., Astron. & Astrophys. 145,339(1985)
42. Beer, H., Walter, G., Macklin, R.L., 5th Int. Symp. on Capt. γ-Ray Spectr. and Related Topics (AIP Conf. Proc. No.125, NY 1985) p. 778
43. Beer, H., Käppeler, F., Yokoi, K., Takahashi, K., Ap. J. 278,388(1984)
44. Beer, H., Macklin, R.L., Ap. J. (to be published)
45. Beer, H., Käppeler, F., 4th Int. Symp. on Neutron Capt. γ-Ray Spectr. and Related Topics (The Inst. of Physics, Bristol, London 1982, Conf. Series No.62)p. 588
46. Walter, G., Beer, H., Käppeler, F., Penzhorn, R.-D., Astron. & Astrophys. 155,247(1986)
47. Penzhorn, R.D., Walter, G., Beer, H., Z. Naturforsch. 38a,712(1983)
48. Käppeler, F., Naqvi, A.A., Al-Ohali, M., Phys. Rev. C35,936(1987)
49. Ratynski, W., Käppeler, F., Contribution to the 6th Int. Symp. on Capture γ-Ray Spectr. and Related Topic, Leuven 1987
50. Wisshak, K., Käppeler, F., Müller, H., Nucl. Instr. & Methods A251,101(1986)

Observations of s-Process Indicators in Barium Stars*

Robert. A. Malaney
W.K. Kellogg Radiation Laboratory
California Institute of Technology
Pasadena, CA91125
U.S.A.

David L. Lambert
Department of Astronomy
University of Texas, Austin,
Texas 78712
U.S.A.

ABSTRACT

The relative abundances of the magnesium isotopes and the abundance of atomic rubidium observed in a stellar atmosphere are known to be indicators of the *s*-process neutron source and neutron density, respectively. This paper reports observations of these key *s*-process indicators in the cool Ba II star, HD178717, using molecular MgH lines at 5101Å and 5107Å, and the Rb I line at 7800Å. It is determined that the $^{22}Ne(\alpha, n)^{25}Mg$ source, believed to operate in intermediate-mass AGB stars, is unlikely to be responsible for the *s*-process in this star. A low neutron density of $\sim 2 \times 10^7 \text{cm}^{-3}$ at the *s*-process site is determined from the rubidium abundance. The consequences of these observations on our understanding of the origin of the *s*-process enhancements in Ba II stars are discussed.

*Based on observations obtained at the European Southern Observatory, La Silla.

I. INTRODUCTION

The origin of the s-process enhancements observed in the classical Barium (Ba II) stars[1-3] remains one of the most fundamental challenges in stellar nucleosynthesis theory. An understanding of this phenomenon would lead to an improvement in our knowledge of both s-process systematics and mixing processes occuring during the late phases of stellar evolution.[4] A determination of two crucial aspects of the s-process site, namely the neutron source and the neutron density, would be a significant advance towards this goal. Knowledge of these two parameters would allow strong constraints to be placed on any evolutionary hypothesis purporting to explain the Ba II star phenomenon. Hitherto, the neutron source has been analysed in only two Ba II stars,[5-6] and the neutron density also in only two.[5-7] In order to extend such studies to other members of this important stellar class, spectroscopic observations of a large number of both northern and southern hemisphere Ba II stars were obtained. In this paper, which reports the first results of this study, we discuss our determination of the neutron source and neutron density in the cool Ba II star HD178717 (spectral type K4[8]). Many abundance characteristics of this star have been previously determined,[9-10] and it is known to possess s-process enhancements of ~ 1 dex.

The two most likely neutron producing reactions in a stellar interior are the $^{13}C(\alpha,n)^{16}O$ and the $^{22}Ne(\alpha,n)^{25}Mg$ reactions.[11-12] It is well known[13-15] that the operation of the $^{22}Ne(\alpha,n)^{25}Mg$ neutron source should lead to an observable distortion in the relative abundances of the magnesium isotopes from their solar system distribution of ^{24}Mg:^{25}Mg:^{26}Mg = 79:10:11. In order to obtain information on the source of neutrons, the magnesium isotopic mixture in HD178717 was determined from observations of the molecular MgH (0,0) band lines at 5101Å and 5107Å. Also, since the observed rubidium abundance is known to be an indicator of the neutron density at the s-process site,[5),7),15] the abundance of this element in HD178717 was determined from observations of the Rb I line at 7800Å. In order to minimize non-LTE effects,[5),16-17] this rubidium abundance determination was carried out differentially with respect to the standard K3 giant μ Aql. This star has been previously found to possess a solar-like composition of the heavy elements.[18)]

The consequences of our results with regard to the origin of the Ba II stars are discussed. Specifically, we compare our results with the abundances predicted if mass transfer from an evolved asymptotic giant branch (AGB) star has occurred. This scenario for the origin of the s-process enhancements in Ba II stars has received a great deal of theoretical and observational attention in recent years.[19–21)]

II. OBSERVATIONS

The observations discussed here were obtained with the Reticon-equipped coudé echelle spectrometer of the 1.4m Coudé Auxiliary Telescope at the European Southern Observatory, La Silla. Spectra of HD178717 centered at 5100Å and 7800Å, and of μ Aql centered at 7800Å were obtained. The length and the resolution of the spectra were 40Å and 0.05Å, respectively, for the 5100Å centered spectrum, and 60Å and 0.08Å, respectively, for the 7800Å centered spectra. The signal to noise ratio in the continiuum exceeded 100 in all of the obtained spectra.

III. ANALYSIS AND RESULTS

In order to determine the magnesium isotopic distribution and the rubidium abundance of our stars, the observed spectra were compared with synthetic spectra calculated using an LTE spectral synthesis program.[22)] The required input for the synthesis program consists of the wavelengths, lower excitation potentials and gf values of the relevant spectral lines. The line parameters used in this study have been previously used elsewhere,[5–6)] and the reader is directed to these references for further discussion. Also required as input is the relevant stellar models. These were obtained using interpolations from a grid of published models.[23)] The atmospheric parameters, namely, the surface temperature T_{eff}, the surface gravity log g and the microturbulent velocity ξ, of our stars have been previously determined[7–9] and are given in table 1. A composite of the rotational, macroturbulent and instrumental broadening was applied to the synthetic spectra. This broadening was determined by matching the profile of an apparently unblended line of similar strength to the MgH lines.

Table 1. Atmospheric Parameters

Star	$T_{\rm eff}$ (K)	log g	ξ (km s^{-1})
HD178717	4100	1.0	2.2
μ Aql	4500	2.1	1.9

a) Isotopic Magnesium.

In order to determine the relative abundances of the magnesium isotopes a comparison of the observed MgH lines of the (0,0) band at 5101Å and 5107Å are made with synthetic spectra. The MgH lines at 5101Å have the advantage of a large isotopic splitting (~0.14Å) . Contamination by lines from the C_2(0,0) and C_2(1,1) bands, however, leads to significant blending in this spectral region. Although the MgH lines at 5107Å have the disadvantage of a smaller separation (~0.1Å), the ^{25}MgH and ^{26}MgH lines are unaffected by C_2 blends. Before proceeding it was necessary to determine the contribution of C_2 in the wavelength region under investigation. The strength of C_2 was estimated by fitting the profile of the C_2 line at 5086Å. It should be noted, however, that due to blending with nearby unidentified lines, caution must be used in order to avoid an overestimate of the C_2 strength.

Figure 1 displays a comparison of the observed MgH spectra of HD178717 in the 5101Å region with synthetic spectra calculated for different distributions of the magnesium isotopes. The calculation with the magnesium isotopes in the ratio ^{24}Mg:^{25}Mg:^{26}Mg = 60:25:15 clearly gives the best fit to the observed data. The fact that this fit is not exact could be the result of a number of error sources, such as poor fitting of the continuum, poorly determined gf values or stellar parameters,

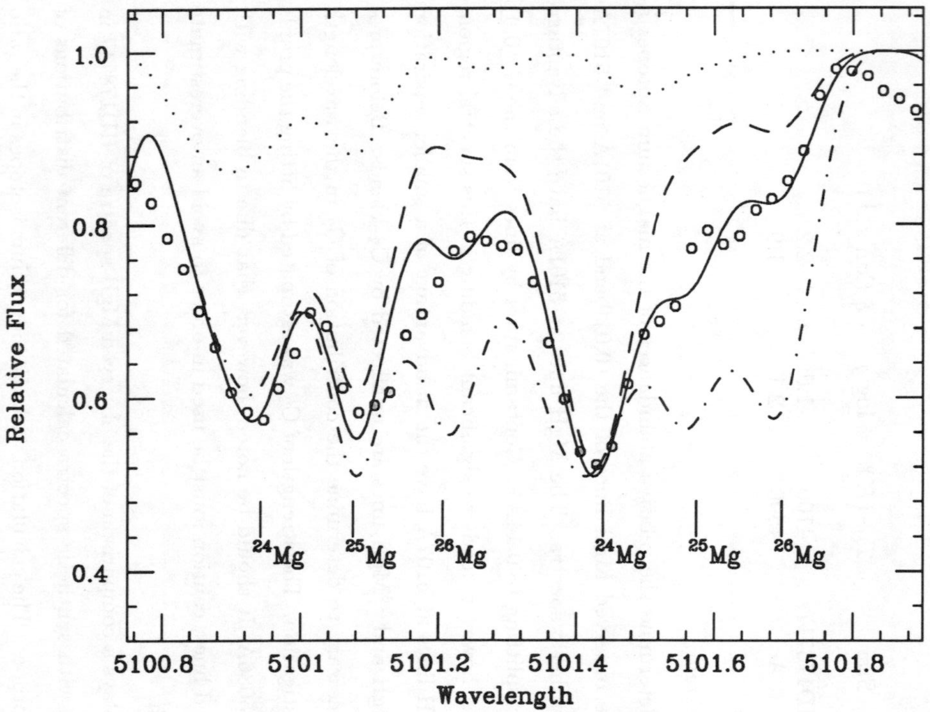

Figure 1. Comparison of the observed MgH feature at 5101Å with synthetic spectra. The synthetic spectra are calculated assuming a C_2 contribution alone (dotted curve); a solar distribution of magnesium isotopes (dashed curve); a best fit distribution of magnesium isotopes where ^{24}Mg:^{25}Mg:^{26}Mg = 60:25:15 (solid curve); and an equal distribution of magnesium isotopes (dot-dashed curve). The circles represent the observed data points. The principal lines of the spectrum are indicated.

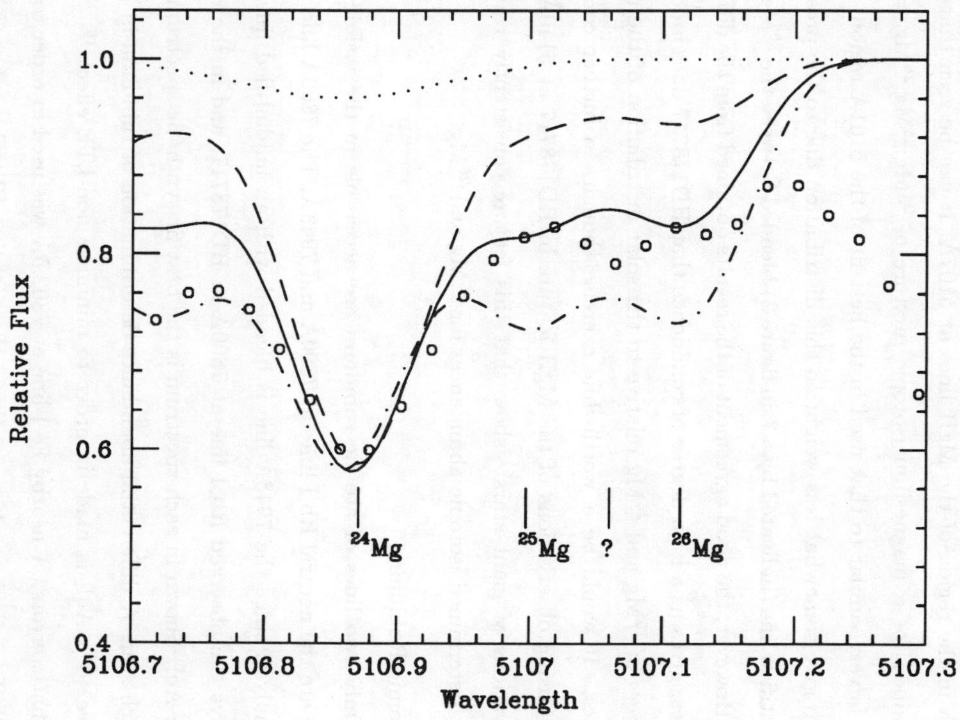

Figure 2. As in figure 1 except the best fit for the 5107Å MgH feature is ^{24}Mg:^{25}Mg:^{26}Mg = 56:22:22.

a poor determination of the C_2 strength, or the presence of unknown blends. The latter source of error is the most likely in view of the cool nature of the star under study. In light of these uncertainties it is important to check these results using another portion of the stellar spectrum. Figure 2 shows the observed and synthetic spectra in the region of the MgH lines at 5107Å. It can be seen that a good fit is obtained for a magnesium isotopic mixture of ^{24}Mg:^{25}Mg:^{26}Mg = 56:22:22, which is very similar to that used in the best fit of the 5101Å region. We would tend to give somewhat less weight to this distribution due to the presence of an unidentified line (indicated by a ? in figure 2) blended between the ^{25}Mg and ^{26}Mg lines. However, the good agreement of the results obtained from the different MgH features, gives us a high degree of confidence that HD178717 does indeed possess an excess of ^{25}Mg and ^{26}Mg relative to the solar distribution of the magnesium isotopes. It would be a worthwhile exercise though, to further confirm these results using observations of the MgH feature in HD178717 at 5140Å. Previous observations of giant stars[6)] show that this feature consistently results in the lowest determined isotopic abundances for ^{25}Mg and ^{26}Mg.

b) Atomic Rubidium.

Only two lines of atomic rubidium are accessible in the visible spectrum. These are the neutral Rb I lines at 7800Å and 7948Å. The 7800Å line was chosen for analysis since the 7948Å line is blended with an unidintified line. Figure 3 displays the observed Rb I line at 7800Å in HD178717, and in the comparison star μ Aql. Shown in each spectrum is the best fit synthetic spectrum computed for each star. A differential abundance determination of rubidium in HD178717, relative to μ Aql, is made in order to minimise non-LTE effects[16−17)] and any systematic errors. A nearby Fe I line at 7802.5Å was used in order to determine the Rb/Fe ratio in each star. Using the notation [Rb/Fe] = log $(\mathrm{Rb/Fe})_{\mathrm{BaII}}$ - log$(\mathrm{Rb/Fe})_{\mu \mathrm{Aql}}$, we find [Rb/Fe] equal to +0.2 in HD178717. Knowing the dilution factor of the irradiated material, the Rb/Fe ratio of the irradiated material can then be estimated in a simple manner.[5)] The change in [Rb/Fe] caused by varying

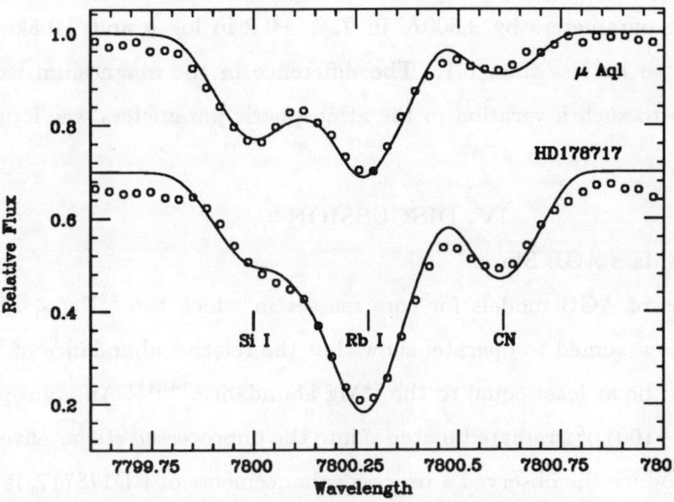

Figure 3 The observed and best-fit synthesized spectra for μ Aql (top) and HD178717 (bottom; a constant is subtracted from this spectrum) in the region of the Rb I 7800Å line.

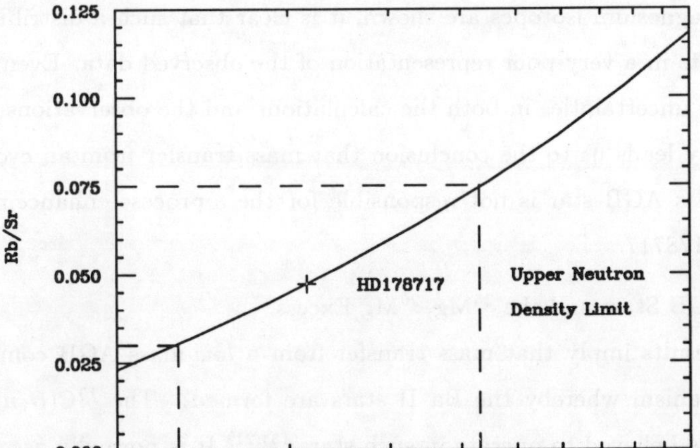

Figure 4 The calculated Rb/Sr ratio as a function of the neutron density (solid curve). Also indicated (cross) is the measured value of Rb/Sr in HD178717. The horizontal dashed lines correspond to our estimated error of this ratio . The vertical dashed lines are the resultant error limits on the neutron density.

the atmospheric parameters by $\pm 300K$ in $T_{\rm eff}$, ± 0.2 in log g and ± 0.5km s^{-1} in ξ was found to be less than 0.1. The difference in the magnesium isotope abundances due to such a varation in the atmospheric parameters was found to be negligible.

IV. DISCUSSION

a) Intermediate-Mass AGB Stars.

Calculations of AGB models for core masses in which the ^{22}Ne$(\alpha, n)^{25}$Mg neutron source is assumed to operate, show that the relative abundance of ^{25}Mg and ^{26}Mg should be at least equal to the ^{24}Mg abundance.[14),15)] An appropriate dilution factor (~100) of irradiated material into the unprocessed stellar envelope, required to reproduce the observed *s*-process enhancements of HD178717, is used in these calculations. For example, calculations for an AGB model possessing a core mass of $1.16M_\odot$[15)] predict an observed ^{26}Mg/^{24}Mg ratio of ~3. From figures 1 and 2 where calculations of synthetic spectra assuming an equal distribution of the magnesium isotopes are shown, it is clear that such a distribution of isotopes results in a very poor representation of the observed data. Even taking into account uncertainties in both the calculations and the observations, this large discrepancy leads us to the conclusion that mass transfer from an evolved intermediate-mass AGB star is not responsible for the *s*-process enhancements observed in HD178717.

b) Low-Mass AGB Stars and the ^{25}Mg–^{26}Mg Excess.

Do these results imply that mass transfer from a *low*-mass AGB companion is the mechanism whereby the Ba II stars are formed? The ^{13}C$(\alpha, n)^{16}$O neutron source is believed to operate in such stars.[15),21)] It is normally assumed that the ^{13}C$(\alpha, n)^{16}$O neutron source does not lead to any significant alteration of the magnesium isotopes due to the small neutron absorption cross section[24)] of ^{16}O. However, recent calculations[25)] have shown that it may be possible, in a limited parameter space, to produce observable anomalies in the magnesium isotopic mixture if it is assumed that all the CNO isotopes in the intershell region of the AGB star have been transformed into ^{22}Ne. In such circumstances, neutron absorption reactions on ^{22}Ne and up through the magnesium isotopes, could lead to a small observable excess of ^{25}Mg and ^{26}Mg similar to that observed in HD178717 (other calculations[15] in which a solar ^{22}Ne initial intershell abundance is assumed, produce no observable alteration of the magnesium isotopes once dilution of the processed material is taken into account). The question as to what fraction of the CNO isotopes are transformed into ^{22}Ne will be determined by the temperature history of the stellar interior, and the competition between the ^{22}Ne producing reaction sequence $^{14}N(\alpha,\gamma)^{18}F(\beta)^{18}O(\alpha,\gamma)^{22}Ne$, and destruction reactions such as $^{18}O(p,\alpha)^{15}N$ (the $^{18}O/^{22}Ne$ ratio can also have an important effect on the neutron density[26]). The fact that some Ba II stars show normal[5–6] (i.e. solar) isotopic magnesium mixtures, whereas some do not, could simply be an indication of the different temperature histories of the different stars. Since the temperature structure of the star depends sensitively on the stellar mass, it is plausible that this is the stellar parameter determining any distortion of the magnesium isotopes.

In the calculations presented here, it was implicitly assumed that the initial abundances of the magnesium isotopes were given by their solar system distribution. Of course it is not impossible that the ^{25}Mg–^{26}Mg excess observed in HD121447 is simply a result of the star's initial composition. However, this would seem implausible due to the difficulty in explaining such an initial composition, and also since observations[27] of a number of both metal-poor and metal-rich stars, show no excess of ^{25}Mg or ^{26}Mg .

c) The Rb/Sr Ratio and AGB Stars.

Regardless of the stellar mass, the observed rubidium abundances of the Ba II stars either pose a serious problem for current AGB models, or else for the proposed mass transfer scenario. This can be seen from figure 4 where a calculation[15] showing the Rb/Sr ratio as a function of neutron density is shown. The Rb/Sr ratio determined for HD178717 is also indicated. This Rb/Sr ratio is calculated assuming the previously determined[9] [Sr/Fe] ratios of HD178717, and assuming a solar composition for μ Aql.[18] The horizontal dashed lines in figure 4 correspond to the maximum and minimum value of the Rb/Sr ratio assuming a conservative error of $\pm$ 0.2 dex. Current models of the nucleosynthesis occurring in AGB stars,[15] irrespective of the neutron source or core mass assumed, result in Rb/Sr ratios of at least 0.5. It is important to point out here, that, these models take into account the time dependence of the neutron density.[28] This points out, that, the Rb/Sr ratio determined in HD178717 is a measure of the neutron density under which most of the nucleosynthesis took place, and is not simply a measure of the low neutron density occurring at the close of the nucleosynthesis episode. It can be seen from figure 4 that the observed Rb/Sr ratio in HD178717 indicates that the neutron density at the *s*-process site cannot be greater than $\sim 5\times10^7\,\mathrm{cm}^{-3}$. This is comparable with the neutron density deduced from Rb I observations in two other Ba II stars.[5),7),29] In contrast, the neutron densities of AGB stars[15),26),28] are typically in the range of 10^9-$10^{12}\,\mathrm{cm}^{-3}$. This is true even in low-mass AGB stars as a result of the high reaction rate[30] of $^{13}\mathrm{C}(\alpha,n)^{16}\mathrm{O}$ relative to $^{22}\mathrm{Ne}(\alpha,n)^{25}\mathrm{Mg}$. If we assume an AGB mass transfer mechanism as the correct interpretation of the Ba II star phenomenon, then the observed Rb/Sr ratio of HD178717 would appear to be incompatible with current low-mass AGB models. In order to remedy this situation, temperatures at the intershell base of low-mass AGB stars are required to be higher than presently predicted.[31–32] It is interesting to note that recent stellar evolution calculations indicate such a trend.[33–34]

V. CONCLUSIONS

We have investigated the relative abundances of the magnesium isotopes in the cool Ba II star HD178717. We find that a distribution of ^{26}Mg:^{25}Mg:^{26}Mg equal to 60:25:15 gives a good fit to the observed data. This distribution would appear to rule out the operation of the $^{22}\mathrm{Ne}(\alpha,n)^{25}\mathrm{Mg}$ neutron source, and therefore any relationship of this Ba II star to intermediate-mass AGB stars. The excess of

^{25}Mg and ^{26}Mg observed can be explained by the operation of some other neutron source ($^{13}C(\alpha,n)^{16}O$ being the most likely candidate) in a region where a large abundance of ^{22}Ne is initially present. We have also determined the rubidium abundance in HD178717, and deduce an [Rb/Fe] value of +0.2. This ratio indicates a neutron density of $\sim 2\times 10^7 \mathrm{cm}^{-3}$ at the s-process site. If an AGB mass transfer mechanism is responsible for the s-process enhancements of Ba II stars, then, this low neutron density implies that significant revisions of present low-mass AGB models are required.

The authors would like to thank A. McWilliam for his valuable assistance in the calculations of the synthetic spectra. They are also grateful to V. V. Smith and J. Tomkin for their contribution to this study. This work was partly funded by National Science Foundation Grants PHY85-05682 and AST86-16635 (U.S.). DLL acknowledges the Robert A. Welch Foundation of Houston, Texas. RAM acknowledges travel support to La Silla from the SERC and the financial support of an SERC/NATO Fellowship (U.K.).

REFERENCES

1) Bidelman, W. P., and Keenan, P. C., 1951, *Ap.J.*, **114**, 473.
2) Warner, B., 1965, *Mon.Not.R.Ast.Soc.*, **129**, 263.
3) Lambert, D. L., 1985, in *Cool Stars with Excesses of Heavy Elements*, eds. M. Jascheck and P. C. Keenan (Dordrecht: Reidel), p191.
4) Scalo, J. M., 1982, in *Physical Processes in Red Giants*, eds. I. Iben, Jr. and A. Renzini (Dordrecht:Reidel), p77.
5) Tomkin, J., and Lambert, D. L., 1979, *Ap. J.*, **227**, 209.
6) McWilliam, A. and Lambert, D. L., 1987, *Mon.Not.R.Ast.Soc.*, in press.
7) Smith, V. V., and Lambert, D. L., 1984, *Pub. A. S. P.*, **96**, 226.
8) Lü, P. K., Dawson, D. W., Upgren, A. R. and Wies, E. W., 1983, *Ap. J. Supp.*, **52**, 169.
9) Smith, V. V., 1984, *Astron. Astrophys.* **132**, 326.
10) Pinsonneault, M. H., Sneden, C., and Smith, V. V., 1984, *Pub. A. S. P.*, **96**, 239.
11) Cameron, A. G. W., 1955, *Ap. J.*, **121**, 144.
12) Cameron, A. G. W., 1960, *Astr. J.*, **65**, 485.
13) Truran, J. W., and Iben, I., Jr., 1977, *Ap. J.*, **216**, 797.
14) Scalo, J. M., 1978, *Ap. J.*, **221**, 627.
15) Malaney, R. A., 1987, *Ap. J.*, **321**, 832.

[16] Ruland, F., Holweger, H., Griffin. R., Griffin. R. and Biehl, D., 1980, *Astron. Astrophys.*, **92**, 70.

[17] Brown, J. A., Tomkin, J. and Lambert, D. L., 1983, *Ap. J. Letters*, **265**, L93.

[18] Helfer, H. L. and Wallerstein, G., 1968, *Ap. J. Supp.*, **16**, 1.

[19] McLure, R. D., 1985, in *Cool Stars with Excesses of Heavy Elements*, eds. M. Jaschek and P. C. Keenan (Dordrecht: Reidel), p315.

[20] Webbink, R. F., 1986, to be published in *Critical Observations vs. Physical Models for Close Binary Systems*, eds. K.-C., Leung and D.-S. Zhai, (New York: Gordon and Breach).

[21] Iben, I., Jr., and Tutukov, A. V., 1985, *Ap. J. Supp.*, **58**, 661.

[22] Sneden, C., 1974, PhD Thesis, Univ. of Texas.

[23] Bell, R. A., Eriksson, K., Gustafsson, B. and Nordlund, Å., 1976, *Astron. Astrophys. Supp.*, **23**, 37.

[24] Bao, Z. Y., and Käppeler, F., 1986, *At. Data Nuc. Data Tables*, **36**, 411.

[25] Arnould, M. and Jorissen, A., 1986, in *Advances in Nuclear Astrophysics*, eds. E. Vangioni-Flam, J. Audouze, M. Chasse, J. P. Chieze and J. Tran Thanh Van, (Editions Frontieres: France), p419.

[26] Malaney, R. A., 1986, *Mon.Not.R.Ast.Soc.*, **223**, 683

[27] Barbury, B., Spite, F. and Spite, M., 1987, *Astron. Astrophys.*, **178**, 199.

[28] Cosner, K., Iben, I., Jr., and Truran, J. W., 1980, *Ap. J. (Letters)*, **238**, L91.

[29] Beer, H. and Walter, G., 1985, in *Cool Stars with Excesses of Heavy Elements*, eds. M. Jaschek and P. C. Keenan (Dordrecht: Reidel), p373.

[30] Caughlan, G. R., Fowler, W. A., Harris, M. J. and Zimmerman, B. A., 1985, *At. Data Nuc. Data Tables*, **32**, 197.

[31] Mathews, G. J., Ward, R. A. Takahashi, K., and Howard, W. M., 1986, in *Nucleosynthesis and its Implications on Nuclear and Particle Physics*, ed. J. Audouze and N. Mathieu (Reidel:Dordrecht), p277.

[32] Malaney, R. A. and Boothroyd, A. I., 1987, *Ap. J.*, **320**, 866.

[33] Becker, S. A., 1986, private communication.

[34] Boothroyd, A. I., 1987, PhD Thesis, California Institute of Technology.

ZIRCONIUM ISOTOPIC RATIOS AND TECHNETIUM ABUNDANCES IN EVOLVED RED GIANTS

Verne V. Smith
Department of Astronomy
The University of Texas
Austin, TX 78712 USA

ABSTRACT: Relative abundances of six isotopes of Zirconium have been measured in two S stars: R And and R Gem. The isotopic ratios of Zr are sensitive to the neutron densities during s-process nucleosynthesis, and our results suggest that the typical s-process neutron density in these stars is $\lesssim 10^8$ cm^{-3}. Also, abundances of the unstable element Technetium, relative to neighboring s-process elements, have been estimated for eight MS and S stars. Timescales up to ~ 10^6 yr. are inferred for the thermally pulsing phase of AGB evolution for these stars.

I. INTRODUCTION

One of the major observational discoveries connecting both the fields of stellar evolution and the nucleosynthesis of the chemical elements was the observation of unstable technetium in the atmospheres of red giants, primarily of spectral type S, by Merrill (1952). His discovery provided crucial evidence of recent, heavy-element nucleosynthesis within these stars and a stimulus to the theory of stellar nucleosynthesis. The study of stellar nucleosynthesis took a major step forward five years later with the pioneering work of Burbidge, Burbidge, Fowler, and Hoyle (1957-B^{2}FH). A quantitative understanding of the neutron captures which produce heavy elements, such as Tc, via the s-process was provided by B^{2}FH. Since that time, much work (both theoretical and observational) has gone into attempts at understanding more about the astrophysical site of the s-process. Analysis of the solar system s-process abundances provides some estimates of conditions under which they were synthesized. For example, Howard *et al.* (1986) find that solar system s-process nuclei were formed at a site with a mean neutron density of $N_n = 1.1 \times 10^8$ cm^{-3} and temperature of 2.7×10^8 K.

Both of the s-process elements Zr and Tc carry useful diagnostic information on 1) conditions at the site of the s-process, and 2) the stellar evolutionary timescales over which the s-process occurs. In Figure 1, we show a portion of the nuclides-plane with increasing neutron number running along the horizontal axis and increasing proton number along the vertical axis. Stable nuclei are indicated by cross-hatching and half-lives are shown for the unstable nuclei. Successive neutron captures are indicated by the arrowed lines starting with ^{90}Zr. Zirconium-95, with a β-decay half-life of about 65 days, represents an interesting branch point in the main-line nuclear flow, as shown by the diverging solid and dashed arrowed lines. Recent reaction-network calculations by Malaney (1987) show that

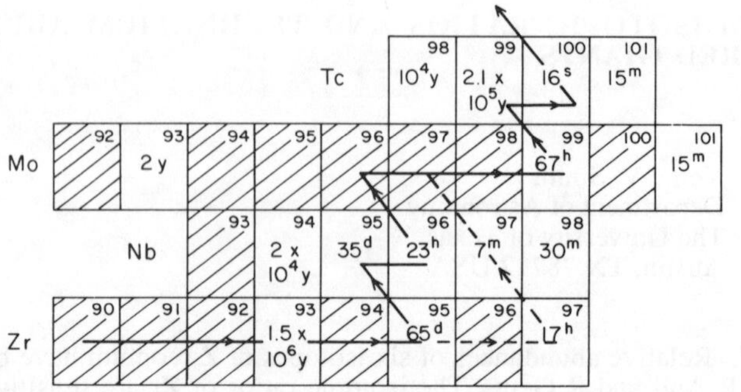

Fig. 1. The main-s-process flow through the Zr-Tc region of nuclei. Stable isotopes are indicated by cross-hatching.

at neutron densities of 10^8 cm^{-3} all of the ^{95}Zr decays away before neutron captures can occur to produce ^{96}Zr, and hence, virtually no ^{96}Zr is produced at these densities. This density-path is shown by the solid arrowed line. At much higher densities, such as 10^{11} cm^{-3}, ^{95}Zr does not have time to decay between neutron captures, and significant amounts of ^{96}Zr are produced at these densities. This path is shown by the dashed line in Figure 1. Thus, the amount of ^{96}Zr present, relative to its fellow Zr-isotopes, can be used to probe the s-process neutron density in a red giant.

Estimating the isotopic abundance of Zr in cool, red-giant atmospheres is possible by exploiting the isotopic splitting of the ZrO bands which are typically very strong spectral features in stars of type MS and S (Keenan 1954). In an earlier study, Zook (1978) provided estimates of the relative abundances of Zr-isotopes in two S stars: R Cyg and V Cnc. Her results were in agreement with an s-process origin to the excess Zr-abundances seen in S stars, including the detection of ^{93}ZrO ($t_{1/2}$ = 1.5 x 10^6 yr), which indicated recent nucleosynthesis. With the high quantum efficiency detectors now available, which can provide high S/N spectra, and recent theoretical predictions from s-process calculations, a new attack on ZrO in evolved, s-process enhanced red giants is of interest, and results for two such stars will be presented in this study.

In addition, as can be seen from Figure 1, ^{99}Tc, with a half-life of 2.1 x 10^5 yr, lies on the main s-process path and is expected to be produced at both low and high neutron densities. With its relatively short half-life, a quantitative estimate of the abundance of Tc relative to other s-process elements can be useful as an age indicator for the s-processing lifetimes of stars on the red-giant branch. Quantitative Tc-abundances in two stars of type SC were provided by Smith and Wallerstein (1983) which Mathews *et al.* (1986) used, along with an analytical model of Tc-production and decay, to deduce timescales of ~ 3-15 x 10^5 yr for the lifetimes of stars in the s-processing phase of evolution. Further quantitative Tc-abundances have been estimated by Dominy and Wallerstein (1986) and Smith and Lambert (1986) for several stars of type M, MS, and S. In this study, we will summarize Tc/Zr and Tc/Y abundance ratios for eight stars of type MS or S.

2. RECENT RESULTS FOR ZR-ISOTOPES

Observations of ZrO have been obtained for a number of S stars in the wavelength region of 6925 Å using the McDonald Observatory's 2.7m telescope and the coudé spectrograph with echelle grating and Reticon detector. The resolution of these spectra are 0.08 Å, and all program stars were ratioed with a featureless hot star to divide out telluric H_2O lines found in this region. The primary ZrO feature in this bandpass is the 0-1 band of the B $^1\Pi$ - X $^1\Sigma^+$ system with the R-branch bandhead at 6931.7 Å for ^{90}ZrO. This is the same band used by Zook (1978) in her analysis, and the total separation from the ^{90}ZrO to ^{96}ZrO bands here is ≈ 2.5 Å. In addition, a blanket of lines from the $\Delta v = -1$ sequence of the γ-system (A $^3\Phi$ - X' $^3\Delta$) contaminate this region. We have found that in some stars with very strong ZrO, the 1-2 band's R_1-branch bandhead of the $^3\Phi$ - $^3\Delta$ system at 6923.4 Å (^{90}ZrO) to be quite strong and an additional isotope indicator.

Line frequencies were computed by Dr. David L. Lambert using molecular constants from Phillips and Davis (1976) for the $^1\Pi$ - $^1\Sigma$ system and Phillips and Davis (1979) for the $^3\Phi$ - $^3\Delta$ system. In addition, Dr. Lambert computed estimated f-values from the data of Littleton and Davis (1985). The final ZrO line list contained the isotopes 90, 91,92, 93, 94, and 96 from the $^1\Pi$ - $^1\Sigma$ 0-1 band and the $^3\Phi$ - $^3\Delta$ bands 0-1 to 9-10 within the wavelength interval 6915 - 6950 Å. This amounts to some 3500 individual molecular lines. In addition, all atomic lines within this bandpass from the Kurucz and Peytremann (1976) line-list were included, and solar gf-values were substituted for all atomic lines visible in the solar spectrum.

Up to now, two stars have been analyzed using the McDonald spectra and this line-list: R And and R Gem, both S stars. The analysis of the spectra utilizes spectrum synthesis with a modified version of an LTE-synthesis program first written by Sneden (1973) and model atmospheres from Johnson, Bernat, and Krupp (1980) and Johnson (1982). Both R And and R Gem are long-period variables, and determination of their atmospheric parameters presents a problem. Fortunately, the relative isotopic abundances derived from a single species, such as ZrO, are quite insensitive to the adopted model T_{eff}, Log g, or C/O ratio. For both stars, a model atmosphere from Johnson (1982) was used with C/O = 0.95, T_{eff} = 3500 K, and Log g = 0.0. Other analyses were done using models with differing C/O-ratios, T_{eff}'s, and Log g's, with the result that no significant differences are found for the relative Zr-isotopic abundances. The overall strength of the synthetic ZrO bands is very sensitive to the model-atmosphere C/O-ratio; however, the relative depths of the various isotopic bandheads yield consistent isotopic-mixtures. In Figure 2, we show the spectrum of R And near the 1-2 band of the γ-system (filled circles) and three synthetic spectra with differing isotopic mixtures. For this small region of spectrum dominated by the ZrO bands, the overall fit to the observed spectrum is quite good, although as mentioned above, the overall Zr-abundance is not reliable due to the great sensitivity of the ZrO strength to the model atmosphere's C/O ratio. The crucial result to note is the definite weakness of any ^{96}ZrO bandhead near 6121 Å. The one feature visible near 6121 Å is actually dominated by ^{90}ZrO from the γ(5-6) bandhead. Even a few-percent mixture of ^{96}Zr will substantially strengthen any absorption near 6121 Å, and this is not observed. In the case of R And, both the γ(1-2) and 0-1 $^1\Pi$ - $^1\Sigma$ bands were used to estimate the fractional Zr-isotopic mixture, whereas in R Gem, the ZrO was much weaker and the γ(1-2) band was too weak to be used reliably. In Table I we present the results for the relative abundances derived from the ZrO bands in R And and R Gem. The results for R And from the two band systems provide reasonably consistent results, and R Gem appears similar to R And in its Zr-isotopic mixture. In both stars, ^{93}ZrO is observed, and this is consistent with Merrill's (1952) observation of Tc in R And and R Gem. Attempts to use ^{93}Zr's half-life of 1.5 x 10^6 yr as a chronometer here are thwarted by our lack of knowledge of the

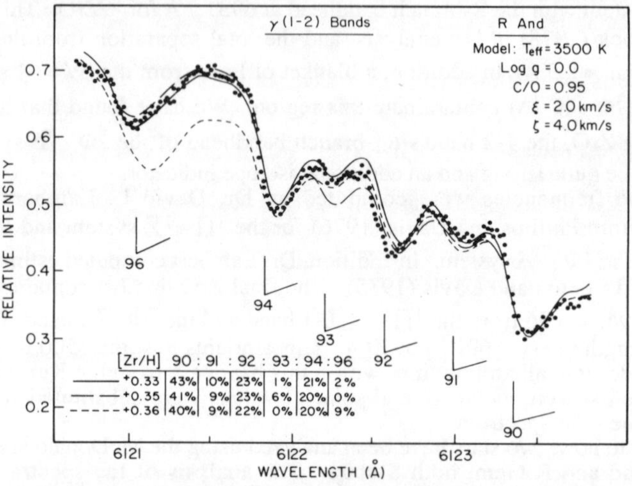

Fig. 2. The observed spectrum of R And (filled circles) and three synthetic spectra with differing Zr-isotopic mixtures for the γ(1-2) isotopic ZrO bandheads.

Table I. Isotopic Zr-Abundances by Fraction

Star and Bandhead	Zr:	90	91	92	93	94	96
R And (γ(1-2))		.42	.10	.22	.02	.21	.02
R And (0-1 $^1\Pi$-$^1\Sigma$)		.32	.18	.14	.05	.24	.07
R Gem (0-1 $^1\Pi$-$^1\Sigma$)		.35	.20	.13	.02	.26	.04

absolute abundance enhancement of recently produced Zr in these stellar atmospheres. The observed Zr is presumably a combination of the isotopic mixture from which the stars formed and some amount of recently s-processed Zr injected into the atmosphere. However, some conclusions can be drawn from the very modest fractions of ^{96}Zr observed in these stars. Boesgaard (1970) and Smith and Lambert (1985, 1986) find substantial enhancements of [Zr/Ti] in MS and S stars (typically + 0.3 - 1.0 dex), and if such enhancements of internally s-processed Zr contained substantial fractions of ^{96}Zr, the observed ^{96}Zr-abundances in these, presumably similar S stars R And and R Gem, would also be enhanced over an assumed initial solar-like fraction of 0.03. No significant quantity of ^{96}Zr-rich material has been added to the atmospheres of either R And or R Gem. Malaney's (1987) models predict that at neutron densities of 10^{11} cm^{-3} about 1/3 of the zirconium is ^{96}Zr. At the rather large s-process enhancements observed in S stars, such a large fraction of ^{96}Zr would be observed as an increase in the ^{96}ZrO band-strength, yet no enhancements are observed over the initial, presumed solar-like, fraction of 0.03. No excess ^{96}Zr has been produced in these stars, and this restricts the s-process neutron densities inside of these stars to not much larger than 10^8 cm^{-3}. Zook (1978) derived similar results for ^{96}Zr in the S stars V Cnc and R Cyg.

3. TECHNETIUM ABUNDANCES IN MS AND S STARS

The abundance of Tc, relative to other s-process elements, can be used as a chronometer of the elapsed time since s-processing has occurred. Quantitative dating is complicated by the current picture of the s-process occurring over repeated cycles in the intershell regions of thermally pulsing Asymptotic Giant Branch (AGB) stars (for a review, see Iben and Renzini 1983). After each s-process cycle and subsequent dredge-up, Tc will decay while

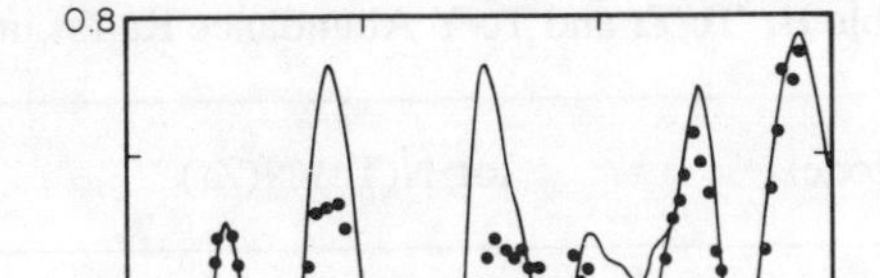

Fig. 3. The observed spectrum of HR 4647 near the 4262 Å Tc I line and two synthetic spectra; one with no Tc and one with log N(Tc) = -0.2 (where log N(H) = 12.0).

the stable s-process nuclei will continually be built up over several pulses. If the s-processing lifetime of an AGB star is much longer than the half-life of Tc (taken to be the s-process isotope-99), however, the amount of Tc-decay, relative to its fellow s-process elements, will be crudely related to the time since s-process dredge-up began on the AGB. Technetium abundances have been estimated in a sample of MS and S stars, all of which are low-amplitude variable stars, using the 4262 Å Tc I resonance line. Observations were carried out with the McDonald Observatory's 2.1m telescope and coudé spectrograph with a CCD detector. Spectral resolutions of 0.2 Å were obtained. The violet regions of cool, red-giant stars are extremely crowded and the region from 4255 - 4265 Å was synthesized in a sample of stars. The line-list included all lines visible in the sun, along with all other lines from the list of Kurucz and Peytremann (1976). This method is described in Smith and Lambert (1986). In order to match the synthetic spectra to the stellar spectra, a normal M-giant, without Tc (but with similar T_{eff} and gravity to the MS and S stars), is chosen as a standard, and the gf-values in the line-list are altered until a good fit to the standard M-giant is obtained. The average amount that the gf-values have to be changed is then applied to the gf-value of the Tc I line (taken from Garstang, 1981). The necessity of "fudging" gf-values introduces an uncertainty in our final Tc-abundances of probably a factor 2-3, but is brought about by the very complicated nature of the violet spectra of cool stars, along with their very low flux-levels in this spectral region. In Figure 3, we illustrate a portion of the spectrum of the MS-star HR 4647 near the 4262 Å Tc I line (filled circles) along with two synthetic spectra: one with no Tc and one with log N(Tc) = -0.2 (on a scale with log N(H) = 12.0). Keeping in mind that gf-values were altered to fit a standard M-giant, the qualitative fit of the synthetic to real spectrum is fairly good.

Table II. Tc/Zr and Tc/Y Abundance Ratios in MS/S Stars

Star or s-process Model	log N(Tc)/N(Zr)	log N(Tc)/N(Y)
o^1 Ori	-3.4	-2.9
HR 4647	-2.7	-1.8
HD 64332	-2.6	-2.3
RS Cnc	-3.9	-3.5
Y Lynx	-4.0	-3.5
HR 6702	-3.9	-3.6
HR 8062	-3.0	-2.8
HR 8714	-3.2	-3.0
Model with $\tau_o = 0.1$ mb^{-1}	-2.5	-2.3
Model with $\tau_o = 0.6$ mb^{-1}	-2.2	-1.7

Results are presented for eight stars in Table II in the form of Tc/Zr and Tc/Y ratios, where the Zr and Y-abundances are derived from near-IR lines near 7500 Å and then corrected for the observed mixture of initial envelope abundance and s-processed material currently in the stellar atmosphere (Smith and Lambert 1986). Also shown in Table II are the predicted ratios of Tc to both Zr and Y from the models of Cowley and Downs (1980) for two, very different, neutron exposures. The simple interpretation of Table II is that a few stars (HR 4647, HD 64332, and HR 8062) are within about 0.5 dex of the Tc-abundance ratios expected for s-processing, while the remaining five stars are well below these abundance ratios. These low Tc/Zr and Tc/Y ratios, if due to Tc-decay, would indicate timescales of $\approx 10^6$ yr for the current s-process lifetimes for these stars. This would agree with Iben's (1983) 1 $M_\odot$ AGB-model's lifetime of $\approx 2 \times 10^6$ yr. These results must be taken with caution, however, as the model atmospheres plus LTE-synthesis are an inadequate representation of the real spectra. It is encouraging that none of the Tc-abundance ratios are wildly larger than predicted by theory; however, more stars will be analyzed in the near future, and along with new predictions from improved s-process reaction networks, the limitations of this technique as applied to the violet Tc I lines will be explored. Another spectral region to consider is 5925 Å where the weaker, intercombination line of Tc I is found and has been identified and measured in some SC stars (where C/O = 1.0 and molecular line-blanketing in this region in these stars is negligible) by Smith and Wallerstein (1983), and more recently, by Kipper (1987).

4. SUMMARY

The results from an analysis of ZrO in the two S stars R And and R Gem reveal that virtually no ^{96}Zr has been added to the atmospheres of these AGB stars through the s-process. This lack of ^{96}Zr enhancement suggests strongly that the neutron density at the s-process site within these stars was not much greater than 10^8 cm^{-3}.

Estimates of the Tc-abundances in eight MS and S stars indicate a large spread in Tc/Zr and Tc/Y ratios ranging from values near those expected from very recent s-processing to values typical of what would be found ~ 10^6 yr after s-processing first began. Timescales on the AGB of 10^6 yr would suggest that many of these stars are of lower mass (M ~ 1 $M_\odot$; Iben 1983).

I thank Dr. J. Tomkin for obtaining some of the spectra used here, and Drs. D. L. Lambert and G. Wallerstein for valuable collaboration. This research is supported in part by the National Science Foundation (grants AST 83-16635 and AST 86-14423) and the Robert A. Welch Foundation of Houston, Texas.

REFERENCES

Boesgaard, A. M. 1970, *Ap. J.*, **161**, 163.
Burbidge, G. R., Burbidge, E. M., Fowler, W. A., and Hoyle, F. 1957, *Rev. Mod. Phys.*, **29**, 547.
Garstang, R. H. 1981, *Pub. A. S. P.*, **93**, 641.
Howard, W. M., Mathews, G. J., Takahashi, K., and Ward, R. A. 1986, *Ap. J.*, **309**, 633.
Iben, I., Jr. 1983, *Ap. J. (Letters)*, **275**, L65.
Iben, I., Jr., and Renzini, A. 1983, *Ann. Rev. Astr. Ap.*, **21**, 271.
Johnson, H. R. 1982, *Ap. J.*, **260**, 254.
Johnson, H. R., Bernat, A. P., and Krupp, B. M. 1980, *Ap. J. Suppl.*, **42**, 501.
Keenan, P. C. 1954, *Ap. J.*, **120**, 484.
Kipper, T. A. 1987, preprint.
Kurucz, R. L., and Peytremann, E. 1976, *Smithsonian Ap. Obs. Spec. Rept.*, No. 362.
Littleton, J. E., and Davis, S. P. 1985, *Ap. J.*, **296**, 152.
Malaney, R. A. 1987, preprint.
Mathews, G. J., Takahashi, K., Ward, R. A., and Howard, W. M. 1986, *Ap. J.*, **302**, 410.
Merrill, P. W. 1952, *Ap. J.*, **116**, 21.
Phillips, J. G., and Davis, S. P. 1976, *Ap. J. Suppl.*, **32**, 537.
Phillips, J. G., and Davis. S. P. 1979, *Ap. J.*, **229**, 867.
Smith, V. V., and Lambert, D. L. 1985, *Ap. J.*, **294**, 326.
Smith, V. V., and Lambert, D. L. 1986, *Ap. J.*, **311**, 843.
Smith, V. V., and Wallerstein, G. 1983, *Ap. J.*, **273**, 742.
Sneden, C. 1973, Ph.D. Thesis, University of Texas, Austin.
Zook, A. C. 1978, *Ap. J. (Letters)*, **221**, L113.

Nucleosynthesis of Heavy Elements
—— Nuclear Physics Aspects

KOHJI TAKAHASHI

Department of Physics
The Ohio State University
Columbus, OH 43210
USA

ABSTRACT

The current status of nuclear data in heavy-element nucleosynthesis and related nucleo-cosmochronology calculations is reviewed.

INTRODUCTION

Since B^2FH [1], various scenarios and model calculations have been presented of the s-process [2,3], the r-process [3-5], the p-process [6], and the related chronology [5,7]. (See [8] for a review on nucleosynthesis in general.) On the one hand, phenomenological approaches with simple models (parametrized with astrophysical variables such as temperature, density etc.) have been successful in providing us with some idea about the astrophysical conditions most appropriate for explaining the solar abundance curve. However, it is generally hard to find astrophysical environments that fulfill such optimal conditions. On the other hand, a few scenarios have dealt with astrophysical sites at which heavy-element nucleosyntheses are somehow inescapable, although the agreements between the computed and solar abundances are not necessarily impressive enough to convince us that those are <u>the</u> very sites.

The credibility of nucleosynthesis calculations or related nucleo-cosmochronology depends on, among others, the quality of input nuclear data which have been accumulating steadily, if not as rapidly as we desire. A few lists of missing nuclear data of specific importance in those studies have been offered [9].

Before presenting a short list of that sort, let me discuss the credibility of nuclear data currently in use. I concentrate mostly on nuclear mass and ß-decay related quantities, as the following talk by Truran [10] will deal with thermonuclear reaction rates.

2. THEORY VERSUS EXPERIMENT

2.1. Nuclear Masses

Masses of very neutron rich nuclei, necessary for r-process calculations, are still unknown [11]. A detailed study of the predictive powers of various mass formulae has been made [12], a part of which is reproduced in part a) of Table 1. The first five mass formulae are from [13]. Myers and v.Groote/Hilf/Takahashi are based on the droplet model with an empirical shell correction. Seeger/Howard incorporates a Nilsson-Strutinsky-type shell correction into a liquid-drop model. Möller/Nix [14] falls into this category. Liran/Zeldes is a semi-empirical shell model, while Comay/Kelson is representing several Garvey-Kelson-type mass relations. Uno/Yamada [15] is a liquid-drop model with an empirical shell correction. The number of parameters is given as #para. RMS stands for the root-mean-square deviation in MeV from experimental masses at the time of the fit, while RMS* is that from the masses newly measured since that time [12]. Not too surprisingly, those formulae with a fewer number of parameters are seen to have somewhat higher stability with respect to fits.

In part b), the mass excess values (in MeV) predicted for neutron-rich Rb isotopes are compared with the experimental values (Wapstra/Boss [13], Wapstra/Audi [11]). An underline indicates that that isotope was included in the fit. [It may be of some interest to note that the 1976 formulae [13] (except CK, of course) had trouble to reproduce/predict the ^{92}Rb mass of WB, which has been revised

Table I

auth.	M	GHT	SH	LZ	CK	MN	UY	exper. WB	exper. WA
#para	16	56	9	178	-	13	≈500		
a) overall fit									
RMS	1.3	0.7	0.7	0.3	0.3	0.8	0.4		
RMS*	1.4	1.1	1.0	0.6	1.3	1.0	1.1		
b) now-known neutron-rich Rb									
A= 91	-77.2	-77.8	-77.5	-78.0	-78.2	-76.2	-77.9	-78.0	-77.8
A= 92	-73.8	-74.4	-74.1	-74.7	-75.1	-72.6	-75.2	-75.2	-74.8
A= 93	-72.6	-73.0	-72.5	-72.8	-73.4	-71.0	-73.1		-72.7
A= 94	-69.3	-69.5	-68.4	-68.9	-69.5	-67.1	-69.3		-68.5
A= 95	-67.9	-68.0	-66.9	-66.7	-67.3	-65.5	-66.3		-65.8
A= 96	-64.3	-64.2	-62.8	-62.3	-62.8	-61.5	-62.0		-61.1
A= 97	-62.5	-62.3	-60.7	-59.8	-60.1	-59.2	-59.6		-58.3
A= 98	-58.4	-58.2	-55.9	-55.0	-55.0	-54.7	-54.9		-54.0
S_n	4.0	3.9	3.2	3.3	2.9	3.5	3.4		3.9
Q_β	10.9	10.9	11.9	12.4	12.6	12.2	12.4		12.4
A= 99	-56.1	-55.8	-53.5	-52.2	-52.0	-51.7	-52.3		-50.9
c) unknown Rb									
A=109	-11.5	-10.6	-6.4	-6.8	-3.5	-	-6.6		
A=110	-5.1	-4.1	0.5	-0.7	3.1	-	-0.8		
S_n	1.7	1.6	1.2	2.0	1.5	-	2.2		
Q_β	17.4	17.5	18.6	18.3	19.1	-	18.2		

in WA.] For ^{98}Rb, the neutron separation energy, S_n, and β-decay Q-value, Q_β, are also given in MeV. Even in these specific examples, it can be easily seen that mass extrapolations beyond a few mass units from known nuclei might lead to much larger deviations than suggested by RMS.

Generally, both the experimental and theoretical mass surfaces are rather smooth. Because of this, predictive powers appear stronger for differential quantities (such as S_n or Q_β) which are the main concern in astrophysical applications. This can be observed in the ^{98}Rb case for which the deviations are large in the masses but relatively small in S_n and Q_β. The mass excess values predicted for very neutron-rich Rb isotopes that are near a canonical r-process path are compared in part c). To see the possible effects of a 1 MeV uncertaintiy in S_n, let us consider the classical r-process model [1] which gives $S_n^o \approx T_9/5(34-\underline{\log} n_n)$ for the odd-N isotope neighboring the (even-N) "waiting-point"(which defines the path for a given Z) isotope, where T_9 is the temperature in 10^9 K and n_n is the neutron number density per cm^3. Assuming $S_n^o \approx 2$ MeV, then LZ and SH would predict ^{109}Rb and ^{105}Rb, respectively, to be the waiting-point nucleus. A shift in calculated abundance curves by as much as four mass units can be generally realized in their comparison with the "observed" one. Alternatively, let the r-process path pass through ^{109}Rb and $T_9 \approx 1$, for example, then the appropriate n_n values calculated with LZ and SH differ four orders of magnitude, exceeding the permissible uncertainty of n_n in a realistic r-process model.

Recently, a Hartree-Fock (energy-density) mass formula has been developed [16], preliminary results of which have been applied to a calculation [17] of nuclear statistical equilibrium in certain supernova. It has also become known [18] that a deformed Hartree-Fock model could explain (and might predict) a sudden change of nuclear shapes such as the one found in neutron-rich Na. The feasibility (and necessity) of such sophisticated models for astrophysical calculations (r-process, in particular) remains to be seen.

2.2. Beta-Decay Rates

A common practice in r-process calculations is to assume that ß-decay half-lives of very neutron rich nuclei in stellar environments would not be far from their terrestrial (ground-state) values. Two sets of predicted ß half-lives, $T_{½,th}$, have been available for some time: one from the gross theory [19], the other from a shell model in the Tamm-Dankoff approximation [20]. Their predictive powers are rather poor as can be seen in Fig.1, where some $T_{½,th}$ values (TYK(MS): [19], KMO: [20]) are compared with <u>newly-measured</u> values, $T_{½,exp}$. Most of part a) and c) are taken

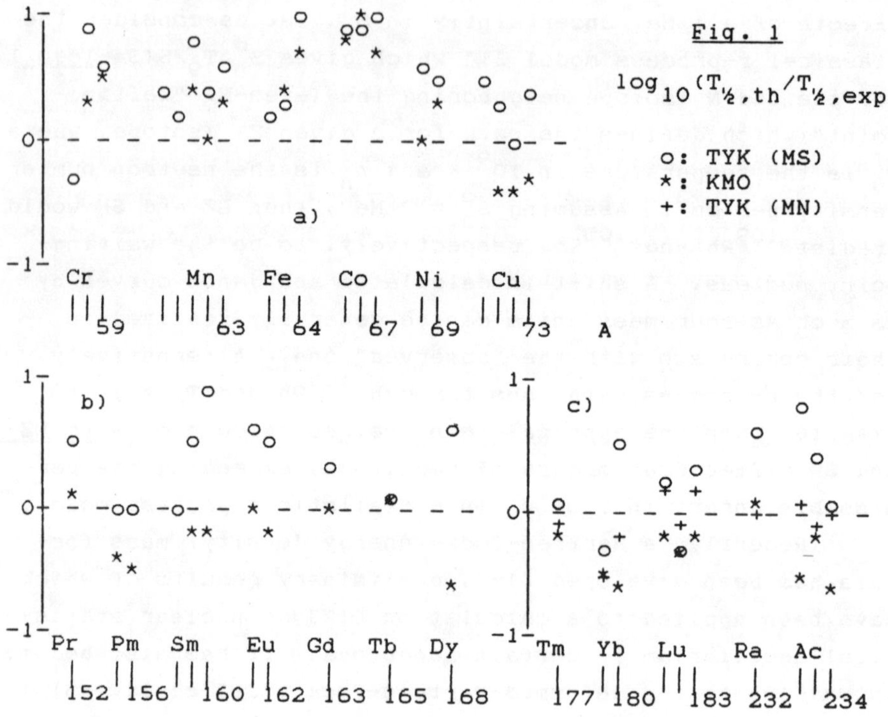

from [21] and part b) from [22]. The mass numbers attached are those for the most neutron rich isotopes. TYK(MS) is seen to generally overestimate the half-lives, while KMO tends to underestimate for heavier nuclei. Such systematic errors could cause appreciable effects on computed r-process abundance curves (cf. [23]). In part c), TYK(MN) stands for the gross-theory values calculated with Q-values from Möller/Nix mass formula [14] rather than those from Myers and Swiatecki (liquid-drop) formula [24]; they show apparent improvements from TYK(MS) in some cases.

A few attempts have been made to predict ß-decay half-lives of heavy nuclei from more elaborate models. One example was a large-scale shell-model calculation [25] for ^{130}Cd which is supposedly on the r-process path, predicting a half-life of 80 ms (with a Gamow-Teller quenching q=0.5) compared with the earlier predictions of 560 ms [19] and 244 ms (with no quenching: q=1) [20]. Recently, the experimental value of 195±35 ms has been reported [26] together with an RPA shell-model [27] value (with q=0.5?) of 300 ms. The ß-decay half-life of ^{80}Zn which is also presumably on the r-process path has been measured to be ≈540 ms [28] in agreement with the KMO [20] prediction (490 ms).

The ongoing vigorous studies of nuclei far from stability [29] would in the near future be able to fix the ß-decay half-lives along the r-process path within a factor of two or less, so that they could impose a stringent constraint upon the dynamical r-process time-scale. It should be kept in mind that the thermal excitations of low-lying states at presumably high r-process temperatures might occasionally lead to an effective half-life significantly different from the ground-state value. Such cases might concern some long-lived isomers possible near magic nuclei.

To calculate ß-decay rates under s-process conditions is not always straightforward since heavy nuclei are expected to be highly ionized in a stellar environment, in which case their terrestrial ß-decay lifetimes may not apply even for their ground-states. Indeed, a detailed treatment [30] of ionization is inescapable in some cases, as has been demonstrated in a recent compilation [31] in comparison with the previous one [32]. Another difficulty concerns the evaluation of ft-values (inverse of the nuclear matrix elements squared) of unknown ß-transitions from thermally-populated excited states, which are often obtained from systematics. This introduces an uncertainty, expected to be a factor of two or so, into the effective decay rate. Although this is probably acceptable in most s-process studies, there exist a few cases, ^{99}Tc decay being one (see Sect.3.1), in which a higher accuracy is required. Recent systematic studies of Gamow-Teller strengths via (p,n) reactions at forward angles give us some hope for improved evaluations of stellar ß-decay rates. For example, a recent (p,n) experiment [33] on ^{81}Br, a possible solar-neutrino detector, indicates a log<u>ft</u> of $4.7^{+0.2}_{-0.1}$ for the (inverse) ^{81}Kr(1/2$^-$;190 keV)→^{81}Br(3/2$^-$; g.s.) transition, which is consistent with the value, $5.0^{+0.2}_{-0.1}$, obtained from direct measurements of electron-captures [34]. (The value predicted from systematics [31] was 5.0.)

2.3. Delayed Neutron/Fission Probabilities

Possibly important roles played by ß-delayed neutron-emissions and fissions during the r-process and in the post r-process cascades have been discussed for many years. To study their effects on the abundance curve or on the Th-U-Pu chronology requires, first of all, the evaluations of $Q_ß$ and S_n or the fission barrier heights [35]. Secondly, ß-strength functions [36], which can sometimes be quite whimsical, must be given for the precursors.

Recently, a systematic study of heaviest neutron-rich progenitor nuclei for those chronometers has been performed [37] within a deformed (Nilsson) shell-model (with RPA for ß-decays [27]). The results imply that the delayed-fission probabilities might have been considerably overestimated in a previous work [38] (based on a spherical shell-model for ß-decays).

Since those nuclei involved are virtually unknown by experiment, and we know little about the astrophysical aspects of the r-process, any claim with respect to the Th-U-Pu production ratios should be taken with great care for some time to come.

2.4. Level Schemes

A detailed nuclear level scheme including radiative transition rates is occasionally required. A typical example concerns the question whether or not a long-lived isomer is in thermal equilibrium with the ground state under s-process conditions [39]. Seemingly, the more the detailed nuclear structure became known, the more it became probable that in most cases the isomers would be quickly equilibrated at least under typical s-process temperatures of ≈ 3×10^8 K. This is the case with the potentially interesting isomers $^{176}Lu^m$ [40] and $^{180}Ta^m$ [41]. Presumably, as astrophysical models mature, detailed nuclear structual information will become more necessary.

2.4. Thermonuclear Reaction Rates

A compilation of radiative neutron-capture cross sections necessary for s-process studies has recently been presented [42], which includes most recent experimental data supplemented by theoretical predictions for unknown cases. It is seen that the reliability of a simple Hauser-Feschbach calculation [43] is rather high. In some limited cases, however, an accurate determination of capture rates under stellar conditions is required. Probably the most important example concerns $^{187}Os^*$(9.8 keV) in conjunction with the Re-Os chronology [44], for which not only experimental study [45] but also a reliable theory should be desirable.

Cross sections for radiative neutron-capture on neutron-rich nuclei which are probably needed in r-process calculations are vitually unknown, although some efforts have been made to utilize the available data on delayed neutrons [46]. Some ingredients of statistical model (Hauser-Feschbach) calculations for nuclei far from stability, such as level densities, optical potentials etc., have been scrutinized theoretically [47]. (See [10] for more discussions on recent developments of statistical model calculations.) Meanwhile, the possible importance of the direct capture process in near-neutron-magic nuclei rather than the compound (statistical) process has been demonstrated [48].

Various reaction rates are needed in p-process studies [6,8] such as those of photoreactions, proton-captures, α-captures etc., depending on the scenarios. (In some cases, ß-decay rates would also play a role [49].) Little attention has been paid in recent years on improving those input data for the p-process. Recently, an experimental study of some (α,p) reaction rates which might be important in certain p-process scenarios has been initiated [50].

3. EXPERIMENTAL PROSPECTS

While experimental and theoretical prospects of improving our understanding of nuclei far from stability will be discussed elsewhere [29], I present here a few examples of experiments possible in the near future, which would determine the quantities much wanted in nucleosynthesis and cosmo-chronological studies.

3.1. (n,p) Reactions on ^{99}Ru and ^{181}Ta

An exciting development in nucleosynthesis (and, in particular, s-process) studies in recent years has been the <u>quantative</u> determination [51] of Tc (very probably ^{99}Tc) and other elemental abundances at the surfaces of several stars. As these surface abundances apparently reflect those in the material which has gone through rather recent s-processings and has been dredged up [52], it is worth attempting [53] to decipher what those data are telling us about <u>an</u> s-process indeed occurring in real stars. A dilemma, classically known as "Tc problem" [54], is that ^{99}Tc would quicky ß-decay at the expectedly high s-process temperatures and hence might not survive, in apparent contradiction with observation. The predicted ß half-lives [31,32,55] at a temperature of $3x10^8$ K ranges from about 1 yr to 20 yr. The enormous enhancements from the terrestrial value of $2x10^5$ yr reflect the expected fast Gamow-Teller transitions from the thermally populated $7/2^+$(140 keV) and $5/2^+$(181keV) excited states in ^{99}Tc to the $5/2^+$ ground-state (and, from the latter, $3/2^+$(89 keV) state) of ^{99}Ru. An experimental determination of the Gamow-Teller strengths for the transitions to the ground-state ^{99}Ru via the possible (see below) ^{99}Ru(n,p)^{99}Tc reaction would therefore reduce one of the main uncertainties in attempts to solve the problem.

The second example concerns a claim [56] that ^{182}Hf is a potential chronometer for the early solar system. The underlying assumption is that ^{182}Hf is produced solely by the r-process and not by the s-process. This requires that the ^{181}Hf half-life should be much enhanced under s-process conditions from its terrestrial value of 42 d, an assumption claimed to be questionable [31]. A ^{181}Ta(n,p)^{181}Hf experiment would determine the Gamow-Teller strength for the ^{181}Hf($9/2^+$,68 keV) → ^{181}Ta($7/2^+$,g.s.) transition, which is a candidate for such an enhancement.

The physics involved in determing Gamow-Teller strengths via (n,p) reactions is much the same as that via (p,n) reactions [57] which have been extensively studied at the Indiana cyclotron. Indeed, the experiments such as ^{54}Fe(n,p)^{54}Mn [58] would help to establish electron-capture rates [59](see also [17]) needed in supernova calculations. Recently a ^{48}Ti(n,p)^{48}Sc experiment has been performed [60] in connection with the ^{48}Ca double-ß decay. A difficulty associated with (n,p) reactions at the moment, which should be overcome in the near future, is obtaining mono-energetic neutrons with high enough intensity. In contrast to the case of (p,n) reactions, furthermore, it is seemingly difficult to find a strong known ß-decay with which to calibrate (n,p) cross sections.

3.2. ß-Decays of $^{187}Re^{75+}$ and $^{205}Tl^{81+}$

The expected [61] considerable enhancement of ^{187}Re ß-decay rates in the interior of stars and its implications to the ^{187}Re-^{187}Os chronology have been discussed in detail elesewhere [62]. The most important (experimentally unknown) quantity is the ß-decay nuclear matrix elements of the ^{187}Re($5/2^+$, g.s.) → ^{187}Os($3/2^-$, 9.8 keV) transition [30] which occurs (as a bound-state ß-decay process [63]) when Re atoms are highly ionized. Similarly, the bound-state ß decay of ^{205}Tl($1/2^+$, g.s.) to ^{205}Pb($1/2^-$, 2.3 keV) has been discussed [64], resurrecting the possible chronometric virtue [65] of this pair in the study of the early solar system (c.f. [66]). This transition is also of interest with respect to a planned ^{205}Tl solar-neutrino detector.

Since these transitions are first-forbidden, it would be rather difficult to deduce the matrix elements from (p,n) reaction data. A better possibility would be to detect bound-state ß-decays directly. To date, no bound-state ß-decay process has been confirmed experimentally. Recently, an experimental setup has been proposed [67], which would lead to a clean measurement of $^3H^+$ bound-state ß-decays. The predicted half-lives of fully ionized ^{187}Re and ^{205}Tl are 14 yr and 0.3 yr [67]. [The half-life of neutral ^{187}Re is $4x10^{10}$ yr, while ^{205}Tl is terrestrially stable.] The prospects of measuring such ß-decays of fully stripped heavy ions have been discussed [67,68].

ACKNOWLEDGMENTS

I should like to thank Richard N. Boyd, David Krofcheck and Evan Sugarbaker for helpful discussions. This work has been supported by the National Science Foundation under grant PHY-8600749 and by the Ohio State Univeristy Office of the Vice President for Research and Graduate Studies.

REFERENCES

[1] Burbidge,E.M., Burbidge,G.R., Fowler,W.A. and Hoyle, F., Rev.Mod.Phys. 29, 547 (1957).

[2] Ulrich,R.K., Essays in Nuclear Astrophysics, eds. C.A.Barnes, D.D.Clayton and D.N.Schramm (Cambridge Univ. Press, 1982), p. 301.

[3] Mathews,G.J. and Ward,R.A., Rep.Prog.Phys. 48, 1317 (1985).

[4] Hillebrandt,W., Space Sci.Rev. 21, 639 (1978).

[5] Schramm,D.N., Essays in Nuclear Astrophysics, op. cit., p. 325 .

[6] Woosley,S.E. and Howard,W.M., Astrophys.J.Suppl. 36, 285 (1978) and references therein.

[7] Fowler, W.A. and Meisel, C.C., Symp. on Cosmogonical Processes, Boulder, 1985.

[8] Truran,J.W., Ann.Rev.Nucl.Part.Sci. 34, 53 (1984).

[9] Mathews,G.J., BNL Report BNL-51778, 1984, p. 485; Arnould, M., TRIUMF Report TRI-85-1, 1985, p. 29.

[10] Truran,J.W., this proceedings.

[11] Wapstra,H. and Audi,G., Nucl.Phys. A432, 1 (1984).

[12] Haustein,P.E., Proc. 7th Int.Conf. on Nuclear Masses and Fundamental Constants, Darmstadt, 1984 (TH Darmstadt, 1985), p. 413.

[13] Maripuu,S., At.Dat.Nucl.Dat.Tables, 17, 411 (1976).

[14] Möller,P. and Nix,R.J., At.Dat.Nucl.Dat.Tables 26, 165 (1981).

[15] Uno,M. and Yamada,M., INS Report INS-NUMA-40, 1982.

[16] Tondeur,F., CERN Report CERN-81-09, 1981, p. 81.

[17] Takahashi,K. and Yokoi,K., CERN Report CERN-81-09, 1981, p. 351.

[18] Epherre,M., Audi,G. and Campi,X., CERN Report CERN-81-09, 1981, p. 62.

[19] Takahashi,K., Yamada,M. and Kondoh,T., At.Dat.Nucl. Dat.Tables 12, 101 (1973).

[20] Klapdor,H.V., Metzinger,J. and Oda,T., At.Dat.Nucl. Dat.Tables 31, 81 (1984).

[21] Rykaczewski,K. et al., XXIII Int. Winter Meeting on Nuclear Physics, Bormio, 1985.

[22] Mach,H. et al., Phys.Rev.Lett. 56, 1547 (1986).

[23] Klapdor, H.V. et al., CERN Report CERN-81-09, 1981, p. 341.

[24] Myers,W.D. and Swiatecki,W.J., Nucl.Phys. 81, 1 1966).

[25] Takahashi,K., Mathews,G.J. and Bloom,S.D., ACS Symp. Ser. No.324, 1986, p. 145.

[26] Kratz,K.L. et al., Z.Phys. A325, 489 (1986).

[27] Krumlinde,J. and Möller,P., Nucl.Phys.A417, 419(1984).

[28] Ekström,B. et al., Phys.Scrip. 34, 614 (1986); Gill,R. L. et al., Phys.Rev.Lett. 56, 1874 (1986).

[29] Proc. 5th Int.Conf. on Nuclei far from Stability, Rosseau Lake, Sept. 1987, to appear.

[30] Takahashi,K. and Yokoi,K., Nucl.Phys. A404, 578(1983).

[31] Takahashi,K. and Yokoi,K., At.Dat.Nucl.Dat.Tables 36, 375 (1987).

[32] Cosner,K. and Truran,J.W., Astrophys.Space Sci. 78, 85 (1981).

[33] Krofcheck,D. et al., Phys.Lett. 189, 299 (1987).

[34] Davids,C,N. et al., Phys.Rev. C35, 1114 (1987); Lowry,M.M. et al., Phys.Rev. C35, 1950 (1987).

[35] Howard,W.M. and Möller,P., At.Dat.Nucl.Dat.Tables 25, 219 (1980).

[36] Klapdor,H.V., Prog.Part.Nucl.Phys. 10, 131 (1983); Takahashi,K., BNL Report BNL-51778, 1983, p. 157.; Hamilton,J.H., Hansen,P.G. and Zganjar,E.F., Rep.Prog. Phys. 48, 631 (1985).

[37] Meyer,B.S. et al., ACS Symp.Ser. No. 324, 1986, p.149.

[38] Thilemann,F.K., Mezinger,J. and Klapdor,H.V., Astron. Astrophys. 123, 162 (1983).

[39] Ward,R.A., Astrophys.J. 216, 540 (1977).

[40] Beer,H., Käppeler,F., Wisshak,K. and Ward,R.A., Astrophys.J.Suppl. 46, 295 (1981); Norman,E.B. et al., Astrophys.J. 291, 834 (1985).

[41] Norman,E.B. et al., Astrophys.J. 281, 360 (1984).

[42] Bao,Z.Y. and Käppeler,F., At.Dat.Nucl.Dat.Tables 36, 411 (1987).

[43] Holmes,J.A., Woosley,S.E., Fowler,W.A. and Zimmerman, B.A., At.Dat.Nucl.Dat.Tables 18, 305 (1976).

[44] Woosley,S.E. and Fowler,W.A., Astrophys.J. 233, 411 (1979).

[45] Hershberger,R.L. et al., Phys.Rev. C28, 2249 (1983).

[46] Kratz,K.L, Ziergert,W., Hillebrandt,W. and Thielemann, F.K., Astrophys.J. 125, 381 (1983); Wiescher,M. et al., AIP Conf.Proc.No.125, 1985, p.908.

[47] Arnould,M. and Tondeur,F., CERN Report CERN-81-09, 1981, p. 229; Thielemann,F.K., Arnould,M. and Truran, J.W., MPI Preprint MPA-262, 1986.

[48] Mathews,G.J., Mangoni,A., Thielemann, F.K. and Fowler, W.A., Astrophys.J. 270, 740 (1983).

[49] Arnould,M., Astron.Astrophys. 46, 117 (1976)

[50] Boyd,R.N., private communication, 1987.

[51] Smith,V.V. and Wallerstein,G., Astrophys.J. 273, 742 (1983).

[52] Iben,I.Jr. and Renzini,A., Ann.Rev.Astron.Astrophys. 21, 271 (1983).

[53] Mathews,G.J., Takahashi,K., Ward,R.A. and Howard,W.M., Astrophys.J. 302, 410 (1986); Takahashi,K., Mathews, G.J., Ward,R.A. and Becker,S.A., Nucleosynthesis and its Implications to Nuclear and Particle Physics (Reidel, 1986), p. 285.

[54] Cameron,A.G.W., Astrophys.J. 130, 452 (1959).

[55] Takahashi,K., Mathews,G.J. and Bloom,S.D., Phys.Rev. C33, 296 (1986).

[56] Norman,E.B. and Schramm,D.N., Nature 304, 515 (1983).

[57] Goodman,C.D., Comments Nucl.Part.Phys. 10, 117 (1981).

[58] Vetterli,M.C. et al., Phys.Rev.Lett 59, 439 (1987).

[59] Fuller,G.M., Fowler,W.A. and Newman,M.J., Astrophys.J. 293, 1 (1985); Bloom,S.D. and Fuller,G.M., Nucl.Phys. A440, 511 (1985); Muto,K., Nucl.Phys. A451, 481(1986).

[60] Alford,W.P., private communication, 1987.

[61] Calyton,D.D., Nature 224, 56 (1969).

[62] Yokoi,K., Takahashi,K. and Arnould,M., Astron. Astrophys. 177, 65 (1983).

[63] Daudel,R., Benoist,P., Jaques,R. and Jean,M., C.R. Acad.Sci. 224, 1427 (1947).

[64] Yokoi,K., Takahashi,K. and Arnould,M., Astron. Astrophys. 145, 339 (1985).

[65] Blake,J.B., Lee,T. and Schramm,D.N., Nature Phys.Sci. 242, 98 (1973).

[66] Clayton,D.D., Space Sci.Rev. 24, 147 (1979).

[67] Takahashi,K., Boyd,R.N., Mathews,G.J. and Yokoi,K., Phys.Rev. C36, in press.

[68] Kienle,P., Nucl.Phys. A447, 419c (1985); Boyd,R.N., this proceedings.

R-PROCESS NUCLEOSYNTHESIS AND COSMOCHRONOLOGY

JOHN J. COWAN
Department of Physics and Astronomy, University of Oklahoma
Norman, OK 73019
and
Harvard-Smithsonian Center for Astrophysics, Cambridge, MA 02138

F.-K. THIELEMANN
Department of Astronomy, Harvard University, Cambridge, MA 02138
and
J. W. TRURAN
Department of Astronomy, University of Illinois, Urbana, IL 61801

ABSTRACT

While the site (or sites) for the r-process is still not known, recent work has helped understand much about how the r-process nuclei are produced and the types of astrophysical environments that might be suitable formation sites. Using a specific model for r-process nucleosynthesis which predicts an overall abundance pattern in good agreement with the distribution of abundances of the r-process nuclei in solar system matter, we have calculated the production ratios of the important nuclear cosmochronometers. An r-process network calculation was performed, utilizing the best available determinations of the rates of neutron capture and photodisintegration, beta decay, beta-delayed neutron emission, and beta-delayed fission. The calculated production ratios were found to be $^{232}Th/^{238}U = 1.60$, $^{235}U/^{238}U = 1.16$, and $^{244}Pu/^{238}U = 0.40$. Using an exponential model for galactic evolution, these ratios give an age range for the Galaxy of $12.4 - 14.7$ billion years depending upon the initial galactic disk enrichment. Nuclear uncertainties result in an additional spread of approximately 2 billion years in the galactic age, while more complex galactic evolutionary models might increase the age by as much as 2 - 3 billion years. The chronometric age for the Galaxy found in this paper is compared to other chronometric determinations, as well as age estimates for the Galaxy and the universe obtained by other means.

1. INTRODUCTION

1.1 Nucleocosmochronology

From studies of the decay of long-lived radioactive nuclei, nucleocosmochronology allows the determination of the age of the Galaxy and hence the universe. Particularly useful for these studies are the long-lived nuclear chronometers: ^{187}Re ($\tau_{\frac{1}{2}} = 4.5 \times 10^{10}$y), ^{232}Th ($\tau_{\frac{1}{2}} = 1.405 \times 10^{10}$y), ^{238}U ($\tau_{\frac{1}{2}} = 4.46 \times 10^{9}$y), and ^{235}U ($\tau_{\frac{1}{2}} = 7.038 \times 10^{8}$y). Nuclei with shorter half lives, like ^{244}Pu ($\tau_{\frac{1}{2}} = 8.26 \times 10^{7}$y) and ^{129}I ($\tau_{\frac{1}{2}} = 1.57 \times 10^{7}$y), can be used to obtain information concerning the rate of nucleosynthesis in the Galaxy occurring close to the time of formation of the solar system.

Age determinations from cosmochronology, utilizing the r-process chronometer pairs $^{232}Th/^{238}U$, $^{235}U/^{238}U$ and $^{244}Pu/^{238}U$, have been made by a number of workers[1-16]. Other age determinations have been made using the parent-daughter ratios $^{187}Re/^{187}Os$, $^{87}Rb/^{87}Sr$, and $^{207}Pb/^{235}U$ rather than the actinide ratios[17-22]. In most studies of nuclear cosmochronology, an exponential model for the time dependence of the nucleosynthesis production in the Galaxy has been assumed, based on the original work by Fowler and Hoyle[1]. Recently, more detailed models of chemical evolution have also been introduced to calculate the age of the Galaxy[12,13,23]. An alternative "model independent" approach considered by Schramm and Wasserburg[2] allows the determination of the mean age of the elements, but ultimately itself becomes somewhat model dependent in its determination of a total galactic age[24,25]. Since a detailed knowledge of the variations of the nucleosynthesis rate over the course of Galactic history is not available and for purposes of comparison, the exponential model of chemical evolution has been adopted in this paper.

Age determinations resulting from these various models are found to be sensitive to : (1) the production ratios for the radioactive isotopes of interest in their appropriate nucleosynthesis sites; (2) the time dependence of the nucleosynthesis rate over the course of galactic evolution (including specifically the level of initial enrichment of the gas from pre-disk stellar populations); and (3) the isotopic ratios in the matter which condensed to form meteorites at the time of formation of the solar system, 4.6 billion years ago. The uncertainties associated with estimates of the production ratios introduce large uncertainties into the age determinations. One of the aims of this paper is to address, specifically, this issue. We will attempt to identify the critical nuclear and astrophysical constraints upon the production ratios and we will provide values of the production ratios in the context of one specific model. In this regard, it is important to note that the cosmochronologically important radioactive isotopes ^{187}Re, ^{232}Th, ^{235}U, ^{238}U, and ^{244}Pu are formed in a single astrophysical process: the rapid neutron capture process (r-process) which synthesizes heavy elements from the mass region A$\sim$70 through the actinides, and perhaps beyond. This feature alone helps substantially to reduce the uncertainties.

1.2 The Search for the Site of the r-Process

There are also uncertainties in all predictions of nucleocosmochronology and Galactic chemical evolution due to the fact that the astrophysical site of r-process nucleosynthesis has not yet been identified[26]. A number of sites have been suggested. (See also Truran[27] for a review of existing theoretical models for the operation of the r-process).

1.2.1 Sites with expansion and cooling of neutronized matter

Early work[28], noting the large neutron fluxes required to produce the r-process nuclei, suggested that high temperature and explosive environments such as the neutronized cores of supernovae were the site for the r-process. Detailed models of the regions just outside of the neutronized cores of supernovae were studied by a number of groups[29–32]. It was also suggested that the r-process nuclei could be produced in "bubbles" or "jets" of neutronized matter ejected from the collapse of rotating magnetized stellar cores[33–36]. Other more exotic models such as neutronized matter resulting from the collision between a neutron star and a black hole were suggested as r-process sites[37].

None of the supernova core models, however, were able to reproduce accurately all of the solar system abundances under physically plausible conditions. One of the problems (particularly with the early models) was that nuclear mass formulas at the time were not reliable in predicting properties of nuclei far from the valley of beta stability. More recent mass formulas have been much more successful in this regard. In addition to the uncertainties in nuclear physics inherent in many of the earlier supernova core models, there were great uncertainties in the detailed astrophysical conditions that occurred during a supernova explosion. Many of these models had the further problem of not being able to provide a natural (or consistent) mechanism for the ejection of the synthesized material into the interstellar medium.

1.2.2 Sites with neutron captures on pre-existing iron-peak nuclei

Since none of the supernova core models was successful in uniquely defining the site for the r-process, other astrophysical environments were suggested as possible sites. One suggested site was the helium zone of a star undergoing a supernova explosion[30,38,40]. As a result of the passage of the supernova shock, high temperature and density conditions will occur in the helium zone of the pre-supernova star. Neutrons would be produced in the helium zone by $^{13}C(\alpha,n)^{16}O$ reactions. The ^{13}C would presumably be produced by the mixing of hydrogen into the helium zone and the subsequent capture of protons by the preexisting ^{12}C. The production of the r-process nuclei would occur as a result of the capture of neutrons by the iron-peak "seed nuclei" already present in the helium zone[41–44].

A number of dynamic (or non-equilibrium) calculations were performed to examine the type of conditions that might produce the solar system r-process abundances[45]. It was found that the solar system r-process abundance curve could be reproduced under conditions that might be typical in a helium zone of a star undergoing a supernova explosion[45]. This model also overcame one of the problems encountered by some earlier supernova core models. That is, it provided a natural ejection mechanism for the synthesized neutron-rich matter. The total r-process abundances in this model depend critically upon the amount of neutron source (*i.e.*^{13}C). The amount of ^{13}C required to produce the total solar system r-process abundances was found, however, to be exceedingly large - approximately 5% by mass of the helium zone, and thus it was concluded that the helium zones cannot be the site for the bulk of the solar system r-process abundances[45].

Another suggested site that also depended upon a preexisting seed was the carbon zone of a star undergoing a supernova explosion. As in the case of the helium zone, the passage of the supernova shock will produce high temperature and density conditions in the carbon zone of the presupernova star. The ^{22}Ne in the carbon zone (produced by subsequent alpha captures on ^{14}N) could then produce neutrons principally by the $^{22}Ne(\alpha,n)^{25}Mg$ reaction. The neutrons produced could then be captured by the iron-peak nuclei (*i.e.* "seed nuclei"), and in this manner the r-process nuclei could be formed[46].

This carbon zone model, like the helium zone site, provided a natural ejection mechanism for getting the synthesized material into the interstellar medium. Its major problem was a lack of sufficient neutrons to produce the total solar system r-process abundances. In particular, the products of ^{22}Ne burning (*e.g.* ^{25}Mg) are very effective at capturing the neutrons and thus preventing a "build-up" to heavier nuclei. Thus, while some neutron-rich isotopes may be produced[47], this site also cannot be responsible for producing the bulk of the r-process nuclei.

The helium core of low mass stars was also suggested as a possible site[48,49]. There have been many suggestions that as a result of helium core flashes in these stars, proton-rich material might be convectively mixed into the core[50]. As a result, ^{13}C would form from the capture of protons by ^{12}C already in the core. The heavier "seed nuclei" already in the core could then form neutron-rich heavy nuclei by capturing neutrons produced by ^{13}C(α,n) reactions in a manner similar to that in the helium zones in supernovae. While the ejection of such synthesized material from the cores might be more difficult than in the supernovae models, this model had the advantage of requiring very little synthesized material per star to produce the total r-process galactic abundances[48,49]. Using the present values for the beta decay rates and neutron capture cross sections, exceedingly large amounts of ^{13}C are required to produce the total solar system r-process abundances. Such large amounts of ^{13}C have not been found to exist in the helium core in any current stellar evolutionary model, and it was again concluded that this site is unlikely to produce the bulk of the r-process nuclei[45].

Thus, although much has been learned about the production of neutron-rich nuclei, the site for r-process nucleosynthesis has not been determined. It is also not clear whether there are several sites with, for example, the heavy r-process nuclei made in the cores of supernovae and light nuclei in helium zones of supernovae. Recent improvements in the nuclear mass formulae needed for predicting the properties of neutron-rich nuclei and increasingly sophisticated astrophysical models (particularly for supernovae) offer hope that the site or sites for the r-process will be identified.

1.3 Overview

Although the site for the r-process has not been identified, it is still possible to use r-process calculations to determine the age of the Galaxy. The specific conditions considered in the present paper are those associated with dynamic helium burning environments[45]. Although there exist strong arguments against this representing the dominant astrophysical site of the r-process (as indicated above), we note that the theoretical calculations which have been performed predict an abundance pattern which reproduces the gross features of the solar system r-process curve. This indicates that the r-process capture path is in approximately the right position in the N-Z plane and that the different mass numbers are produced in roughly the correct proportions.

Our aim in this paper is to determine the production ratios by calculating an r-process distribution with a complete non-equilibrium reaction network which we know provides a reasonable fit to the solar r-process. These calculations employ the best available estimates of the relevant and critical input nuclear parameters: the rates of neutron capture and photodisintegration reactions; the lifetimes of beta decay reactions; and the rates of beta-delayed neutron emission and beta-delayed fission, the effects of which have frequently been ignored in earlier calculations of the production ratios. The input nuclear parameters which provide the basis for this study, including particularly the fission properties of the neutron-rich progenitors of the important nuclear chronometers, are reviewed in section 2. Our results obtained for the production ratios, and their implications for nucleocosmochronology are presented in section 3. Discussion of the results and comparisons of chronometric and other age determinations for the Galaxy and the universe are contained in section 4.

2. CALCULATIONS

The calculations were performed using a code composed of coupled charged–particle and neutron-capture beta-decay networks. This code is a nonequilibrium or dynamic code that does not assume $(n,\gamma)\rightleftarrows(\gamma,n)$ equilibrium (also known as the waiting point approximation). Thus, the r-process path was determined directly by competition between neutron-capture and beta decay, and there are no artificial waiting points. (For a more complete discussion of the nature of the code and the mathematical treatment used in solving the numerical equations, see Cowan *et al.* [42]).

The nonequilibrium calculations depend critically upon the nuclear data, particularly neuron-capture and beta-decay rates for the nuclei. The number of nuclei in the r-process network is determined by the position of the neutron-drip line and thus by the choice of the mass formula. Using the mass formula of Truran, Cameron, and Hilf[51)], there are a total of 6033 nuclei (from Z = 14 to Z = 114) included in our r-process reaction network. Wherever possible experimental values are used, but most of the nuclei in the network do not have experimentally determined values.

The neutron capture cross sections have been calculated using the standard (Hauser Feshbach) methods as described by Truran[52)]. The neutron binding energies and the shell and pairing energy corrections (upon which the nuclear level density systematics are based) were determined using the mass formula of Truran, Cameron and Hilf[51)]. The level density prescription described by Gilbert and Cameron[53)] was adopted, and modified by the use of a revised relation between the level density parameter *a* and the mass formula shell corrections[51)]. This same mass formula was also used to calculate the Q values needed in determining the photodisintegration rates.

Measured beta decay rates have been used wherever possible, but for most of the nuclei measured rates are not available. The beta decay rates for these nuclei were calculated by Klapdor *et al.*[44)]. Some of our early r-process studies[39−41)] employed beta-decay rates calculated using the "gross theory"[54)]. The rates determined by Klapdor *et al.*[44)] were found to be on average 2-5 times faster than the rates determined using the "gross theory"[55)]. While it is not yet clear which set of theoretical rates provides the most consistent agreement with all measured values, some recent experimental determinations[56)] agree well with the rates of Klapdor *et al.*[44)].

The percentages of fission and neutron emission after β-decay (β-delayed fission and neutron emission) for nuclei with proton numbers $75 \leq Z \leq 100$ were calculated by Thielemann, Metzinger, and Klapdor[57)], with the β-strength function taken from Klapdor *et al.*[44)]. The fission barrier heights and nuclear masses were taken from Howard and Möller[58)] and correspond to the results shown in Figure 5c of Thielemann *et al.*[57)].

In order to ensure that the shape of the r-process abundance curve does not reflect artifacts of different sources of nuclear data, it is desirable that nuclear parameters be taken from one consistent source describing all properties (β-decay, masses, fission barriers, etc.) for nuclei far from stability. Particularly for β-delayed fission and neutron emission, where the Q values come from a mass formula (liquid drop or droplet model), it is important that the fission barrier heights are calculated with the same mass formula parameters. Thielemann, Metzinger, and Klapdor[57] performed calculations for β-delayed fission and neutron emission using three different mass formulae : Hilf, von Groote, and Takahashi[59]; von Groote, Hilf, and Takahashi[60]; and Howard and Möller[58]. Using the fission barrier heights calculated with the Howard and Möller[58] mass formula while taking the β-decay Q values of the Hilf *et al.*[59] mass formula, led to the prediction of very large percentages of β-delayed fission which now seem to be overestimates.

There is some indication that mass formulae like those of Myers[61], von Groote, Hilf, and Takahashi[60] and Howard and Möller[58] tend to deviate from experimental masses far from stability[62]. This led Thielemann *et al.*[11,57] to use a different mass formula[59] to perform their r-process calculations and predictions for chronometer ratios. However, based on the analysis by Hoff[63] who compared the abundances (after beta decays) produced as a result of thermonuclear explosions with theoretical predictions, it now seems to be more important to perform a consistent calculation with fission barriers and Q-values from the same source. (This is especially true since presently there are no fission barrier calculations available for mass formulae which more accurately predict experimental masses far from stability.) As noted previously, the neutron capture and photodisintegration rates needed in the r-process study were calculated with Q-values from the mass formula of Truran, Cameron, and Hilf[51]. The macroscopic parameter and mass predictions of this formula are very close to those of the mass formulae of Myers[61] and Howard and Möller[58]. Therefore, the additional inclusion in our present work of β-decay effects based on the Howard and Möller[58] mass formula should not introduce any artificial behavior.

3. RESULTS

3.1 R-Process Production Ratios

Using the nuclear input data for neutron capture, photodisintegration, beta decay rates, beta-delayed neutron emission and beta-delayed fission described in the previous section, a number of nonequilibrium (or dynamic) r-process calculations were performed. Figure 1 shows the model calculation that best fits the solar system r-process abundances. The top curve in the figure shows the solar system abundances from Cameron[64]. The lower theoretical curve (with arbitrary vertical normalization) was produced by assuming conditions of temperature and density that might occur in the helium layer of a star undergoing a supernova explosion. Those conditions were an initial temperature, T, of 4 x 10^8 K and an initial density, ρ, of 10^4 gcm^{-3}. The neutron source was $^{13}C(\alpha,n)^{16}O$ and the ^{13}C was assumed to be approximately 6% by mass in the helium zone. This mass fraction is extremely large and has not been found to exist in any stellar evolutionary models. The requirement of such a large mass fraction for ^{13}C makes it unlikely that the total solar system r-process abundances could have been produced in explosive helium burning reactions[15,45,65]. The purpose of the calculations described in this paper, however, is specifically to determine the production ratios of the nuclear chronometers and hence the age of the Galaxy. A calculation which reproduces the solar r-process abundances perfectly, regardless of its astrophysical relevance, will identify the correct location of the r-process path, as indicated by the presence of the major abundance peaks; and by revealing even the substructures in the r-process curve, this calculation will also determine the correct abundance pattern within the r- process path. It follows that such a calculation would also correctly predict the abundances of the nuclear chronometers produced in the r-process.

Even though the overall shape of the theoretical curve shown in Figure 1 is similar to those resulting from some of our previous calculations[45], it is important to note that this is the first time we have explicitly included beta delayed fission in our calculations. Also, the rates for beta−delayed neutron emission calculated as described above differ from the rates used previously in our calculations. Finally, we note that the theoretical curve shown in Figure 1 has been dashed for mass numbers above A $\sim$ 280. Those nuclei were produced by an r-process path with initially high neutron abundances which bypassed areas of almost complete beta-delayed fission for $Z \leq 100$. Since the present calculations did not include any fission effects for $Z > 100$ and since nuclei with $N > 186$ and $Z > 100$ will also undergo spontaneous or induced fission[66−68], the abundances of nuclei with masses greater than A$\sim$ 280 are uncertain. The abundances of those nuclei will not, however, affect the calculated abundances of the nuclear chronometers.

The r-process calculations illustrated in Figure 1 reproduce the major features of the solar system r-process curve, and therefore allow the abundances of the isotopes in the uranium region to then be determined. The abundances of the chronometers ^{232}Th, ^{235}U, ^{238}U, and ^{244}Pu were calculated by summing over the abundances of the short-lived progenitors in alpha decay chains. The ratios of the nuclear chronometers determined from the calculations shown in Figure 1 are $^{232}\mathrm{Th}/^{238}\mathrm{U} = 1.60$, $^{235}\mathrm{U}/^{238}\mathrm{U} = 1.16$ and $^{244}\mathrm{Pu}/^{238}\mathrm{U} = 0.40$. These values are listed and compared to other determinations of the production ratios in Table 1.

The beta-delayed neutron emission and fission rates for the calculations were calculated using the same mass formula[58)] that was used to determine the fission barrier heights. (Masses from the Howard and Möller mass formula are not available for nuclei with $Z < 76$. For those nuclei, the mass formula of von Groote *et al.*[60)] was used.) As noted above, experimental evidence suggests that this is the most consistent approach for assembling the disparate sources of nuclear data. To determine the sensitivity of the production ratios to the nuclear input data, several other calculations were performed. The beta-delayed neutron emission and the beta-delayed fission rates were recalculated using the mass formula of von Groote *et al.*[60)] rather than that of Howard and Möller[58)]. Thus, in this case the mass formula used to determine the fission barrier heights was not the formula used in determining the beta-delayed neutron emission and fission rates. Using this " less consistent" set of nuclear data and the same astrophysical conditions described above that produced a good fit to the solar system r-process abundances, dynamical r-process calculations were again performed. The general nature of the theoretical r−process curve was unchanged and the fit to the solar system r-process abundance curve was good. The nuclear chronometer ratios were also recalculated and found to be $^{232}\mathrm{Th}/^{238}\mathrm{U} = 1.70$, and $^{244}\mathrm{Pu}/^{238}\mathrm{U} = 0.35$ which were only slightly different than found for the case shown in Figure 1. The ratio of $^{235}\mathrm{U}/^{238}\mathrm{U}$ was again found to be 1.16.

While the fit to the solar system curve shown in Figure 1 is good in the sense that all features (major peaks and substructures) are reproduced, problems are still evident. In particular, we note the displacement of some of the features as a function of A. This might reflect imperfections in the r−process scenario (*i.e.* an incorrect neutron flux) as well as imperfections in the nuclear mass formula. In order to estimate the uncertainties involved in determining the chronometer production ratios as a result of deviations between the theoretical fit and the solar system r-process curve, additional calculations were made under different astrophysical conditions which resulted in different r-process abundance patterns. These calculations were made using the "most consistent" set of nuclear data (*i.e.* using the Howard and Möller[58] mass formula for all the beta-delayed rates). It was found that for calculations which gave a very poor fit to the solar system abundances, the chronometer production ratios can differ by up to approximately 0.30. This leads us to the conclusion that abundance distributions which still give relatively good fits to the solar r-process abundances can result in errors in the chronometer ratios of the order of 0.1, comparable to the errors originating from nuclear uncertainties.

3.2 Galactic Age Determination

The ratio of the abundances of two nuclei A and B at the time of the formation of the solar system, i.e. after a nucleosynthesis epoch of duration $t = \Delta$, can be written as (see Cowan *et al.*[73] for details)

$$\frac{N_A(\Delta)}{N_B(\Delta)} = \frac{P_A}{P_B} f(\lambda_A, \lambda_B, S_0, \lambda_R, \Delta) \tag{1}$$

where P_A/P_B are the production ratios of nuclei A and B, λ_A and λ_B the decay rates of radioactive nuclei A and B, S_0 indicates the fraction of stable r-process material present at solar system formation resulting from the initial enrichment, and λ_R is the time dependence of the effective nucleosynthesis rate.

In order to fulfill observational constraints for the metallicity distribution of stars[74] and the age metallicity relation[75], it is necessary to assume an initial enrichment in metals at the beginning of the galactic disk evolution of the order $Z_0/Z_{sol} = 0.1 - 0.3$[19,24,76]. By assuming an instantaneous initial disk enrichment, it should be noted that our results give lower limits to the age of the Galaxy, because the duration of the initial burst of star formation is not included in the age determination. While this duration time might be as short as $\sim 2 \times 10^8$ years, it could be as long as $(2-3) \times 10^9 y$. This is shown in the more elaborate, but still analytically solvable, models of Clayton[12,13].

Using the simple exponential model, where λ_R is treated as a constant, the duration of galactic nucleosynthesis (Δ) before the formation of the meteorites and the solar system $4.6 \times 10^9 y$ ago has been determined. Eq.(1) was solved for the chronometer pairs $^{232}Th/^{238}U$ and $^{235}U/^{238}U$ (the very short half-life of

^{244}Pu, which can only provide information about the last few $10^8 y$ before the formation of the solar system, makes the ratio of $^{244}Pu/^{238}U$ relatively useless for age determinations), with the production ratios P_A/P_B given above and listed in Table 1, the meteoritic abundance ratios $N_A(\Delta)/N_B(\Delta)$, and by varying the initial disk metallicity enrichment, S_0, between 0.1 and 0.3. This results in values for λ_R and Δ. The meteoritic abundance ratios used here are 0.317 for $^{235}U/^{238}U$ and 2.32 for $^{232}Th/^{238}U$[77].

Table 2 gives the results for the standard case, the r-process calculation with an abundance distribution which best fits the solar r-process abundances (*i.e.* the calculation shown in Figure 1). The values range from 7.8 $\times$ $10^9 y$ (for an initial disk enrichment of 0.29 or 29%) to 10.1 $\times$ $10^9 y$ (for $S_0 = 0.11$) for Δ. This translates into Galactic ages of 12.4 $\times$ $10^9 y$ to 14.7 $\times$ $10^9 y$. These ages do not include the duration of Galactic evolution which produced the initial enrichment. As discussed above, more complex evolution models, which start with a small gas mass in the disk and have an increasing star formation rate during the early infall period, can result in a duration of galactic nucleosynthesis which is larger by as much as approximately $(2-3) \times 10^9 y$.

Age determinations were also made using the production ratios calculated with somewhat less consistent nuclear data. The results are an age for the Galaxy of 10.3 $\times$ $10^9 y$ to 11.2 $\times$ $10^9 y$ for initial enrichments, S_0, of .30 to 0.14 respectively. We see that the nuclear data can introduce an uncertainty in the galactic age of between two and three billion years, in addition to the uncertainty due to the rate of initial enrichment.

4. DISCUSSION

Using the most consistent nuclear data and a reasonable model for chemical evolution, our results indicate a Galactic age of 12.4 $\times$ $10^9 y$ to 14.7 $\times$ $10^9 y$, depending upon the amount of initial disk enrichment. It is clear, however, from the previous section that uncertainties remain. These uncertainties prevent us from ruling out Galactic ages as young as 10.3 $\times$ $10^9 y$ (based on nuclear uncertainties) or as old as approximately 16.7 $\times$ $10^9 y$ (based on uncertainties in the model of chemical evolution of the Galaxy). Much more work needs to be done to eliminate these and other uncertainties in the age determination.

Recognizing the uncertainties, it is possible to use the Galactic age determined from nucleocosmochronology to estimate the age of the universe. To do so requires adding the elapsed time between the big bang and the formation of the Galaxy. While this exact time is not well known, a frequently adopted value is $10^9 y$ or 1 Gyr[16,79]. Thus, we will use this value for purposes of comparison, but note that longer times between the big bang and the formation of the Galaxy have also been suggested[80]. Our best determination for the Galactic age, as listed in section 3, therefore, suggests an age for the universe of

$$13.4\ Gyr \leq\ t_0\ \leq 15.7\ Gyr. \tag{2}$$

We note again that our uncertainties do not let us rule out ages for the universe as young as 11.3 Gyr or as old as approximately 17.7 Gyr.

There have been other recent chronometric determinations of the age of the Galaxy and hence the age of the universe. Using production ratios different than ours (see Table 1), Fowler and Meisl[72] and Fowler [16] find a relatively low age for the universe of $t_0 = 11.0 \pm 1.6$ Gyr. Schramm and Meyer[81] put limits on the age of the Galaxy of 12.0 Gyr $\lesssim T_{GAL} \lesssim$ 19.8 Gyr which leads to 13.0 Gyr $\lesssim t_0 \lesssim$ 20.8 Gyr. They note, however, that uncertainties in the cosmochronological data give an even larger spread in their estimates of the age of the Galaxy and the universe. Butcher[82], using observations of the ratio of Th/Nd, finds a very low upper limit to the age of the Galaxy of 9.6 Gyr. Mathews and Schramm[83], however, argue that this result is model dependent and suggest that an upper limit of 15 Gyr is consistent with the Th/Nd observations.

The very low estimates for the age of the universe do not seem consistent with most estimates of the age of the Galaxy based on the study of globular clusters. Sandage[80], for example, finds ages for the globular clusters of 17 ± 2 Gyr which give an age to the universe of $t_0 = 18 \pm 2$ Gyr. Vanden Berg[84] finds similar ages for the globular clusters of 15-18 Gyr or $16 \leq t_0 \leq 19$ Gyr. Recent stellar evolution calculations, which assume compositions with enhanced [O/Fe] and [α/Fe] ratios, however, suggest that the maximum age of globular clusters is 13-14 Gyr[85]. These recent calculations therefore suggest 14 Gyr $\leq t_0 \leq$ 15 Gyr. This new age estimate for the globular clusters is still significantly larger than the age estimate for the universe of $t_0 = 10.3 \pm 2.2$ Gyr, based on the cooling of Galactic white dwarfs[79].

These age determinations for the universe can also be compared with inverse Hubble times, T_0. There is, of course, no current agreement on the value for the Hubble constant, H_0. Recent estimates of the Hubble constant, using the Tully-Fischer relation[86], give $H_0 \sim 90$ km $s^{-1} Mpc^{-1}$. The inverse Hubble time would then be $\sim$ 11 Gyr. On the other hand, Sandage and Tammann[87] and Tammann[88] find $H_0 = 50 \pm 10$ km $s^{-1} Mpc^{-1}$, using redshift-distance relations. This gives $T_0 = 20$ Gyr. The inverse Hubble time is related to the age of the universe by $t_0 = \alpha T_0$, where $\alpha = 2/3$ for $\Omega = 1$ and $\alpha \approx 1$ for $\Omega << 1$. Therefore, for $\Omega = 1$, a Hubble constant of 50 suggests an age for the universe of $\approx$ 13.3 Gyr. Still other determinations of the Hubble constant[89,90] find a value of 60 ± 15 km$s^{-1} Mpc^{-1}$, suggesting $t_0 \approx 16.7$ Gyr for $\Omega << 1$ and $t_0 \approx 11$ Gyr for $\Omega = 1$. We note here that the inflationary model of the universe suggests $\Omega = 1$.

We conclude that the uncertainty in the age of the Galaxy and the universe is still relatively large. Our current results from nuclear cosmochronology and the globular cluster ages, with a more adequate treatment of composition (*i.e.* an increased [O/Fe] ratio for low metallicities), both fall into a range of 13 - 16 Gyr and therefore agree quite well. This would argue for an intermediate value of the Hubble constant for $\Omega << 1$ or a relatively low value of $H_0 \lesssim 50$ km $s^{-1} Mpc^{-1}$ for the case of $\Omega = 1$.

5. ACKNOWLEDGEMENTS

We thank W. A. Fowler for providing us with his latest results, and we thank W. A. Fowler, D. D. Clayton, and B. Pagel for helpful comments. This work has been supported in part by the National Science Foundation (AST 82-14964 and AST 85-21705 at the University of Oklahoma, and AST 83-14415 and AST 86-11500 at the University of Illinois). J.J.C. gratefully acknowledges support from NASA grant NGR 22-007-272 and Smithsonian Institutional Funds. J.W.T. wishes to express his thanks to the Alexander von Humboldt Foundation for support by a U.S. Senior Scientist Award during his stay at the Max-Planck-Institut fur Astrophysik, Garching bei Munchen.

REFERENCES

1) Fowler, W.A. and Hoyle, F., *Ann. Phys.*, **10**, 280 (1960).
2) Schramm, D.N. and Wasserburg, G.T., *Ap. J.*, **162**, 57 (1970).
3) Truran, J.W., and Cameron, A.G.W., *Ap. Space Sci.*, **14**, 179 (1971).
4) Fowler, W.A., in *Cosmology, Fusion and Other Matters* ed. F. Reines (Boulder:Colorado Associated University Press), p. 67 (1972).

5) Fowler, W.A., *Proceedings of the Welch Foundation Conferences on Chemical Research* **XXI**, ed. W.D. Milligan (Houston:Robert A. Welch Foundation) p. 61 (1978).

6) Tinsley, B.M., *Ap. J.*, **198**, 145 (1975).

7) Tinsley, B.M., *Ap. J.*, **216**, 548 (1977).

8) Hainebach, K.L., and Schramm, D.N., *Ap. J.*, **212**, 347 (1977).

9) Reeves, H., *Ap. J.*, **237**, 229 (1979).

10) Symbalisty, E.M.D., and Schramm, D.N., *Rep. Prog. Phys.*, **44**, 293 (1981).

11) Thielemann, F.-K., Metzinger, J., and Klapdor, H.V. , *Astron. Ap.*, **123**, 162 (1983).

12) Clayton, D.D., *Ap. J.*, **285**, 411 (1984).

13) Clayton, D.D., in *Nucleosynthesis: Challenges and New Developments*, eds. D. Arnett and J.W. Truran (Chicago:Chicago Press), p. 65 (1985).

14) Thielemann, F.-K., and Truran, J.W., in *The Extragalactic Distance Scale and Deviations from the Hubble Expansion* eds. B. Mador and B. Tully (Dordrecht:Reidel), p. 185 (1986).

15) Cowan, J.J., Thielemann, F.-K., and Truran, J.W., in *Advances in Nuclear Astrophysics*, eds. E. Vangioni-Flam, J. Audouze, M. Casse, J.-P. Chieze, and J. Tran Thanh Van, (Gif sur Yvette:Editions Frontieres), p. 477 (1986).

16) Fowler, W.A., *Quart.J.R.A.S.* **28**, 87 (1987).

17) Clayton, D.D., *Ap. J.*, **139**, 637 (1964).

18) Woosley, S.E., and Fowler, W.A., *Ap. J.*, **233**, 411 (1979).

19) Yokoi, K., Takahashi, K., and Arnould, M. *Astron. Ap.* **117**, 65 (1983).

20) Arnould, M., Takahashi, K., and Yokoi, K., *Astr. Ap.*, **137**, 51 (1984).

21) Beer, H., and Walter, G., *Ap. Space Sci.*, **100**, 243 (1984).

22) Beer, H., and Macklin, R.L., *Phys. Rev.*, **C32**, 738 (1985).

23) Arnould, M., in *Nucleosynthesis and its Implications for Nuclear and Particle Physics*, eds. J. Audouze and N. Mathieu (Dordrecht:Reidel), p. 363 (1986).

24) Tinsley, B.M., *Fund. Cosmic Phys.*, **5**, 287 (1980).

25) Meyer, B.S. and Schramm, D.N., *Ap. J.*, **311**, 406 (1986).

26) Cowan, J.J., Cameron, A.G.W., Truran, J.W., and Sneden, C., in *Advances in Nuclear Astrophysics*, eds. E. Vangioni-Flam, J. Audouze, M. Casse,

J.-P. Chieze, and J. Tran Thanh Van, (Gif sur Yvette:Editions Frontieres), p. 429 (1986).

27) Truran, J.W., *Ann. Rev. Nucl. Part. Sci.*, **34**, 53 (1984).

28) Burbidge, E.M., Burbidge, G.R., Fowler, W.A., and Hoyle, F., *Rev. Mod. Phys.* **29**, 547 (1957).

29) Seeger, P.A., Fowler, W.A., and Clayton, D.D., *Ap.J. Suppl.* **11**, 121 (1965).

30) Cameron, A.G.W., Delano, M.D., and Truran, J.W., in *The Properties of Nuclei far from the Region of Beta Stability* **Vol. 2** (Geneva:CERN), p.735 (1970).

31) Schramm, D.N., *Ap.J.* **185**, 293 (1973).

32) Hillebrandt, W., Takahashi, K., and Kodama, T., *Astron. Ap.* **52**, 63 (1976).

33) LeBlanc, J.M. and Wilson, J.R., *Ap.J.* **161**, 541 (1970).

34) Schramm, D.N. and Barkat, Z., *Ap.J.* **173**, 195 (1972).

35) Meier, D.L., Epstein, R.I., Arnett, W.D., and Schramm, D.N., *Ap. J.* **204**, 869 (1976).

36) Symbalisty, E.M.D., Schramm, D.N., and Wilson, J.R., *Ap. J. (Letters)* **291**, L59 (1985).

37) Lattimer, J.M., and Schramm, D.N., *Ap. J. (Letters)* **192**, L145 (1974).

38) Hillebrandt, W. and Thielemann, F.K., *Mitt. Astron. Gesellschaft* **43**, 224 (1977).

39) Truran, J.W., Cowan, J.J., and Cameron, A.G.W., *Ap. J. (Letters)* **222**, L63 (1978).

40) Truran, J.W., Cowan, J.J., and Cameron, A.G.W. , in Proceedings of the International Symposium on Superheavy Elements, ed. M.A.K. Lodhi (New York: Pergamon Press), p. 515 (1978).

41) Cowan, J.J., Cameron, A.G.W., and Truran, J.W., *Ap.J.* **241**, 1090 (1980).

42) Cowan, J.J., Cameron, A.G.W., and Truran, J.W., *Ap.J.* **265**, 429 (1983).

43) Blake, J.B., Woosley, S.E., Weaver, T.A., and Schramm, D.N., *Ap.J.* **248**, 315 (1981).

44) Klapdor, H.V., Oda, T., Metzinger, J., Hillebrandt, W., and Thielemann, F.K., *Z. Phys. A - Atoms and Nuclei* **299**, 213 (1981).

45) Cowan, J.J., Cameron, A.G.W., and Truran, J.W., *Ap.J.* **294**, 656 (1985).

46) Lee, T., Schramm, D.N., Wefel, J.P., and Blake, J.B., *Ap.J.* **232**, 854 (1979).

47) Wefel, J.P., Schramm, D.N., Blake, J.B., and Pridmore-Brown, D., *Ap. J. (Suppl.)* **45**, 565 (1981).

48) Cowan, J.J., Cameron, A.G.W., and Truran, J.W., *Ap.J.* **252**, 348 (1982).

49) Cowan, J.J., Cameron, A.G.W., and Truran, J.W., in "Stellar Nucleosynthesis," eds. C. Chiosi and A. Renzini (Dordrecht: Reidel), p. 151 (1984).

50) Deupree, R.G. and Cole, P.W., *Ap.J.* **269**, 676 (1983).

51) Truran, J.W., Cameron, A.G.W., and Hilf, E., *CERN* **70-30**, p. 275 (1970).

52) Truran, J.W., *Ap. Space Sci.* **18**, 308 (1972).

53) Gilbert, A. and Cameron, A.G.W., *Can. J. Phys.* **43**, 1446 (1965).

54) Takahashi, K. and Yamada, M., *Progr. Theor. Phys.* **41**, 1470 (1969).

55) Cameron, A.G.W., Cowan, J.J., Klapdor, H.V., Metzinger, J., Oda, T., and Truran, J.W., *Ap. Space Sci.* **91**, 221 (1983).

56) Gill, R.L., Casten, R.F., Warner, D.D., Piotrowski, A., Mach, H., Hill, J.C., Wohn, F.K., Winger, J.A., and Moreh, R., *Phys. Rev. Letters* **56**, 1874 (1986).

57) Thielemann, F.-K., Metzinger, J., and Klapdor, H.V., *Z. Phys.*, **A309**, 301 (1983).

58) Howard, W.M. and Möller, P., *At. Data Nucl. Data Tables*, **25**, 219 (1980).

59) Hilf, E.R., von Groote, H. and Takahashi, K., in *Proc. Third International Conference on Nuclei Far From Stability* (Geneva: CERN 76-13), p. 142 (1976).

60) von Groote, H., Hilf, E.R., and Takahashi, K., *At. Data Nucl. Data Tables*, **17**, 418 (1976).

61) Myers, W.D., *At. Data Nucl. Data Tables*, **17**, 411 (1976).

62) Haustein, P.E., *Proc. 7th Intl. Conference on Atomic Masses (AMCO -7)*, ed. O. Klepper, *THD Schriftenreihe fur Wissenschaft und Technik* **26**, p.413 (1984).

63) Hoff, R., in *Weak and Electromagnetic Interactions in Nuclei*, ed. H.V. Klapdor, (Heidelberg:Springer), p. 207 (1986).

64) Cameron, A.G.W., in *Essays in Nuclear Astrophysics*, eds. C. A. Barnes, D. D. Clayton, and D. N. Schramm (Cambridge Univ. Press) p. 23 (1982).

65) Cameron, A.G.W., Cowan, J.J., and Truran, J.W., in *Nucleosynthesis: Challenges and New Developments*, eds. D. Arnett and J.W. Truran (Chicago: Chicago Press), p. 190 (1985).

66) Howard, W.M., *Physica Scripta*, **10A**, 138 (1974).

67) Bengtsson, R., Bolev, R., Larsson, S.E., and Randrup, J., *Physica Scripta*, **10A**, 142 (1974).

68) Meldner, H.W., Nuckolls, J., and Wood, L., *Physica Scripta*, **10A**, 149 (1974).

69) Seeger, P.A., and Schramm, D.N., *Ap. J. (Letters)*, **160**, L157 (1970).

70) Wene, C.O., and Johansson, S.A.E., *Proc. 3rd Intl. Conference on Nuclei far from Stability* **CERN 76-13**, p. 584 (1976).

71) Krumlinde, J. Möller, P., Wene, C.O., and Howard, W.M., *Proc. 4th Intl. Conference on Nuclei far from Stability* **CERN 81-09**, p. 260 (1981).

72) Fowler, W.A., and Meisl, C.C., in *Cosmogonical Processes*, eds. W.D. Arnett, C.J. Hansen, J.W. Truran, and S. Tsuruta, (Singapore:VNU Press), p. 83 (1986).

73) Cowan, J.J., Thielemann, F.-K., and Truran, J.W., *Ap. J.*, in press (1987).

74) Pagel, B.E.J. and Patchett, B.E., *M.N.R.A.S.*, **172**, 13 (1975).

75) Twarog, B.A., *Ap. J.*, **242**, 242 (1980).

76) Lynden-Bell, D., *Vis. in Astron.*, **19**, 299 (1975).

78) Anders, E. and Ebihara, M., *Geochim. Cosmochim. Acta*, **46**, 2363 (1982).

79) Winget, D.E., Hansen, C.J., Liebert, J., Van Horn, H.M., Fontaine, G., Nather, R.E., Kepler, S.O., and Lamb, D.Q., *Ap. J. (Letters)*, **315**, L77 (1987).

80) Sandage, A., *Ap. J.*, **252**, 553 (1982).

81) Schramm, D.N. and Meyer, B.S., in *Advances in Nuclear Astrophysics*, eds. E. Vangioni-Flam, J. Audouze, M. Casse, J.-P. Chieze, and J. Tran Thanh Van, (Gif sur Yvette:Editions Frontieres), p. 441 (1986).

82) Butcher, H.R., *Nature* **328**, 127 (1987).

83) Mathews, G.J. and Schramm, D.N., preprint (1987).

84) VandenBerg, D.A., *Ap. J. Suppl.* **51**, 29 (1983).

85) VandenBerg, D.A., *IAU Colloquium 126*, in press (1987).

86) Huchra, J.P., in *13th Texas Symposium on Relativistic Astrophysics*, ed. M. P. Ulmer, (Singapore:World Scientific), p. 1 (1987).

87) Sandage, A. and Tammann, G.A., *Nature* **307**, 326.

88) Tammann, G.A., in *13th Texas Symposium on Relativistic Astrophysics*, ed. M. P. Ulmer, (Singapore:World Scientific), p. 8 (1987).

89) Arnett, W.D., Branch, D., and Wheeler, J.C., *Nature* **314**, 337 (1985).
90) Branch, D., *Ap. J. (Letters)* **320**, L23 (1987).

TABLE 1

PRODUCTION RATIOS FROM R-PROCESS CALCULATIONS

$^{232}Th/^{238}U$	$^{235}U/^{238}U$	$^{244}Pu/^{238}U$	REFERENCE
1.65	1.42	0.90	Seeger *et al.*[29]
			Fowler[4]
1.90	1.89	0.96	Seeger and Schramm[69]
1.70	0.89	0.53	Wene and Johansson[70]
1.80	1.42	0.90	Fowler[5]
1.90	1.50	0.90	Symbalisty and Schramm[10]
1.50	1.10	0.40	Krumlinde *et al.*[71]
1.40	1.24	0.12	Thielemann *et al.*[11,57]
1.71	1.34	0.66	Fowler and Meisl[72]
			Fowler[16]
1.60	1.16	0.40	This work

TABLE 2

THE AGE OF THE GALAXY, t_G, FOR VARIOUS VALUES OF THE INITIAL DISK ENRICHMENT, S_0.[a]

S_0	λ_R (10^{-10}Years^{-1})	Δ (Years)	t_G (Years)
0.29	-2.00	7.8 x 10^9	12.4 x 10^9
0.26	-1.88	8.0 x 10^9	12.6 x 10^9
0.23	-1.75	8.4 x 10^9	13.0 x 10^9
0.18	-1.63	9.0 x 10^9	13.6 x 10^9
0.11	-1.50	10.1 x 10^9	14.7 x 10^9

[a] Calculated from the duration Δ of nucleosynthesis before the formation of the solar system, using best-fit case with "most consistent" nuclear data.

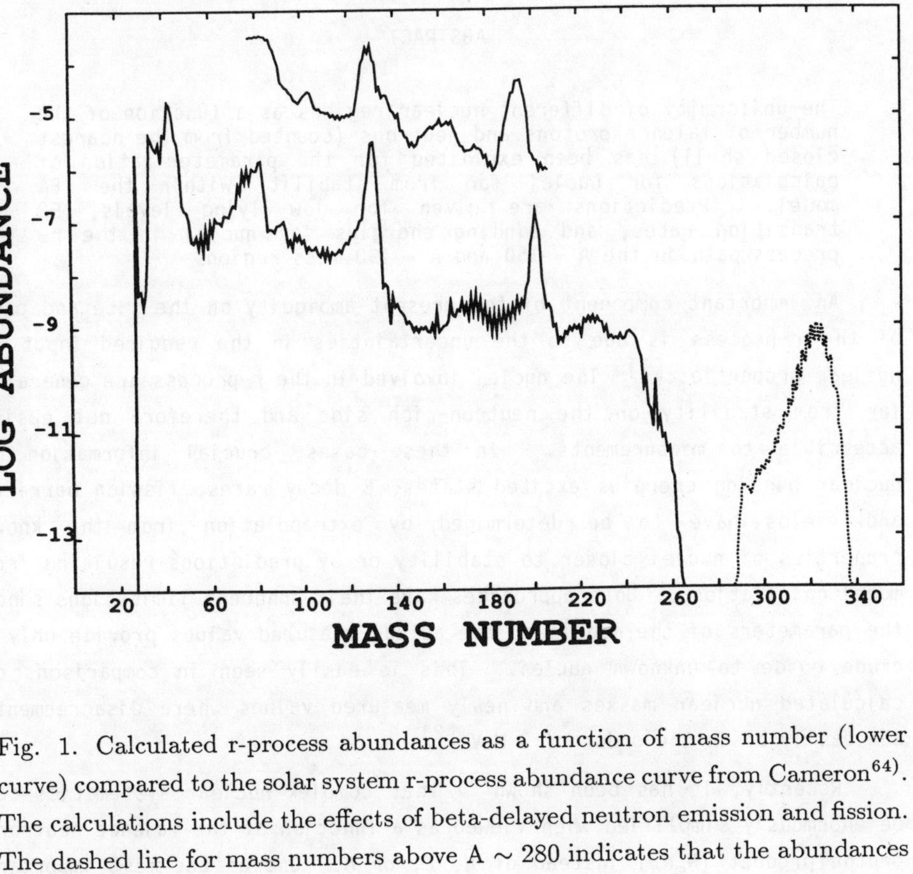

Fig. 1. Calculated r-process abundances as a function of mass number (lower curve) compared to the solar system r-process abundance curve from Cameron[64]. The calculations include the effects of beta-delayed neutron emission and fission. The dashed line for mass numbers above A ~ 280 indicates that the abundances of those nuclei are uncertain.

Nuclear Properties Along the r-Process Path

A. APRAHAMIAN
Lawrence Livermore National Laboratory
Livermore, California 94550
USA

ABSTRACT

The uniformity of different nuclear regions as a function of the number of valence protons and neutrons (counted from the nearest closed shell) has been exploited for the parameterization of calculations for nuclei far from stability within the IBA model. Predictions are given for low lying levels, E2 transition rates, and binding energies for nuclei in the r-process path in the A = 150 and A = 190 mass regions

An important component of the present ambiguity on the site and path of the r-process is due to the uncertainties in the required input of nuclear properties.[1] The nuclei involved in the r-process are generally far from stability on the neutron-rich side and therefore not easily accessible to measurements. In these cases, crucial information on nuclear binding energies excited states, β decay rates, fission barriers and yields have to be determined by extrapolation from the known properties of nuclei closer to stability or by predictions resulting from model calculations. Both approaches have their inherent limitations since the parameters of the models or the actual measured values provide only a crude guide to unknown nuclei. This is easily seen in comparisons of calculated nuclear masses and newly measured values where disagreements can easily be in the order of 3 MeV.[2]

Recently, it has been shown[3] that complex nuclear systematics can be enormously simplified when viewed as a function of the valence neutron-proton product (N_pN_n) instead of N, Z, or A. One of the most important features of the N_pN_n scheme is the conversion of the extrapolation process into a more reliable one of interpolation. The purpose of this paper is to exploit this idea in parametrizing collective model calculations by N_pN_n and thereby predict the properties of nuclei along the r-process path near the abundance peaks at A=150 and 190.

The N_pN_n scheme is based on the role of the neutron-proton interaction in determining nuclear structure in heavy nuclei. A number of recent studies[3,4,5] have shown that the n-p interaction is the controlling agent for the onset of deformation, the development of collectivity, and the structure of nuclear phase transitions. A simple illustration of the idea is shown in Fig. 1. A traditional plot of the first 2^+_1 energies for a series of nuclei in the A = 150 region are shown to the left sides of the figure. What

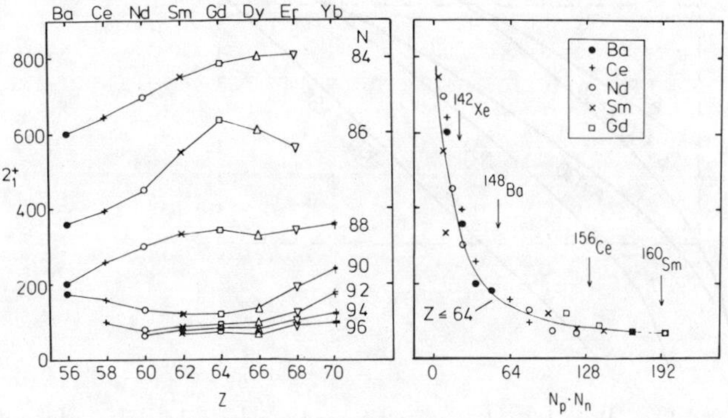

Fig. 1 Systematics of the $E_{2_1^+}$ ratios in the A = 150 region.

is observed is a complicated function of Z. The right side of Fig. 1 shows the same data as one smooth function of N_pN_n, the product of the number of valence protons and neutrons (counted as holes past midshells or midsubshells where appropriate). Similar simplifications also occur in the A = 100, A = 130, A = 190, and A = 230 nuclear regions. Furthermore, comparisons of N_pN_n plots for all of the above transition regions behave in a similar way. A collection of the curves for five mass regions is shown in Fig. 2 for the $E_{4_1^+/2_1^+}$ ratios. There are only small differences in slope, otherwise the curves are nearly identical. These results have two very significant implications.

First, the existence of a single smooth curve for a given region greatly facilitates making predictions even if only a few points are known. Often, the N_pN_n values of many nuclei far off stability are less than those of some well known nuclei close to stability. For example the ^{166}Ba nucleus is 28 neutrons away from the last Ba isotope but it has an N_pN_n value of 96 which lies in well defined regions of both N_pN_n curves shown in Figs. 1 and 2. In this way, the properties of unknown nuclei far off stability can be determined by interpolation of known values.

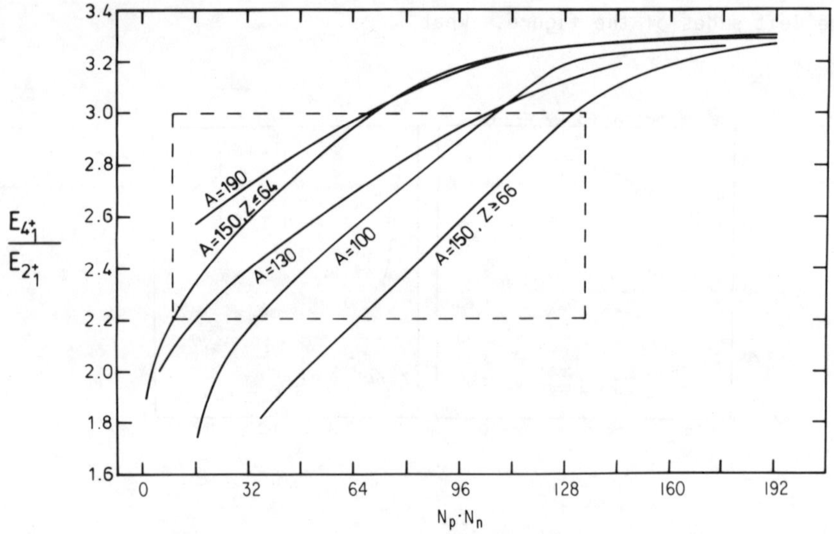

Fig. 2 Plot of the five N_pN_n curves associated with the labelled amss regions.

The second effect of the N_pN_n approach is in the parameterization of model calculations. If a nuclear region can be described so simply, then the parameters of a given model may be written as a function of N_pN_n and allow a unified description of a whole region. This idea was demonstrated in the calculations of 70 known nuclei in the A = 150 mass region within the IBA[4] framework with a set of five fixed parameters with excellent agreement.

Figure 3 shows the empirical and calculated vales for the $E_{4_1^+}/2_1^+$ ratio in the A = 150 region. The calculations clearly reproduce the empirical behavior within the region.

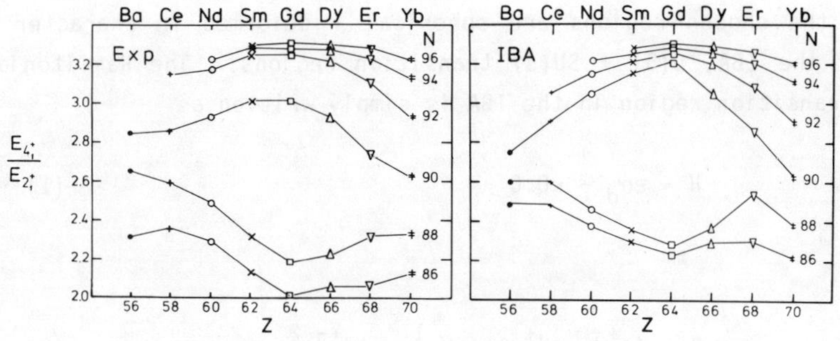

Fig. 3 Empirical and calculated $E_{4_1^+}/E_{2_1^+}$ ratios for the A = 150 region.

Therefore, for select regions of nuclei which are well reproduced by these calculations, the extrapolation-interpolation inversion can be used to yield predictions for unknown nuclei. These calculations have been carried out within the framework of the Interacting Boson Approximation for approximately 50 nuclei in the regions near the r-process abundance peaks at A = 150 and A = 190. Figure 4 shows the sequences of unknown nuclei that have been calculated.

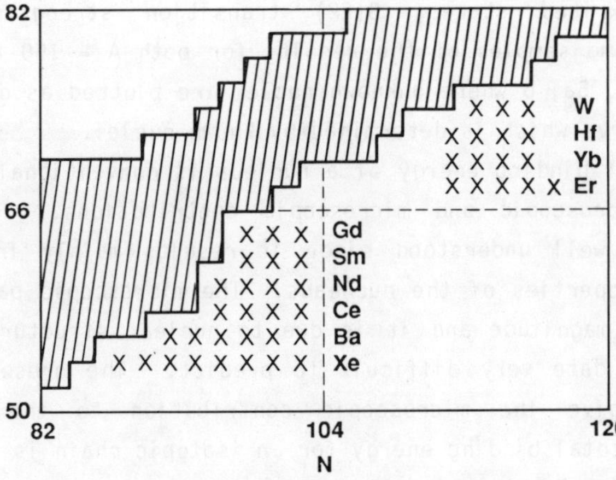

Fig. 4 The relevant portion of the N-Z plane. The shaded parts indicate known nuclei. The x's show the calculated nuclei.

Both of the chosen regions are spherical → deformed in character or in terms of the IBA, U(5) → SU(3) transition regions. The Hamiltonian for such a transition region in the IBA is simply written as

$$H = \varepsilon n_d - \kappa Q.Q \tag{1}$$

where

$$Q = (s^+\tilde{d} + d^+s) + \sqrt{\frac{1}{5}}\, \chi\, (d^+\tilde{d})^2. \tag{2}$$

The U(5) or vibrational limit results if $\kappa = 0$ and the SU(3) or rotor limit results when $\varepsilon = 0$. The parameterization of the ε as a function of N_pN_n is written in the following way[4]

$$\varepsilon = \varepsilon_0\, e^{-\theta(\frac{N_pN_n}{4} - N_0)} \,. \tag{3}$$

with the parameters ε, Θ, and N_0. In each region, the calculations were performed by keeping all of the 5 parameter (κ, χ, ε_0, Θ, and N_0) constant and simply varying the ε as a function of N_pN_n.

The predicted results yield the complete spectrum of energies and spins of excited states, B(E2) transition strengths and binding energies. Some samples of the results for both A = 150 and A = 190 are shown in Figs. 5a, b where unknown nuclei are plotted as data points on a solid N_pN_n curve which is determined by known nuclei.

The total binding energy of a nucleus is conventionally expressed as a sum of macroscopic and microscopic contributions. The macroscopic component is well understood since it results mainly from the "liquid drop-like" properties of the nucleaus. The microscopic part is typically 10-15 MeV in magnitude and it is due to nuclear structure effects which have been to date very difficult to predict. The present calculations essentially give the microscopic contribution to the total binding energy. The total binding energy for an isotopic chain is written as,

$$\text{Binding Energy} = C + AN_n + BE_{IBA}$$

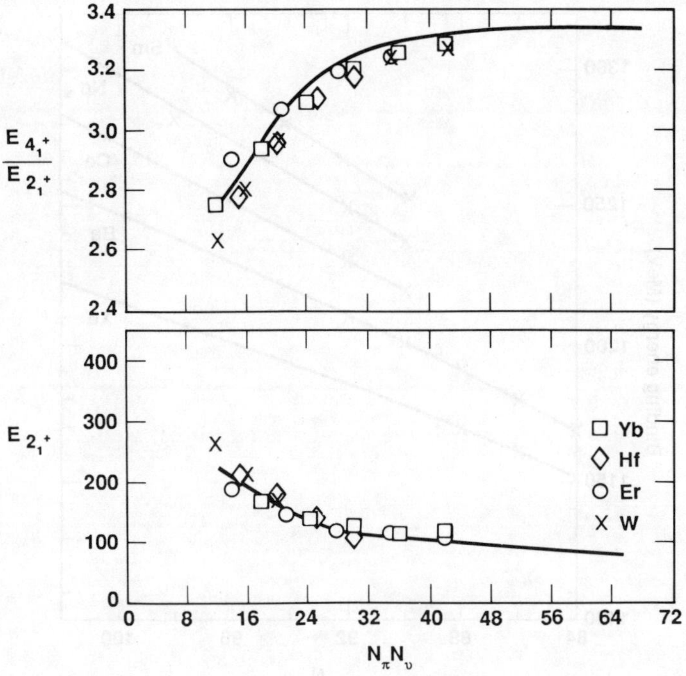

Fig. 5a,b The calculated $E_{2_1^+}$ energies and $E_{4_1^+}/E_{2_1^+}$ ratios for unknown nuclei in the A=190 region. The solid lines represent the N_pN_n systematics of known nuclei in the same region.

where C and A are constants, N_n is the number of valence neutrons and BE_{IBA} is the calculated structure contribution. The parameters C and A are determined from measured binding energies and then used along with the calculated BE_{IBA} to yield the binding energies of isotopes far from stability. The results for A = 150 region are shown in Fig. 6.

The calculated and measured values for this region are in excellent agreement. These predictions however represent only a preliminary attempt at calculating crucial nuclear properties along the r-process path. The N_pN_n approach is worthy of further investigation since it represents a unique tool for parametrization of nuclear models and therefore a reliable

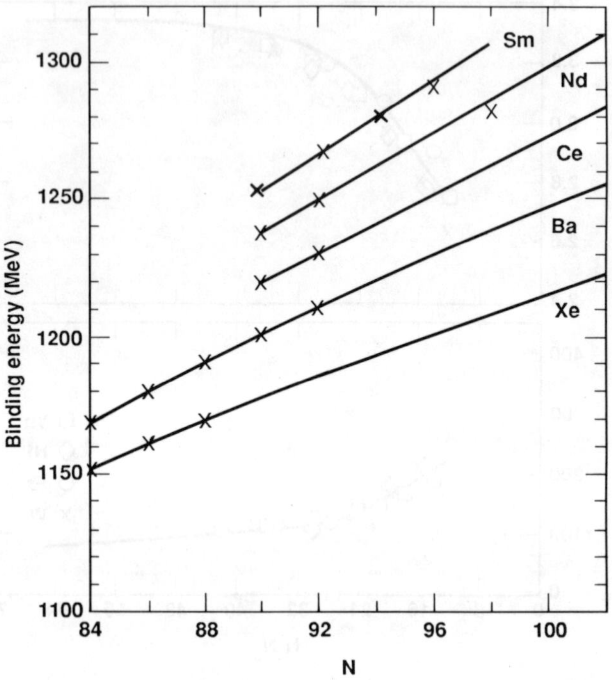

Fig. 6 The total binding energies for select nuclei from the A=150 region. The x's indicated measured values while the solid lines are the predictions.

method for predictions of nuclear properties far from stability. Also, the use of the IBA model allows calculations of the complete nuclear excitation structure, and decay strengths as as well as the binding energies.

The author would like to gratefully acknowledge discussions with R. F. Casten and E. Henry.

REFERENCES

1. Mathews, G.J., and Ward, R.A., Rep. Prog. Phys. 48, 1371 (1985).
2. Brenner, D. S. et al., Phys. Rev. C. 26, 2166 (1982).
3. Casten, R.F., Phys. Lett. 152B, 145 (1985).
4. Casten, R.F., Nucl. Phys. A444, 133, (1985).
5. Casten, R.F., Phys. Rev. C 33 1819 (1986).
6. Arima, A., and Iachello, F., Ann. Phys. 99, 253 (1976]; Ann. Phys. 111, 201 (1978); and Ann. Phys. 123, 468 (1979).

New Nuclear Structure Calculations and the Astrophysical r-Process

W. MICHAEL HOWARD, PETER MÖLLER*, GRANT J. MATHEWS
and BRADLEY S. MEYER

Lawrence Livermore National Laboratory
Livermore, CA 94550

ABSTRACT

Recent developments in nuclear-structure calculations used to calculate the nuclear properties of neutron-rich heavy elements can have a dramatic effect on astrophysical r-process calculations. In particular, in the context of the macroscopic-microscopic method, studies of the effect of the finite range of the nuclear force on the surface energy, the shape dependence of the Wigner term, new shape degrees of freedom near scission, and improved single particle potentials for neutron-rich heavy elements have led to improved calculations of the properties of heavy elements. We will discuss preliminary applications of these new models to nuclei of interest in the astrophysical r-process and the implications for nuclear cosmochronology and the existence of superheavy elements in nature.

INTRODUCTION

The most recent extensive calculation of nuclear structure quantities that are used as input into studies of the decay from the astrophysical r-process path to the line of β-stability in the actinide region and cosmochronology is the work by Howard and Möller[1]. In this study fission barriers and nuclear ground-state masses are calculated for all nuclei with $76 \leq Z \leq 100$ and $140 \leq N \leq 184$.

An important question when nuclear structure models are applied to problems of interest in astrophysics is: are the models at all reliable

* Permenent Address: Department of Mathematical Physics, Lund Institute of Technology, Lund, Sweden.

so far from the region of nuclei where the models were studied and where the model parameters were adjusted? Until recently there was not much basis for answering this question; one could only hope that the models were reliable. In the calculation of Howard and Möller a model of the macroscopic-microscopic type was employed. A Nilsson modified-oscillator single-particle potential was used to calculate the microscopic shell corrections. The droplet model of Myers and Swiatecki[2] was used to calculate the macroscopic part. It is now well-known that the droplet model gives too low mass excesses for neutron-rich heavy elments. It is also well-known that it is somewhat difficult to find appropriate parameters for the modified oscillator single-particle model for new regions of nuclei where the parameters have to be determined by extrapolation. For example, Figure 1 gives[3] the residual error, as a function of the number of neutrons from stability, for nuclei with a wide range of A, as calculated with the droplet model. As one can see, there is a systematic effect for the droplet model to predict an overbinding of neutron-rich nuclei.

YUKAWA-PLUS-EXPONENTIAL MACROSCOPIC MODEL

However, now there are some studies with models of the macroscopic-microscopic type that indicate that at least some of those models are fairly reliable far from stability and for new regions of nuclei. In particular, extensive and systematic studies with the folded-Yukawa macroscopic-microscopic model, where the macroscopic energy is calculated by a Yukawa-plus-exponential model and the microscopic energy is calculated with a folded-Yukawa single-particle model, have been carried out over the past seven years[4,5,6,7,8,9]. Evaluation of the results obtained have shown that the error in the mass model does not seem to increase as a function of increasing neutron excess. The major inclusion of new physics in the Yukawa-plus-exponential macroscopic model is the inclusion of the effect of the finite range of the nuclear force on the nuclear surface potential energy. In addition, improvements in the treatment of pairing and in the shape dependence of the Wigner term are also under stduy, but are not included in the present calculations.

Figure 2 is a corresponding plot of the residual error as a

function of the number of neutrons away from stability using the Yukawa-plus-exponential macroscopic energy and folded-Yukawa single-particle energy levels[10]. For this model there are no systematic errors for increasing neutron excess and the predicted masses in the neutron-rich region agree very well. In addition, in the very heavy-element region evidence for a rock of stability was present in the calculated results already in 1981[5,6]. This rock was studied further later[7], and its existence has been studied experimentally by Armbruster and collaborators at the GSI. The agreement between calculated and experimentally determined alpha decay energies for Z = 100 through 105 is also excellent, including the trend as a function of neutron number. Another achievement of the Yukawa-plus-exponential model is a more complete and consistent understanding of the fission properties in the heavy-element region[11,14] in the vicinity of ^{258}Fm. An interesting prediction of the Yukawa-plus-exponential model[7] is that the predicted alpha decay half-life (the shortest decay mode) of the superheavy elements ($^{110}X^{288}$ and $^{110}X^{290}$) are only about 200 days.

For astrophysical r-process calculations, a matter of prime importance is the position of the neutron-drip line. The position of this line determines the r-process path. The observed abundance peaks in the distribution of r-process nuclei in nature provides a constraint on the r-process path. The r-process path must cross the neutron closed shells N = 50, 82, and 126 at the values of A = 80, 130, and 195 to reproduce the r-process peaks. It has been known for some time that the droplet model predicts neutron-rich nuclei that are too bound. That is, the neutron binding energy falls off too slowly with neutron number. A success of the new nuclear structure models that include the effects of the finite range of the nuclear force is that it predicts that the neutron binding energies decrease more rapidly with increasing neutron number, than previous models.

FINITE-RANGE DROPLET MODEL

The effects of the finite range of the nuclear force have also been introduced into the droplet model, producing a model called the finite-range droplet model[9]. Figure 3 shows the line of β-stability and the neutron drip model as given by the droplet model and the finite range droplet model. As one can see, the finite range droplet model decreases the position of the neutron drip line by about 10 neutrons at Z = 90, relative to that given by the droplet model. Because of the change in the position of the neutron-drip line, this new model allowss for a more consistent calculation of the r-process than was possible with the droplet model.

In addition to the improvements in the macroscopic models, improvements in the microscopic models are important as well. Because of some fundamental differences between models with an oscillator single-particle potential and models with a more correct description of the nuclear surface diffuseness, we have come to realize that for extrapolations to new regions the folded-Yukawa single-particle model is much more reliable than the modified-oscillator Nilsson single-particle model.

NUCLEAR SURFACE ENERGY

For nuclear cosmochronology studies, we are mainly concerned with the fissility of neutron-rich heavy elements[12]. An important quantity in determining the fissility of neutron-rich nuclei is how rapidly the nuclear surface energy decreases with increasing neutron excess. Consider the surface energy as calculated in the macroscopic model. In the liquid drop model[13] the surface energy is given as

$$E_s(Z,\ N,\ \text{shape}) = a_s\,(1 - \kappa_s I^2)\,A^{2/3} \cdot B_s \qquad (1)$$

where

$$I \equiv \frac{N-Z}{N+Z} \qquad (2)$$

is the relative neutron excess, κ_s the surface-asymmetry coefficient and a_s is the surface-energy coefficient. The quantity B_s is the surface area of the nucleus relative to that for a sphere. The Yukawa-plus-exponential macroscopic energy uses the same form for the surface energy, except that B_s now represents the relative generalized surface energy in a model that accounts for the effect of the finite range of the nuclear force. For this model, the quantity B_s is given by[8]

$$B_s = \frac{A^{-2/3}}{8\pi r_o^2 a^4} \int_V \left(2 - \frac{|\vec{r}-\vec{r}'|}{a}\right) \frac{e^{-|\vec{r}-\vec{r}'|/a}}{|\vec{r}-\vec{r}'|/a} d^3\vec{r}\, d^3\vec{r}' \quad (3)$$

where r_o is the nuclear radius constant and a is the range of the Yukawa-plus-exponential potential. In this model r_o has the value 1.16 fm and a has the value 0.68 fm[8]. The normalization constants in front of the integral are chosen such that for spherical shape, in the limit of the parameter a approaching zero, B_s approaches unity. If one integrates Eq. (3) over a sphere of radius $r_o A^{1/3}$, then one obtains[8]

$$B_s = 1 - \frac{3}{x_o^2} + (1+x_o)\,(2+3/x_o + 3/x_o^2)\, e^{-2x_o} \quad (4)$$

where

$$x_o = \frac{r_o A^{1/3}}{a} \quad (5)$$

Since x_o is a rather large number for r-process nuclei (x_o = 10), one would think that this correction is small for these nuclei and that the finite range effects would be negligible. That is in fact the case. However, the finite range effects are important for light nuclei and it is their inclusion in the mass formulae that effects the overall physical coefficients, particularly the value for a_s amd κ_s. It is the fact that a_s and κ_s are different from their values in other models, that changes the calculated properties of the neutron-rich heavy elements. The coefficients a_s and κ_s are primarily determined by the fit to the fission barriers. Fission barriers are sensitive to the ratio of the surface energy to the

Coulomb energy. The Coulomb energy is inversely proportional to r_o, which can be accurately determined from electon scattering experiments. The finite range of the nuclear force reduces the surface energy, particularly for saddle-point shapes with well developed necks. Such shapes exist for fission saddle-point shapes for relatively light nuclei as well as for some heavy nuclei such as ^{258}Fm. A careful adjustment of a_s and κ_s to a wide range of fission barriers, using the folded-Yukawa model has allowed a more accurate determination of these quantities than in previous models.

In the droplet model the shape dependence of the nuclear potential energy is more complicated than in the liquid drop model and the constant κ no longer enters. In the droplet model[2] the surface energy is given as
energy is given as

$$E_s(Z,\ N,\ \text{shape}) = (a_s + \frac{9}{4}\ [\frac{J^2}{Q}]\ \bar{\delta}^{\ 2})\ B_s\ A^{2/3} \qquad (6)$$

where J is a symmetry energy constant, Q is an effective surface stiffness constant, δ is a mean deviation of the nucleus from its bulk density, and B_s is the relative surface term of the liquid-drop model. The term δ has a complicated shape dependence involving the relative surface term B_s and a term representing a Coulomb redistribution of protons on the nuclear surface. However, by neglecting the relatively small influence of the this shape-dependent term, the more complicated dependence of the droplet model on the surface and coulomb energies can be described approximately in terms of an effective value of κ which would yield the same saddle point as the liquid-drop model. The result is[14]

$$\kappa_{eff} = \frac{9}{4}(J^2/Q) \Big/ a_s\ \left(1 + \frac{9K\hat{B}_s}{4QA^{1/3}}\right)^2 \qquad (7)$$

where $\hat{B}_s$ is the relative surface term of the saddle-point shape. Thus, unlike in the liquid-drop model, this effective κ depends on the mass number, the saddle point shape, and the surface energy coefficient.

Table I gives the effective κ_s for the neutron-rich nucleus ^{260}U, for the various models, including the liquid drop and droplet model. In the fitting of the coefficients, the effect of the finite range dependence is to increase the effective surface stiffness coefficient, Q, by about a factor 2. The first obvious result of the inclusion of the the finite range of the nuclear force into the mass model is to reduce the effective surface asymmetry coefficient. This decreases the surface energy less with an incrase in neutron excess and raises the fission barrier for neutron-rich heavy elements relative to that predicted by previous models. However, there are other effects as well and we will discuss them in the next section.

Table I

Surface Asymmetry Coefficient κ_S for ^{260}U

Liquid Drop Model (1967)	1.7826
Droplet Model (1973)	2.8
Folded Yukawa Model (1981)	2.3
Finite-Range Droplet Model (1984)	1.8

SOME COMPARISONS BETWEEN THE DROPLET MODEL AND THE YUKAWA-PLUS-EXPONENTIAL MODEL FOR R-PROCESS NUCLEI.

We have perform some preliminary calculations for neutron-rich heavy elements with the Yukawa-plus-exponential macroscopic model and the folded Yukawa single-particle microscopic model and compare them with the results of Howard and Möller. These calculations are preliminary in the sense that the single-particle energies have been scaled from the single-particle levels from one nucleus (^{240}Fm). A more definitive calculation would interpolate the single particle energies from the calculated single-particle levels from a grid of nuclei. Figs. 4, 5, and 6 give a comparison between the calculation with the droplet model and the Yukawa-plus-exponential model for ^{236}U, ^{262}U, and ^{276}U. As one can see there are some dramatic differences between the old and the new models. The first difference is that the overall barriers are higher, primarily because of the decreased value of the surface-asymmetry coefficient. However, another dramatic difference (for ^{262}U) is that the ground-state shape is almost spherical or even slightly oblate in the new model, rather than the very prolate as in the droplet model. This change in ground state shape is due in part to the higher macroscopic barrier. Also contributing to the different result is that we use a folded-Yukawa single-particle potential. This type of potential is believed to extrapolate more reliably to neutron-rich elements, because of the more predictable behavior of the spin-orbit coupling term and the diffuseness parameter as a function of mass number. This change in ground state shape will have several effects of the fissility of neutron-rich heavy elements. The fact that the ground state shapes are close to spherical will mean that the spontaneous fission probability will be greatly reduced because of the increased width of the barrier that must be penetrated, as well as the effect of the increased height of the barrier.

The ground state shape also influences the rates of beta delayed fission, that are important in the decay from the r-process line to the valley of β-stability. Previous calculations of beta strength functions for neutron-rich nuclei with large prolate ground states, demonstrate that the Gamow-Teller strength is smeared out over a relative large range in energy due to the relatively regular spacing of the single particle levels for large deformations. If the ground states are nearly spherical, then the beta strength will be concentrated in relatively few highly degenerate levels. For this case, there may be relatively less beta decay rate available above the fission barrier. In addition, the detailed beta decay rates will be much more dependent upon detailed nuclear structure effects, rather than upon systematic macroscopic effects.

Considering only the effect of the higher fission barriers, it appears that the Yukawa-plus-exponential model will tend to give even less beta-delayed fission than the droplet model, and thus beta-delayed fission may be of little consequence for calculations of the production of actinide nuclei in the r-process. However, microscopic effects on the beta decay rates may change this conclusion somewhat. The termination of the r-process by neutron-induced fission will also be affected, but since this occurs near the closed shell at N = 184, the overall effect is less clear, except to say that it will probably occur at a higher mass number than in the droplet model. In regards to the question of the production of superheavy nuclei by the astrophysical r-process, suffice it to say that it will probably still prove difficult. Moreover, the Yukawa-plus-exponential model predicts lifetimes of order of 200 days, which is not a time span of astrophysical consequence.

One should notice that the model changes most severely effect very neutron-rich nuclei, rather than experimentally known nuclei such as ^{236}U. One would expect this since both the droplet model and the Yukawa-plus-exponential model coefficients were chosen to reproduce the fission properties of nuclei on the valley of β-stability. The differences in the predictons of the two models highlight the difficulty in studying the nuclear properties of r-process nuclei.

CONCLUSIONS

Calculations of the astrophysical r-process require knowledge of the binding energies and fission properties of neutron-rich heavy elements, that are not accessible to experimental study. In recent years, substantial progress has been made in improving the reliability of the prediction of the nuclear structure quantities of neutron-rich heavy elements. Although the extrapolation power of the models has improved, r-process nuclei are still many neutrons removed from known nuclei, so that the extrapolations can be considered rather uncertain. However, the more reliable models predict that the r-porcess nuclei are probably less susceptible to both beta-delay and neutron-induced fission than previously thought. The exact details and astrophysical consequences remain to be worked out. However, the study of the astrophysical r-process remains a rich environment for the study of challenging problems in nuclear physics.

*Work performed under the auspices of the U.S. Department of Energy by the Lawrence Livermore National Laboratory under contract number W-7405-ENG-48.

REFERENCES

1] Howard, W. M. and Möller, P., At. Nucl. Dat. Tables, 25, 219, (1980).

2] Myers, W. D., and Swiatecki, W. J., Ann. Physics (N.Y.), 55, 395, (1969).

3] Haustein, P. E., Nuclei Off the Line of Stability, (R. A. Meyer and D. S. Brenner, eds., American Chemical Society, Washington, D. C.), 126, (1986).

4] Möller, P. and Nix, J. R., Nucl. Phys. A361, 117 (1981).

5] Möller, P. and Nix, J. R., At. Nucl. Dat. Tables 26, 165, (1981).

6] Bengtsson, R., Möller, P. Nix, J. R. and Zhang, Jing-ye Phys. Scr., 29, 402, (1984).

7] Möller, P., Leander, G. A. and Nix, J. R., Z. Phys. A323, 41, (1986).

8] Möller, P. and Nix, J. R., Preprint LA-UR-86-3983, Los Alamos National Laboratory, At. Nucl. Data Tables, in press, (1986).

9] Möller, P. , Myers, W. D., Swiatecki, W. J. and Treiner, J., Preprint LA-UR-86-4149, Los Alamos National Laboratory, At. Nucl. Data Tables, in press, (1986).

10] Haustein, P. E. , private communication, (1987).

11] Möller, P., Nix, J. R. and Swiatecki, W. J., Nucl. Phys. A, in press, (1987).

12] Meyer, B. S., Howard, W. M., Mathews, G. J., Takahashi, K., Möller, P., and Leander, G. A., Lawrence Livermore National Laboratory preprint UCRL-96754, (1987).

13] Myers, W. D. amd Swiatecki, W. J., Ark. Fys., 36, 343,(1967).

14] Howard, W. M. and Nix, J. R., Phys. and Chem. of Fission (IAEA, Vienna), 1, 145, (1974).

Figure Captions

Figure 1 Relative error (in MeV) in calculated masses as a function of neutrons from stability for droplet model.[2]

Figure 2 Relative error (in MeV) in the calculated masses as a function of neutrons from stability for the finite-range model.[4]

Figure 3 A comparison of the neutron drip line for the droplet model[2] (dashed line) and the finite-range droplet model[9] (dot-dashed line).

Figure 4 A comparison of the fission barrier for ^{236}U as calculated by Howard and Möller[1] and as calculated with the finite-range model.[4] The fission barriers are calculated as a function of ϵ_2 and include mass symmetric and mass asymmetric deformations.

Figure 5 Same as Fig. 4, except for ^{262}U.

Figure 6 Same as Fig. 4, except for ^{276}U.

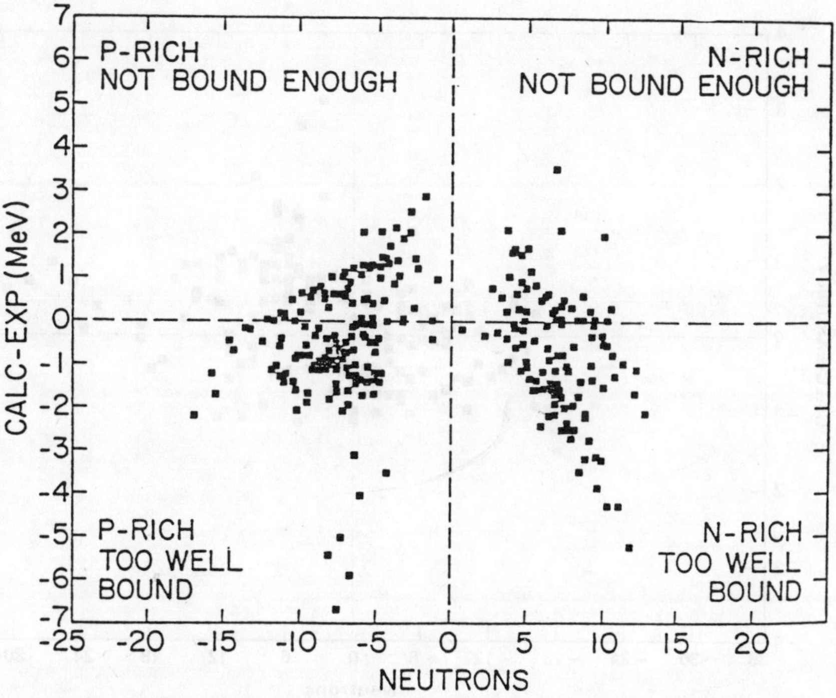
P-RICH
NOT BOUND ENOUGH
N-RICH
NOT BOUND ENOUGH
P-RICH
TOO WELL
BOUND
N-RICH
TOO WELL
BOUND
CALC-EXP (MeV)
NEUTRONS
7
6
5
4
3
2
1
0
-1
-2
-3
-4
-5
-6
-7
-25
-20
-15
-10
-5
0
5
10
15
20

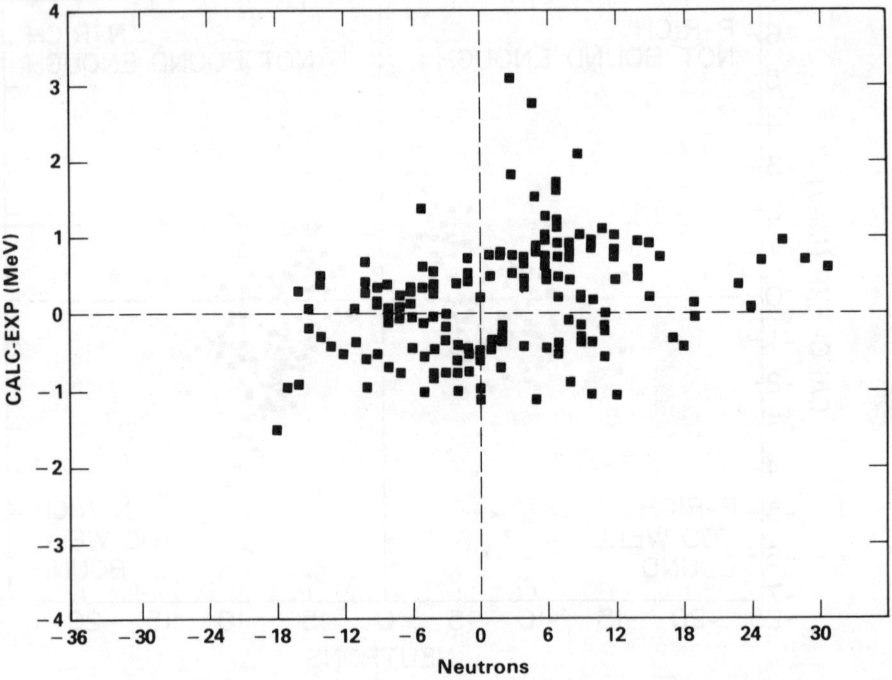

Figure 2

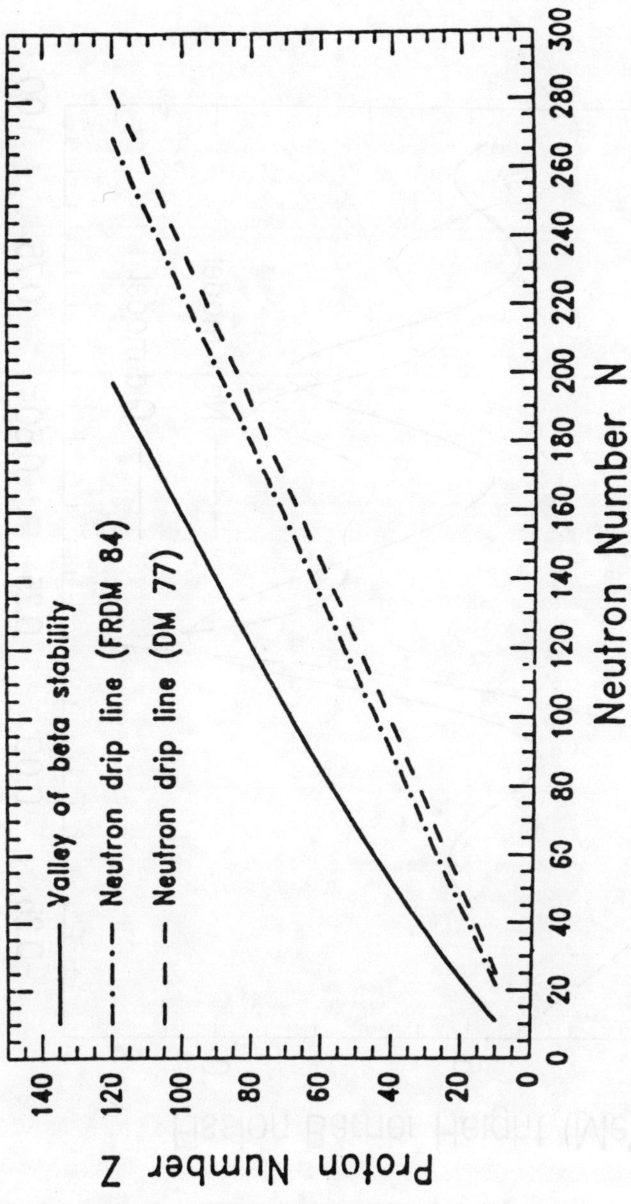

Valley of beta stability
Neutron drip line (FRDM 84)
Neutron drip line (DM 77)
Proton Number Z
Neutron Number N
0
20
40
60
80
100
120
140
160
180
200
220
240
260
280
300

Figure 3

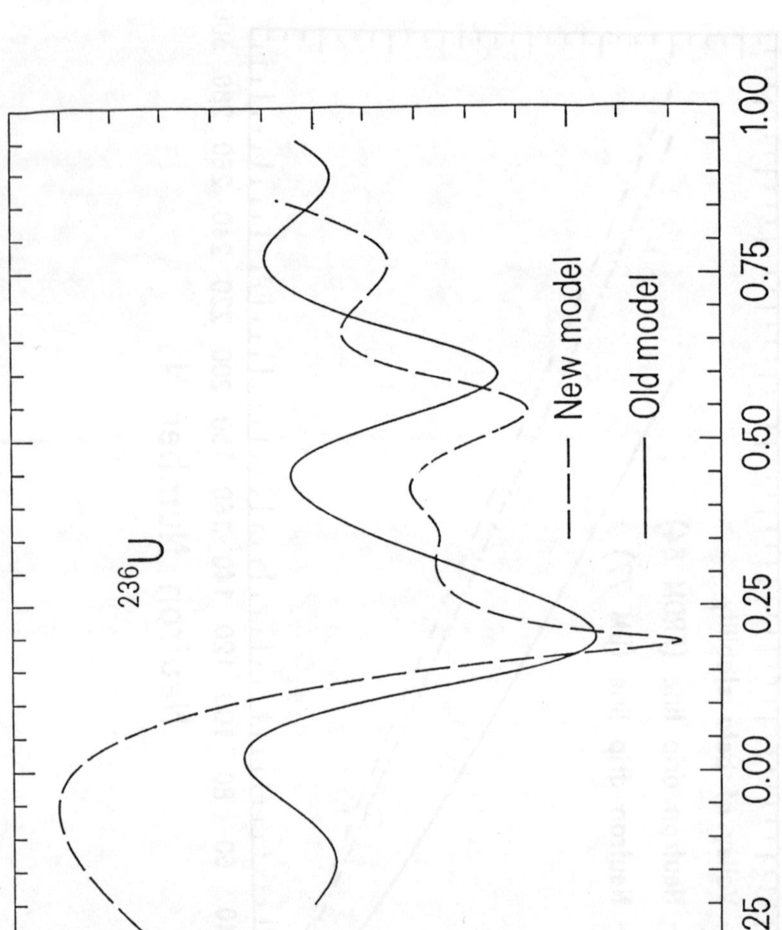

Fission Barrier Height (MeV)
10
5
0
^{236}U
New model
Old model
−0.25
0.00
0.25
0.50
0.75
1.00
ε_2

Figure 4

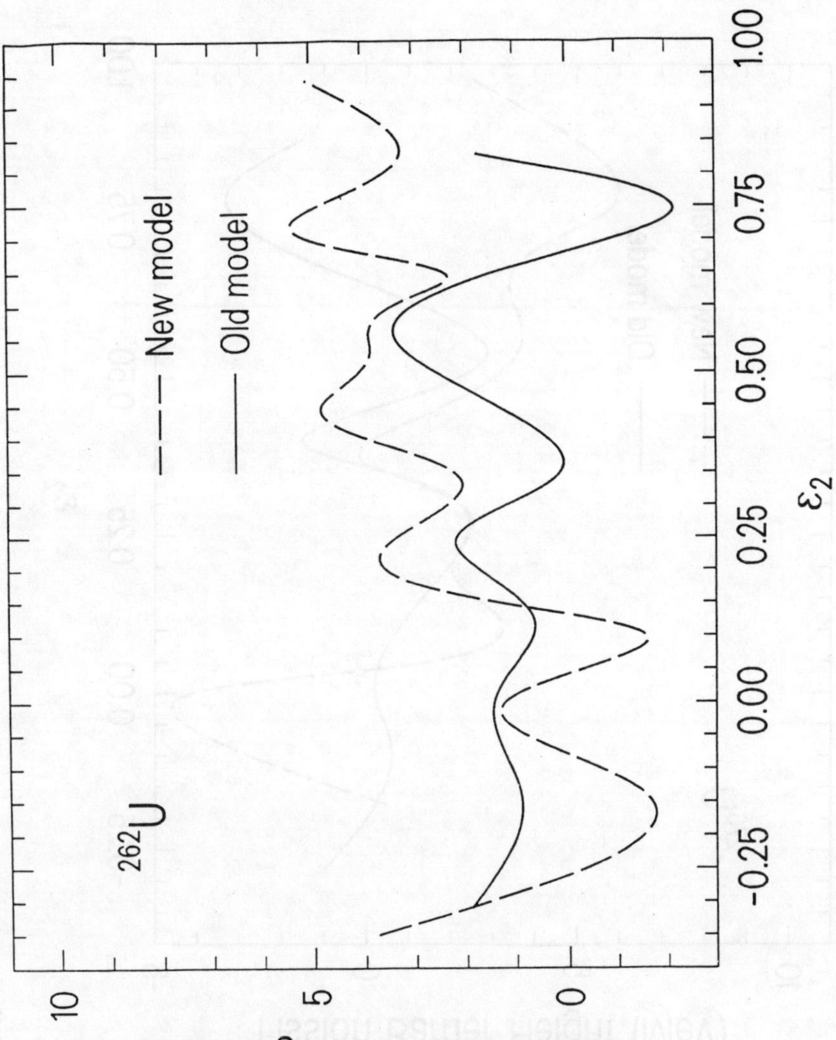

^{262}U
New model
Old model
Fission Barrier Height (MeV)
10
5
0
-0.25
0.00
0.25
0.50
0.75
1.00
ε_2

Figure 5

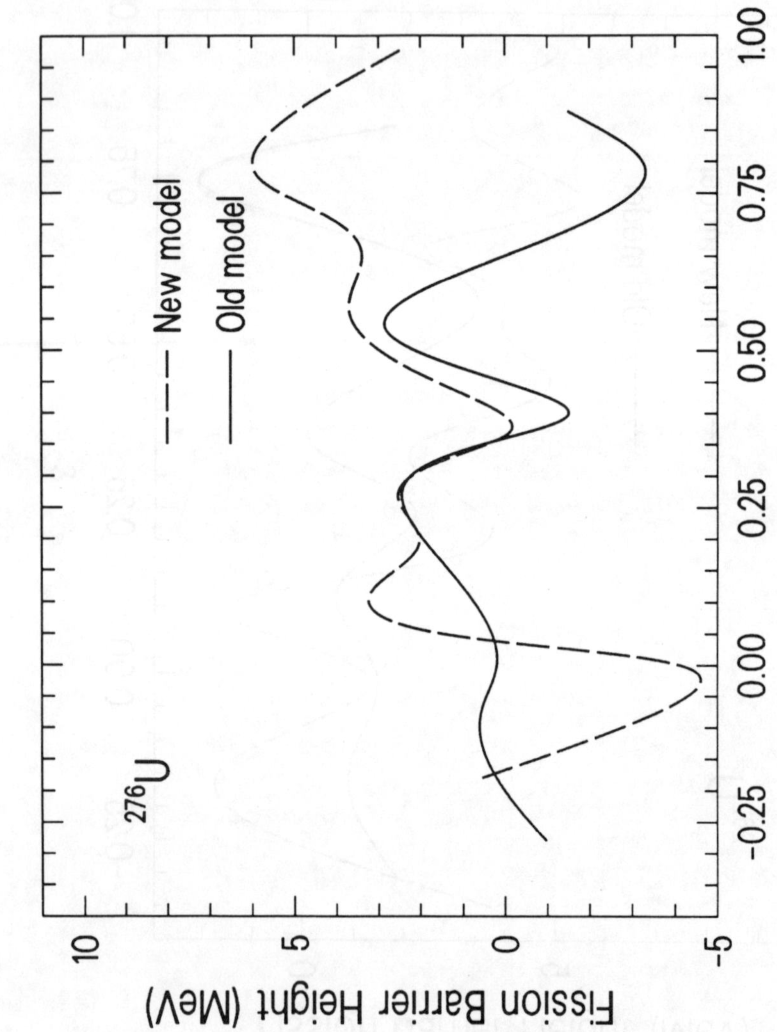

^{276}U
New model
Old model
Fission Barrier Height (MeV)
10
5
0
-5
-0.25
0.00
0.25
0.50
0.75
1.00
ε_2

Figure 6

A PARAMETER STUDY OF THE R PROCESS

K. E. Sale and G. J. Mathews

Lawrence Livermore National Laboratory
University of California
Livermore, CA 94550
USA

ABSTRACT

We consider the r-process production of actinide nuclei from a high temperature nucleon gas in a model which is most favorable for the r process. In this way, we can place upper limits on the value of the free electron fraction, Y_e, of the material which experiences the r process. We find that for a broad range of initial nucleon densities, Y_e must be small (close to that for equilibrium neutron star matter).

1. INTRODUCTION

Observations of old metal-deficient halo stars [1,2] indicate that abundances of r-process elements, such as Eu, are nearly constant relative to Fe. This suggests that r-process nucleosysthesis occurs along with the production of Fe group seed material. This limits the viability of possible r-process sites which require the prior production of enhanced heavy element seed such as the He and C burning zones of supernovae [3,4]. Nevertheless, there is still a wide variety of proposed sites which do not require prior processing. The possibilities include neutronized material ejected from supernova cores [5,6] and material ejected in the collisions of a neutron star with another compact object [7,8]. Both of these remaining scenarios (supernova cores and neutron star collisions) would probably begin at sufficiently high temperatures for nuclei to be completely photodisassociated in the ejecta which later cool to form Fe group and r-process elements.

A distinction might be made between these scenarios, however, in the degree of deleptonization of the ejecta. The time scale for the ejection of material from a supernova may be slow enough that the deleptonized matter establishes a new equilibrium. Neutron star collisions, however, may allow for a more rapid mass ejection which does not permit equilibration at lower density. This can be parameterized by the free electron fraction, Y_e. During a core bounce supernova Y_e of the ejecta could be higher ($Y_e > 0.3$) than for cold neutron star matter in equilibrium ($Y_e = 0.125$). The r-process yields will be sensitive to Y_e. A larger value of Y_e will ultimately mean a lower number of neutrons available per seed nucleus. Thus, sites with large values of Y_e may not be able to produce r-process elements before the supply of neutrons is exhausted.

In what follows, we make a parameter study of r-process element production to determine which values of Y_e are compatible with a successful r process as a function of initial matter density. Limits on the ranges of temperature and density have been set by Norman and Schramm [9]. A successful r process is defined, in this context, by the production of actinides from a photodisassociated nucleon gas. We use a model defined below which is optimum for r-process production, and therefore provides upper limits on the values for Y_e. From this study we conclude that the r-process environment must have been significantly deleptonized (as may be the case in neutron star coalescence).

2. THE MODEL

The calculations we have done are based on the approximations of the classical r process. Our first assumption is that of (n,γ)-(γ,n) equilibrium, i.e., the abundances of the different isotopes of each element are in statistical equilibrium. The relative abundances of the isotopes of each element are then determined by only the neutron number density and the temperature at any time (given the neutron separation energy for each isotope, which we estimate using the droplet model). The validity of this assumption requires that the neutron capture and photo-neutron lifetimes are short compared to the other time scales in the problem. In this sense, the model we consider is a best possible case for the r process in that we do not include the time scale for neutron captures.

The second assumption is that the seed nuclei for the r process (taken to be ^{56}Fe) are either formed very rapidly at the start or are present at the beginning of the r-process event. This also makes a best case for the r process in that we do not consider the time delay for the formation of Fe seed material. Thirdly, we assume that the nucleosynthetic process halts abruptly, followed by beta decays back to the valley of stability. The effects of beta-delayed neutron emission are included in the decay back phase.

In our model we have assumed that the total mass density varies as $\rho \propto t^{-2}$ (constant velocity expansion of a spherical shell), with a time scale set by free fall:

$$\tau = \frac{1}{\sqrt{24\pi G \rho_o}} = \frac{466}{\sqrt{\rho_o}} \tag{1}$$

This time scale is an upper limit [10,11]. Taking this slowest permitted time scale gives the greatest opportunity to create the actinide elements. The model we use allows for the presence of neutrons, seed nuclei and r-process product nuclei only. This model will give the largest possible r-process yields for given initial conditions. We take the expansion to be adiabatic with $\gamma = 4/3$ and account for the reheat of material from the nuclear reactions and beta decays. The initial temperature is taken to be 5.0×10^9 K (the photodisintegration temperature of ^{56}Fe). The effects of fission are simulated by the breakup of any nuclei with Z = 100 (Fm) into nuclei of Z = 62 (Sm) and Z = 28 (Ni). The energy and neutrons released during fission and neutron number equilibration are taken into account.

3. RESULTS

Three example mass fraction curves from our calculations are shown in Fig. 1. Figure 1a shows a case in which all of the available neutrons were absorbed without the formation of significant amounts of high mass nuclei, i.e., the r process runs out of neutrons. The second curve (Fig. 1b) is from a case in which the yield is similar to that which is observed. Figure 1c is the result from a case with a large abundance of neutrons at the start. In this last case, the heavy elements are seen to be overproduced. Again, in this model the production of heavy nuclei is overestimated.

We summarize the results of our calculations in Fig. 2. The initial conditions for which we have computed r-process yields are marked in the Y_e-ρ plane. Points denoted by open circles indicate cases in which the heavy elements were underproduced. Filled circles indicate acceptable or excessive production of high mass nuclei.

We conclude that for the r process to produce heavy elements at the observed abundances, the initial conditions must be such that

$$Y_e \leq \frac{\log \rho}{36} + 0.1 \tag{2}$$

over the range $10^3 < \rho < 10^8$ g/cm^3. This constraint is at best marginally consistent with supernova ejecta, but is possibly compatible with ejecta from binary neutron star coalescence.

REFERENCES

1. Spite, M. and Spite, F., Astron. Astrophys. 67, 23 (1978).
2. Sneden, C. and Parthasarathy, M., Ap. J. 267, 757 (1983).
3. Lee, T. et al., Ap. J. 232, 854 (1979).
4. Cowan, C. R., Cameron, A. G. and Truran, J. W., Ap. J. 265, 429 (1983).
5. Delano, M. D. and Cameron, A. G., Astrophys. Sp. Sci. 10, 203 (1971).
6. Hillebrandt, W., Nomoto, K. and Wolff, R. G., Astron. Astrophys. 133, 175 (1984).
7. Lattimer, J. and Schramm, D. N., Ap. J. 192L, 145 (1974).
8. Symbalisty, E. and Schramm, D. N., Astrophys. Lett. 22, 143 (1982).
9. Norman, E. B. and Schramm, D. N., Ap. J. 228, 881 (1979).
10. Sato, K., Prog. Theor. Phys. 51, 726 (1974).
11. Evans, C. R. and Mathews, G. J., this conference proceedings.

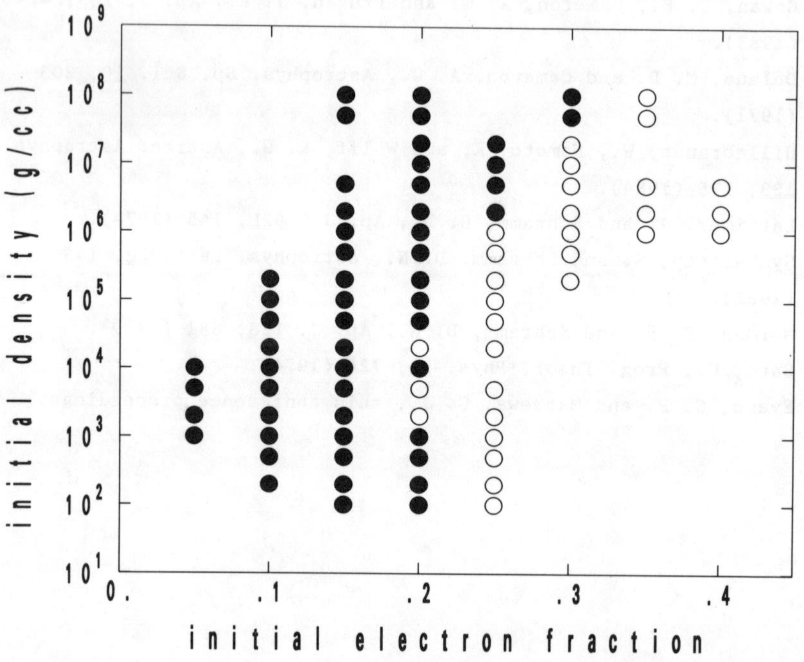

figure 2.

Results of r-process calculations.
Filled circles indicate successful r processes,
open ones mark failures.

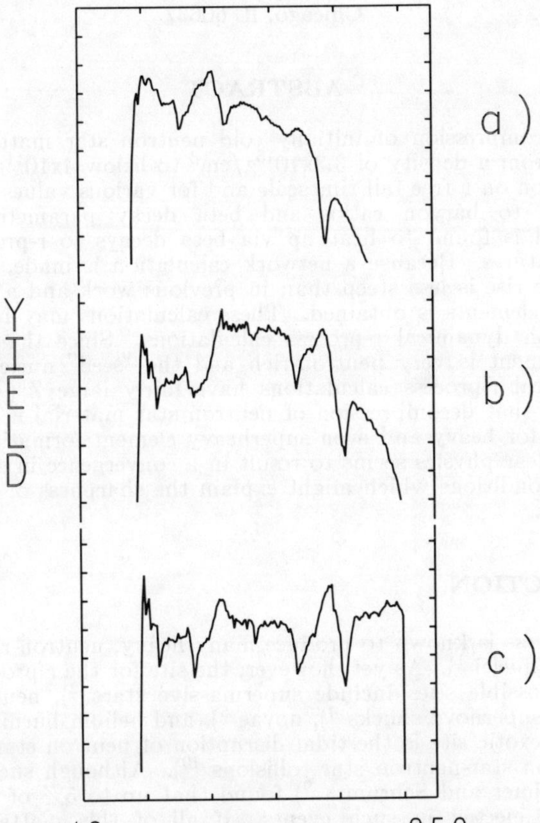

figure 1.

R-process yields for the cases described in the text.

Vertical scale is one decade per tic mark.

NUCLEOSYNTHESIS FROM THE DECOMPRESSION OF INITIALLY COLD NEUTRON STAR MATTER

B. S. Meyer and D. N. Schramm
University of Chicago
Chicago, IL 60637

ABSTRACT

The decompression of initially cold neutron star matter is followed from a density of $3.3 \times 10^{13} g/cm^3$ to below $4 \times 10^{11} g/cm^3$ for expansion on a free fall timescale and for various values of initial electron to baryon ratios and beta decay parameters. The material is found to heat up via beta decays to r-process like temperatures. Because a network calculation is made, the temperature rise is less steep than in previous work and a distribution of elements is obtained. These calculations may be continued with dynamical r-process calculations. Since the resulting environment is very neutron-rich and the "seed" nuclei for the dynamical r-process calculations have fairly large Z ($Z \approx 60$), it appears that decompression of neutron star material is a plausible site for heavy and even superheavy element formation. Also the nuclear physics seems to result in a convergence in seeds and initial conditions which might explain the sharpness of r-process features.

I. INTRODUCTION

The r process is known to produce many heavy, neutron-rich nuclei as well as the actinide nuclei [1]). As yet, however, the site for the r process has not been determined. Possible sites include supermassive stars [2]), neutronized cores of supernovae [3]), supernova shocks [4]), novae [5]), and helium burning in supernovae [6,7,8]). A more exotic site is the tidal disruption of neutron stars by black holes [9]) or by neutron star-neutron star collisions [10]). Although such sites are probably rare, Lattimer and Schramm [9]) found that up to 5% of the neutron star mass could be ejected in such events. If all of this matter were r-process material, such sites could produce much of the Galaxy's r-process matter.

In this paper we will argue that neutron star disruption by a black hole is a plausible site at which to make heavy and superheavy elements. The material forms into very heavy, neutron-rich nuclei which work their way down to the beta-stable line as the material decompresses. Furthermore, because the nuclei start at fairly large proton number Z, it is easier for them to reach the heavy and superheavy element regions.

Another interesting advantage for such a scenario is that it may explain the sharpness of r-process features. Because a number of r-process events have contributed to the solar system elements, some mechanism must force r-process material to follow a fairly narrow path in the N-Z plane of the nuclides. Apart from neutron star disruption, the nucleosynthesis from the above mentioned events all depend explicitly on the hydrodynamics in the problem. Since the hydrodynamics vary from event to event, it is hard to see how a number of events could have similiar enough conditions that they give sharp r-process peaks. Decompressing neutron star matter, on the other hand, will tend to converge upon the neutron drip line with r-process like temperatures and neutron densities. This occurs under a variety of conditions, as long as the temperature is not too large. Because the material converges on the drip line, this would be a case of the nuclear physics in the problem imposing the mechanism to force the material into a narrow path in the N-Z plane, independently of the hydrodynamics.

In another paper in this volume [11]), ejection of hot, shocked neutron star matter is discussed as a possible element production site. Hot ejected matter will reach nuclear statistical equilibrium and will lose the special advantages cold neutron star matter has to make very heavy elements. We discuss instead the decompression of initially cold neutron star matter. For this we require a nuclear model appropriate for densities above neutron drip ($\rho \geq 4\text{x}10^{11}\text{g/cm}^3$). A network type of calculation is also required to allow the decompression results to merge consistently into dynamical r-processes. In this paper we present some results of such calculations and some discussion. For more results and details, see ref. 12.

II. CALCULATIONS

The nuclear model employed for the decompression is based on that of Lattimer et al. [13]) with modifications to include the effects of nuclear deformation, nuclear shells, and coulomb thickness. The model treats nuclei as compressible liquid drops in Wigner-Seitz cells filled with free neutrons and electrons. The free energy of a cell is given as a sum of bulk, surface, coulomb, translation, lepton, and shell terms [13,12]) and is minimized to give the equilibrium configuration of the nuclei.

For very low temperatures ($T \ll 1$ MeV), only one species of nuclei will be present. At higher temperatures, more than one species will be present, however, due to interactions of the nuclei with the external particles. We thus need a Saha equation to describe the configuration of the nuclei. The regular Saha equation used in r-process calculations [1]) is not appropriate since at high densities the sea of neutrons outside nuclei may be degenerate. A nuclear Saha equation that takes a degenerate external neutron sea into account is derived in ref. 12 and is used throughout the present calculations.

The decompression of initially cold neutron star matter proceeds as follows. The material, at T=0, expands. Because T=0, only one species of nuclei is present, namely, the equilibrium species. Once the electron density becomes low enough, however, the degenerate sea of electrons no longer blocks beta decays, and beta decays can occur. These beta decays release energy which causes the temperature of the material to rise. Furthermore, the beta decays drive the material away from a one species composition both because of the higher temperature and because nuclei of different proton number Z result. It is clear, then, that, unless a one species description is maintained by forcing all nuclei to beta decay at the same time so that all nuclei have the same Z and by not allowing for an isotopic spread, we require some sort of network calculation.

Such a network description is discussed in ref. 12. It is analogous to dynamical (n—γ) equilibrium r-process calculations [14]) but employs the degenerate neutron sea Saha equation discussed previously and the nuclear model appropriate for material above neutron drip. Because both the nuclear model and the Saha equation go over to forms used in normal r-process calculations, these calculations become dynamical r-process calculations below neutron drip.

The essential nature of the calculations is as follows. At each timestep, an initial guess external neutron density is determined. Nuclear abundances are determined from the Saha equation. A new external neutron density is determined by subtracting the number density of neutrons in nuclei from the known total neutron density. The new neutron density is compared with the old. If they do not agree, the new neutron density is used to compute new nuclear abundances and hence another new neutron density. This process is repeated until the neutron density converges. Then beta decay rates are computed for the nuclei as in ref. 14. Beta decays are allowed to occur and the material is allowed to expand. The temperature change is computed [12]) from the entropy change and from cooling due to expansion and to neutrino processes [15,16]). This process is repeated for subsequent timesteps.

The calculations presented here are for material expanding on a timescale $446\alpha/\rho^{1/2}$ sec, where ρ is the density in g/cm^3 and α measures the deviation from the free fall timescale (for these calculations, α=1.0 throughout). Beta decays depend on the parameter $<\rho_L/ft>$ [14]). The present calculations treat the cases where log $<\rho_L/ft>$ is -3.5 and -4.5. The initial values of the electron to baryon ratio Y_e are taken to be 0.03, 0.05, 0.10, and 0.15. These values are chosen to explore a possible range of ejected matter, perhaps from varying depths inside a neutron star. Also, the wide range allows for consideration of uncertainties in neutron star composition. We begin to follow the decompression at a density of $3.3 \times 10^{13} g/cm^3$. This density is low enough that there is no difficulty presented by a possible bubbles phase [17,13]) but high enough that beta decays are still blocked. For all four initial Y_e, the nuclei begin at Z=50.

Although the calculations may be continued well below neutron drip since they become in essence dynamical r-process calculations, they are cut-off for $\rho \leq 4 \times 10^{11} g/cm^3$ (neutron drip) and $T \geq 0.0689$ MeV ($T = 8 \times 10^8$ K). The reason for this is that the computations become extremely time-consuming as the nuclei work their way up the network since each nuclear mass must be calculated anew at each timestep. Fortunately, as discussed above, the nuclear model goes over to a normal incompressible liquid drop model below neutron drip; thus, the results from the decompression calculations could be used in a very consistent way as input for normal dynamical r-process calculations.

III. RESULTS

Figures 1 and 2 show the temperature of the expanding matter versus the mass density for the two cases log $<\rho_L/ft> = -3.5$, α=1.0 and log $<\rho_L/ft> =$ -4.5, α=1.0. In both figures, the curves are labeled a,b,c, and d for $(Y_e)_i = 0.03$, 0.05, 0.10, and 0.15, respectively.

The first feature to notice is that the curves rise from zero temperature at different densities. This is due to the blocking of beta decays by the degenerate

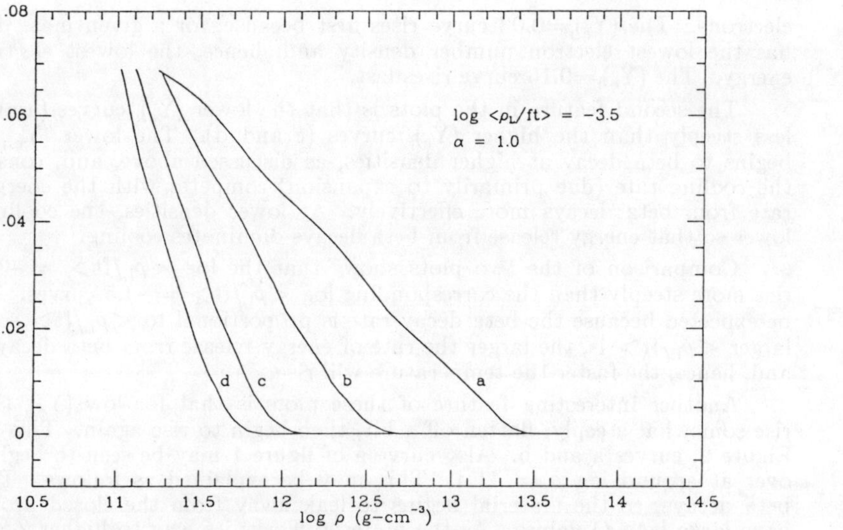

Figure 1. Temperature vs. Density for log $<\rho_L/\mathrm{ft}>$= -3.5, α=1.0, and $(Y_e)_i$= (a) 0.03, (b) 0.05, (c) 0.10, and (d) 0.15.

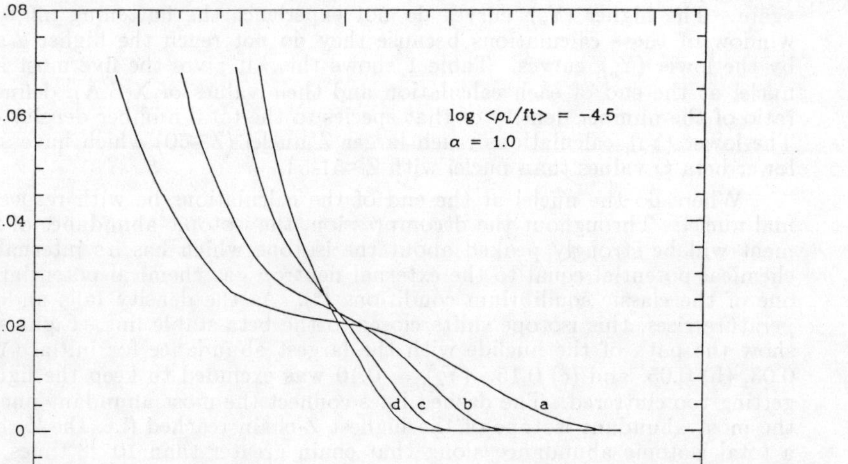

Figure 2. Temperature vs. Density for log $<\rho_L/\mathrm{ft}>$= -4.5, α=1.0, and $(Y_e)_i$= (a) 0.03, (b) 0.05, (c) 0.10, and (d) 0.15.

electrons. The $(Y_e)_i$=0.03 curve rises first because, for a given mass density, it has the lowest electron number density and, hence, the lowest electron fermi energy. The $(Y_e)_i$=0.15 curve rises last.

The second feature in the plots is that the lower $(Y_e)_i$ curves (a and b) rise less steeply than the higher $(Y_e)_i$ curves (c and d). The lower $(Y_e)_i$ material begins to beta decay at higher densities, as discussed above, and, consequently, the cooling rate (due primarily to expansion) competes with the energy release rate from beta decays more effectively. At lower densities, the cooling rate is lower so that energy release from beta decays dominates cooling.

Comparison of the two plots shows that the log $<\rho_L/ft>$ = -3.5 curves rise more steeply than the corresponding log $<\rho_L/ft>$ = -4.5 curves. This is to be expected because the beta decay rate is proportional to $<\rho_L/ft>$ so that the larger $<\rho_L/ft>$ is, the larger the rate of energy release from beta decays will be, and, hence, the faster the temperature will rise.

Another interesting feature of these plots is that for low $(Y_e)_i$ the curves rise somewhat steeply, flatten off a bit, then begin to rise again. This is seen in Figure 2, curves a and b. Also curve a of figure 1 may be seen to begin to turn over at around log ρ = 11.4. This may be explained as follows. The initial beta decays, as the material begins to leak away from the closed proton shell, have large beta Q values. As the material works its way to higher Z and away from the first few proton numbers above the closed shell, the beta Q values fall off somewhat, thereby leading to smaller release of energy from beta decays and a less rapid rise in the temperature. Once the density becomes small enough, the cooling from expansion falls off so that the temperature rises more sharply again. The higher $(Y_e)_i$ curves do not experience the flattening off within the window of these calculations because they do not reach the higher Z's reached by the lower $(Y_e)_i$ curves. Table 1 shows this. It gives the five most abundant nuclei at the end of each calculation and their values of X(Z,A), defined as the ratio of the number density of that species to the total number density of nuclei. The lower $(Y_e)_i$ calculations reach larger Z nuclei ($Z\approx60$) which have somewhat lower beta Q values than nuclei with $Z\approx51$-54.

Where do the nuclei at the end of the calculations lie with respect to normal nuclei? Throughout the decompression, the isotopic abundance of some element will be strongly peaked about the isotope which has its internal neutron chemical potential equal to the external neutron gas chemical potential. This is one of the classic equilibrium conditions [18]). As the density falls and the temperature rises, this isotope shifts closer to the beta-stable line. Figures 3 and 4 show the path of the nuclide with the largest abundance for initial $(Y_e)_i$ = (a) 0.03, (b) 0.05, and (c) 0.15. $(Y_e)_i$ = 0.10 was excluded to keep the figures from getting too cluttered. The dashed lines connect the most abundant nucleus with the most abundant isotope of the highest Z-chain reached (i.e. the Z-chain with a total isotopic abundance along that chain greater than 10^{-10} times the total isotopic abundance along the Z-chain with the greatest abundance. Other nuclei fall on the dashed line at the end of each calculations. Interestingly, all of the nuclei for all of the calculations tend to converge on the same path. The "seed" nuclei for the continuation r-process calculations thus all start at roughly the same path. This may be of significance in that it may help to explain the sharpness of the r-process peaks.

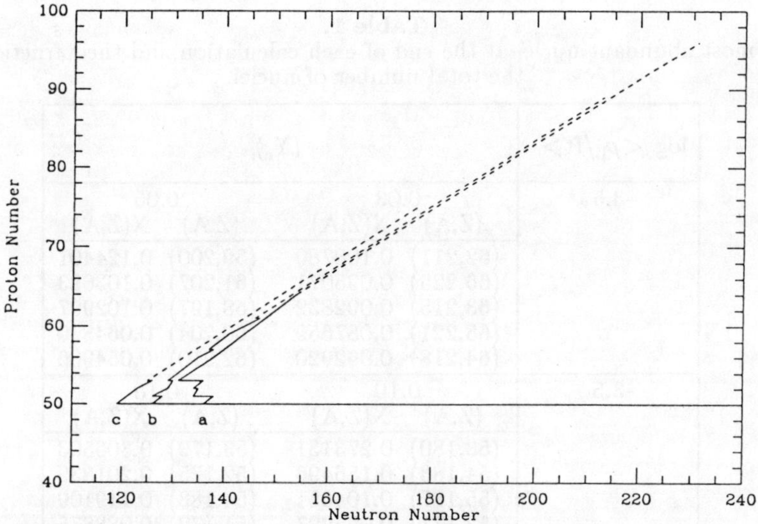

Figure 3. Path of the most abundant nucleus in the N-Z plane (solid line) for $\log <\rho_L/ft> = -3.5$, $\alpha = 1.0$, and for $(Y_e)_i$ = (a) 0.03, (b) 0.05, and (c) 0.15. For an explanation of the dashed lines, see the text.

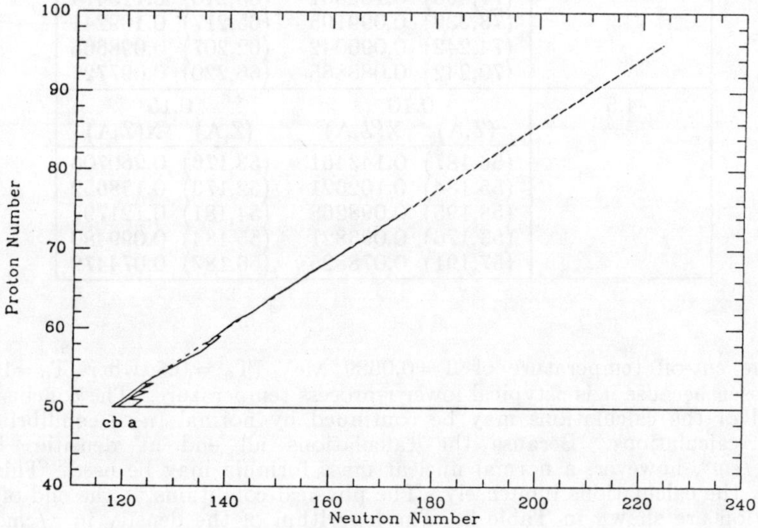

Figure 4. Path of the most abundant nucleus in the N-Z plane (solid line) for $\log <\rho_L/ft> = -4.5$, $\alpha = 1.0$, and for $(Y_e)_i$ = (a) 0.03, (b) 0.05, and (c) 0.15. For an explanation of the dashed lines, see the text.

Table 1.
Five most abundant nuclei at the end of each calculation and their fractions of the total number of nuclei.

$\log \langle \rho_L/ft \rangle$	$(Y_e)_i$			
-3.5	0.03		0.05	
	(Z,A)	X(Z,A)	(Z,A)	X(Z,A)
	(62,211)	0.108780	(59,200)	0.124401
	(66,225)	0.095071	(61,207)	0.103623
	(63,215)	0.092832	(58,197)	0.102997
	(65,221)	0.087659	(60,204)	0.064870
	(64,218)	0.062920	(62,210)	0.054906
-3.5	0.10		0.15	
	(Z,A)	X(Z,A)	(Z,A)	X(Z,A)
	(53,180)	0.273131	(53,179)	0.306503
	(54,183)	0.156496	(52,175)	0.201839
	(55,187)	0.104651	(54,183)	0.119109
	(52,177)	0.095997	(51,172)	0.088875
	(56,189)	0.065318	(55,186)	0.083627
-4.5	0.03		0.05	
	(Z,A)	X(Z,A)	(Z,A)	X(Z,A)
	(72,237)	0.104995	(64,213)	0.121017
	(71,235)	0.104561	(63,210)	0.113474
	(73,239)	0.099195	(65,217)	0.109241
	(74,242)	0.090042	(62,207)	0.098665
	(70,242)	0.088855	(66,220)	0.097727
-4.5	0.10		0.15	
	(Z,A)	X(Z,A)	(Z,A)	X(Z,A)
	(56,187)	0.143461	(53,176)	0.260706
	(55,184)	0.102621	(52,173)	0.158652
	(58,195)	0.098268	(54,181)	0.121797
	(53,176)	0.095821	(55,184)	0.099480
	(57,191)	0.078825	(56,187)	0.074472

The cut-off temperature of T=0.0689 MeV (T_9 = 0.8, where $T_9 = 10^9$ K) was chosen because it is a typical lower r-process temperature. The conclusion is that all of the calculations may be continued by normal (n-γ) equilibrium r-process calculations. Because the calculations all end at densities below 4×10^{11} g/cm^3, however, a normal nuclear mass formula may be used. This will simplify the calculations immensely. The physical conditions at the end of each calculation are shown in Table 2. The logarithm of the density in g/cm^3 and free neutron density in cm^{-3} are given, as well as the temperature in T_9, the final Y_e, and the highest Z-chain reached by beta decays such that $n_{Z,max} > 10^{-10} n_{Z,O}$, where $n_{Z,O}$ is the Z-chain with the largest abundance.

Table 2.
Physical quantities at the end of the calculations.

log $<\rho_L/ft>$	Quantities at End	$(Y_e)_i$ 0.03	0.05	0.10	0.15
-3.5	log ρ	11.24	11.26	11.11	11.02
	log n_n	34.96	34.94	34.69	34.46
	T_9	0.80	0.80	0.80	0.80
	Y_e	0.0381	0.0572	0.1084	0.1594
	Z_{max}	96	89	77	72
-4.5	log ρ	8.76	9.42	9.95	10.17
	log n_n	32.42	33.08	33.52	33.62
	T_9	0.80	0.80	0.80	0.80
	Y_e	0.0378	0.0667	0.1132	0.1606
	Z_{max}	96	96	83	75

IV. CONCLUSIONS

Qualitative differences of these results with previous work [19]) result from the network approach in this work. Because expansion is always allowed to compete with beta decays, the temperature rises more slowly. Also fewer beta decays occur because only a fraction of the nuclei decay during each timestep, not all of them; thus, although the highest Z reached is roughly comparable to results obtained previously [19]) at the cut-off densities in this work, the dominant Z-chains are lower. Because a distribution of nuclei is obtained, the present results would meld more naturally into a dynamical r-process calculation.

The main conclusion of considerations of the decompression of initially cold neutron star matter remains that an r-process like environment can result. Because the "seed" nuclei for the continuation r-process calculation are very neutron-rich and have a larger Z ($Z \approx 60$) than usual r-process calculations ($Z \approx 26$), it is likely that heavy or even superheavy elements will be produced. Also, because the "seed" nuclei begin nearly on the same path, it is possible that the sharpness of r-process features could be explained. The only caveat is that if temperatures that are too large are reached ($T_9 > 3$) in the continuation r-process calculations, the material will tend to nuclear statistical equilibrium and will destroy the unique ability decompression of cold neutron star matter has to make very heavy elements. Calculations are under way to determine whether or not this occurs.

ACKNOWLEGEMENTS

One of us (B.S.M.) acknowledges the support of a National Science Foundation Graduate Fellowship.

REFERENCES

1. Burbidge, E. M., Burbidge, G. R., Fowler, W. A., and Hoyle, F., Rev. Mod. Phys. 29, 547 (1957).
2. Seeger, P. A., Fowler, W. A., and Clayton, D. D., Ap. J. Suppl. 97, 121 (1965).
3. Cameron, A. G. W., Delano, M. D., and Truran, J. W., in Proc. on the Properties of Nuclei Far from the Region of Beta-Stability. Stern Rep. 70-30, Leyson (1970).
4. Colgate, S. A., Ap. J. 163, 221 (1971).
5. Hoyle, F. and Clayton, D. D., Ap. J. 191, 705 (1974).
6. Hillebrandt, W. and Thielemann, F.-K., Astron. Ap. 58, 357 (1977).
7. Truran, J. W., Cowan, J. J., and Cameron, A. G. W., Ap. J. Lett. 222, L63 (1978).
8. Blake, J. B., Woosley, S. E., Weaver, T. A., and Schramm, D. N., Ap. J. 248, 315 (1981).
9. Lattimer, J. M. and Schramm, D. N., Ap. J. 210, 549 (1976).
10. Symbalisty, E. and Schramm, D. N., Astrophysical Letters 22, 143, (1982).
11. Evans, C. R. and Mathews, G. J., this volume.
12. Meyer, B. S., in preparation.
13. Lattimer, J. M., Pethick, C. J., Ravenhall, D. G., and Lamb, D. Q., Nucl. Phys. A432, 646 (1985).
14. Sato, K., Prog. Theor. Phys. 51, 726 (1974).
15. Bahcall, J. N. and Wolf, R. A., Phys. Rev. 140, B1452 (1965).
16. Beaudet, G., Petrosian, V., and Salpeter, E. E., Ap. J. 150, 979 (1967).
17. Lamb, D. Q., Lattimer, J. M., Pethick, C. J., and Ravenhall, D. G., Nucl. Phys. A411, 449 (1983).
18. Baym, G., Bethe, H. A., and Pethick, C. J., Nucl. Phys. A175, 225 (1971).
19. Lattimer, J. M., Mackie, F., Ravenhall, D. G., and Schramm, D. N., Ap. J. 213, 225 (1977).

Shock Driven Neutronized Wind from Binary Neutron Star Coalescence as a Site for the r-process

Charles R. Evans

Theoretical Astrophysics, 130 - 33, California Institute of Technology, Pasadena, CA 91125

Grant J. Mathews

University of California, Lawrence Livermore National Laboratory, Livermore, CA 94550

ABSTRACT

Several issues are raised in connection with r-process nucleosynthesis occurring in material ejected during the coalescence of binary neutron stars. An example of such an event is the binary pulsar system PSR 1913+16 which will coalesce in some 3.5×10^8 years due to gravitational radiation induced orbital separation shrinkage. We discuss a mechanism by which hot, high n/p matter is ejected from the coalescing cores. Several points are stressed: (1) the ejected material likely rises to a high initial temperature and therefore begins as a photo-dissociated free nucleon gas; (2) the ejection occurs as a result of the repeated passage of shock waves generated by the rapidly rotating, newly merged core; (3) this process leads to an energy excess in the ejecta that brings the material to a relatively high terminal velocity and with an associated fast expansion time scale. We show results of a preliminary r-process calculation that incorporates a decompression track implied by this ejection process.

The r-process has long been associated with supernovae and stellar collapse ever since the recognition[1,2] that a high neutron flux is needed to synthesize the abundance peaks at $A = 80, 130, 195$ as well as the abundance of naturally occurring elements heavier than Pb. Various sites within supernovae (high density inner mass cuts, low density envelope, etc.) have been examined (see Mathews and Ward[3] for a recent review), however the astrophysical site of the r-process is still unknown. In light of this uncertainty, various authors have proposed as alternatives r-process production in material ejected from neutron stars during black hole - neutron star encounters[4] or during neutron star - neutron star coalescence.[5,6] These exotic sites have the apparent advantages of initiating the r-process with material having a significantly higher n/p ratio and possible high mass-ejection efficiency per event, but suffer the disadvantage of highly uncertain event rates. However, even if relatively rare, this source may provide a significant contribution to the production of actinides, which are otherwise difficult to produce.[7] In this paper we begin examining how material can be ejected from the vicinity of two merging neutron stars and, given the associated time scales, whether it will undergo subsequent r-process.

Consider two $1.4M_{\odot}$ neutron stars and assume their radii are $\approx 12\,km$. Equal mass stars will reach tidal-lobe overflow at about the time their surfaces are brought into contact, when the separation becomes roughly $24\,km$. The orbital period of such a system is then only $\approx 1.2\,ms$ with stellar orbital velocities of $\approx 6.2\times10^{9}\,cm\,s^{-1}$. Matter released from the stars will have an initial binding energy of order $130\,MeV$ per nucleon. It is a nontrivial question as to how such tight binding can be overcome. Lattimer and Schramm[4] in considering a neutron star spiraling into a more massive black hole ($M_{BH} \gtrsim 5M_{\odot}$) determined that tidal disruption of the neutron star would lead to about 5 % of the material reaching infinity in a so-called "tube-of-toothpaste" effect. In considering neutron star - neutron star coalescence, Clark and Eardley[5] suggested that $\gtrsim 0.1M_{\odot}$ was ejected. However, Symbalisty and Schramm[6] pointed out that some other attribute of the collision besides tidal disruption (e.g., shocks) might be necessary to unbind significant amounts of material. Another idea, proposed in a different context, is that material may be driven off a hot, neutrino-radiating neutron star in a radiative wind.[8]

Lattimer et al.[9] previously considered the decompression of cold neutron star matter and assumed material would initially return to the form of nuclear clusters in a degenerate electron sea. It is likely, however, that any ejection mechanism will impart some fraction of the energy required to free material to infinity to thermal energy. This can easily exceed the temperature $T \approx 10^{10}\,K$ necessary to photo-dissociate nuclei to a free nucleon gas.

Shock-driven ejection in the form of a wind may be a promising mechanism given the ubiquity of shocks expected in a binary merger. We are currently examining whether shocks can drive a significant wind in such a scenario. A recent hydrodynamic simulation[10] of the *head-on* collision of two compact stars provided further impetus for thinking about the binary merger scenario. The axisymmetric calculation was intended to investigate gravitational radiation emission and accordingly a polytropic equation of state was assumed. Despite the idealized nature of the head-on assumption, some important insights were gained. After the initial coalescence of the two stars, the new single core is surrounded by a hot tenuous medium of released stellar material. The core is in a state of violent, large amplitude oscillation and circulation. This motion continually drives shocks out into the surrounding material, which is then steadily driven outward. Figure 1 shows contour levels of pressure from the axisymmetric calculation during the post-merger, oscillation phase. A complex pattern of shocks is evident in the low density surrounding material. The calculation exhibited a high rate of mass being transported outward; in this idealized calculation, a total of 8 % of the initial stellar material was driven to large radius in a few core oscillation periods.

While the merging of a circular orbit binary occurs differently, its hydrodynamic effects on the liberated fluid may be quite similar. Assuming the system continues to evolve as Newtonian point masses after first contact gives a timescale for merging of only

$$\tau = \frac{5}{64}\frac{c^5}{G^3}\frac{a^4}{M^3} = 1.2\,(M/2.8)^{-3}(a/24.0km)^4\,ms\,, \qquad (1)$$

Figure 1. Contours of pressure after axisymmetric collision and merger. Collision axis is vertical. Outer boundary has a radius of 80 km. Approximately 85 % of the material is within an inner core of radius $\sim 20\ km$. Shocks driven by large amplitude oscillations and circulation within this inner, denser core are evident in the surrounding hot envelope.

or about one orbital period. It is apparent the stars merge into a single core quite rapidly. The time scale to reach an axisymmetric state though is significantly longer. The immediate coalescence leaves a central, rapidly rotating ($P \approx 1\ ms$) triaxial core. Once the equatorial ellipticity reaches $\varepsilon \approx 0.2$ the decay time due to gravitational radiation emission is on the order of a second. During the abrupt coalescence, a substantial but unknown amount of neutron star matter will become unbound and go into forming a circumstellar envelope. Only a three-dimensional hydrodynamic calculation could identify the amount of mass deposited in the envelope.

If we assume $0.3\ M_\odot$ goes into the envelope, the density at the base of the envelope will be on the order of $\approx 10^{12}\ g\ cm^{-3}$. The flow of this envelope material will be dominated by a pair of spiral shocks generated by the nonspherical rapidly rotating inner core. The shock speed at the base is equal to the equatorial surface velocity of the core. The density of semi-degenerate matter at the base of the envelope is sufficiently high that initial shock strengths will be moderate ($p_1/p_0 \lesssim 5$). As the shocks descend the density gradient in the envelope, they rapidly strengthen.

Each region of the envelope will encounter a shock once every 0.6 ms. Between shocks the flow will be adiabatic if the temperature is below several MeV (or dense enough to trap neutrinos). As material is repeatedly shocked it will gain total specific energy and rise against the gravitational potential. As material moves upward, at some height (as yet undetermined) a final shock will give the material net positive specific energy, launching a wind. Ejected matter reaches a terminal velocity v_t, which may be high depending on the strength of the last shock. The decompression behavior can be quite different from (reverse) free-fall[9]; for example a wind will satisfy $\rho t^2 \sim \dot{M}/(4\pi v_t{}^3)$, while free-fall gives $\rho t^2 \sim (6\pi G)^{-1}$. In most cases the value of ρt^2 for a wind will be significantly less than that for reverse free-fall, indicating a more rapid dilution of the gas.

In the region below the base of the wind, repeated shocks will tend to push the matter temperature up, while the upward flow and adiabatic expansion will tend to cool the material. If the temperature rises significantly above 1 MeV, pairs are produced and degeneracy is broken. A potential problem with our scenario is that matter must traverse any such high temperature region sufficiently rapidly to avoid coming into beta equilibrium due to the reactions

$$p + e^- \rightarrow n + \nu_e, \tag{2a}$$

$$n + e^+ \rightarrow p + \bar{\nu}_e. \tag{2b}$$

In high density regions these reactions may remain blocked if neutrinos are trapped. Otherwise the n/p ratio is quickly pushed down toward unity, and an r-process would be less likely. However, Arnett[11] suggests that even a n/p slightly greater than unity may serve to give an r-process if a large fraction of free protons quickly lock up into He. We are investigating the possibility of modeling significant aspects of the wind generation with 1D or 2D hydrodynamic simulations

As a further study of the plausibility of this shock driven neutron star wind as a site for r-process nucleosynthesis, we have calculated the r-process yields in the framework of a schematic classical r-process model similar to that discussed elsewhere in this volume.[12] In this shock-driven wind, the density diminishes as $\rho \sim r^{-2} \sim t^{-2}$. The temperature diminishes adiabatically along with the density $T \sim \rho^{1/3}$. For the purposes of this schematic illustration, we will ignore the release of internal energy from the r-process nuclear reactions, although this must eventually be considered. Omitting this contribution shifts the r-process toward more highly neutron rich isotopes, so in this sense this is an optimistic scenario. In addition to the density decrease along a streamline in the wind, the free neutron number in a comoving volume rapidly drops due to capture on nuclei.

For initial conditions we assume a total mass for the coalescing neutron star system of $2.8 M_\odot$, and a mass loss rate of $0.02\, M\ s^{-1}$ (sufficient to give a ~10 % loss in 10 s). The terminal velocity of the wind (specific energy excess) can be high depending on the shock strength and height of the critical point. We assume for a first calculation that this gives a high terminal velocity of $\sim 7.5\times10^9\ cm\ s^{-1}$. These conditions specify the temperature and density in the wind at any distance from the source. We begin the r-process calculation at $T = 1.5\times10^9 K$ ($\rho = 10^4\ g\ cm^{-3}$) somewhat below the photo-dissociation temperature of Fe, so that at this point all of the available protons are assumed, as a most optimist scenario, to be incorporated into Fe seed. The n/p ratio is taken to be 8, assumed typical for equilibrium material released from the neutron stars.

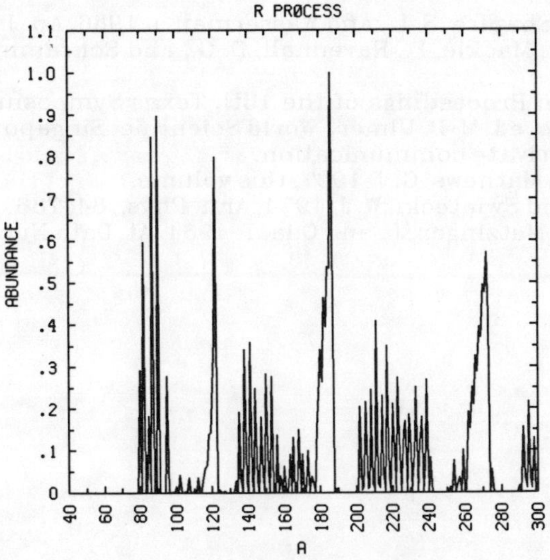

Figure 2. Calculated r-process abundances versus A, prior to decay back to stability.

Figure 2 shows the resultant classical dynamical r-process abundances before decay back to stability. These calculations utilized the droplet model[13)] for estimating the nuclear masses away from stability, and the Klapdor beta decay rates.[14)] It appears the calculated r-process is too good in that the abundance peaks occur for mass numbers below the corresponding solar system peaks. This suggests that a more realistic model might produce an abundance distribution very similar to solar system r-process material. This is a calculation we are now pursuing.

ACKNOWLEDGMENTS

This research is supported in part by NSF grants PHY83-08826 and AST85-14911 and U.S. Department of Energy contract number W-7405-ENG-48. Numerical computations were performed on the Cray X-MP at the National Center for Supercomputing Applications, University of Illinois, Urbana, Illinois.

REFERENCES

[1] Burbidge, E. M., Burbidge, G. R., Fowler, W. A., and Hoyle, F. 1957, Rev. Mod. Phys., **29**, 547.
[2] Cameron, A. G. W. 1957, Chalk River Report CRL-41.
[3] Mathews, G. J., and Ward, R. A. 1985, Rep. Prog. Phys., **48**, 1371.

[4] Lattimer, J., and Schramm, D. N. 1974, Ap. J. Lett., **192,** L145.
[5] Clark, J. P. A., and Eardley, D. M. 1977, Ap. J., **215,** 311.
[6] Symbalisty, E. M. D., and Schramm, D. N. 1982, Ap. Letters, **22,** 143.
[7] Blake, J. B., Weaver, T., Woosley, S. E., and Schramm, D. N. 1981, Ap. J., **248,** 315.
[8] Duncan, R. C., Shapiro, S. L., and Wasserman, I. 1986, Ap. J., **309,** 141.
[9] Lattimer, J. M., Mackie, F., Ravenhall, D. G., and Schramm, D. N. 1977, **213,** 225.
[10] Evans, C. R., in Proceedings of the 13th Texas Symposium on Relativistic Astrophysics, ed. M. P. Ulmer. World Scientific: Singapore (1987).
[11] Arnett, W. D., private communication.
[12] Sale, K. E., and Mathews, G. J. 1987, this volume.
[13] Myers, W. D., and Swiatecki, W. J. 1974, Ann. Phys., 84, 786.
[14] Klapdor, H. V., Metzinger, J., and Oda, T. 1984, At. Data Nucl. Data Tables, 31, 81.

THE p-PROCESS REVISITED

M. Rayet[1], N. Prantzos[2], and M. Arnould[1]
[1] Institut d'Astronomie, d'Astrophysique et de Géophysique,
Université Libre de Bruxelles, Belgium
[2] Institut d'Astrophysique de Paris, France

ABSTRACT

We propose a scenario for the synthesis of p-nuclei in the oxygen burning layers of massive exploding stars, where appropriate conditions are expected, according to recent models.

The results of some test computations performed with updated nuclear reaction rates and an extended reaction network are reported.

1. INTRODUCTION

The stable neutron-deficient isotopes of the elements with charge number Z≥34 are classically referred to as the p-nuclei. Those nuclides are observed only in the solar system, where they represent no more than 1% to 0.1% (with increasing Z) of the bulk Z≥34 elemental abundances, made predominantly of the more neutron-rich s- and r-nuclei. However, different mixtures of the p-, s- and r-isotopes are found in certain meteoritic inclusions for some elements, like Kr, Sr, Xe and Sm (e.g. Papanastassiou 1985; Anders 1987). In some cases, a p-nuclei deficiency with respect to the sr-component is observed, while an excess is derived in the other ones. In addition, certain data are more consistent with a decoupling between the p- and sr-components, whereas the famed Xe-HL exhibits a strong correlation between its lightest (p-type) and heaviest (r-type) isotopes (there has also been some claim that Te exhibits a similar pattern, but these data are disputed; e.g. Smith and De Laeter 1986). Let us also note that the ^{142}Nd excess observed in some meteoritic inclusions is sometimes attributed to the in-situ decay of the short-lived neutron-deficient radionuclide ^{146}Sm (Lugmair et al. 1983).

The stellar mechanism(s) responsible for the synthesis of the p-nuclei, referred to as the p-process, has(have) to be entirely distinct from the slow and rapid neutron capture processes leading to the production of the s- and r-nuclei. In general, it is considered that radiative proton captures and (γ,n) photodisintegrations on preexisting more neutron-rich species play a leading role in the p-process. The relative importance of those two types of reactions, as well as the nature and extent of their complement of nuclear transformations depend sensitively upon the envisioned stellar site.

The astrophysically most plausible site for the p-process is the deep interior of highly evolved massive stars. More specifically, Arnould (1976) envisions the oxygen/neon layers at the presupernova stage, while Woosley and Howard (1978) consider the same types of layers in Type II supernovae. Those two models do not necessarily contradict each other, but might instead be complementary in a natural sequence of stellar events. These and other proposed p-process models are reviewed elsewhere (Rayet 1987).

In spite of the success encountered by some models in their attempt to interpret the bulk solar system abundance distribution of the p-nuclei, they also have their difficulties and shortcomings, and have at best to be considered as first order approximations with respect to both their astrophysical and nuclear physics content.

In particular, the limited agreement between theory and observation is obtained in the framework of more or less highly parametrized and schematic stellar models which demonstrate the high sensitivity of the p-nuclei abundances to the temperature and time scale conditions, and emphasize the need for extremely fine-tuned combinations of such conditions. The astrophysical plausibility of some (if not all) of the physical conditions considered so far has to be demonstrated, or even questioned. This is namely the case for the post-shock Type II supernova H-rich envelope model proposed by Burbidge et al. (1957), and studied most thoroughly by Audouze and Truran (1975).

On the nuclear physics side, the thermonuclear reaction rates relevant to the p-process have been calculated up to now with very schematic models. In addition, use has been made of more or less restricted networks. In particular, the network adopted by Woosley and Howard (1978) includes photodisintegrations only. On the other hand, in order to handle their more extended networks, Audouze and Truran (1975) and Arnould (1976) solve a number of subnetworks separately, and couple them afterwards in an approximate way.

In this paper, we propose a new model which avoids many of the above mentioned astrophysical and nuclear physics shortcomings. The envisioned astrophysical scenario is discussed in Sect. 2, while the adopted nuclear reaction network is described in Sect. 3. Sect. 4 presents some test cases making use of parametrized astrophysical conditions already adopted in previous works, instead of those derived from the detailed astrophysical models described in Sect. 2. This allows us to make some comparison with previous works. Some conclusions are drawn in Sect. 5. The results of the fully self-consistent integration of the p-process network into the stellar model of Sect. 2 will be presented in detail elsewhere.

2. THE ASTROPHYSICAL SCENARIO

As a continuation and extension of the work of Arnould (1976) and Woosley and Howard (1978), we contemplate the possibility of development of the p-process in the deeply buried oxygen-neon layers of an evolved star. In many instances, the simplest and most reliable scenario in which such zones can be implied involves the post-He-burning evolution of very massive ($M_{ZAMS} \geq 80\ M_{\odot}$) mass losing stars (VMS). The evolution of such stars through the Wolf-Rayet (WR) stage up to the end of central He burning has been studied in detail by various authors (e.g. Prantzos et al. 1986; Langer and El Eid 1986). Of special interest for the future evolution of such objects is the development at He exhaustion of an almost pure O core with mass $M_O \geq 45\ M_{\odot}$. The concomitant nucleosynthesis has also been followed with great care (Prantzos et al. 1986, 1987). In particular, the core abundances of all the stable species between O and Bi have been calculated with the aid of a detailed reaction network.

As shown by e.g. El Eid and Langer (1986), the VMS post-He burning evolution could terminate with a dynamical collapse due to the e^+e^- pair instability. As a consequence, the explosive burning of oxygen would be triggered in the innermost stellar regions after just a very brief quasi-hydrostatic phase of gravitational contraction and of burning of the small available amounts of C and Ne. Complete explosive disruption might ensue, at least for WR stars with $M_0 \gtrsim 60\ M_\odot$ (see also El Eid, this workshop).

Those calculations provide detailed temperature and density profiles for different mass zones between He exhaustion and core disruption. The concomitant nucleosynthesis can thus be followed in detail, starting with the abundances of all the stable nuclides between H and Bi throughout the whole star calculated at the end of central He burning.

The post-He burning of the exploding WR stars most likely involves some level of p-processing in a more or less large fraction of their mass. For example, the 61 $M_\odot$ core calculated by El Eid and Langer (1986) exhibits peak temperatures at the time of explosion ranging from $T \gtrsim 2 \times 10^9$ K at the outskirts of the O core up to $\approx 4 \times 10^9$ K at the center, where T remains in excess of $\approx 10^9$ K for about 80 s (El Eid 1987). As shown by previous works, such physical conditions span those of interest for the development of a p-process.

In summary, the VMS model described above could be an ideal framework in order to develop for the first time a fully consistent p-process model built into a realistic stellar model, and making use of self-consistent initial abundances for the seed s- and r-nuclei.

3. THE NUCLEAR PHYSICS INPUT

The remaining essential ingredient which is required in order to follow the details of the p-processing in the scenario described in Sect. 2 is the nuclear reaction network. It has to be large enough for avoiding artificial boundary effects on the reaction fluxes, and has also to make use of as reliable thermonuclear reaction rates as possible.

The network adopted in the $Z \geq 32$ range is displayed in Fig. 1. It involves all the possible seed s- and r-nuclei, and extends deeply into the neutron-deficient region, even going beyond the last observed isotope of some elements. The network also extends below Zn down to C, and includes neutrons, protons, and α-particles as well. It thus permits a detailed calculation of the concentrations of those light background particles in a self-consistent way.

The network includes all possible reactions on each of the considered nuclides, viz. (n,p), (n,α), (n,γ), (p,α), (p,γ), (α,γ), and their reverse. The special 3α reaction, and the various astrophysically relevant $^{12}C+^{12}C$, $^{12}C+^{16}O$ and $^{16}O+^{16}O$ channels are also included, and make possible a consistency check between the energy production rates predicted by the abbreviated network used in the stellar model calculations and by our full nucleosynthesis network. All in all, the constructed network involves ≈1000 nuclear species linked by ≈10260 reactions, among which 10% relate to the Mg to Ni region of interest for oxygen burning.

Up to Si, the reaction rates are taken from the compilation by Caughlan et al. (1985), or, when necessary, from earlier versions of it. Above Si, all reaction rates are predicted theoretically. The experimental information is indeed very scarce in that mass region, and concerns only some stable targets in their ground state. Particle-induced reaction cross sections are calculated in the framework of the Hauser-Feshbach model, following the prescription of Thielemann et al. (1986). This prescription reproduces very satisfactorily proton-induced reaction data in medium-weight nuclei, as well as recent measurements of the reaction $^{90}Zr(p,\gamma)^{91}Nb$ (Laird et al. 1987). The neutron-induced reaction data are also very well accounted for in the whole $40 \leq A \leq 209$ mass range. This success gives us some confidence in the adopted prescription for the reaction rate calculations. Of course, the reliability of the method cannot be fully ascertained when dealing with very neutron-deficient nuclei. Photodisintegration rates are calculated from the particle-induced reaction rates using the reversibility theorem.

The β-decay rates of the nuclides included in our network have been calculated by Arnould (1977) with the gross theory of β-decay for allowed and first-order forbidden transitions (e.g. Takahashi et al. 1973) extended in order to account for continuum electron captures (see also Egawa et al. 1975; Takahashi et al. 1978). The gross theory is particularly well suited for the description of the high Q-value transitions (involved in the decay of highly unstable nuclei) which are the only ones to be rapid enough to have a chance to compete with particle captures and photodisintegrations in the p-process context. Low Q-value transitions are indeed slow enough for being completely negligible with respect to those nuclear reactions at temperatures which are high enough for the latter ones to be operative. If the nuclear reactions are frozen out, due to the absence of light particles and/or to low temperatures, it is largely sufficient to assume that the stable species are fed by the decay of the unstable isobars following the laboratory decay pattern.

Those β-decays have not been included in the test cases to be reported in Sect 4. This is justified by the fact that nuclear processing time scales no longer than ≈1 second are considered there. This is shorter than the β-decay time scales we have calculated for the nuclides belonging to our network. The role of the β-decays will be examined in detail when the p-process calculations will be performed on longer time scales appropriate to (quasi-)hydrostatic conditions, or to the VMS framework described in Sect. 2.

4. SOME TEST CASES

We present here some p-process calculations based on the network and input nuclear physics described in Sect. 3. These are performed in the framework of a "canonical" parametrized explosion model which has been used already in previous works. This allows us to make some comparison with older results obtained in particular with more limited networks and more schematic nuclear reaction models.

More specifically, we assume that the explosion can be simulated by a sudden increase of temperature and density to peak values T_m and ρ_m, respectively, followed by an adiabatic decrease described by $T(t) = T_m \exp(-t/\tau_T)$ and $\rho(t) = \rho_m \exp(-3t/\tau_T)$. Inspired by previous p-process calculations, we adopt $2.2 \leq T_m \leq 3.2 \times 10^9$ K and $\rho_m = 10^6$ gcm^{-3}, which may be appropriate for neon/oxygen supernova layers. For the sake of comparison, we select $\tau_T = 1$ s, as in Woosley and Howard (1978).

As initial composition, we adopt the neon burning yields presented by Woosley (1986) for the $Z \leq 20$ elements, and the abundances at He exhaustion in a 60 $M_\odot$ star core calculated by Prantzos et al. (1987) with a full s-process network for the Z>20 elements up to Bi[1].

4.1. *Typical Nuclear Flows*

The main characteristics of the nuclear flows in the neon/oxygen zones considered above have already been discussed in some detail by Arnould (1976) and Woosley and Howard (1978), and need not be reexamined here thoroughly.

Fig. 1 presents two snapshots of the flow patterns in the $Z \geq 32$ range at times ~0.03 s and $\approx 2\ 10^{-6}$ s after an explosion with peak temperatures $T_{m9} = 2.2$ and 3.2, respectively (T_9 is the temperature in units of 10^9 K). Those times correspond in each case to maximum neutron concentrations.

In that Z range, it appears that the seed s- and r-material flows to the neutron-deficient region as a result of (γ,n) reactions. That migration is impeded and eventually stopped or deflected by a variety of reactions. The precise location in the (N,Z) plane where that happens, as well as the nature and relative importance of the reactions which are responsible, depend sensitively upon the astrophysical conditions, especially the temperature and the background neutron concentration. They also depend critically upon the nuclear properties, which are intimately related to the location of the nuclides in the (N,Z) plane.

First of all, some isotopic flows associated to (γ,n) chains appear to be possibly hindered by (n,γ) reactions. This is especially illustrated by the $T_{m9}=3.2$ case of Fig. 1, and still more so below the

---

[1]As shown by Prantzos et al. (1987), the core $Z \geq 20$ abundance patterns at He exhaustion are almost insensitive to M_{ZAMS} in the whole $50 \leq M_{ZAMS}/M_\odot \leq 100$ range

Figure 1. Schematic diagram in the N-Z plane showing the isotopes involved in the reaction network. The diagram has been split up in order to show the relevant information on reaction fluxes for two different peak temperatures (cases 1 and 2).

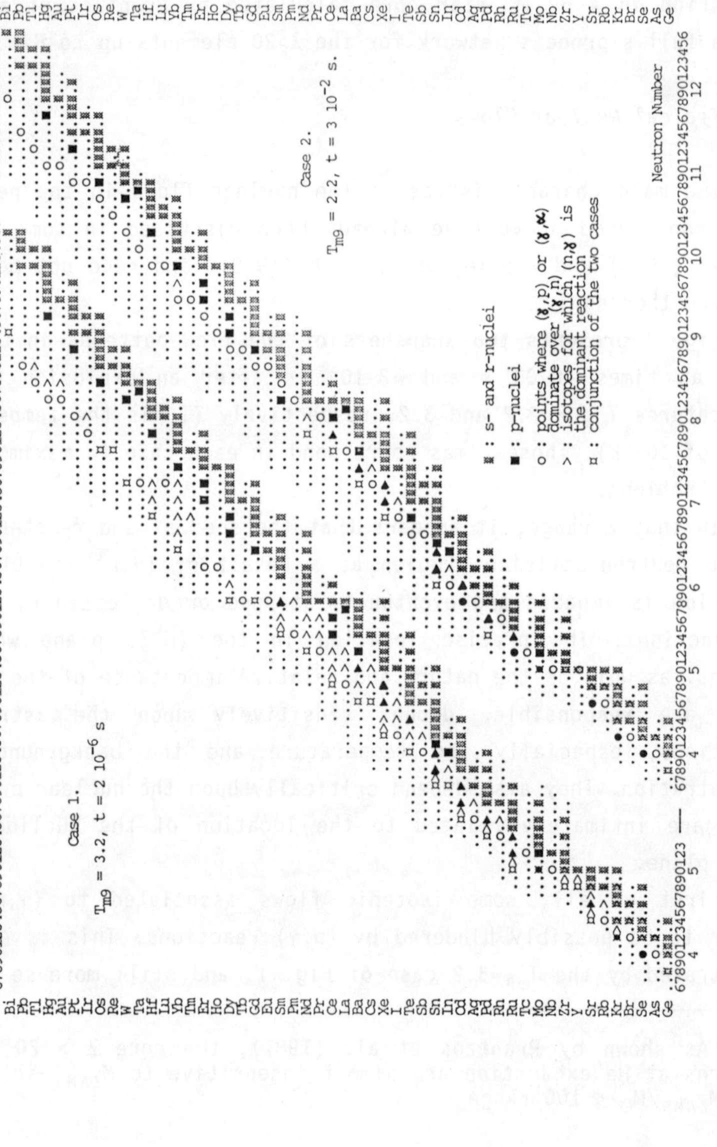

N=82 shell closure. At the time of the snapshots of Fig. 1, the neutron mass fractions have reached their maximum values of $\approx 3.4\times10^{-10}$ and 5.7×10^{-18} for $T_{m9}=3.2$ and 2.2, respectively. Those values are attained after a very rapid increase, and before a relatively slow decrease, which is steeper with increasing temperatures. It has to be remarked that Woosley and Howard (1978) estimate neutron mass fractions of at most $\approx 10^{-12}$ and $\approx 10^{-15}$ during explosive oxygen burning with $T_{m9}=3.0$ and 2.5, respectively. They use those values to conclude from qualitative considerations that neutron captures are unimportant in oxygen zones with $2.2\lesssim T_{m9}\lesssim 3.2$, except in the $A\lesssim 90$ region at $T_{m9}\gtrsim 2.5$. In spite of that restriction, their network neglects all the (n,γ) reactions. In such conditions, their predicted yields for the lightest p-nuclei (which they find to be produced in substantial amounts at $T_{m9}\gtrsim 3$ only; see also Sect. 4.2) are unreliable. In fact, we find that the neglect of the (n,γ) reactions is still less justified than anticipated by Woosley and Howard (1978), particularly in view of the higher neutron concentrations derived from our full network calculations. For example, we find maximum neutron mass fractions ≈ 50 times larger than Woosley and Howard's predictions at $T_{m9}=3$.

In contrast, proton captures are never found to be the dominant transformation path for the $Z\geq 32$ network nuclides. This is true not only for the cases illustrated in Fig. 1, but also for all the other considered conditions (maximum proton mass fractions from $\approx 10^{-15}$ for $T_{m9}=2.2$ to $\approx 7\times10^{-9}$ for $T_{m9}=3.2$ are obtained with the p-process network). In such conditions, the nuclear flows from one Z value to the other are due only to (γ,p) and (γ,α) photodisintegrations, which provide a leak to lower Z values out of the (γ,n)-(n,γ) chains of isotopic transformations. As illustrated in Fig. 1, the main part of that leak is located closer to the proton drip line at higher temperatures. The (γ,n) (as the other photodisintegration) rates are indeed extremely rapidly increasing with T. Such an increase of the (γ,n) flow cannot be compensated by the larger (n,γ) flow resulting from the higher neutron concentrations obtained at higher temperatures.

In summary, material in the considered temperature range has a tendency to flow from high Z-values downward along a more or less neutron-deficient path, the stream being fed by (γ,n) transformations of the seed nuclei. That flow is impeded by neutron captures and by the relatively high neutron, proton and α-particle separation energies of certain nuclei. For long enough time scales at a given temperature, essentially all the material heavier than the iron group would thus be transformed back into iron peak nuclei. Consequently, rapid enough freezing of the nuclear reactions is of key importance for the production of the p-nuclei, either directly, or through the β^+-decay of unstable isobars after freezing.

4.2. Resulting Abundances of the p-Nuclei

Table 1 summarizes the results of our nucleosynthesis calculations for 6 different T_m values. Listed are the overproductions with respect to solar (Cameron 1982) of the nuclides which are classified by Cameron (1982) as pure p-nuclei, or which are at least considered as having a dominant p-process character. Of course, the expected relative importance of the possible p-, s- and even r-process contributions to some nuclides may depend more or less critically upon the considered astrophysical scenarios, and also upon the nuclear physics content of the models.

At the three lowest considered T_m values, the heaviest p-nuclei appear to be more or less largely overproduced, while the lightest ones are not synthesized in any significant amount. The opposite situation holds true for the three highest temperatures. This typical pattern of results has already been found by Woosley and Howard (1978), and is clearly the signature of the photodisintegration flow from Pb to lower and lower Z values described briefly in Sect. 4.1. During the τ_T=1 s time scale adopted in all cases, the flow at the lowest temperatures accumulates material essentially in the N≥82 region, and has no time to leak significantly out of that shell closure barrier. This contrasts with the situation encountered at the highest temperatures, where the photodisintegrations have time to empty the N≥82 pool mainly in favor of the 50≤N≤82 range for T_{m9}=2.8 and 3.2, and even the N≤50 range at

Table 1. Overproduction factors for the p-nuclei, at various explosion peak temperatures. Factors smaller than 2 are not shown. Cases where at least 50% of the abundances are produced as unstable nuclei are underlined.

	T_{m9}					
	2.2	2.4	2.6	2.8	3.0	3.2
^{74}Se	---	---	---	5.39E+1	7.65E+2	7.54E+3
^{78}Kr	---	---	---	4.39E+0	3.98E+1	3.98E+3
^{84}Sr	---	---	---	2.22E+0	3.58E+0	4.77E+2
^{92}Mo	---	---	---	1.59E+1	3.24E+1	5.14E+0
^{94}Mo	---	1.39E+1	4.10E+1	3.62E+1	6.14E+0	---
^{96}Ru	---	---	---	1.57E+1	4.92E+1	2.58E+0
^{98}Ru	---	3.03E+1	9.52E+1	4.37E+2	1.80E+2	---
^{102}Pd	---	---	2.79E+1	8.39E+2	1.19E+3	1.45E+1
^{106}Cd	---	---	---	1.15E+2	4.77E+2	3.89E+0
^{108}Cd	6.07E+1	6.09E+1	1.52E+2	9.72E+2	5.38E+2	---
^{113}In	6.16E+1	6.55E+1	1.17E+2	4.02E+1	7.86E+1	---
^{112}Sn	---	---	---	1.82E+2	7.80E+2	---
^{114}Sn	5.12E+0	7.45E+0	7.14E+1	9.13E+2	3.96E+2	---
^{115}Sn	3.63E+0	---	2.39E+0	1.16E+1	2.72E+0	---
^{120}Te	---	---	1.05E+2	5.84E+2	9.19E+1	---
^{124}Xe	---	---	9.07E+0	7.60E+2	1.59E+3	---
^{126}Xe	---	2.07E+0	2.31E+2	3.31E+2	2.43E+2	---
^{130}Ba	---	---	2.60E+2	4.49E+3	7.99E+2	---
^{132}Ba	---	7.05E+0	2.00E+3	3.32E+2	1.13E+2	---
^{138}La	5.29E+0	3.25E+1	2.91E+1	---	---	---
^{136}Ce	---	---	1.34E+2	1.20E+3	7.65E+1	---
^{138}Ce	---	1.94E+1	7.83E+2	1.23E+2	3.96E+2	---
^{144}Sm	---	6.02E+1	1.46E+2	2.46E+2	3.09E+2	---
^{152}Gd	8.48E+1	6.94E+1	---	---	---	---
^{156}Dy	---	1.90E+2	1.70E+2	---	---	---
^{158}Dy	1.52E+1	5.86E+2	5.92E+2	---	---	---
^{162}Er	4.94E+0	7.64E+2	1.61E+2	---	---	---
^{164}Er	3.52E+1	9.09E+1	8.99E+1	---	---	---
^{168}Yb	7.07E+1	1.69E+3	1.42E+2	---	---	---
^{174}Hf	3.68E+2	8.32E+2	7.72E+1	---	---	---
^{180}Ta	2.71E+1	---	---	---	---	---
^{180}W	6.57E+2	9.70E+1	3.43E+0	---	---	---
^{184}Os	7.70E+2	1.67E+3	---	---	---	---
^{190}Pt	9.21E+1	1.92E+3	1.09E+4	---	---	---
^{196}Hg	9.51E+1	1.76E+3	2.67E+4	---	---	---

T_{m9}=3.2.

Even if the general pattern of overproductions just described agrees qualitatively with Woosley and Howard's (1978) results, some more or less local and large deviations are encountered, which are most likely due to differences in the calculated reaction rates, and possibly also to our consideration of (n,γ) reactions. Roughly speaking, our overproduction factors have a tendency to be higher than theirs in the $N \lesssim 82$ range, and lower at larger N values. Some exceptions to that general rule are, however, noticeable. Among others, this is namely the case for ^{84}Sr at $T_{m9} \gtrsim 2.8$ [which is not likely to be predicted reliably by Woosley and Howard (1978), as discussed in Sect. 4.1], and for ^{196}Hg at T_{m9}=2.6.

As already emphasized by Arnould (1976) and Woosley and Howard (1978), no single set of astrophysical conditions relevant to the oxygen/neon zones can explain the solar system p-nuclei distribution. This is why we calculate for the p-nucleus j a weighted overproduction factor

$$\langle F \rangle_j = \Sigma_i C_i F_{ji}, \qquad (1)$$

where the sum runs over the 6 explosion conditions with overproduction factors F_{ji} given in Table 1, the coefficients C_i of the linear combination being such that $\Sigma C_i = 1$. The simplest case to be considered in our test calculations corresponds to $C_i = 1/6$ for all i. The derived $\langle F \rangle_j$ values normalized to $F_0 = \Sigma_j \langle F \rangle_j$ are given in Fig. 2(a).

In that schematic model, especially large overproductions are found for the lightest (^{74}Se, ^{78}Kr) and heaviest (^{190}Pt, ^{196}Hg) p-nuclei, while $^{92,94}Mo$ and ^{96}Ru are most dramatically underproduced, the other largely underabundant species (^{113}In, ^{138}La, ^{152}Gd, ^{180}Ta) being possibly produced by other stellar or non-stellar processes (e.g. Rayet 1987 for a brief review). Neglecting those last four cases, ≈70% of the p-nuclei have average overproductions in the $0.3 \lesssim \langle F \rangle / F_0 \lesssim 3$ range. At this point, we want to emphasize that the agreement with the solar system distribution of p-nuclei, or the discrepancies with respect to it, have not to be taken too seriously in the schematic test calculations leading to Fig. 2(a). This also holds true for the calculations by Woosley and Howard (1978). It has indeed to be remembered that their

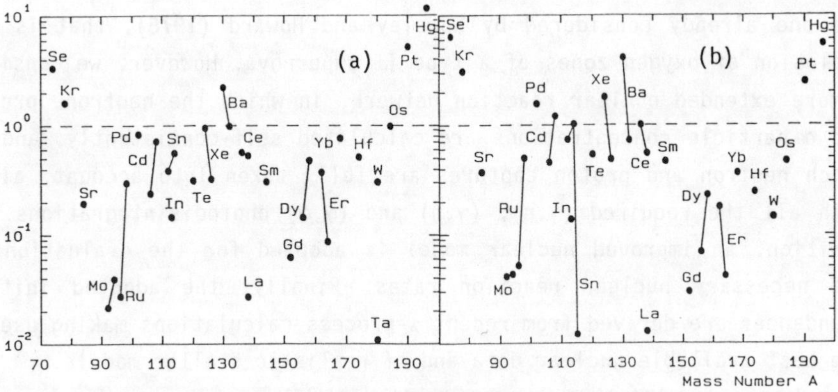

Figure 2. Overproduction factors for the p-nuclei obtained with two different weighted combinations of contributions from the six peak temperatures listed in Table 1 (see text)

adopted model, as well as the one leading to Fig. 2(a), contain a quite rich variety of parameters, like T_m, τ_T, and the various C_i that have to be selected in order to calculate <F> from Eq. (1), some of which have a drastic influence on the predicted p-process yields. In order to illustrate the influence of the choice of the C_i values, Fig. 2(b) is constructed similarly to Fig. 2(a), but with the adoption of C_i=1/15 for the T_{m9}=2.2, 2.4 and 2.6 layers (where most of the heavy p-nuclei are produced), and C_i=4/15 for the T_{m9}=2.8, 3.0 and 3.2 cases (from which the light p-nuclei emerge). That particular choice allows ^{74}Se and ^{196}Hg to be overproduced in roughly equal amounts. As expected, some A≳150 p-nuclei now appear to be more or less largely underproduced, while the problem raised especially by the ^{92,94}Mo and ^{96}Ru underproduction remains unsolved.

5. CONCLUSIONS

This paper describes some results of test calculations that have to be considered as a first step towards the elaboration of more complete and self-consistent p-process models.

Those test calculations use the same astrophysical scenarios as the one already considered by Woosley and Howard (1978), that is the explosion of oxygen zones of a type II supernova. However, we consider a more extended nuclear reaction network, in which the neutron, proton and α-particle concentrations are calculated self-consistently, and in which neutron and proton captures are fully taken into account, along with all the required (γ,n), (γ,p) and (γ,α) photodisintegrations. In addition, an improved nuclear model is adopted for the evaluation of the necessary nuclear reaction rates. Finally, the adopted initial abundances are derived from recent s-process calculations making use of the best available nuclear data and of realistic stellar models.

In spite of those improvements with respect to Woosley and Howard's (1978) calculations, we feel premature to elaborate upon a detailed comparison between our predicted p-process yields and the solar system data. We instead consider our calculations as a first step towards a more complete p-process model to be developed self-consistently in the framework of realistic models for exploding very massive stars. Such stellar explosions are obviously not very frequent in the Galaxy, and their net yields of p-nuclei may turn out not to be sufficient to account for the p-nuclei at the solar level *if it is truly representative of the whole galactic disk*. However, it has to be remembered that no information is available on the p-nuclei outside the solar system. In such conditions, one may speculate about the more or less local nature of the solar system p-nuclei content. We will see!

Acknowledgments. We are grateful to M. El Eid for many informative discussions about the VMS models, and to F.-K. Thielemann for his collaboration on the nuclear reaction rate calculations. This work has been supported in part by the PICS (Programme International de Collaboration Scientifique) nr. 18 between F.N.R.S. (Belgium) and C.N.R.S. (France). M.A. and M.R. are Chercheurs Qualifiés F.N.R.S.

REFERENCES

Anders, E., preprint (1987)
Arnould, M., *Astron. Astrophys.* 8, 436 (1976)
Arnould, M., unpublished (1977)

Audouze, J. and Truran, J.W. *Astrophys.J.* **202**, 204 (1975)
Burbidge, E.M., Burbidge, G.R., Fowler, W.A., and Hoyle, F., *Rev. Mod. Phys.* **29**, 547 (1957)
Cameron, A.G.W., in *Essays in Nuclear Astrophysics*, eds. C. Barnes, D.D. Clayton, and D.N. Schramm (Cambridge University Press), p.23 (1982)
Caughlan, G.R., Fowler, W.A., Harris, M.J., and Zimmerman, B.A., *Atomic Data Nucl. Data Tables* **32**, 197 (1985)
Egawa, Y., Yokoi, K., and Yamada, M., *Prog. Theor. Phys.* **54**, 1339 (1975)
El Eid, M.F., private communication (1987)
El Eid, M.F. and Langer, N., *Astron. Astrophys.* **167**, 274 (1986)
Laird, C.E., Flynn, D., Hershberger, R.L., and Gabbard, F., *Phys. Rev. C* **35**, 1265 (1987)
Langer, N. and El Eid, M.F., *Astron. Astrophys.* **167**, 265 (1986)
Lugmair, G.W., Shimamura, T., Lewis, R.S., and Anders, E., *Science* **222**, 1015 (1983)
Papanastassiou, D.A., in *Isotopic Ratios in the Solar System*, ed. Centre National d'Etudes Spatiales (Toulouse: Cepadues-Editions), p. 77 (1985)
Prantzos, N., Doom, C., Arnould, M., and de Loore, C., *Astrophys.J.* **304**, 695 (1986)
Prantzos, N., Arnould, M., and Arcoragi, J.-P., *Astrophys.J.* **315**, 209 (1987)
Rayet, M., in *Proceedings of the 4th Workshop on Nuclear Astrophysics*, Tegernsee (to appear in *Lecture Notes in Physics*, Springer-Verlag) (1987)
Smith, C.L. and De Laeter, J.R., *Meteoritics* **21**, 133 (1986)
Takahashi, K., Yamada, M., and Kondoh, T., *Atomic Data Nucl. Data Tables* **12**, 101 (1973)
Takahashi, K., El Eid, M.F., and Hillebrandt, W., *Astron. Astrophys.* **67**, 185 (1978)
Thielemann, F.-K., Arnould, M., and Truran, J.W., in *Advances in Nuclear Astrophysics*, eds. E. Vangioni-Flam, J. Audouze, M. Cassé, J.-P. Chièze, and J. Tran Thanh Van (Gif-sur-Yvette: Editions Frontières), p. 525 (1986)
Woosley, S.E., in *Nucleosynthesis and Chemical Evolution*, eds. B. Hauck, A. Maeder, and G. Meynet (Observatoire de Genève), p. 1 (1986)
Woosley, S.E. and Howard, W.M., *Astrophys.J. Suppl.* **36**, 285 (1978)

CHAPTER VI

INPUT NUCLEAR DATA

"It takes two to speak the truth, – one to speak, and another to hear."

Henry David Thoreau
A Week on the Concord and
Merrimack Rivers (1849)
Wednesday

Thermonuclear Reaction Rates For Astrophysical Studies

JAMES W. TRURAN
Dept. of Astronomy, University of Illinois
Urbana, Illinois 61801 USA

FRIEDRICH-KARL THIELEMANN
Department of Astronomy, Harvard University
Cambridge, MA 02138 USA

MARCEL ARNOULD
Institut d'Astronomie et d'Astrophysique
Universite Libre de Bruxelles, Belgium

ABSTRACT

We report the results of statistical model calculations of thermonuclear reaction rates which include provision for the self-consistent treatment of the effects of electron screening. A significant improvement over earlier work has been achieved with the use of both realistic optical model potentials for neutron, proton, and alpha particle interactions and a revised prescription for the determination of nuclear level densities as a function of energy. Theoretical cross section predictions for charged particle reactions and for neutron capture reactions are found to be generally consistent with experimental values to within a factor of two. General expressions are also provided for corrections due to the effects of election screening in dense plasmas on thermonuclear reaction rates, both for particle capture and photodisintegration reactions and for reactions involving charged particles in both the incoming and outgoing channels. These results provide a unified and consistent procedure for the determination of thermonuclear reaction rates as a function of temperature and density.

1. INTRODUCTION

The rates of thermonuclear reactions as a function of temperature and density represent essential input to astrophysical studies. They are required for numerical calculations of both the hydrostatic phases of stellar evolution[9] and the dynamic phases of stellar collapse and supernova explosions[36], as well as of the nova[29] and X-ray burst[20] phenomena. The wide range of physical conditions which are sampled in these astrophysical environments naturally allow the synthesis of rare nuclear species, ranging substantially all the way from the proton drip line to the neutron drip line and from the light element region through to transuranic elements. Studies of hydrogen thermonuclear runaways on white dwarfs, leading to classical nova outbursts[29], and on neutron stars, relevant to X-ray burst models[20], involve the operation of the CNO cycles at high temperature and the possible breakout from these cycles[32,34] and thus require a knowledge of the reaction properties of proton-rich isotopes in the vicinity of the Z=N line through at least ^{60}Zn. Astrophysical processes in which stable and unstable proton-rich isotopes are produced also include the formation of the iron-peak nuclei in explosive nucleosynthesis in supernovae (including the ^{56}Ni which powers the light curve) and the p-process, which forms the proton-rich isotopes of elements in the mass range $A > 70$. On the neutron-rich side, the r-process of neutron capture is characterized by flows which extend deep into the neutron-rich regions, even to the neutron drip line, while the s-process forms heavy nuclei $A \gtrsim 60$ along the valley of beta stability. The significant progress which has been achieved in recent years in our understanding of these nucleosynthesis processes is discussed in recent review articles[30,36].

These diverse processes of nuclear energy generation and nucleosynthesis probe the limits of stable masses on both the proton-rich and neutron-rich sides of the valley of beta stability. Nuclear physics requirements for these studies include a knowledge of the rates of reactions involving the interactions of neutrons, protons, alpha particles, and photons with these sometimes exotic heavy nuclei. Since it is clearly impossible to obtain direct experimental cross section information for most of these cases, it is necessary to develop reliable theoretical methods for the determination of thermonuclear reaction rates. Substantial success has been achieved with the adoption of a Hauser-Feshbach[7] or equivalent expression for the energy-averaged cross section. Statistical model based calculations of thermonuclear reaction rates for astrophysical applications have previously been performed by Truran[27], Truran et al.[31], Michaud et al[17], Michaud and Fowler[16], Truran[28], Arnould[1], Thielemann[21], Holmes et al.[8], and Woosley et al.[35]. Our own recent calculations[23,24] are reviewed and discussed in the next section.

The treatment of thermonuclear reactions in high density astrophysical environments requires modification due to the effects of electron screening[10,11,18,19]. Such conditions are typically encountered, for example, in calculations of supernova core collapse leading to Type II supernovae, in studies of central carbon ignition at high densities in accreting carbon oxygen white dwarfs, and in calculations of thermonuclear runaways on neutron stars or white dwarfs. Screening enhancements by substantial factors are possible for some reactions in very dense plasmas. In general, these plasma effects can be described by a screening potential which must be added to the nuclear and pure Coulomb potential. A general procedure[25,26] for the treatment of screening for particle-capture and photodisintegration reactions and for reactions involving charged particles in both the incoming and outgoing channels is outlined in Section 3.

2. THERMONUCLEAR REACTION RATE CALCULATIONS

We restrict ourselves to reactions involving nuclei in the mass range A > 20, for which it can generally be assumed that there exists a sufficiently high density of excited states in the compound nucleus at the optimum bombarding energy to allow cross section calculations based upon statistical model (Hauser-Feshbach) assumptions. This model assumes that, for each angular momentum of the projectile, there exists a state in the compound nucleus at the bombarding energy with appropriate spin and parity. Assuming (for convenience in notation only) the target or final nucleus to exist or be produced only in its ground state, the cross section for a particle-induced reaction $i(j,k)\ell$ can be written

$$\sigma(E) = \frac{\pi\hbar^2/2\mu_{ij}E}{(2J_i+1)(2J_j+1)} \sum_{J,\pi} (2J+1) \frac{T_j(E,J,\pi)T_k(Q_k+E,J,\pi)}{\sum_n T_n(Q_n+E,J,\pi)} \tag{1}$$

where the $T_i(E,J,\pi)$ are the appropriate transmission coefficients for the particle and photon channels. The astrophysical thermonuclear reaction rate is given by the integral over a Maxwellian distribution

$$<\sigma v> = \left(\frac{8}{\mu_{ij}\pi}\right)^{1/2} (kT)^{-3/2} \int_0^\infty E\sigma(E) \ \exp \ (-E/kT) \ dE. \qquad (2)$$

The general expression for the thermonuclear rate can then be written in the form

$$<\sigma v> = c \ (kT)^{-3/2} \int_0^\infty \sum_{J,\pi} (2J+1) \ \frac{T_j (E,J,\pi) \ T_k(Q_k+E,J,\pi)}{\sum_n T_n(Q_n+E,J,\pi)} \ \exp \ (-E/kT) dE \qquad (3)$$

The challenge here is then to provide accurate estimates of the required neutron, proton, alpha particle, and photon transmission coefficients from statistical model assumptions. A knowledge of the nuclear level density as a function of energy represents critical input in this regard. Additionally, the transmission coefficients must of course be appropriately modified in the presence of electron screening.

Previous calculations of thermonuclear reaction rates for astrophysical studies used optical square well potentials (black nucleus model) for the particle channels, for all but perhaps s-wave and p-wave neutrons[28]. We have achieved a significant improvement over earlier work by solving the Schrödinger equation with a more realistic optical potential for the appropriate particle-nucleus combination. For protons and neutrons, we employ the optical potential derived by Jeukenne et al.[12], with the correction for the imaginary part as discussed in Fantoni et al.[5] and Mahaux[13]. For alpha particles, we used the phenomenological Woods-Saxon potential derived by Mann[14], based on the extensive data of McFadden and Satchler[15].

For the gamma ray transmission coefficients, both electric and magnetic dipole transitions (E1 and M1) are included in the calculation of the total photon width. M1 transitions were treated as discussed in Holmes et al.[8]. The more important E1 contributions were calculated on the basis of the Lorentzian representation of the Giant Dipole Resonance, utilizing the phenomenological model for the resonance width proposed by Thielemann and Arnould[22].

Calculations of these transmission coefficients require as well a knowledge of the nuclear level densities. Our treatment of nuclear level densities closely parallels that of Gilbert and Cameron[6] and combines the use of the Bethe[3] formula, based on the Fermi gas model, at high energies

$$\rho(U) = \frac{1}{12\sqrt{2}\,\sigma\, a^{1/4}} \frac{\exp(2\sqrt{aU})}{U^{5/4}} \tag{4a}$$

and the empirically based Ericson[4] formula at low energies

$$\rho(E) = T^{-1} \exp[(E-E_o)/T] \tag{4b}$$

The parameter σ is the spin cutoff factor. The energy shift of the first excited state due to the necessity of breaking up a proton or neutron pair before exciting nucleons to higher states, not included in the independent particle Fermi gas model, is accounted for in the "back-shifted" Fermi gas formalism by adopting $U = E-\delta$ in Eq. (4), where δ is the pairing correction. We also identify $E_o \equiv \delta$ in the Ericson formula. The level density at high densities is thus determined by the single parameter a, while fitting of the observed levels at low energies gives T. Following Gilbert and Cameron[6], our empirical prescription for the level density parameter a is again built upon a correlation with the shell correction energies of a semi-empirical mass formula. Further details of these procedures are presented elsewhere[23,24].

A measure of the level of success achieveable with the use of these procedures is provided by comparisons of our cross section predictions with experiment. In Figure 1, our calculated cross sections as a function of energy for charged particle reactions are compared with experiment for representative intermediate mass nuclei which participate in stellar silicon burning. We generally find that the energy dependences and absolute values of charged particle cross sections are quite well reproduced, including specifically the magnitudes of cusps at the threshold energies of channel openings. The improvements achieved and the importance of experimental cross section information are both

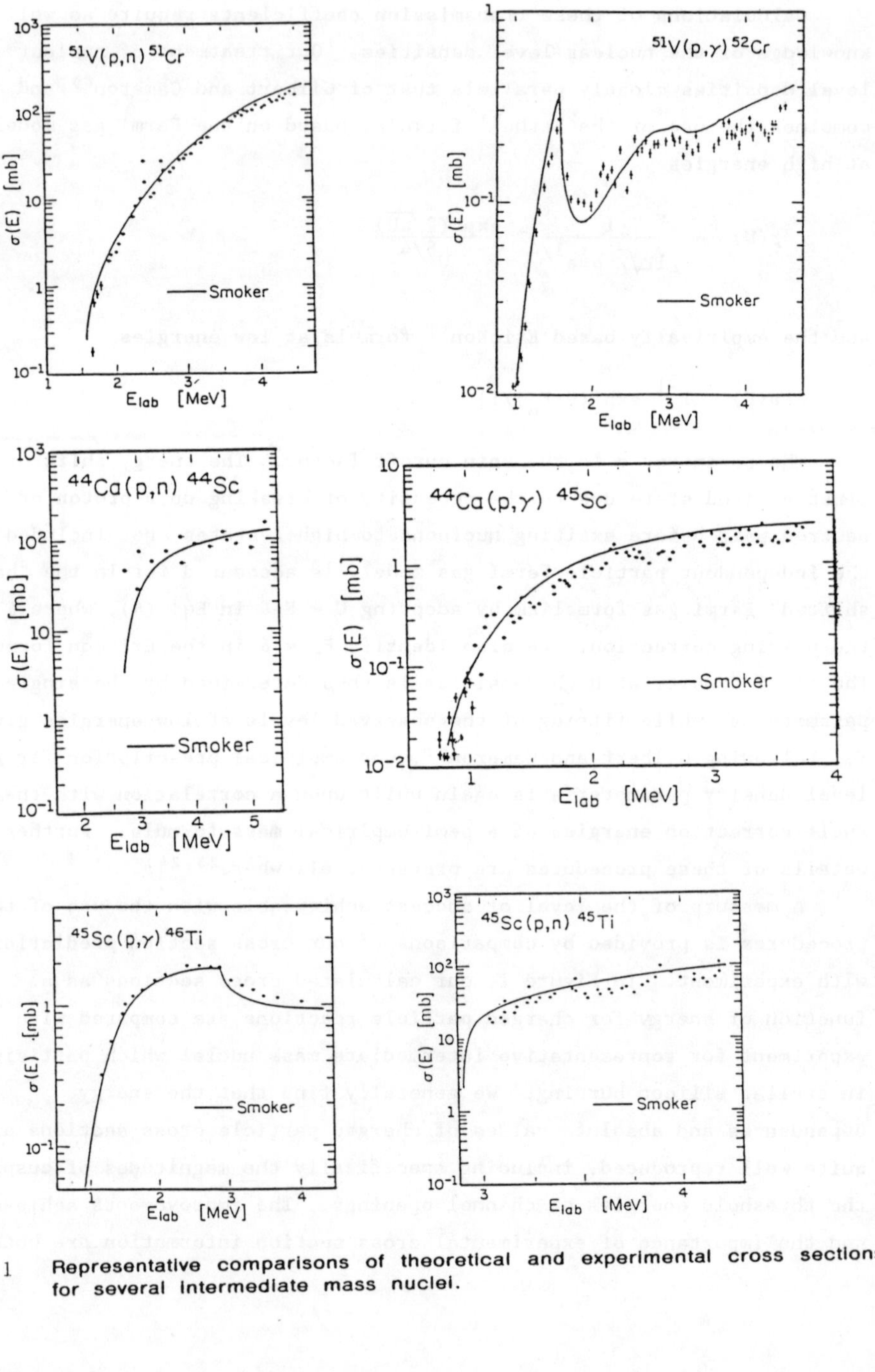

Fig. 1 **Representative comparisons of theoretical and experimental cross sections for several intermediate mass nuclei.**

evident from the comparison of calculated and experimental neutron capture cross sections presented in Figure 2. The upper figure shows the ratios of the cross sections calculated by Truran[28] to the experimental cross sections as a function of neutron number. The points corresponding to isotopes of identified elements are connected by solid lines, revealing systematic variations in the vicinities of closed neutron shells. These trends arise from the inadequacies of systematic treatments of nuclear level densities at low excitations, particularly in the vicinity of closed shells. The lower figure, which shows a similar comparison of our recent calculations[23,24] with the experimental data compiled by Bao and Käppeler[2], reveals that the situation has been considerably improved with the use of our revised prescriptions for nuclear level densities and radiation widths. Cross section predictions obtained within this framework are expected to lie safely within a factor of two of experimental values.

3. ELECTRON SCREENING CORRECTIONS

In order to be able to provide compilations of thermonuclear reaction rates as a function of both temperature and density, we have sought to provide a simple but quite general expression for transmission coefficients in an environment where screening effects are important[25,26]. We first note that, given the nuclear and pure Coulomb potential and the orbital quantum number ℓ, the transmission coefficients are a function only of the kinetic energy at infinity T(E). The Schrödinger equation in the presence of screening is

$$\frac{d^2u_\ell(r)}{dr^2} + [k^2 - \frac{\ell(\ell+1)}{r^2} - \frac{2\mu}{\hbar^2}\,(V_{nucl}(r) + V_{coul}(r) + V_{scr}(r))]\,u_\ell(r) = 0 \tag{5}$$

where we generally take

$$V_{scr}(r) = -U_o + f(r) \quad . \tag{6}$$

We consider barrier penetration in the WKB approximation, which yields

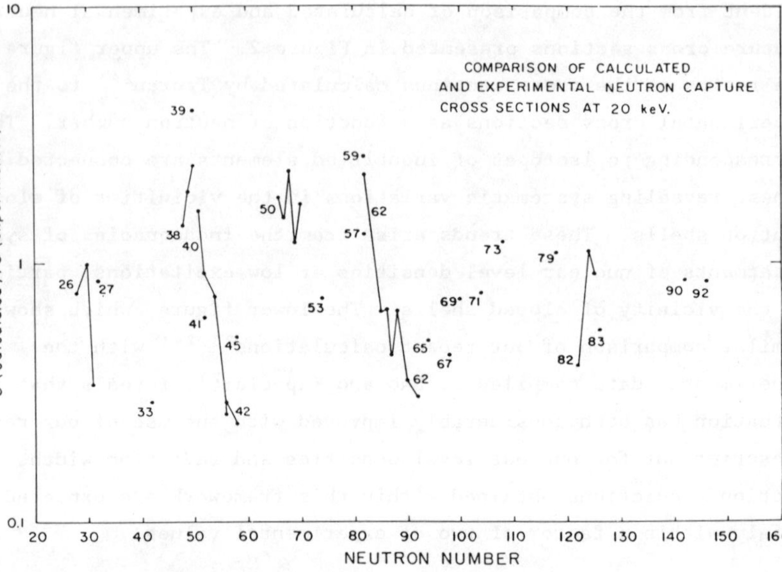

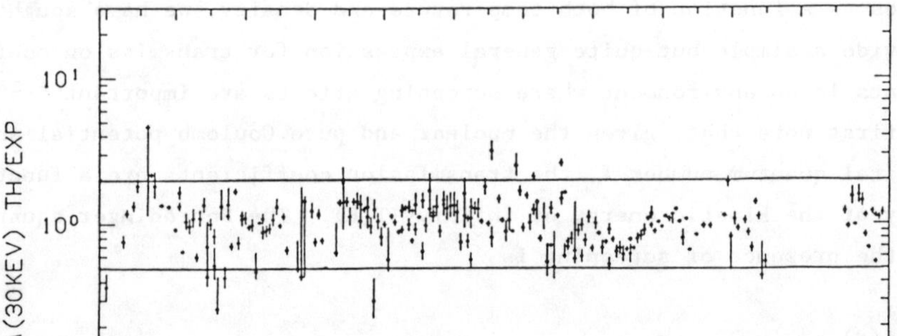

Fig. 2 Ratios of theoretical to experimental 30keV neutron capture cross sections versus A.

the equation for the transmission coefficient

$$T(E) = \exp\{-2/\hbar \int_{x_1}^{x_2} [2\mu(V(r)-E)] \, dr\} \tag{7}$$

where x_1 and x_2 are the classical turning points. In the region $r > R$, and for the case $\ell = 0$, we have $V(r) = V_{coul}(r) + V_{scr}(r)$ and, for our assumed form of $V_{scr}(r)$, it follows that

$$T_{scr}(E) = T(E+U_o)\ \xi\,(E+U_o) \tag{8}$$

where $T_{scr}(E)$ is the screened transmission coefficient, $T(E+U_o)$ is the usual transmission coefficient evaluated at the elevated energy $(E+U_o)$, and $\xi(E+U_o)$ is a correction factor which arises due to the radial dependence of the screening potential $V_{scr}(r)$.

In the presence of screening, our integral expression for $\langle\sigma v\rangle$ in Eq. (3) involves a sum of integrals of the form

$$\int_o^\infty \frac{T_{scr,j}(E,J,\pi)\ T_{scr,k}(Q_{scr,k}+E,J,\pi)}{\sum_n T_{scr,n}(Q_{scr,n}+E,J,\pi)} \exp(-E/kT)\, dE \tag{9}$$

where account is taken of the modifications of the Q-values for individual channels in the presence of screening. In particular, the nuclear potential in the incoming channel is lowered by an amount $U_{o,j}$ at $r = 0$ and thus a state with higher intrinsic excitation energy is produced. However, the nuclear potential in the outgoing channel is lowered by an amount $U_{o,n}$ at $r = 0$ and thus the available kinetic energy at infinity must be reduced accordingly. The Q-values of individual channels in the presence of screening may thus be written

$$Q_{scr,n} = Q_n + U_{o,j} - U_{o,n}\ . \tag{10}$$

Our integral expression Eq. (9) may then be written as

$$\int_o^\infty \frac{\xi_j(E+U_{o,j})T_j(E+U_{o,j})\ \xi_k(Q_k+E+U_{o,j})T_k(Q_k+E+U_{o,j})}{\sum_n \xi_n(Q_n+E+U_{o,j})T_n(Q_n+E+U_{o,j})} \exp(-E/kT)dE \tag{11}$$

We now make the usual variable transformation $E' = E+U_{o,j}$, which yields a factor $\exp(U_{o,j}/kT)$ in front of the integral similar to the screening factor in the absence of a radial dependence of the screening potential. We also approximate the integrand by evaluating the factors $\xi_n(Q_n+E+U_{o,j})$ at the maximum of the integrand E'_{max} (Gamow peak energy) and arrive at the expression.

$$\exp(U_{o,j}/kT) \frac{\xi_j(E'_{max})\xi_k(Q_k+E'_{max})}{\xi_{n,dom}(Q_n+E'_{max})} \int_{U_{o,j}}^{\infty} \frac{T_j(E')T_k(Q_k+E')}{\sum_n T_n(Q_n+E')} \exp(-E'/kT)dE'. \tag{12}$$

Here, $\xi_{n,dom}(Q_n+E'_{max})$ denotes the correction factor in the dominant reaction channel at the maximum of the integrand. This expression holds for each integral in the sum and, assuming a weak J dependence of E'_{max}, for the screened thermonuclear rate in general. The screening factor for the reaction $i(j,k)\ell$ can thus be written

$$\exp(H_{ij,k}) = \exp(U_{o,j}/kT) \quad \frac{\xi(E'_{max})\xi_k(Q_k+E'_{max})}{\xi_{n,dom}(Q_n+E'_{max})} \tag{13}$$

the analogous expression for the screening factor for the case of a photodisintegration reaction $(\lambda_{\gamma,j})$ is given by[25,26]

$$\exp(H_{\ell\gamma,j}) = \frac{\xi(Q_j+E'_{max})}{\xi_{n,dom}(Q_n+E'_{max})} . \tag{14}$$

In both instances, the determination of the screening correction due to the existence of a radial dependence of the screening potential $V_{scr}(r)$ requires a knowledge of which channel is dominant at the energy corresponding to the peak of the integrand $(Q_n+E'_{max})$. This must be provided either from experiment or from the predictions of theoretical calculations of thermonuclear rates. Future compilations of theoretical rates[24] will provide this information and E'_{max}.

4. SUMMARY AND CONCLUSION

Our use of realistic optical model potentials in the determination of particle transmission coefficients and of revised and improved prescriptions for the determination of nuclear level densities and of electric dipole photon transmission coefficients has given rise to significantly improved estimates of thermonuclear reaction rates. The statistical model calculations described in this paper have been found specifically to predict thermonuclear reaction rates for both charged particle and neutron induced reactions which are generally in agreement with experimental determinations to within a factor of two.

The establishment of a general relation between particle transmission coefficients in screened and unscreened environments has also allowed us to derive a general expression for the screening enhancements of thermonuclear reaction rates, including screening of reactions involving charged particles in both the incoming and outgoing channels. The screening enhancement factors derived in the previous section can also be shown to be entirely consistent with the distribution of abundances in a nuclear statistical equilibrium, when proper account is taken of the changes of the chemical potentials for all species due to Coulomb interactions[26].

We note in conclusion that the combined studies outlined in this paper[23-26] allow for the first time a consistent and straightforward determination of reaction rates as a function of both temperature and density. The thermonuclear reaction rate dependence on the temperature follows from the appropriate integral over a Maxwellian distribution. Additionally, the approximate expressions for the screening enhancements derived in the preceding section may be applied if one knows only both the energy E'_{max} of the maximum of the integrand and the dominant reaction channel at that energy. The forthcoming compilation of thermonuclear reaction rates[24], calculated for nuclei in the mass range $20 \leq A \leq 70$ with the SMOKER program, will include this information.

5. ACKNOWLEDGEMENTS

This research was supported in part by the United States National

Science Foundation under grant AST 86-11500 at the University of Illinois.

6. REFERENCES

(1) Arnould, M., Astron. Astrophys. 19, 82 (1972).

(2) Bao, Z.Y. and Käppeler, F., At. Data Nucl. Data Tables, in press (1987).

(3) Bethe, H., Phys. Rev. 50, 352 (1936).

(4) Ericson, T., Nucl. Phys. 11, 481 (1959).

(5) Fantomi, S., Friman, B.L., and Pandharipande, V.R., Phys. Lett. 104B, 89 (1981).

(6) Gilbert, A. and Cameron, A.G.W., Can. J. Phys. 43, 1446 (1965).

(7) Hauser, W. and Feshbach, H., Phys. Rev. 87, 366 (1952).

(8) Holmes, J.A., Woosley, S.E., Fowler, W.A., and Zimmerman, B.A., At. Data Nucl. Data Tables 18, 306 (1976).

(9) Iben, I., Jr. and Renzini, A., Ann. Rev. Astron. Astrophys. 21, 271 (1983).

(10) Ichimaru, S. and Utsumi, K., Ap. J. (Letters) 269, L51 (1983).

(11) Ichimaru, S. and Utsumi, K., Ap. J. 286, 363 (1984).

(12) Jeukenne, J.P., Lejeune, A., and Mahaux, C., Phys. Rev. C16, 80 (1977).

(13) Mahaux, C., Phys. Rev. C82, 1848 (1982).

(14) Mann, F.M., Hauser 5, A. Computer Code to Calculate Nuclear Cross Sections, Hanford Engineering: HEDL-TME 78-83 (1978).

(15) McFadden, L. and Satchler, G.R., Nucl. Phys. 84, 177 (1966).

(16) Michaud, G. and Fowler, W.A., Phys. Rev. C2, 2041 (1970).

(17) Michaud, G., Scherk, L., and Vogt, E., Phys. Rev. C1, 864 (1970).

(18) Mochkovitch, R. and Nomoto, K., Astron. Astrophys. 154, 115 (1986).

(19) Salpeter, E.E., Austral. J. Phys. 7, 373 (1954).

(20) Taam, R.E., Ann. Rev. Nucl. Part. Sci. 35, 1 (1985).

(21) Thielemann, F.-K., Thesis, Technische Hochschule Darmstadt (1980).

(22) Thielemann, F.-K., and Arnould, M., Proc. Int. Conf. on Nuclear Data for Science and Technology, Antwerpen, ed. K. Böckhoff, p. 762 (1983).

(23) Thielemann, F.-K., Arnould, M., and Truran, J.W., in Advances in Nuclear Astrophysics, eds. E. Vangioni-Flam, J. Audouze, M. Casse, J.-P. Chieze, and J. Tran Thanh Van (Gif-sur-Yvette: Editions Frontieres), p. 525 (1987).

(24) Thielemann, F.-K., Arnould, M., and Truran, J.W., in preparation (1987).

(25) Thielemann, F.-K., Truran, J.W., in Advances in Nuclear Astrophysics, eds. E. Vangioni-Flam, J. Audouze, M. Casse, J.-P. Chieze, and J. Tran Thanh Van (Gif-sur-Yvette: Editions Frontieres), p. 541 (1987).

(26) Thielemann, F.-K. and Truran, J.W., in preparation (1987).

(27) Truran, J.W., Ph.D. thesis, Yale University (1966).

(28) Truran, J.W., Astrophys. Space Sci. 18, 306 (1972).

(29) Truran, J.W., in Essays in Nuclear Astrophysics, eds. C.A. Barnes, D.D. Clayton, and D.N. Schramm (Cambridge: Cambridge Univ. Press), p. 467 (1982).

(30) Truran, J.W., Ann. Rev. Nucl. Part. Sci. 34, 53 (1984).

(31) Truran, J.W., Hansen, C.J., Cameron, A.G.W., and Gilbert, A., Can. J. Phys. 44, 151 (1966).

(32) Wallace, R.K. and Woosley, S.E., Astrophys. J. Suppl. 45, 389 (1981).

(33) Weigmann, H. and Rohr, G., Reactor Centrum Nederland, Report 203, p. 194 (1973).

(34) Weischer, M., Görres, J., Thielemann, F.-K., and Ritter, H., Astron. Astrophys. 160, 56 (1986).

(35) Woosley, S.E., Fowler, W.A., Holmes, J.A., and Zimmermann, B.A., At. Data Nucl. Data Tables 22, 371 (1979).

(36) Woosley, S.E. and Weaver, T.A., Ann. Rev. Astron. Astrophys. 24, 205 (1986).

INFLUENCES OF THE ASTROPHYSICAL ENVIRONMENT ON NUCLEAR DECAY RATES

Eric B. Norman
Nuclear Science Division, Lawrence Berkeley Laboratory
University of California
Berkeley, CA 94720

ABSTRACT

In many astronomical environments, physical conditions are so extreme that nuclear decay rates can be significantly altered from their laboratory values. Such effects are relevant to a number of current problems in nuclear astrophysics. Experiments related to these problems are now being pursued, and will be described in this talk.

1. INTRODUCTION

The decay rates of atomic nuclei are usually thought to be determined soley by the interplay of the strong, electromagnetic, and weak interactions between the neutrons and protons in the nucleus. As a result, it is generally taken for granted that these decay rates are unaffected by environmental factors. Under terrestrial conditions, this is an extremely good approximation. In many astronomical environments, however, physical conditions are so extreme that nuclear decay rates can be significantly altered from their laboratory values. In order to understand the nucleosynthesis and history of the chemical elements that we observe, it is necessary to know what the actual half-lives of nuclei are under the conditions of temperature and density that they have experienced.

At the high temperatures found in stellar interiors, nuclei are continuously irradiated by large fluxes of high-energy photons. These photons can be absorbed by a nucleus with the effect that the nucleus is excited to a higher energy level. In several cases of astrophysical interest, the beta decay of the ground state is a forbidden and hence very slow transition, while those of the excited states are allowed transitions. Thus, if these excited states are populated, significant increases in the effective decay rate of the nucleus can result.

Another effect of the high temperatures found in stellar interiors is that during nucleosynthesis, matter is essentially fully ionized. For nuclei which beta-minus decay, this leads to the possibility of bound-state beta decay. In this process, the beta particle is emitted directly into a previously unoccupied bound atomic orbital. For decays with low endpoint energies, the increase in decay energy provided by the atomic electron binding energy can lead to large increases in the nuclear decay rate.

A final effect that also results from the absence of atomic electrons is of importance in cosmic-ray physics. Some of the species present in the cosmic rays are radioactive nuclei which, under terrestrial conditions, decay by electron capture. High-energy cosmic rays are fully stripped of their atomic electrons in passing through the interstellar medium. Thus, because there are no bound atomic electrons to be captured, dramatic decreases in the decay rates occur for such cosmic-ray nuclei. For some species, the result is that they become "stable" , while for others, terrestrially unobserved beta-minus or beta-plus decays will determine their decay rates.

Effects such as these are relevant to a number of current problems in nuclear astrophysics. Nuclear physics experiments related to these problems are now being pursued and will be described in this talk.

2. THE S-PROCESS BRANCH AT ^{148}PM

While it is generally true that in the s-process the neutron-capture rates are low compared to typical beta-decay rates, there are places along the s-process path where the half-lives are sufficiently long that neutron capture on these unstable nuclei competes with beta decay. If one knows the relevant isotopic abundances, neutron-capture cross sections, and beta-decay half-lives, such "branch points" allow one to infer the neutron density during the s-process.

Through recent measurements of the neutron-capture cross sections of ^{148,149,150}Sm, Winters et al.[1] have shown that such a branch in the s-process path occurs at the odd-odd nucleus ^{148}Pm. The path of the s-process near A=148 is shown in Figure 1.

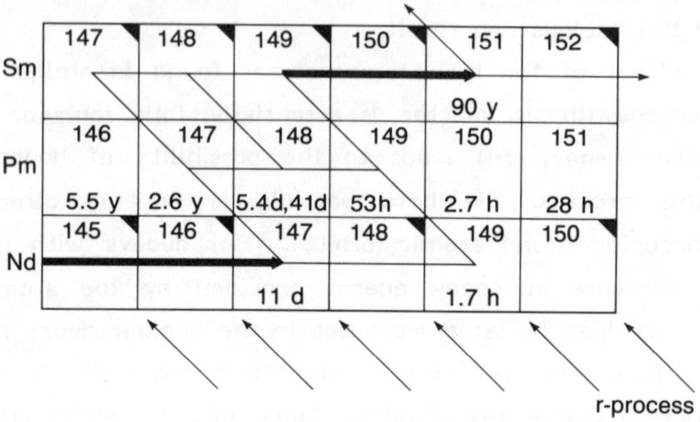

Figure 1. The path of the s-process near A=148.

However, the usefulness of this branch cannot yet be fully exploited because of the lack of information on the nuclear structure of ^{148}Pm. Until recently, only three levels in ^{148}Pm were known.[2] The ground state, ^{148}Pm^{g}, is a $J^{\pi} = 1^-$ level with a 5.37-day beta-decay half-life. The first excited state at 76 keV is a $J^{\pi} = 2^-$ level that gamma decays to the ground state. The second excited state at 137 keV is a $J^{\pi} = 6^-$ isomer, ^{148}Pm^{m}, that decays almost entirely by beta-minus decay with a 41.3-day half life. It has been estimated[1] that during the s-process, the ^{147}Pm(n,γ) reaction produces roughly equal amounts of ^{148}Pm^{g} and ^{148}Pm^{m}. However, at the high temperatures at which the s-process is believed to occur, (1-4) x 10^8 K, the question arises as to whether this population will be preserved or whether ^{148}Pm^{g} and ^{148}Pm^{m} could come into thermal equilibrium. Winters et al.[1] have shown that the neutron density inferred from this s-process branch point is a factor of three larger if equilibrium is reached than if the initial equal mixture of ^{148}Pm^{g} and ^{148}Pm^{m} is preserved.

There are many potential mechanisms by which ^{148}Pm^{g,m} could reach thermal equilibrium during the s-process. One of the most important is undoubtably photoexcitation. In the hot stellar environment at which the s-process is thought to occur, there is an enormous flux of high energy photons. Thus it is possible that a nucleus, initially in the isomeric state, could absorb one of these photons and be excited to a higher-lying level which subsequently decays to the ground state. This process is illustrated in Figure 2.

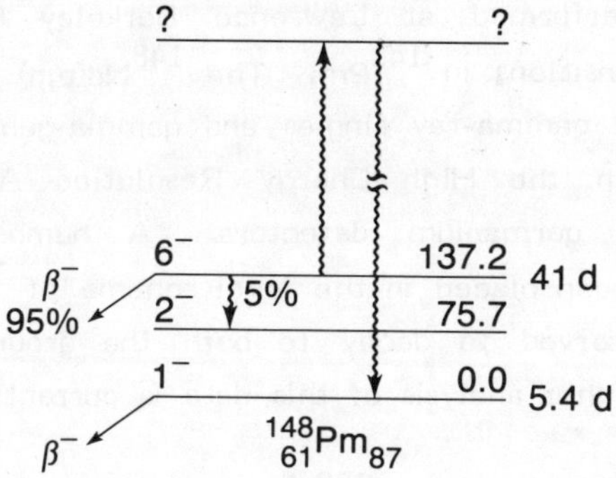

Figure 2. Schematic illustration of the process of photoexcitation of $^{148}Pm^m$ to a higher lying state that subsequently decays to $^{148}Pm^g$.

In order for the timescale for equilibration under s-process conditions to be shorter than the half-life of $^{148}Pm^g$, such mediating levels must lie below approximately 1 MeV excitation energy. Thus, to decide if this actually happens during the s-process, the positions and gamma-decay properties of levels in ^{148}Pm must be known.

As a first step toward answering the question of whether $^{148}Pm^{g,m}$ reach thermal equilibrium in the s-process, an experiment was carried out to locate low-lying levels in ^{148}Pm. The $^{149}Sm(d,{}^3He)^{148}Pm$ reaction was performed using a deuteron beam from the Princeton University cyclotron. The reaction ^3He particles were momentum analyzed with a QDDD spectrometer and were detected at the focal surface with a position-sensitive proportional counter followed by a plastic scintillator. From these measurements, the positions of twenty new states below 1-MeV excitation energy in ^{148}Pm were determined.

An example of the type of state that could mediate transitions between $^{148}Pm^g$ and $^{148}Pm^m$ is one of the $J^\pi = 4^-$ levels that are expected to be located at low excitation energy. Such a state should have sizable decay branches to both the $J^\pi = 2^-$ level at 76 keV (which decays to the ground state) and to the isomer. To determine which, if any, of the levels observed in the present investigation serve as such a mediator during the s-process, the gamma-decay properties of these levels must be known. An experiment was performed at Lawrence Berkeley Laboratory to study electromagnetic transitions in ^{148}Pm. The ^{148}Nd(p,n) reaction was used to produce ^{148}Pm and gamma-ray singles and gamma-gamma coincidence data were acquired with the High Energy Resolution Array (HERA) of 20 Compton-surpressed germanium detectors. A number of the observed gamma rays have been placed in the level scheme of ^{148}Pm, but so far no level has been observed to decay to both the ground state and to the isomeric level. Further analysis of this data is currently in progress.

3. THE NUCLEOSYNTHESIS OF ^{180}TA

One of the continuing puzzles in nuclear astrophysics is the origin of the naturally occurring $J^{\pi} = 9^{-}$ isomer $^{180}Ta^{m}$, which appears as 0.012% of natural tantalum.[2] Even this small abundance cannot be explained in terms of the conventional s- and r- processes. The presence of the stable $^{176-180}$Hf isotopes allows the standard s-process flow to bypass ^{180}Ta. The beta decays following r-process neutron captures would terminate at the first stable A=180 nuclide, ^{180}Hf. Recently, however, it has been suggested that $^{180}Ta^{m}$ may be synthesized via small beta-decay branches off the main s- and r-process paths.[3,4] Beer and Ward[3] have proposed a model in which $^{180}Ta^{m}$ is produced through the beta decay of $^{180}Hf^{m}$. These authors also pointed out that the r-process could contribute to the observed abundance of $^{180}Ta^{m}$ if a fraction of the beta decays of ^{180}Lu lead to the population of $^{180}Hf^{m}$, a fraction of which subsequently decays to $^{180}Ta^{m}$. Figure 3 summarizes the basic ideas of this model.

Searches for these small beta decay branches have been conducted. The beta decay of $^{180}Hf^{m}$ to $^{180}Ta^{m}$ has been observed,[5,6] but the measured branching ratio is too small to allow the s-process to be the major source of $^{180}Ta^{m}$. Searches by two different groups for the beta decay of ^{180}Lu to $^{180}Hf^{m}$ have given conflicting results.[7,8,9]

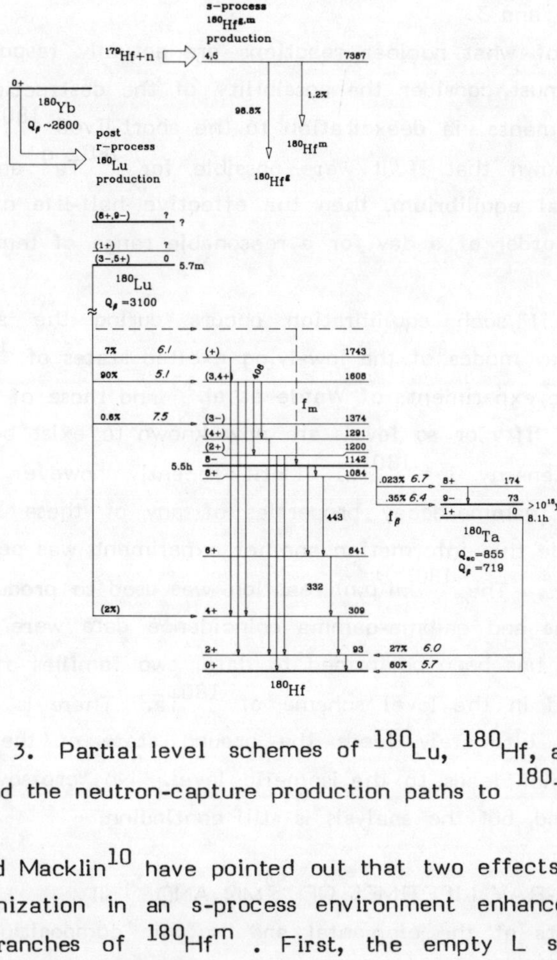

Figure 3. Partial level schemes of ^{180}Lu, ^{180}Hf, and ^{180}Ta and the neutron-capture production paths to ^{180}Ta^m.

Beer and Macklin[10] have pointed out that two effects due to the high degree of ionization in an s-process environment enhance the laboratory beta-decay branches of ^{180}Hf^m. First, the empty L shell inhibits the internal conversion of the 57.5-keV transition. This effect alone increases the half-life of ^{180}Hf^m against electromagnetic decay by a factor of 1.4. Second, the nearly empty K shell opens the door for bound-state beta decay which significantly supplements beta-decay strength into the continuum. Takahashi and Yokoi[11] have calculated the ratio of the rates for bound-state decay to that of decay into the continuum. For the ^{180}Hf^m decay to the $J^\pi = 8^+$ level at 174-keV excitation energy in ^{180}Ta, they find that this ratio ranges between 1.5 and 11 for plausible s-process conditions. A similar calculation for the direct decay to the $J^\pi = 9^-$ ^{180}Ta^m results in a ratio between 0.5 and 3.

Regardless of what nuclear reactions are actually responsible for its production, one must consider the possibility of the destruction of $^{180}Ta^{m}$ in stellar environments via deexcitation to the short-lived $^{180}Ta^{g}$. Norman et al.[12] have shown that if it were possible for $^{180}Ta^{g}$ and $^{180}Ta^{m}$ to come into thermal equilibrium, then the effective half-life of the nucleus would be on the order of a day for a reasonable range of temperatures and densities.

To decide if such equilibration occurs during the s-process, the positions and decay modes of the low-lying excited states of ^{180}Ta must be known. From the experiments of Warde et al.[13] and those of Dewberry and Naumann[14] some fifty or so levels are now known to exist below about 1-MeV excitation energy in ^{180}Ta. Until recently however, nothing was known about the gamma-decay properties of any of these levels. In an attempt to provide this information another experiment was performed using the HERA at LBL. The $^{180}Hf(p,n)$ reaction was used to produce ^{180}Ta and gamma-ray singles and gamma-gamma coincidence data were acquired. In the analysis that has been performed to date, two families of gamma rays have been placed in the level scheme of ^{180}Ta. There is one group of transitions which ultimately feeds the ground state of the nucleus and another group which leads to the isomeric level. No "crossover" transition has yet been found, but the analysis is still continuing.

4. THE COSMIC-RAY LIFETIMES OF ^{54}MN AND ^{56}NI

Measurements of the elemental and isotopic composition and energy spectra of the cosmic-ray nuclei just below iron have indicated the presence of the radioactive nucleus ^{54}Mn in the cosmic rays.[15,16] As can be seen from Figure 4, in the laboratory ^{54}Mn decays via electron capture with a half-life of 312 days. As a high-energy cosmic ray, ^{54}Mn would be stripped of its atomic electrons through interactions with the interstellar medium. Thus cosmic-ray ^{54}Mn would be stable were it not for the possibility of both beta-minus and beta-plus decay. Such decays are energetically allowed, although hindered because of the large spin changes required. The most likely decay of a bare ^{54}Mn nucleus is the beta-minus decay to the ground state of ^{54}Fe, and the partial half-life for this branch has been estimated to be (0.06 - 10) x 10^6 years.[17] The measured fluxes of cosmic-ray ^{54}Mn could be used to determine the mean age of cosmic-ray secondary nuclei if this half-life were known.[18]

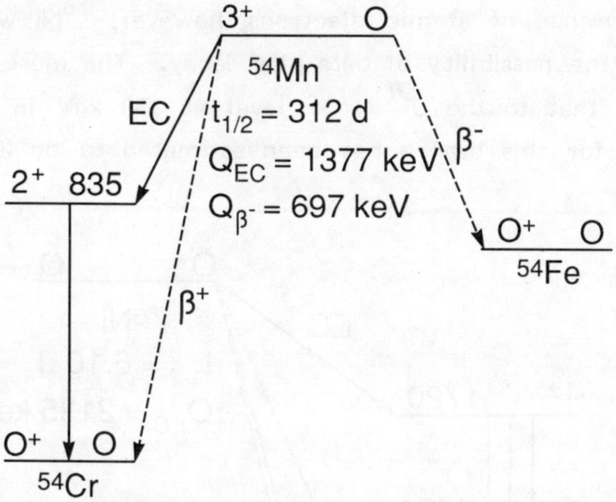

Figure 4. Decay scheme of ^{54}Mn.

We have recently begun a project to measure the beta-minus decay lifetime of ^{54}Mn. A source of ^{54}Mn is mounted in close geometry to a silicon surface-barrier detector telescope. This assembly is then placed at the center of a large annular sodium iodide detector. The well of the annulus is then filled with two sidecap sodium iodide detectors. Searches are then made for events in the silicon detector telescope that are in anticoincidence with signals in the sodium iodide array. Such events will include the sought-for beta-minus particles. In a series of preliminary experiments of this kind, we found that electrons produced by the 2.5 x 10^{-4} internal conversion decay branch of ^{54}Mn are a problem for our measurement. Such electrons can backscatter off the silicon detectors and thus deposit only a fraction of their total energy. This produces a large background that at present hides the sought-for beta particles. However, from analysis of the data taken thus far, we can set an upper limit on the beta-minus decay branching ratio of ^{54}Mn of 5 x 10^{-4} . Further improvements of our system will soon be made in order for the beta-minus decay lifetime of ^{54}Mn to be measured.

A major product of charged-particle induced nucleosynthesis in stars is the doubly-magic nucleus ^{56}Ni. As can be seen in Figure 5, in the laboratory, this isotope decays via electron capture with a half-life of 6.1 days. In the absence of atomic electrons, however, ^{56}Ni would be stable were it not for the possibility of beta-plus decay. The most likey beta-plus decay branch is that to the $J^{\pi} = 3^{+}$ level at 158 keV in ^{56}Co, and the partial half-life for this branch has been estimated to be $(0.1 - 6) \times 10^{6}$ years.[17]

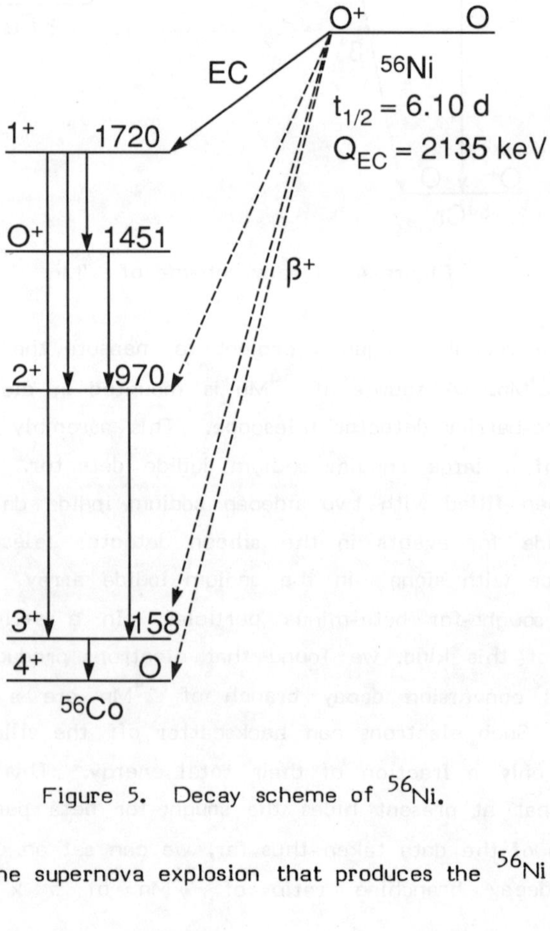

Figure 5. Decay scheme of ^{56}Ni.

Soon after the supernova explosion that produces the ^{56}Ni takes place, temperatures are low enough that electrons are bound to nuclei and neutral atoms can exist. Thus, the half-life of ^{56}Ni would be its laboratory value. However, it is thought that supernovae may be sites of cosmic-ray production and acceleration. If ^{56}Ni produced in such explosions were accelerated to high energies then it too would be quickly stripped of its electrons. Thus, the cosmic-ray lifetime of ^{56}Ni would be determined by its beta-plus decay half-life. Future observations of ^{56}Ni in the cosmic rays could be used to determine the time interval between nucleosynthesis and cosmic-ray acceleration if this half-life were known.[19]

We plan to measure the beta-plus decay lifetime of ^{56}Ni using a four-fold coincidence technique. ^{56}Ni is produced by bombarding iron foils with a beam of ^{3}He ions provided by LBL's 88-Inch Cyclotron, and is then chemically separated from the target. To detect ^{56}Ni beta-plus decay, the activity is placed at the center of a large sodium iodide annular detector that is split into two optically isolated halves. Located inside the annulus on one side of the source is a germanium gamma-ray detector, and on the other side of the source is a silicon surface-barrier detector. This system is being used to search for events in which a positron is detected in the silicon detector, a 511-keV annihilation gamma ray is recorded in each half of the annulus, and a 158-keV gamma ray is recorded in the germanium detector. In a series of preliminary tests of this scheme, a large number of such four-fold coincident events have been observed. If these were naively assumed to be due to the sought-for decay, the inferred branching ratio would be on the order of 10^{-3} which is much greater than the expected value of 10^{-7} to 10^{-6}. The origin of these events is now being investigated.

5. CONCLUSIONS

It has been shown that under the physical conditions that occur in a variety of astronomical environments, several different processes can lead to the alteration of nuclear decay rates from their terrestrial values. Careful studies of specific nuclear properties can provide the quantitative information needed to properly incorporate these effects into nucleosynthesis calculations. These required measurements provide real challenges for experimenters.

6. ACKNOWLEDGEMENTS

The experiments described in this talk have involved the the efforts of a number of people. I wish to thank my collaborators on these projects, K. T. Lesko, R. M. Larimer, R. M. Diamond, F. S. Stephens, M. A. Deleplanque, E. M. Beck, J. C. Bacelar, A. E. Champagne, S. E. Kellogg, and E. Horch. This work is supported by the U. S. Department of Energy under Contract No. DE-AC03-76SF00098.

7. REFERENCES

1. R. R. Winters et al., Astrophys. J. 300 , 41 (1986).
2. C. M. Lederer and V. S. Shirley, Table of Isotopes , 7th ed. (Wiley, New York, 1978).
3. H. Beer and R. A. Ward, Nature 291 , 308 (1981).
4. K. Yokoi and K. Takahashi, Nature 305 , 198 (1983).
5. S. E. Kellogg and E. B. Norman, Phys. Rev. C 31 , 1505 (1985).
6. S. E. Kellogg, Ph. D. thesis, Univ. of Washington (1987).
7. S. E. Kellogg and E. B. Norman, Phys. Rev. C 34 , 2248 (1986).
8. W. Eschner et al., Z. Phys. A 317 , 281 (1984).
9. K. T. Lesko et al., Phys. Rev C 34 , 2256 (1986).
10. H. Beer and R. L. Macklin, Phys. Rev. C 26 , 1404 (1982).
11. K. Takahashi and K. Yokoi, Nucl. Phys. A 404 , 578 (1983), and private communication.
12. E. B. Norman et al., Astrophys. J. 281 , 360 (1984).
13. E. Warde et al., Phys. Rev. C 27 , 98 (1983).
14. R. A. Dewberry and R. A. Naumann, Phys. Rev. C 28 , 2259 (1983).
15. G. Tarlé et al., Astrophys. J. 230 , 607 (1979).
16. L. Koch et al., Astron. Astrophys. 77 , L9 (1981).
17. L. Wilson, Ph. D. thesis, Univ. of California (1978), LBL-7723.
18. M. Cassé, Astrophys. J. 180 , 623 (1973).
19. M. Cassé, Proc. 13th Int. Conf. Cosmic Rays, Vol. 1, p.546 (1973).

EXPERIMENTS INVOLVING SHORT-LIVED RADIOACTIVE NUCLEI AND THEIR IMPACT ON NUCLEOSYNTHESIS AND COSMOLOGY

by

Richard N. Boyd

Department of Physics, Department of Astronomy

The Ohio State University, Columbus, OH 43210

and

Lawrence Livermore National Laboratory

Livermore, CA 94550

ABSTRACT: Radioactive ion beams will allow for the study of a variety of nuclear reactions of interest to astrophysics. Furthermore, such beams in heavy ion accelerator storage rings will allow for measurements of lifetimes of highly ionized nuclei, some of which are important to cosmology.

1. INTRODUCTION

The radioactive ion beams (RIBs) I will be discussing generally involve nuclei with halflives less than or the order of a minute. Such short-lived nuclei would be impossible to make into targets, since a typical time for an experimental run to measure a nuclear cross section is an hour. Thus if such nuclei are to be used in reaction studies, they must be made into beams, and the nucleus which usually forms the projectile must then become the target.

The primary interest in low energy RIBs is astrophysical[1-3]), although there are some nuclear physics questions which can be addressed with RIBs, and other research spinoffs[1-3]) exist as well. I will, however, concentrate on astrophysical aspects of RIB uses. The reactions I will discuss constitute a sample of a large number of reactions we will ultimately try to observe. Unfortunately, I will not be presenting any data. Since we are presently installing our RIB facility at the new FN tandem van de Graaff facility at Lawrence Livermore National Laboratory (LLNL), we are not yet able to run. The first experiment we will try, the ^{7}Be(p,γ) reaction, is of interest to the solar neutrino problem; we have discussed our plans to do that reaction previously[4,5]). We did obtain a ^7Be beam on target of $3x10^5$ ions/s with the old LLNL accelerator, as well as some preliminary data for that reaction, last fall. However, we are still concerned about backgrounds, so we have made some improvements, and will rerun the experiment this fall.

2. THE HCNO CYCLE

The Hot CNO (HCNO) cycle provides the first reaction I will discuss. That cycle has been studied theoretically[6]); a subset of the processses considered is shown in Fig. 1. Its understanding requires information about several nuclear reactions involving short-lived nuclei, e.g., $^{13}N(p,\gamma)^{14}O$ (^{13}N halflife = 10.0 min.), $^{14}O(\alpha,p)^{17}F$ (^{14}O halflife = 70.6 sec.), and $^{14}O(\alpha,\gamma)^{18}Ne$. While all of these reactions are of interest, Fig. 1 suggests that one of them, $^{14}O(\alpha,p)^{17}F$, is pivotal to the whole process. At low stellar temperatures, tens of millions K, the only process available to ^{14}O (if it is formed at all) is β-decay to ^{14}N, since proton capture on ^{14}O is not energetically allowed. The processes in this route are indicated in Fig. 1 by the double coupling lines: they comprise the "β-delayed CNO cycle"; as the speed at which it can proceed is limited by the lifetimes of the β-decaying nuclides, ^{14}O and ^{15}O. At higher temperatures, the ^4He-^{14}O Coulomb barrier can be overcome sufficiently to allow the $^{14}O(\alpha,p)$ and $^{14}O(\alpha,\gamma)$ reactions to proceed. Theoretical studies[7,8]) suggest that, of these reactions, the former will dominate. Thus the relationship between the thermonuclear reaction rate, which is determined by the cross section, for $^{14}O(\alpha,p)$ and the ^{14}O lifetime is crucial to understanding the mode in which this cycle, if indeed it even is a cycle (it will not be at high temperatures), can operate.

The setup for measuring the $^{14}O(\alpha,p)$ reaction is shown in general form in Fig. 2. A high intensity beam of 25 MeV ^{12}C ions will first be produced by the FN tandem van de Graaff accelerator at LLNL, and then brought to bear on the Ohio State University (OSU) ^3He gas cell. This ^3He target system is necessarily complicated, as it must be dense and localized, and must recirculate the ^3He gas; its details have been described elsewhere[9]). The $^{12}C(^3He,n)^{14}O$ reaction is known to have a large cross section; when it is reversed as $^3He(^{12}C,^{14}O)n$ the ^{14}O yield receives a large enhancement in the laboratory system, as the reaction products are all scattered into a 6.8° cone about the incident ^{12}C beam direction. This results in a secondary ^{14}O beam, after energy loss in various foils is taken into account, of about 18 MeV (or about 4 MeV in the center-of-mass system) at the entrance to the

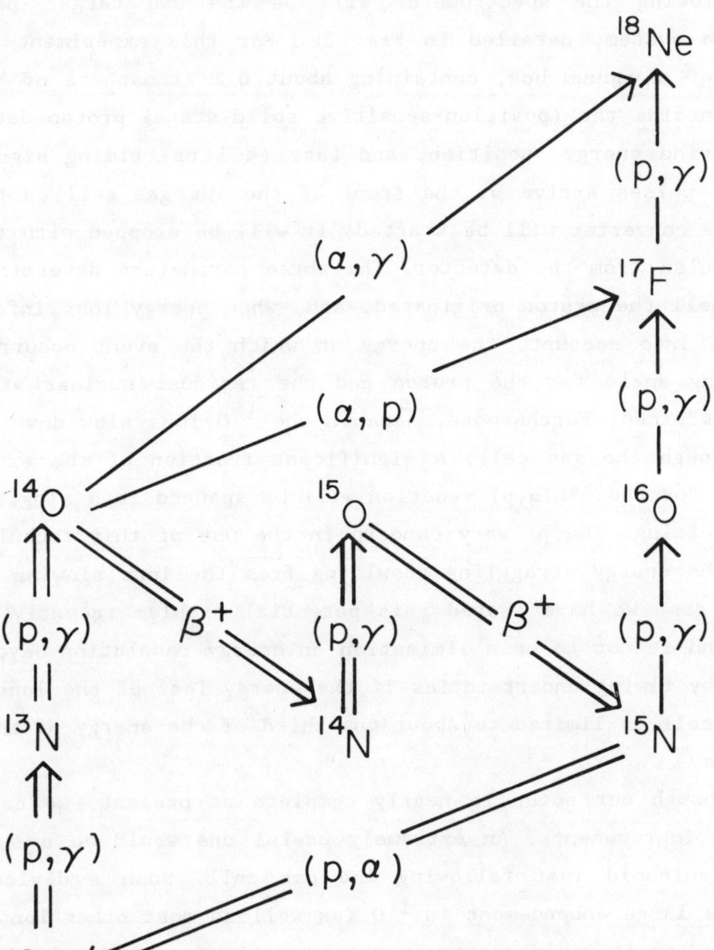

Fig. 1. Some of the reaction processes of the HCNO cycle.

(downstream) ^{4}He gas cell.

The ^{3}He gas cell will be followed by the LLNL double-triplet magnetic quadrupole spectrometer, a device which consists of six quadrupole elements, has an inner diameter of 20 cm at maximum, and is, in total, about 3 m long. In addition to the quadrupole elements, it has an electrostatic deflector between the two quadrupole triplets. Further details are given in previous publications[4,5]).

Following the spectrometer will be the OSU target--particle-detection system, detailed in Fig. 2b. For this experiment it will consist of a vacuum box, containing about 0.1 atmosphere of ^{4}He gas, which contains the (position-sensitive solid-state) proton detectors; they provide energy, position, and fast (< 1 ns) timing signals. As the beam pulses arrive at the front of the ^{4}He gas cell, a time-to-amplitude converter will be started; it will be stopped with the fast timing pulse from the detector. The three parameters determine where in the cell the proton originated, and, when energy loss information is taken into account, the energy at which the event occurred. The scattering angle for the proton and the residual nuclear state are also determined. Furthermore, because the ^{14}O ions slow down as they move through the gas cell, a significant fraction of the excitation function for the ^{14}O(α,p) reaction will be spanned in a single accelerator setting. The primary concern in the use of this technique involves the energy straggling resulting from the ions slowing down in the ^{4}He gas. We have tested this potential problem recently[10]), and have found it not to be a limitation on energy resolution beyond that imposed by timing uncertainties if the energy loss of the ions in the ^{4}He gas cell is limited to about one-third of the energy of the incident ions.

Although our setup is nearly complete at present, we can still dream of improvements. An extremely useful one would be a superconducting solenoid just following the gas cell. Such a device would provide a large enhancement in ^{14}O (as well as most other ion's) beam intensity, as it would constrain the heavy ions emerging from the ^{3}He cell to a small divergence angle, thus making it considerably easier to get them through the spectrometer. In the context of upgrades, I should note that radically different schemes from the one I am discussing also have been proposed for producing RIBs[11,12]).

3. THE GALACTIC ^{26}Al ABUNDANCE

A second experiment also involves the use of the basic components of the LLNL-OSU facility, but illustrates additional degrees of complexity with which one must deal in some RIB experiments. The experi-

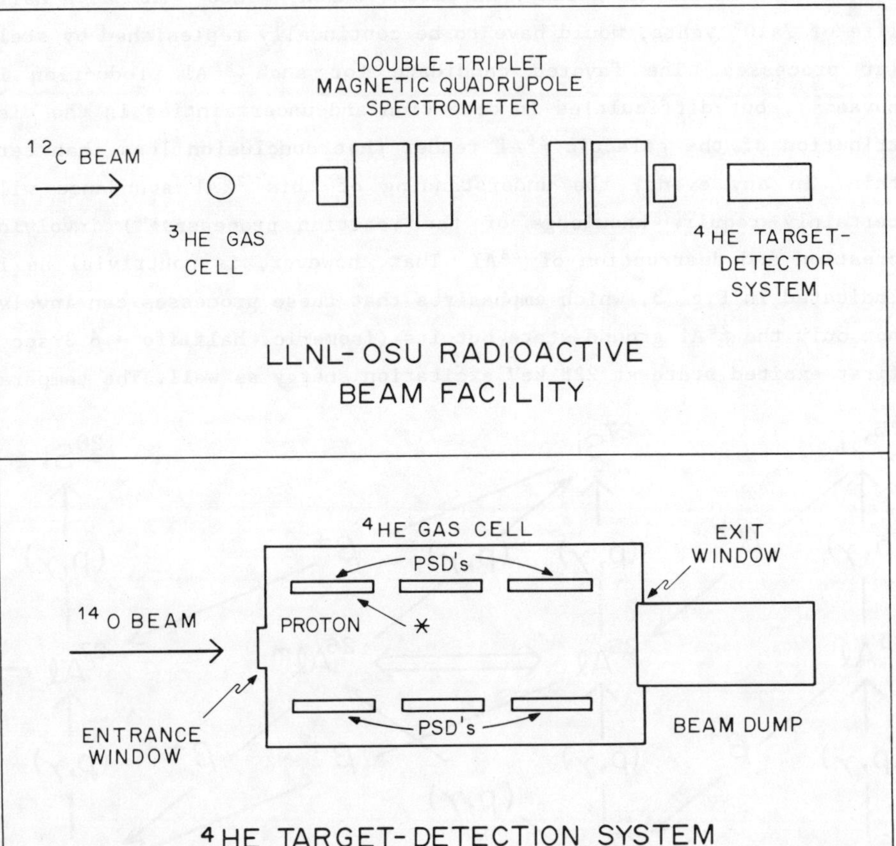

Fig. 2. Upper half: Basic components of the LLNL-OSU Radioactive Beam Facility. Lower half: Details of the OSU target--particle-detection system.

ment, the $^{26}Al(p,\gamma)^{27}Si$ reaction, is of importance to understanding the ongoing processes of nucleosynthesis in our galaxy, dramatic confirmation of which was obtained from the detection of γ-rays from the decay of ^{26}Al by the HEAO-3 γ-ray observatory[13]). The intensity level of these γ-rays (actually they result from the decay from the first excited state of ^{26}Mg, which is populated in the β-decay of ^{26}Al, to the ^{26}Mg ground state) suggests[14]) that there are about three solar masses of ^{26}Al in the galaxy, an amount which, since ^{26}Al has a half-life of $7x10^5$ years, would have to be continually replenished by stellar processes. The favored candidate for such ^{26}Al production is novae[15]), but difficulties in detection and uncertainties in the distribution of the galactic ^{26}Al render that conclusion less than certain. In any event, the understanding of this ^{26}Al abundance will certainly require knowledge of the reaction processes[16]) involving creation and destruction of ^{26}Al. That, however, is nontrivial as is indicated in Fig. 3, which emphasizes that these processes can involve not only the ^{26}Al ground state but its (isomeric, halflife = 6.3 sec.) first excited state at 228 keV excitation energy as well. The tempera-

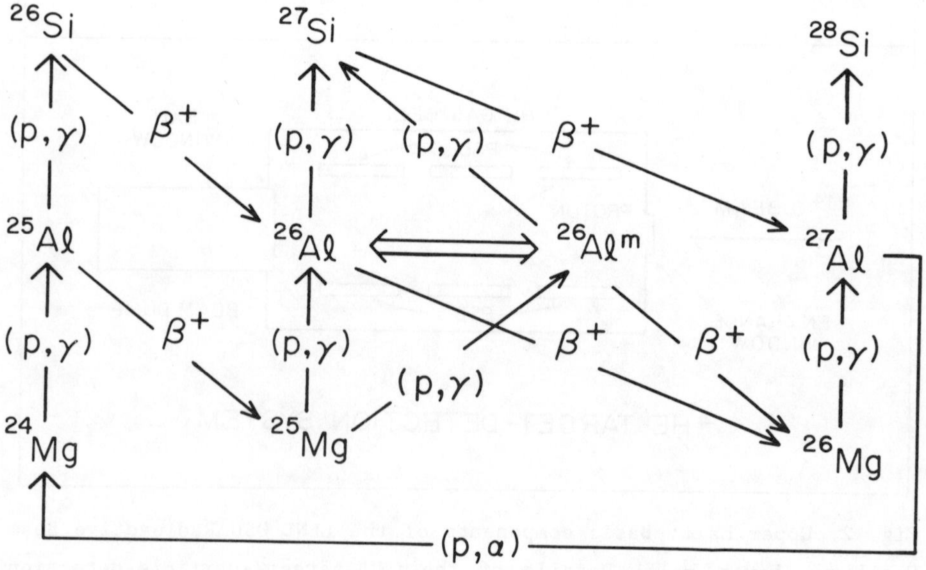

Fig. 3. Processes of stellar creation and destruction of ^{26}Al.

ture of the stellar environment is crucial here: the establishment of thermal equilibrium at high temperatures will populate the first excited state. That will cause a rapid depletion in ^{26}Al abundance, since that state β-decays quickly to ^{26}Mg.

Thus understanding of the processes which produce ^{26}Al requires not only measurements of reactions such as ^{26}Al(p,γ)^{27}Si, a difficult (but not impossible) study because of the radioactivity of ^{26}Al, but of ^{26}Al^{m}(p,γ)^{27}Si as well. This latter reaction cannot possibly be done without making a ^{26}Al^{m} beam, which we hope to do. Such a study requires one to do two experiments, the first primarily with a beam of ^{26}Al and the second of ^{26}Al^{m}, then subtract to get the ^{26}Al(p,γ) and ^{26}Al^{m}(p,γ) cross sections. Fortunately reactions exist which preferentially populate ^{26}Al or ^{26}Al^{m}. The ^{24}Mg(^{3}He,p)^{26}Al reaction would, at low energies, be expected to populate states which can be reached by low angular momentum transfer, e.g., the first excited ^{26}Al^{m}(0^{+}) state, considerably more strongly than those populated by a large angular momentum transfer, e.g., the ground (5^{+}) state. However the ^{27}Al(^{3}He,α) reaction should, at low energies, populate both the ground and first excited states, since both would be reached by pickup of a $d_{5/2}$ neutron. By judicious selection of the scattering angle in the ^{3}He(^{27}Al,^{26}Al)^{4}He reaction, a beam can probably be produced predominately in the ^{26}Al ground state. The reactions produced by this beam on an H_2 target should, then, differ appreciably from those produced by the (primarily) ^{26}Al^{m} beam from the ^{3}He(^{24}Mg,^{26}Al)p reaction. Studies done to date tend to confirm our predictions about the above two production reactions, although we need to perform lower energy measurements with ^{3}He beams on ^{24}Mg and ^{27}Al targets before we know what yields of ^{26}Al and ^{26}Al^{m} RIBs to anticipate.

Proton radiative capture studies also require a different detection scheme than that discussed above for the ^{14}O(α,p)^{17}F experiment. Fortunately the γ-rays produced in the ^{26}Al(p,γ) reaction are about 7 MeV, at which energy backgrounds can be virtually eliminated. We will detect the γ-rays by surrounding a gas cell of H_2 with six 10 cm x 10 cm x 40 cm NaI detectors, which are in turn surrounded with plastic anticoincidence detectors. This system, the OSU target--γ-ray detection system, has been assembled and tested, and has indeed been found to exhibit vanishingly low backgrounds at the energies of interest. Furthermore the large size of the NaI detectors should allow a fairly high efficiency for detection of the γ-rays, even when they are as energetic as 7 MeV.

4. MEASUREMENTS OF LIFETIMES OF FULLY STRIPPED IONS

Finally I will describe another kind of RIB experiment which is of great interest to astrophysics. It involves observation of β-decay lifetimes of highly stripped nuclei. To understand why such information is astrophysically important, consider ^{187}Re, a nucleus used as a chronometer[17]) in determining the age of the universe from abundances of radionuclides. One can relate the present day abundances of ^{187}Re and ^{187}Os to those at the time of the earliest nucleosynthesis using their halflives, our knowledge of the s-process, and reasonable assumptions about their history throughout the chemical evolution of our galaxy. The latter, however, involves stellar environments at high enough temperatures that the ions may be in high states of ionization. For most neutron rich nuclides, the number of bound electrons will have little effect on the lifetime. However, the ionization state can profoundly effect those nuclides which, in a state in which they have most of their bound electrons, are unstable, but which, in a nearly fully stripped state, might have much longer lifetimes for normal β-decay. In such cases, "bound state β-decay," a process in which the electron emitted is captured into an atomic orbit, can become the dominant decay mode. This process, though known for some time[18]), has recently been given an improved theoretical treatment[19]). It has, however, never been observed experimentally. In the present discussion, I have ignored effects which are important to an accurate description of ^{187}Re as a chronometer; these have recently been described in great detail[21]). But such effects do not change the need for an accurate description of bound-state β-decay, a process which has been estimated to have a 20% effect on the effective lifetime of ^{187}Re, and hence on the age of the universe determined from that chronometer.

Therefore a test of the theoretical description of bound-state β-

decay for fully stripped ions becomes crucial. Experiments to perform such measurements using accelerator storage rings have recently been proposed[19,20]. The detailed experiment discussed[19] utilized a stored ^{3}H beam but, as was noted, similar experiments using the heavy-ion storage ring at GSI Darmstadt, can be performed for nearly fully stripped heavy ions. Basically the experiment could be performed as shown in Fig. 4, using ^{163}Dy as an example. Any normal β-decays which occur along the straight sections of the storage ring, e.g., from C′ to D, will change the ionic charge by one unit, so can be detected in a detector located, as shown, after the next bending magnet. A bound-state decay, however, will not change the ionic charge state, so will (aside from tiny recoil effects) not affect the ion's trajectory at all. Thus such ions would remain in the storage ring.

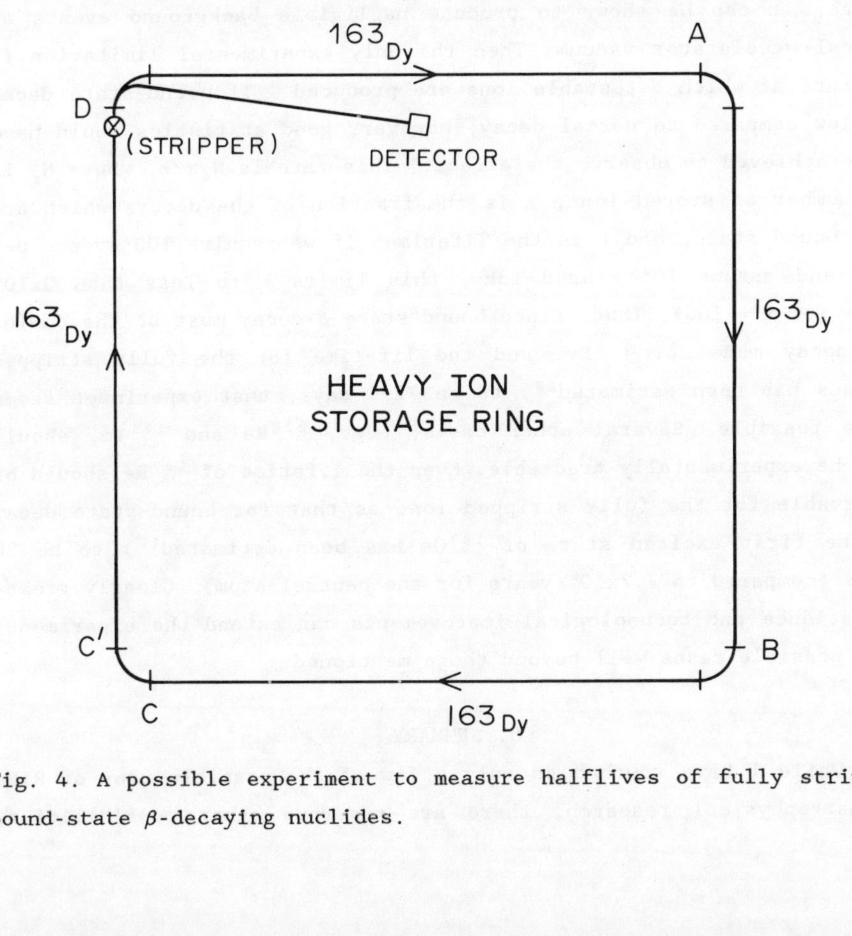

Fig. 4. A possible experiment to measure halflives of fully stripped bound-state β-decaying nuclides.

However, introduction of a stripper just before one of the magnets would result in the ions produced in the bound-state decay mode being bent out of the storage ring by the magnet immediately following the stripper, where they would be detected together with the ions produced in normal β-decay in the last straight section preceeding the magnet. However, the ions produced in bound-state decay since the last pass through the stripper, i.e., throughout one entire loop, would be detected. Backgrounds in such an experiment, if fully stripped ions were used, would be virtually nonexistent. Such would not be the case, however, if less than fully stripped ions were used, since the residual gas atoms in the storage ring could produce electron charge exchange reactions which would simulate normal decays. But increase of the nuclear charge by one unit is only possible via nuclear charge exchange; it can be shown to produce negligible background events at typical accelerator vacuum. Then the only experimental limitation is the rate at which detectable ions are produced. (If bound-state decay is slow compared to normal decay, however, good statistics would have to be achieved to observe the effect.) This rate is $N_s x/\tau$, where N_s is the number of stored ions, x is the fraction of the decays which are to a bound state, and τ is the lifetime. If we require 100 events per day, and assume 10^8 stored ions, this limits τ to less than 3×10^4 years for x = 100%. Thus, since bound state β-decay must be the dominant decay mode for ^{163}Dy, and the lifetime for the fully stripped nucleus has been estimated[19]) to be 72.3 days, that experiment seems quite feasible. Several other cases, e.g., ^{228}Ra and ^{171}Tm, should also be experimentally tractable. Even the lifetime of ^{187}Re should be observable for the fully stripped ion, as that for bound-state decay to the first excited state of ^{187}Os has been estimated[19]) to be 20 years (compared to 7.2×10^9 years for the neutral atom). Clearly greater patience and technological improvements can extend the experimentally possible cases well beyond those mentioned.

5. SUMMARY

While I have touched on only a few of the possible uses of RIBs in astrophysical research, there are numerous other possibilities, both in astrophysics and in other research disciplines. Furthermore we expect to see developments in the technology of radioactive beams in the future; this will open up even more possibilities for their use.

ACKNOWLEDGEMENTS

This work has been supported by National Science Foundation grant 86-00749 and by Lawrence Livermore National Laboratory. I also acknowledge numerous useful discussions with G.J. Mathews, K.E. Sale, and K. Takahashi.

REFERENCES

1. Proc. Workshop on Radioactive Ion Beams and Small Cross Section Measurements, ed. by R.N. Boyd, Burr Oak, OH, 1981.

2. Proc. Workshop on Prospects for Research with Radioactive Beams from Heavy Ion Accelerators, ed., J.M Nitschke (Lawrence Berkeley Lab. Report LBL-18187), 1984.

3. Proc. Accel. Radioactive Beams Workshop, ed. L. Buchmann and J.M. D'Auria, TRIUMF Report TRI-85-1, 1985.

4. Sale, K.E., Bauer, R.W., Boyd, R.N., Mathews, G.J., Haight, R.C., and Corn, P.B., Nucl Instr. and Methods in Phys. Res. B24/25, 490 (1987).

5. Haight, R.C., Mathews, G.J., White, R.J., Aviles, L.A., and Woodard, S.E., IEEE Trans on Nucl. Sci, NS-30, 1160 (1983); Haight, R.C., Mathews, G.J., and Bauer, R.W., Nucl. Instr. and Methods in Phys. Res. B10/11, 361 (1985).

6. Wallace, R.K. and Woosley, S.E., Astrophys. J. Suppl. Series 45, 389 (1981); Caughlan, G.R., CNO Isotopes in Astrophysics, ed. by J. Audouze, D. Reidel, Boston, 1977, p. 121.

7. Langanke, K., private communication, 1987.

8. Wiescher, M., Harms, V., Gorres, J., Thielemann, F-K. and Rybarcyk, L.J., Astrophys. J. 316, 162 (1987).

9. Boyd, R.N., Rybarcyk, L., Wiescher, M., and Hausman, H.J., IEEE Trans on Nucl. Sci. NS-30, 1387 (1983); Boyd, R.N., Rybarcyk, L., Hausman, H.J., Kim, W., and Schmalbrock, P., Nucl. Instr. Methods in Phys. Res. B10/11, 378 (1985).

10. Islam, M.S., Corn, P.B., Boyd, R.N., and Norbeck, E., private communication, 1987.

11. Buchmann, L., Proc. Accel. Radioactive Beams Workshop, ed. L. Buchmann and J.M. D'Auria, TRIUMF Report TRI-85-1, 1985, p. 149.

12. Nitschke, J.M., Proc. Workshop on Prospects for Research with Radioactive Beams from Heavy Ion Accelerators, ed. J.M. Nitschke (Lawrence Berkeley Lab. Report LBL-18187), 1984, p. 48.

13. Mahoney, W.A., Ling, J.C., and Jacobson, A.S., Astrophys. J. 262, 742 (1982).

14. Mahoney, W.A., Ling, J.C., Wheaton, W.A., and Jacobson, A.S., Astrophys. J. 286, 578 (1984).

15. Clayton, D.D., Astrophys. J. 280, 144 (1984); Fowler, W.A., Rev. Mod. Phys. 56, 137 (1984).

16. Ward, R.A. and Fowler, W.A., Astrophys. J. 238, 266 (1980).

17. See, e.g., Meyer, B.S. and Schramm, D.N., Astrophys. J. 311, 406 (1986); Fowler, W.A. and Meisel, C.C., Proc. Symp. on Cosmogonical Processes, Boulder, 1985.

18. Daudel, R., Benoist, P., Jacaques, R., and Jean, M., Compt. Rend. 224, 1427 (1947); Daudel, R., Jean, M., and Lecoin, M., J. Phys. Radium 8, 238 (1947); Compt. Rend. 225, 290 (1947).

19. Takahashi, K., Boyd, R.N., Mathews, G.J., and Yokoi, K., to be published in Phys. Rev., 1987.

20. Kienle, P., Nucl. Phys. A447, 419c (1986).

21. Yokoi, K., Takahashi, K., and Arnould, M., Astron. Astrophys. 117, 65 (1983); Arnould, M., Takahashi, K., and Yokoi, K., Astron. Astrophys. 137, 51 (1984).

NEW LABORATORY APPROACHES IN STUDYING NUCLEAR REACTIONS IMPORTANT FOR NUCLEOSYNTHESIS

H.W.Becker, A.Krauss, C.Rolfs and H.P.Trautvetter
Universität Münster, Münster, West Germany

1. INTRODUCTION

For the understanding of the origin and distribution of the elements it is important to study those nuclear reactions, which play a crucial role in the nucleosynthesis during the early phase of the universe as well as the evolution of stars [1,2]. In spite of the large number of experiments carried out over the last 40 years involving the light elements, there are still substantial uncertainties in the rates of several important reactions.

To improve the situation one has to study experimentally the nuclear cross sections over a wide range of energies, together with a detailed study of the individual level parameters of the nuclei involved, in order to facilitate reliable theoretical extrapolations. But most useful is the attempt to pursue the experimental studies to energies as low as possible, with good statistical accuracy. Two examples are presented: the $^{3}He(^{3}He,2p)^{4}He$ reaction, which is of relevance to the solar neutrino problem, and the $^{27}Al(p,\alpha)^{24}Mg$ reaction, which closes the MgAl cycle.

The $^{14}N(n,p)^{14}C$ reaction at helium burning temperatures is related to questions concerning the stellar neutron production and may act as a neutron poison. The influence of a possible change in the reaction rate compared with literature is discussed in [4] for BaII-stars.

At elevated stellar temperatures during explosive burning the normal reaction sequence $^{14}N(\alpha,\gamma)^{18}F(\beta^{+}\nu)^{18}O(\alpha,\gamma)^{22}Ne(\alpha,n)^{25}Mg$ will not produce enough neutrons due to the relatively long halflive of ^{18}F (110 min). The radioactive ^{18}F nucleus acts as a target for the competing $^{18}F(\alpha,p)^{21}Ne$ reaction. However, the branching between both nuclear processes, beta-decay versus (α,p), is not well known since experiments on radioactive nuclei are yet rare. In order to check predictions of statistical-model calculations for $^{18}F(\alpha,p)^{21}Ne$, we have studied the reverse reaction $^{21}Ne(p,\alpha)^{18}F$.

Finally, future experimental studies of reactions involving radioactive nuclei, which are of importance for the understanding of the explosive burning stages in stars, require e.g. novel detection techniques. A status-report on the technical development of a detector, insensitive to γ-radiation below 2.2 MeV, will be given.

2. CHARGED-PARTICLE-INDUCED REACTIONS AT LOW ENERGIES

2.1. The $^3He(^3He,2p)^4He$ Reaction

It has been suggested [5,6] that the solar neutrino problem might be solved by a hitherto undetected resonance close to the threshold of the $^3He(^3He,2p)^4He$ reaction, which is the major competing reaction (86%) to the $^3He(\alpha,\gamma)^7Be$ branch (14%) in the p-p chains [7]. The latter branch leads ultimately to the ^8B-neutrinos to which the Davis experiment [7] is sensitive. Direct measurements of the $^3He(^3He,2p)^4He$ reaction have been carried out with energies as low as $E_{c.m.}$ = 30 keV [8], in search for such a resonance, with no positive evidence for it for energies below 40 keV [8]. For this reason we have attempted to pursue the investigation of this reaction down to the thermal energy region of the sun: $E_0 = 21.9 \pm 6.2$ keV for a central solar temperature of $T_6 = 15.5$. At such low energies, the cross section for the $^3He+^3He$ reaction can be estimated assuming a constant S-factor value of 5 MeV b [8] to yield $\sigma(22\text{ keV}) \approx 10^{-36}\text{ cm}^2$. With typical experimental conditions one expects a few events per day. The experimental signal can be obscured by the $^3He(d,p)^4He$ background reaction, by cosmic-ray background, and by high-frequency noise (induced in the electronic circuits). Details of the technical aspects of this experiment are given in [9] and only the major results are discussed below.

The $^3He(d,p)^4He$ background problem was solved by the proton-proton coincidence technique for identifying uniquely the $^3He(^3He,2p)^4He$ events. The cosmic-ray problem could be reduced to a level of less than 1 event per 200 h with the aid of an active plastic scintillator shielding, and the high-frequency noise was reduced to less than 1 event per week by other means [9]. With this setup the energetically lowest data point was obtained at E = 24.5 keV, after 6 days of running or an accumulated charge of about

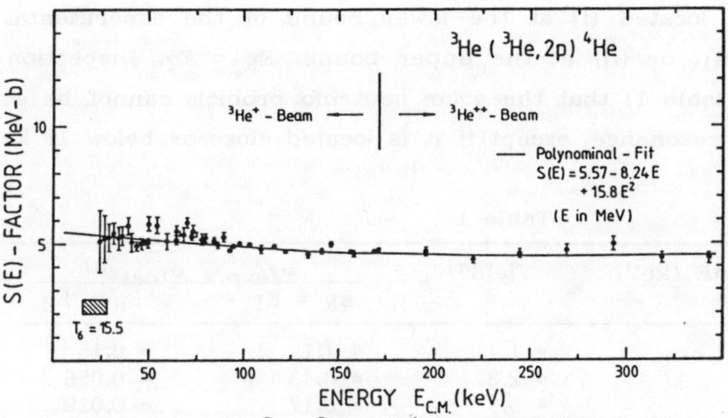

Fig. 1: S(E)-factor for the ^{3}He(^{3}He,2p)^{4}He reaction.

180 Coulombs. This point lies well within the thermal energy region of the sun. The resulting S-factor curve is shown in Fig. 1 together with a polynomial fit to the data yielding S(0) = 5.57 ± 0.31 MeV b, in excellent agreement with the value 5.5 MeV b recommended in [10]. The present result does not support S(0) = 4.7 MeV b used in the calculation of the solar neutrino flux [11]: N_ν(theo) = 5.8 ± 2.2 SNU. With the relation $N_\nu \propto S(0)^{0.42}$ and the present value of S(0), the predicted neutrino flux reduces to 5.4 SNU, leaving still a sizable discrepancy with observation, N_ν = 2.1 ± 0.3 SNU [11]. However, the solar neutrino problem could be solved if the resonance was located below 24.5 keV. With the relation $N_\nu \propto \langle\sigma v\rangle^{-0.42}$, where $\langle\sigma v\rangle = \langle\sigma v\rangle_{NR} + \langle\sigma v\rangle_R$ is the total reaction rate consisting of the sum of non-resonant (NR) and resonant (R) contributions, one can estimate [9] that the solar neutrino problem would be solved for a ratio $\lambda = \langle\sigma v\rangle_R/\langle\sigma v\rangle_{NR} \simeq 9$. This ratio depends on the resonance energy E_R and on the resonance strength $\omega\gamma_R$, where $\omega\gamma_R$ is proportional to the thick-target yield for a narrow resonance [2]. We have therefore divided the Gamow peak width ΔE_0 = 12.5 keV into 4 energy intervals with a target thickness of $\Delta \simeq 3$ keV and determined experimentally upper limits on the reaction yield (table 1). Also given in table 1 is the ratio of the deduced experimental upper limit for the resonance strength $\omega\gamma_{exp}$ to the calculated strength $\omega\gamma_{cal}$ fulfilling the condition $\lambda \simeq 9$ for two cases: if the assumed

resonance was located (i) at the lower bound of the experimental energy region, $E_R = E_1$, or (ii) at the upper bound, $E_R = E_2$. Inspection of this ratio reveals (table 1) that the solar neutrino problem cannot be solved by the suggested resonance, except if it is located close or below 16 keV.

Table 1

Energy range ΔE (keV) $E_1 \rightarrow E_2$	Yield[a]	$\omega\gamma_{exp} / \omega\gamma_{cal}$ [b] $E_R = E_1$	$E_R = E_2$
16.2 → 18.9	≤ 1.1	≤ 1.1	≤ 0.15
18.8 → 21.5	≤ 2.3	≤ 0.43	≤ 0.056
21.4 → 24.4	≤ 5.5	≤ 0.17	≤ 0.019
24.0 → 27.1	≤ 17	≤ 0.084	≤ 0.008

[a] In units of 10^{-18} reactions per incident particle.
[b] For $\lambda \simeq 9$ (see text).

2.2. The $^{27}Al(p,\alpha)^{24}Mg$ Reaction

The $^{27}Al(p,\alpha)^{24}Mg$ reaction is another example, where we have still large uncertainties concerning the reaction rates: below $T_9 \simeq 0.6$ ($E_p \leq 0.5$ MeV) no experimental information exists. This reaction closes the MgAl cycle, which takes place during hydrogen burning and which has been identified as a possible production mechanism for ^{26}Al. Various astrophysical models [12 and references therein] such as novae, supernovae, red giants, Wolf-Rayet stars and supermassive stars have been suggested for the production site of ^{26}Al, all of which are confronted with the γ-ray observation [13] resulting in the presence of about 4 solar masses of ^{26}Al in our Galaxy [14] as well as with the observation of the ^{26}Mg isotopic enrichments (extinct ^{26}Al) in carbonaceous meteorites [15]. At present no definite decision can be made on the dominant production mechanism for ^{26}Al, but the most popular scenarios appear to be associated with novae and Wolf-Rayet stars, where ^{26}Al nuelosynthesis occurs near $T_9 = 0.1$. Due to the uncertainties in the rates of $^{27}Al(p,\alpha)^{24}Mg$, it is not clear how much catalytic material will be lost via the competing $^{27}Al(p,\gamma)^{28}Si$ reaction.

Details of the experiment are given in [16] and only the results are discussed below. The precision of the measurements was limited by the

$^{11}B(p,3\alpha)$ reaction arising mainly from boron contaminants in the target or backing, which could not be reduced below a level of about 10^{13} atoms/cm^2.

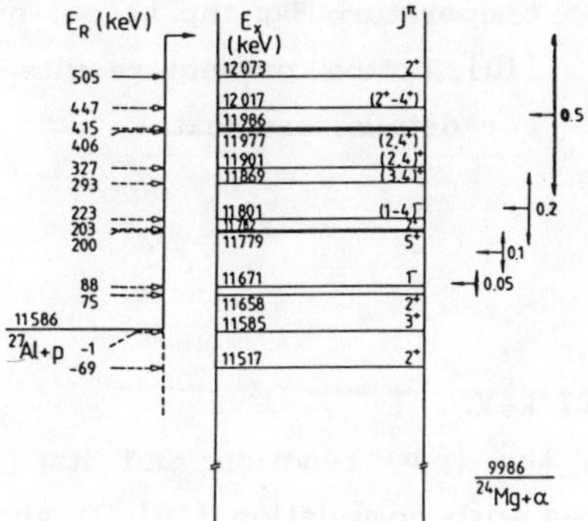

Fig. 2: Level-diagram for ^{28}Si. Indicated by the arrows are the expected resonances for $^{27}Al(p,\alpha)^{24}Mg$. Also shown on the right side are the Gamow-peak regions for various temperatures T_9.

The experimental situation before the present work is illustated in Fig. 2. The arrows indicate the position of the expected resonances (the resonance at E_R = 505 keV is the lowest previous observation, solid arrow) together with two possible subthreshold resonances at -1 and -69 keV. New upper limits for the strength of the E_R = 203, 223, 293 and 327 keV resonances could be determined: they are about a factor 1000 lower than previously assumed [17,18]. For example, in the determination of the reaction rates, the upper limit of $\omega\gamma(E_R = 223 \text{ keV}) \leq 8.3 \times 10^{-4}$ eV was taken previously [18] as the actual value for this resonance, while the present work leads to an upper limit of 4.5×10^{-7} eV. The influence of the subthreshold resonances was estimated on the basis of new spectroscopic information [17] and found to be negligible. The contribution of the low-energy wings of high-energy resonances to the total rate is smaller than that using the new upper limits of the resonances between 203 and 327 keV. The influence of the E_R = 75 and 88 keV resonances is still uncertain; its rate contribution was either omitted (curve labelled "low" in Fig. 3) or was estimated as an upper limit given by the full reduced particle width for the corresponding levels in ^{28}Si (Fig. 2) (curve labelled "high" in Fig. 3). In summary, the major changes are due to the new upper

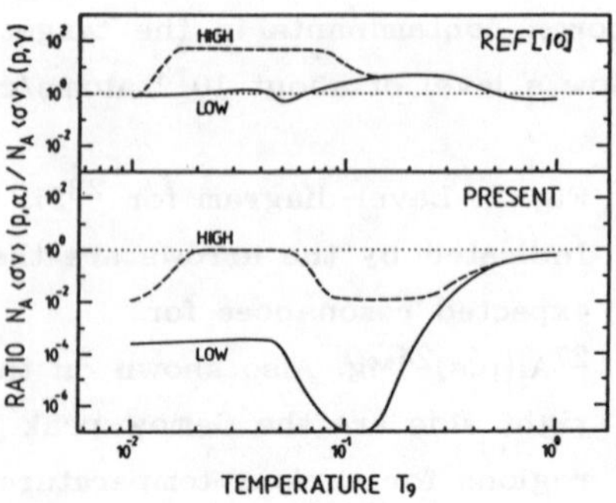

Fig. 3: Ratio of the reaction rates for $^{27}Al(p,\alpha)^{24}Mg$ to $^{27}Al(p,\gamma)^{28}Si$ versus temperature T_9: top as out of [10], bottom present results. For details, see text.

limits of the resonances at E_R = 203 to 327 keV.

The ratio for the rates between the (p,α) reaction and its (p,γ) competitor is shown in Fig. 3 and compared with compilation [18]. It should be noted that the (p,γ) reaction rate also was changed [16] in the present work due to new spectroscopic information obtained for the E_R = 75 and 88 keV resonances [19]. The new results presented in Fig. 3 show that the cycle is not closed at all relevant astropyhsical temperatures. This result requires new network calculations in the various models for ^{26}Al nucleosynthesis and may also change somewhat previous conclusions.

3. REACTIONS RELATED TO NEUTRON PRODUCTION

3.1. The $^{14}N(n,p)^{14}C$ Reaction

It has been argued [4] that in sites where $^{13}C(\alpha,n)^{16}O$ is the neutron source, the $^{13}C(p,\gamma)^{14}N$ reaction might lead to a significant ^{14}N production. However, this ^{14}N may constitute an important neutron poison through $^{14}N(n,p)^{14}C$ [4]. Due to such a possible neutron "hold-up", the neutron production by $^{13}C(\alpha,n)^{16}O$, required for the heavy element synthesis at the observed level in BaII-stars, must be so large that some ^{16}O overproduction is unavoidable [4].

The $^{14}N(n,p)^{14}C$ cross section has been measured previously for neutron energies down to 150 keV [20] and was then extrapolated by the 1/v-law to thermal energies [21]. We have determined experimentally Maxwell-averaged cross sections for this reaction for kT = 25 and 52 keV with a method similar to that given in [22]. The experiment is described in detail in [23]. We find that the average cross sections at both kT values are about a factor 3 lower than the values given in [21]: $\sigma(kT = 25\ keV) = 0.74 \pm 0.05$ mb and $\sigma(kT = 52\ keV) = 0.58 \pm 0.07$ mb. This means that the neutron poison through $^{14}N(n,p)^{14}C$ is not as effective as previously assumed [4].

3.2. The $^{21}Ne(p,\alpha)^{18}F$ Reaction

The significance of the $^{18}F(\alpha,p)^{21}Ne$ reaction for neutron production was pointed out in [24]. As discussed above, this reaction was studied via the reverse reaction $^{21}Ne(p,\alpha)^{18}F$. Two methods were employed [25]: (i) direct spectroscopic observation of the reaction α-particles and (ii) determination of the number of ^{18}F nuclei produced via the ^{18}F activity.

(i) For the direct spectroscopy a supersonic recirculating gas jet target [26] was used in conjunction with thin (15-20 μm) surface barrier detectors positioned at several angles relative to the beam axis. With the choice of the thin detectors, the α-particles are observed at higher pulse heights than the elastically scattered protons and can be clearly identified. The $^{22}Ne(p,\alpha)^{19}F$ background reaction produces α-particles with similar energies as for the $^{21}Ne(p,\alpha)^{18}F$ reaction due to similar Q-values. This background was unavoidable due to the 90% enrichment of our ^{21}Ne gas. Therefore, another excitation curve was obtained for $^{22}Ne(p,\alpha)^{19}F$ (using 99.9% ^{22}Ne) to be subtracted from the $^{21}Ne(p,\alpha)^{18}F$ data. This fact limited the accuracy of the results. An example of an excitation curve for the 90° detector is shown in Fig. 4. It can be seen that individual resonances are present, for which no spin assignments are known. Therefore, this curve does not represent angle-integrated cross sections.

(ii) The second method, measuring the ^{18}F activity, has the advantage not being sensitive to the $^{22}Ne(p,\alpha)^{19}F$ background reaction and also yielding angle-integrated results. However, since a closed cell was used, it was unavoidable to produce also ^{18}F on the oxydation layers of the entrance foil as well as on the beam stop. This contribution was determined separately and corrected for. The results are shown in Fig. 4 together with the energy regions covered by the chosen target thickness. It can be seen

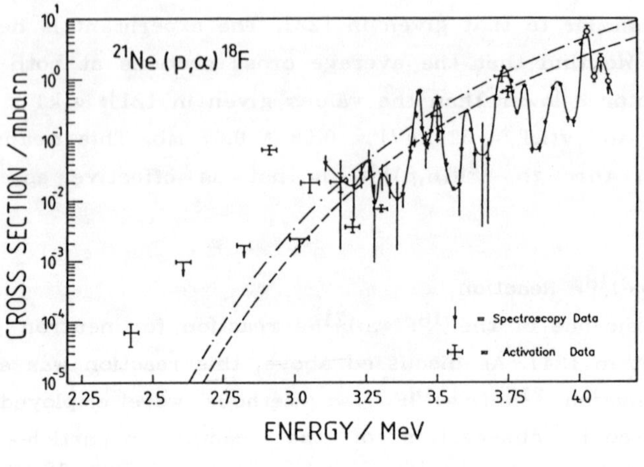

Fig. 4: Excitation curve for the $^{21}Ne(p,\alpha)^{18}F$ reaction. The solid line through the α-spectroscopic data points is to guide the eye. For details, see text.

that for E_R = 3.49 and 3.73 MeV the activation results represent the mean value over the resonance structure. Fig. 4 also shows two model predictions for the $^{21}Ne(p,\alpha)^{18}F$ reaction cross sections (dash-dotted line = [27], dashed line = [28]). Fair agreement is noted with [27] and somewhat better agreement with [28]. The two model predictions differ by a factor of 2.4. The low-energy activation points, E = 2.4 and 2.9 MeV, lie considerably higher than both model predictions and indicate that the resonance structure continues towards low energies. Thus, additional experiments are needed to verify this interpretation and to establish the reaction rates at the temperature region important for nuclear atrophysics.

4. GAMMA-RAY DETECTOR FOR HOT RADIOACTIVE TARGETS

The situation discussed in section 3.2, where the β-decay of ^{18}F takes too long and nuclear reactions on the radioactive nucleus will overtake, is also typical for many (p,γ) ad (α,γ) fusion reactions in hot explosive astrophysical scenarios. Direct observation of the γ-rays using a radioactive target is hampered by the presence of high radioactivity. To avoid the sensitivity of a detector to such relatively low energy γ-rays, one could take advantage of the photodisintegration of deuterium with the threshold at 2.23 MeV and use the liberated neutrons as a signal. Most (p,γ) and (α,γ) reactions have relatively large Q-values to yield γ-rays with energies above this threshold. A prototype of such a detector has been developed [29] and the construction of a full-sized detector consisting of 242 *l* D_2O and 30 ^{3}He-filled proportional neutron counters has now been completed [30]. The heavy water serves two purposes: (i) high-density deuterium target for photodisintegration and (ii) moderator for the liberated neutrons. Fig. 5 shows a cross section through this detector with the D_2O tank of cylindrical shape (inner radius = 34 cm, length = 70 cm).

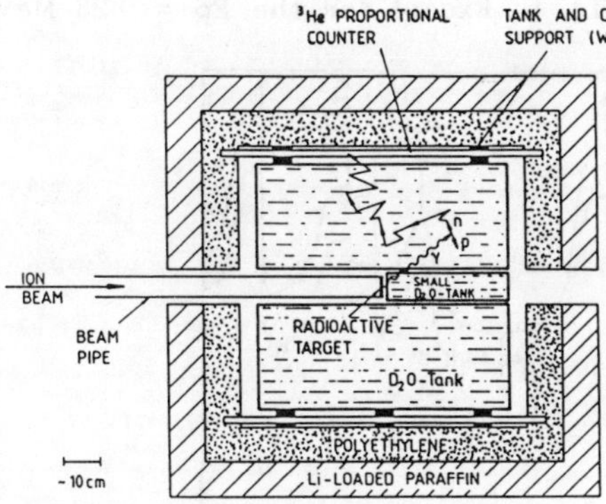

Fig. 5: Cross section through the D_2O detector. For details, see text.

This size was primarily given by the amount of heavy water available [30], but should represent a reasonable size since the mean free path length of 8 MeV γ-rays in D_2O is 42 cm and the root-mean-square distance reached by neutrons of thermal age is about 30 cm in 99.8% D_2O. The tank is surrounded by 18 ^{3}He counters of 1 m length (Fig. 5) and 6 ^{3}He counters of 88 cm length at each the front and rear side of the tank. This structure is packed in 10 cm thick polyethylene [30], which serves as an additional moderator as well as a reflector for neutrons, which escaped detection by the ^{3}He counters. Finally, the detector was shielded by 10 cm thick

Li-loaded paraffin against cosmic-ray neutrons [(p,n) reactions after pion-induced showers].

The neutron detection efficiency of this system was determined with a calibrated ^{252}Cf source placed at the center of the detector to yield ε_n = 0.18. The background count rate of the detector was measured to be typically 0.7 counts/sec, the origin of which is not well understood at present. This number limits the sensitivity in the detection of nuclear resonances to strengths above $\omega\gamma \simeq 0.3$ meV for reasonable beam currents, which is, however, sufficient for most applications.

For a test of the operation of this new detector, excitation curves of the well-known [17] $^{27}Al(p,\gamma)^{28}Si$ reaction were obtained with the D_2O detector and a NaI(Tl) crystal (placed in the center hole (Fig. 5) close to the target). The results are shown in Fig. 6. Except for the E_R = 0.23 MeV

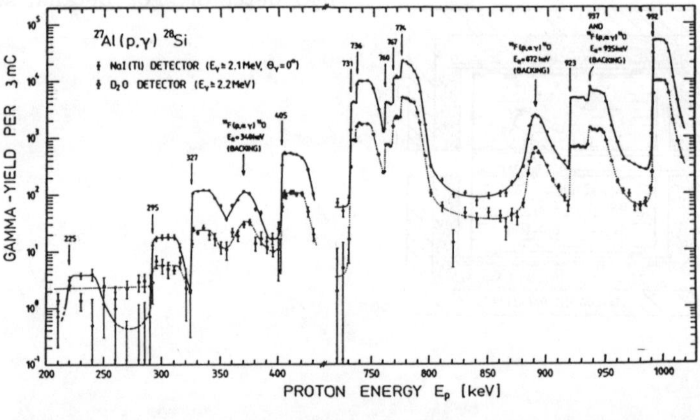

Fig. 6: Excitation curve for the known resonances in $^{27}Al(p,\gamma)^{28}Si$ [17]. The dotted line (D_2O detector) and the solid line (NaI(Tl) crystal) are to guide the eye.

resonance with $\omega\gamma$ = 43 μeV [17], all resonances have been observed with the new detector. This detector gives no information on the γ-ray energy observed (except for the threshold condition $E_\gamma \geq 2.22$ MeV) and hence multiple counting of cascading γ-rays is unavoidable. This problem needs special treatment. On the other hand the new detector is practically a 4π detector and gives therefore angle-integrated results.

With the aid of similar measurements on known [17] resonances , the γ-ray efficiency of the detector was determined and a value of $\varepsilon_\gamma = 5x10^{-4}$ was obtained at E_γ = 4.4 MeV.

The detector is ready for its first use in measurements involving radioactive targets: $^{22}Na(p,\gamma)^{23}Mg$. The preparation of the ^{22}Na target is underway and will be described by J.M.D'Auria in these proceedings.

REFERENCES

[1] W.A.Fowler, Rev.Mod.Phys. 56(1984)149
[2] C.Rolfs, H.P.Trautvetter and W.S.Rodney, Rep.Progr.Phys. 50(1987)3
[3] B.Fillipone, these proceedings
[4] A.Jorrison and M.Arnould: Advances in Nuclear Astrophysics, ed. E.Vangioni-Flam et al. (Editions Frontieres, Gif Sur Yvette, 1986) p.419
[5] V.N.Fetisov and Y.S.Kopysov, Nucl.Phys. A239(1975)511
[6] W.A.Fowler, Nature 238(1972)24
[7] J.N.Bahcall and R.Davis, Essays in Nuclear Astrophysics, ed.C.A.Barnes et al (Cambridge Press, 1982) p. 243
[8] M.R.Dwarakanath, Phys.Rev. C9(1974)805
[9] A.Krauss et al., Nucl.Phys. A467(1987)273
[10] M.J.Harris et al., Ann.Rev.Astr.Ap. 21(1983)165
[11] J.N.Bahcall et al., Rev.Mod.Phys. 54(1982)767
[12] N.Prantzos and M.Casse, Ap.J. 307(1986)324
[13] W.A.Mahoney et al., Ap.J. 262(1982)742
[14] D.D.Clayton, Ap.J. 280(1984)144
[15] G.J.Wasserburg and D.A.Papanastassiou, Essays in Nuclear Astrophysics, ed. C.A.Barnes et al. (Cambridge Press, 1982) p.77
[16] R.Timmermann, Diplomarbeit, Universität Münster (1987) and Nucl.Phys. (submitted)
[17] P.M.Endt and C. van der Leun, Nucl.Phys. A310(1978)1
[18] W.A.Fowler, G.R.Caughlan and B.A.Zimmerman, Ann.Rev.Astr.Ap. 13(1975)69
[19] A.E.Champagne et al., Nucl.Phys. A459(1986)239
[20] C.H.Johnson and H.H.Barschall, Phys.Rev. 80(1950)818
[21] W.E.Alley and R.M.Lessler, Nucl.Data Tables A11(1972)621
[22] H.P.Trautvetter et al., Z.Phys. A323(1986)1
[23] K.Brehm, Diplomarbeit, Universität Münster (1987); to be published
[24] W.Hillebrandt et al., Proc. XXI Rencontre de Moriond, Les Arcs (1986)
[25] U.Giesen, Diplomarbeit, Universität Münster (1987); to be published
[26] H.W.Becker et al., Nucl.Instr.Meth. 198(1982)277
[27] S.E.Woosley et al., Atomic Dat.Nucl.Dat.Tab. 22(1978)371
[28] F.Thielemann, private communication
[29] C.Rolfs and R.W.Kavanagh, Nucl.Instr.Meth. B26(1987)165
[30] S.Seuthe et al., Universität Münster, in collaboration with R.W.Kavanagh (Caltech) and F.B.Waanders (Potchefstroom); to be published

Status of Stellar Helium Burning Reaction Rates

B. W. Filippone
W. K. Kellogg Radiation Laboratory
California Institute of Technology
Pasadena, California, 91125

ABSTRACT

Recent experimental and theoretical work has lead to modifications of the best-estimates for the stellar cross sections and S-factors for the $^{16}O(\alpha,\gamma)^{20}Ne$ and $^{12}C(\alpha,\gamma)^{16}O$ reactions. These new results are discussed.

The two reactions $3\alpha \rightarrow {}^{12}C$ and $^{12}C(\alpha,\gamma)^{16}O$ are critically important in determining both the elemental abundance distributions and the evolutionary path of medium-mass stars following helium burning[1]. The $^{16}O(\alpha,\gamma)^{20}Ne$ reaction is expected to be less important because of the absence of any resonance structure in ^{20}Ne at the energies relevent to helium burning ($E_{cm} \lesssim 300$ keV). While the $3\alpha \rightarrow {}^{12}C$ reaction appears to be fairly well understood[2] (stellar rate known to $\sim 15\%$), the $^{12}C(\alpha,\gamma)^{16}O$ reaction rate at helium burning energies still posesses considerable uncertainty. Also, while the $^{16}O(\alpha,\gamma)^{20}Ne$ reaction rate is thought to be relatively unimportant for helium burning, this is based almost entirely on theoretical estimates of the direct capture cross section. In this contribution I will discuss new results for the latter two reactions with particular emphasis on new experimental work at Caltech.

Looking first at the $^{16}O(\alpha,\gamma)^{20}Ne$ reaction; the stellar reaction rate, as mentioned above, is expected to be dominated by non-resonant direct capture. A measurement at higher energies (where the cross section is accessible to experiment) would allow a determination of the normalization of the cross section which is the most difficult quantity to determine theoretically. We have performed such a measurement using a new apparatus which detects both particles in the final state (γ-ray and ^{20}Ne recoil), providing high sensitivity for small cross sections. The apparatus, shown in Fig. 1, was designed specifically to study the

$^{12}C(\alpha,\gamma)^{16}O$ reaction but because of the similarity in the kinematics (with the heavy-ion as the incident particle) the $^{16}O(\alpha,\gamma)^{20}Ne$ reaction can also be investigated. Briefly, by using the heavy-ion as the incident beam, the recoil ion is kinematically constrained to a small forward cone with an opening angle $\lesssim 3°$. The reaction γ rays are detected with large NaI(Tl) crystals while the recoil ions are separated from the main beam by an ExB velocity filter and detected by an EΔE ionization detector in time-of-flight coincidence with the γ ray.

Fig. 2 shows the time-of-flight spectra obtained at both the highest and lowest yield points in the measurement, demonstrating that the coincidence requirement provides an essentially background-free signature of the reaction. By measuring the cross section in the energy range $E_{cm} = 1.7-2.4$ MeV the S-factor for capture into the ^{20}Ne ground state could be determined. The measurements are shown in Fig. 3 along with a prediction of the S-factor, made before the experiment, by Langanke [3]. The new measurements, along with theoretical calculation of the energy dependence of the cross section[3,4] and γ-ray branching ratios lead to a best estimate for the S-factor for $^{16}O(\alpha,\gamma)^{20}Ne$ of 0.7 ± 0.3 MeV-b at $E_{cm} = 300$ keV. Details of this experiment will be published shortly[5].

The apparatus described above has also been used to study the $^{12}C(\alpha,\gamma)^{16}O$ reaction. Previous measurements[6,7,8] have relied on detection of the capture γ ray alone and have therfore been confronted with considerable background from beam-induced γ rays as well as from cosmic rays and natural radioactivities. In the present experiment the electric dipole (E1) part of the cross section has been determined by using the velocity dispersion of the velocity filter described above to preferentially select recoils which are associated with E1 γ rays. The E1 cross section from the present experiment is compared with previous measurements in Fig. 4. The agreement between the experiments is reasonable, indicating that the cross section is well understood in the energy range of $E_{cm} = 1.0-3.0$ MeV. However, the interesting question for astrophysics is how well do we know the reaction rate at stellar energies ($E_{cm} = 300$ keV).

The reaction rate at stellar energies is expected to be dominated by the contribution of a (1^-) bound state in ^{16}O lying just below zero relative energy for $^{12}C(\alpha,\gamma)^{16}O$. Thus the challenge is to use the cross section measured at higher energies to constrain the paramters (in particular the α width) of the bound state. This is possible in principle by observing the interference of this state with the broad resonance seen at E_{cm} = 2.4 MeV (see Fig. 4). However there is also likely to be some capture amplitude due to 1^- states at even higher excitation energies. In order to gain some information on these various contributions we have, following the previous measurements[6,8], performed a three level R-matrix analysis[9]. The three levels correspond to the $E^* = 7.12$ MeV bound state, the 9.5 MeV unbound state (the dominant feature in Fig. 4), and a level at higher excitation energy (called the "background" level) to account for all higher lying 1^- states.

Three parameters are required to define each level: an α width, a γ width, and an eigenenergy. Two of these parameters can be expressed in terms of the others[10], and the eigenenergy of the third state is also fixed. This leaves six parameters to be determined from the data. Some of these same parameters can be constrained by calculating the $l = 1$ phase shift for α-^{12}C elastic scattering and comparing with the measurements. In fact, new data has recently become available for the elastic scattering phase shifts[11]. By fitting the parameters of the 3 level R-matrix analysis to both the $^{12}C(\alpha,\alpha)^{12}C$ data[11,12,13] and the $^{12}C(\alpha,\gamma)^{16}O$ data from the present measurement simultaneously, we obtain an allowed range for the S-factor at 300 keV of 0.00 - 0.14 MeV-b. The χ^2 for the fit is actually minimized at a value of 0.01 MeV-b, but the χ^2 dependence on S(300) is quite flat for the above range. The allowed range was deduced by observing a 30% increase in χ^2 from its minimum value of 1.3 per data point, where χ^2 is calculated for both the elastic scattering data as well as the present data. This value can be compared with two previous determinations of the E1 S-factor at 0.3 MeV using a similar analysis which yielded ranges of 0.02 - 0.28

MeV-b[6] and 0.02 - 0.48 MeV-b[8]. In all cases the large range in the allowed values results from the correlation of the strength of the bound state with that of the background level. A large value for the background strength and a small value for the bound state strength as well as the converse give a reasonable description of the measurements.

Several attempts have been made to characterize the background strength in order to eliminate this parameter and thus reduce the uncertainty in the extrapolation. In the hybrid R-Matrix[14] analysis the 9.5 MeV state and the background strength are both described by a single potential. Barker[15] has argued that the contribution to the $^{12}C(\alpha,\gamma)^{16}O$ reaction from the background state is negligible in the energy range of the measurements (i.e $\Gamma_\gamma = 0$ for this state) also eliminating this as a free parameter. For comparison, these analyses yield best fit values (fitting both the elastic scattering and the α capture data from the present experiment simultaneously) for $S_{E1}(300)$ of 0.08 MeV-b for the hybrid calculation[16] and 0.14 MeV-b for a three level R-matrix fit with $\Gamma_\gamma = 0$ for the background state. These fits are shown in Fig. 5 together with the data for the present experiment. In addition there is likely to be a contribution at stellar energies from E2 capture due primarily to a 2^- bound state near zero energy in $^{12}C(\alpha,\gamma)^{16}O$. Data taken at higher energies[8] indicates that this contribution may amount to $\simeq 50\%$ of the E1 cross section, but with considerable uncertainty.

A combined analysis of the various experiments is underway[17] using the techniques described above in order to better limit the allowed range in $S_{E1}(300)$. New measurements of the $^{12}C(\alpha,\gamma)^{16}O$ reaction at *higher* energies may also prove useful as a means of experimentally characterizing the background strength and thus reducing the uncertainty in $S_{E1}(300)$.

ACKNOWLEDGMENTS

I would like to acknowledge my experimental collegues in this work: C. A. Barnes, K. H. Chang, T. R. Donoghue, H. C. Evans, K. H. Hahn, R. M. Kremer, and L. W. Mitchell; as well as helpful discussions with W. A. Fowler, S. E.

Koonin, and K. Langanke.

REFERENCES

[1] See for example: W. D. Arnett and F.-K. Thielemann, *Astrophys. J.* **295**, 589 (1985); F.-K. Thielemann and W. D. Arnett, *Astrophys. J.* **295**, 604 (1985); S. E. Woosley, *Proc. Accelerated Radioactive Beams Workshop, TRI-85-1*, L. Buchman and J. M. d'Auria eds. Vancouver:TRIUMF (1985)

[2] C. A. Barnes, *Essays in Nuclear Astrophysics* C. A. Barnes, D. D. Clayton, and D. N. Schramm eds. Cambridge: Cambridge Univ. (1982)

[3] K. Langanke, *Z. Phys.* **A317**, 325 (1984)

[4] P. Descouvemont and D. Baye, *Phys. Lett.* **127B**, 286 (1983)

[5] K. Hahn, et al., *Phys. Rev.* **C** in press (1987)

[6] P. Dyer C. A. Barnes, *Nucl. Phys.* **A233**, 495 1974

[7] K. U. Kettner, et al., *Z. Phys.* **A308**, 73 1982

[8] A. Redder et al., *Nucl. Phys.* **A462**, 385 (1987)

[9] A. M. Lane and R. G. Thomas, *Rev. Mod. Phys.* **30**, 269 (1958)

[10] D. C. Weisser et al., *Nucl. Phys.* **A235**, 460 (1974)

[11] R. Plaga et al., *Nucl. Phys.* **A465**, 291 (1987)

[12] C. M. Jones, et al., *Nucl. Phys.* **37**, 1 (1962)

[13] G. J. Clark, et al. *Nucl. Phys.* **A110**, 481 (1968)

[14] S. E. Koonin, et al., *Nucl. Phys.* **A220**, 221 (1974)

[15] F. C. Barker *Aust. J. Phys.* **40**, 25 (1987)

[16] K. Langanke, Private Communication

[17] K. Langanke and B. W. Filippone, to be published

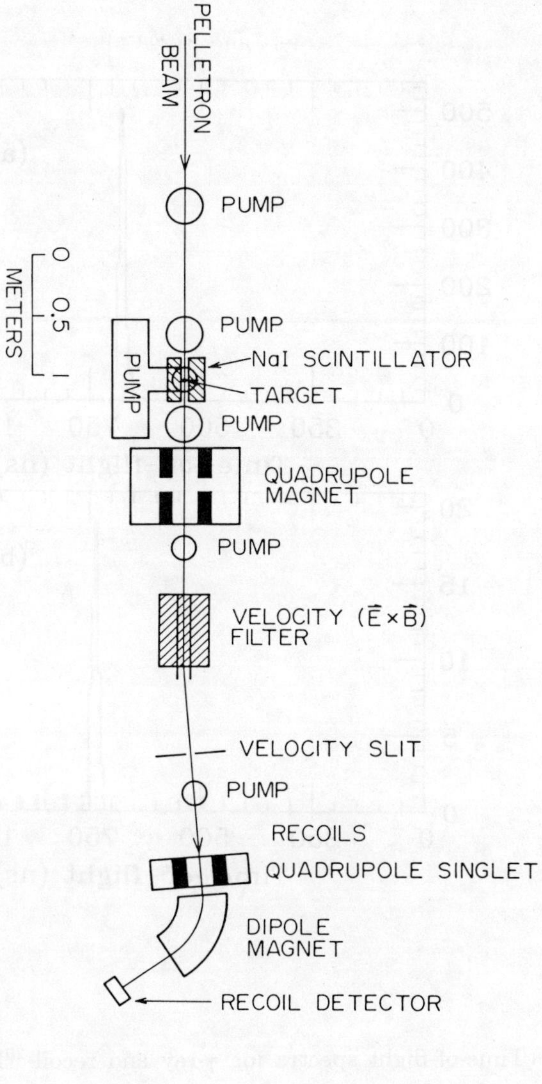

Fig. 1. Schematic diagram of the experimental apparatus.

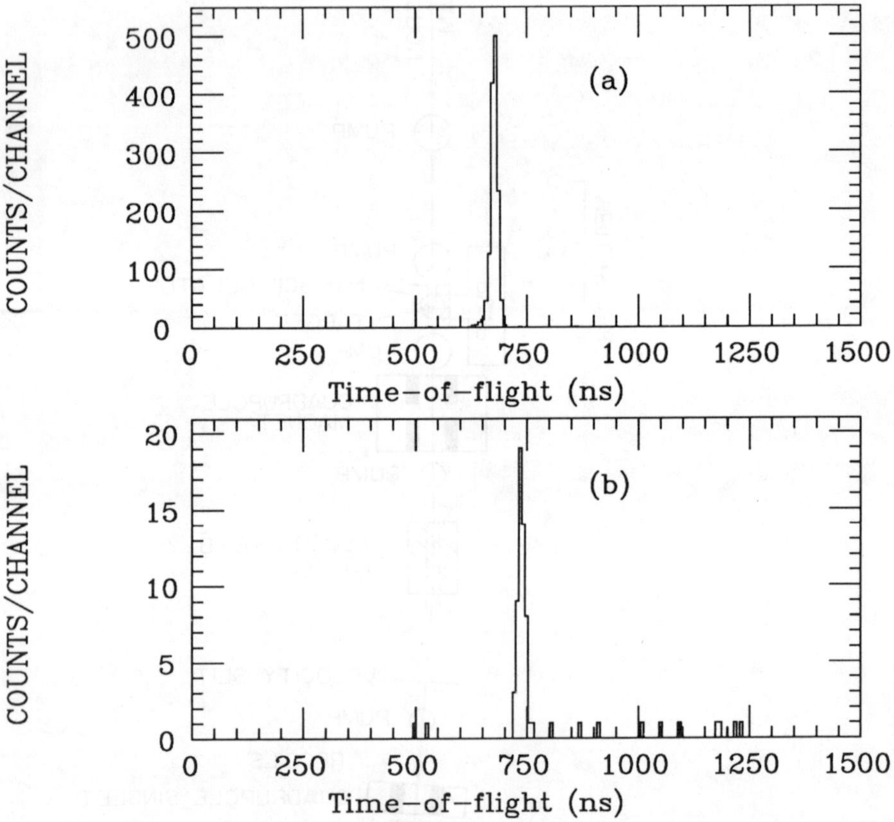

Fig. 2. Time-of-flight spectra for γ-ray and recoil-^{20}Ne coincidence from the $^{16}O(\alpha,\gamma)^{20}Ne$ reaction: (a) E_{cm} = 1.99 MeV; (b) E_{cm} = 1.70 MeV.

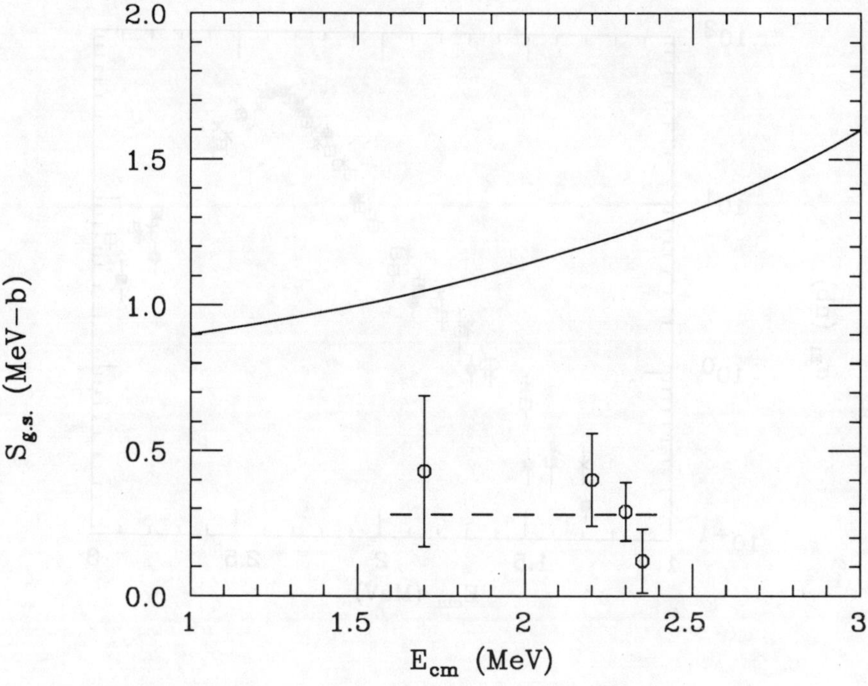

Fig. 3. S-factor for capture into the ^{20}Ne ground state from $^{16}O(\alpha, \gamma)^{20}Ne$. The solid curve is the prediction of Langanke[3] and the dashed curve is the weighted average of the measured values.

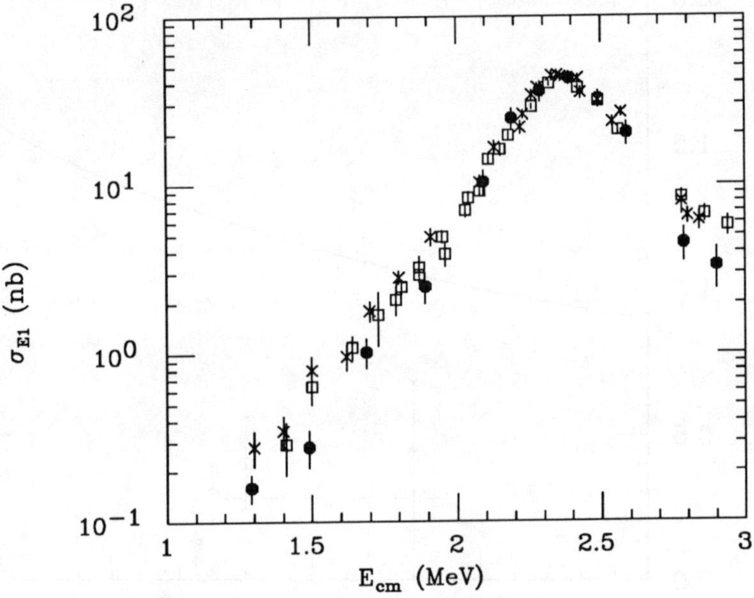

Fig. 4. E1 cross sections for the $^{12}C(\alpha,\gamma)^{16}O$ reaction for γ decay to the ^{16}O ground state. The open squares are from Ref. 6, the crosses from Ref. 8 and the solid points from the present experiment.

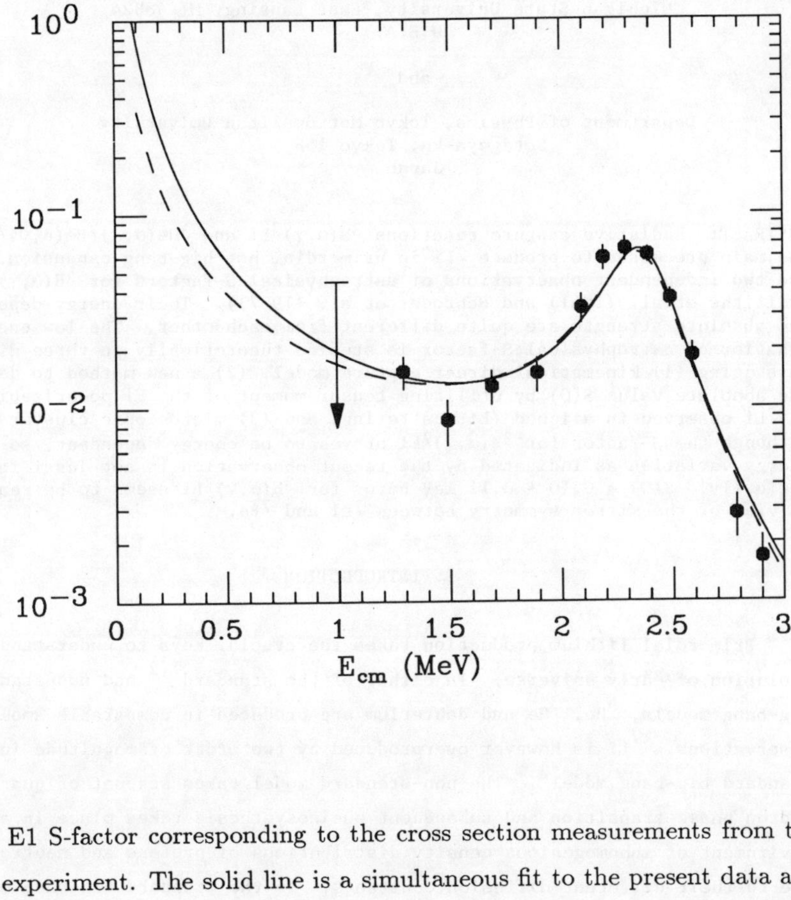

Fig. 5. E1 S-factor corresponding to the cross section measurements from the present experiment. The solid line is a simultaneous fit to the present data and the elastic scattering measurements of Refs. 11,12,13 using the procedure of Barker[15]. The dashed line is the fit using the hybrid R-matrix analysis discussed in the text.

ASTROPHYSICAL S-FACTOR FOR RADIATIVE CAPTURE REACTIONS --- LITHIUM AND BERYLLIUM FORMATION ---

Toshitaka Kajino

National Superconducting Cyclotron Laboratory
Michigan State University, East Lansing, MI 48824
U.S.A.

and

Department of Physics, Tokyo Metropolitan University
Setagaya-ku, Tokyo 158
Japan

ABSTRACT: Radiative capture reactions $^3H(\alpha,\gamma)^7Li$ and $^3He(\alpha,\gamma)^7Be(e,\nu)^7Li$ are the main processes to produce ^7Li in primordial hot big-bang expansion. There are two independent observations of astrophysical S-factors for $^3H(\alpha,\gamma)^7Li$ (by Griffiths et al. (1961) and Schroder et al. (1987)). Their energy dependence and absolute strength are quite different from each other. The low energy behavior of astrophysical S-factor is studied theoretically in three different ways using (1) kinematical direct capture model, (2) a new method to determine the absolute value S(0) by utilizing tensor moment of the El polarizability of ^7Li observed in aligned ^7Li scattering, and (3) microscopic cluster model. Although the S-factor for $^3H(\alpha,\gamma)^7Li$ proves to be energy dependent, so strong energy variation as indicated by the recent observation is not justified theoretically. S(0) = 0.10 - 0.11 keV barns for $^3H(\alpha,\gamma)^7Li$ seems to be reasonable in view of the mirror symmetry between ^7Li and ^7Be.

1. INTRODUCTION

Primordial lithium production takes the crucial keys to understand the evolution of early universe. In either of the standard[1,2] and non-standard[3] big-bang models, ^4He, ^3He and deuterium are produced in comparable amounts to observations. ^7Li is however overproduced by two order of magnitude in non-standard big-bang model[3]. The non-standard model takes account of quark hadron phase transition and subsequent nucleosynthesis takes place in an envirnment of inhomogenious density distributions of protons and neutrons due to their different diffusion constants. In the standard big-bang model[2], on the other hand, all these light elements are produced systematically with $\Omega_B \sim 0.1$, assuming homogenious energy and matter density distributions. ($\Omega_B = \rho_B/\rho_C$, $\rho_C = 3H_0^2/8\pi G$ and $\rho_C = 5.7\ 10^{-30}$ g/cm^3. ρ_B = baryon mass density.)

There are a few essential parameters in the big-bang model. They are the baryon density ρ_B and the number of light neutrino flavors N_ν (and, in addition, the well-known neutron half life). Cosmological interest lies in constraining these parameters[2-4].

The radiative capture reaction of triton and alpha particles mainly mediates the ^{7}Li production at the lower baryon densities $\rho_B < 10^{-31}$ g/cm^3 in the standard scinario. At the higher baryon densities 10^{-31} g/cm$^3 < \rho_B$, the mirror radiative capture reaction of alpha and ^{3}He followed by electron capture dominates the ^{7}Li production. In the non-standard big-bang scinario, only the production sites differ from the standard model; $^{3}H(\alpha,\gamma)^{7}Li$ takes place mostly in neutron-rich region and $^{3}He(\alpha,\gamma)^{7}Be(e,\nu)^{7}Li$ in proton-rich region. The calculated primordial ^{7}Li abundances depend on these reaction rates strongly, and it is necessary to know their accurate values in order to constrain the baryon density and the number of neutrino flavors precisely.

Two independent nuclear measurements[5,6] of the reaction rate for $^{3}H(\alpha,\gamma)^{7}Li$ indicate about a factor of two difference from each other, leading to a factor of two uncertainty[7,8] in primordial ^{7}Li production. Although the previous data of Griffiths et al.[5] had large error bars and their central values of S-factor look like energy independent, the recent measurement[6] has found a relatively strong energy variation as the incident energy goes lower. As a result, new S(0) value is twice as large as that of Griffiths' data. The energy dependence is a very important finding because such a behavior of S-factor has been predicted theoretically[9,10]. However, the observed energy dependence is too strong, compared with theoretical calculations in the microscopic cluster model[10-12]. The new data are in agreement with kinematical model calculation done by Williams and Koonin[9], and this theoretical curve was used in extrapolating the data to zero energy. However, their theoretical calculation has proved to be wrong[13].

I have hence decided to study the astrophysical S-factor for the two radiative capture reactions theoretically. The purpose of this study is to remove uncertainty from the nuclear reaction data. This paper is organized as follows. First, in the next section, the energy dependence of S at the low energies is calculated in the kinematical direct capture model[9,13]. Although this model is extended to any radiative capture processes leading to weakly bound systems, I discuss only the $^{3}H(\alpha,\gamma)^{7}Li$ and $^{3}He(\alpha,\gamma)^{7}Be$ reactions in this article. Absolute strength of S is not predicted in the kinematical model. In sect 3, I propose a new method[14] to estimate the absolute strength of the S-factor for $^{3}H(\alpha,\gamma)^{7}Li$. This should be compared with the previous direct capture γ-ray measurements. In this new method the electric dipole polarizability observed in aligned ^{7}Li scattering is used to determine the S-factor. A theoretical calculation in the microscopic cluster model[10-12] also is helpful as another way to estimate the absolute strength of S. This model is essentially free from any fitting parameter such as an overall normalization of the calculated S-factor to observation. In sect. 4, both of the energy dependence and absolute strength of the S-factor are discussed in the microscopic cluster model. Finally, in sect. 5, the conclusion of the present study is shown.

2. ENERGY DEPENDENCE AT LOW ENERGIES[13]

Astrophysical S-factor is defined by

$$S(E_{cm}) = E_{cm}\exp(2\pi\eta)\,\sigma(E_{cm}), \qquad (1)$$

where $\eta = Z_1Z_2e^2/\hbar v$ and $\sigma(E_{cm})$ is the radiative capture cross section at the incident energy E_{cm}. Since the electric dipole transition dominates the low energy (E_{cm} < 2 MeV) reactions $^3H(\alpha,\gamma)^7Li$ and $^3He(\alpha,\gamma)^7Be$ by more than 98%[10], the total capture cross section is approximated to be

$$\sigma(E_{cm}) = \frac{1}{2}\sum_{J_iL_iJ_f}\frac{16\pi}{9}\frac{1}{\hbar}\left\{\frac{E_{cm}+E_{th}(J_f)}{\hbar c}\right\}^3 \left|\langle\Psi_{J_f}(A{=}7)\,\|\,M(E1)\,\|\,\Psi_{J_i}(\alpha{+}(A{=}3))\rangle\right|^2, \qquad (2)$$

where M(E1) is the electric dipole operator. In this study, we have assumed the spin-orbit coupling scheme for the nuclear wave functions in the cluster model, which is a good approximation[10-12,15] to the seven nucleon systems $^3H + \alpha$ and $^3He + \alpha$;

$$\Psi_{JM} = \sqrt{\frac{4!3!}{7!}}\,A\,\{[\phi(\xi_\alpha)\,\phi^{(\frac{1}{2})}(\xi_3)]\otimes i^L Y^{(L)}(\hat{r})\chi_{JL}(r)\}^{(J)}_M . \qquad (3)$$

A is the antisymmetrizer of nucleons, ϕ_α and ϕ_3 are the intrinsic wave functions of two interacting clusters alpha and triton (or ^3He), ξ_α and ξ_3 are the corresponding internal coordinates and spin-isospin variables, and $\chi(r)$ is the relative wave function with respect to the intercluster distance r. At astrophysical low energies E_{cm} < 300 keV, two charged particles alpha and triton (or ^3He) have a small probability to find each other in the interaction region due to the Coulomb barrier. Therefore, the radiative capture process takes place in the external region[12] 5 fm < r < 30 fm far beyond the range of nuclear force r ∿ 3 fm. In this region the antisymmetrizer A is approximated to be unity[†] and only the kinematical condition describes the energy dependence of incident scattering waves. Since the nuclear phase shifts for the $^3H + \alpha$ and $^3He + \alpha$ scatterings are known[16] to be described by scattering from a charged hard sphere of radius $r_0 = 2.8$ fm for s and d waves at low energies $E_{cm} < 5$ MeV, the scattering waves are written as

$$\chi_{JL}(k,r) \sim \frac{1}{\sqrt{v}} \{F_L(k,r) - \frac{F_L(k,r_0)}{G_L(k,r_0)} G_L(k,r)\}/kr, \tag{4}$$

in the external region, where F_L and G_L are the regular and irregular Coulomb functions.

There are two sources having an energy dependence of astrophysical S-factor (1). They are the photon phase space factor $(E_{cm} + E_{th})^3$ and the initial scattring waves. The trivial factor $E_{cm}\exp(2\pi\eta)$ in eq. (1) cancels out the penetrability of scattering waves (4) as will be shown later.

The Coulomb functions are expanded analytically in terms of non-dimensional energy dependent factor,

$$\eta^2 = E_0/E_{cm}, \tag{5}$$

where

$$E_0 = \frac{\mu}{2}\{\frac{Z_1 Z_2 e^2}{\hbar}\}^2, \tag{6}$$

is introduced[13] as a natural energy scaling. $E_0 = 170$ keV and 680 keV for the $^3H(\alpha,\gamma)^7Li$ and $^3He(\alpha,\gamma)^7Be$ reactions. Among many expansion formulae, we adopt the following two methods: The first is the power series expansion[18]

$$\begin{Bmatrix} F_0 \\ G_0 \end{Bmatrix}(k,r) = \sqrt{\frac{2\pi\eta}{e^{2\pi\eta}-1}} \sum_{m=0} \rho \begin{Bmatrix} f_m(\eta\rho) \\ g_m(\eta\rho)/C_0^2(\eta) \end{Bmatrix} (1/\eta^2)^m, \tag{7}$$

---

† This is not true for the bound states of ^7Li and ^7Be because the absolute asymptotic amplitudes of the $^3H + \alpha$ and $^3He + \alpha$ clustering components are affected strongly by the antisymmetrizer in the interaction region. However, once we solve the Schrodinger eq. of motion for seven nucleon systems as in the resonating group method, A is approximated to be a norm kernel the eigenvalues of which are slightly different from unity. Since we are interested in the energy dependence of incident wave functions alone, this approximation is justified.

where f_m is a polynomial expression of $\eta\rho$, and g_m is a function of $\ln(2\eta\rho)$ and a power series of $\eta\rho$. Note that $\eta\rho = \eta \times kr$ is energy independent. The second is the expansion[19] in terms of modified Bessel-Clifford functions I_k and K_k as

$$\begin{Bmatrix} F_0 \\ G_0 \end{Bmatrix}(k,r) = \sqrt{\frac{2\pi\eta}{e^{2\pi\eta}-1}} \sum_{m=0} \rho \sum_k \begin{Bmatrix} \tilde{f}_{km}(2\sqrt{2\eta\rho})\ I_k(2\sqrt{2\eta\rho}) \\ \tilde{g}_{km}(2\sqrt{2\eta\rho})\ K_k(2\sqrt{2\eta\rho})/C_0^2(\eta)\rho \end{Bmatrix} (1/\eta^2)^m, \qquad (8)$$

where $\tilde{f}_{km}$ and $\tilde{g}_{km}$ are functions of $2\sqrt{2\eta\rho}$. Extension to higher partial waves is easy. $e^{2\pi\eta} \gg 1$ at astrophysical energy, and the factor $C_0(\eta) = \sqrt{2\pi\eta/(e^{2\pi\eta}-1)} \cong (2\pi\eta/e^{2\pi\eta})^{\frac{1}{2}}$ together with wave function normalization $1/\sqrt{v}$ appearing in eq. (4) countervalance the trivial energy dependence $E_{cm}\cdot\exp(2\pi\eta)$ in eq. (1). A singular factor $1/C_0^2(\eta)$ in the expression of $G_0(k,r)$ does not matter in the present parametrization of the scattering waves (4). The energy dependence of the scattering waves is carried only by $(1/\eta^2)^m = (E_{cm}/E_0)^m$. This is a simple form making it easy to estimate the energy dependence of S-factor in terms of energy power series. It is to be noted that the present expansion is applied to the lower energies than the natural energy scaling, $E_{cm} < E_0$. The final bound state wave functions are approximated to be Whittacker function. Using Whittacker function for $\Psi_{J_f}(A=7)$ and eq.(8) for $\Psi_{J_i}(\alpha+(A=3))$, the reduced matrix element in eq. (2) proves to converge to all orders m of energy expansion. To ensure this convergence (regarding an

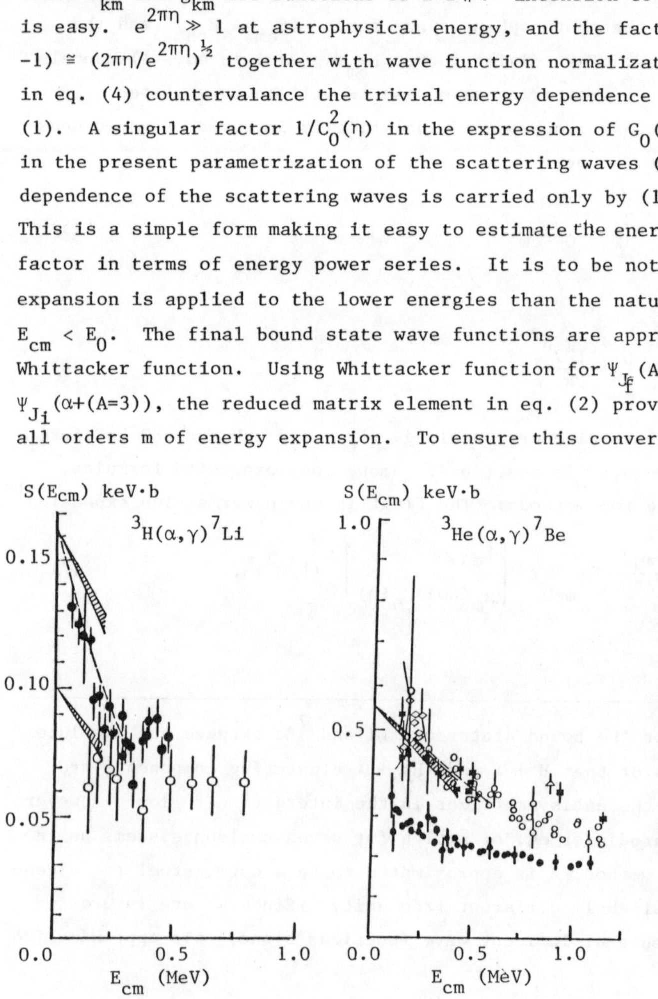

Fig. 1 Energy dependence of astrophysical S-factor for $^3H(\alpha,\gamma)^7Li$ and $^3He(\alpha,\gamma)^7Be$. Solid shaded area are the present results, and dashed lines are the results of Williams and Koonin. Open and closed circles are from refs. 5 and 6 for $^3H(\alpha,\gamma)^7Li$. Open and closed circles, open and closed squares, triangle and cross are from Parker and Kavanagh, Krawinkel et al., Nagatani et al., Osborne et al., Robertson et al. and Alexander et al. in ref. 10 for $^3He(\alpha,\gamma)^7Be$.

integral with respect to r), we have performed the numerical integration over $\eta\rho_0 < \eta\rho < 5$ for $^3H(\alpha,\gamma)^7Li$ and $\eta\rho_0 < \eta\rho < 13$ for $^3He(\alpha,\gamma)^7Be$. Here, $\rho_0 = kr_0$, and the difference of upper limit of the argument $\eta\rho$ comes from weaker binding of the $^3He + \alpha = {}^7Be$ system than the $^3H + \alpha = {}^7Li$ system from the cluster thresholds.

The calculated energy dependence of the S-factor is shown in fig. 1 and compared with observations. In order to see the effects of r_0 on the variation of $S(E_{cm})$, the core radius r_0 is varied as $r_0 = 2.5 - 3.0$ fm, corresponding to the shaded region. Two expansion methods (7) and (8) give exactly the same result for the both radiative capture reactions. The normalization is arbitrary. Although the present result[13] for the $^3He(\alpha,\gamma)^7Be$ reaction agrees marginally with calculation of Williams and Koonin[9], the results for the mirror $^3H(\alpha,\gamma)^7Li$ reaction differ remarkably from each other. When one expresses the energy variation of astrophysical S-factor as

$$\begin{aligned} S(E_{cm}) &= S_{3/2^-}(E_{cm}) + S_{1/2^-}(E_{cm}) \\ &= S_{3/2^-}(0)[(1 + E_{cm}/E_{th}(3/2^-))^3(1 + a_1E_{cm} + a_2E_{cm}^2 + \cdots) \\ &\quad + R(1 + E_{cm}/E_{th}(1/2^-))^3(1 + b_1E_{cm} + b_2E_{cm}^2 + \cdots)], \end{aligned} \tag{9}$$

the logarithmic derivative of $S(E_{cm})$ is

$$\lim_{E_{cm}\to 0} \frac{d}{dE_{cm}} \ln S(E_{cm}) = \frac{1}{1 + R}[(a_1 + Rb_1) + (3/E_{th}(3/2^-) + R\cdot 3/E_{th}(1/2^-))], \tag{10}$$

where $R = S_{1/2^-}(0)/S_{3/2^-}(0)$ is the branching ratio at $E_{cm} = 0$ keV. Excluding the trivial photon phase space factor $(3/E_{th}(3/2^-) + R\cdot 3/E_{th}(1/2^-))/(1 + R)$ from eq. (10), the quantity $|(a_1 + Rb_1)/(1 + R)|$ stands for an average gradient of $S(E_{cm})$ which is a pure contribution from the reduced matrix elements in eq. (2).

$$|(a_1 + Rb_1)/(1 + R)|_{^3H(\alpha,\gamma)^7Li} < |(a_1 + Rb_1)/(1 + R)|_{^3He(\alpha,\gamma)^7Be} \tag{11}$$

should hold on this quantity because of the weaker binding of the $^3He + \alpha = {}^7Be$ system than the $^3H + \alpha = {}^7Li$ system. As summarized in table 1, the present calculation satisfies this relation, but Williams and Koonin's calculation does not. It is not advisable to use Williams and Koonin's calculation in order to extrapolate the observed $S(E_{cm})$ values towards zero energy for the $^3H(\alpha,\gamma)^7Li$ reaction.

The present result of energy dependence in fig. 1 is in reasonable agreement with the prediction in the microscopic cluster model[10-12,31] and potential model[16,32] calculations.

Table 1. Logarithmic derivative of $S(E_{cm})$ at E_{cm} = 0 keV, excluding photon phase space factor;

$$(a_1 + Rb_1)/(1 + R) \quad MeV^{-1}$$

r_0 = 2.8 fm is used. See text for detail.

	$^3H(\alpha,\gamma)^7Li$	$^3He(\alpha,\gamma)^7Be$
Williams and Koonin[9]	-3.34	-2.67
Kajino and Bertsch[13]	-2.45	-2.85

3. ABSOLUTE STRENGTH --- A NEW METHOD TO DETERMINE S(0) [14]

A recent high precision measurement[20] of tensor analysing powers T_{20} for aligned ^7Li scattering provides with a determination of the tensor moment of E1 polarizability τ_{11}.

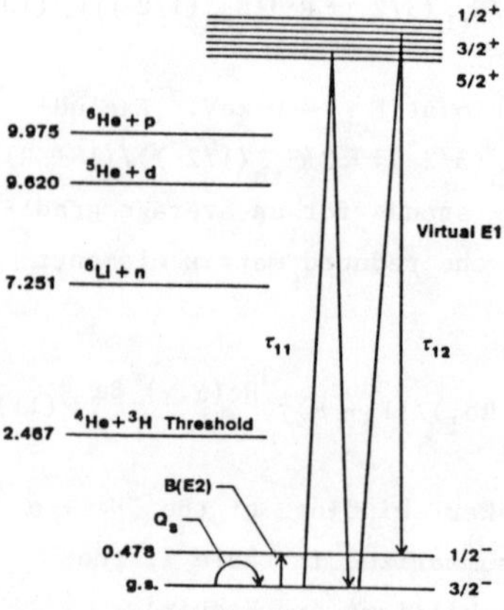

Fig. 2 Electric dipole and quadrupole couplings of ^7Li considered in the present study. See text for definition of τ_{if}.

The E1 polarizability τ_{11} of ^7Li is defined by

$$\tau_{11} = \frac{8}{9}\sqrt{\frac{10}{3}} \sum_n W(1\ 1\ 3/2\ 3/2;\ 2\ I_n) \frac{\langle 3/2^- \| M(E1) \| I_n^+ \rangle \langle I_n^+ \| M(E1) \| 3/2^- \rangle}{E_n - E_{3/2^-}}, \quad (12)$$

where $W(1\ 1\ 3/2\ 3/2;\ 2\ I_n)$ is the Racah coefficient, and $|3/2^-\rangle$ and $|I_n^+\rangle$ are the ground and excited states of ^7Li. There is no clear giant dipole resonance observed in ^7Li [21-23]. In addition, the E1 transition to the continuum breakup channels (fig. 2) at the threshold energy region are so strong as to exhaust molecular dipole sum rule[24]. The breakups to the continuum channels of ^{7}Li are therefore presumed to make a major contribution to the E1 polarizability;

$$\tau_{11} = \sum_c \tau_{11}^c, \tag{13}$$

where c spans over all physical continuum channels as shown in fig. 2, and the partial strength for the ^{7}Li(γ,t)α channel of astrophysical interest is expressed in terms of astrophysical E1 S-factor as[14]

$$\tau_{11}^{\alpha t} = \sqrt{\frac{10}{3}}\, \frac{(\hbar c)(\pi c^2)}{\pi^2} \sum_{I_n} (-)^{I_n - 3/2}\, W(1\ 1\ 3/2\ 3/2;\ 2\ I_n^+)$$
$$\int_0^\infty S(E_{cm}, E1;\ I_n^+ \to 3/2^-)\, \exp(-2\pi\eta)/(E_{cm} + E_{th})^4\, dE_{cm}. \tag{14}$$

$S(E_{cm},E1;I_n^+ \to 3/2^-)$ is defined by eqs. (1) and (2), and $E_n - E_{3/2^-} = E_{cm} + E_{th}$. Knowing $\tau_{11}^{\alpha t}$ with eq. (14), one can estimate the absolute strength of $S(E_{cm})$.

In order to estimate the partial strength $\tau_{11}^{\alpha t}$ from total τ_{11}, it is useful to consider the inverse energy-weighted photo-disintegration cross section $\sigma_{-2}^{\alpha t}$.

$$\sigma_{-2}^{\alpha t} = \int_{E_{th}}^\infty \left[\frac{\sigma_{dis}^{\alpha t}(1/2^+)}{E^2} + \frac{\sigma_{dis}^{\alpha t}(3/2^+)}{E^2} + \frac{\sigma_{dis}^{\alpha t}(5/2^+)}{E^2}\right] dE, \tag{15}$$

where $E = E_{cm} + E_{th}$ is the photon energy. Using the definition of photo-disintegration cross section $\sigma_{dis}^{\alpha t}$, $\tau_{11}^{\alpha t}$ is rewritten in terms of $\sigma_{dis}^{\alpha t}$ as

$$\tau_{11}^{\alpha t} = -\frac{\sqrt{5}}{3}\,\frac{\hbar c}{\pi^2} \int_{E_{th}}^\infty \left[\frac{\sigma_{dis}^{\alpha t}(1/2^+)}{E^2} + \frac{\sigma_{dis}^{\alpha t}(3/2^+)}{E^2}\right] dE, \tag{16}$$

in the LS-scheme. Two quantities have a strong similarity to each other except for an overall normalization and sign†. Photo-disintegration cross sections are fortunately observed[21-23] for all breakup channels of ^{7}Li. Hence, assuming the proportionality relation

---

†Integrand in the right hand side of eq. (15) has a braod maximum just above the threshold as shown in fig. 3. This is a general trend for any breakup channels of ^{7}Li, and one confirms the significance of the continuum contribution to the E1 polarizability, eq. (16).

$$\tau_{11}^{\alpha t} / \sum_c \tau_{11}^c \cong [\sigma_{-2}^{\alpha t} / \sum_c \sigma_{-2}^c]_{exp}, \quad (17)$$

and knowing the total E1 polarizability $\tau_{11} = \sum_c \tau_{11}^c$, one can estimate the partial strength $\tau_{11}^{\alpha t}$.

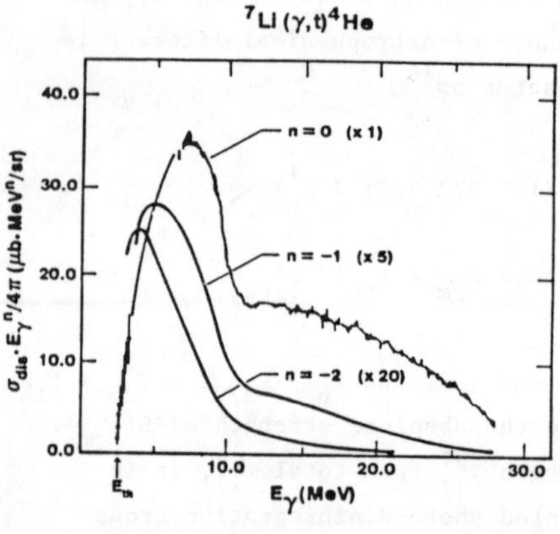

Fig. 3 Energy-weighted photo-disintegration cross sections for $^7Li(\gamma, t)^4He$;
$\sigma_{dis} \cdot E_\gamma^n / 4\pi$ (b·MeV^n/sr).
E_{th} = 2.467 MeV.

Let us go to the calculation of tensor analysing powers to determine the total τ_{11}. We have taken account of the polarization potential[25] of the second rank in the calculation of T_{20} for $^{120}Sn(^7\vec{Li}, ^7Li)$ and $^{58}Ni(^7\vec{Li}, ^7Li)$ at sub-Coulomb energies;

$$<I_i M_i | v_{pol}^{(2)} | I_f M_f>$$
$$= \{\frac{4\pi}{5} \frac{Z_T e}{r^3} <I_i \| M(E2) \| I_f>$$
$$- \sqrt{\frac{9\pi}{5}} \frac{(Z_T e)^2}{r^4} \tau_{if}\} \, g Y_\mu^{(\lambda)*}(\hat{s} \cdot \hat{r}). \quad (18)$$

$|IM>$ is the substate of ^7Li with total spin I = 3/2 or 1/2 and its projection M, and M(E2) is the electric quadrupole operator. τ_{if} is the E1 polarizability defined by eq. (12) representing the second order effects of virtual E1 transitions. $g = (I_f \; M_f \; 2 \; \mu | \; I_i \; M_i)$. Large electric quadrupole deformation of ^7Li dominates T_{20}, but the effects of E1 polarizability also contribute at the 15% level. The effects of nuclear force, higher multipoles, couplings with higher excited states of ^7Li, vacuum polarization and the relativistic correction are negligibly small at sub-Coulomb energies.[20,26,27] There are four parameters in the potential (18). They are the electric quadrupole moment eQ_s of the ground state, B(E2) strength for the E2 transition between the ground spin-doublets of ^7Li, and E1 polarizabilities τ_{11} and τ_{12}. The electric quadrupole moment[28] -4.06 (± 0.04) efm^2 is used to fix eQ_s. This value is determined by using molecular beam spectroscopy, and I believe that the inferred value is the most precise among many observations of Q_s. In the LS-coupling scheme, which has succeeded[10-12] in explaining the electric properties of ^7Li

and ^{7}Be, the following relations constrain the four independent parameters[14];

$$B(E2) = (1+\varepsilon)\frac{25}{16\pi}(eQ_s)^2, \qquad (19\text{-a})$$

$$\tau_{12} = (1+\eta)\,\tau_{11}, \qquad (19\text{-b})$$

and τ_{11} is negative. Here, $\varepsilon = 0.085$ and $\eta = 0.073$ are the correction terms arising from the threshold effects. Since the measured B(E2) values scatter depending on the experiments, eq. (19-a) is adopted to fix B(E2) together with $eQ_s = -4.06$ efm^2. With the relation (19-b), there remains only one parameter τ_{11} to be determined.

First, I calculated T_{20} by turning off the effects of El polarizability, i.e. $\tau_{11} = \tau_{12} = 0.0$. (Dotted curve in fig. 4.) The theoretical calculation overestimates the data. Putting in negative El polarizability $\tau_{11} = -0.269$ fm^3, the data are explained well by a destructive effects of the virtual El transitions of ^{7}Li to the continuum states. (Solid curve in fig. 4.)

Eq. (17) and the inferred value $\tau_{11} = -0.269$ fm^3 lead to

$$\tau_{11}^{\alpha t} = -0.088 \pm 0.035 \text{ fm}^3, \qquad (20)$$

as the partial strength of the El polarizability. Error comes from the uncertainty of the observed σ_{-2}^{c} in eq. (17). Inserting (20) into eq. (14) and assuming only the well established energy dependence of S-factor, the relative s/d-waves ratio and the branching ratio calculated in the microscopic cluster model, one obtains the astrophysical S-factor for ^{3}H$(\alpha,\gamma)^7$Li as shown in fig. 5 by solid curve. The S(0) value turns out to be

$$S(0) = 0.097 \pm 0.038 \text{ keV barns.} \qquad (21)$$

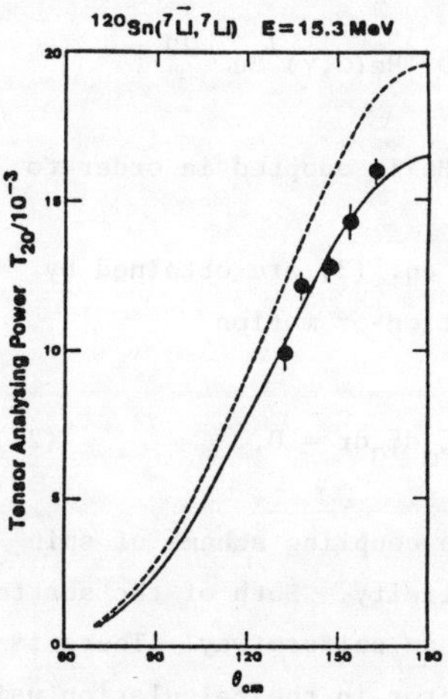

Fig. 4 Tensor analysing powers T_{20} for the elastic scattering of ^{7}Li from ^{120}Sn. Experimental data are from ref. 20. Solid and dotted curves are the calculated results with and without the effects of El polarizabilities of ^{7}Li, respectively.

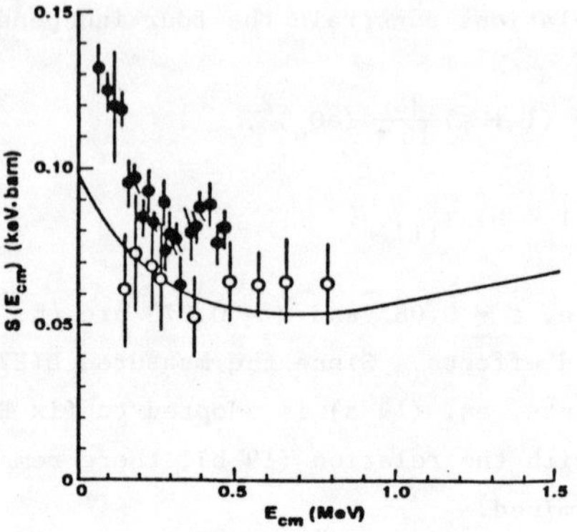

Fig. 5 Astrophysical S-factor for the $^3H(\alpha,\gamma)^7Li$ reaction. Open and closed circles are the experimental data respectively from refs. 5 and 6. Solid curve is the calculated result.

4. MIRROR SYMMETRY BETWEEN $^3H(\alpha,\gamma)^7Li$ AND $^3He(\alpha,\gamma)^7Be$ [29]

The microscopic cluster model for ^7Li and ^7Be is adopted in order to discuss the mirror symmetry between these nuclei.

The intercluster relative wave functions in eq. (3) are obtained by solving dynamically an integro-differential equation of motion

$$\int \phi^*(\xi_\alpha)\ \phi^*(\xi_3)\ \{H - E\}\ \ A\{\phi(\xi_\alpha)\ \phi(\xi_3)\ \chi(\vec{r})\}\ d\xi_\alpha d\xi_3 d\hat{r} = 0, \tag{22}$$

in the resonating group method (RGM)[30]. Here the coupling scheme of spin and orbital angular momentum is dropped for simplicity. Both of the scattering and bound state wave functions are solved in an uniform way. There is no artificial parameter to fit the observed S-factor in the calculation using these microscopic cluster model wave functions. Their success is reported in refs. 10-12.

The calculated absolute S(0) values in this model depend somehow on adopted effective nuclear forces in the Hamiltonian. However, the ratio of S(0)'s for the two mirror reactions is almost unchanged independently of interactions used;

$$S(0)^{^3H(\alpha,\gamma)^7Li}\ /\ S(0)^{^3He(\alpha,\gamma)^7Be} = 0.202 \pm 0.007. \tag{23}$$

This fact shows an existence of a good mirror symmetry between ^{7}Be and ^{7}Li.† As for the ^{3}He(α,γ)^{7}Be reaction, many established S-factors[17] observed in the different methods agree with one another (except for the measurement of Krawinkel et al.[17]). Their weighted average[4] excluding Krawinkel's data is

$$S(0)^{\,^{3}\mathrm{He}(\alpha,\gamma)^{7}\mathrm{Be}} = 0.56 \pm 0.03 \text{ keV barns}, \qquad (24)$$

and consistent with another independent estimate 0.54 ± 0.02 keV barns by Parker[37] and Bahcall[35]. Eqs. (23) and (24) lead to

$$S(0)^{\,^{3}\mathrm{H}(\alpha,\gamma)^{7}\mathrm{Li}} = 0.11 \pm 0.03 \text{ keV barns}. \qquad (25)$$

Schroder et al. have reported that the observed branching ratio $R \cong 0.3$ is smaller than the Griffiths' data $R \cong 0.4$ and almost all existing calculations[10-12,31,32] which are consistent with Griffiths' data. $R \cong 0.3$ is too small from the theoretical viewpoint[13]. I would like to discuss this problem briefly here.

When one extrapolates the observed $S_{3/2^-}(E_{cm})$ and $S_{1/2^-}(E_{cm})$ values to zero energy by using theoretical calculation of the energy dependence of S-factor, reduced widths $\theta^2_{3/2}$ and $\theta^2_{1/2}$ of the final bound states are used as free fitting parameters. Based on the semi-empirical potential model calculation, Tombrello and Parker[16] have succeeded in explaining the ^{3}He(α,γ)^{7}Be and Griffiths' ^{3}H(α,γ)^{7}Li data with the use of common reduced widths to both nuclei, i.e. $\theta^2_J(^{7}\mathrm{Be}) = \theta^2_J(^{7}\mathrm{Li})^{\dagger\dagger}$. This is a consequence of good mirror symmetry between ^{7}Be and ^{7}Li, being the reason why the theoretical calculations in the cluster model also can explain these data very well. If one tries to fit recent Schroder's ^{3}H(α,γ)^{7}Li and established ^{3}He(α,γ)^{7}Be data simultaneously, this symmetry has to be broken.

---

† The small breaking of mirror symmetry between ^{7}Li and ^{7}Be is taken into consideration[12] by using slightly different effective nuclear forces separately for the two nuclei in order to reproduce the observed cluster separation energies correctly. Coulomb force and such symmetry breaking force explain almost all differences observed between the ^{3}He + α and ^{3}H + α systems.

†† We are not interested in each magnitude of θ^2_J because it depends on the adopted potential model and the condition.

Alpha spectroscopic factors[36] $C^2S^{\alpha}_J$ also are used† as free fitting parameters instead of reduced widths θ^2_J. Schroder et al. have extracted quite different alpha spectroscopic factors from each other for the ground and first excited states of ^{7}Li in the analysis of their small branching ratio $R \cong 0.3$; $C^2S^{\alpha}_{1/2}/C^2S^{\alpha}_{3/2} = 0.75 \pm 0.15$. On the other hand, they have also extracted $C^2S^{\alpha}_{1/2}/C^2S^{\alpha}_{3/2} \sim 1.0$ from Griffiths' data $R \cong 0.4$, which is consistent with the potential model and microscopic cluster model analyses. The difference of $C^2S^{\alpha}_J$, if any, originates from the spin-orbit interaction that is responsible for the energy separation of the ground and first excited states of ^{7}Li. It is to be stressed that the theoretical models that are consistent with large branching ratio $R \cong 0.4$ describe the p-wave phase shifts $\delta_{3/2^-}$ and $\delta_{1/2^-}$ and their differences at low energies very well.

5. DISCUSSIONS AND CONCLUSION

The energy dependence of astrophysical S-factor for $^3H(\alpha,\gamma)^7Li$ calculated in the kinematical direct capture model[13] agrees well with the microscopic cluster model[10-12,31] and potential model[16,32] calculations at low energies. Williams and Koonin's calculation[9] is known to predict stronger energy variation than the others for this reaction. However, it is not justified theoretically.

There are two independent observations of the S-factor for $^3H(\alpha,\gamma)^7Li$. They are the data of Schroder et al.[6] and Griffiths et al.[5] Although Schroder's measurement indicates a similar energy dependence to Williams and Koonin's calculation at low energies $E_{cm} < 200$ keV and leads to a large value $S(0) = 0.162 \pm 0.024$ keV barns[6], this value is ruled out because the basis theoretical calculation in extrapolation is not reliable. Another extrapolation using the energy dependence predicted in the microscopic model leads to $S(0) = 0.134 \pm 0.020$ keV barns.[6] Griffiths' data lead to $S(0) = 0.100 \pm 0.025$ keV barns in the same extrapolation method.[4]

A new method[14] was proposed in order to determine the absolute streng-

† Although the ratios $\theta^2_{1/2}/\theta^2_{3/2}$ and $C^2S^{\alpha}_{1/2}/C^2S^{\alpha}_{3/2}$ extracted from the same data using different potential model calculations are compared with each other for some time, thses two quantities are not equivalent.

th of S-factor, which utilizes the tensor moment of E1 polarizability of ^{7}Li measured in sub-Coulomb $^7\vec{\mathrm{Li}}$ scattering. This method is quite different from the previous direct capture γ-ray measurements[5,6,17]. An absolute strength S(0) = 0.097 ± 0.038 keV barns is obtained in this method, being in a reasonable agreement with Griffiths' S(0) value.

Theoretical calculation in the microscopic cluster model[12] predicts the ratio of S-factors for the two mirror reactions ^{3}H(α,γ)^{7}Li and ^{3}He(α,γ)^{7}Be fairly precisely. Connecting this with the well examined[4] S(0) value[17] for ^{3}He(α,γ)^{7}Be, one obtains S(0) = 0.11 ± 0.03 keV barns for ^{3}H(α,γ)^{7}Li. What will happen in turn when one uses the theoretical ratio (23) of S-factors and three different S(0) values for the ^{3}H(α,γ)^{7}Li reaction, which were separately determined from Griffiths' data and Schroder's data using different extrapolation to zero energy? Griffiths' S(0) = 0.100 ± 0.025 keV barns for ^{3}H(α,γ)^{7}Li leads to S(0) = 0.50 ± 0.03 keV barns for ^{3}He(α,γ)^{7}Be. Schroder's S(0) = 0.162 ± 0.024 keV barns and 0.134 ± 0.020 keV barns for ^{3}H(α,γ)^{7}Li lead respectively to S(0) = 0.80 ± 0.025 keV barns and 0.66 ± 0.02 keV barns for ^{3}He(α,γ)^{7}Be. If the latter two values were used in the calculation of solar neutrino counting rate in the ^{37}Cl-detector, the discrepancy between theoretical (10.7 SNU and 9.2 SNU corresponding to S(0) = 0.80 and 0.66 keV barns for ^{3}He(α,γ)^{7}Be) and observed (2.0 ± 0.3 SNU)[33] neutrino counting rates might be even larger. The current theoretical value is 7.6 ± 1.1 SNU (ref. 34) or 6.5 ± 0.9 SNU (ref. 4). (A new Z/X value leads to 7.9 ± 0.87 SNU in ref. 35.)

To summarize, a finite energy variation of S-factor for the ^{3}H(α,γ)^{7}Li reaction is established both experimentally and theoretically. However, it is hard to theoretically explain so strong energy dependence at low energies E_{cm} < 150 keV as indicated by the recent measurement of Schroder et al. In view of the mirror symmetry between ^{7}Li and ^{7}Be, S-values S(0) = 0.10 - 0.11 keV barns for ^{3}H(α,γ)^{7}Li seem to be consistent with S(0) = 0.50 - 0.55 keV barns for the mirror reaction ^{3}He(α,γ)^{7}Be.

I acknowledge George F. Bertsch, Ken-ichi Kubo, Sam M. Austin, Fredelick C. Barker, and Hiroshi Toki for their stimulating discussions. This work is based on the collaborations with G.F. Bertsch (ref. 13) and with K. Kubo and G.F. Bertsch (ref. 14).

REFERENCES

1) R.V. Wagoner, Astrophys. J. 178 (1973) 343.
2) J. Yang et al., Astrophys. J. 281 (1984) 493.
3) C. Alcock, G.M. Fuller and G.J. Mathews, Astrophys. J. 320 (1987) 439; G.M. Fuller, G.J. Mathews and C. Alcock, preprint UCRL-97316 (1987).
4) T. Kajino, H. Toki and S.M. Austin, Astrophys. J. 319 (1987) 531.
5) G.M. Griffiths et al., Can. J. Phys. 39 (1961) 1387.
6) U. Schroder et al., Phys. Lett. B192 (1987) 55.
7) B.E.J. Pagel, Phil. Trans. R. Soc. Lond. A320 (1986) 557.
8) L. Kawano, D.N. Schramm and G. Steigman, (1987) preprint.
9) R.D. Williams and S.E. Koonin, Phys. Rev. C23 (1981) 2773.
10) T. Kajino and A. Arima, Phys. Rev. Lett. 52 (1984) 739; T. Kajino, Suppl. J. Phys. Soc. Japan 54 (1985) 321; T. Kajino, Lecture Notes in Physics 219 (Springer Verlag, New York, 1985) 85.
11) H. Walliser, H. Kanada and Y.C. Tang, Nucl. Phys. A419 (1984) 133.
12) T. Kajino, Nucl. Phys. A460 (1986) 559.
13) T. Kajino and G.F. Bertsch, to be published.
14) T. Kajino, K. Kubo and G.F. Bertsch, to be published in Phys. Rev. C.
15) T. Kajino, T. Matsuse and A. Arima, Nucl. Phys. A413 (1984) 323 and Nucl. Phys. A414 (1984) 185.
16) T.A. Tombrello and P.D. Parker, Phys. Rev. 131 (1963) 2582.
17) P.D. Parker and R.W. Kavanagh, Phys. Rev. 131 (1963) 2578; K. Nagatani, M.R. Dwarakanath and S. Ashery, Nucl. Phys. A128 (1969) 325; J.L. Osborne et al., Phys. Rev. Lett. 48 (1982) 1664 and Nucl. Phys. A149 (1984) 115; R.G.H. Robertson et al., Phys. Rev. C27 (1983) 11; H. Volk et al., Z. Phys. A310 (1983) 91; T. K. Alexander et al., Nucl. Phys. A427 (1984) 526; H. Krawinkel et al., Z. Phys. A304 (1982) 307.
18) M. Abramowitz, J. Math. Phys. 33 (1954) 111.
19) M. Abramowitz, J. Math. Phys. 29 (1950) 303; C. Froberg, Rev. Mod. Phys. 27 (1955) 399.
20) A. Weller et al., Phys. Rev. Lett. 55 (1985) 480.
21) G. Junghans et al., Z. Phys. A291 (1979) 353.
22) V.P. Denisov and L.K. Kul'chitskii, Sov. J. Nucl. Phys. 5 (1967) 344.
23) M.K. Leung et al., Can. J. Phys. 55 (1977) 252; D.M. Skopik et al., Phys. Rev. C20 (1979) 2025.
24) Y. Alhassid, M. Gai and G.F. Bertsch, Phys. Rev. Lett. 49 (1982) 1482.
25) K. Alder and A. Winther, Electromagnetic Excitation (North Holland, Amsterdam, 1975).
26) O. Hausser et al., Nucl. Phys. A212 (1973) 613 and Phys. Lett. 38B (1972) 75.
27) W.G. Lynch et al., Phys. Rev. Lett. 48 (1982) 979.
28) D. Sundholm et al., Chem. Phys. Lett. 112 (1984) 1.
29) T. Kajino, to be published.
30) J.A. Wheeler, Phys. Rev. 52 (1937) 1083 and 1107.
31) T. Mertelmeier and H.M. Hofmann, Nucl. Phys. A459 (1986) 387.
32) K. Langanke, Nucl. Phys. A457 (1986) 351.
33) R Davis. Jr., Proc. Informal Conf. on Status and Future of Solar Neutrino Research, ed. by G. Friedlander (Brookhaven National Laboratory Report No. 50879) Vol.1, p.1.
34) J.N. Bahcall et al., Rev. Mod. Phys. 54 (1982) 767.
35) J.N. Bahcall and R.K. Ulrich, (1987) preprint submitted to Rev. Mod. Phys.
36) C. Rolfs, Nucl. Phys. A217 (1973) 29.
37) P.D. Parker, Physics of the Sun, ed. by P.A. Sturrock (New York, D. Reidel Publ. Co., 1986) Vol.1, p.15.

NEW EXPERIMENTAL RESULTS FOR NUCLEAR REACTIONS IN EXPLOSIVE HYDROGEN BURNING

J. Görres and M. Wiescher

Department of Physics
University of Notre Dame
Notre Dame, IN 46556

ABSTRACT

We shall describe new experiments to derive the nuclear reaction rates of $^{22}Na(p,\gamma)^{23}Mg$ and $^{22}Mg(p,\gamma)^{23}Al$, important to the understanding of ^{22}Na nucleosynthesis in explosive hydrogen burning processes.

The understanding of rp-process nucleosynthesis [1] in the Ne-Na-Mg region is of particular interest for the interpretation of ^{22}Ne anomalies in meteoritic inclusions [2] as decay product of ^{22}Na ($T_{1/2}$=2.6y) after the freeze out period of the explosive event. Several reaction sequences, depending on the temperature and density conditions, may have lead to the formation of ^{22}Na. Most favorable is the Ne-Na cycle [3]

$$^{20}Ne(p,\gamma)^{21}Na(\beta^{+}\nu)^{21}Ne(p,\gamma)^{22}Na$$

where ^{22}Na is formed by proton capture reactions on Ne-isotopes, which are well known [4,5]. The abundance of ^{22}Na, however, depends also on the rates of possible depletion reactions, in this case $^{22}Na(p,\gamma)^{23}Mg$. Only theoretical estimates exist for this rate [1,6,7]. Another mechanism for producing ^{22}Na, more likely to occur at high temperature and density conditions [1,6], is

$$^{21}Na(p,\gamma)^{22}Mg(\beta^{+}\nu)^{22}Na.$$

In this case the production rate depends strongly also on the rate of the competing proton capture reaction on ^{22}Mg, $^{22}Mg(p,\gamma)^{23}Al$, which is not known as well [6].

In the following we will desribe attempts to determine experimentally the reaction rate of $^{22}Na(p,\gamma)^{23}Mg$ directly and, in an more indirect way, the rate of $^{22}Mg(p,\gamma)^{23}Al$ by studying the structure of the compound nucleus ^{23}Al.

I. A direct study of the capture reaction $^{22}Na(p,\gamma)^{23}Mg$

Little was known about the structure of the compound nucleus ^{23}Mg above the proton threshold at Q = 7.58 MeV when the first estimate of the rate was published [1]. Therefore considerable effort was raised to study the proton unbound levels in ^{23}Mg by $^{25}Mg(p,t)^{23}Mg$ [8] and $^{24}Mg(^{3}He,\alpha)^{23}Mg$ [9]. The experimental results are displayed in figure 1. Hovewer, the data do not allow a unique spin and parity assignment for the levels, making it difficult to derive a reasonable estimate for the $^{22}Na(p,\gamma)$ - reaction rate. Despite the lack of sufficient input information the rate was estimated [6,7], but due to the discussed uncertainties [7] an experimental verification is highly desirable.

The difficulty in measuring the $^{22}Na(p,\gamma)^{23}Mg$ reaction is caused by the extremly large γ-background due to the β^{+}-decay of ^{22}Na to the first excited state (98%) in ^{22}Ne. This creates γ-radiation from the annihilation of the positrons (0.511 MeV) and from the subsequent

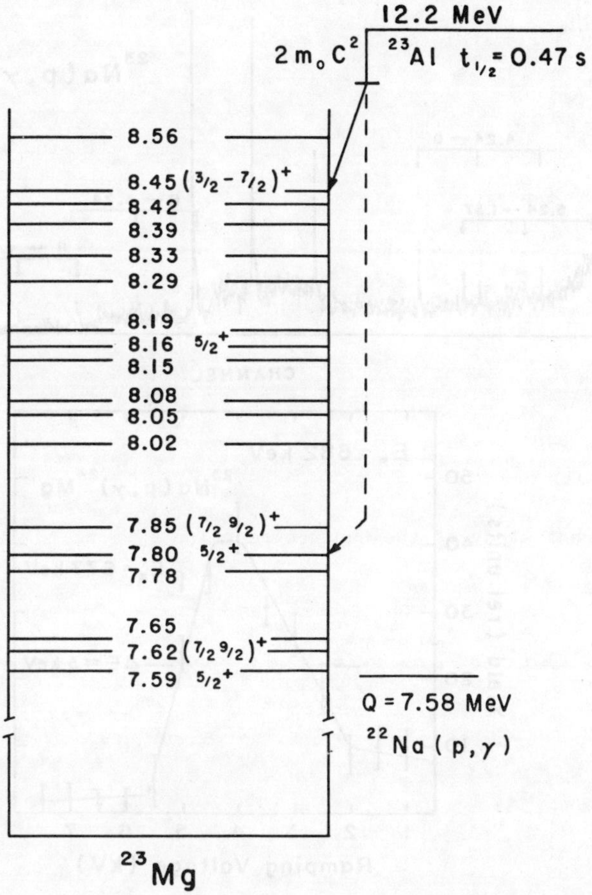

Fig.1. Level scheme for the proton unbound states in ^{23}Mg

γ-decay of the first excited state in ^{22}Ne to the ground state (1.275 MeV). The target material was commercially obtained (Amersham Corp) and the total target activity was 60 μCi, which corresponds to

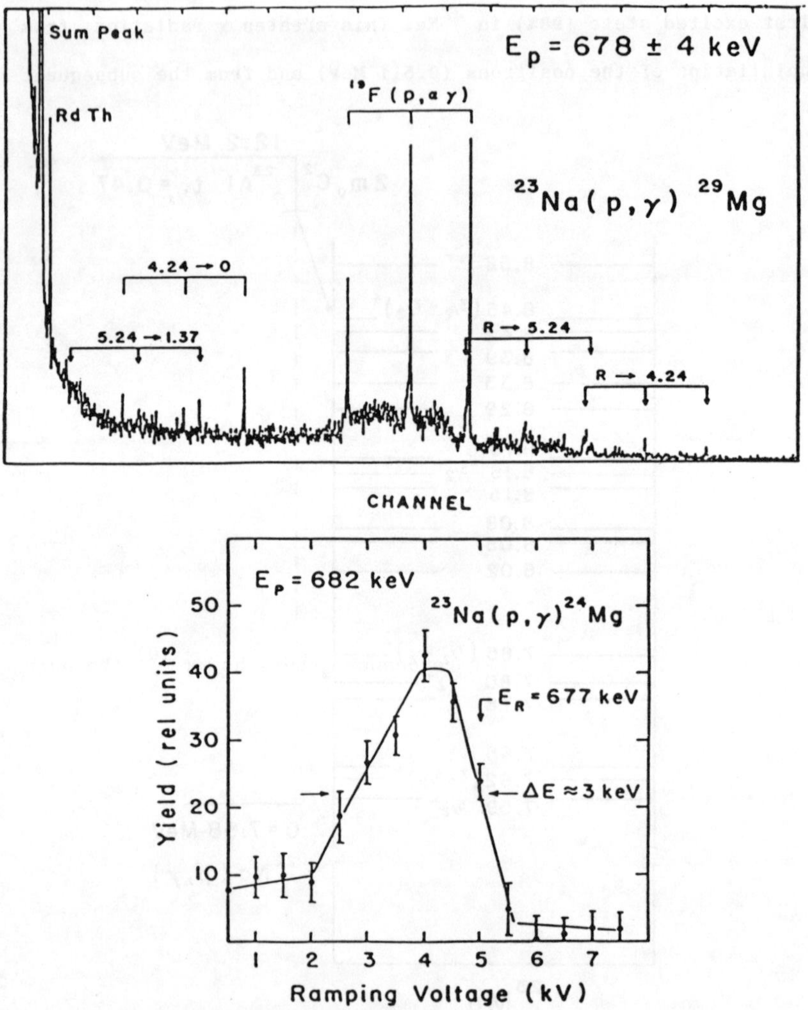

Fig.2. The lower part shows the yield curve over the 677 keV $^{23}Na(p,\gamma)^{24}Mg$ resonance. The associated γ-spectrum is shown above.

$1\cdot10^{15}$ ^{22}Na-atoms/cm^2. Scans of the target activity verified a homogeneous distribution of the ^{22}Na over the target area. Test experiments showed that the target could withstand beam currents of 10 μA (E_p = 700 keV) without noticable deterioration (<10%).

The experiments were performed with the 2MV Pelletron at the California Institute of Technology. We investigated the reaction in the bombarding energy range E_p = 0.40 - 1.25 MeV with beam currents of 8 - 30 μA. The γ-radiation was measured at 0° with a large volume Ge-detector shielded with 2.2cm of lead against the target activity. A low frequency ramping voltage of 8 kV on the target allowed us to cover an excitation range of 8 keV per run. Significant impurities in the target material (^{23}Na, Cl, Mg, Si) could be observed by measuring $^{23}Na(p,\gamma)$-, $^{35}Cl(p,\gamma)$-, $^{26}Mg(p,\gamma)$-, $^{29}Si(p,\gamma)$- and $^{30}Si(p,\gamma)$-resonances. From the known resonance strengths $\omega\gamma$ [10] the number of atoms for the particular impurities could be determined: $1\cdot10^{15}$ ^{23}Na, $1.2\cdot10^{15}$ Cl, $3.2\cdot10^{15}$ Mg and $1.1\cdot10^{17}$ Si atoms/cm^2. Figure 2 shows the yield curve and the γ-spectrum of the $^{23}Na(p,\gamma)^{24}Mg$ resonance at E_r = 0.677 MeV ($\omega\gamma$ = 1 eV). This allows the derivation of the target thickness and the stopping power since ^{23}Na in the target material behaves chemically as ^{22}Na.

The excitation curve was measured in energy steps of 8 keV collecting a charge of 0.1 - 0.2 Cb per run. The obtained γ-spectra were analyzed in terms of all possible γ-transitions to bound states in ^{23}Mg below E_x = 4.4 MeV. Figure 3 shows two typical spectra, obtained at the energy E_p = 0.614 MeV, where a strong 1 eV resonance was expected [7], and at E_p = 0.796 MeV as background run. Indicated

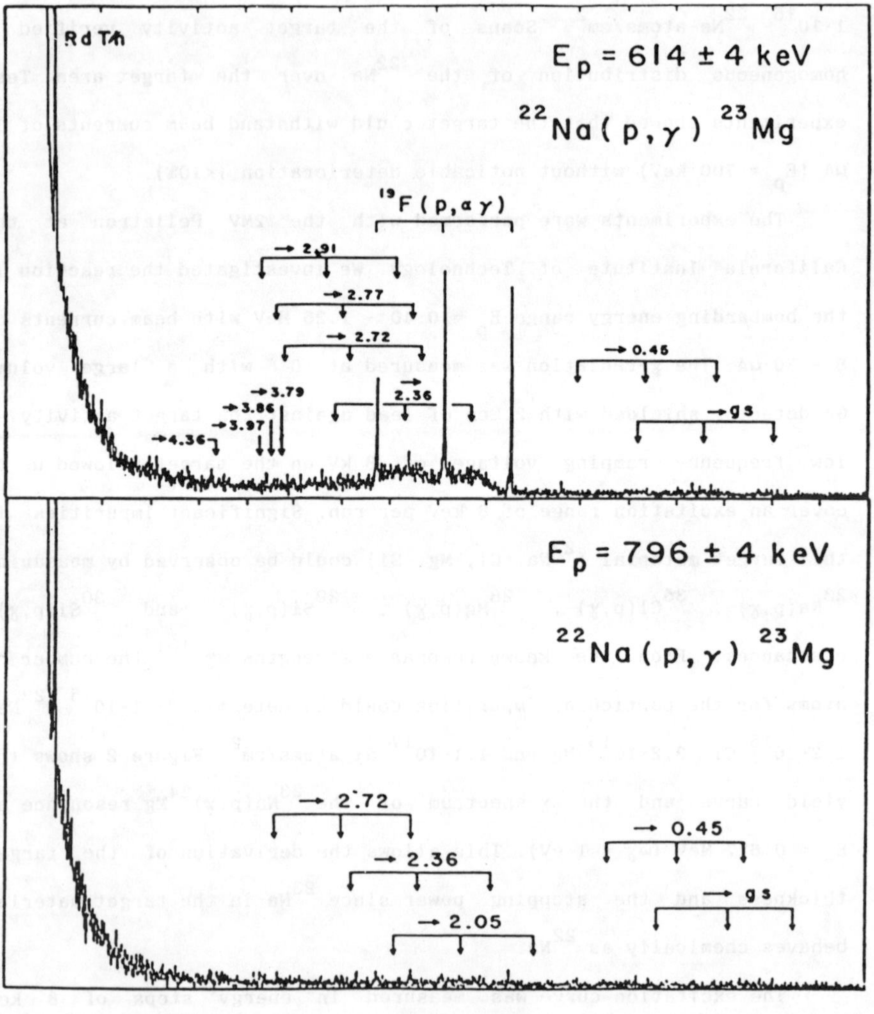

Fig.3. Typical γ-spectrum obtained at E_p=614 and 796 keV in search of resonances in ^{22}Na(p,γ)^{23}Mg.

are the positions of possible γ-transitions in $^{22}Na(p,\gamma)^{23}Mg$. No obvious transition could be observed. From the derived yield an upper limit for the strength of possible γ-transitions can be determined by assuming the same target composition for ^{22}Na as for ^{23}Na. Figure 4 shows the resulting upper limits (1σ) as a function of the proton energy for the transitions to the ground state ($3/2^+$) in ^{23}Mg. The derived upper limits are between 30 meV and 200 meV, depending on the reaction background and are significantly lower than the predicted values [7].

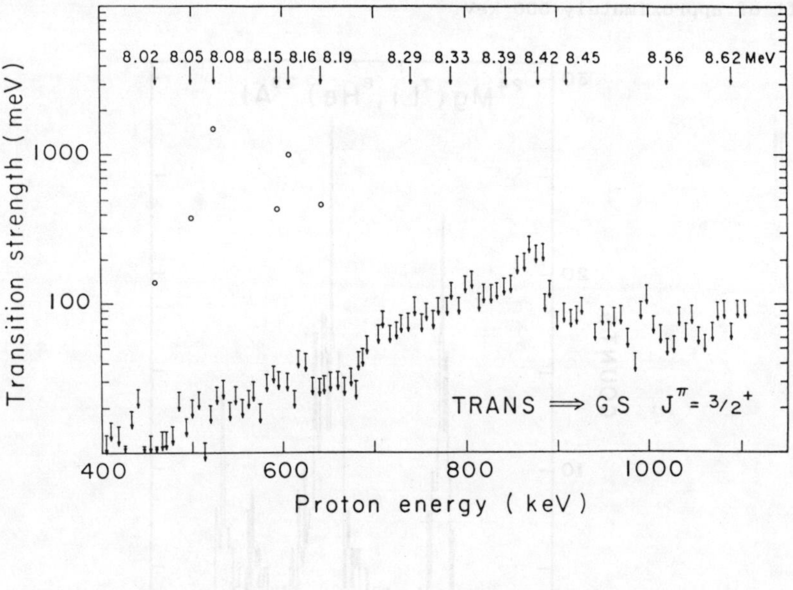

Fig.4. Upper limit for the resonance strength in the ground state transition of $^{22}Na(p,\gamma)^{23}Mg$. The squares indicate the estimated resonance strength in ref. [7].

II. Search for the first excited state in ^{23}Al and its influence for the rate of $^{22}Mg(p,\gamma)^{23}Al$.

Due to the low proton threshold (Q = 0.127 MeV) in the compound nucleus ^{23}Al the reaction rate of $^{22}Mg(p,\gamma)^{23}Al$ will be determined by the nonresonant direct capture to the ground state and by the s-wave resonant capture to the first excited state in ^{23}Al [6]. The resonant part of the rate depends exponentially on the unknown resonance energy and linearily on its resonance strength, neither of which is known. The mirror level in ^{23}Ne has an excitation energy of 1.02 MeV, its pronounced single particle structure suggests a large Thomas-Ehrman shift of approximately 500 keV.

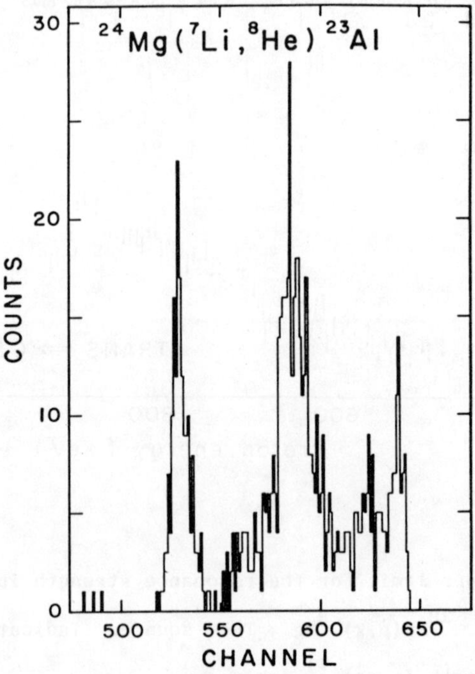

Fig.5. A spectrum from the $^{24}Mg(^{7}Li,^{8}He)^{23}Al$ reaction at 200 MeV.

We measured the reaction $^{24}Mg(^{7}Li,^{8}He)^{23}Al$ to determine the level energy. The experiments were performed at the NSCL Cyclotron Lab at MSU. A natural Mg target was bombarded with 200 MeV ^{7}Li beam. The reaction products were detected with an energy resolution of 300 keV using the S-320 Spectrograph. Figure 5 shows the obtained ^{8}He spectrum. Peakshape and resolution were determined by measuring the ground state transitions in $^{40}Ca(^{7}Li,^{8}He)$ as shown in figure 6.

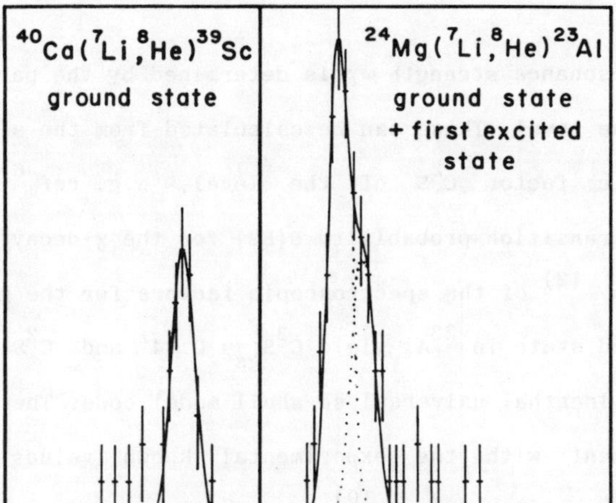

Fig.6. Peak shape analysis for the transitions to the ground state and the first excited state in ^{23}Al, using the fit parameters from the ground state transition in $^{40}Ca(^{7}Li,^{8}He)^{39}Sc$.

The groundstate in ^{23}Al ($J^{\pi}=5/2^{+}$) is strongly populated and the first excited state ($J^{\pi}=1/2^{+}$) can be identified as a pronounced shoulder on its low energy side. Standard peak fitting analysis results in an excitation energy of $E_x = 460\pm60$ keV well in agreement with predictions from Thomas-Ehrman shift calculations. This agrees also with a study of ^{23}Al by $^{28}Si(^3He,^8Li)^{23}Al$ at bombarding energies of 76.3 MeV [11]. Reanalysis of the measured spectrum yields an excitation energy for the weakly populated first excited state of $E_x = 475\pm50$ keV. The average of the two results yields $E_x = 470 \pm 40$ KeV.

This excitation energy corresponds to a resonance energy of $E_p^{lab} = 360 \pm 40$ keV, which is significantly lower than previously suggested [6].

The resonance strength $\omega\gamma$ is determined by the partial widths Γ_p and Γ_γ of the level. These can be calculated from the single particle spectroscopic factor C^2S of the level, e.g. ref. [6], and from the reduced E2 transition probability B(E2) for the γ-decay of the state. Calculations [12] of the spectroscopic factors for the ground- and the first excited state in ^{23}Al yield $C^2S_{gs} = 0.34$ and $C^2S_1 = 0.66$, using the Brown-Wildenthal universal sd-shell model code. The results are in good agreement with the experimental known values for the mirror states in ^{23}Ne, 0.22 and 0.7 [10], respectively. Using the same code the reduced transition probability was calculated to B(E2)= 20.4 e^2fm^4. The resulting widths, $\Gamma_p = 32$ eV and $\Gamma_\gamma = 2\cdot10^{-7}$ eV allow to determine the resonance strength to $\omega\gamma = 2\cdot10^{-7}$ eV. The reaction contribution of the direct capture to the ground state depends on its spectroscopic factor. The shell model calculations suggest $C^2S = 0.34$, which is slightly higher than the previously adopted value of 0.22 [6] from the mirror state in ^{23}Ne. The reaction rate for $^{22}Mg(p,\gamma)^{23}Al$ is calculated [13] and shown in figure 7.

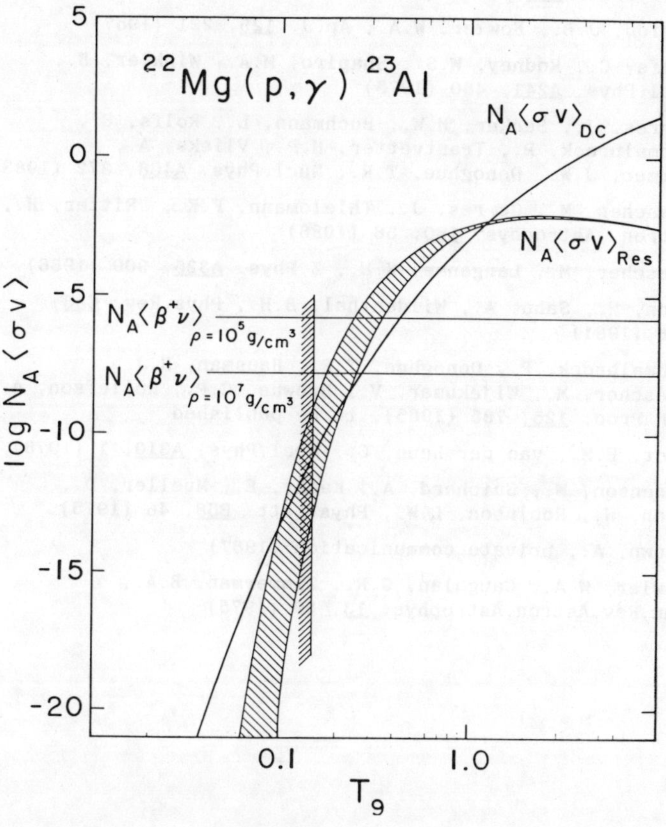

Fig.7. Resonant and nonresonant contributions to the reaction rate of $^{22}Mg(p,\gamma)^{23}Al$. The rates of the competing β-decay at different densities are shown for comparison. The vertical line indicates the temperature at which a significant amount of ^{22}Mg is produced [7].

References

1) Wallace, R.K., Woosley, S.E., Ap.J.Suppl. 45, 389 (1981)

2) Eberhardt, P., Jungck, M.H.A., Meier, I.O., Niederer, F., Ap.J.Lett. 234, L169 (1979)

3) Marion, J.B., Fowler, W.A., Ap.J. 125, 221 (1957)

4) Rolfs, C., Rodney, W.S., Shapiro, M.A., Winkler, H., Nucl.Phys. A241, 460 (1975)

5) Görres, J., Becker, H.W., Buchmann, L., Rolfs, C., Schmalbrock, P., Trautvetter, H.P., Vlieks, A., Hammer, J.W., Donoghue, T.R., Nucl.Phys. A408, 372 (1983)

6) Wiescher, M., Görres, J., Thielemann, F.K., Ritter, H., Astron. Astrophys. 160, 56 (1986)

7) Wiescher, M., Langanke, K.H., Z.Phys. A325, 309 (1986)

8) Nann, H., Saha, A., Wildenthal, B.H., Phys.Rev. C23, 606 (1981)

9) Schmalbrock, P., Donoghue, T.R., Hausman, H.J., Wiescher, M., Wijekumar, V., Browne, C.P., Rollefson, A.A., AIP Proc. 125, 785 (1985), to be published

10) Endt, P.M., van der Leun, C., Nucl.Phys. A310, 1 (1978)

11) Benenson, W., Guichard, A., Kashy, E., Mueller, D., Nann, H., Robinson, L.W., Phys.Lett. B58, 46 (1975)

12) Brown, A., private communication (1987)

13) Fowler, W.A., Caughlan, G.R., Zimmerman, B.A., Ann.Rev.Astron.Astrophys. 13, 69 (1975)

EXPERIMENTAL DETERMINATION OF REACTION RATES IN THE HOT-CNO-CYCLE AND THE rp-PROCESS*

Peter Parker
A.W. Wright Nuclear Structure Laboratory
Yale University, New Haven, CT 06511, U.S.A.

ABSTRACT

For resonances involving short-lived radioactive nuclei [e.g., the $^{13}N(p,\gamma)$ and $^{15}O(\alpha,\gamma)$ reactions] it is often possible to determine the relevant reaction rates by making use of standard nuclear spectroscopy techniques together with stable targets and beams. More generally, nuclear spectroscopy techniques can also play a very valuable role in determining the rates of **very** low energy resonances (resonances which are so far below the Coulomb barrier that they can not presently be studied directly) even when both the beam and target nuclei are readily available; examples relating to the $^{25}Mg(p,\gamma)$ and $^{26}Al(p,\gamma)$ reactions will be discussed.

In explosive nucleosynthetic events the temperatures and densities involved are sufficiently high so that proton and alpha-particle induced nuclear reactions can compete with beta decay [e.g., the Hot-CNO-cycle[1)] and the rp process[2)]; Figure 1]. In order to study these reactions **directly** would require the use of short-lived radioactive targets or beams; however, it is often possible to use standard nuclear spectroscopy techniques to determine the relevant nuclear reaction rates **indirectly** (without the necessity for radioactive targets or beams) by making use of other nuclear reactions (transfer reactions, charge-exchange reactions, etc) to form the relevant resonant state and to study its decay into particular channels. In this way, for example, the $^{13}N(p,\gamma)^{14}O$ reaction rate can be determined by using a reaction such as $^{14}N(^{3}He,t)$ or $^{12}C(^{3}He,n)$ to form a specific $^{13}N(p,\gamma)$

resonant state in ^{14}O and by then measuring the $^{14}O^* \rightarrow {}^{13}N+p$ and $^{14}O^* \rightarrow {}^{14}O+\gamma$ decay rates (Section A), and the $^{15}O(\alpha,\gamma)^{19}Ne$ reaction rate can be determined by using a reaction such as $^{20}Ne(^{3}He,\alpha)$ or $^{19}F(^{3}He,t)$ to form a specific $^{15}O(\alpha,\gamma)$ resonant state in ^{19}Ne and by then measuring the $^{19}Ne^* \rightarrow {}^{15}O+\alpha$ and $^{19}Ne^* \rightarrow {}^{19}Ne+\gamma$ decay rates (Section B). Furthermore, in cases where the relevant target nuclei are readily available, the use of nuclear spectroscopy techniques combined with appropriate transfer reactions [e.g., the use of the (^{3}He,d) reaction as a substitute for the (p,γ) reaction] makes it possible to determine the rates of resonances which are so far below the Coulomb barrier that they can not presently be studied directly; examples are given below of the identification of such **very** low energy resonances for the $^{25}Mg(p,\gamma)$ reaction (Section C) and the $^{26}Al(p,\gamma)$ reaction (Section D).

A. THE $^{13}N(p,\gamma)^{14}O$ REACTION

The role of the Hot-CNO-cycle is determined by a comparison of the rate of ^{13}N positron decay ($t_{1/2}$=9.96 minutes) and the rate of the competing $^{13}N(p,\gamma)^{14}O$ reaction. At the temperatures of interest, $T < 10^9$ K ($T_9 < 1$), the rate of the $^{13}N(p,\gamma)^{14}O$ reaction is determined by the properties of the low energy ($E_{c.m.}$=541 keV), s-wave resonance corresponding to the first excited state of ^{14}O, as shown in Figure 2, (ref. 3). Recently, we have measured E_x = 5168.5 ± 1.8 keV and Γ_{TOT}= 38.1 ± 1.8 keV for this state using the $^{14}N(^{3}He,t)^{14}O$ reaction in conjunction with the Princeton Q3D magnetic spectrograph[4]. Three separate sets of experiments are currently underway to measure $\Gamma_\gamma/\Gamma_{TOT}$ for this state and to thereby determine $\omega\gamma$ for this resonance.

At Yale we have made a preliminary set of measurements[5] using the $^{14}N(^{3}He,t)^{14}O^*(^{14}O)\gamma$ reaction in which coincidences are measured between (a) the tritons corresponding to the formation of the 5.17-MeV first excited state of ^{14}O and (b) the recoiling ^{14}O

nuclei following the gamma decay of this state. An important aspect of this coincidence experiment is the fact that these recoiling residual ^{14}O nuclei can be detected with ≈ 100% efficiency by positioning the detector at the kinematically required angle for the initial $^{14}O^*$ recoil; the angular spread due to the gamma decay is only ± 1/2°. Greater than 99.9% of the time, the $^{14}O^*$(5.17-MeV) state decays by breaking up into the ^{13}N+p channel, and therefore the "recoil" detector must be able to separate the recoiling residual ^{14}O nuclei from the ^{13}N recoils, etc. In the initial measurements this separation was provided by a ΔExE detector telescope. The experiment is shown schematically in Figure 3, and the kinematic separation of these two decay channels is shown in Figure 4. A built-in calibration and check on this method is provided (a) by the detection of coincident recoils from the $^{14}N(^3He,t_0)^{14}O_{g.s.}$ reaction at its kinematically required angle and (b) by the detection of $^{15}N(^3He,t)^{15}O^*(^{15}O)\gamma$ coincidences corresponding to the bound excited states of ^{15}O.

The results of these initial measurements[5)] are an upper limit of $\Gamma_\gamma/\Gamma \leq 1/2500$. These measurements were limited by the sensitivity and count-rate capability of the recoil detector; a new set of measurements at Yale, using the newly installed Enge split-pole spectrograph for the recoil detector, should be able to improve this limit by at least a factor of 10. [Model predictions for Γ_γ/Γ for this state are 1/15,000 (ref. 6) and 1/25,000 (ref. 7)].

At the same time, experiments have been carried out at Seattle[8)] using the $^{12}C(^3He,n)^{14}O^*(\gamma)^{14}O$ reaction in which coincidences have been measured between the neutrons corresponding to the formation of the $^{14}O^*$(5.17-MeV) state and its decay gamma rays. Their current upper limit is $\Gamma_\gamma/\Gamma \leq 1/6000$, limited by their ability to perform a background subtraction of $^{14}N(^3He,n)^{16}O^*(\gamma)^{16}O$ coincidences due to ^{14}N contamination in their targets.

Very recently, at ANL a third set of measurements has been

started in which the $^{1}H(^{14}N,^{14}O)n\gamma$ reaction is studied using a 175-MeV ^{14}N beam from the ATLAS accelerator[9]. The kinematics of this reaction require that **all** the ^{14}O recoils corresponding to the 5.17-MeV state be within a $\pm 1^{o}$ cone centered at 0°. A magnetic spectrograph and its associated detector system are used to measure the momentum, time-of-flight, ΔE, and E of the reaction products.

Within the next 12 months, refinements in each of these three experiments should be able to produce a positive measurement of $\omega\gamma$ for this $^{13}N(p,\gamma)^{14}O$ resonance.

B. THE $^{15}O(\alpha,\gamma)^{19}Ne$ REACTION

At still higher temperatures and pressures, the Hot-CNO-cycle can be broken by the $^{15}O(\alpha,\gamma)^{19}Ne$ reaction, leading to the rp-process which converts the CNO nuclei to heavier proton-rich nuclei up to ^{56}Ni. (See Figure 1.) Calculations of the rates of the $^{15}O(\alpha,\gamma)^{19}Ne$ reaction assume that the reaction proceeds through resonances in ^{19}Ne just above the alpha particle threshold, and hence it is necessary to determine the properties of these states (E_x, J^{π}, Γ_{α}, Γ_{γ} and Γ_{TOT}) as accurately as possible. The contribution of the resonances to the reaction rate in the temperature range $0.1 < T_9 < 10$ can be estimated using the sharp resonance formula[10],

$$\lambda_{\alpha\gamma} = \left(\frac{2\pi}{MkT}\right)^{3/2} \hbar^2 \sum_r (\omega\gamma)_r \exp\left(-E_R/kT\right), \qquad (1)$$

where $\lambda_{\alpha\gamma}$ is the rate for the $^{15}O(\alpha,\gamma)^{19}Ne$ reaction, $\sum_r$ is a sum over resonances, $\omega = (2J_r+1)/[(2I_1+1)(2I_2+1)]$, $\gamma = \Gamma_\gamma\Gamma_\alpha/\Gamma_{TOT}$, and M is the reduced mass of the system. For the $^{15}O(\alpha,\gamma)^{19}Ne$ reaction there are no $J^{\pi} = 1/2^{-}$ states (corresponding to $\ell=0$ resonances) near the $^{15}O + \alpha$ threshold; however, four states have been identified which could correspond to low angular momentum resonances ($\ell \leqslant 3$); see Figure 5.

A recent measurement[12] of the thick target $^{15}N(\alpha,\gamma)^{19}F$

yields corresponding to the $E_{c.m.}$ = 536 and 542 keV resonances (E_x=4.550 and 4.556 MeV) can be used to determine Γ_α for each of these resonances; based on the fact that $\Gamma_\gamma \gg \Gamma_\alpha$ for these two resonances, Γ_α=(3.2±0.7) x 10^{-5} eV and $\Gamma_\alpha < 5 \times 10^{-6}$ eV, respectively. The α-widths of the corresponding states in ^{19}Ne (E_x=4.600 and 4.549 MeV, respectively) can be estimated by assuming that mirror states have the same dimensionless reduced α-particle widths ($\theta^2_{\alpha\ell}$) defined by[2)]

$$\Gamma_\alpha = \theta^2_{\alpha\ell} E_R f_\alpha P_\ell(R_1, E_\alpha). \tag{2}$$

Here $E_R = \hbar^2/2MR_1^2$, f_α is the reflection factor[2)] and is taken to be 4.8, R_1 is the interaction radius and is taken to be 4.99 fm, and $P_\ell = kR/(F_\ell^2+G_\ell^2)$ where F_ℓ and G_ℓ are the regular and irregular Coulomb functions.

Using this equation to correct for the different barrier penetrabilities of the ^{19}Ne and ^{19}F systems, we can extract values for Γ_α for the two states in ^{19}Ne, and by combining these values for Γ_α with the values for Γ_γ measured in ^{19}F we can determine ωγ for these two ^{15}O(α,γ)^{19}Ne resonances.

Resonance Parameters for ^{15}O(α,γ)^{19}Ne

$E_{c.m.}$ (MeV)	E_x (MeV)	J^π	Γ_α (eV)	Γ_γ (eV)	ωγ(exp) (eV)	ωγ(LWFG) (eV)
1.021	4.549	$3/2^-$	$<3.6\times10^{-3}$	$<3.8\times10^{-2}$	$<7.3\times10^{-3}$	1.19×10^{-1}
1.076	4.600	$5/2^+$	8.8×10^{-2}	$>1.3\times10^{-2}$	$>3.4\times10^{-2}$ $<26.4\times10^{-2}$	2.45×10^{-2}

A comparison of the reaction rates determined by these measured parameters and by the parameters calculated by Langanke **et al.**[13)] (LWFG) is given in Figure 6. For the 4.600-MeV state the LWFG calculation (corresponding to their assumption that $\theta^2_{\alpha\ell}$= 0.02) is consistent with our determination of ωγ, but their calculation for the 4.549 MeV state is 16 times larger than our experimental upper limit. Compared to the LWFG calculations, our elimination of the 4.549 MeV state in ^{19}Ne as an important $\ell = 2$ resonance ($E_{c.m.}$=1021 keV) reduces the rate of this reaction by a factor of ≈ 3 for temperatures $T_9 \approx 1$, as shown in Figure 6; in the LWFG calculation this state had accounted for 2/3 of the reaction rate at $T_9 = 1$.

The values of Γ_α and $\omega\gamma$ have not yet been measured directly for the 505 keV (ℓ=1) and 851 keV (ℓ=3) resonances in ^{19}Ne or for their isospin mirror states in ^{19}F. Measurements of these two states as resonances in the $^{15}N(\alpha,\gamma)^{19}F$ reaction are not practical because the corresponding $7/2^+$ state in ^{19}F is too far below the Coulomb barrier and the corresponding $3/2^+$ state is bound; for the $3/2^+$ state a value of $\theta_\alpha^2 = 0.06$ has been reported[14] in ^{19}F on the basis of a DWBA analysis of the $^{15}N(^6Li,d)^{19}F$ α-stripping reaction. The two $\ell = 4$ resonances (E_x=4.140 and 4.197 MeV) do not make significant contributions (≤5%) to the $^{15}O(\alpha,\gamma)$ reaction because of their large angular momentum barriers.

In order to determine the contributions of other resonances to the $^{15}O(\alpha,\gamma)^{19}Ne$ reaction [and to better determine the contributions of the 1021-keV and 1076-keV resonances] it is necessary to measure the ratio $\Gamma_\gamma/\Gamma_\alpha$ for each state together with the magnitude of Γ_γ, Γ_α, or Γ_{TOT} for each state. [On the basis of model calculations[13], the ratio $\Gamma_\gamma/\Gamma_\alpha$ is expected to fall in the range from ≈ 1 to 10^4 for these states.]

One way to carry out a measurement of $\Gamma_\gamma/\Gamma_\alpha$ is to make use of the $^{19}F(^3He,t)^{19}Ne^*(^{15}O)\alpha$ or $^{19}F(^3He,t)^{19}Ne^*(^{19}Ne)\gamma$ reactions to measure coincidences between the formation of these states and their various decay modes. Figure 7 shows the corresponding $^{19}F(^3He,t)$ spectrum for the ^{19}Ne excited states just above the $^{15}O + \alpha$ threshold. For temperatures below $T_9 \approx 0.8$ the rate of the $^{15}O(\alpha,\gamma)$ reaction is predicted to be dominated by the resonance corresponding to the 4.033-MeV state (ℓ=1; $E_{c.m.}$=505 keV). The experimental geometry and kinematics for the $^{19}F(^3He,t)^{19}Ne^*(^{15}O)\alpha$ and $^{19}F(^3He,t)^{19}Ne^*(^{19}Ne)\gamma$ coincidence reactions are shown in

Figures 8 and 9 for the ^{19}Ne state at 4.033 MeV. Using a magnetic spectrograph to separate and analyze the ^{15}O and ^{19}Ne recoil nuclei, this experiment will have the sensitivity to measure these decay branching ratios down to the level of $< 10^{-4}$ which, when combined with lifetime measurements of Γ_γ for these states or their analogs in ^{19}F, will provide an experimental determination of the rate of the $^{15}O(\alpha,\gamma)^{19}Ne$ reaction as a function temperature.

C. THE $^{25}Mg(p,\gamma)^{26}Al$ REACTION

The discovery[15) of fossil ^{26}Al (as excess ^{26}Mg) in meteorites and particularly the discovery[16) of live ^{26}Al in the interstellar medium [on the basis of the observation by the HEAO-3 satellite of the 1809-keV gamma rays from the decay of ^{26}Al to the first excited state of ^{26}Mg] suggest[17) that substantially more ^{26}Al is being produced in our galaxy than can be readily accounted for in high-temperature sites such as supernovae; this has motivated a number of studies looking for possible low-energy resonances in the $^{25}Mg(p,\gamma)^{26}Al$ reaction. Direct $^{25}Mg(p,\gamma)$ measurements have been carried out down to the 190-keV resonance (Figure 10). In order to search for resonances at even lower energies and to determine their resonance strengths, Champagne[18) has developed a method utilizing the (^{3}He,d) proton-transfer reaction coupled to a normalization based on a comparison between (^{3}He,d) and (p,γ) measurements at higher-energy, directly-observable resonances. This method uses an R-matrix parameterization to determine the relative resonance widths of two states provided they (1) are formed by the same partial wave, (2) have similar spectroscopic strengths, and (3) have similar excitation energies (ΔE≤500 keV). Such states will have virtually identical potentials, and therefore the difference in their resonance widths will depend only on the difference in their spectroscopic strengths and on the energy dependence of their penetrabilities,

$$\Gamma_{p1}/\Gamma_{p2} = [S_1/S_2]\,[P_\ell(E_1)/P_\ell(E_2)] \qquad (3)$$

where $S_1/S_2 = \gamma^2_{p1}/\gamma^2_{p2}$.

Because this method makes use of the **ratio** of the penetrabilities, it is quite insensitive to the shape, depth, and size of the nuclear potential, as long as the **same** potential is used for both resonances. Applying this method to the $^{25}Mg(p,\gamma)$ reaction, an $\ell=0$ resonance has been found[19] at $E_{c.m.}$ = 57.5 keV (Figure 11); based on a comparison[19] with the well-known $\ell=0$ (p,γ) resonance at $E_{c.m.}$ = 374 keV, the 57.5-keV resonance has a resonance strength of $\omega=(1.6\pm0.6)\times10^{-13}$ eV which increases the rate of the $^{25}Mg(p,\gamma)^{26}Al$ reaction very significantly for temperatures below $T_9 = 0.1$, thereby indicating that red giants may play a major role in the production of the ^{26}Al which is observed in the interstellar medium. The 57.5-keV resonance, when combined with known, higher energy resonances yields the following analytic expression for the $^{25}Mg(p,\gamma)^{26}Al_{g.s.}$ reaction rate (in units of $cm^3\ mole^{-1}\ sec^{-1}$) as a function of temperature, $T_9(10^9$ K):

$$\begin{aligned} N_A\langle\sigma v\rangle = {} & 1.90\times10^{-8}\ T_9^{-3/2}\ \exp(-0.668/T_9) \\ & +0.0689T_9^{-3/2}\ \exp(-2.230/T_9) \\ & +2.84\times10^{3}\ T_9^{-3/2}\ \exp(-3.528/T_9) \\ & +2.60\times10^{4}\ \exp(-4.045/T_9). \end{aligned} \tag{4}$$

The first three terms in this expression represent the contributions of the $E_{c.m.}$ = 57.54 keV, 192.2 keV and 304.0 keV resonances, while the final term describes the contribution of other resonances up to $E_{c.m.}$ = 1.66 MeV. This expression for the reaction rate is plotted as a function of temperature in Figure 12.

D. THE $^{26}Al(p,\gamma)^{27}Si$ REACTION

A second example of the importance of this method [using transfer reactions to measure the properties of very low energy resonances] can be seen in the problem of determining the rate at which the ^{26}Al, produced via the $^{25}Mg(p,\gamma)$ reaction above, will be destroyed by the $^{26}Al(p,\gamma)$ reaction. Using ^{26}Al targets on thick backings, two groups (at Munster[20] and at Caltech[21]) have carried out direct (p,γ) studies, locating and measuring a number of resonances, the two lowest of which lie at $E_{c.m.}$ = 277 keV and 368 keV. Meanwhile, a Notre Dame-OSU group[22] and a Princeton-Yale group[5,23,24] have studied the level structure of ^{27}Si in the region of the ^{26}Al+p threshold using the ^{28}Si(^{3}He,α)^{27}Si and ^{27}Al(^{3}He,t)^{27}Si reactions. (See, for example, Figure 13.) These results show (Figure 14) that there are at least six ^{27}Si states between the ^{26}Al+p threshold and the lowest directly observed (p,γ) resonance; a comparison with mirror states in ^{27}Al (studied via the ^{26}Mg(^{3}He,d) and ^{26}Mg(α,t) reactions[25]) suggests that at least one of these six states probably has spin $J \geq 7/2$ and therefore would be a candidate for an $\ell = 0$ or $\ell = 1$ ^{26}Al(p,γ) resonance. Neither the ^{27}Al(^{3}He,t)^{27}Si reaction nor the ^{28}Si(^{3}He,α)^{27}Si reaction (its measured angular distributions for these states are more characteristic of a compound-nucleus reaction than a direct neutron-pick-up reaction[24]) is able to determine the J^{π}'s or the resonance strengths (proton widths or spectroscopic factors) for these states.

Very recently, however, Vogelaar and Kavanagh have succeeded in making a 0.6 μg/cm^2 ^{26}Al target (in a matrix of 9.6 μg/cm^2 ^{27}Al) mounted on a 27 μg/cm^2 carbon backing, suitable for charged-particle spectroscopy measurements using a magnetic spectrograph. A preliminary result of the initial ^{26}Al(^{3}He,d)^{27}Si measurements[26] with this target by a Caltech-Princeton-Yale collaboration using the Princeton Q3D is shown in Figure 15. Clear evidence is seen in these spectra for (a) the ^{27}Si state at 7741 keV corresponding to the lowest directly observed (p,γ) resonance at $E_{c.m.}$ = 277 keV and (b) the ^{27}Si state at 7651 keV which would correspond to a previously unobserved (p,γ) resonance at $E_{c.m.}$ = 187 keV. These are the **only** two peaks which could be separated cleanly from the contaminant peaks and identified as corresponding to states in ^{27}Si in the range $7300 \leq E_x \leq 8300$ keV. While additional analyses of these data and of similar data measured at six other angles, will be necessary in order to make a quantitative evaluation of the role of the 187-keV resonance in determining the rate of the $^{26}Al(p,\gamma)^{27}Si$ reaction, the strength of its peak in the (^{3}He,d) spectrum suggests, at least qualitatively, that this resonance could make an important contribution to the rate of this reaction at low temperatures. No similar strength was found for the ^{27}Si state at E_x = 7589 keV ($E_{c.m.}$=125 keV), even though its peak was located in a region of the spectrum that was much less obscured by peaks due to ^{27}Al and other target contaminants. Additional measurements are planned following further refinements to the target in order to remove more of the contaminants, possibly including the use of an isotope separator in order to improve the $^{26}Al/^{27}Al$ ratio.

CONCLUSION

In conclusion, we have demonstrated that traditional nuclear spectroscopy methods can be used to indirectly determine nuclear reaction rates which can not be measured directly either because one of the reactants is too short-lived or because the resonance of interest is too far below the Coulomb barrier. Using this method it is expected that within the next year definitive measurements will be completed to determine the rates of the $^{13}N(p,\gamma)^{14}O$, $^{15}O(\alpha,\gamma)^{19}Ne$, and $^{26}Al(p,\gamma)^{27}Si$ reactions.

ACKNOWLEDGEMENTS

It is a pleasure to acknowledge the following collaborators for their contributions to various aspects of this work: at Princeton, A.E. Champagne, T. Chupp, and A.B. McDonald; at Yale, A. Caraley, A.J. Howard, P.V. Magnus, M.S. Smith, and T.F. Wang; at Toronto, R.E. Azuma, C. Campbell, J.D. King, and J. Vise; at Caltech, B. Vogelaar and R.W. Kavanagh; and at Argonne, T.F. Wang, C.N. Davids, S.J. Sanders, and E. Rehm.

REFERENCES

*Work supported in part by U.S.D.O.E. Contract No. DE-AC02-76ER03074

1. Caughlin, G.R. and Fowler, W.A., Nature Phys. Sci. **238,** 23 (1972).
2. Wallace, R.K. and Woosley, S.E., Ap. J. Supp. **45,** 389 (1981).
3. Ajzenberg-Selove, F., Nucl. Phys. **A449,** 1 (1986).
4. Chupp, T.E., Kouzes, R.T., McDonald, A.B., Parker, P.D., Wang, T.F., and Howard, A.J., Phys. Rev. **C31,** 1023 (1985).
5. Wang, T.F., Ph.D. Thesis, Yale University (1985).
6. Mathews, G.J. and Dietrich, F.S., Ap. J. **287,** 969 (1984).
7. Langanke, K., van Roosmalen, O., and Fowler, W.A., Nucl. Phys. **A435,** 657 (1985).
8. Fernandez, P.B., Adelberger, E.G., Khandaker, M., and Hindi, M.M., B.A.P.S. **31,** 1209 (1986); and private communication (1987).
9. Wang, T.F., **et al.,** ATLAS Experiment #202.
10. e.g., Clayton, D.D., **Principles of Stellar Evolution and Nucleosynthesis** (McGraw Hill, NY) (1968).
11. Ajzenberg-Selove, F., Nucl. Phys. **A392,** 1 (1983).
12. Magnus, P.V., Smith, M.S., Parker, P.D., Azuma, R.E., Campbell, C., King, J.D., and Vise, J., Nucl. Phys. **A470,** 206 (1987).
13. Langanke, K., Wiescher, M., Fowler, W.A., and Gorres, J., Ap. J. **301,** 629 (1986).
14. Fortune, H.T., private communication (1986).
15. e.g., Lee, T., Papanastassiou, D.A., and Wasserburg, G.J., Ap. J. **211**, L107 (1977).
16. Mahoney, W.A., Ling, J.C., Wheaton, W.A., and Jacobson, A.S., Ap. J. **286,** 578 (1984).
17. e.g., Clayton, D.D., Ap. J. **280,** 144 (1984).
18. Champagne, A.E., Howard, A.J., and Parker, P.D., Nucl. Phys. **A402,** 179 (1983).

19. Champagne, A.E., McDonald, A.B., Wang, T.F., Howard, A.J., Magnus, P.V., and Parker, P.D., Nucl. Phys. **A451,** 498 (1986).

20. Buchmann, L., Hilgemeier, M., Krauss, A., Redder, A., Rolfs, C., and Trautvetter, H.P., Nucl. Phys. **A415,** 93 (1984).

21. Vogelaar, B., and Kavanagh, R.W., private communication (1987).

22. Schmalbrock, P., Donoghue, T.R., Weischer, M., Wijekumar, V., Browne, C.P., Rollefson, A.A., Rolfs, C., and Vlieks, A., Nucl. Phys. **A457,** 182 (1986).

23. Wang, T.F., Magnus, P.V., Hadden, J.D., Howard, A.J., Parker, P.D., and Champagne, A.E., B.A.P.S. **31,** 786 (1986).

24. Hadden, J.D., Champagne, A.E., **et al.,** B.A.P.S. **32** (1987) Rutgers APS-DNP Meeting.

25. Howard, A.J., Champagne, A.E., **et al.,** private communication (1987).

26. Vogelaar, B., **et al.,** B.A.P.S. **32** (1987) Rutgers APS-DNP Meeting.

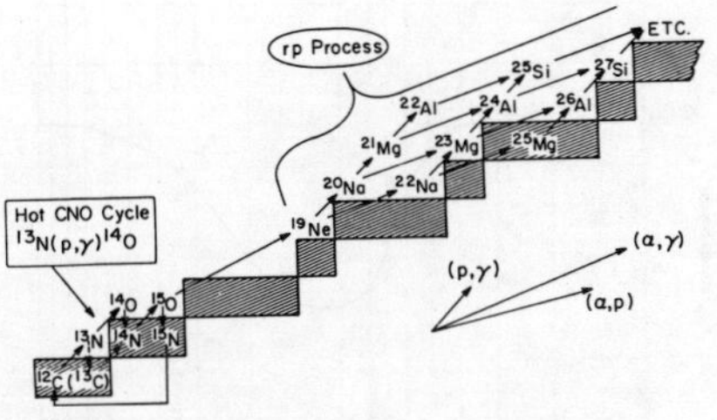

Figure 1: Explosive Hydrogen Burning; the Hot-CNO-Cycle and the rp Process.

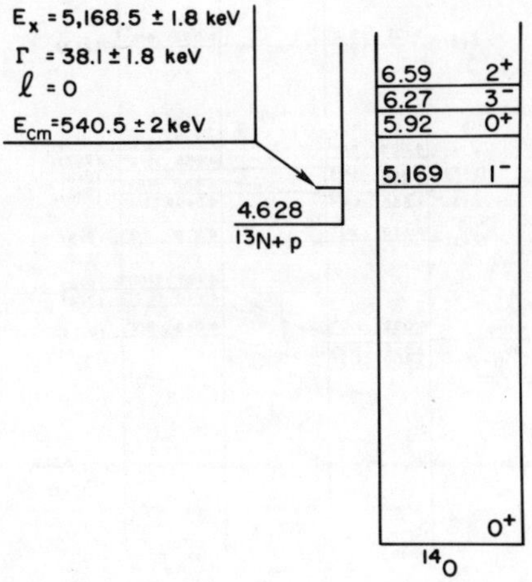

Figure 2: Level structure of ^{14}O

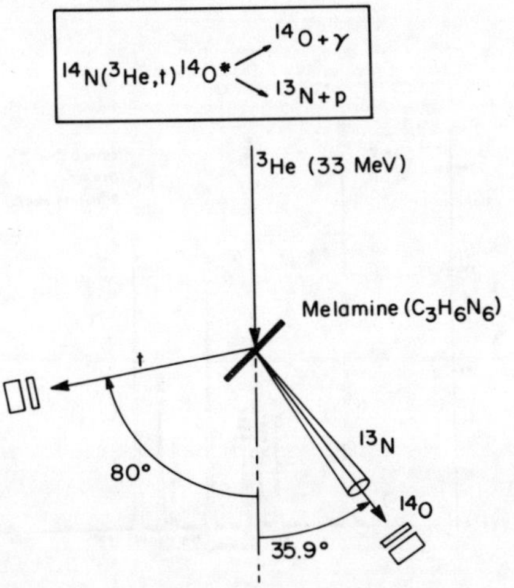

Figure 3: Schematic layout of the coincidence experiment to measure the $^{14}N(^3He,t)^{14}O^*(^{14}O)\gamma$ and $^{14}N(^3He,t)^{14}O^*(^{13}N)p$ reactions.

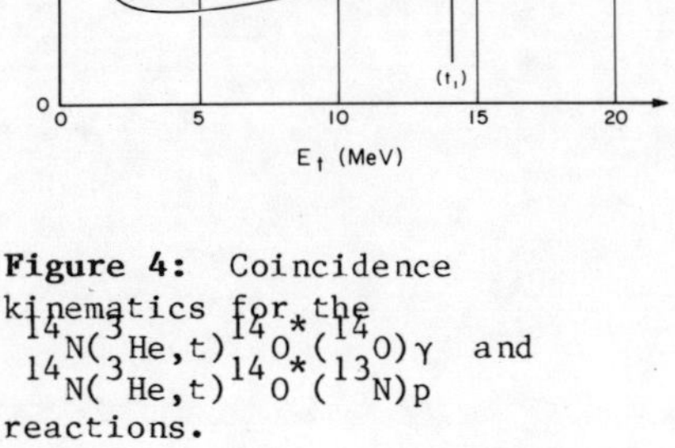

Figure 4: Coincidence kinematics for the $^{14}N(^3He,t)^{14}O^*(^{14}O)\gamma$ and $^{14}N(^3He,t)^{14}O^*(^{13}N)p$ reactions.

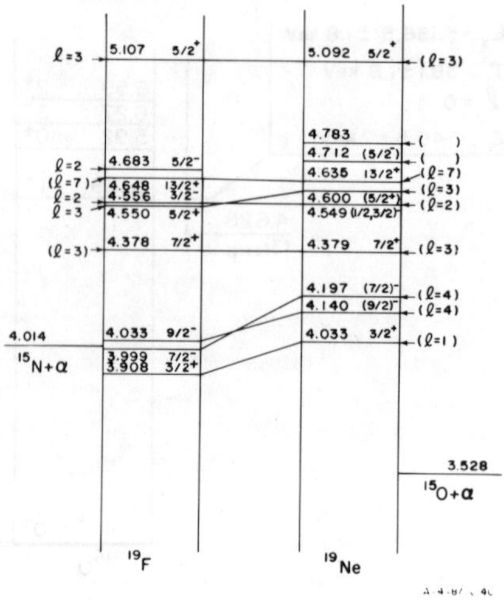

Figure 5: Level structure of ^{19}F and ^{19}Ne in the neighborhood of the $^{15}N+\alpha$ and $^{15}O+\alpha$ thresholds, respectively. (The next lower state in each of these nuclei is at E_x=2.8 MeV.) The ^{19}F states at 4.550, 4.556, 4.683, and 5.107 MeV have been studied directly as $^{15}N(\alpha,\gamma)^{19}F$ resonances. The ^{19}Ne state at 4.033 MeV is predicted to be the dominant resonance for the $^{15}O(\alpha,\gamma)^{19}Ne$ reaction at temperatures $T_9 \leq 0.8$ (ref. 13).

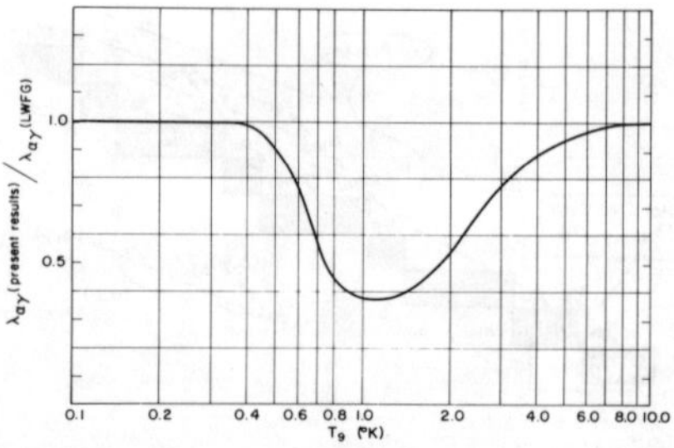

Figure 6: Impact of the recent $^{15}N(\alpha,\gamma)^{19}F$ measurements (ref. 12) on the previous model calculations (ref. 13) of the rate of the $^{15}O(\alpha,\gamma)^{19}Ne$ reaction.

Figure 7: Spectrum of ^{19}Ne states in the region just above the $^{15}O+\alpha$ threshold (3.528 MeV).

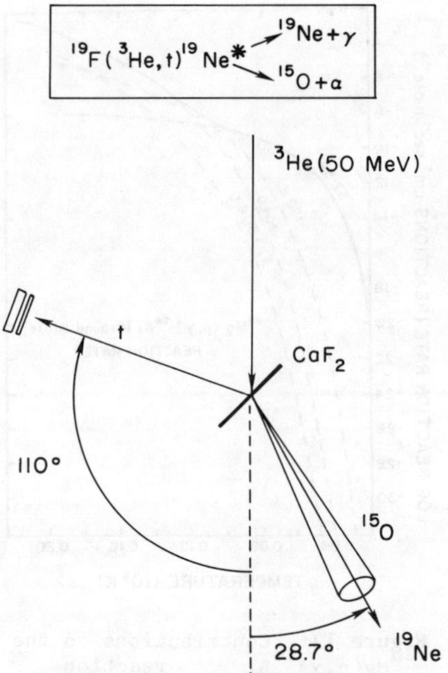

Figure 8: Schematic layout of one proposed coincidence experiment to measure the $^{19}F(^3He,t)^{19}Ne^*(^{19}Ne)\gamma$ and $^{19}F(^3He,t)^{19}Ne^*(^{15}O)\alpha$ reactions.

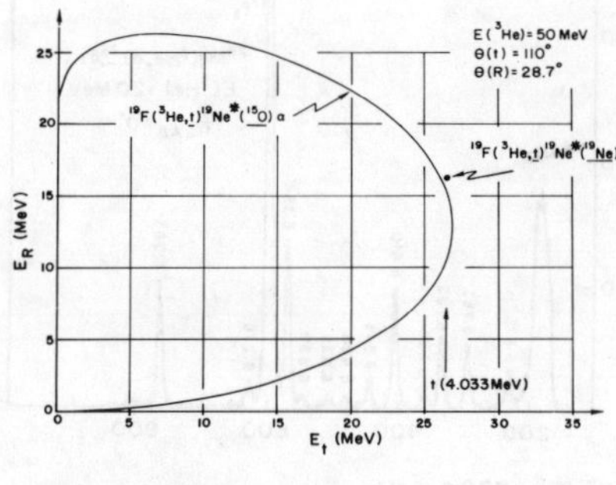

Figure 9: Coincidence kinematics for the $^{19}F(^3He,t)^{19}Ne^*(^{19}Ne)\gamma$ and $^{19}F(^3He,t)^{19}Ne^*(^{15}O)\alpha$ reactions.

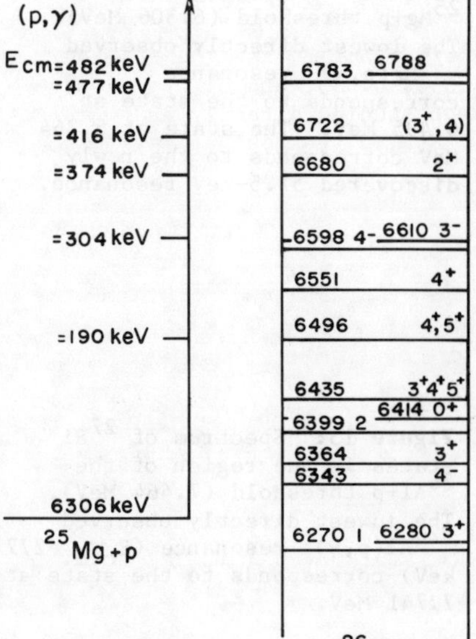

Figure 10: Level structure of ^{26}Al in the region of the ^{25}Mg+p threshold. The lowest directly observed $^{25}Mg(p,\gamma)$ resonances are indicated. Using the $^{25}Mg(^3He,d)^{26}Al$ reaction, an important $\ell=0$ (p,γ) resonance has been identified at $E_{c.m.}$=57.5 keV, corresponding to the ^{26}Al state of 6.364 keV.

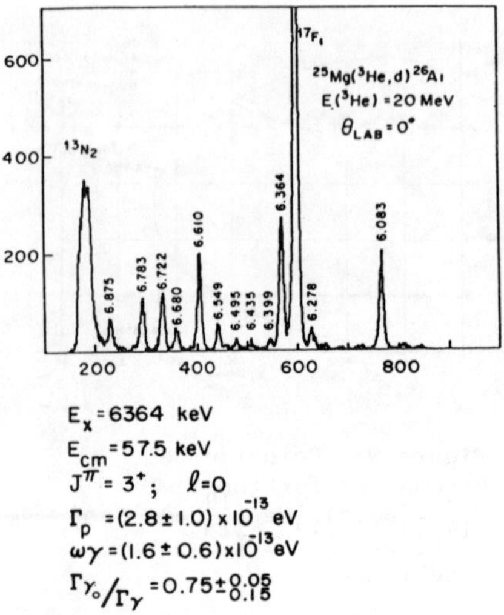

Figure 11: The $^{25}Mg(^3He,d)^{26}Al$ spectrum in the region of the ^{25}Mg+p threshold (6.306 MeV). The lowest directly observed $^{25}Mg(p,\gamma)$ resonance corresponds to the state at 6.495 MeV. The state at 6.364 MeV corresponds to the newly discovered 57.5-keV resonance.

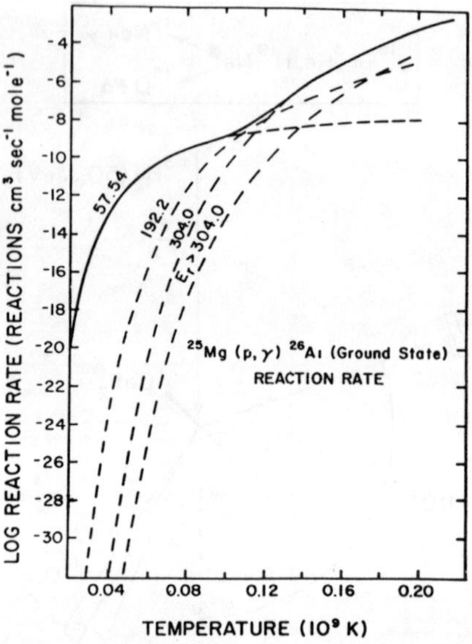

Figure 12: Contributions to the $^{25}Mg(p,\gamma)^{26}Al_{g.s.}$ reaction rate from its most important low-energy resonances.

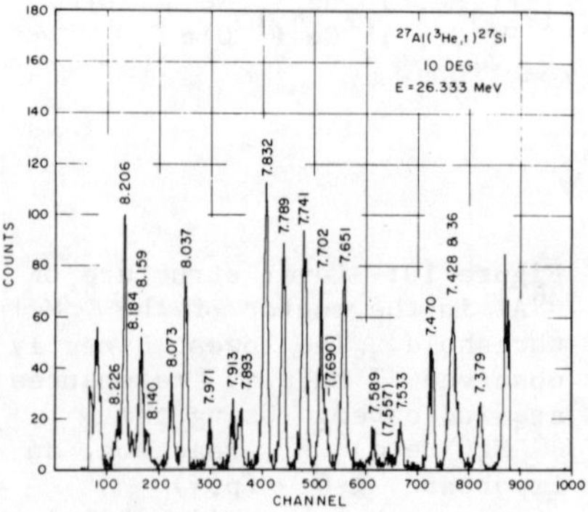

Figure 13: Spectrum of ^{27}Si states in the region of the ^{26}Al+p threshold (7.464 MeV). The lowest directly observed $^{26}Al(p,\gamma)$ resonance ($E_{c.m.}$ =277 keV) corresponds to the state at 7.741 MeV.

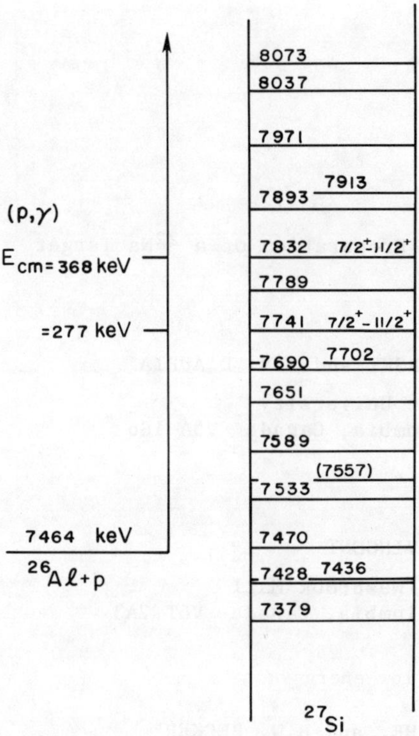

Figure 14: Level structure of ^{27}Si in the region of the ^{26}Al+p threshold. The lowest directly observed ^{26}Al(p,γ) resonances are indicated. Using the ^{26}Al(^{3}He,d)^{27}Si reaction (Fig. 15) peaks are identified corresponding to ^{27}Si states with E_x=7741 keV (the 277-keV resonance) and with E_x=7651 keV (which would correspond to a previously unobserved 187-keV resonance).

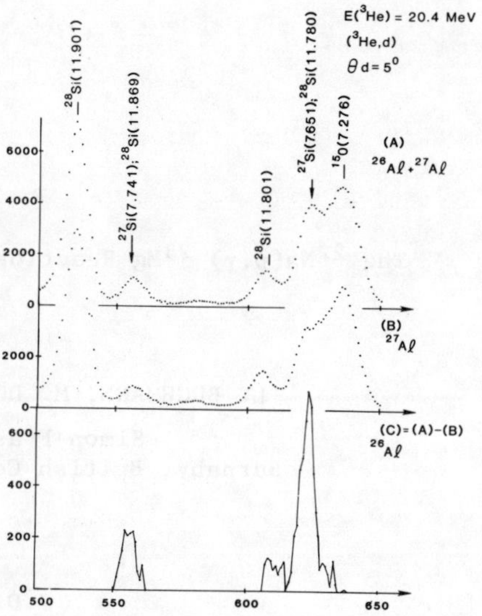

Figure 15: The ^{26}Al(^{3}He,d)^{27}Si Reaction. [E(^{3}He)=20.4 MeV; θ_d=5°] (A):(^{3}He,d) spectrum measured using a [0.6 μg/cm^{2} ^{26}Al + 9.6 μg/cm^{2} ^{27}Al] target. (B):(^{3}He,d) spectrum measured using an ^{27}Al target, normalized to spectrum (A). (C): Net ^{26}Al(^{3}He,d)^{27}Si spectrum; (C)=(B)-(A). The two peaks observed in (C) correspond to the lowest directly observed ^{26}Al(p,γ) resonance [E_x=7.741 MeV, $E_{c.m}$=277 keV] and to a still lower ^{27}Si state [E_x=7.651 MeV] which would correspond to a previously unobserved ^{26}Al(p,γ) resonance at $E_{c.m.}$=187 keV.

The $^{22}Na(p,\gamma)$ ^{23}Mg Reaction: Preparation of a ^{22}Na Target

L. BUCHMANN, M. DOMBSKY and J.M. D'AURIA
Simon Fraser University
Burnaby, British Columbia, Canada V5A 1S6

J.D. VINCENT
TRIUMF, 4004 Wesbrook Mall
Vancouver, British Columbia, Canada V6T 2A3

C. ROLFS, S. SEUTHE, and H.W. BECKER
Institut für Kernphysik, Universität Münster, Münster, West Germany

R. KAVANAGH
California Institute of Technology
Pasadena, CA 91125, U.S.A.

1. INTRODUCTION

Recently[1,2] it was discovered that so-called Ne-E inclusions are present within carbonaceous chondrite meteorites. Since these inclusions are pure ^{22}Ne, within experimental limits, it has been suggested that these inclusions arose from the decay of ^{22}Na ($T_{1/2}$=2.6 a). The ^{22}Na in such a case would have been formed in explosive stellar events such as supernovae or more likely novae, ejected and later, while the stellar ejecta were cooling, condensed and incorporated into micrometer sized grains. With the formation of the solar system such grains could have been coalesced in primitive meteorites without changing their internal structure. The pure ^{22}Ne observed in the mass spectroscopy of parts of the meteorite then shows a picture integrated over many of these tiny grains.

With the observation of a βγ decay line from the long lived, ^{26}Al ($T_{1/2}$=716000 a), which is believed to be formed in similar stellar scenarios as ^{22}Na, it also seems to be feasible, given more sensitive γ-ray observing satellites, to detect the βγ decay line of ^{22}Na (1275 keV). This would prove the ongoing production of this radioisotope.

To calculate the production of ^{22}Na in stellar environments and thus to discriminate between several production scenarios, it is necessary to know the nuclear stellar rates for the production and destruction of ^{22}Na. Such information would also allow one to calculate the galactic abundance of ^{22}Na and stable products formed in the same process. Thus by observing ^{22}Na, conclusions could be formulated on stable isotope abundances as has been done in the case of $^{26}Al/^{27}Al$.

Due to the vulnerability of ^{22}Na to the $^{22}Na(n,p)^{22}Ne$ reaction, which has a positive Q-Value, ^{22}Na is generally believed to be formed in proton rich, neutron free environments at temperatures moderately low, i.e. some hundred million degrees or below. One possible nuclear reaction chain leading to ^{22}Na is the NeNa cycle as displayed in Figure 1. This cycle can have some leakage into the

MgAl cycle (at higher temperatures) in which ^{26}Al may be produced. The cycle starts with the stable and rather abundant isotopes of neon and propagates via proton capture and subsequent beta decay to ^{23}Na, at which point the cycle is closed by a (p,α) reaction back into ^{20}Ne. ^{22}Na is formed in this cycle by the well studied $^{21}Ne(p,\gamma)^{22}Na$ reaction[3]), while even at moderate temperatures it is destroyed by proton capture ($^{22}Na(p,\gamma)^{23}Mg$) instead of the competing β^+ decay. Thus it is essential to know the $^{22}Na(p,\gamma)^{23}Mg$ reaction rate in order to calculate the ^{22}Na concentration in such a cycle or any proton rich environment.

Unfortunately, for any measurement of the $^{22}Na(p,\gamma)^{23}Mg$ reaction, a ^{22}Na target exhibits some nasty properties. These include extremely high $\beta\gamma$ fluxes for any useful target (10^{17} atoms exhibit activities up to about 20mC) and reactive chemical properties for sodium which make preparation of suitable targets difficult. The approach chosen here to overcome these problems was both to develop a detector only sensitive to the high energy γ's of the proton capture and to produce a target by ion implantation which shows stability and a high degree of purity to match the experimental conditions required by the detector.

The detector is described in detail elsewhere[4]). Briefly, it is based on the photodisintegration of the deuteron, which has a threshold of 2.2 MeV. The target is placed inside the detector and is then surrounded by a tank containing about 200 ℓ of heavy water. A photon above the disintegration threshold can then split the deuteron into its constituents: a proton and a neutron. To obtain high detector efficiency the proton is not detected but rather the neutron is moderated and detected in about 20 ^{3}He counters surrounding the heavy water tank. Some polyethylene is also used outside the detector for a further moderation of the neutrons produced. Thus the entire detector is only sensitive to the number of high energy photons escaping the target while it is bombarded with protons. The total efficiency of the disintegration detector is expected to be of the order of 10^{-3}, comparable with normal Ge(Li) gamma spectrometers.

Given such a threshold detector, which does not exhibit energy resolution for the reaction gamma-rays, the target has to be of reasonably high purity. Therefore it has been decided to implant a mass separated ^{22}Na ion beam into a suitable backing. The decision to use an implanted target was also made feasible by the high ionization efficiencies demonstrated by surface ionization sources. The subsequent sections will describe the developments made towards an implanted ^{22}Na target.

2. THE DUAL OVEN ION SOURCE

In the implantation process to be tried here the ^{22}Na is produced by activating a thick piece of aluminum (~ grams). Experiments with such samples have shown that radiosodium is released rapidly after the aluminum has become molten (660°C)[5]. Thus it is planned to melt such an activated piece of aluminum in a special oven which is connected to the ion source.

Given the small amount of ^{22}Na available coupled with its high activity, it was considered necessary to set up the mass analyser (as described in the subsequent section) using a stable beam (e.g. ^{39}K) prior to releasing the ^{22}Na. Thus, the separator can be operated for non-radioactive stable ion beams before the implantation of ^{22}Na is tried. To avoid any significant effects or changes on the source during switching between beams, a dual oven system connected to a surface ionization source was developed. This is shown in Figure 2 and described below.

The system has two side ovens connected to a centre oven, and each side oven may contain either the activated piece of aluminum or a mixture of sodium/potassium/rubidium compounds for the production of stable beams (1+). The ovens can be heated up alternately. To connect both of them to the source, the side ovens as well as the surface ionization source are connected to the middle/centre oven, which stays heated continuously when the source is in operation. The ion source, consisting of a rhenium ionizer, is heated continuously while the source is in operation. All heaters consist of 0.3-0.5 mm thick graphite, through which high currents (typically some hundred amperes) are conducted. Thermocouples inserted in the side ovens as well as the ionizer measure the temperatures associated with the system. The ovens and the source can operate at a high potential of 18 kV, required for extraction of the ion (1+) beam.

To protect the side ovens (when not operational) from the heat produced by the rest of the system, they have been separated as far as possible, within the available geometry, from the rest of the source while being surrounded by heat absorbing copper shields. Since the side oven not in operation is a heat sink to the vapour (to be ionized) produced by the alternate oven, the side ovens are separated by boron nitride disks containing only a small (~1 mm) throughput hole from the middle oven. The ratio of the transport resistivity of this hole over the one of the ionizer hole essentially determines the vapour loss rate of the system. The rhenium ionizer is insulated from its respective oven by a boron nitride tube.

The high current feedthroughs are directly water-cooled by transporting the water through their centres to the lowest, hottest point and having the outside flow cooling the upper parts. The outer sides of the side ovens are clamped into copper pegs brassed to the respective feedthroughs to transport heat away as efficiently as possible while not in use. The middle oven as well as the inner part of the side ovens is clamped into graphite pegs to avoid heat sinks when both sides are in operation.

Given the work function of the rhenium, and relating to similar experiences with tungsten surface ionizer[6)], the sodium ions diffusing through the rhenium tube (diameter about 2 mm) should be ionized with an efficiency of about 50% with the ionizer at a temperature of near 1000°C. The ions then will be extracted by a grounded 4 mm extraction electrode; its distance to the ionizer can be varied externally.

3. THE MASS SEPARATOR

The new TISOL mass separator facility at TRIUMF will be used to produce the implanted ^{22}Na target. This facility is designed to handle highly radioactive sources and will become contaminated as it is used in an on-line mode. The TISOL facility is described in detail elsewhere[7,8], so only a brief summary will be given here.

The entire facility is displayed in Figure 3. The Dual Oven Source is inserted into the TIS (Target Ion Source) box and connected by insulated cables to the power supplies and other services located at a Faraday cage 15 m away; all of which must be capable of operating at high tensions up to 20,000 V. The walls of the TIS box are protected from the heat radiating from the source by directly cooled copper shields. The movable extraction electrode is inserted via a side flange. A cryopump is used to obtain pressures of typically 10^{-6} Torr in the TIS box.

Following extraction, the ion beam is focussed by an einzel-lense (5 cm in diameter) at 10 cm distance from the extraction electrode to gain higher transmission through to the magnetic quadrupoles to follow. These quadrupoles, located about a metre from the ion source, focus the beam in the non-bend as well as the bend plane (in addition to the mass-separating dipole) into the collector device situated on top of the focus box. At about 3 m distance from the source, the ions enter the dipole and become mass separated with a typical mass resolution of $M/\Delta M$ of 1000 to 2000, depending on the source and the transmission. The curvature radius of the dipole is 1.25 m. The collector is situated close to the focus of the dipole, about 4 m from the exit of the dipole.

Two Faraday cups, positioned after the TIS box and the magnetic quadrupoles, are available as beam analysing elements, as well as the collector system, which is described in the subsequent section. Both cups as well as the entire vacuum system, and the power supplies, are remotely controlled by an OPTUMUX/IBM clone system from the control platform located above the shielding blocks. The vacuum system for the beam line from the TIS box through the dipole contains two turbo (450 ℓ) pumps and produces pressures typically in the 10^{-6} Torr range. The three heater power supplies for the ovens described earlier are able to deliver 500 A/10 V, 500 A/10 V, 100 A/10 V, respectively, and are also controllable by the microcomputer system, even at high tension.

4. THE COLLECTOR DEVICE

The collector device, which contains the targetbacking and the targetholder for the subsequent proton studies is displayed in Figure 4 and described below.

The ion (^{22}Na) beam will be stopped in a nickel layer which is evaporated onto a 0.2 mm thick tantalum sheet. In preliminary studies nickel has proven to be the material which shows the smallest diffusion of sodium ions, when bombarded with intense beams (~200 μA) of low energy (~300 keV) protons. The tantalum sheet is directly sealed into the targetholder for the later proton bombardment. The tantalum sheet can be directly water-cooled if necessary.

To ease the exchange of the target and to keep the cold trap under vacuum, two gate valves separate the beam line vacuum from the target. This allows one to remove the "hot" target while maintaining the beam line vacuum. A port between the two gate valves allows for roughing and venting.

The ion beam will be transported through a 1" copper pipe, which is kept at liquid nitrogen temperatures (cold trap), to avoid carbon deposits on the target surface. Since the amount of carbon deposited on the target surface critically depends on the distance of the cold trap to the target, the gate valves, the roughing flange and the targetholder are made movable by a bellow connection to slip the copper pipe through the gate valves to a distance of about 1 mm from the target surface. Expanding the bellow (guided by linear bearings) allows one to retract the target and to close the gate valves.

Both the copper pipe and the targetholder serve as Faraday cups. This allows measurement of the current to minimize beam losses in the pipe. In addition, the copper pipe is on a negative potential (300 V) relative to the targetholder, suppressing secondary electron emission from the target surface. A shield mounted at the entrance of the copper pipe serves as collimator for mass separation as well as suppressing contamination of most of the rest of the system by the radioactive ^{22}Na beam.

5. STATUS

At present the Dual Oven Ion Source and the Collection Device are being tested and modified, using stable (non-radioactive) beams. Studies are planned using irradiated (at TRIUMF) aluminum pieces with essentially only ^{24}Na ($T_{1/2}$=15 h) to determine the overall transmission efficiency of this method, along with discovering potential "hot" spots (places where high deposits are observed). Fortunately, this radioisotope does not lead to long-term contamination of TISOL. Following a successful series of studies with ^{24}Na, irradiated pieces of aluminum, obtained from Los Alamos and with up to 100 mCi of ^{22}Na present, are then available for the preparation of a ^{22}Na target with optimally of the order of 20 mCi (or about 10^{17} g/cm^2).

REFERENCES

1) Eberhardt, P., Jungk, M.H.A., Meier, F.O. and Niederer, F.K., Geochim. Cosmochim. Acta 45, 1515 (1981).

2) Arnould, M. in: Proceedings of the Accelerated Radioactive Beams Workshop, ed. by Buchmann, L. and D'Auria, J.M., TRI-85-1, p.29 (1985).

3) Goerres, J. Becker, H.W., Buchmann, L., Rolfs, C., Schmalbrock, P., Trautvetter, H.P. and Vlieks, A., Nucl. Phys., A408, 372 (1983).

4) Rolfs, C. and Kavanaugh, R.W., NIM, B26, 165 (1987).

5) Ravn, H., CERN-ISOLDE, private communication (1987).

6) Ravn, H., in reference 2), p.9 (1985).

7) Oxorn, K., Crawford, J.E., Dautet, H., Lee, J.K.P., Moore, R.B., Nikinnen, L., Buchmann, L., D'Auria, J., Kokke, R., Otter, A.J., Sprenger, H. and Vincent, J., NIM, B26, 143 (1987).

8) D'Auria, J. et al., to be submitted to NIM.

FIGURE CAPTIONS

Figure 1. The NeNa and the MgAl cycles as examples of nuclear networks in a proton rich stellar environment leading to the long-lived isotopes, ^{22}Na and ^{26}Al.

Figure 2. The Dual Oven Ion Source to be used for the implantation of ^{22}Na. The ^{22}Na is contained in pieces of activated aluminum, which, when molten, should release it. A rhenium ionizer and subsequent extraction, produces the singly charged ion beam of ^{22}Na. For further details, see text.

Figure 3. A schematic, elevation view of the TISOL on-line mass separator at TRIUMF. For further details see text as well as references 7) and 8).

Figure 4. The ^{22}Na collector device in an expanded (out) position. For implantation, the gate valves are opened and the target holder placed close to the cooled copper pipe (compressed position). For further details, see text.

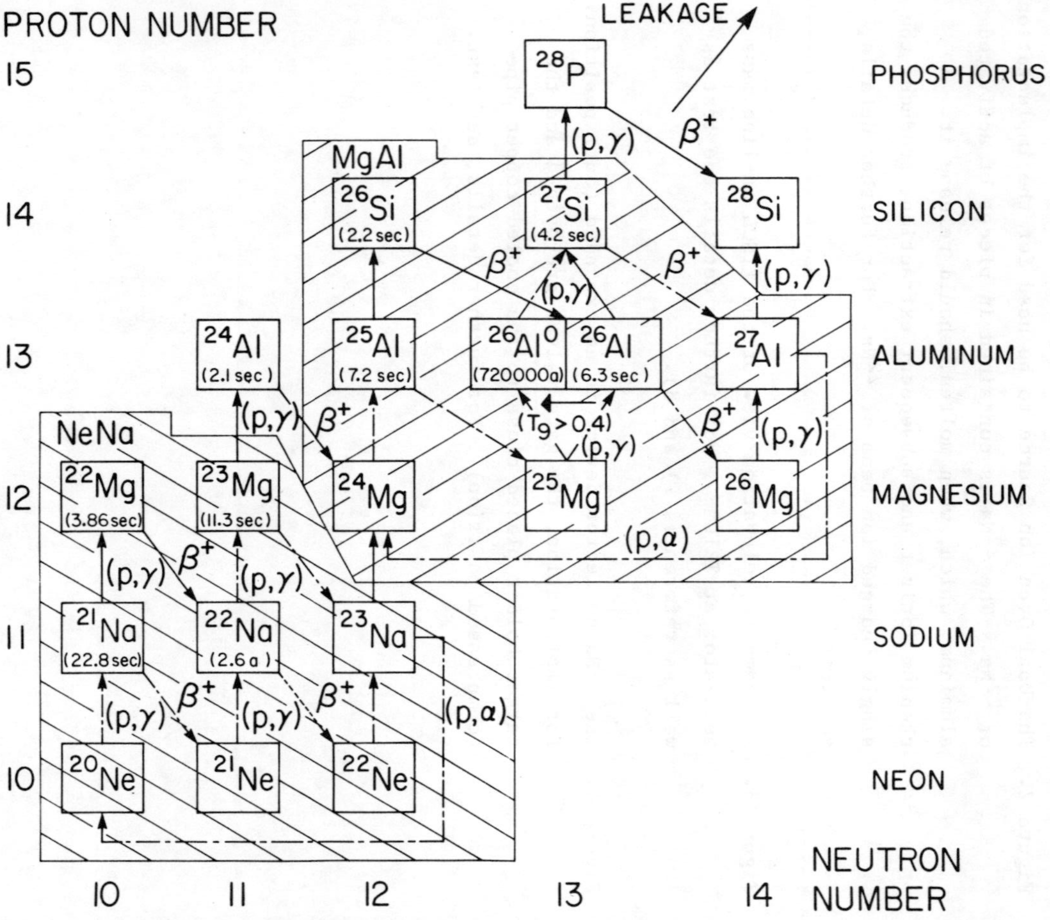
PROTON NUMBER
15
14
13
12
11
10
LEAKAGE
MgAl
NeNa
^{28}P
^{26}Si (2.2 sec)
^{27}Si (4.2 sec)
^{28}Si
^{24}Al (2.1 sec)
^{25}Al (7.2 sec)
^{26}Al^{0} (720000a)
^{26}Al (6.3 sec)
^{27}Al
(T_9 > 0.4)
^{22}Mg (3.86 sec)
^{23}Mg (11.3 sec)
^{24}Mg
^{25}Mg
^{26}Mg
^{21}Na (22.8 sec)
^{22}Na (2.6 a)
^{23}Na
^{20}Ne
^{21}Ne
^{22}Ne
(p,γ)
β+
(p,α)
PHOSPHORUS
SILICON
ALUMINUM
MAGNESIUM
SODIUM
NEON
10
11
12
13
14
NEUTRON NUMBER

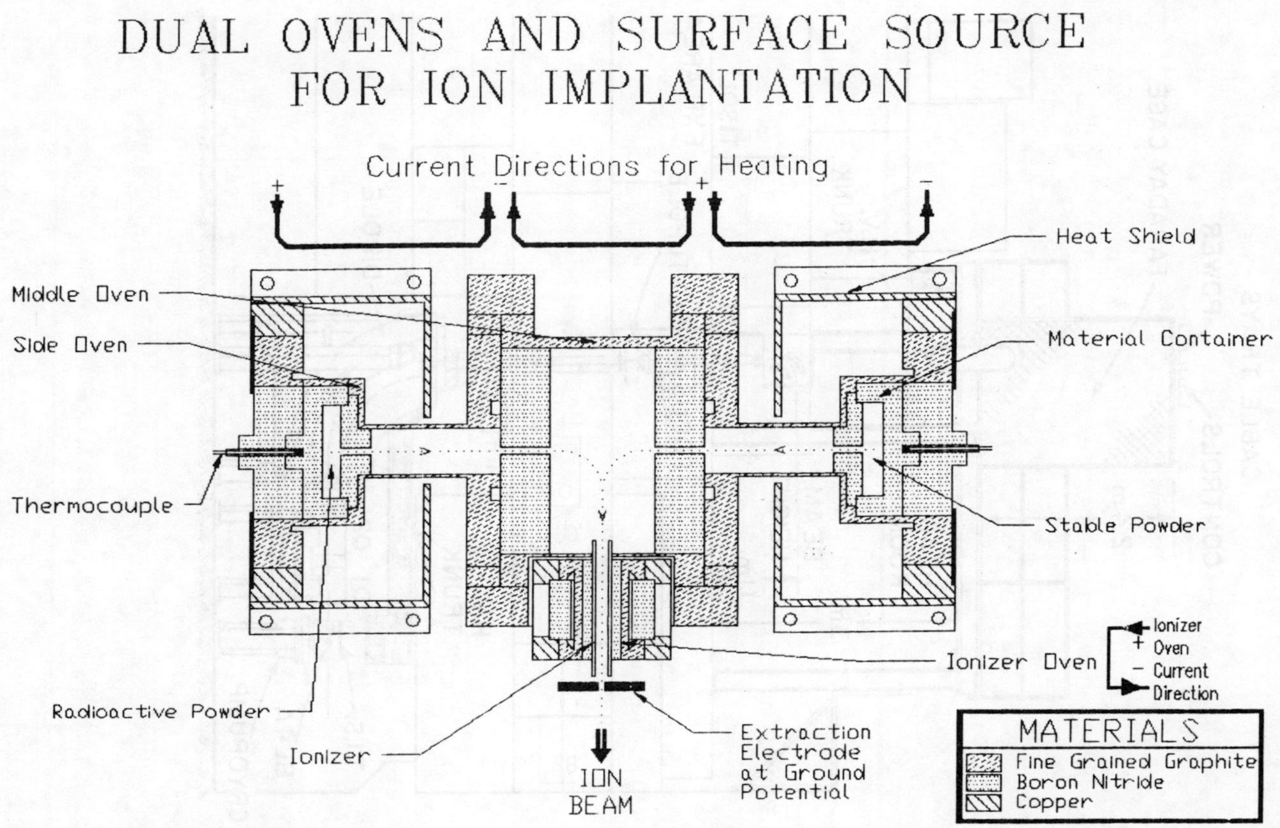
DUAL OVENS AND SURFACE SOURCE
FOR ION IMPLANTATION
Current Directions for Heating
+
-
+
-
Heat Shield
Middle Oven
Side Oven
Material Container
Thermocouple
Stable Powder
Ionizer Oven
Ionizer
+ Oven
- Current
Direction
Radioactive Powder
Ionizer
Extraction
Electrode
at Ground
Potential
ION
BEAM
MATERIALS
Fine Grained Graphite
Boron Nitride
Copper

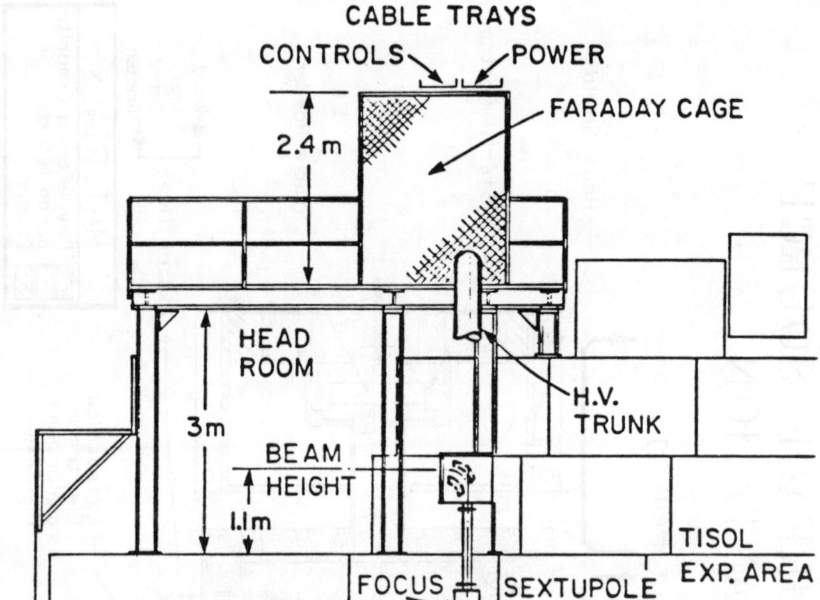
CABLE TRAYS
CONTROLS
POWER
FARADAY CAGE
2.4 m
HEAD ROOM
3m
H.V. TRUNK
8.5 m
7.3 m
BEAM HEIGHT
1.1m
TISOL EXP. AREA
6.1 m
FOCUS
SEXTUPOLE
4.8 m
3.6 m
T.P.
H.V. TRUNK
2.4 m
DIPOLE
T.I.S.
Q1
Q2
1.2 m
BL4A
CRYOPUMP

IMPLANTATION CATCHER UNIT

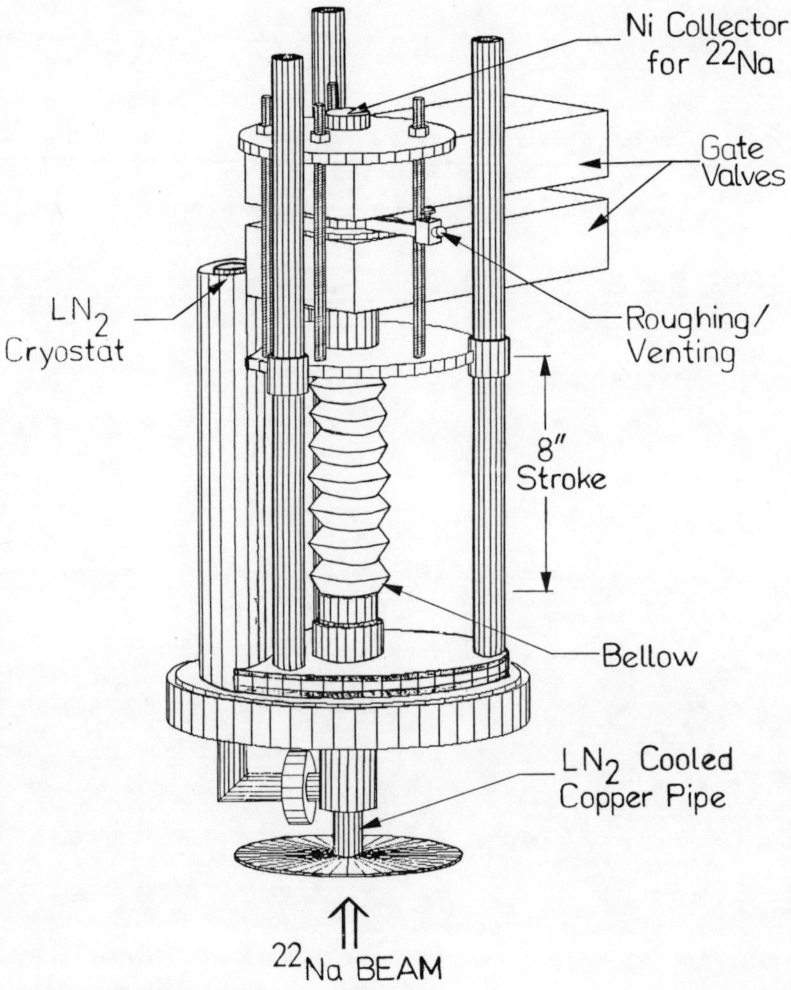

EPILOGUE

"And thus without a wing
Or service of a keel
Our summer made her light escape
Into the beautiful."

Emily Dickenson
As Imperceptibly as Grief

THE END

SYMPOSIUM ON THE ORIGIN AND DISTRIBUTION OF THE ELEMENTS

List of Speakers

Dr. C. Alcock, L-413
University of California
Institute of Geophysics and Planetary Physics
Lawrence Livermore National Laboratory
Livermore, CA 94550

Dr. E. Anders
Enrico Fermi Institute
5640 Ellis Ave.
University of Chicago
Chicago, IL 60637

Dr. A. Aprahamian
University of California
Lawrence Livermore National Laboratory
Livermore, CA 94550

Prof. W. D. Arnett
Dept. of Astronomy and Astrophysics
University of Chicago
5640 South Ellis Avenue
Chicago, IL 60637

Dr. Jean Audouze
Institut d'Astrophysique de Paris
98 bis Bd ARAGO
F-75014 Paris, France

Dr. S. A. Becker
Los Alamos National Laboratory
MS B-220
Los Alamos, NM 87545

Dr. H. Beer
Kernforschungszentrum Karlsruhe
IK-III
D-7500, Karlsruhe, West Germany

Prof. R. N. Boyd
OSU Van de Graaff Laboratory
1302 Kinnear Rd.
Columbus, OH 43212

Dr. W. M. Brunish
Los Alamos National Laboratory
ESS-5 Earth and Space Science Division
MS-F665
Los Alamos, NM 87545

Prof. L. Buchmann
Simon Fraser University
Burnaby, British Columbia
Canada V5A 1S6

Prof. B. J. Carr
Queen Mary College
University of London
England

Dr. J. Cooperstein
Dept. of Physics
Brookhaven National Laboratory
Upton, NY 11973

Prof. J. J. Cowan
Department of Physics and Astronomy
University of Oklahoma
Norman, OK 73019

Dr. I. J. Danziger
European Southern Observatory
Karl Schwarzschild Str. 2
D-8046 Garching bei Munchen
West Germany

Dr. M. F. El Eid
Universität Sternwarle Gottinger
D-3400 Gottingen,
W. Germany

Dr. C. R. Evans
Department of Physics
California Institute of Technology
Pasadena, CA 91125

Prof. B. W. Filippone
W. K. Kellogg Radiation Laboratory
106-38
California Institute of Technology
Pasadena, CA 91125

Dr. J. Frieman
Stanford Linear Accelerator Centers
Stanford University
Stanford, CA 94305

Dr. G. M. Fuller
University of California
Institute of Geophysics and Planetary Physics
Lawrence Livermore National Laboratory
Livermore, CA 94550

Dr. J. Görres
Dept. of Physics
University of Notre Dame
Notre Dame, IN 46556

Dr. L. J. Hall
Lawrence Berkeley Laboratory
University of california
Berkeley, CA 94720

Dr. D. Hartman
Lick Observatory
University of California
Santa Cruz, CA 95064

Dr. Isabel Hawkins
University of California
Institute of Geophysics and Planetary Physics
Lawrence Livermore National Laboratory
Livermore, CA 94550

Dr. W. H. Howard
University of California
Institute of Geophysics and Planetary Physics
Lawrence Livermore National Laboratory
Livermore, CA 94550

Prof. Icko Iben, Jr.
University of Illinois
341 Astronomy Bldg.
Urbana, IL 61801

Dr. Toshitaka Kajino
Michigan State University
Cyclotron Laboratory
East Lansing, MI 48824-1321

Dr. E. W. Kolb
Theoretical Astrophysics
Fermi National Accelerator Laboratory
MS-209
Batavia, IL 60510

J. C. Lattanzio
University of California
Institute of Geophysics and Planetary Physics
Lawrence Livermore National Laboratory
Livermore, CA 94550

Prof. J. M. Lattimer
Department of Earth and Space Science
SUNY at Stony Brook
Stony Brook, NY 11794

Dr. R. A. Malaney
W. K. Kellogg Radiation Laboratory
106-38
California Institute of Technology
Pasadena, CA 91125

Dr. G. J. Mathews
University of California
Lawrence Livermore National Laboratory
Livermore, CA 94550

Dr. R. W. Mayle, L-35
University of California
Lawrence Livermore National Laboratory
Livermore, CA 94550

Dr. F. Matteucci
European Southern Observatory
Karl Schwarzschilde Str 2
D-8046 Garching bei Munchen
West Germany

Dr. B. S. Meyer
Department of Astronomy and Astrophysics
University of Chicago
5640 S. Ellis
Chicago, IL 60637

Prof. Richard Matzner
Physics Dept.
Univ. of Texas
Austin, TX 78712

Dr. J.-P. Meyer
Service d'Astrophysique
DPHG/SAP
91191 Gif sur Yvette Cedex
CEN-Saclay
France

Dr. S. Niemeyer, L-232
University of California
Lawrence Livermore National Laboratory
Livermore, CA 94550

Dr. E. B. Norman
Lawrence Berkeley Laboratory
Bldg. 88
Berkeley, CA

Dr. B. E. J. Pagel
Royal Greenwich Observatory
Herstmonceux Castle
Hailsham, Sussex, BN27 1RP
England

Prof. Peter D. Parker
Physics Department
Yale University
New Haven, CT 06520

Dr. N. Prantzos
Institut d'Astrophysique de Paris
98 bis Bd Arago
F-75014 Paris, France

Dr. K. E. Sale
University of California
Lawrence Livermore National Laboratory
Livermore, CA 94550

Prof. D. N. Schramm
Department of Astronomy and Astrophysics
University of Chicago
5640 S. Ellis
Chicago, IL 606370

Dr. Verne V. Smith
Dept. of Astronomy
University of Texas
Austin, TX 78712

Dr. N. J. Snyderman
University of California
Lawrence Livermore National Laboratory
Livermore, CA 94550

Prof. G. Steigman
Department of Physics
Ohio State University
Columbus, OH 43212

Dr. G. Stringfellow
Lick Observatory
University of California
Santa Cruz, CA 95064

Dr. K. Takahashi
Physics Department
Ohio State University
Columbus, OH 43212

Dr. H. P. Trautvetter
Institut für Kernphysik
Der Universität Münster
D-4400 Münster
Federal Republic of Germany

Prof. V. Trimble
Department of Physics
University of California
Irvine, CA 92664

Prof. J. W. Truran
Department of Astronomy
University of Illinois
1011 W. Springfield Avenue
Urbana, IL 61801

Prof. V. E. Viola
Indiana University Cyclotron Facility
2401 Milo B. Sampson Lane
Bloomington, IN 41405

Prof. C. J. Waddington
School of Physics and Astronomy
University of Minnesota
116 Church St. S.E.
Minneapolis, MN 55455

Dr. R. K. Wallace
LANL
X-Division
MS-B218
Los Alamos, NM 87545

Prof. G. Wallerstein
Astronomy FM-20
University of Washington
Seattle, WA 98195

Dr. Rosemary F. G. Wyse
Astronomy Dept.
U.C. Berkeley
Berkeley, CA 94720

AUTHOR INDEX